Lecture Notes in Computer Science   13944

Founding Editors

Gerhard Goos
Juris Hartmanis

The series Lecture Notes in Computer Science (LNCS), including its subseries Lecture Notes in Artificial Intelligence (LNAI) and Lecture Notes in Bioinformatics (LNBI), has established itself as a medium for the publication of new developments in computer science and information technology research, teaching, and education.

LNCS enjoys close cooperation with the computer science R & D community, the series counts many renowned academics among its volume editors and paper authors, and collaborates with prestigious societies. Its mission is to serve this international community by providing an invaluable service, mainly focused on the publication of conference and workshop proceedings and postproceedings. LNCS commenced publication in 1973.

Xin Wang · Maria Luisa Sapino ·
Wook-Shin Han · Amr El Abbadi · Gill Dobbie ·
Zhiyong Feng · Yingxiao Shao · Hongzhi Yin
Editors

Database Systems for Advanced Applications

28th International Conference, DASFAA 2023
Tianjin, China, April 17–20, 2023
Proceedings, Part II

 Springer

Editors
Xin Wang (iD)
Tianjin University
Tianjin, China

Wook-Shin Han
POSTECH
Pohang, Korea (Republic of)

Gill Dobbie (iD)
University of Auckland
Auckland, New Zealand

Yingxiao Shao (iD)
Beijing University of Posts
and Telecommunications
Beijing, China

Maria Luisa Sapino (iD)
University of Torino
Turin, Italy

Amr El Abbadi
University of California Santa Barbara
Santa Barbara, CA, USA

Zhiyong Feng
Tianjin University
Tianjin, China

Hongzhi Yin (iD)
The University of Queensland
Brisbane, QLD, Australia

ISSN 0302-9743   ISSN 1611-3349 (electronic)
Lecture Notes in Computer Science
ISBN 978-3-031-30671-6   ISBN 978-3-031-30672-3 (eBook)
https://doi.org/10.1007/978-3-031-30672-3

This Springer imprint is published by the registered company Springer Nature Switzerland AG
The registered company address is: Gewerbestrasse 11, 6330 Cham, Switzerland

Preface

It is our great pleasure to present the proceedings of the 28th International Conference on Database Systems for Advanced Applications (DASFAA 2023), organized by Tianjin University and held during April 17–20, 2023, in Tianjin, China. DASFAA is an annual international database conference which showcases state-of-the-art R&D activities in database systems and advanced applications. It provides a premier international forum for technical presentations and discussions among database researchers, developers, and users from both academia and industry.

This year we received a record high number of 652 research paper submissions. We conducted a double-blind review following the tradition of DASFAA, and constructed a large committee consisting of 31 Senior Program Committee (SPC) members and 254 Program Committee (PC) members. Each valid submission was reviewed by at least three PC members and meta-reviewed by one SPC member, who also led the discussion with the PC members. We, the PC co-chairs, considered the recommendations from the SPC members and investigated each submission as well as its reviews to make the final decisions. As a result, 125 full papers (acceptance ratio of 19.2%) and 66 short papers (acceptance ratio of 29.3%) were accepted. The review process was supported by the Microsoft CMT system. During the three main conference days, these 191 papers were presented in 17 research sessions. The dominant keywords for the accepted papers included model, graph, learning, performance, knowledge, time, recommendation, representation, attention, prediction, and network. In addition, we included 15 industry papers, 15 demo papers, 5 PhD consortium papers, and 7 tutorials in the program. Finally, to shed light on the direction in which the database field is headed, the conference program included four invited keynote presentations by Sihem Amer-Yahia (CNRS, France), Kyuseok Shim (Seoul National University, South Korea), Angela Bonifati (Lyon 1 University, France), and Jianliang Xu (Hong Kong Baptist University, China).

Four workshops were selected by the workshop co-chairs to be held in conjunction with DASFAA 2023, which were the 9th International Workshop on Big Data Management and Service (BDMS 2023), the 8th International Workshop on Big Data Quality Management (BDQM 2023), the 7th International Workshop on Graph Data Management and Analysis (GDMA 2023), and the 1st International Workshop on Bundle-based Recommendation Systems (BundleRS 2023). The workshop papers are included in a separate volume of the proceedings also published by Springer in its Lecture Notes in Computer Science series.

We are grateful to the general chairs, Amr El Abbadi, UCSB, USA, Gill Dobbie, University of Auckland, New Zealand, and Zhiyong Feng, Tianjin University, China, all SPC members, PC members, and external reviewers who contributed their time and expertise to the DASFAA 2023 paper-reviewing process. We would like to thank all the members of the Organizing Committee, and the many volunteers, for their great support in the conference organization. Lastly, many thanks to the authors who submitted their papers to the conference.

March 2023   Xin Wang
                                                                              Maria Luisa Sapino
                                                                                Wook-Shin Han

Organization

Steering Committee Members

Chair

Lei Chen   Hong Kong University of Science and Technology (Guangzhou), China

Vice Chair

Stéphane Bressan   National University of Singapore, Singapore

Treasurer

Yasushi Sakurai   Osaka University, Japan

Secretary

Kyuseok Shim   Seoul National University, South Korea

Members

Zhiyong Peng	Wuhan University of China, China
Zhanhuai Li	Northwestern Polytechnical University, China
Krishna Reddy	IIIT Hyderabad, India
Yunmook Nah	DKU, South Korea
Wenjia Zhang	University of New South Wales, Australia
Zi Huang	University of Queensland, Australia
Guoliang Li	Tsinghua University, China
Sourav Bhowmick	Nanyang Technological University, Singapore
Atsuyuki Morishima	University of Tsukuba, Japan
Sang-Won Lee	SKKU, South Korea
Yang-Sae Moon	Kangwon National University, South Korea

Organizing Committee

Honorary Chairs

Christian S. Jensen   Aalborg University, Denmark
Keqiu Li   Tianjin University, China

General Chairs

Amr El Abbadi   UCSB, USA
Gill Dobbie   University of Auckland, New Zealand
Zhiyong Feng   Tianjin University, China

Program Committee Chairs

Xin Wang   Tianjin University, China
Maria Luisa Sapino   University of Torino, Italy
Wook-Shin Han   POSTECH, South Korea

Industry Program Chairs

Jiannan Wang   Simon Fraser University, Canada
Jinwei Zhu   Huawei, China

Tutorial Chairs

Jianxin Li   Deakin University, Australia
Herodotos Herodotou   Cyprus University of Technology, Cyprus

Demo Chairs

Zhifeng Bao   RMIT, Australia
Yasushi Sakurai   Osaka University, Japan
Xiaoli Wang   Xiamen University, China

Workshop Chairs

Lu Chen   Zhejiang University, China
Xiaohui Tao   University of Southern Queensland, Australia

Panel Chairs

Lei Chen   Hong Kong University of Science and
                                Technology (Guangzhou), China
Xiaochun Yang   Northeastern University, China

PhD Consortium Chairs

Leong Hou U.   University of Macau, China
Panagiotis Karras   Aarhus University, Denmark

Publicity Chairs

Yueguo Chen   Renmin University of China, China
Kyuseok Shim   Seoul National University, South Korea
Yoshiharu Ishikawa   Nagoya University, Japan
Arnab Bhattacharya   IIT Kanpur, India

Publication Chairs

Yingxiao Shao   Beijing University of Posts and
                                Telecommunications, China
Hongzhi Yin   University of Queensland, Australia

DASFAA Steering Committee Liaison

Lei Chen   Hong Kong University of Science and
                                Technology (Guangzhou), China

Local Arrangement Committee

Xiaowang Zhang   Tianjin University, China
Guozheng Rao   Tianjin University, China
Yajun Yang   Tianjin University, China
Shizhan Chen   Tianjin University, China
Xueli Liu   Tianjin University, China
Xiaofei Wang   Tianjin University, China
Chao Qiu   Tianjin University, China
Dong Han   Tianjin Academy of Fine Arts, China
Ying Guo   Tianjin University, China

Hui Jiang   Tianjin Ren'ai College, China
Kun Liang   Tianjin University of Science and Technology,
                                        China

Web Master

Zirui Chen   Tianjin University, China

Program Committee Chairs

Xin Wang   Tianjin University, China
Maria Luisa Sapino   University of Torino, Italy
Wook-Shin Han   POSTECH, South Korea

Senior Program Committee (SPC) Members

Baihua Zheng   Singapore Management University, Singapore
Bin Cui   Peking University, China
Bingsheng He   National University of Singapore, Singapore
Chee-Yong Chan   National University of Singapore, Singapore
Chengfei Liu   Swinburne University of Technology, Australia
Haofen Wang   Tongji University, China
Hong Gao   Harbin Institute of Technology, China
Hongzhi Yin   University of Queensland, Australia
Jiaheng Lu   University of Helsinki, Finland
Jianliang Xu   Hong Kong Baptist University, China
Jianyong Wang   Tsinghua University, China
K. Selçuk Candan   Arizona State University, USA
Kyuseok Shim   Seoul National University, South Korea
Lei Li   Hong Kong University of Science and
                                        Technology (Guangzhou), China
Lina Yao   University of New South Wales, Australia
Ling Liu   Georgia Institute of Technology, USA
Nikos Bikakis   Athena Research Center, Greece
Qiang Zhu   University of Michigan-Dearborn, USA
Reynold Cheng   University of Hong Kong, China
Ronghua Li   Beijing Institute of Technology, China
Vana Kalogeraki   Athens University of Economics and Business,
                                        Greece
Vincent Tseng   National Yang Ming Chiao Tung University,
                                        Taiwan
Wang-Chien Lee   Pennsylvania State University, USA

Xiang Zhao   National University of Defense Technology,
                              China
Xiaoyong Du   Renmin University of China, China
Ye Yuan   Beijing Institute of Technology, China
Yongxin Tong   Beihang University, China
Yoshiharu Ishikawa   Nagoya University, Japan
Yufei Tao   Chinese University of Hong Kong, China
Yunjun Gao   Zhejiang University, China
Zhiyong Peng   Wuhan University, China

Program Committee (PC) Members

Alexander Zhou   Hong Kong University of Science and
                              Technology, China
Alkis Simitsis   Athena Research Center, Greece
Amr Ebaid   Google, USA
An Liu   Soochow University, China
Anne Laurent   University of Montpellier, France
Antonio Corral   University of Almería, Spain
Baoning Niu   Taiyuan University of Technology, China
Barbara Catania   University of Genoa, Italy
Bin Cui   Peking University, China
Bin Wang   Northeastern University, China
Bing Li   Institute of High Performance Computing,
                              Singapore
Bohan Li   Nanjing University of Aeronautics and
                              Astronautics, China
Changdong Wang   SYSU, China
Chao Huang   University of Notre Dame, USA
Chao Zhang   Tsinghua University, China
Chaokun Wang   Tsinghua University, China
Chenyang Wang   Aalto University, Finland
Cheqing Jin   East China Normal University, China
Chih-Ya Shen   National Tsing Hua University, Taiwan
Christos Doulkeridis   University of Pireaus, Greece
Chuan Ma   Zhejiang Lab, China
Chuan Xiao   Osaka University and Nagoya University, Japan
Chuanyu Zong   Shenyang Aerospace University, China
Chunbin Lin   Amazon AWS, USA
Cindy Chen   UMass Lowell, USA
Claudio Schifanella   University of Torino, Italy
Cuiping Li   Renmin University of China, China

Damiani Ernesto	University of Milan, Italy
Dan He	University of Queensland, Australia
De-Nian Yang	Academia Sinica, Taiwan
Derong Shen	Northeastern University, China
Dhaval Patel	IBM Research, USA
Dian Ouyang	Guangzhou University, China
Dieter Pfoser	George Mason University, USA
Dimitris Kotzinos	ETIS, France
Dong Wen	University of New South Wales, Australia
Dongxiang Zhang	Zhejiang University, China
Dongxiao He	Tianjin University, China
Faming Li	Northeastern University, USA
Ge Yu	Northeastern University, China
Goce Trajcevski	Iowa State University, USA
Gong Cheng	Nanjing University, China
Guandong Xu	University of Technology Sydney, Australia
Guanhua Ye	University of Queensland, Australia
Guoliang Li	Tsinghua University, China
Haida Zhang	WorldQuant, USA
Hailong Liu	Northwestern Polytechnical University, China
Haiwei Zhang	Nankai University, China
Hantao Zhao	ETH, Switzerland
Hao Peng	Beihang University, China
Hiroaki Shiokawa	University of Tsukuba, Japan
Hongbin Pei	Xi'an Jiaotong University, China
Hongxu Chen	Commonwealth Bank of Australia, Australia
Hongzhi Wang	Harbin Institute of Technology, China
Hongzhi Yin	University of Queensland, Australia
Huaijie Zhu	Sun Yat-sen University, China
Hui Li	Xidian University, China
Huiqi Hu	East China Normal University, China
Hye-Young Paik	University of New South Wales, Australia
Ioannis Konstantinou	University of Thessaly, Greece
Ismail Hakki Toroslu	METU, Turkey
Jagat Sesh Challa	BITS Pilani, India
Ji Zhang	University of Southern Queensland, Australia
Jia Xu	Guangxi University, China
Jiali Mao	East China Normal University, China
Jianbin Qin	Shenzhen Institute of Computing Sciences, China
Jianmin Wang	Tsinghua University, China
Jianqiu Xu	Nanjing University of Aeronautics and Astronautics, China

Jianxin Li	Deakin University, Australia
Jianye Yang	Guangzhou University, China
Jiawei Jiang	Wuhan University, China
Jie Shao	University of Electronic Science and Technology of China, China
Jilian Zhang	Jinan University, China
Jilin Hu	Aalborg University, Denmark
Jin Wang	Megagon Labs, USA
Jing Tang	Hong Kong University of Science and Technology, China
Jithin Vachery	NUS, Singapore
Jongik Kim	Chungnam National University, South Korea
Ju Fan	Renmin University of China, China
Jun Gao	Peking University, China
Jun Miyazaki	Tokyo Institute of Technology, Japan
Junhu Wang	Griffith University, Australia
Junhua Zhang	University of New South Wales, Australia
Junliang Yu	The University of Queensland, Australia
Kai Wang	Shanghai Jiao Tong University, China
Kai Zheng	University of Electronic Science and Technology of China, China
Kangfei Zhao	The Chinese University of Hong Kong, China
Kesheng Wu	LBNL, USA
Kristian Torp	Aalborg University, Denmark
Kun Yue	School of Information and Engineering, China
Kyoung-Sook Kim	National Institute of Advanced Industrial Science and Technology, Japan
Ladjel Bellatreche	ISAE-ENSMA, France
Latifur Khan	University of Texas at Dallas, USA
Lei Cao	MIT, USA
Lei Duan	Sichuan University, China
Lei Guo	Shandong Normal University, China
Leong Hou U.	University of Macau, China
Liang Hong	Wuhan University, China
Libin Zheng	Sun Yat-sen University, China
Lidan Shou	Zhejiang University, China
Lijun Chang	University of Sydney, Australia
Lin Li	Wuhan University of Technology, China
Lizhen Cui	Shandong University, China
Long Yuan	Nanjing University of Science and Technology, China
Lu Chen	Swinburne University of Technology, Australia

Lu Chen	Zhejiang University, China
Makoto Onizuka	Osaka University, Japan
Manish Kesarwani	IBM Research, India
Manolis Koubarakis	University of Athens, Greece
Markus Schneider	University of Florida, USA
Meihui Zhang	Beijing Institute of Technology, China
Meng Wang	Southeast University, China
Meng-Fen Chiang	University of Auckland, New Zealand
Ming Zhong	Wuhan University, China
Minghe Yu	Northeastern University, China
Mizuho Iwaihara	Waseda University, Japan
Mo Li	Liaoning University, China
Ning Wang	Beijing Jiaotong University, China
Ningning Cui	Anhui University, China
Norio Katayama	National Institute of Informatics, Japan
Noseong Park	George Mason University, USA
Panagiotis Bouros	Johannes Gutenberg University Mainz, Germany
Peiquan Jin	University of Science and Technology of China, China
Peng Cheng	East China Normal University, China
Peng Peng	Hunan University, China
Pengpeng Zhao	Soochow University, China
Ping Lu	Beihang University, China
Pinghui Wang	Xi'an Jiaotong University, China
Qiang Yin	Shanghai Jiao Tong University, China
Qianzhen Zhang	National University of Defense Technology, China
Qing Liao	Harbin Institute of Technology (Shenzhen), China
Qing Liu	CSIRO, Australia
Qingpeng Zhang	City University of Hong Kong, China
Qingqing Ye	Hong Kong Polytechnic University, China
Quanqing Xu	A*STAR, Singapore
Rong Zhu	Alibaba Group, China
Rui Zhou	Swinburne University of Technology, Australia
Rui Zhu	Shenyang Aerospace University, China
Ruihong Qiu	University of Queensland, Australia
Ruixuan Li	Huazhong University of Science and Technology, China
Ruiyuan Li	Chongqing University, China
Sai Wu	Zhejiang University, China
Sanghyun Park	Yonsei University, South Korea

Sanjay Kumar Madria	Missouri University of Science & Technology, USA
Sebastian Link	University of Auckland, New Zealand
Sen Wang	University of Queensland, Australia
Shaoxu Song	Tsinghua University, China
Sheng Wang	Wuhan University, China
Shijie Zhang	Tencent, China
Shiyu Yang	Guangzhou University, China
Shuhao Zhang	Singapore University of Technology and Design, Singapore
Shuiqiao Yang	UNSW, Australia
Shuyuan Li	Beihang University, China
Sibo Wang	Chinese University of Hong Kong, China
Silvestro Roberto Poccia	University of Turin, Italy
Tao Qiu	Shenyang Aerospace University, China
Tao Zhao	National University of Defense Technology, China
Taotao Cai	Macquarie University, Australia
Thanh Tam Nguyen	Griffith University, Australia
Theodoros Chondrogiannis	University of Konstanz, Germany
Tieke He	State Key Laboratory for Novel Software Technology, China
Tieyun Qian	Wuhan University, China
Tiezheng Nie	Northeastern University, China
Tsz Nam (Edison) Chan	Hong Kong Baptist University, China
Uday Kiran Rage	University of Aizu, Japan
Verena Kantere	National Technical University of Athens, Greece
Wei Hu	Nanjing University, China
Wei Li	Harbin Engineering University, China
Wei Lu	RUC, China
Wei Shen	Nankai University, China
Wei Song	Wuhan University, China
Wei Wang	Hong Kong University of Science and Technology (Guangzhou), China
Wei Zhang	ECNU, China
Wei Emma Zhang	The University of Adelaide, Australia
Weiguo Zheng	Fudan University, China
Weijun Wang	University of Göttingen, Germany
Weiren Yu	University of Warwick, UK
Weitong Chen	Adelaide University, Australia
Weiwei Sun	Fudan University, China
Weixiong Rao	Tongji University, China

Wen Hua	Hong Kong Polytechnic University, China
Wenchao Zhou	Georgetown University, USA
Wentao Li	University of Technology Sydney, Australia
Wentao Zhang	Mila, Canada
Werner Nutt	Free University of Bozen-Bolzano, Italy
Wolf-Tilo Balke	TU Braunschweig, Germany
Wookey Lee	Inha University, South Korea
Xi Guo	University of Science and Technology Beijing, China
Xiang Ao	Institute of Computing Technology, CAS, China
Xiang Lian	Kent State University, USA
Xiang Zhao	National University of Defense Technology, China
Xiangguo Sun	Chinese University of Hong Kong, China
Xiangmin Zhou	RMIT University, Australia
Xiangyu Song	Swinburne University of Technology, Australia
Xiao Pan	Shijiazhuang Tiedao University, China
Xiao Fan Liu	City University of Hong Kong, China
Xiaochun Yang	Northeastern University, China
Xiaofeng Gao	Shanghai Jiaotong University, China
Xiaoling Wang	East China Normal University, China
Xiaowang Zhang	Tianjin University, China
Xiaoyang Wang	University of New South Wales, Australia
Ximing Li	Jilin University, China
Xin Cao	University of New South Wales, Australia
Xin Huang	Hong Kong Baptist University, China
Xin Wang	Southwest Petroleum University, China
Xinqiang Xie	Neusoft, China
Xiuhua Li	Chongqing University, China
Xiulong Liu	Tianjin University, China
Xu Zhou	Hunan University, China
Xuequn Shang	Northwestern Polytechnical University, China
Xupeng Miao	Carnegie Mellon University, USA
Xuyun Zhang	Macquarie University, Australia
Yajun Yang	Tianjin University, China
Yan Zhang	Peking University, China
Yanfeng Zhang	Northeastern University, China, and Macquarie University, Australia
Yang Cao	Hokkaido University, Japan
Yang Chen	Fudan University, China
Yang-Sae Moon	Kangwon National University, South Korea
Yanjie Fu	University of Central Florida, USA

Yanlong Wen	Nankai University, China
Ye Yuan	Beijing Institute of Technology, China
Yexuan Shi	Beihang University, China
Yi Cai	South China University of Technology, China
Ying Zhang	Nankai University, China
Yingxia Shao	BUPT, China
Yiru Chen	Columbia University, USA
Yixiang Fang	Chinese University of Hong Kong, Shenzhen, China
Yong Tang	South China Normal University, China
Yong Zhang	Tsinghua University, China
Yongchao Liu	Ant Group, China
Yongpan Sheng	Southwest University, China
Yongxin Tong	Beihang University, China
You Peng	University of New South Wales, Australia
Yu Gu	Northeastern University, China
Yu Yang	Hong Kong Polytechnic University, China
Yu Yang	City University of Hong Kong, China
Yuanyuan Zhu	Wuhan University, China
Yue Kou	Northeastern University, China
Yunpeng Chai	Renmin University of China, China
Yunyan Guo	Tsinghua University, China
Yunzhang Huo	Hong Kong Polytechnic University, China
Yurong Cheng	Beijing Institute of Technology, China
Yuxiang Zeng	Hong Kong University of Science and Technology, China
Zeke Wang	Zhejiang University, China
Zhaojing Luo	National University of Singapore, Singapore
Zhaonian Zou	Harbin Institute of Technology, China
Zheng Liu	Nanjing University of Posts and Telecommunications, China
Zhengyi Yang	University of New South Wales, Australia
Zhenya Huang	University of Science and Technology of China, China
Zhenying He	Fudan University, China
Zhipeng Zhang	Alibaba, China
Zhiwei Zhang	Beijing Institute of Technology, China
Zhixu Li	Fudan University, China
Zhongnan Zhang	Xiamen University, China

Industry Program Chairs

Jiannan Wang	Simon Fraser University, Canada
Jinwei Zhu	Huawei, China

Industry Program Committee Members

Bohan Li	Nanjing University of Aeronautics and Astronautics, China
Changbo Qu	Simon Fraser University, Canada
Chengliang Chai	Tsinghua University, China
Denis Ponomaryov	The Institute of Informatics Systems of the Siberian Division of Russian Academy of Sciences, Russia
Hongzhi Wang	Harbin Institute of Technology, China
Jianhua Yin	Shandong University, China
Jiannan Wang	Simon Fraser University, Canada
Jinglin Peng	Simon Fraser University, Canada
Jinwei Zhu	Huawei Technologies Co. Ltd., China
Ju Fan	Renmin University of China, China
Minghe Yu	Northeastern University, China
Nikos Ntarmos	Huawei Technologies R&D (UK) Ltd., UK
Sheng Wang	Alibaba Group, China
Wei Zhang	East China Normal University, China
Weiyuan Wu	Simon Fraser University, Canada
Xiang Li	East China Normal University, China
Xiaofeng Gao	Shanghai Jiaotong University, China
Xiaoou Ding	Harbin Institute of Technology, China
Yang Ren	Huawei, China
Yinan Mei	Tsinghua University, China
Yongxin Tong	Beihang University, China

Demo Track Program Chairs

Zhifeng Bao	RMIT, Australia
Yasushi Sakurai	Osaka University, Japan
Xiaoli Wang	Xiamen University, China

Demo Track Program Committee Members

Benyou Wang	Chinese University of Hong Kong, Shenzhen, China
Changchang Sun	Illinois Institute of Technology, USA
Chen Lin	Xiamen University, China
Chengliang Chai	Tsinghua University, China
Chenhao Ma	Chinese University of Hong Kong, Shenzhen, China
Dario Garigliotti	Aalborg University, Denmark
Ergute Bao	National University of Singapore, Singapore
Jianzhong Qi	The University of Melbourne, Australia
Jiayuan He	RMIT University, Australia
Kaiping Zheng	National University of Singapore, Singapore
Kajal Kansal	NUS, Singapore
Lei Cao	MIT, USA
Liang Zhang	WPI, USA
Lu Chen	Swinburne University of Technology, Australia
Meihui Zhang	Beijing Institute of Technology, China
Mengfan Tang	University of California, Irvine, USA
Na Zheng	National University of Singapore, Singapore
Pinghui Wang	Xi'an Jiaotong University, China
Qing Xie	Wuhan University of Technology, China
Ruihong Qiu	University of Queensland, Australia
Tong Chen	University of Queensland, Australia
Yile Chen	Nanyang Technological University, Singapore
Yuya Sasaki	Osaka University, Japan
Yuyu Luo	Tsinghua University, China
Zhanhao Zhao	Renmin University of China, China
Zheng Wang	Huawei Singapore Research Center, Singapore
Zhuo Zhang	University of Melbourne, Australia

PhD Consortium Track Program Chairs

Leong Hou U.	University of Macau, China
Panagiotis Karras	Aarhus University, Denmark

PhD Consortium Track Program Committee Members

Anton Tsitsulin	Google, USA
Bo Tang	Southern University of Science and Technology, China

Hao Wang   Wuhan University, China
Jieming Shi   Hong Kong Polytechnic University, China
Tsz Nam (Edison) Chan   Hong Kong Baptist University, China
Xiaowei Wu   State Key Lab of IoT for Smart City, University of
                                     Macau, China

Contents – Part II

Knowledge Graph

Recommendation

KRec-C2: A Knowledge Graph Enhanced Recommendation with Context Awareness and Contrastive Learning

Yingtao Peng[1], Zhendong Zhao[2], Aishan Maoliniyazi[1], and Xiaofeng Meng[1(✉)]

[1] School of Information, Renmin University of China, Beijing, China
{yingtaopeng,aishan,xfmeng}@ruc.edu.cn
[2] Institute of Information Engineering, Chinese Academy of Sciences, Beijing, China
zhaozhendong@iie.ac.cn

Abstract. Knowledge graph (KG) has been widely utilized in recommendation system to its rich semantic information. There are two main challenges in real-world applications: high-quality knowledge graphs and modeling user-item relationships. However, existing methods try to solve the above challenges by adopting unified relational rules and simple node aggregation, which cannot cope with complex structured graph data. In this paper, we propose a **K**nowledge graph enhanced **Re**commendation with **C**ontext awareness and **C**ontrastive learning (KRec-C2) to overcome the issue. Specifically, we design an category-level contrastive learning module to model underlying node relationships from noisy real-world graph data. Furthermore, we propose a sequential context-based information aggregation module to accurately learn item-level relation features from a knowledge graph. Extensive experiments conducted on three real-world datasets demonstrate the superiority of our KRec-C2 model over existing state-of-the-art methods.

Keywords: Recommendation System · Knowledge Graph · Contrastive Learning · Context Awareness

1 Introduction

Nowadays, we are suffering from a severe information overload issue in all fields as never before. The recommendation system (RS), aim to filter relevant information from massive amounts of data, which makes it easy for the user to obtain high-value information from modern big data platform such as e-commerce [16,19], multimedia [33], and online news [14]. As one of the most important technologies in the current big data era, the recommendation system brings great convenience to users, and attracts great attention from academia and industry.

Among the existing research from recommendation communities, the Collaborative Filtering (CF) [9,13,15] framework assumes that users with similar behaviors may have similar interests, and is the most commonly used technique for recommendation systems to predict user preferences. Arguably, most existing CF-based models can be roughly divided into two categories: (1) The traditional

X. Wang et al. (Eds.): DASFAA 2023, LNCS 13944, pp. 3–20, 2023.
https://doi.org/10.1007/978-3-031-30672-3_1

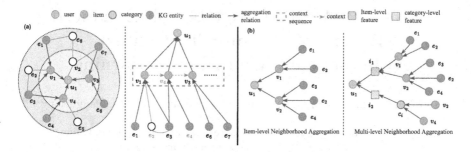

Fig. 1. (a) is an example of user representation on context awareness, where an arrow is the contextual relation within a user's interaction sequence. (b) is an example of the category-level signal fusion schemes, where i_2 represent the category-level signal.

CF approaches. These models find a certain similarity measure through the behavior of groups and make recommendations for users [13]. (2) The neural network-based CF approaches. These methods represent each user and item as a low-dimensional vector, and then evaluate the degree of association according to defined vector operations [4]. For example, some works obtain higher-order feature vectors by combining the non-linearity of a neural network to improve the performance of vector representation [9,10]. In contrast, we design a novel context-aware fusion network to capture the user's representation in this paper.

In the real world, the user's ratings are mostly empty, which leads to the extremely sparse interactive data we collect. In addition, due to the limited attention of the user, an item without interaction may also be a great favorite. However, because the interaction data is sparse in real-world situations and existing neural network-based CF models are over-reliant on the interaction data, these approaches perform unsatisfactorily in universal scenarios [36]. To address this problem, some research works [2,25] introduce knowledge graph (KG) as auxiliary information in recommender systems, which can better integrate different information into the overall network to learn better embedding representations. The KG allows the developer to model relationships between users and items with the entity and relationship information and enhance the semantic association of nodes by embedding [22,36].

Although the KG-based RS has been applied in several areas, its effectiveness remained over-reliant on the quality of knowledge graphs. Mostly, the KG is often noisy and redundancy in real-world scenes, which contains topic-irrelevant connections between entities and longtail entity distribution, leading to poor performance of the model on feature learning [17,21]. Even though existing self-supervised learning-based RS methods [32,37] try to alleviate the issue by graph-based augmentation techniques (such as edge dropout, node dropout and subgraph sampling) [32,34], there is fall short in three factors: i) **Ignoring the context feature**. The context feature is a relationship dependence between items within a interaction sequence [20] in the recommendation system, which helps us to obtain potential relationships between the interaction items.

Unfortunately, existing KG-based works do not account for this crucial information, leading to the insufficient representation performance of user and item. Taking the right of Fig. 1 as an example, arrows are the potantial contextual relations within a user's clicked sequence. On the contrary, ignoring contextual features limits the discovery of contextual intention in the left Fig. 1 (a). ii) **Overlooking the category-level signal**. We define a category-level signal as the common characteristics of similar items in the same catalog, such as function, classification, etc., which provide a solution to capture the category-level features for similar items. Regrettably, existing KG-based models focus on the node-based aggregation scheme [26] at the item level, leading to missing the category-level feature. As shown in the right side of Fig. 1 (b), the representation of u_1 fusions the feature of item i_1 and the category-level feature i_2. By comparison, u_i cannot capture the category-level feature on the left side of Fig. 1 (b). iii) **Shortage of the self-supervised signal**. The lack of self-supervised signals leads to the insufficient ability of the model to deal with noise in a real-world scenario. Self-supervised learning generally learns a feature representation for downstream tasks by extracting information and transferable knowledge. Existing KG-based models with self-supervised learning ignore the significant category-level view, leading to a lower ability to alleviate the sparse data and noisy issue.

To solve the above problems, we propose a knowledge graph enhanced recommendation with context awareness and contrastive learning (KRec-C2). Our model consists of three components: (1) **Item-level context awareness module (ICAM)**. Each user-item interaction is enriched with the underlying intents for the user. Although we can express these items as vectors, their contextual semantics within a sequence are opaque to understand. Hence, we combine the intent and the context based on node aggregation representation. Technically, context awareness is an attentive design of relation embeddings, where the important intent is assigned with a larger weight factor. (2) **Category-level intents fusion module (CIFM)**. The node-based aggregation scheme focuses on the receptive field of the item to obtain an embedded representation for the item-self. In contrast, CIFM provides a solution to capture the common intrinsic properties on the category level between items. To overcome the technical challenge, we first obtain the embedded vector for item-self with node-based aggregation and capture the intrinsic features on the category level for items by a higher-order aggregation scheme. Then we fusion each signal into a representation vector. (3) **Contrastive learning module (CLM)**. We carefully design a contrastive learning (CL) optimization objective for capturing the relevance of the category view to alleviate the noise and sparse issues. Specifically, contrastive learning takes inspiration from the KG-based representation and self-supervised data augmentation to guide our model in refining the category-level representations with the contrastive view on the category level. The major novelty lies in achieving the optimization objective with mutual information maximization at the training stage.

We summarize the contributions of this work as: (1) We propose a knowledge graph enhanced recommendation with context awareness and contrastive learning (KRec-C2) to enhance the recommendation performance. (2) We design three novel modules in the KRec-C2 framework for KG-based recommendation, including ICAM, CIFM, and CLM. Based on these modules, context features and the category-level signal have been effectively utilized and integrated into the KG-based recommender. (3) Extensive experiments conducted on three real-world datasets demonstrate the effectiveness of our proposed approach.

2   Preliminaries

We first define our recommendation task, and then present necessary notations and structural data, such as user-item interactions and KGs.

Task Description. Assume that we have the interaction data $\mathbf{Y}$ and the KG $\mathcal{G}$. Our goal is to learn the probability $\hat{y}_{uv} = \sigma(\mathbf{u} \cdot \mathbf{v})$ that a user u adopt a candidate item v, where σ is activation function, $\mathbf{u}$ and $\mathbf{v}$ are vectors representation of an user u and a candidate item v.

Interaction Data. In our recommendation task, we focus on the user-item interactions (e.g., browse, click, purchase and comment, etc.). Here we let $\mathcal{U} = \{u_1, u_2, u_3, \cdots\}$ be a set of users and $\mathcal{V} = \{v_1, v_2, v_3, \cdots\}$ be a set of items. We define $\mathbf{Y} = \{y_{uv} | u \in \mathcal{U}, v \in \mathcal{V}\}$ as the user-item interaction matrix.

Knowledge Graph. In application scenes of the RS, we can obtain the auxiliary property information of items (e.g., categories, and attributes), which make up various real-world entities and numerous relations in KG. We organize knowledge graph $\mathcal{G}$ in the form of triple, let $\mathcal{G} = \{(h, r, t) | h, t \in \mathcal{E}, r \in \mathcal{R}\}$, where its relations are composed of (item-property-item) triples, and denote that a relation r from head node h to tail node t. For example, (Titanic, love movie, Flipped) describes that Titanic and Flipped both belong to love movies.

3   Methodology

In this section, we first introduce our model framework and then discuss each module of KRec-C2 in detail.

3.1   Framework

The framework of our model is illustrated in Fig. 2, where we innovatively model context, category-level signals, and self-supervised features by three modules to improve the recommendation effect. KRec-C2 inputs interaction data and KG, and outputs the probability that user u would interact item v. Specifically, our model consists of three components: (1) ICAM. First, we aggregate neighbor's information of items by KGCN. Then, the context awareness layer captures implicit contextual relationships among the interaction sequence items and predicts the user's potential preferences. (2) CIFM. The category-aware aggregation module propagates embeddings from a category node's neighbors to update its representation. Then, we fuse contextual features and category-level signals as user feature representations during the training process. (3) CLM. We design a

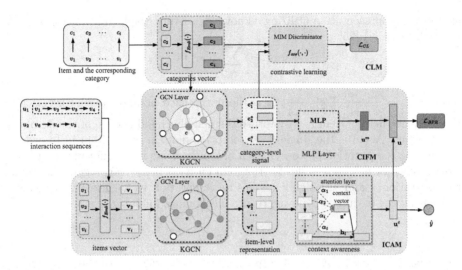

Fig. 2. An overview of the proposed KRec-C2 framework.

contrastive learning module on a category-level view to capture the relevance of the category for alleviating noise and sparse issues. Finally, on the prediction layer, we predict the probability of a user will interact a candidate item based on contextual features and item vector.

3.2 Item-Level Context Awareness Module

Previous GNN-based research only aggregates the information of neighbor nodes [25,26,30], while our goal is to capture the contextual relationship between inter-acted items and predict users' latent intentions. First, we utilize the KGCN layer to obtain item-level aggregation features, which contain the associated informa-tion of items in the KG. For example, in the movie's case, the item-level aggrega-tion feature includes various relationships from directors, actors, locations, etc. Then, we can obtain potential intent from the user's interaction sequence. Intu-itively, user interactions contain potential user preferences, which are essential to features. It is necessary to characterize users by extracting personalized fea-tures to improve the recommendation ability. This idea motivates us to model user-item relationships with context awareness.

Item-Level Aggregation Representation. The neighbor information aggre-gation captures the local proximity structure and stores representation in each entity to reflect the user's personalized interests better. For example, in the real world, a goods node has many rich relations with neighbor nodes in KG, such as brand, size, color, etc. We use $\mathcal{N}(v) = \{e_1, e_2, \cdots, e_n\}$ to denote the entities' set for the item's neighbors in KG. Then, we learn the current item's high-order aggregation representation utilizing a knowledge-aware graph convolution neural network (KGCN) [25].

$$\mathbf{v}_{\mathcal{N}(v)} = \sum_{e \in \mathcal{N}(v)} \zeta_{r_{v,e}} \mathbf{e}, \tag{1}$$

where $\zeta_{r_{v,e}} = \dfrac{\exp\left(\tilde{\zeta}_{r_{v,e}}\right)}{\sum_{e \in \mathcal{N}(v)} \exp\left(\tilde{\zeta}_{r_{v,e}}\right)}$, ζ_r is the normalized node-relation score $\tilde{\zeta}_r$. In the receptive field of item v, we capture the neighborhood feature of the item as $\mathbf{v}_{\mathcal{N}(v)}$, where $\mathcal{N}(v)$ is the single-layer receptive domain of the item v. In order to keep the computation pattern of each batch fixed, we first uniformly sample a fixed-size set of neighbor nodes for the item entity. Then we aggregate the item representation $\mathbf{v}$ and its neighbor nodes representation $\mathbf{v}_{\mathcal{N}(v)}$ into a vector. For calculating the feature $\mathbf{v}^a$, our aggregate methods as follows:

$$\mathbf{v}^a = \sigma \left(W \cdot f_{agg}\left(\cdot\right) + b\right) \tag{2}$$

where $f_{agg}\left(\cdot\right)$ notes the aggregate operator, we implement it with three methods: sum operator $f_{agg}^{sum} = \mathbf{v} + \mathbf{v}_{\mathcal{N}(v)}$, concat operator $f_{agg}^{concat} = concat\left(\mathbf{v}, \mathbf{v}_{\mathcal{N}(v)}\right)$, and neighbor operator $f_{agg}^{neighbor} = \mathbf{v}_{\mathcal{N}(v)}$. W and b denote transformation weight and bias, σ denotes the nonlinear function.

Context Awareness Representation. To capture the user's history interest and predict potential intention, we designed the context-awareness module, composed of a bidirectional gate recurrent unit (Bi-GRU) and a self-attention network based on the output of KGCN. The module inputs a sequence of the user's interaction item, which is the output result of item-level aggregation representation. Our goal is to capture the contextual relationship and output the user's potential intention.

Based on item-level aggregation representation, we obtain the item-level representations $\mathcal{V}_l = \{\mathbf{v}_1^a, \mathbf{v}_2^a, \cdots, \mathbf{v}_l^a\}$ of user's interaction sequence $\{v_1, v_2, \cdots, v_l\}$. In this module, we adopt a Bi-GRU based self-attention network composed of a bidirectional gate recurrent unit layer and a self-attention layer. In i-step of Bi-GRU, our input information are the i-step item-level representation $\mathbf{v}_i^a$ and the hidden representation of $i - 1$-step of Bi-GRU, and our output information are an intermediate feature $\mathbf{h}_i \in R^m$, where m denotes the dimension of the hidden representation of Bi-GRU, the formula is as follows:

$$\mathbf{h}_i = f_{BiGRU}\left(\mathbf{h}_{i-1}, \mathbf{v}_i^a; \Theta_{gru}\right), \tag{3}$$

where $\mathbf{h}_i$ denotes intermediate feature of GRU's i-step, Θ_{gru} denotes all the related parameters of the Bi-GRU network.

In the Bi-GRU layer, the last hidden state $\mathbf{h}_i$ denotes the sequential representation of the input sequence. In order to explore the latent relationship of user interaction sequences, we consider that the previous hidden states may play different role for user's latent intention, and then we can capture the user's potential feature by self-attention mechanism. Therefore, we take the hidden embedding vector $H = [\mathbf{h}_1, \mathbf{h}_2, \cdots, \mathbf{h}_l] \in R^{m \times n}$ as input, and exploit self-attention mechanism to capture the user's contextual interest feature, the formula is as follows:

$$Z^* = V \; softmax \left(\frac{K^T Q}{\sqrt{D}} \right), \tag{4}$$

where $softmax\,(\cdot)$ denotes the normalization function. $Q = W_q H$, $K = W_k H$, and $V = W_v H$ are mapped vectors by H, and W_q, W_k, W_v are the parameter matrices of the linear mapping. D is the dimension of the input vector H.

Finally, we obtain item's contextual representation $\mathbf{u}^c = [\mathbf{z}^*; \mathbf{h}_l]$ by merging the user's the contextual representation $\mathbf{z}^*$ and the prediction feature $\mathbf{h}_l$.

3.3   Category-Level Intents Fusion Module

Existing methods learn item representation by node-level aggregation scheme, which ignores the category-level signal, resulting in a limited effect. In this section, we aim to capture the category-level signals and incorporate them into user feature representations in the stage of training. Technically, we learn the category node and its neighbor representations as the category-level features, and fuse them with item-level user representations. Based on the above analysis, we model interactions at the category level.

Category-Level Aggregation Representation. Mostly, items in the same category have similar attributes in recommended scenarios. Existing works only capture the single relations with item by the category node such as (Titanic, classification, love movie), missing salient common feature between items in the same catalog (e.g., region, director, etc.). To alleviate this issue, we construct the category-item graph to capture the common feature as a category-level signal.

In this part, we obtain a set of triples $\mathcal{T}$ from the items and their categories, where $\mathcal{T} = \{(c_1, r_1, v_1), (c_2, r_2, v_2), (c_2, r_2, v_3), \cdots\}$, r_i notes relation between item v_i and category c_i, such as (sports, golf, PGA Tour winners). Similar to item-level aggregate representation, we use $\mathcal{N}(c) = (e_1, e_2, \cdots, e_n)$ to denote the set of entities directly connected to the category c. The items' similarities attributes in a same category can regard as the common preferences of users. Similarly, we adopt a knowledge-based graph convolution neural network (KGCN) to capture the category-level representation. Specifically, in a single GCN layer, the formula is as follows:

$$\mathbf{H}^{l+1} = \sigma \left(\tilde{\mathbf{D}}^{-\frac{1}{2}} \tilde{\mathbf{A}} \tilde{\mathbf{D}}^{-\frac{1}{2}} \mathbf{H}^l W \right), \tag{5}$$

where $\mathbf{H}^{l+1}$ is output feature, σ is a nonlinear activation function. W is the linear transformation matrix parameter. $\tilde{\mathbf{A}} = \mathbf{A} + \mathbf{I}$ is a self connected adjacency matrix of $\mathbf{A}$, we construct adjacency matrix $\mathbf{A}$ by our KG. $\tilde{\mathbf{D}}_{ii} = \sum_j \tilde{\mathbf{A}}_{ij}$ is a degree matrix.

Then, in the category's receptive field, we compute the neighborhood representation of the category c as $\mathbf{c}_{\mathcal{N}(c)}$. Similarly, we calculate the category representation $\mathbf{c}$ by sum aggregator $f_{agg}^{sum} = \mathbf{c} + \mathbf{c}_{\mathcal{N}(c)}$, because we found that the sum aggregator performs better, which is specified in Sect. 4.5. The formula is as follows:

$$\mathbf{c}^a = \sigma\left(W \cdot \left(\mathbf{c} + \mathbf{c}_{\mathcal{N}(c)}\right) + b\right), \tag{6}$$

where $\mathbf{c}$ is the category embedding vector. Based on the category graph, we obtain the category-level aggregation representation $\mathbf{c}^a$.

Category signals fusion Layer. We obtain the user's context feature representation $\mathbf{u}^c$ from the ICAM module at Sect. 3.2. From the category-level aggregation representation layer, we obtain the category-level signal vector $\mathbf{c}^a$ of the user's interaction item. To extract feature and align the dimension with $\mathbf{u}^c$, we utilize the MLP layer to acquire the the category-level feature $\mathbf{u}^m$, $\mathbf{u}^m = f_{MLP}(\mathbf{c}^a)$, where f_{MLP} is a MLP neural network. Then, we fuse the two parts of the features through the connect operation to obtain fusion feature $\mathbf{u}$ in the training stage. The formula is as follows:

$$\mathbf{u} = \sigma\left(concat\left(\mathbf{u}^c, \mathbf{u}^m\right)\right), \tag{7}$$

where σ is a activation function.

3.4   Contrastive Learning Module

In general, knowledge graphs contain redundancy and noise, and low-quality KG affects the performance of recommender systems. Generally, items of the same category include similar user preference information. Therefore, inspired by recent advances in self-supervised learning and contrastive learning techniques [5, 28], we built a contrastive learning module to improve the model's performance. Specifically, we take the aggregated features of category nodes as a view and make them closer to the category-level signal during the training process. Through contrastive learning, we can alleviate the redundancy and noise issues and enhance the robustness of the model.

We use a contrastive learning framework to guide self-supervised learning of class-level features to maximize the mutual information [11] between the category information and the category-level signal. First, the category is one of the essential components of an item's attributes. For an item, we obtain the vector of its category by embedding function $f_{Emb}(\cdot; \cdot)$, $\mathbf{c} = f_{Emb}(c; \Phi)$, where $\mathbf{c}$ denotes the category initialization vector, and Φ is parameters of the network f_{Emb}.

Then, the neighbor nodes of a category contain its fine-grained common characteristic such as function, classification, etc. We fuse highlight attributes as the category aggregation signal by modeling feature correlation to inject the attribute information into the category representation. Technically, we obtain the category-level signal vector $\mathbf{c}_i^a$ through the category-level aggregation representation layer. We treat a category vector $\mathbf{c}_i$ and its category-level signal feature $\mathbf{c}_i^a$ as two different views. More formally, we minimize the associated feature prediction loss by:

$$\mathcal{L}_{CL} = \sum_{c \in \mathcal{C}} -\log \frac{\exp\left(f_{cos}\left(\mathbf{c}_i, \mathbf{c}_i^a\right)/\gamma\right)}{\sum_{c \in \mathcal{C}} \exp\left(f_{cos}\left(\mathbf{c}_i, \mathbf{c}_{i'}\right)/\gamma\right)}, \tag{8}$$

Algorithm 1. KRec-C2 Learning algorithm.

Require:
    Interaction matrix $\mathbf{Y}$; knowledge graph $\mathcal{G}$
Ensure:
    Model parameters Θ
1: Randomly initialize neural parameters Θ
2: Constructe adjacency matrix of entities $\mathbf{A}_e$ and adjacency matrix of relations $\mathbf{A}_r$
    from $\mathcal{G}$
3: **while** An epoch is not end **do**
4:    Sample minibatch of interactions from Interaction matrix $\mathbf{Y}$
5:    Compute the loss $\mathcal{L}_{BPR}(\Theta)\,(Eq.\,(10))$
6:    Compute the loss $\mathcal{L}_{CL}(\Theta)\,(Eq.\,(8))$
7:    $\mathcal{L}(\Theta) \leftarrow \mathcal{L}_{BPR}(\Theta) + \lambda_1 \mathcal{L}_{CL}(\Theta)$
8:    Update neural parameters
9: **end while**
10: **return** Θ

---

where $f_{cos}(\cdot,\cdot)$ is the cosine similarity function here, $i \neq i^{'}$, γ is the hyper-parameter to the temperature in softmax function. $\mathcal{C}$ is set of the category node and $\mathcal{C} = \{c_1, c_2, \cdots, c_n\}$.

3.5 Prediction Module

In prediction layer, we calculate the relevance probability by a user's contextual feature $\mathbf{u}_i^c$ and vectors $\mathbf{v}$ of candidate items as follow: $\hat{y}_{uv} = \sigma(\mathbf{u}_i^c \cdot \mathbf{v})$, where σ is a sigmoid activation function, $\mathbf{u}_i^c$ is obtained from Sect. 3.2, $\mathbf{v}$ notes the embedding representation of items.

3.6 Model Optimization

To optimize the recommendation model, we minimize the following objective function to learn the model parameter by combining the BPR loss [27] and the independence loss:

$$\mathcal{L} = \mathcal{L}_{BPR} + \lambda_1 \mathcal{L}_{CL} + \lambda_2 \|\Theta\|_2^2, \tag{9}$$

where Θ is the set of model parameters; λ_1 and λ_2 are two hyper-parameters to control the independence loss $\mathcal{L}_{CL}$ and L_2 regularization term, respectively.

In particular, we assumes that observed interactions indicate more user preferences and should be given higher predictive values than unobserved interactions:

$$\mathcal{L}_{BPR} = \sum_{(u,v^+,v^-)\in\mathcal{O}} -\ln\sigma\left(\hat{y}_{u,v^+}^{\mathcal{L}_{BPR}} - \hat{y}_{u,v^-}^{\mathcal{L}_{BPR}}\right), \tag{10}$$

where $\hat{y}_{uv}^{\mathcal{L}_{BPR}} = \sigma(\mathbf{u}\cdot\mathbf{v})$, $O = \{(u,v_+,v_-)\,|\,(u,v_+)\in\mathcal{R}^+, (u,v_-)\in\mathcal{R}^-\}$ denotes the pairwise training data, $\mathcal{R}^+$ denotes the positive samples, $\mathcal{R}^-$ denotes the negative samples, v_+ is user clicked item, v_- is no click item. $\hat{y}_{u,v}^{\mathcal{L}_{BPR}}$ indicates the calculated score. Furthermore, the parameters Θ in our model are jointly optimized. The training procedure of our model is illustrated in Algorithm 1.

Table 1. Statistics of the three evaluation datasets.

Dataset	User	Items	Interactions	Entities	Relation	Triplets
Last-FM	1,872	3,846	42,346	18,165	62	15,518
MIND	299,999	47,034	20,000	57,434	62	793,304
Douban-movie	1,883	57,018	1,563,754	101,408	6	319,918

4 Experiments

In this section, we firstly describe our interaction datasets and KG. Then, we introduce the baselines and experiment setup. Finally, we present the experiment results and discuss the influence of hyper-parameters.

4.1 Experimental Settings

Datasets Description. We utilize the three common datasets in our experiments. (1) **MIND**[1] was constructed from the Microsoft News. (2) **Last-FM**[2] was collected from the online music system Last.fm. (3) **Douban-Movie**[3] was collected from Douban Movies.

We preprocess these datasets and use the user and item ID embeddings as raw input. For the Last-FM dataset, we follow the data processing method released by KGIN [28] and obtain get classifications of artists using the tag data. We follow the knowledge graph construction method with Microsoft Satori[4] released by KGCN [25]. Then, we construct the category-item graph utilizing the item triples as (category, type.object.name, item). For the MIND dataset, we follow the data processing method published by KGCL [34] and use the catalogs as the category label. We follow the pre-processing data strategy in KGCL to construct the knowledge graph based on the spacy-entity-linker tool[5] and Wikidata[6]. We construct the category-relation-item triples using catalog information. For the Douban-movie dataset, we remove genres with fewer than 16 items to guarantee the data quality, and extract attributes to construct the KG utilizing genres, actors, directors, etc. The statistics of the data set are shown in Table 1.

Evaluation Metrics. To evaluate the performance, we adopt two universal metrics for Top-K recommendation: Recall@K, and Normalized Discounted Cumulative Gain (NDCG@K) [3,10].

Baselines Models. We compare our proposed approach with the following baseline methods: **FM** [18] is a bechmark factorization model, which models contextual information to provide context-aware rating predictions. **NFM** [9] combines

---

[1] https://msnews.github.io/.

[2] https://grouplens.org/datasets/hetrec-2011/.

[3] https://movie.douban.com/.

[4] https://searchengineland.com/library/bing/bing-satori.

[5] https://github.com/egerber/spaCy-entity-linker.

[6] https://query.wikidata.org/.

Table 2. Performance comparison of all methods

Dataset	Last-FM		MIND		Douban-movie	
Model	Recall	NDCG	Recall	NDCG	Recall	NDCG
FM	0.0591	0.0577	0.0651	0.0303	0.0291	0.0091
NFM	0.0743	0.0661	0.0749	0.0335	0.0307	0.0107
CKE	0.0732	0.0631	0.0820	0.0384	0.0316	0.0116
RippleNet	0.0785	0.0702	0.0768	0.0401	0.0439	0.0139
KGCN	0.0881	0.0744	0.0799	0.0406	0.0454	0.0154
KGAT	0.0943	0.0812	0.0849	0.0431	0.0481	0.0171
CKAN	0.0882	0.0796	0.0887	0.0468	0.0458	0.0158
KGIN	0.1298	0.0948	<u>0.1008</u>	<u>0.0493</u>	0.0517	0.0317
KGCL	<u>0.1473</u>	<u>0.1019</u>	0.0983	0.0481	<u>0.0558</u>	<u>0.0358</u>
KRec-C2	**0.1675**	**0.1130**	**0.1209**	**0.0626**	**0.0657**	**0.0478**

the linearity of FM and the non-linearity of neural network in modelling higher-order feature interactions. **CKE** [36] leverages the heterogeneous information in a knowledge base to improve the quality of recommender systems. **RippleNet** [22] stimulates the propagation of user preferences over the set of knowledge entities by automatically and iteratively extending a user's potential interests along links in the knowledge graph. **KGCN** [25] is an end-to-end framework utilizing high-order structure information and semantic information to capture inter-item relatedness effectively. **KGAT** [26] models the high-order connectivities in KG with an attention mechanism to discriminate the importance of the neighbors. **KGIN** [28] mainly performs relational path-aware aggregation of user intent-item and KG triples to enhanced learning. **CKAN** [30] introduces a heterogeneous propagation mechanism and combines collaborative filtering representation features with knowledge graph embeddings. **KGCL** [34] adopts a knowledge graph augmentation schema to suppress KG noise in information aggregation.

Parameter Settings. We set the size d of the ID initialization embedding to 16, and the batch size s of the method is 128 for all methods. The learning rate η is adjusted in $\{10^{-2}, 10^{-3}, 5^{-3}, 10^{-4}, 5^{-4}\}$, and the $L2$ regularization coefficient in the method is searched in $\{10^{-5}, 10^{-4}, \cdots, 10^{-1}\}$. In addition, we set the receptive neighborhood size of KGCN to 16 and adjust the number of layers L to be $\{1, 2, 3\}$. We discuss other parameters in Sect. 4.5.

4.2   Performance Comparison

We report the overall performance evaluation of all methods in Table 2, where we set $K=20$ following the settings of most methods [28,34], and the strongest baseline is underlined. From the table, we summarize the following observations: KRec-C2 consistently outperforms other baselines across three datasets. Specifically, our model achieves a significant experimental effect compared with the

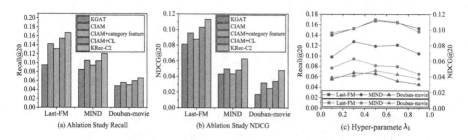

Fig. 3. (a) and (b) are results of ablation study. (c) is result of hyperparameter λ_1

strongest baselines in this paper, where the result of Recall@20 are 16.75%, 12.08%, and 6.57% in Last-FM, MIND, and Douban-movie, respectively. We obtain the best result at recall@20 on Last-FM data, which is about 2.02% higher than the KGCL. While, we obtain the minimum value at recall@20 on the Douban-movie data, which is still about 1.01% higher than the KGCL. The above results verify the effectiveness of our KRec-C2 model. Overall, the improvements obtained from KRec-C2 can be attributed to three modules: (1) CIAM. In contrast, all baselines ignore the context and only model user-item latent features such as the node-based aggregation method. (2) CIFM. Compared with baselines (such as KGAT, CKAN, and KGIN), it can learn the common category-level features from the KG. (3) CLM. Self-supervised signals relieve the data noise in KG and enhance the model's robustness. Further, we also observe that the difference between our model with other baselines is more stable through horizontal comparison.

The Table 2 shows that most KG-based recommender systems (such as KGAT, KGIN, and CKAN) achieve better performance than FM, NFM, and CKE. Detailed, KGIN obtains the best result at recall@20 on Last-FM data, which is about 5.66% higher than the CKE. Specifically, KGIN introduces the relational path embedding user intent, enhances user features by aggregation scheme. However, such methods ignore the contextual relationship and latent intent between items within a interaction sequence. On the other hand, the above table results show that the KGCL model is better than other models on Last-FM and Douban-movie. In the worst case, KGCL has a lower value of 0.0983 at Recall@20 on MIND, within spitting distance of the KGIN result. Excitingly, our model is 1.505% higher than KGCL on average from the above table. The excellent performance demonstrates that our model effectively combines context awareness and the category-level self-supervised signal to improve representation.

4.3 Ablation Study of KRec-C2 Model

In this section, we conduct ablation studies to verify the effectiveness of our three core components. The results are shown in Fig. 3 (a) and (b). First, compared with KGAT, the effect of CIAM has been a noticeable improvement. Specifi-

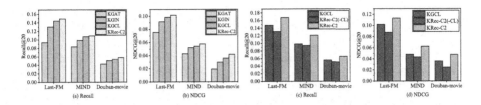

Fig. 4. Performance in data sparsity and noise.

cally, the maximum improvement value is 4.78% at Recall@20 on Last-FM, and the minimum improvement value is 0.66% at NDCG@20 on MIND. The above results prove that applying context awareness to user-item relational modeling is vital. Second, the average result of CIAM+category feature decreases by 0.81% compared with ICAM. Fusing the category feature enhances the model's generalization performance, leading to poor effects. Finally, after adding the contrastive learning of the category-level view, compared with ICAM+CL, the maximum value of improving is 1.64% at Recall@20 on MIND, and the minimum value of improving is 0.61% at Recall@20 on Douban-movie. The above results show the effectiveness of the CIFM and CLM.

4.4   The Effects of Alleviating Data Sparsity and Noisy

In this subsection, we evaluate the performance of KRec-C2 on noisy and sparse data. To be specific, noise data refers to the error and redundancy in KG. Sparse data means few interactions between users and items.

First, to verify the robustness of our KRec-C2 in handling a few interactions, we follow similar settings in KGCL to generate sparse user sets for MIND data, 20 interactions for Last-FM data, and 32 interactions for Douban-movie data. From the Fig. 4 (a) and (b), we observe that KRec-C2 is slightly higher than the KGCL. Specifically, the maximum improvement efficiency was improved up to 0.6% at NDCG@20 on Douban-movie, and the minimum was improved up to 0.18% at Recall@20 on MIND. Moreover, the performance of KRec-C2 is significantly superior to KGAT and KGIN. Then, we compare the results of KRec-C2 with KGCL and KRec-C2(-CL) in the Fig. 4 (c) and (d). Specifically, the effects of Recall@20 rose by 2.02 Technically, we designed a CLM component to verify our hypothesis that CLM is an essential component in alleviating the noise issue and highlighting the category-level signal. Figure 4 (c) and (d) show that the result of KRec-C2(-CL) is lower than KGCL and KRec-C2.

4.5   Hyper-Parameter Sensitivity

To evaluate the effect of the hyperparameter λ_1, aggregator, and interaction sequence size, we conduct experiments on three datasets by varying their values.

(1) The hyperparameter λ_1. As shown in Fig. 3 (c), the performance of KRec-C2 first increases and then decreases with the increase of λ_1. The three

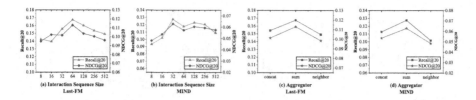

Fig. 5. Impact of interaction sequence size and aggregator on Last-FM and MIND, separately.

datasets of Last-FM, MIND, and Douban-movie achieve the best results when λ=0.5, 0.3, and 0.5, respectively. The results show that the model can enhance knowledge representation and improve robustness by introducing self-supervised learning features. **(2) Interaction sequence size.** As shown in Fig. 5 (a) and (b), Last-FM and MIND usually get the best performance when L=64 and 32 respectively. While aggregating more features may cause poor performance. The reason for the above result may be related to the training of the discriminator. More hidden layers need more parameters to be trained so that the discriminator hardly reaches a steady state. **(3) Aggregator.** We choose different aggregators from *sum, concat* and *neighbor* to study the impact of perception in KRec-C2. The results are shown in Fig. 5 (c) and (d), which indicates that KRec-C2 is more sensitive to the sum aggregator. The outperforms of the sum aggregator surpass the other two aggregators.

5   Related Work

Knowledge Graph-Based Recommendation. Existing KG-enhanced works for recommendation fall into three categories: embedding-based, path-based, and joint models. i) The embedding-based models generally obtain vector representations of products, users, and their relationships by the knowledge graph and apply these representations to similarity calculation [23,24,36]. ii) The Path-based models commonly enhance recommendation effects through user-product connection similarity, where path similarity is used to measure the similarity of connection entities in KG [12,29,35]. iii) The joint models usually capture the correlation between entities by mining attributes and discovering the high-level structural and potential information in KG [6,22,25,26,30]. Although many effective models have been proposed, the primary problem is that these methods (such as KGAT [26], KGIN [28], and CG-KGR [6]) only focus on aggregating potential features in the KG, and ignore the context information in the interaction sequence. This may lead to insufficient profiling for both users and items, and then the capability of recommendation system may be suppressed. To address these issues, we proposed ICAM in our model.

Contrastive Learning for Recommendation. Similar to works in NLP [8] and CV [5], contrastive learning aims to learn self-supervised representations by comparing positive and negative samples from different views in RS [1, 7, 37]. Under the scenario of KG-based RS, most contrastive learning methods generate two views by uniform data augmentation schemes [31, 32, 34]. Other works focus on exploring contrastive learning with a novel multi-level interactive view [38]. For instance, MCCLK [38] performs contrastive learning across three views on both local and global levels, mining comprehensive graph feature and structure information in a self-supervised manner. However, these works ignore the learning of the category-level signals, which are essential for acquiring salient features. To alleviate this problem, we proposed CML by the category-level view.

6 Conclusion

In this paper, we explore context-awareness, category-level intent fusion, self-supervised learning, and implement joint modeling for KG-based RS. We propose a novel framework KRec-C2, which enhances the performance with an end-to-end fashion under the paradigm of KG-based RS. Specifically, our method consists of three modules, namely ICAM, CIFM, and CLM. ICAM captures the context feature from an interaction sequence to enhance the user's representation. CIFM fuses the category-level signal in the training stage to obtain the category-level intents. CLM captures the relevance of the category-level view for alleviating noise issues. Our experiments on three real-world datasets demonstrate the rationality and effectiveness. This work explores the potential of context awareness and contrastive learning in KG-based recommendations. In the future, we will optimize the performance of recommendation system from the model structure.

Acknowledgements. This work was supported in part by the NSFC (Natural Science Foundation of China) grants (62172423 and 91846204).

References

1. Bian, S., Zhao, W.X., Zhou, K., et al.: Contrastive curriculum learning for sequential user behavior modeling via data augmentation. In: The 30th ACM International Conference on Information and Knowledge Management (CIKM) (2021)
2. Cao, Y., Wang, X., He, X., et al.: Unifying knowledge graph learning and recommendation: Towards a better understanding of user preferences. In: The World Wide Web Conference (WWW), pp. 151–161 (2019)
3. Chen, C., Zhang, M., Ma, W., et al.: Jointly non-sampling learning for knowledge graph enhanced recommendation. In: Proceedings of the 43rd International ACM SIGIR Conference on Research and Development in Information Retrieval (SIGIR), pp. 189–198 (2020)
4. Chen, J., Zhang, H., He, X., et al.: Attentive collaborative filtering: multimedia recommendation with item-and component-level attention. In: Proceedings of the 40th International ACM SIGIR Conference on Research and Development in Information Retrieval (SIGIR) (2017)
5. Chen, T., Kornblith, S., Norouzi, M., et al.: A simple framework for contrastive learning of visual representations. In: International Conference on Machine Learning (ICML), pp. 1597–1607 (2020)

6. Chen, Y., Yang, Y., Wang, Y., et al.: Attentive knowledge-aware graph convolutional networks with collaborative guidance for personalized recommendation. In: 2022 IEEE 38th International Conference on Data Engineering (ICDE), pp. 299–311. IEEE (2022)
7. Chen, Y., Liu, Z., Li, J., et al.: Intent contrastive learning for sequential recommendation. In: Proceedings of the ACM Web Conference (WWW), pp. 2172–2182 (2022)
8. Fu, H., Zhou, S., Yang, Q., et al.: LRC-BERT: latent-representation contrastive knowledge distillation for natural language understanding. In: Proceedings of the AAAI Conference on Artificial Intelligence (AAAI) (2021)
9. He, X., Chua, T.S.: Neural factorization machines for sparse predictive analytics. In: Proceedings of the 40th International ACM SIGIR Conference on Research and Development in Information Retrieval (SIGIR), pp. 355–364 (2017)
10. He, X., Deng, K., Wang, X., et al.: LightGCN: simplifying and powering graph convolution network for recommendation. In: Proceedings of the 43rd International ACM SIGIR Conference on Research and Development in Information Retrieval (SIGIR), pp. 639–648 (2020)
11. Hjelm, R.D., Fedorov, A., Lavoie-Marchildon, S., et al.: Learning deep representations by mutual information estimation and maximization. arXiv preprint arXiv:1808.06670 (2018)
12. Hu, B., Shi, C., Zhao, W.X., et al.: Leveraging meta-path based context for top-n recommendation with a neural co-attention model. In: Proceedings of the 24th ACM SIGKDD International Conference on Knowledge Discovery & Data Mining (SIGKDD) (2018)
13. Koren, Y., Bell, R., Volinsky, C.: Matrix factorization techniques for recommender systems. Computer **42**(8), 30–37 (2009)
14. Lee, D., Oh, B., Seo, S., et al.: News recommendation with topic-enriched knowledge graphs. In: Proceedings of the 29th ACM International Conference on Information & Knowledge Management (CIKM), pp. 695–704 (2020)
15. Liang, D., Krishnan, R.G., Hoffman, M.D., et al.: Variational autoencoders for collaborative filtering. In: Proceedings of the 2018 World Wide Web Conference (WWW), pp. 689–698 (2018)
16. Lin, T.H., Gao, C., Li, Y.: CROSS: cross-platform recommendation for social E-commerce. In: The 42nd International ACM SIGIR Conference (SIGIR), pp. 515–524 (2019)
17. Pujara, J., Augustine, E., Getoor, L.: Sparsity and noise: where knowledge graph embeddings fall short. In: Proceedings of the 2017 Conference on Empirical Methods in Natural Language Processing (EMNLP), pp. 1751–1756 (2017)
18. Rendle, S., Gantner, Z., Freudenthaler, C., et al.: Fast context-aware recommendations with factorization machines. In: Proceedings of the 34th international ACM SIGIR Conference on Research and Development in Information Retrieval (SIGIR), pp. 635–644 (2011)
19. Schafer, J.B., Konstan, J.A., Riedl, J.: E-commerce recommendation applications. Data Min. Knowl. Disc. **5**, 115–153 (2001)
20. Smirnova, E., Vasile, F.: Contextual sequence modeling for recommendation with recurrent neural networks. In: Proceedings of the 2nd Workshop on Deep Learning for Recommender Systems (RecSys), pp. 2–9 (2017)
21. Wang, G., Zhang, W., Wang, R., et al.: Label-free distant supervision for relation extraction via knowledge graph embedding. In: Proceedings of the 2018 Conference on Empirical Methods in Natural Language Processing (EMNLP), pp. 2246–2255 (2018)

22. Wang, H., Zhang, F., Wang, J., et al.: Ripplenet: propagating user preferences on the knowledge graph for recommender systems. In: Proceedings of the 27th ACM International Conference on Information and Knowledge Management (CIKM), pp. 417–426 (2018)
23. Wang, H., Zhang, F., Xie, X., et al.: DKN: deep knowledge-aware network for news recommendation. In: Proceedings of the 2018 World Wide Web Conference (WWW), pp. 1835–1844 (2018)
24. Wang, H., Zhang, F., Zhao, M., et al.: Multi-task feature learning for knowledge graph enhanced recommendation. In: The World Wide Web Conference (WWW), pp. 2000–2010 (2019)
25. Wang, H., Zhao, M., Xie, X., et al.: Knowledge graph convolutional networks for recommender systems. In: The World Wide Web Conference (WWW), pp. 3307–3313 (2019)
26. Wang, X., He, X., Cao, Y., et al.: KGAT: knowledge graph attention network for recommendation. In: Proceedings of the 25th ACM SIGKDD International Conference on Knowledge Discovery & Data Mining (SIGKDD), pp. 950–958 (2019)
27. Wang, X., He, X., Wang, M., et al.: Neural graph collaborative filtering. In: Proceedings of the 42nd International ACM SIGIR Conference on Research and Development in Information Retrieval (SIGIR), pp. 165–174 (2019)
28. Wang, X., Huang, T., Wang, D., et al.: Learning intents behind interactions with knowledge graph for recommendation. In: Proceedings of the Web Conference (WWW), pp. 878–887 (2021)
29. Wang, X., Wang, D., Xu, C., et al.: Explainable reasoning over knowledge graphs for recommendation. In: Proceedings of the AAAI Conference on Artificial Intelligence (AAAI). vol. 33, pp. 5329–5336 (2019)
30. Wang, Z., Lin, G., Tan, H., et al.: CKAN: collaborative knowledge-aware attentive network for recommender systems. In: Proceedings of the 43rd International ACM SIGIR Conference on Research and Development in Information Retrieval (SIGIR), pp. 219–228 (2020)
31. Wei, W., Huang, C., Xia, L., et al.: Contrastive meta learning with behavior multiplicity for recommendation. In: Proceedings of the Fifteenth ACM International Conference on Web Search and Data Mining (WSDM), pp. 1120–1128 (2022)
32. Wu, J., Wang, X., Feng, F., et al.: Self-supervised graph learning for recommendation. In: Proceedings of the 44th International ACM SIGIR Conference on Research and Development in Information Retrieval (SIGIR), pp. 726–735 (2021)
33. Xu, Q., Shen, F., Liu, L., et al.: GraphCAR: content-aware multimedia recommendation with graph autoencoder. In: The 41st International ACM SIGIR Conference on Research & Development in Information Retrieval (SIGIR), pp. 981–984 (2018)
34. Yang, Y., Huang, C., Xia, L., et al.: Knowledge graph contrastive learning for recommendation. In: Proceedings of the 45th International ACM SIGIR Conference on Research and Development in Information Retrieval (SIGIR), pp. 1434–1443 (2022)
35. Yu, X., Ren, X., Sun, Y., et al.: Personalized entity recommendation: a heterogeneous information network approach. In: Proceedings of the 7th ACM International Conference on Web Search and Data Mining (WSDM), pp. 283–292 (2014)
36. Zhang, F., Yuan, N.J., Lian, D., et al.: Collaborative knowledge base embedding for recommender systems. In: Proceedings of the 22nd ACM SIGKDD International Conference on Knowledge Discovery and Data Mining (SIGKDD), pp. 353–362 (2016)

37. Zhou, K., Wang, H., Zhao, W.X., et al.: S3-Rec: self-supervised learning for sequential recommendation with mutual information maximization. In: Proceedings of the 29th ACM International Conference on Information & Knowledge Management (CIKM), pp. 1893–1902 (2020)
38. Zou, D., Wei, W., Mao, X.L., et al.: Multi-level cross-view contrastive learning for knowledge-aware recommender system. In: Proceedings of the 45th International ACM SIGIR Conference on Research and Development in Information Retrieval (SIGIR), pp. 1358–1368 (2022)

HIT: Learning a Hierarchical Tree-Based Model with Variable-Length Layers for Recommendation Systems

Anran Xu[1], Shuo Yang[2], Shuai Li[2], Zhenzhe Zheng[1(✉)], LingLing Yao[2], Fan Wu[1], Guihai Chen[1], and Jie Jiang[2]

[1] Shanghai Jiao Tong University, Minhang, China
{xuanran,zhengzhenzhe,fwu}@cs.sjtu.edu.cn, gchen@cs.sjtu.edu.cn
[2] Tencent, Shenzhen, China
{rafaelyang,shawnshli,zeus,vincentyao}@tencent.com

Abstract. Large-scale industrial recommendation systems (RS) usually confront computational problems due to the enormous corpus size. Hence, an efficient indexing structure is a practical solution to retrieve and recommend the most relevant items within a limited response time. The existing approaches that adopted embedding or tree-based index structures cannot handle the long-tail phenomenon. To address this issue, we propose a **HI**erarchical **T**ree-based model with variable-length layers (HIT) for recommendation systems. HIT consists of a hierarchical tree index structure and a user preference prediction model. It can fully exploit all the training data by dynamically adjusting the lengths of layers in its tree index structure, which can effectively alleviate the long-tail problem. To assess the models' resistance against the long-tail problem, we further define two types of equilibrium under our index structure. To satisfy the equilibrium, we propose a corresponding hierarchical tree learning algorithm. Furthermore, for those items with a rare appearance in the training data, on which the learning algorithm would fail, we design a dedicated bandit layer to solve them. Extensive experiments on three large-scale real-world datasets show that HIT can significantly outperform the existing methods in terms of efficient recommendations on items with different frequencies.

Keywords: Recommendation Systems · Long-tail Phenomenon · Hierarchical Index Structure

1 Introduction

With the explosive growth of information on the internet, recommendation systems (RS) are used to help users efficiently discover the content that matches their interests. Industrial RS usually deals with hundreds of millions of heterogeneous items and delivers recommendations within milliseconds [21]. To serve users' highly personalized content within a limited latency, industrial RS is required to balance effectiveness and efficiency. It follows a cascade item retrieval procedure: matching/retrieval module (*i.e.*, candidate generation) and then fine-grained ranking module [5]. The retrieval module aims to efficiently retrieve a small subset of relevant items (usually hundreds of) from the entire corpus, while the ranking module aims to precisely evaluate the retrieved items by the sophisticated learning-based algorithm [22].

The retrieval module is critical to the overall performance of recommendation systems, as it determines the performance upper bound of the subsequent modules. The vector embedding methods have been widely adopted for the retrieval module, such as DSSM [10], FM [18], DeepFM [7], etc. However, as the number of items increases, the cost of brute force computation of the inner products for all items becomes less practical. Thus, efficient algorithms like approximate nearest neighbors (ANN), such as PQ [12], HNSW [16], and maximum inner product search, such as LSH [20], are proposed to retrieve top relevant items in such scenarios. Recently, tree-based models [27–29] and KD matrix-based model [6] are designed to further improve the performance by leveraging the power of specific index structures. However, these fixed index structures still seriously suffer from the long-tail phenomenon. The improvement of the recommendation performance often comes from recommending too many head items.

The long-tail phenomenon is prevalent in recommendation systems [17,24]: a small fraction of head items account for most user-item interactions. When trained on such a biased data set, various learning-based prediction models for performance indicators in RS, such as deep interest network [26] for click-through-rate (CTR) and entire space multi-task model [15] for conversion-rate (CVR), will suffer from inferior performance. They usually give higher scores to head items than their ideal values, while predicting large variances for the long-tail items. As a result, head items are recommended even more frequently than their original popularity exhibited in the dataset, leading to the popularity bias [2]. Most current works focusing on long-tail problems in RS apply constraint and regularization to its prediction model to increase the exposure of the long-tail items [2,11,19]. Previous work tends to focus on some advanced methods and complex models, which are often applied in the ranking module. However, increasing the learning ability while keeping the efficiency in the retrieval module is also important for RS. Currently, the retrieved candidates still suffer severe bias in RS's initial (retrieval) module under the fixed index structure, which will seriously affect the follow-up recommendation performance. Yet, few efforts have explored the effective index structure in the retrieval module.

To address the challenges above, we propose a novel **HI**erarchical **T**ree-based model with variable-length layers for recommendation systems (HIT). HIT aims to alleviate the popularity and loop bias in the retrieval module from the perspective of index learning. To the best of our knowledge, HIT is the first framework that can jointly learn index structure and business goals to fully exploit all the training data and relief the long-tail problem in the retrieval phase.

The main contributions of this paper are:

- We propose HIT, which consists of a variable-length hierarchical tree index and user preference prediction model. HIT can relieve the long-tail problem in the retrieval phase and improve the accuracy of user preference prediction in different parts of the items.
- We propose a multi-layered training scheme based on HIT. Items with different layers will have different scopes that affect the parameters of each layer. Furthermore, we define two types of equilibrium tree structures and prove their ability to solve the

long-tail problem. Then a tree learning algorithm is proposed for learning such an equilibrium tree.

- For the items with a rare appearance in the training data, on which the user preference prediction model would fail, we expand the linUCB framework and propose a bandit layer to achieve the exploration of these items.
- We conduct extensive experiments on three large-scale real-world datasets, which show that HIT outperforms existing methods in terms of different parts of items.

2 Related Work

The existing approaches adopted embedding-based or tree-based index structures with approximate nearest neighbors (ANN) or beam search retrieval styles. Among numerous large-scale recommendation algorithms, the closest works to ours are the layer-structure methods, including tree-based deep model [27–29], and matrix-based model DR [6]. They all learn an objective function for the structure and the model parameters jointly. Tree-based methods map each item to one leaf node in a tree-based structure model. DR extends this structure. Each item corresponds to multiple paths, and each path can contain various items. These architectures are all fixed index structures. They encode items with the same number of layers, as shown in Fig. 1(a). The traditional fixed-length layers model was affected by the severe long-tail phenomenon. Head items with more training samples and play a leading role in both structures' model training process and prediction. Training data collected from users' feedback is very sparse for most items, and thus causes model predictions to have large variances for long-tail content. In a word, they both fail to relieve the long-tail problem on the retrieval module.

Various techniques have been proposed to resolve the issue of popularity bias. Some [19] proposed the graph-based model or hybrid algorithm combining multiple models, to optimize accuracy and diversity at the same time. Hurley *et al.* [11] formulated the trade-off between diversity and retrieval quality as a binary optimization problem and used a parameter to control it. Kamishima *et al.* [13] applied the mean-match regularizer in their information-neutral RS to correct popularity bias. Chen *et al.* [3] devised a process-oriented regularization term, proposing the Entire Space Adaptation Model. There are some other methods to reduce popularity bias, such as re-ranking [1] and decreasing the influence of head items to model training [23]. Some other methods using counterfactual inference and causal graph can also be found in [9,25]. Most of the above methods are for a specific model or ranking module. In this paper, we target to alleviate the long-tail problem by learning an effective index structure (HIT) in the retrieval module, which has not been addressed by the above methods.

3 Hierarchical Tree-Based Model

In the hierarchical tree-based model, we use the variable-length hierarchical tree index instead of the fixed-length layers structure. The item assigned structure in HIT is shown in Fig 1 (b). Let $\mathcal{V} = \{1, \ldots, V\}$ be the set of all items. Then the mapping of an item v to node is $\pi : v \rightarrow [K]^l, v \in \mathcal{V}, l \in \{1, ..., L\}$. Different from the traditional fixed-length layer, we can assign items on the node at any level of the tree. Given a pair of training sample (x, y), where x is a user and y is the corresponding item. If item y is assigned

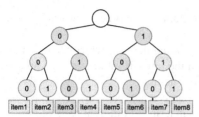 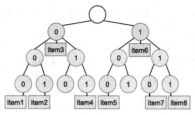

(a) Fixed-length layers: map all items to the same layer(leaf nodes)

(b) Variable-length layers: map items to different layers

Fig. 1. Different types of index mapping under a tree structure with ary $K = 2$ and layer $L = 3$. Red means popular and blue means unpopular. (Color figure online)

to node n_l, $\pi(y) = n_l$. Then the node path, *i.e.*, the nodes associated with the item y are $n_1, \ldots, n_l$. For example, in Fig. 1(b), item 1 is assigned to [0, 0 ,0]. Then the nodes associated with item 1 are [0] in layer 1, [0, 0] in layer 2, and [0, 0, 0] in layer 3. item 3 is assigned to [0], which means that the node associated with item 3 is only [0] in layer 1. Under a specific π, we can train a user preference prediction model with parameters θ in which we express the relationship between nodes as a probability multiplication. It is worth noting that in the case of variable-length layers, the number of nodes to be multiplied is related to the layer of the assigned node. As shown in Fig. 2, the first layer takes the user embedding emb(x) as input, and outputs a probability $p(n_1 \mid x, \theta_1)$ over the K nodes of the first layer. For the intermediate layer l, we concatenate the user embedding emb(x) and the embeddings of all the predecessor layers emb(n_{l-1}) as the input of network in this layer, which outputs probability $p(n_l|x, n_1, ..., n_{l-1}, \theta_l)$ over the K nodes of layer l. The probability of the assigned node n_l in layer l given user x is the product of the probabilities of all the predecessor layers' outputs:

$$p(n_l \mid x, \theta) = \prod_{i=1}^{l} p\left(n_i \mid x, n_1, \ldots, n_{i-1}, \theta_i\right). \tag{1}$$

We use the probabilistic multiplication model for the following reasons: concatenating embeddings of all the predecessor nodes can pass down their information to further increase the learning ability of the successor layer. Through such a hierarchical probability model, successor items can learn the characteristics of predecessor items to enhance their learning ability and improve results. For example, in Fig. 1 we maintain the information of the predecessor node (item 3) to help generate a candidate set (items 1, 2, 4). Suppose the user prefers item 1, and we can further learn item 1 in the candidate set during the successor layers' network training, making item 1 wins in the layer 3. Then for M training interactions (x_m, y_m), we can write the objective as:

$$Q(\theta, \pi) = \sum_{i=1}^{M} \log p\left(\pi\left(y_m\right) \mid x_m, \theta\right)$$

$$= \sum_{v=1}^{V} \sum_{m:y_m=v} \sum_{i=1}^{l} \log p\left(n_{m,i} \mid x_m, n_{m,1}, ..., n_{m,i-1}, \theta_i\right), \tag{2}$$

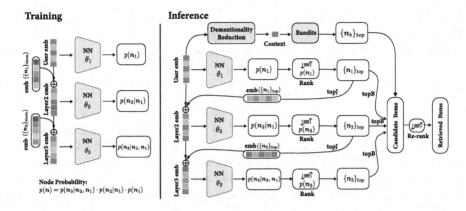

Fig. 2. Training and inference process of the user preference prediction model in HIT.

where v is assigned to n_l and l is the layer. Under such an objective, only the interactions that include the layer l's node can participate in the training process of θ_l.

Next, we show the reasons behind this newly proposed variable-length tree structure to enhance the learning ability and prediction performance for the long-tail items. First, for the traditional fixed-length layers model, as shown in Fig. 1 (a), all items are in the same layer. If one head item (such as item 3) has much larger training samples than long-tail items, it will impact all layers' parameters θ and guide the network to select its corresponding node ([0] in layer 1, [0, 1] in layer 2, and [0, 1, 0] in layer 3). However, in the proposed variable-length layers structure, supposing v is a head item, we find that the item v only acts in $\theta_1, \ldots, \theta_l$, *i.e.*, the $\theta_i (i > l)$ will be not influenced by v by the objective in Eq. (2). For example, item 3 can only impact the parameter of layer 1, without changing the parameters of layer 2 and 3. These layers' parameters that are not affected by head items provide the possibility for further learning on long-tail data.

It is worth mentioning that, even if we use the single node assignment to explain the main ideas of HIT for simplicity, HIT can be easily extended to the multi-node scenario, in which an item can be assigned to multiple (J) nodes in the same layer.

4   Joint Offline Learning Framework

Our model follows a joint learning framework with the optimization goal in Eq. (2), in which the parameters θ and the mapping π are optimized alternatively. We randomly generate a hierarchical tree with fixed-length layers, assigning all items to the last layer L. Then we can get the initial scores of each layer's nodes. Following the objective in Eq. (2), the parameters θ can be optimized by any gradient-based optimizer. Then, the joint learning framework iteratively repeats the following steps:

Model Training: For a fixed mapping $\pi^{(t-1)}$, optimize parameters θ using a gradient-based optimizer to maximize the objective $Q\left(\theta, \pi^{(t-1)}\right)$, enhancing the learning performance on the long-tail items.

Tree Learning: For fixed parameters θ, update mapping $\pi^{(t)}$ to maximize the same objective. Based on the inequality $\sum_{m=1}^{M_v} \log p_m \leq M_v(\log \sum_{m=1}^{M_v} p_m - \log M_v)$ as in [6], we reorganize the M_v training data from the outside to inside according to the layer, node, and item in the tree structure (*e.g.* layer 1, node [0] and item 3 in Fig. 3 (c)). In such a way, we can write the original objective as:

$$Q(\theta, \pi) = \sum_{l=1}^{L} \sum_{n \in \mathcal{N}_l} \sum_{v \in \mathcal{V}_n} s(v, n), \tag{3}$$

where $s(v, n) = M_v (\log \sum_{m:y_m=v} p(n \mid x_m, \theta) - \log M_v)$, M_v is the amount of inter-actions of the item v, $\mathcal{N}_l$ is the set of nodes in layer l, and $\mathcal{V}_n$ is the set of items assigned to node n. The term $s(v, n)$ can be viewed as the impact of assigning item v to the node n on tree learning. However, directly optimizing this objective is infeasible for the following reasons. First, the upper layer will have the highest probability for each item, as

$$p(n_1 \mid x, \theta_1) \geq \prod_{i=1}^{l} p(n_i \mid x, n_1, .., n_{i-1}.\theta_i). \tag{4}$$

As a result, all items tend to be assigned to the first layer. Also, items are apt to be assigned for the same node to increase the probability. Suppose an extreme case in which all items are allocated to a single node at the first layer. Then the probability of retrieving this single node for any user is 1, which catastrophically compromises the retrieval's efficiency. Therefore, directly optimizing the original objective in 3 is less promising. Hence, before we design an algorithm for tree learning (Sect. 4.2), we need to first introduce constraints (*i.e.*, two kinds of equilibrium) to prevent the model from collapsing and deal with the long-tail phenomenon (Sect. 4.1).

4.1   Equilibrium in HIT

As Eq. (3) shows, a small number of nodes $n \in \mathcal{N}_l$ can occupy most of the samples in the same layer l, which we call the *external long-tail phenomenon*. Similarly, a small number of items $v \in \mathcal{V}_n$ can occupy most of the samples in the same node n, which we call the *internal long-tail phenomenon*. To relieve these two phenomena, we define the equilibrium tree with the following two properties:

– **External equilibrium**: To make that each node in the same layer plays a similar role in the model, we maximize the objective: min $\sum_{v \in \mathcal{V}_n} s(v, n)$ in each layer, making $\sum_{v \in \mathcal{V}_n} s(v, n)$ is similar for all the nodes $n \in \mathcal{N}_l$ in the same layer l.
– **Internal equilibrium**: To make that each item v on the same node plays a similar role in the model, we maximize the objective: min $s(v, n)$ for each node, making $s(v, n)$ similar for all the items $v \in \mathcal{V}_n$ on the same node n.

As Fig. 3(a) shows, in the case that there is only one item in each node, the model will be guaranteed by the internal equilibrium, but no external equilibrium. The popular nodes (the nodes associated with item 3 and item 6, *i.e.* purple nodes in Fig. 3(a)) have a more significant impact on the network parameters during the training process, while the other nodes will get insufficient training under such equilibrium. The DR model will fall into this situation initially, where long-tail items are hardly retrieved. On the contrary, as Fig. 3(b) shows, this tree structure guarantees the model an external equilibrium, as there are only two paths remaining in the tree. All nodes associated with items (*i.e.* purple

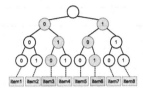
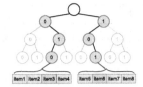
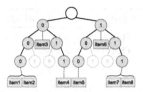

(a) Structure with internal equilibrium but without external equilibrium.

(b) Structure with external equilibrium but without internal equilibrium.

(c) Structure with internal and external equilibrium at the same time.

Fig. 3. Comparison of equilibrium in different situations.

nodes in the Fig. 3(b)) can get adequate training. However, the number of samples for each item on the same node would be quite different, resulting in a long-tail phenomenon of items within that node. For example, for item 1–4 in path 1, the training process and the node choice are almost determined by item 3. In such case, long-tail items (item 1, 2, 4) are retrieved directly without careful evaluation. When items assemble after multiple training rounds, the DR model will fall into this situation, where the long-tail items are retrieved along with the popular items without sufficient learning.

Neither of the above two cases can ensure the performance for retrievaling results for the long-tail items. Thus, instead of the original joint learning objective, the goal of our tree learning algorithm is to optimize an equilibrious hierarchical tree with variable-length layers. By advancing item 3 and item 6 to the predecessor nodes in Fig. 3 (c), the score of long-tail nodes, *i.e.* node [0, 0, 0], [0, 1, 1], [1, 0, 0] and [1, 1, 1], will raise to a similar level, keeping external equilibrium. Furthermore, the scores of long-tail items are more balanced in a node *e.g.*, item 1 and item 2 in node [0,0,0], keeping internal equilibrium.

4.2 Hierarchical Tree Learning Algorithm

Now considering these equilibriums raised above, we propose a hierarchical tree learning Algorithm. First, combining the equilibrium and the original objective, the process of tree learning can be formulated as such optimization problem:

$$\max_{\pi} \begin{bmatrix} Q(\theta, \pi) & (1) \\ \min \sum_{v \in \mathcal{V}_n} s(v, n) \text{ for each } l & (2) \\ \min s(v, n) \text{ for each } n & (3) \end{bmatrix}, \tag{5}$$

where (1) is the original objective, (2) is to satisfy external equilibrium, and (3) is to satisfy internal equilibrium in Sect. 4.1. It is difficult to directly optimize this multi-objective combinational optimization problem. Instead, we adopt a heuristic method.

We consider solving this problem from the outside in, addressing it separately at different levels. We can assign items layer by layer to satisfy the internal and external equilibrium simultaneously. We first fix items in each layer and then assign items to different nodes on each layer. We set the constraints as ϵ, which is a set of constants to control the size of items in each layer by arranging the set of items $\mathcal{V}_l$ to the layer l.

Considering the optimization problem of the original objective, we can show that the head item with a large number of training samples, and thus having a higher score,

---

Algorithm 1: Node assignment

Input: Score functions $s[v, n]$, ϵ, α
Output: Node assignments π

1 **for** *items* $v \in \mathcal{V}$ **do**
2 $\quad \lfloor$ assign v to $\mathcal{V}_l$ by ϵ

3 **for** $l = 1$ *to* L **do**
4 $\quad$ **for** $t = 1$ *to* T **do**
5 $\quad\quad$ **for** *items* $v \in \mathcal{V}_l$ **do**
6 $\quad\quad\quad$ **if** $t > 1$ **then**
7 $\quad\quad\quad\quad \lfloor \ |\pi^{(t-1)}(v)| \leftarrow |\pi^{(t-1)}(v)| - 1$
8 $\quad\quad\quad$ **for** *candidate node* n *of* v **do**
9 $\quad\quad\quad\quad$ $\Delta s[v, n] =$
$\quad\quad\quad\quad\quad \lfloor \ \log(sum + s[v, n]) - \log(sum) - \alpha_l \left(f(|\mathcal{V}_n| + 1) - f(|\mathcal{V}_n|) \right)$
10 $\quad\quad\quad$ $\pi^{(t)}(v) \leftarrow \arg\max_n \Delta s[v, n]$, $\mathrm{sum} \leftarrow \mathrm{sum} + s\left[v, \pi_{(t)}(v)\right]$
11 $\quad\quad\quad \lfloor \ |\pi^{(t)}(v)| \leftarrow |\pi^{(t)}(v)| + 1$

12 Return Node assignments π

---

should be placed at the predecessor layer, giving larger values for the original objective. Thus, the metric of the constants to control the items' assignment in each layer, ϵ, can be set as the popularity of training samples for each item v, *i.e.* M_v. Furthermore, to satisfy the internal equilibrium, we need to keep M_v of each item in the same layer similar. If there is no long tail phenomenon in each layer, then for each node of the current layer, its internal items must satisfy the internal equilibrium. Using the two principles above, for example, we set $\epsilon = \{0.01, 0.2, 1\}$, then items with top 1% interactions are in the layer 1, items with top 20% interactions are in the layer 2, and the remaining items are in the layer 3.

Next, we satisfy the external equilibrium at each layer and control the number of items contained in each node by penalizing the number of items that each node contains on the original objective. For each layer l,

$$Q(\theta, \pi) = \sum_{v \in \mathcal{V}_l} M_v \left(\log \sum_{m : y_m = v} p(n \mid x_m, \theta) - \log M_v \right) - \alpha_l(\pi), \qquad (6)$$

where α_l is the penalty coefficient in layer l. As $M_v \log M_v$ does not affect the objective in Eq. (6), and after allocating layer according to M_v, each M_v is similar as we have already satisfied the internal equilibrium by ϵ. Thus, for each layer l,

$$Q(\theta, \pi) = \sum_{v \in \mathcal{V}_l} \sum_{m : y_m = v} p(n \mid x_m, \theta) - \alpha_l(\pi). \qquad (7)$$

Then in each layer l, we can follow the coordinate descent algorithm in [6] to solve this problem. At step T, the incremental gain of the objective function by choosing $n = \pi_T(v)$, $n \in \mathcal{N}_l$ is given as follows:

$$\left(\sum_{t=1}^{T-1} s\left[v, \pi_t(v)\right] + s[v, n] \right) - \left(\sum_{t=1}^{T-1} s\left[v, \pi_t(v)\right] \right) - \alpha_l(f(|\mathcal{V}_n| + 1) - f(|\mathcal{V}_n|)), \qquad (8)$$

where we reuse $s[v, n]$ to denote $\log \sum_{m:y_m=v} p(n \mid x_m, \theta)$ for simplicity. $f(x) = x^4/4$. If the current node's total score is higher, it will have more punishment by $f(x)$. Thus, it will tend to assign a new item to a node less score to keep the external equilibrium. A more detailed description is shown in Algorithm 1. What's more, considering an extreme situation, if the interaction amount of the items is all the same, there is no long-tail phenomenon. Then under the node capacity constraint, all items will not be fixed in the predecessor layer but will move to the same successor layer. Then the hierarchical tree degenerates into a fixed-length layers tree, just like DR.

5   Online Learning and Serving

In practice, if the long-tail problem is rather severe, the networks' training will have difficulty in convergence at the successor layers, as it is hard to optimize the neural network on the successor nodes that are associated with insufficient training samples. Under such high uncertainty of the network, instead of learning from the historical behavior, we model it to a bandit problem to explore items with tiny training samples. In this way, we can improve performance through online learning during inference.

5.1   Bandit Layer

linUCB [14] is a typical bandit method in RS, in which they use ridge regression to overcome the data sparsity. However, we cannot directly use it on our HIT framework, as many candidate items are enormous. It is impossible to treat each item as an arm and do the bandit. The computational complexity is linear to the number of arms, and the algorithm can only run efficiently when the arm's size cannot be too large. Also, only one item can be retrieved in the LinUCB framework, but the items in the retrieval system are multiple.

Based on the idea of linUCB and the characteristics in the retrieval module, here we propose a bandit layer to explore the new potential items. In the bandit layer, firstly, we follow the previous approach [4, 14] to make the dimensionality reduction. We denote the user feature as ϕ_u, the item feature as ϕ_i, and suppose a weight matrix W, st. the probability of user clicking on item i is: $P = \phi_u^T W \phi_i$. After using logistic regression to get W, we cluster the projected feature vector $\phi_u^T W$ as the context, which can be seen as mapping the user to the item dimension. As in our HIT framework, we can directly regard the user embeddings as $\phi_u^T W$ and the probability of the user clicking on the item is just the inner product of the item embedding and the user embedding. Thus, we use the following process to make the dimensionality reduction.

We use K-means to group users into five clusters by the well-trained user embeddings and get five centroids $o_k, k = 1, ..., 5$ as the common setting [4, 14]. Then the user's features x_t at trail t is a six-vector: five member features $x_k, k = 1...5$ (normalized so that they sum up to unity) corresponded to that user's membership in these 5 clusters, and the sixth was a constant 1. The user's member feature x_k can be determined by the inner product between the user embedding and the centroids of clusters,

$$x_k = <\text{emb}(x), o_k> . \tag{9}$$

We define the arms as the candidate nodes in the bandit layer. Let $\mathbf{D}_n$ be a design matrix of dimension $m \times d$ at trial t, whose rows correspond to m training inputs (*e.g.*,

---

Algorithm 2: Layer-wise Beam Search Retrieval Algorithm in Prediction

---

Input: user x, model parameter θ, beam size B, top size T, node to item π, bandit parameters $\mathbf{A}, \mathbf{b}$

Output: The set of recommended items

1 Result set $\mathcal{A} \leftarrow \emptyset$
2 $\mathcal{N}_1$ be top B nodes of $\{p(n_1 \mid x, \theta)\}$ in layer 1
3 $\mathcal{A} \leftarrow \mathcal{A} \cup \pi(\mathcal{N}_1)$
4 **for** $l = 2$ *to* L **do**
5 $\quad$ Calculate the node probability $p(n_l) =$
   $\quad$ $p(n_1, \ldots, n_{l-1} \mid x, \theta) p(n_l \mid x, n_1, \ldots, n_{l-1}, \theta)$ for each remaining node n_l
6 $\quad$ Sort nodes and derive the set of top B nodes as $\mathcal{N}_l$
7 $\quad$ $\mathcal{A} \leftarrow \mathcal{A} \cup \pi(\mathcal{N}_l)$
8 Calculate the features of current user $\mathbf{x}_t$
9 **foreach** *remaining node $n \in \mathcal{N}_b$ in bandit layer* **do**
10 $\quad$ $\hat{\theta}_n \leftarrow \mathbf{A}_n^{-1} \mathbf{b}_n$, $p_{t,n} \leftarrow \mathbf{x}_t^{\top} \hat{\theta}_n + \gamma \sqrt{\mathbf{x}_t^{\top} \mathbf{A}_n^{-1} \mathbf{x}_t}$
11 Choose node $\mathcal{N}_t = \arg \max B_{n \in \mathcal{N}_b} p_{t,n}$
12 $\mathcal{A} \leftarrow Top_T(\mathcal{A}) \cup \pi(\mathcal{N}_t)$
13 Return $\mathcal{A}$.

---

m contexts that are observed previously for node n), and $\mathbf{c}_n \in \mathbb{R}^m$ be the corresponding response vector (*e.g.*, the corresponding m click/no-click user feedback). Applying ridge regression to the training data $(\mathbf{D}_n, \mathbf{c}_n)$ gives an estimate of the coefficients:

$$\hat{\theta}_n = \left(\mathbf{D}_n^{\top} \mathbf{D}_n + \mathbf{I}_d\right)^{-1} \mathbf{D}_n^{\top} \mathbf{c}_n, \tag{10}$$

where $\mathbf{I}_d$ is the $d \times d$ identity matrix. When components in $\mathbf{c}_n$ are independent conditioned on corresponding rows in $\mathbf{D}_n$. Then, if we want to get top B nodes, the arm-selection strategy in the bandit layer can be set as:

$$\mathcal{N}_t = \arg \max B_{n \in N_b} \left(\mathbf{x}_t^{\top} \hat{\theta}_n + \gamma \sqrt{\mathbf{x}_t^{\top} \mathbf{A}_n^{-1} \mathbf{x}_t}\right), \tag{11}$$

where $\mathbf{A}_n = \mathbf{D}_n^{\top} \mathbf{D}_n + \mathbf{I}_d$, $\mathbf{b}_n = \mathbf{D}_n^{\top} \mathbf{c}_n$, and γ is a constant. Then the parameters can be updated by:

$$\mathbf{A}_n = \mathbf{A}_n + \mathbf{x}_t \mathbf{x}_t^{\mathrm{T}}, \quad \mathbf{b}_n = \mathbf{b}_n + r_n \mathbf{x}_t, \tag{12}$$

where r_n is the corresponding payoff, $r_n = \sum_{v \in n} \mathbb{1}(v), n \in \mathcal{N}_t$.

5.2 Beam Search for Inference

Given user embeddings as input, we want to retrieve items from the HIT model in the inference stage. To improve retrieval efficiency, we use the beam search algorithm to retrieve the most probable nodes. In each layer, the algorithm selects the top B nodes from all the successors of the selected nodes from the predecessor. When $B = 1$, this becomes the greedy search. In each layer, choosing the top B from $B \times K$ candidates has a time complexity of $O(KB \log B)$. The total complexity is $O(LKB \log B)$, which is sublinear to the total number of items $\mathcal{V}$. The whole online serving algorithm, combined

Table 1. Dimensions of the three datasets after pre-processing. One record is a user-item pair that represents user feedback.

	#of users	#of items	#of interactions	Top items	Popular items	Long-tail items
AmazonMusic	13,159	131,878	277,847	0%–1%	1%–20%	20%–100%
MovieLens	129,797	20,709	9,939,873	0%–1%	1%–20%	20%–100%
IndustrialAds	269,414	3,274,379	16,404,625	0%–0.02%	0.02%–1%	1%–100%

by bandit layer and beam search inference, is shown in Algorithm 2. The final output combines the predictions using the user preference prediction model in Sect. 4 and the bandit layer in Sect. 5.1. The inference stage is illustrated in the right part of Fig. 2 with the results of the bandit layer.

6   Experimental Study

We study the performance of the proposed HIT in this section. Experimental results on MovieLens-20M [8], Amazon Digital Music, and a real industrial dataset are presented. In the experiments, we compare the performance of HIT with the state-of-the-art (SOTA) ANN model (using DSSM [10] + HNSW [16]), SOTA index structure model (DR [6]), and Brute-force algorithm (using DSSM for all items) to show the effectiveness of the model. What's more, we show that DR is a special case of HIT when we set $\epsilon = \{0, 0, 1\}$, *i.e.* not using the proposed component. Thus the HIT without equilibrium ("HITWOE") is the same as DR. HITWOE (DR) provides the ablation study at the same time. We also conduct empirical studies to show how the tree-based model and tree learning algorithm work, which further show the relationship between the model performance and the selection of ϵ and B. At the end of this section, we demonstrate the ability to retrieve potential items of HIT and the prediction efficiency of HIT.

6.1   Datasets and Metrics

AmazonMusic. This dataset contains user reviews of music from Amazon. Each user-music interaction contains a user-id, a music-id, and a timestamp. We only keep users with at least ten reviews. Then we randomly sample 1,000 users and corresponding records to construct the validation set, another 1,000 users to construct the test set, and the remaining users to construct the training set.

MovieLens-20M. This dataset contains movie ratings scored by users between 1995 and 2015. Each user-movie interaction contains a used-id, a movie-id, a rating between 1.0 to 5.0, and a timestamp. We follow the same data pre-processing procedure as [6, 28]. We only keep records with ratings higher or equal to 4.0 and users with at least ten reviews. Then we randomly sample 1,000 users and corresponding records to construct the validation set, another 1,000 users to construct the test set, and the remaining users to construct the training set.

IndustrialAds. This dataset is a subset of real industrial ads-clicking information till September 15, 2021. Each user-ad interaction contains a user-id, an ad-id, and a timestamp. We only keep users with at least 30 clicks. Then we randomly sample 5,000

Table 2. Comparison of performance for ANN, HITWOE, DR, HIT, and Brute-force retrieval on the whole AmazonMusic(@50), MovieLens-20M(@10) and IndustrialAds(@200).

	AmazonMusic			MovieLens-20M			IndustrialAds		
	P	R	F1	P	R	F1	P	R	F1
ANN	0.25%	1.44%	0.43%	19.23%	11.53%	14.41%	1.21%	8.01%	2.10%
HITWOE(DR)	0.13%	0.73%	0.22%	20.52%	12.21%	15.31%	1.39%	9.22%	2.42%
HIT	**0.27%**	**1.62%**	**0.46%**	**20.70%**	**12.29%**	**15.43%**	**1.68%**	**11.17%**	**2.92%**
Brute-force	0.26%	1.48%	0.45%	21.11%	12.50%	15.70%	1.47%	9.77%	2.56%

users and corresponding records to construct the validation set, another 5,000 users to construct the test set, and the remaining users to construct the training set.

To evaluate the results under different data distributions, we choose datasets with different data distribution characteristics. AmazonMusic is a more sparse dataset than MovieLens-20M. The number of users interacting with each item is less, which is more similar to real industrial scenarios (*i.e.*, IndustrialAds). On the contrary, MovieLens-20M has denser data interactions, where the number of users in the dataset is significantly more than the number of items. In these three datasets, for each user, the first half of the interactions, according to the timestamp, are historical behavior features, and the latter half is the ground truths to be predicted.

Metrics. To evaluate the model performance in different parts of the items, we divide the items into three parts by the frequency of interactions for each dataset, shown in Table 1. We calculate the metrics by retrieving the top 50 items in different parts of music on AmazonMusic, the top 10 items in different parts of movies for each user on MovieLens-20M, and the top 200 items in different parts of advertisements on IndustrialAds. Besides, to evaluate the bandit layer's performance, we set Bottom items as $80\% - 100\%$ frequency items with few interactions. Also, we choose the top 10 movies on the MovieLens-20M (and the top 50 music on AmazonMusic), which we downsample in training data, making them fall into the bottom items, to see whether the bandit layer can retrieve these top items without the historical training data. Further, we evaluate it respectively in the whole dataset and each part of the dataset. We use Precision@M, Recall@M, and F1-Measure@M (P, R, F1, respectively, in Table 2 and Table 3 as metrics to evaluate the performance under different algorithms. Following the same setting as DR, the metrics are calculated for each user individually and averaged without weight across users. We denote the retrieved set of items for a user x as $\mathcal{P}_{x,t}$, where t is the type of items, including long-tail, popular, top, potential, and all items, respectively. The user's ground truth set of each corresponding part is $\mathcal{G}_{x,t}$. Precision@M, Recall@M and F1-Measure@M are:

$$\text{Precision@M}\,(x,t) = \frac{|\mathcal{P}_{x,t} \cap \mathcal{G}_{x,t}|}{|\mathcal{P}_{x,t}|}, \text{ Recall@M}\,(x,t) = \frac{|\mathcal{P}_{x,t} \cap \mathcal{G}_{x,t}|}{|\mathcal{G}_{x,t}|},$$
$$\text{F1-Measure@M}(x,t) = \frac{2 * \text{Precision@M}\,(x,t) * \text{Recall @M}(x,t)}{\text{Precision@M}\,(x,t) + \text{Recall@M}\,(x,t)}. \quad (13)$$

6.2 Empirical Results

HIT Results. We compare the performance of HIT with the following algorithms: ANN, HIT without equilibrium and Brute-force. We independently train the same

Table 3. Comparison of performance of different models on long-tail items, popular items, and top items of AmazonMusic(@50), MovieLens-20M(@10) and IndustrialAds(@200).

Algorithm-Dataset	Long-tail items			Popular items			Top items		
	P	R	F1	P	R	F1	P	R	F1
ANN-AM	0.002%	0.012%	0.003%	0.04%	0.43%	0.07%	0.24%	4.72%	0.46%
HITWOE(DR)-AM	0.004%	0.056%	0.007%	0.04%	0.35%	0.07%	0.19%	3.44%	0.39%
HIT-AM	**0.008%**	**0.095%**	**0.015%**	**0.07%**	**0.70%**	**0.13%**	**0.28%**	**4.45%**	**0.52%**
Brute-force	0.002%	0.009%	0.003%	0.04%	0.45%	0.08%	0.28%	5.10%	0.52%
ANN-ML	0.64%	1.43%	0.88%	11.96%	12.64%	12.29%	18.31%	25.32%	21.25%
HITWOE(DR)-ML	0.63%	1.95%	0.96%	12.71%	12.32%	12.51%	19.95%	26.74%	22.85%
HIT-ML	**0.89%**	**2.63%**	**1.33%**	**13.23%**	**13.00%**	**13.11%**	**20.03%**	**27.04%**	**23.01%**
Brute-force	0.95%	3.03%	1.45%	13.59%	13.22%	13.40%	20.47%	27.71%	23.55%
ANN-IA	0.03%	0.39%	0.05%	0.31%	5.16%	0.58%	1.25%	54.82%	2.45%
HITWOE(DR)-IA	0.08%	1.29%	0.14%	0.74%	12.31%	1.40%	1.13%	40.37%	2.19%
HIT-IA	**0.08%**	**1.30%**	**0.16%**	**0.76%**	**12.64%**	**1.43%**	**1.59%**	**67.35%**	**3.10%**
Brute-force	0.05%	0.77%	0.09%	0.60%	10.11%	1.14%	1.50%	63.34%	2.93%

model 5 times and compute the mean of each metric. We show the overall performance of different algorithms on AmazonMusic, MovieLens-20M and IndustrialAds in Table 2. The results indicate that the proposed HIT outperforms other methods, including the ANN, DR and HIT without equilibrium. Moreover, The performance of HIT can even exceed Brute-force on AmazonMusic and IndustrialAds because HIT can generate a better candidate set than the whole original items. It also indicates that the learning ability of HIT on sparse data is better, while the traditional inner-product method is better on more dense data sets.

As for the different parts of the items, we show the results in Table 3. First, on MovieLens-20M, the performances of HIT, top items, popular items, and long-tail items are all not much different from Brute-force. Compared with ANN, DR and HITWOE, the improvement of HIT on long-tail items is particularly significant. While on AmazonMusic, HIT can perform better than other algorithms, even the Brute-force on popular and long-tail items. Because different datasets have different data distribution types thus, we can find that for sparse datasets, HIT can significantly improve the learning effect on the middle and tail items, while for dense datasets, HIT can significantly increase the learning effect only on the tail. Last, on IndustrialAds, a sparser dataset, we find that HIT mainly significantly improves the performance of the top and popular items. It shows that over the datasets with different sparsity, the long-tail phenomenon is more prominent, and the affected sample range is larger for sparse datasets. HIT can help solve this phenomenon. Second, we found that for the items short of training data, although there is a slight improvement using HIT, it cannot be improved to a very high level due to the small amount of data. This requires us to use new methods to explore. Therefore, HIT mainly helps items with enough training samples and can support the network's training but are massive enough to become the head items. Affected by the long tail effect, these items are difficult to win in the prediction process in the traditional methods, but HIT can help them. Moreover, if there is no long-tail effect and the data set is sufficient, HIT is no need to do the equilibrium, then it will degenerate into HIT-WOE(DR). If the amount of data is very scarce in the whole items, HIT will degenerate into a simple bandit tree.

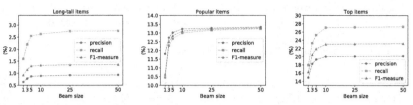

(a) Comparison of precision@10, Recall@10 and F1-Measure@10 on MovieLens-20M under different part of items and beam size B

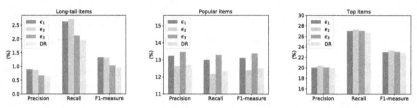

(b) Comparison of precision @10, recall @10 and F1-measure @10 on MovieLens-20M under different part of items and ϵ.

Fig. 4. Sensitivity of hyper-parameters

Sensitivity of Hyper-parameters. On the MovieLens-20M dataset, we set the parameters of tree structure $K = 50, L = 3, J = 3$. On the AmazonMusic dataset, we set the parameters of tree structure $K = 50, L = 3, J = 3$. For other coefficients to the HIT structure, we set the $\alpha = \{0.3, 0.03, 0.03\}$, $\epsilon = \{0.01, 0.2, 1\}$ and beam size $B = 10$ on MovieLens-20M; $\alpha = \{3, 0.03, 0.03\}$, $\epsilon = \{0.01, 0.2, 1\}$ and beam size $B = 10$ on AmazonMusic; $\alpha = \{0.3, 0.3, 3\}$, $\epsilon = \{0.0002, 0.01, 1\}$ and beam size $B = 50$ on IndustrialAds. Beam size B controls the number of candidate nodes to be retrieved. Larger B leads to better performance and heavier computation in the inference stage. As a result, Fig. 4 (a) shows, for MovieLens-20M, 5 or 10 is the better choice. Thus we set the beam size as 10 for MovieLens-20M. Also, as the beam size increases, the performance gets closer to the Brute-force search. Basically, when $B = 25$, it can be close to stability on the MovieLens-20M. We also compare the performance under different parts of data and different settings of ϵ in the tree learning stage on the MovieLens-20M. We set $\epsilon_1 = \{0.01, 0.2, 1\}$, $\epsilon_2 = \{0, 0.2, 1\}$, and $\epsilon_3 = \{0.01, 0.01, 1\}$, respectively. We can also treat DR as a special HIT, whose $\epsilon = 0, 0, 1$, *i.e.* all items are assigned to nodes in the last layer. The results are shown in Fig. 4 (b): On long-tail items, ϵ_1 and ϵ_2 can get better performance as the long-tail items are in the separate layer, will not affect by the popularity bias of top items and popular items. As for ϵ_3, long-tail items are influenced by the popular items, and in DR, long-tail items are influenced by the popular items and top items simultaneously. Thus the performance is worse in them. Meanwhile, on the popular items, ϵ_1 and ϵ_3 can learn more information than ϵ_2 or DR. They will not be affected by the top items. Also, we can see that the performance on top items is nearly the same under different settings due to the large number of top item interactions in MovieLens-20M, which can learn sufficiently regardless of the position.

Explore Results. For the extremely long-tail item, affected by the serious popularity bias, the precision and recall under Brute-force on bottom 20% items are all 0. Under

Table 4. Bandit layer ability.

AmazonMusic		
Precision@50	Recall@50	F1-Measure@50
0.054%	0.878%	0.102%
MovieLens-20M		
Precision@10	Recall@10	F1-Measure@10
5.20%	21.83%	8.0%

different index structures, the situation is also the same. Table 4 shows that under the bandit layer, the HIT could explore the bottom items and have the ability to find the potential hot items. Considering that the actual online advertisement level is at the level of hundreds of thousands, we chose the hyper-parameters settings of $K = 50$ and $B = 10$. Thus datasets AmazonMusic and MovieLens-20M are used. Since it is impossible to obtain whether an item is popular in real industrial scenarios, we take real user interactions (top 10 on MovieLens-20M and top 50 on AmazonMusic) as the correct predictions. Thus the performance on a sparser dataset AmazonMusic is not as good as MovieLens-20M due to the lack of data. Real scenarios will not have this problem. We can get more accurate and complete feedback. Also, when a new item is added to the list, we can directly put it on the bandit layer and gradually move it upwards, leaving the bandit layer to the advanced neural network layer, getting a more accurate prediction performance as the data accumulates.

Prediction Efficiency. As we mentioned, the complexity of the whole prediction process is $O(LKBlogB)$, which is sublinear *w.r.t* the total number of items. In practice, the efficiency of HIT depends on the number of candidate items. Under the real industrial advertising level ($100,000 - 200,000$), HIT has a similar efficiency with DR, about 0.25 milliseconds per user on average (4,000 qps), better than ANN (3,000 qps), and much faster than the Brute-force (which is not acceptable in terms of efficiency under the real industrial advertising level).

7   Conclusion

The retrieval module plays a vital role in the candidate item generation of large-scale industrial RS. In this paper, we have proposed HIT, a novel hierarchical tree-based model with variable-length layers for recommendation systems. Inside HIT, we have proposed a hierarchical equilibrium tree structure to deal with the popularity bias and improve the accuracy of user preference prediction in different parts of the items. Further, we have designed a bandit layer to explore the potential items at the bottom layer. Extensive experiments on large-scale real-world datasets show that HIT can significantly outperform existing methods in terms of efficient long-tail recommendation.

Acknowledgements. This work was supported in part by National Key R&D Program of China No. 2020YFB1707900, in part by China NSF grant No. 62132018, U2268204, 62272307

61902248, 61972254, 61972252, 62025204, 62072303, in part by Shanghai Science and Technology fund 20PJ1407900, in part by Alibaba Group through Alibaba Innovative Research Program, and in part by Tencent Rhino Bird Key Research Project. The opinions, findings, conclusions, and recommendations expressed in this paper are those of the authors and do not necessarily reflect the views of the funding agencies or the government.

References

1. Abdollahpouri, H.: Popularity bias in ranking and recommendation. In: AAAI/ACM Conference on AI, Ethic and Society (AIES), pp. 529–530 (2019)
2. Abdollahpouri, H., Burke, R., Mobasher, B.: Controlling popularity bias in learning-to-rank recommendation. In: 11th ACM Conference on Recommender Systems (RecSys), pp. 42–46 (2017)
3. Chen, Z., Xiao, R., Li, C., Ye, G., Sun, H., Deng, H.: ESAM: discriminative domain adaptation with non-displayed items to improve long-tail performance. In: The 43rd International ACM SIGIR conference on research and development in Information Retrieval (SIGIR), pp. 579–588 (2020)
4. Chu, W., et al.: A case study of behavior-driven conjoint analysis on yahoo! Front page today module. In: Proceedings of the 15th ACM SIGKDD International Conference on Knowledge Discovery and Data Mining (SIGKDD), pp. 1097–1104 (2009)
5. Covington, P., Adams, J., Sargin, E.: Deep neural networks for youtube recommendations. In: Proceedings of the 10th ACM Conference on Recommender Systems (RecSys), pp. 191–198 (2016)
6. Gao, W., et al.: Learning an end-to-end structure for retrieval in large-scale recommendations. In: The 30th ACM International Conference on Information and Knowledge Management (CIKM) (2021)
7. Guo, H., Tang, R., Ye, Y., Li, Z., He, X.: DeepFM: a factorization-machine based neural network for CTR prediction. In: Proceedings of the 26th International Joint Conference on Artificial Intelligence (IJCAI), pp. 1725–1731 (2017)
8. Harper, F.M., Konstan, J.A.: The movielens datasets: history and context. TiiS **5**(4), 1–19 (2015)
9. He, M., Li, C., Hu, X., Chen, X., Wang, J.: Mitigating popularity bias in recommendation via counterfactual inference. In: Database Systems for Advanced Applications (DASFAA), pp. 377–388 (2022)
10. Huang, P.S., He, X., Gao, J., Deng, L., Acero, A., Heck, L.: Learning deep structured semantic models for web search using clickthrough data. In: Proceedings of the 22nd ACM International Conference on Conference on Information and Knowledge Management (CIKM), pp. 2333–2338 (2013)
11. Hurley, N., Zhang, M.: Novelty and diversity in top-N recommendation-analysis and evaluation. TOIT **10**(4), 1–30 (2011)
12. Jegou, H., Douze, M., Schmid, C.: Product quantization for nearest neighbor search. TPAMI **33**(1), 117–128 (2010)
13. Kamishima, T., Akaho, S., Asoh, H., Sakuma, J.: Correcting popularity bias by enhancing recommendation neutrality. In: The 8th ACM Conference on Recommender Systems (RecSys), Posters (2014)
14. Li, L., Chu, W., Langford, J., Schapire, R.E.: A contextual-bandit approach to personalized news article recommendation. In: Proceedings of the 19th International Conference on World Wide Web (WWW), pp. 661–670 (2010)

15. Ma, X., et al.: Entire space multi-task model: An effective approach for estimating post-click conversion rate. In: The 41st International ACM SIGIR Conference on Research & Development in Information Retrieval (SIGIR), pp. 1137–1140 (2018)

16. Malkov, Y.A., Yashunin, D.A.: Efficient and robust approximate nearest neighbor search using hierarchical navigable small world graphs. TPAMI **42**(4), 824–836 (2018)

17. Park, Y.J., Tuzhilin, A.: The long tail of recommender systems and how to leverage it. In: ACM Conference on Recommender Systems (RecSys), pp. 11–18 (2008)

18. Rendle, S.: Factorization machines. In: 2010 IEEE International Conference on Data Mining (ICDM), pp. 995–1000 (2010)

19. Shi, L.: Trading-off among accuracy, similarity, diversity, and long-tail: a graph-based recommendation approach. In: Proceedings of the 7th ACM Conference on Recommender Systems (RecSys), pp. 57–64 (2013)

20. Shrivastava, A., Li, P.: Asymmetric LSH (ALSH) for sublinear time maximum inner product search (MIPS). In: Advances in Neural Information Processing Systems (NeurIPS), pp. 2321–2329 (2014)

21. Wang, R., Fu, B., Fu, G., Wang, M.: Deep & cross network for ad click predictions. In: Proceedings of the ADKDD, pp. 1–7 (2017)

22. Xie, R., Qiu, Z., Rao, J., Liu, Y., Zhang, B., Lin, L.: Internal and contextual attention network for cold-start multi-channel matching in recommendation. In: Proceedings of the 29h International Joint Conference on Artificial Intelligence (IJCAI), pp. 2732–2738 (2020)

23. Yang, L., Cui, Y., Xuan, Y., Wang, C., Belongie, S., Estrin, D.: Unbiased offline recommender evaluation for missing-not-at-random implicit feedback. In: Proceedings of the 12th ACM Conference on Recommender Systems (RecSys), pp. 279–287 (2018)

24. Yin, H., Cui, B., Li, J., Yao, J., Chen, C.: Challenging the long tail recommendation. Proc. VLDB Endow. **5**(9), 896–907 (2012)

25. Zheng, Y., Gao, C., Li, X., He, X., Li, Y., Jin, D.: Disentangling user interest and popularity bias for recommendation with causal embedding. arXiv preprint arXiv:2006.11011 (2020)

26. Zhou, G., et al.: Deep interest network for click-through rate prediction. In: Proceedings of the 24th ACM SIGKDD International Conference on Knowledge Discovery & Data Mining (SIGKDD), pp. 1059–1068 (2018)

27. Zhu, H., et al.: Joint optimization of tree-based index and deep model for recommender systems. In: Advances in Neural Information Processing Systems (NeurIPS), pp. 3971–3980 (2019)

28. Zhu, H., et al.: Learning tree-based deep model for recommender systems. In: Proceedings of the 24th ACM SIGKDD International Conference on Knowledge Discovery & Data Mining (SIGKDD), pp. 1079–1088 (2018)

29. Zhuo, J., Xu, Z., Dai, W., Zhu, H., Li, H., Xu, J., Gai, K.: Learning optimal tree models under beam search. In: International Conference on Machine Learning (ICML), pp. 11650–11659 (2020)

Disentangling User Intention
for Sequential Recommendation
with Dual Intention Decoupling Network

Liangliang Chen and Guang Chen[✉]

Beijing University of Posts and Telecommunications, Beijing, China
{outside,chenguang}@bupt.edu.cn

Abstract. Modern recommender systems often use sequential neural networks to capture users' dynamic and evolving intentions from behavior data. However, a user's different intentions might evolve over time at different speeds. Some user intentions are relatively stable with respect to time (i.e., time-invariant), and simply feeding all behavior data to sequential neural networks might not capture these time-invariant intentions well, since the inductive bias of sequential neural networks could prefer time-varying patterns than time-invariant patterns. In this paper, we propose a novel Dual Intention Decoupling Network (DIDN) framework to model time-invariant patterns and time-varying patterns in users' behavior data separately, thus both types of patterns could be modeled more accurately. To do so, we first introduce a self-attention based model and a tree-based clustering algorithm to model time-varying and time-invariant patterns respectively, and then combine these two models to generate the overall click-through rate prediction. In the self-attention module, we further introduce a candidate item attention mechanism to implicitly decouple a user's mixed intentions. Experimental results on three benchmarks show that our DIDN outperforms the state-of-the-art baselines in the topk sequential recommendation task.

Keywords: Sequential Recommendation · Self-Attention based Sequential Model · Disentangled Representation Learning

1 Introduction

Recommender systems need to catch users' interests based on their feedback (e.g., click, follow) and recommend items that users might be interested in. Users' interests often evolve over time, thus it is necessary to develop automatic approaches to facilitate the sequential recommendation, which has gradually appealed to academia and industry sectors in recent years.

Sequential recommenders regard the user's historical interactions as sequential data and predict the successive item that the user is most likely to interact with. Numerous works have been successfully applied in this area. Markov

This work was partially supported by MoE-CMCC "Artificial Intelligence" Project No. MCM20190701.

chain based methods [2, 17, 25, 35] argued that the user's current behavior is only affected by the last few behaviors. Recent methods often treat users' behavior data as sequential inputs, and use sequential neural encoders (e.g., RNNs) to embed the behavior data into a latent vector [1, 36], and then make recommendations based on the latent vector to achieve state-of-the-art performance [15, 16, 27, 31].

Despite their pioneering performance, we argue that the existing paradigms of sequential neural recommenders are insufficient to model the evolution of users' intentions. We notice that a user might have many different intentions, and they might evolve over time at different speeds. In observed behavior data, all intentions are mixed together, and we reckon this might confuse the sequential neural encoders and lead to suboptimal representations of intentions. Specifically, we hypothesis some user intentions evolve over time at relatively high speeds (i.e., time-varying intentions), while some user intentions evolve at relatively low speeds (i.e., time-invariant intentions). For example, a user may purchase computer speakers soon after buying a skytech archangel gaming computer PC desktop, a gaming monitor, a keyboard and mouse combo, successively under normal circumstances. The user intentions evolving in the above interactions are time-varying, which can be captured by the sequential model effectively [15, 16, 27, 31]. However, the periodic changes of some intentions are relatively slow or even stable, regarded as time-invariant, e.g., the user might prefer a movie just because of the genre or director. This preference may exist in the user's mind for a long time and drive the user's decision making process. Existing sequential neural recommenders usually feed the behavior data into the sequential model, and the inductive bias of the sequential neural networks might suppress the time-invariant patterns in the sequential inputs [37]. Recent pieces of literature have also shown that disentangling the mixed user intentions from different perspectives is critical to learning high-quality representations [9, 10]. Therefore, we argue that time-varying intentions and time-invariant intentions could be decoupled and modeled separately to improve the effectiveness of intention modeling.

In this research, we propose a novel Dual Intention Decoupling Network (DIDN) framework to disentangle time-varying and time-invariant intentions for better modeling of users' intentions. To do so, we explicitly map items into two independent embedding spaces for capturing time-varying and time-invariant intentions, respectively. We use a self-attention sequential encoder to encode time-varying patterns in the behavior data, and design a tree-based clustering module to capture the time-invariant patterns. The final recommendation score is a weighted sum of scores from these two modules. In addition, when modeling the time-varying intentions, we further introduce a candidate item attention mechanism in the self-attention block to implicitly decouple the user's mixed intentions related to the candidate item. The effect of the implicit decoupling could be magnified by stacking more self-attention blocks. Extensive experiments on three benchmarks verify the effectiveness of our proposed DIDN. Ablation studies and visualizations are also conducted to prove the effectiveness and interpretability of key components.

Our contributions are summarized as follows: 1) To our best knowledge, we are the first to distinguish time-varying and time-invariant intentions in the sequential recommendation. 2) Experimental results on three large-scale datasets show that our proposed framework achieves superior performance. 3) We visualize the disentangled representations of our model, which further unveils interesting insights into the recommendation tasks.

2   Related Work

2.1   Sequential Recommendation

The sequential recommender aims to predict the successive desired item according to one's recent interactions, which has always been the focus of academia and industry. The existing works can mainly be divided into four types: MCs-based methods [2,17], RNN based methods [11–13,16], self-attention based methods [15,27], and methods based on the external memory network [31]. [17] utilized Markov Chains to explore the transition probability among items to predict the successive item depending on the last several actions. [2] proposed Factorizing Personalized Markov Chains (FPMC) that modeled MCs via Matrix Factorization (MF). [13] proposed session-based GRU (GRU4Rec) to encode a user's interaction sequence into a vector which is used to make predictions. [11] adopted new loss functions and a novel sampling strategy to improve GRU4Rec. [12] proposed purpose-specific recurrent units (PSRUs) to capture a user's multiple purposes. [16] used two layers of GRU to extract the evolution of user interests. Motivated by the recent success of self-attention based models in natural language processing, some researchers leveraged the self-attention structure for the sequential recommendation [15,27]. [27] applied unidirectional self-attention blocks to encode the user's interaction sequence (SASRec). [15] regarded the time intervals between each pair of interactions to improve SASRec. However, considering a single vector was inefficient to express the user's multiple interest preferences, [31] comprehensively summarized the user's history interactions into multiple interest channel memory networks.

Despite their success, we argue that these sequential recommenders failed to capture the user's current intentions precisely. Since each interaction is driven by multiple and entangled intentions of the user, some of them need to be modeled separately instead of being fed into the sequential model directly. To this end, our proposed framework DIDN models the time-varying and time-invariant intentions separately using two independent embedding spaces and respective modeling paradigms, which are vital to estimating the finer granularity of user intentions.

2.2   Disentangled Representation Learning

Disentangled representation learning aims to disentangle the latent explanatory factors hidden in the observed data which is crucial for improving the interpretability and robustness of the neural network model [28]. The majority of the existing works focus on computer vision [3,4,30]. For example, β-VAE [4] added a KL-divergence term to the loss function to learn disentangled representations from raw pictures in an unsupervised manner. [30] utilized a causal layer to generate causal factors that correspond to causally related concepts in data. Until recently, the application of disentangled representation learning in the recommendation field has not been widely studied [7,9,10]. [7] used Variational Auto-Encoder (VAE) to disentangle user preferences into macro-level and micro-level factors. [9] utilized graph neural networks to learn disentangled representations of different user intentions. In particular, each graph corresponded to a specific user intention. [10] designed a framework with causality-oriented data to disentangle user interest and conformity. ComiRec [20] is a recent representative work for extracting multiple interests, including two variants, ComiRec-DR and ComiRec-SA. SINE [21] maintained a large-scale set of intention prototypes and offers the ability to activate a sparse set of preferred intentions automatically. Our work differs from them in that we attempt to disentangle intentions from the time-varying v.s. time-invariant perspective.

3   Problem Formulation

In sequential recommendation, let $\mathcal{U} = \{u_1, u_2, ..., u_{|\mathcal{U}|}\}$ denotes a set of users, $\mathcal{V} = \{v_1, v_2, ..., v_{|\mathcal{V}|}\}$ denotes a set of items, and $H_u^{all} = [v_1^{(u)}, ..., v_i^{(u)}, ..., v_n^{(u)}]$ represents the historical interaction sequence for user $u \in \mathcal{U}$, where n is the length. In practice, we set the maximum length of the sequence as N to form H_u. Given H_u, the recommender aims to predict the item that the user u would like to interact with at the next moment, which can be expressed as:

$$v_{\hat{k}} = \underset{k}{\mathrm{argmax}} \mathcal{P}\left(v_{n+1}^{(u)} = v_k \mid H_u\right) \tag{1}$$

4   The Proposed Method

In this section, we describe the proposed framework Dual Intention Decoupling Network (DIDN) in detail and explain the reason why we put forward the two-stage structure of dual intention decoupling. In brief, we utilize two independent sets of embeddings to capture the time-varying and time-invariant intentions separately. Afterward, the self-attention based sequential model and a tree routing module estimate both intentions respectively as the first stage of decoupling (i.e., explicit decoupling). Finally, we apply a weighted sum to produce the overall score of every candidate item v:

$$S_{uv} = \sigma\left(w_u\right) S_{uv}^{(tv)} + \left(1 - \sigma\left(w_u\right)\right) S_{uv}^{(ti)} \tag{2}$$

where S_{uv} represents the overall score, $S_{uv}^{(tv)}$ and $S_{uv}^{(ti)}$ are the scores predicted by the user's time-varying and time-invariant intention respectively. σ is the

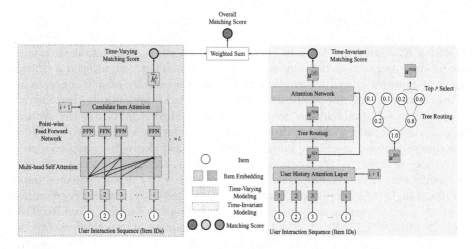

Fig. 1. The arcHRecture of DIDN consists of explicit and implicit decoupling mechanisms. The former disentangles the users' mixed intentions into the time-varying and time-invariant parts and models them on independent embeddings and separate networks. Besides, the latter further leads the candidate item to guide the intention disentanglement adaptively.

nonlinear sigmoid function to constrain the output between 0 and 1, and w_u is the learnable weight specific to u for balancing two scores. In addition, we introduce a novel candidate item attention module in self-attention block to further disentangle users' intentions related to the candidate item adaptively as the second stage of decoupling (i.e., implicit decoupling). The overall framework is shown in Fig. 1.

4.1 Embedding Layer

In our proposed DIDN framework, to capture the user's time-invariant and time-varying intentions respectively, we explicitly use separate embeddings to learn them. Specifically, we use $M_{\mathcal{V}}^{(ti)} = \left[m_1^{(ti)}, ..., m_{|\mathcal{V}|}^{(ti)}\right]^{\top} \in \mathbb{R}^{|\mathcal{V}| \times d}$ to represent the embeddings of time-invariant intentions, and use $M_{\mathcal{V}}^{(tv)} = \left[m_1^{(tv)}, ..., m_{|\mathcal{V}|}^{(tv)}\right]^{\top} \in \mathbb{R}^{|\mathcal{V}| \times d}$ to represent the embeddings of time-varying intentions, where d is the latent dimension.

4.2 Time-Varying Intention Modeling

Previous studies [15,27] have shown that the self-attention block is capable of modeling the users' time-varying intention which evolves over time at relatively high speeds. Hence our time-varying intention modeling module employs the self-attention block [1] as the backbone. For convenience, we stack the time-varying embeddings of the user's interacted items together resulting in a matrix

$M_N^{(tv)} = \left[m_1^{(tv)}, ..., m_N^{(tv)} \right]^\top \in \mathbb{R}^{N \times d}$. As the practice in [15], we add the learnable positional embeddings to $M_N^{(tv)}$ to let DIDN be aware of sequential signal.

Multi-head Self-attention Sub-layer. Here, we linearly project hidden state $\widetilde{H}$ (i.e., $\widetilde{H} = M_N^{(tv)}$) into z subspaces (i.e., heads) to produce the output representations as follows:

$$\bar{H} = \text{MHSA}\left(\widetilde{H}\right) = [head_1; head_2; ...; head_z] \cdot W^O$$

$$head_k = \text{Attention}\left(\widetilde{H} \cdot W_k^Q, \widetilde{H} \cdot W_k^K, \widetilde{H} \cdot W_k^V\right)$$

(3)

where $[;]$ means the concatenation operation, $W_k^Q \in \mathbb{R}^{d \times d/z}$, $W_k^K \in \mathbb{R}^{d \times d/z}$, $W_k^V \in \mathbb{R}^{d \times d/z}$ are trainable parameters in each head, and $W^O \in \mathbb{R}^{d \times d}$ is a trainable parameter too. Since [15] further improved the performance of self-attention based sequential recommender by considering item inputs $M_N^{(tv)}$, the relative time intervals between the interacted items, and the absolute position of each interacted item in Eq. 3. We follow their practice directly.[1]

Position-Wise Feed-Forward Network (PFFN). Though the multi-head self-attention sub-layer can integrate the input $\widetilde{H}$, it is still a linear model. To endow the model with nonlinearity while keeping scale-invariant features, we apply a point-wise feed-forward network on top of the self-attention sub-layer. For each position i:

$$F_i = \text{PFFN}\left(\bar{h}_i\right) = \text{ReLU}\left(\bar{h}_i \cdot W^{(1)} + b^{(1)}\right) \cdot W^{(2)} + b^{(2)}$$

(4)

where $\bar{h}_i$ denotes the output of MHSA at position i, $W^{(1)} \in \mathbb{R}^{d \times 4d}$, $W^{(2)} \in \mathbb{R}^{4d \times d}$, $b^{(1)} \in \mathbb{R}^{4d}$ and $b^{(2)} \in \mathbb{R}^d$ are learnable parameters and shared across all positions, ReLU is a nonlinear function.

Candidate Item Attention. The time-varying intentions themselves could also be entangled. For example, if the recommender wants to predict whether a user likes the items endorsed by Adele or not, it should attend more over the star-style items the user has interacted with instead of the general ones. Among all time-varying intentions, it is desired for our DIDN to focus on those intentions that are related to the candidate item. Thus, we propose a candidate item attention module to further decouple the intentions (i.e., implicit decoupling).

To model such finer-grained relationships between the candidate item and the interacted items, we introduce an attention mechanism on top of PFFN. To be more specific, we adopt the local activation ability of the attention mechanism to

---

[1] In this paper, we focus on the intention decoupling paradigm. Thus we omit more details about the multi-head self-attention sub-layer [15] for brevity.

learn an attention score distribution. This helps model the time-varying intention evolving process relative to the candidate item. It can be expressed as:

$$\alpha_i = \text{MLP} \left(\left[F_i; m_{cand}^{(tv)}; F_i - m_{cand}^{(tv)}; F_i \odot m_{cand}^{(tv)} \right] \right) \tag{5}$$

where $\odot$ denotes the element-wise product, F_i represents the output of PFFN for the i-th position and $m_{cand}^{(tv)}$ denotes time-varying embedding of the candidate item, MLP stands for a multi-layer perception. Afterward, the updated output of PFFN can be formulated as:

$$\alpha_i' = \frac{\exp \alpha_i}{\sum_{k=1}^{N} \exp \alpha_k}$$
$$\hat{F}_i = \alpha_i' \odot F_i \tag{6}$$

It is worth noting to mention that we stack L self-attention blocks to learn more complex representations as shown in Fig. 1.

Time-Varying Matching Score. After L self-attention blocks, we get the final output $\widetilde{H}^L$ for all items of the input sequence. To produce the time-varying score of item v at position i, we employ a latent factor model as follows:

$$S_{uv}^{(tv)} = \widetilde{h}_i^{L\top} \cdot m_v^{(tv)} + b_O^{(tv)} \tag{7}$$

where $m_v^{(tv)} \in M_{\mathcal{V}}^{(tv)}$, $\widetilde{h}_i^L$ denotes the representation of user time-varying intention at position i, and $b_O^{(tv)}$ is a bias term. Here we use the shared embedding matrix $M_{\mathcal{V}}^{(tv)}$ for reducing the model size.

4.3   Time-Invariant Intention Modeling

In this section, we focus on modeling the time-invariant intentions. To do so, we need to invoke a module that is not sensitive to time-varying intentions, thus the time-varying intentions in the behavior data could be somehow "filtered". Motivated by [33,34], clustering could capture slow evolving patterns in the behavior data, which is well-suited for our task. We propose a tree routing module for clustering, which could route users with similar time-invariant intentions to the same leaf nodes automatically. The key insight of the tree structure is that the clustered user categories can be extended from coarse to fine by traversing tree nodes which is essential to improve the robustness and interpretability of the recommendation system. Based on the clustering results, our model generates the user's time-invariant intention.

User History Attention Layer. For each user u and his/her historical interactions H_u, an attention network is introduced to build a vector called user history embedding to represent the user's time-invariant intentions directly:

$$u^{his} = \sum_{j=1}^{N} \beta_j m_j^{(ti)}$$

$$\beta_j = \frac{\exp\left(m_{cand}^{(ti)\top} m_j^{(ti)}\right)}{\sum_{k=1}^{N} \exp\left(m_{cand}^{(ti)\top} m_k^{(ti)}\right)} \tag{8}$$

where β_j presents the relevance between the user's j-th interacted item and the candidate item. The number of interactions is consistent with that in Sect. 4.2.

Tree Routing. We design a novel tree routing module, where each node maintains a diversion vector ν and a representation μ. The node decision weight τ stands for the relevance to the user is not only calculated by the correlation of its diversion vector ν and user history embedding u^{his} but also inherited from its parent node. When the decision weight τ gradually propagates from the root node to the leaf nodes, the user is routed to the most relevant user categories. In summary, the routing algorithm can be expressed as:

$$\tau_{j'+x}^{i+1} = \tau_j^i \frac{\exp\left(u^{his\top} \nu_{j'+x}^{i+1}\right)}{\sum_{y=0}^{c-1} \exp\left(u^{his\top} \nu_{j'+y}^{i+1}\right)}, \quad x = 0,\ldots,c-1 \tag{9}$$

where $\nu_{j'+x}^{i+1}$ is the node diversion vector of $j' + x$-th node in layer $i + 1$. The node decision weight $\tau_{j'+x}^{i+1}$ inherits from its parent node's one τ_j^i, and c is the number of its sibling nodes. Afterward, user routing embedding u^{rou} is formed by the representations of the leaf nodes with the largest P weights as follows:

$$u^{rou} = \frac{1}{\sum_{j=1}^{P} \tau_j} \sum_{j=1}^{P} \tau_j \mu_j \tag{10}$$

where $\{\tau_1,\ldots,\tau_P\}$ and $\{\mu_1,\ldots,\mu_P\}$ present the largest P decision weights and the corresponding representations on the leaf layer, respectively.

Time-Invariant Matching Score. To generate the final representation of user time-invariant intentions, we use an attention network to combine user history embedding u^{his} and user routing embedding u^{rou} in the following steps:

$$\begin{aligned} e_u^{his}, e_u^{rou} &= \text{ReLU}\left(\left[u^{his}; u^{rou}\right] W + b\right) \gamma^T \\ \delta_u^{his}, \delta_u^{rou} &= \text{softmax}\left(e_u^{his}, e_u^{rou}\right) \\ u^{(ti)} &= \delta_u^{his} u^{his} + \delta_u^{rou} u^{rou} \end{aligned} \tag{11}$$

where $W \in \mathbb{R}^{d \times q}, b \in \mathbb{R}^q, \gamma \in \mathbb{R}^q$ are the parameters of the attention network, and q denotes the hidden size of the attention network. Based on $u^{(ti)}$, we produce the time-invariant matching score between the user u and item v as follows:

$$S_{uv}^{(ti)} = u^{(ti)^\top} \cdot m_v^{(ti)} + b_O^{(ti)} \qquad (12)$$

where $m_v^{(ti)} \in M_{\mathcal{V}}^{(ti)}$, $b_O^{(ti)}$ is a bias term. We here also use the shared item embedding matrix $M_{\mathcal{V}}^{(ti)}$ for reducing model size like Sect. 4.2.

4.4 Model Learning

Sample Processing. Like the practices in training the conventional unidirectional sequential recommendation models [14,15,27], we treat each position of the sequence as a sample. In detail, each training sample uses the first i behaviors of u to predict the $(i + 1)$-th behavior, where $i = 1, 2, ..., (N - 1)$. For each ground-truth item v^+, like traditional BPR [19], we randomly sample an item from the user's non-interacted items as the negative item v^- and the loss function of predicting item is defined as:

$$\mathcal{L}_{BPR} = \sum_{(u,v^+,v^-) \in \mathcal{D}} -\ln \sigma \left(S_{uv} \left(v^+ \mid H_u' \right) - S_{uv} \left(v^- \mid H_u' \right) \right) \qquad (13)$$

where $\mathcal{D}$ denotes the set of item pairs for training, H_u' is the truncated version of user historical interaction sequence H_u. We mask the loss of the padding items.

Multi-task Training

$$\mathcal{L}_{discrepancy} = -\ell_2 \left(M_{\mathcal{V}}^{(ti)}, M_{\mathcal{V}}^{(tv)} \right) \qquad (14)$$

We optimize the parameters θ by combining the loss function of both item prediction and disentangled representation learning together so the training objective is as follows:

$$\mathcal{L} = \mathcal{L}_{BPR} + \omega \mathcal{L}_{discrepancy} + \phi ||\theta||^2 \qquad (15)$$

where $||\theta||^2$ is the regularizer term to avoid overfitting, $\omega, \phi \in (0, 1)$ are trade-off parameters in multi-task training.

5 Experiments

In this section, we introduce our experimental settings and report our empirical results. Our experiments are designed to answer the following research questions: RQ1: Can our proposed DIDN outperform state-of-the-art baselines for sequential recommendation tasks? RQ2: How does the dual intention decoupling contribute to the results of DIDN through ablation studies? RQ3: Can the proposed DIDN interpretably disentangle the user's mixed intentions?

Table 1. Statistics of the datasets.

Dataset	#Users	#Items	#Interaction	#Avg.length
MovieLens-1m	6040	3416	0.987 m	163.5
Amazon Movies & TV	40928	37564	1.05 m	25.55
Amazon Video Games	30935	12111	0.2 m	6.46

Table 2. Hyperparameter settings in tree routing module. Tree structure is the number of nodes at each layer.

Dataset	Tree Structure	Top P Selection
MovieLens-1 m	1-2-8	4
Amazon Movies & TV	1-32-256	32
Amazon Video Games	1-16-256	32

5.1 Datasets

We evaluate the proposed framework on three representative datasets from two real-world applications, the Movielens dataset [29], and the Amazon dataset [18]. **MovieLens**: A widely used benchmark dataset for evaluating recommendation algorithms. In this work, we adopt one well-established version, MovieLens-1m. **Amazon**: A series of item review datasets crawled from Amazon.com by [18]. They divide the data into separate datasets according to the categories. In this work, we adopt two categories, "Movies&TV" and "Video Games". To reduce data sparsity and the size of the dataset, we filter out all the users and items that have less than five instances of feedback. Table 1 lists the statistics of datasets.

5.2 Task Settings and Evaluation Metrics

To evaluate the sequential recommendation models, we adopt the leave-one-out evaluation (i.e., next item recommendation) task, which has been widely used in [15,27]. For each user, we order his/her interacted items by timestamp to conduct the personal sequence. And we hold out the recent item as the test data, treat the second recent item as the validation set, and utilize the remaining items for training. For a fair comparison, we follow the common strategy in [15,27], pairing each ground truth item in the test set with 100 randomly sampled negative items that the user has not interacted with.

Evaluation Metrics. To evaluate the ranking result of all the models, we employ two common top-k evaluation metrics, including HR Ratio (HR) and Normalized Discounted Cumulative Gain (NDCG). HR@k counts the rates of the ground truth item that appears in the top k items. NDCG measures the gain of a candidate depending on its position in the ranked result list. Higher values indicate better performance for all metrics.

5.3   Baselines and Implementation Details

We compare DIDN with the following representative baselines: 1) POP: Counts the number of interactions of each item by all users as the popularity of the item in the training set. Recommends items based on popularity. 2) BPR-MF [19]: Optimizes the matrix factorization with implicit negative feedback using a pairwise ranking loss. 3) GRU4Rec+ [11]: An improved version of GRU4Rec with a new class of loss functions and sampling strategy for the session-based recommendation. 4) FPMC [2]: It captures users' long-term preferences as well as their sequential behaviors by combing MF with first-order Markov Chains. 5) TiSASRec [15]: A state-of-the-art model that has been proposed recently. Compared with SASRec, TiSASRec adopts relative interval embeddings to represent time intervals between each pair of the interacted items. 6) Comi-SA [20]: A self-attention network for extracting multiple interests. 7) SINE [21]: A multi-interest framework based on intention prototypes [8].

For TiSASRec, Comi-SA[2], SINE, we use the codebase provided by the corresponding authors and fine-tuned hyper-parameters using the validation set and terminated training if validation performance didn't improve within 20 epochs. For other baselines, we directly use the evaluation metrics reported in [15], where [15] fine-tuned default hyper-parameter settings according to the respective papers. To be fair, we don't consider making a comparison with related sequential recommenders [5,6] that are not open source.

We implement our framework DIDN with TensorFlow. To accelerate our training process, dynamic negative item sampling (DNIS) [26] is used to adaptively select the highest-ranked negative sample. DNIS is considered to be one of the most effective samplers of BPR loss. We set the number of negative samples in DNIS $BPR_{neg} = 4$, the hyper-parameters in the tree routing module as Table 2, the trade-off parameter in Eq. 15 $\omega = 0.01$ across all datasets. For other hyper-parameters, like ϕ, we use the default settings suggested by [15]. For all baselines, we fix the embedding size as 50 according to [15]. While for our DIDN, the embedding size is fixed as 25, since we contain two sets of embeddings. Therefore, the number of item embeddings is the same for all methods to guarantee fair comparison. All the models are trained from scratch without any pre-training on a single Tesla T4 GPU with a batch size of 128.

5.4   Overall Performance Comparison (RQ1)

We report the recommendation performance of all methods in Table 3. It can be observed that: among all the baseline methods, sequential methods (e.g., FPMC and GRU4Rec+) have obviously better performance than non-sequential methods (e.g., POP and BPR-MF) on all datasets consistently. This observation verifies that considering sequential information is vital to improving the performances of recommender systems. Furthermore, TiSASRec [15], which considers

---

[2] After computing the multiple interest embeddings and the corresponding multiple matching scores, we use an argmax operator to choose the final matching score.

Table 3. Performance comparison of different methods on public datasets. The best results are highlighted with the bold fold. * indicates the significant improvement over all baselines with p < 0.05.

Dataset	Metric	Pop	BPR-MF	FPMC	GRU4Rec+	TiSASRec	Comi-SA	SINE	Ours
MovieLens-1m	NDCG@10	0.239	0.342	0.391	0.433	0.571	0.560	0.569	**0.584***
	HR@10	0.439	0.595	0.618	0.652	0.799	0.788	0.790	**0.807**
Amazon Movies & TV	NDCG@10	0.272	0.348	0.338	0.299	0.480	0.472	0.473	**0.498***
	HR@10	0.458	0.554	0.510	0.485	0.694	0.660	0.690	**0.710***
Amazon Video Games	NDCG@10	0.23	0.273	0.341	0.232	0.476	0.466	0.470	**0.492***
	HR@10	0.401	0.442	0.522	0.397	0.696	0.690	0.693	**0.706***

Table 4. Ablation study of DIDN.

Dataset	metrics	DIDN w/o ed	DIDN w/o id	DIDN
MovieLens-1m	NDCG@10	0.583	0.579	0.584
	HR@10	0.799	0.802	0.807
Amazon Movies & TV	NDCG@10	0.485	0.488	0.498
	HR@10	0.695	0.693	0.710
Amazon Video Games	NDCG@10	0.481	0.489	0.492
	HR@10	0.697	0.698	0.706

the relative time intervals and absolute positions among items to predict future interactions, has a better performance compared with the other baselines. However, DIDN improves over the best baseline methods across all datasets with respect to HR@10 and NDCG@10. This demonstrates the superior effectiveness of DIDN. We attribute such improvements to the following aspects, 1) our DIDN takes advantage of disentangled representations and independent modeling processes which can estimate the user's time-varying and time-invariant intentions separately, while the previous methods (e.g., GRU4Rec+, TiSASRec) feed the entangled representation of users' intention into the sequential model. It may bring unexpected noise to original sequential modeling, which can not estimate the user's current intention effectively and precisely; 2) we integrate the candidate item attention into each self-attention block, as discussed in Sect. 4.2. By assigning different attention scores to the historical interactions, our model implicitly decouples the user's intentions related to the candidate item. Moreover, by stacking more than one self-attention block, the effect of implicit decoupling could be magnified block by block.

5.5   Ablation Study (RQ2)

As introduced previously, the dual intention decoupling is at the core of DIDN, which not only adopts two separate embedding spaces and respective modeling processes to distinguish user time-varying and time-invariant intentions but also applies the candidate item attention to assign the higher score to those more relevant interacted items to candidate item. Here we investigate how dual

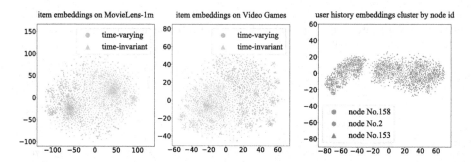

Fig. 2. Visualization of the learned embeddings of DIDN through t-SNE.

intention decoupling affects the performance by ablation experiments. Table 4 summarizes the empirical results on three datasets, where "DIDN w/o ed" represents the variant version that DIDN without explicit decoupling (i.e., two sets of independent embeddings and respective modeling process) while "DIDN w/o id" represents the variant version that DIDN without implicit decoupling (i.e., candidate item attention).

We have several findings: 1) Compared to Amazon Games, the dataset with larger average actions/user (e.g., MovieLens-1m), "DIDN w/o ed" has obviously better performance than TiSARec. We speculate the reason is that when a user has enough historical interactions, the attention score allocation in the candidate item attention mechanism may be more effective. On the contrary, if a user has limited behaviors, the interacted items are not enough to describe the evolution of the user's time-varying intention, which may limit the performance of implicit decoupling. 2) Clearly, "DIDN w/o id" outperforms TiSASRec with respect to metrics across all datasets. This might suggest that the novel embedding structure is capable of endowing the recommender with better representation ability. Together with the respective modeling processes which have avoided the mutual influence, DIDN estimates the user's intentions more precisely.

5.6 Visualization of Dual Intention Decoupling and Tree Routing (RQ3)

In what follows, we conduct experiments to get deep insights into the intention decoupling and tree routing.

Explicit Decoupling. We visualize the learned disentangled item embeddings $M_V^{(tv)}$ and $M_V^{(ti)}$ through t-SNE to get an intuitive understanding. From the left and mid sub-figures of Fig. 2, we observe that the two sets of embeddings are far from each other. Moreover, compared to the wider and more scattered distribution of time-varying embeddings, time-invariant embeddings tend to cluster into several groups. This illustrates the high quality of disentanglement in DIDN and verifies the rationality of the clustering method (i.e., our tree routing) to model the time-invariant intentions.

Tree Routing. As introduced previously, we adopt tree routing as a clustering approach to estimate the user time-invariant intention. The users with similar time-invariant intentions will be routed to the same leaf nodes. Here we take three leaf nodes (e.g., No.2, No.153, and No.158) in Amazon Movies&TV dataset as examples. We collect and visualize the users' history embeddings u^{his} (i.e., the input of the tree routing module) in Eq. 8 from these three leaf nodes. As the right sub-figure of Fig. 2 shows, embeddings of the same node almost belong to one cluster, which shows the clustering property of tree routing clearly, i.e., stores different user intentions in different leaf nodes. Moreover, we observe an interesting phenomenon that the clustering results of sibling nodes are close (i.e., No.153 and No.158), while that of not siblings are far from each other, even can be separated by a line (i.e., No.2 and No.158). We attribute this to the strength of the tree structure, where the clustered user categories can be extended from coarse to fine by traversing tree nodes. That is, the users routed to the siblings are relatively similar. Accordingly, we believe that the obtained clustering results play a crucial role in the interpretability of the prediction for user personalized modeling and recommendation.

6   Conclusion

In this paper, we propose a novel framework called DIDN for disentangling users' mixed intention in the sequential recommendation. DIDN utilizes two independent embedding spaces and their respective modeling paradigms to capture the time-varying and time-invariant intentions separately as explicit decoupling. Besides, a candidate item attention mechanism is further introduced to decouple the relevant intentions corresponding to the candidate item as implicit decoupling. Extensive experiments illustrate that DIDN outperformed state-of-the-art baselines. Moreover, we offer deep insights into DIDN. The visualizations depict that our decoupling results have a strong explanatory, which is of great significance for us to understand the recommendation itself.

References

1. Vaswani, A., et al.: Attention is all you need. In: Advances in Neural Information Processing Systems, pp. 5998–6008 (2017)
2. Rendle, S., Freudenthaler, C., Schmidt-Thieme, L.: Factorizing personalized markov chains for next-basket recommendation. In: Proceedings of the 19th International Conference on World Wide Web, pp. 811–820 (2010)
3. Gidaris, S., Singh, P., Komodakis, N.: Unsupervised representation learning by predicting image rotations. arXiv preprint arXiv:1803.07728 (2018)
4. Higgins, I., et al.: beta-VAE: learning basic visual concepts with a constrained variational framework (2016)
5. Ye, W., Wang, S., Chen, X., Wang, X., Qin, Z., Yin, D.: Time matters: sequential recommendation with complex temporal information. In: Proceedings of the 43rd International ACM SIGIR Conference on Research and Development in Information Retrieval, pp. 1459–1468 (2020)

6. Ma, C., Ma, L., Zhang, Y., Sun, J., Liu, X., Coates, M.: Memory augmented graph neural networks for sequential recommendation. Proc. AAAI Conf. Artif. Intell. **34**(04), 5045–5052 (2020)
7. Ma, J., Zhou, C., Cui, P., Yang, H., Zhu, W.: Learning disentangled representations for recommendation. arXiv preprint arXiv:1910.14238 (2019)
8. Ma, J., Cui, P., Kuang, K., Wang, X., Zhu, W.: Disentangled graph convolutional networks. In: International Conference on Machine Learning (PMLR), pp. 4212–4221 (2019)
9. Wang, X., Jin, H., Zhang, A., He, X., Xu, T., Chua, T.-S.: Disentangled graph collaborative filtering. In: Proceedings of the 43rd International ACM SIGIR Conference on Research and Development in Information Retrieval, pp. 1001–1010 (2020)
10. Zheng, Y., Gao, C., Li, X., He, X., Jin, D., Li, Y.: Disentangling user interest and conformity for recommendation with causal embedding. arXiv preprint arXiv:2006.11011 (2020)
11. Hidasi, B., Karatzoglou, A.: Recurrent neural networks with top-k gains for session-based recommendations. In: Proceedings of the 27th ACM International Conference on Information and Knowledge Management, pp. 843–852 (2018)
12. Wang, S., Hu, L., Wang, Y., Sheng, Q.Z., Orgun, M., Cao, L.: Modeling multi-purpose sessions for next-item recommendations via mixture-channel purpose routing networks. In: International Joint Conference on Artificial Intelligence. International Joint Conferences on Artificial Intelligence (2019)
13. Hidasi, B., Karatzoglou, A., Baltrunas, L., Tikk, D.: Session-based recommendations with recurrent neural networks. arXiv preprint arXiv:1511.06939 (2015)
14. Pi, Q., Bian, W., Zhou, G., Zhu, X., Gai, K.: Practice on long sequential user behavior modeling for click-through rate prediction. In: Proceedings of the 25th ACM SIGKDD International Conference on Knowledge Discovery & Data Mining, pp. 2671–2679 (2019)
15. Li, J., Wang, Y., McAuley, J.: Time interval aware self-attention for sequential recommendation. In: Proceedings of the 13th International Conference on Web Search and Data Mining, pp. 322–330 (2020)
16. Zhou, G., et al.: Deep interest evolution network for click-through rate prediction. Proc. AAAI Conf. Artif. Intell. **33**(01), 5941–5948 (2019)
17. Chen, S., Moore, J.L., Turnbull, D., Joachims, T.: Playlist prediction via metric embedding. In: Proceedings of the 18th ACM SIGKDD International Conference on Knowledge Discovery and Data Mining, pp. 714–722 (2012)
18. McAuley, J., Targett, C., Shi, Q., Van Den Hengel, A.: Image-based recommendations on styles and substitutes. In: Proceedings of the 38th International ACM SIGIR Conference on Research and Development in Information Retrieval, pp. 43–52 (2015)
19. Rendle, S., Freudenthaler, C., Gantner, Z., Schmidt-Thieme, L.: BPR: bayesian personalized ranking from implicit feedback. arXiv preprint arXiv:1205.2618 (2012)
20. Cen, Y., Zhang, J., Zou, X., Zhou, C., Yang, H., Tang, J.: Controllable multi-interest framework for recommendation. In: Proceedings of the 26th ACM SIGKDD International Conference on Knowledge Discovery & Data Mining, pp. 2942–2951 (2020)
21. Tan, Q., et al.: Sparse-interest network for sequential recommendation. In: Proceedings of the 14th ACM International Conference on Web Search and Data Mining, pp. 598–606 (2021)

22. Srivastava, N., Hinton, G., Krizhevsky, A., Sutskever, I., Salakhutdinov, R.: Dropout: a simple way to prevent neural networks from overfitting. J. Mach. Learn. Res. **15**(1), 1929–1958 (2014)
23. He, K., Zhang, X., Ren, S., Sun, J.: Deep residual learning for image recognition. In: Proceedings of the IEEE Conference on Computer Vision and Pattern Recognition, pp. 770–778 (2016)
24. Ba, J.L., Kiros, J.R., Hinton, G.E.: Layer normalization. arXiv preprint arXiv:1607.06450 (2016)
25. He, R., Kang, W.-C., McAuley, J.: Translation-based recommendation. In: Proceedings of the Eleventh ACM Conference on Recommender Systems, pp. 161–169 (2017)
26. Zhang, W., Chen, T., Wang, J., Yu, Y.: Optimizing top-n collaborative filtering via dynamic negative item sampling. In: Proceedings of the 36th International ACM SIGIR Conference on Research and Development in Information Retrieval, pp. 785–788 (2013)
27. Kang, W.-C., McAuley, J.: Self-attentive sequential recommendation. In: 2018 IEEE International Conference on Data Mining (ICDM), pp. 197–206. IEEE (2018)
28. Bengio, Y., Courville, A., Vincent, P.: Representation learning: a review and new perspectives. IEEE Trans. Pattern Anal. Mach. Intell. **35**(8), 1798–1828 (2013)
29. Harper, F.M., Konstan, J.A.: The movielens datasets: history and context. ACM Trans. Interact. Intell. Syst. **5**(4), 1–19 (2015)
30. Yang, M., Liu, F., Chen, Z., Shen, X., Hao, J., Wang, J.: CausalVAE: disentangled representation learning via neural structural causal models. In: Proceedings of the IEEE/CVF Conference on Computer Vision and Pattern Recognition, pp. 9593–9602 (2021)
31. Lian, J., et al.: Multi-interest-aware user modeling for large-scale sequential recommendations. arXiv preprint arXiv:2102.09211 (2021)
32. Liu, H., Lin, H., Chen, G.: TANTP: conversational emotion recognition using tree-based attention networks with transformer pre-training. In: Karlapalem, K. (ed.) PAKDD 2021. LNCS (LNAI), vol. 12713, pp. 730–742. Springer, Cham (2021). https://doi.org/10.1007/978-3-030-75765-6_58
33. Shi, S., Ma, W., Zhang, M., Zhang, Y., Ma, S.: Beyond user embedding matrix: learning to hash for modeling large-scale users in recommendation. In: SIGIR 2020: The 43rd International ACM SIGIR Conference on Research and Development in Information Retrieval (2020)
34. Batmaz, Z., Yurekli, A., Bilge, A., Kaleli, C.: A review on deep learning for recommender systems: challenges and remedies. Artif. Intell. Rev. **52**(1), 1–37 (2019)
35. He, R., McAuley, J.: Fusing similarity models with markov chains for sparse sequential recommendation. In: 2016 IEEE 16th International Conference on Data Mining (ICDM), pp. 191–200. IEEE (2016)
36. Cho, K., van Merriënboer, B., Bahdanau, D., Bengio, Y.: On the properties of neural machine translation: Encoder-decoder approaches. In: Proceedings of SSST-8, Eighth Workshop on Syntax, Semantics and Structure in Statistical Translation, pp. 103–111 (2014)
37. Pascanu, R., Mikolov, T., Bengio, Y.: On the difficulty of training recurrent neural networks. In: International conference on machine learning (PMLR), pp. 1310–1318 (2013)

Thompson Sampling with Time-Varying Reward for Contextual Bandits

Cairong Yan$^{(\boxtimes)}$, Hualu Xu, Haixia Han, Yanting Zhang$^{(\boxtimes)}$, and Zijian Wang

School of Computer Science and Technology, Donghua University, Shanghai, China
{cryan,ytzhang,wang.zijian}@dhu.edu.cn,
{hualuXu,haixiaHan}@mail.dhu.edu.cn

Abstract. Contextual bandits efficiently solve the exploration and exploitation (EE) problem in online recommendation tasks. Most existing contextual bandit algorithms utilize a fixed reward mechanism, which makes it difficult to accurately capture the preference changes of users in non-stationary environments, thus affecting recommendation performance. In this paper, we formalize the online recommendation task as a contextual bandit problem and propose a Thompson sampling algorithm with time-varying reward (TV-TS) that captures user preference changes from three perspectives: (1) forgetting past preferences based on a functional decay method while capturing possible periodic demands, (2) mining fine-grained preference changes from multi-behavioral implicit feedback, and (3) iterating the reward weights adaptively. We also provide theoretical regret analysis to demonstrate the sublinearity of the algorithm. Extensive empirical experiments on two real-world datasets show that our proposed algorithm outperforms state-of-the-art time-varying bandit algorithms. Furthermore, the designed reward mechanism can be flexibly configured to other bandit algorithms to improve them.

Keywords: Thompson sampling · time-varying reward · contextual bandit · personalized recommendation

1 Introduction

Users' feedback on items can help recommender systems uncover their preferences more accurately. These behaviors become an essential basis for optimizing and improving the results of the next recommendation. However, items that are not recommended and thus do not receive any feedback may also be of potential interest to the user, and this type of information is often overlooked. Therefore, there has been a classic challenge in online recommendation tasks: balancing exploration and exploitation, usually referred to as the EE problem [8].

Contextual bandits are inherently able to cope with this problem, so online recommendation tasks are often modelled as contextual bandit problems, where the reward settings for actions are critical to the models' performance. In a contextual bandit, N arms represent the available actions, and the reward for each arm obeys an independent fixed probability distribution. The model is trained using a reward mechanism, i.e., the agent pulls arms, receives feedback from the

X. Wang et al. (Eds.): DASFAA 2023, LNCS 13944, pp. 54–63, 2023.
https://doi.org/10.1007/978-3-031-30672-3_4

interaction with the environment, and then guides the selection of the best subsequent action [14]. When selecting arms, contextual bandits consider the characteristics and feedback of arms pulled in the past. Due to their high efficiency and interpretability, they are widely applied in the E-commerce, advertising, and even medical field.

Most existing contextual bandit models describe user feedback through fixed rewards [1,15], i.e., the probability distribution of rewards does not change over time. However, influenced by the ever-changing surroundings, users' behavior changes at different stages, which may lead to a drift in the rewards for each arm. If the same weight is maintained for the feedback generated by user behavior at different stages, a substantial part of the historical feedback data collected will become noise and interfere with the model's prediction of the user's current preference.

In this paper, we formalize online recommendation as a contextual bandit problem and propose a Thompson sampling algorithm for non-stationary scenarios to cope with changes in user preferences. Our contributions are as follows.

(1) We propose a time-varying reward mechanism (TV-RM). It captures user preference changes in two ways: forgetting past preferences based on a functional decay method while capturing possible periodic demands, and fine-grained mining preference changes from multi-behavioral implicit feedback. We design TV-RM in a modular fashion, and the module can be flexibly configured to other bandit algorithms to help improve performance.
(2) We propose a Thompson sampling algorithm with time-varying rewards(TV-TS). Each arm maintains a reward function with time-decaying properties and iterates the reward weights adaptively. Thus, the algorithm features the same time complexity as the traditional contextual Thompson sampling algorithm. We also prove that the upper bound on the cumulative regret of the algorithm is $O\left(d\sqrt{T\ln\frac{T}{\gamma}}\right)$ with sublinear properties.
(3) Extensive experiments show that the proposed TV-TS algorithm performs better in all four metrics than the stochastic bandits and the state-of-the-art time-varying bandit algorithms. Compared with the Thompson sampling algorithm, it improves the recommendation performance metric F1 by 2.18% and 0.75% on two datasets, IJCAI-15 and UDB, respectively.

2   Related Work

2.1   Adversarial Bandits

In adversarial bandits, rewards are no longer assumed to be obtained from a fixed sample set with a known distribution but are determined by the adversarial environment [2,3,11]. The well-known EXP3 [2] algorithm sets a probability for each arm to be selected, and all arms compete against each other to motivate continuous optimization in each round. Based on this, Vakili et al. [11] constructed a piece-wise stationary model with time-varying rewards for the reward distribution of the arms. Since the adversarial framework is more applicable to the non-stationary environment with significant variability, Besbes et al. [3] weighed the adversarial framework and stochastic contextual bandits and established an integrated model to provide more personalized recommendations.

2.2   Non-stationary Bandits

The sensitivity of non-stationary bandits to reward changes is somewhere between that of stochastic bandits and adversarial bandits. To better capture changes in user interest, two strategies are typically used to calculate rewards: sliding window [4,6,10,13], and discounted reward [5,9].

Sliding Window. Cheung et al. [4], and Xu et al. [13] proposed sliding window-based UCB algorithms, where fresh data are stored in a sliding window to eliminate the effect of obsolete data. Preference changes around the window are compared to detect abrupt reward changes. Trovò et al. [10] and Ghatak et al. [6] also proposed sliding window-based Thomson Sampling algorithms. These methods can cope with the non-stationarity of mutations and smooth environmental changes. Still, They cannot distinguish between fresh or outdated data better, set window sizes and thresholds for mutations.

Discounted Reward. In the bandit algorithms with discounted reward, indicating a smooth forgetting of past information at a rate of decay in a specific function. Russac et al. [9] proposed the D-LinUCB algorithm, which uses exponential weights to smoothly decay past information, providing sufficient theoretical guarantees for policy choice in both slow or abruptly changing environments. Deng et al. [5] leveraged a weighted Gaussian process regression approach to determine the reward weights.

3   Methodology

3.1   Problem Formulation

Suppose a contextual bandit with N arms that make up a set of arms $A = \{a_1, a_2, a_3, \ldots, a_N\}$. Each arm is associated with an d-dimensional feature vector x_t at time t. The reward r_t for each arm is generated by a function $x_t^T \mu_t$, where $\mu_t \in R^d$ is a stationary but unknown parameter sampling from Gaussian distribution $N\left(\hat{\mu}_t, V_t^{-1}\right)$. Here, $\hat{\mu}_t$ and V_t^{-1} can be calculated as

$$\hat{\mu}_t = V_t^{-1} \sum_{\tau=1}^{t-1} x_\tau r_\tau, V_t = I_d + \sum_{\tau=1}^{t-1} x_\tau x_\tau^T. \tag{1}$$

We assume that the prior of the arm a is given by $N\left(\hat{\mu}_t, V_t^{-1}\right)$ at time t, combining prior and likelihood function, the posterior distribution is calculated as

$$p(\hat{\mu}_t | r_t) \propto p\left(\hat{\mu}_t\right) p\left(r_t \mid \hat{\mu}_t\right). \tag{2}$$

Therefore, due to the conjugate property of the Gaussian distribution, the posterior can be rewritten as $N\left(\hat{\mu}_{t+1}, V_{t+1}^{-1}\right)$ at time $t + 1$.

In a contextual bandit problem, the agent chooses to pull the arm with the highest expected reward, observes its actual reward based on the user feedback, and updates its reward distribution before continuing with the next round of recommendations. We define regret $R(t)$ as the difference between the expected reward $x_t^{*T} \mu_t^*$ of the best arm a^* and the actual reward $x_t^T \mu_t$ of the pulled arm. The objective of the agent is to minimize the cumulative regret $R(\mathcal{T})$,

$$R(\mathcal{T}) = \sum_{t=1}^{\mathcal{T}} R(t) = \sum_{t=1}^{\mathcal{T}} \left(x_t^{*T} \mu_t^* - x_t^T \mu_t \right). \tag{3}$$

3.2   Time-Varying Reward Mechanism

To cope with the non-stationary environment, we propose a time-varying reward mechanism (TV-RM). It focuses on three aspects to reveal the judgment and change in user preferences: (1) forgetting past preferences based on a functional decay method while capturing possible periodic demands, (2) mining fine-grained reference value of reward from implicit feedback of multiple behaviors, (3) and adaptively iterating the weights of rewards over time.

Periodic Forgetting Strategy. Users' interest in items tends to fade over time. For example, a user purchased an item a long time ago but hasn't had any interaction with that item since then. Then we assume that his interest in this has waned over that period of time. This phenomenon is known as "smoothly forgetting" [9]. However, once the user interacts with it again, he regains interest and continues to forget on the basic of the new memory slowly, and so on. In such an application scenario, we map the user's interest in the items to the average reward of the arms, while treating the degeneration of interest between two interactions as a cycle, which shows a periodic decay over time.

When setting up dynamic rewards, we consider two aspects. First, we utilize a time-dependent *sigmoid* function to weaken the weight of past rewards to avoid the infinite increase in rewards and to counteract the decay of user attention. Then, we define n_t to indicate the number of times each arm has not been pulled recently. Specifically, at time t, the estimate of n_t for arm a is given by

$$n_t = \varepsilon + \mathbb{1}\,(I_t \neq a) \sum_{\tau=j}^{t} \mathbb{1}\,(I_\tau \neq a), \tag{4}$$

where ε is a value greater than 0 but minimal, j represents the moment when the arm was last pulled, and $\mathbb{1}\,(I_t \neq a)$ indicates whether arm a was pulled at round t. Let $f_t = \sum_{\tau=1}^{t-1} x_\tau r_\tau$ denote a part of the composition of $\hat{\mu}_t$. Then, we get f_{t+1} for arm a:

$$f_{t+1} = f_t \cdot sigmoid\left(n_t^{-1}\right) + x_t r_t. \tag{5}$$

Based on the above definition, n_t is assigned a tiny value indicating that the proportion of historical reward reaches a high point for the arm that has just been pulled. For the unpulled arms, the reward value decreases, where n_t indirectly reflects the periodicity of the reward.

Fine-Grained Preference Mining Strategy. We classify user's behavior into strong and weak interactions, and use them as the basis for reward setting. The strong interaction behavior arms A^s include arms with multiple clicks, favorite, add, or purchase, indicating that the user is interested in A^s. The positive feedback is generated with a value of 1, indicating a successful recommendation. The weak interaction behavior arms A^w are defined as no interaction or a single click without subsequent purchase, indicating that the user is not interested in A^w. Conversely, a negative feedback value of -1 is generated. We use s_t to denote the implicit feedback,

$$s_t = \begin{cases} 1, a \in A^s \\ -1, a \in A^w \end{cases}.$$

(6)

Adaptive Reward Weight Setting Strategy. For the reward at time t, we assign a reward weight $\Delta \bar{w}_t$ to the expected arm a^{opt} such that its reward mean μ_{t+1}^{opt} maintained at time $t+1$ will have the ability to outperform the arm a^* that currently has the best reward mean μ_t^*, facilitating the next round of accurate recommendations. According to the descriptions above, for $\xi > \mu_{t+1}^{opt} - \mu_t^{opt}$, Δw_t is defined as

$$\Delta w_t = (\mu_t^* - \mu_t^{opt}) + \xi.$$

(7)

To avoid storing a large number of historical operations and to accommodate slow changes, we update the weights in a fully recursive manner to correct the latest reward weight $\Delta \bar{w}_t$:

$$\Delta \bar{w}_t = \frac{(t-1)\Delta \bar{w}_{t-1} + \Delta w_t}{t}.$$

(8)

$$\begin{aligned} \Delta w_{t+1} - \Delta w_t &= (\mu_{t+1}^* - \mu_{t+1}^{opt}) - (\mu_t^* - \mu_t^{opt}) \\ &= \mu_t^{opt} - \mu_{t+1}^{opt} + \xi, \end{aligned}$$

(9)

when $\xi > \mu_{t+1}^{opt} - \mu_t^{opt}$, we get $\Delta w_{t+1} > \Delta w_t$, so $\Delta \bar{w}_t$ is a function that increases with time. Therefore, we obtain a weighted time-varying bandit with general nonlinear rewards $\mathbb{r}_t = s_t \Delta \bar{w}_t$. In this way, f_{t+1} is updated as

$$f_{t+1} = f_t \cdot sigmoid\left(n_t^{-1}\right) + x_t \mathbb{r}_t.$$

(10)

Combining the above three points, we apply TV-TM to a complete CMAB. Figure 1 shows the workflow of a contextual bandit configured with the TV-RM in a non-stationary case, detailing the components of the time-varying reward.

When the agent pulls the arm with the highest expected reward and makes recommendations to users, the environment will generate feedback to judge the effectiveness of this round. The mechanism intelligently predicts the payoffs generated and the method by which they should be decayed for past rewards.

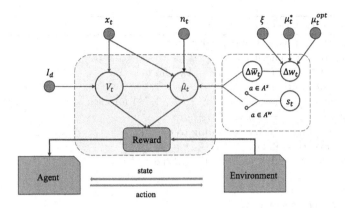

Fig. 1. The workflow of a contextual bandit configured with TV-RM.

3.3 Thompson Sampling Algorithm with Time-Varying Reward

It was shown that contextual bandit has a low cumulative regret value [15]. Therefore, based on the Thompson sampling algorithm for contextual bandit, this paper integrates the TV-RM to capture changes in user interest dynamically. We first build arms for the contextual bandit by referring to the method of [13], each arm represents a cluster of items with the same characteristics, and their rewards obey the corresponding Gaussian distribution. In each round, the agent randomly samples a parameter μ_t from the reward distribution of each arm and observes its associated d-dimensional contextual vector x_t, which is the sequence of the selected arms in such a situation [15]. The payoff obtained by arm a is calculated by $x_t^T \mu_t$. At this time, the agent pulls the arm a_t^* with the highest expected reward to generate a recommendation list, and finally updates the reward parameters according to the TV-RM. Algorithm 1 shows the process of TV-TS with TV-RM.

Regret Analysis of TV-TS . we will prove that the TV-TS algorithm has sublinear properties. First, we give the following definitions to help the subsequent proof.

Definition 1. *Super-martingale [1].* $X_t = \mathbb{I}\left(N_{x_t^*} > 0\right) x_t^{*T} \mu_t^* - \mathbb{I}\left(N_{x_t} > 0\right)$ $x_t^T \mu_t$, $Y_t = \sum_{\tau=1}^t X_\tau$, *where* Y_t *denotes the super-martingale process.*

Definition 2. *Confidence Ellipsoid [7].* Define $\beta_t(\delta) = R\sqrt{d\ln\frac{1+tL/\lambda}{\delta}} + \sqrt{\lambda L}$, *we can get* $\parallel \hat{\mu}_t - \mu_t \parallel_{V_t} \leq \beta_t(\delta)$ *with probability* $1 - \delta$.

The cumulative regret $R(\mathcal{T})$ has been defined in Eq. 3, which depends on the reward distribution $N(x_t^T \mu_t, \sigma^2)$. The whole process is divided into four steps:

Step1: We assume that $0 \leq\parallel x_t \parallel\leq 1$, $0 \leq\parallel \mu_t \parallel\leq 1$ for all arms. Then the inequality $0 \leq x_t^T \mu_t \leq 1$ is true, so $X_t \leq 1$. Then according to the Azuma-Hoeffding inequality, we get that $Y_{\mathcal{T}} \leq 1\sqrt{2\mathcal{T}\ln\frac{2}{\delta}}$ with probability $1 - \delta$.

Algorithm 1: TV-TS algorithm

Input: Set of arms A, Contextual vector x_t
Output: Recommend list L
1 **Init:** $f_0 = 0^d$, $V_0 = I \in R^{d \times d}$, $\hat{\mu}_0 = V_0^{-1} f_0$, $n_0 = \varepsilon$
2 **for** $t = 1$ to T **do**
3   Sampling μ_t from distribution $N\left(\hat{\mu}_t, V_t^{-1}\right)$ for each arm
4   Choose arm $a_t^* = argmax_{a \in A} x_t^T \mu_t$
5   Get recommend list L according to a_t^*
6   **Observe payoff** $\mathbf{r}_t$:
7   Set s_t, Δw_t, $\Delta \bar{w}_t$ as Eq.6, Eq.7, Eq.8
8        $\mathbf{r}_t = s_t \cdot \Delta \bar{w}_t$
9   **for** $a \in A$ **do**
10   **if** $\mathbb{1}\left(I_t \neq a\right)$ **then**
11          $n_t = n_{t-1} + 1$
12   **else**
13          $n_t = \varepsilon$
14   Update V_{t+1}, f_{t+1} and $\hat{\mu}_{t+1}$ as follows:
15        $f_{t+1} = f_t \cdot sigmoid\left(n_t^{-1}\right) + x_t \mathbf{r}_t$
16        $V_{t+1} = V_t + x_t x_t^T$
17        $\hat{\mu}_{t+1} = V_{t+1}^{-1} f_{t+1}$

Step2: According to the Cauchy-Schwarz inequality and Deduction of Confidence Ellipsoid in Definition. 2, for any vector x_t,

$$x_t^T \left(\mu_t - \hat{\mu}_t\right) = x_t^T V_t^{-\frac{1}{2}} V_t^{\frac{1}{2}} \left(\mu_t - \hat{\mu}_t\right) \leq \| x_t \|_{V_t^{-1}} \beta_t\left(\delta\right). \tag{11}$$

Step3: For each round, the regret value can be calculated by

$$\begin{aligned} R\left(t\right) &= X_t + \left(x_t^{*T} \mu_t - x_t^{*T} \hat{\mu}_t\right) - \left(x_t^T \mu_t - x_t^T \hat{\mu}_t\right) \\ &\leq X_t + \| x_t^* \|_{V_t^{-1}} \beta_t\left(\delta\right) + \| x_t \|_{V_t^{-1}} \beta_t\left(\delta\right). \end{aligned} \tag{12}$$

Step4: According to the previous definition of $\beta_t\left(\delta\right)$, Y_T and $\| x_t \|_{V_t^{-1}}$, let $\delta = \frac{\gamma}{6(T+1)^2}$ replace the parameter, we can get $\beta_t\left(\delta\right) = O(\sqrt{d \ln T^3/\gamma})$, $Y_T = O(\sqrt{T \ln \frac{T^2}{\gamma}})$, and $\sum_{t=1}^T \| x_t \|_{V_t^{-1}} = O(\sqrt{dT})$.

$$R\left(T\right) \leq Y_T + \sum_{t=1}^T \left(\| x_t^* \|_{V_t^{-1}} + \| x_t \|_{V_t^{-1}}\right) \beta_t\left(\delta\right) = O\left(d\sqrt{T \ln \frac{T}{\gamma}}\right). \tag{13}$$

To summarize, we prove that the upper bound of the cumulative regret of the algorithm is $O\left(d\sqrt{T \ln \frac{T}{\gamma}}\right)$ with sublinear properties.

4 Experiment and Evaluation

In this section, we conduct extensive experiments to answer the following questions:

RQ1: How does our proposed TV-TS algorithm perform on real-world datasets compared to the baselines for the online recommendation task?

RQ2: Can the proposed novel reward mechanism TV-RM be applied to other classical bandit algorithms and improve their performance?

4.1 Experiment Setting

We evaluate all algorithms on two open e-commerce datasets, IJCAI-15[1] and UBD[2], which provides a large number of real behavior records of random users. We employ precision, recall, F1-measure (F1), and cumulative regret to evaluate the algorithms. Especially, cumulative regret is an essential indicator for measuring algorithms to solve the problem of exploration and exploitation trade-off inherent, which has already been defined in Eq. 3. Lower regret values indicate better results. As for baseline, We have implemented some well-known bandit algorithms for comparison, covering both classic stochastic bandits such as ε-greedy, Upper Confidence Bound (UCB), and Thompson sampling (TS), as well as time-varying bandits such as Exp3.S [11], var-TS [12], and SW-TS [10]. All these algorithms set the same parameters except for the reward mechanism.

4.2 Performance Comparison (RQ1)

We compare our algorithm TV-TS with six baselines, and the different metric values for each algorithm on the two datasets are shown in Table 1.

Table 1. Performance comparison of different bandit algorithms on two datasets.

	algorithm	precision↑	recall↑	F1↑	regret↓
IJCAI-2015	ε-greedy	0.0196	0.0266	0.0226	94153
	UCB	0.0198	0.0265	0.0226	93596
	TS	0.0213	0.0278	0.0241	89739
	Exp3.S	0.0175	0.0133	0.0151	122982
	var-TS	0.0277	0.0258	0.0267	82271
	SW-TS	0.0322	0.0270	0.0293	80597
	TV-TS	**0.0423**	**0.0503**	**0.0459**	**78684**
UDB	ε-greedy	0.0039	0.0055	0.0045	163709
	UCB	0.0052	0.0046	0.0048	168462
	TS	0.0091	0.0084	0.0087	131281
	Exp3.S	0.0053	0.0039	0.0044	163445
	var-TS	0.0094	0.0080	0.0086	137254
	SW-TS	0.0105	0.0102	0.0103	126761
	TV-TS	**0.0166**	**0.0159**	**0.0162**	**105220**

---

[1] https://tianchi.aliyun.com/dataset/dataDetail?dataId=42.

[2] https://tianchi.aliyun.com/dataset/dataDetail?dataId=649.

First, we compare TV-TS with the three classic stochastic bandit algorithms, ε-greedy, UCB, and TS. Our algorithm performs best on all four metrics. Second, we compare TV-TS with the three time-varying bandit algorithms, our algorithm achieves the best results on all three metrics regarding recommendation performance and one metric regarding the EE problem.

4.3   Effect of TV-RM (RQ2)

Setting rewards is a vital part of the bandit algorithm. We design TV-RM to be flexibly configured into the stochastic bandit algorithms. Figure 2 shows the significant improvement in recommendation performance after configuring the TV-RM module on both stochastic bandit algorithms. Compared with TS, TV-TS improves F1 scores by 2.18% and 0.75% on both datasets and decreases cumulative regret values by 11,055 and 26,061, respectively. Similarly, TV-UCB is also better than UCB.

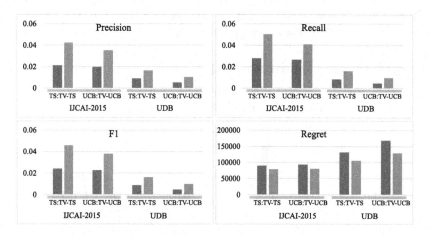

Fig. 2. Effect of TV-RM on two classic bandit algorithms.

5   Conclusion

Reward setting is an important component of the bandit model. In this paper, we propose a time-varying reward mechanism for non-stationary environments that exploits the fine-grained and periodic features of behavior and the time-decaying nature of memory to help obtain more accurate user preferences, respectively, and then consequently propose a time-varying contextual Thompson sampling algorithm. We demonstrate the sublinearity of the proposed algorithm through regret analysis. Extensive experimental results on two public datasets also show that the algorithm outperforms the state-of-the-art time-varying bandit algorithms. Moreover, the proposed reward mechanism can be flexibly configured to other bandit algorithms to help improve the recommendation performance.

Acknowledgements. This work is supported by National Natural Science Foundation of China (No. 62206046) and Shanghai Science and Technology Innovation Action Plan Project (No. 22511100700).

References

1. Agrawal, S., Goyal, N.: Thompson sampling for contextual bandits with linear payoffs. In: Proceedings of the 30th International Conference on Machine Learning, pp. 127–135 (2013)
2. Auer, P., Cesa-Bianchi, N., Freund, Y., Schapire, R.E.: Gambling in a rigged casino: the adversarial multi-armed bandit problem. In: Proceedings of the 36th Annual Foundations of Computer Science, pp. 322–331 (1995)
3. Besbes, O., Gur, Y., Zeevi, A.: Stochastic multi-armed-bandit problem with non-stationary rewards. In: Proceedings of the 28th Conference on Neural Information Processing Systems, pp. 199–207 (2014)
4. Cheung, W.C., Simchi-Levi, D., Zhu, R.: Learning to optimize under non-stationarity. In: Proceedings of the 22nd International Conference on Artificial Intelligence and Statistics, pp. 1079–1087 (2019)
5. Deng, Y., Zhou, X., Kim, B., Tewari, A., Gupta, A., Shroff, N.: Weighted gaussian process bandits for non-stationary environments. In: Proceedings of the International Conference on Artificial Intelligence and Statistics, pp. 6909–6932 (2022)
6. Ghatak, G.: A change-detection-based thompson sampling framework for non-stationary bandits. IEEE Trans. Comput. **70**(10), 1670–1676 (2020)
7. Li, C., Wang, H.: Asynchronous upper confidence bound algorithms for federated linear bandits. In: Proceedings of The 25th International Conference on Artificial Intelligence and Statistics, pp. 6529–6553 (2022)
8. Liu, E.Z., Raghunathan, A., Liang, P., Finn, C.: Decoupling exploration and exploitation for meta-reinforcement learning without sacrifices. In: Proceedings of the 38th International Conference on Machine Learning, pp. 6925–6935 (2021)
9. Russac, Y., Vernade, C., Cappé, O.: Weighted linear bandits for non-stationary environments. In: Advances in Neural Information Processing Systems (2019)
10. Trovo, F., Paladino, S., Restelli, M., Gatti, N.: Sliding-window thompson sampling for non-stationary settings. J. Artif. Intell. Res. **68**, 311–364 (2020)
11. Vakili, S., Zhao, Q., Zhou, Y.: Time-varying stochastic multi-armed bandit problems. In: Proceedings of the 48th Asilomar Conference on Signals, Systems and Computers, pp. 2103–2107 (2014)
12. Xu, L., Jiang, C., Qian, Y., Zhao, Y., Li, J., Ren, Y.: Dynamic privacy pricing: a multi-armed bandit approach with time-variant rewards. IEEE Trans. Inf. Forensics Secur. **12**(2), 271–285 (2016)
13. Xu, X., Dong, F., Li, Y., He, S., Li, X.: Contextual-bandit based personalized recommendation with time-varying user interests. In: Proceedings of the 34th AAAI Conference on Artificial Intelligence, pp. 6518–6525 (2020)
14. Yan, C., Han, H., Zhang, Y., Zhu, D., Wan, Y.: Dynamic clustering based contextual combinatorial multi-armed bandit for online recommendation. Knowl.-Based Syst. **257**, 109927 (2022)
15. Zhu, Z., Huang, L., Xu, H.: Self-accelerated thompson sampling with near-optimal regret upper bound. Neurocomputing **399**, 37–47 (2020)

Cold-Start Based Multi-scenario Ranking Model for Click-Through Rate Prediction

Peilin Chen[1], Hong Wen[1], Jing Zhang[2], Fuyu Lv[1], Zhao Li[3], Qijie Shen[1], Wanjie Tao[1], Ying Zhou[4], and Chao Zhang[3(✉)]

[1] Alibaba Group, Hangzhou, China
{peilin.cpl,qinggan.wh,fuyu.lfy,qijie.sqj,wanjie.twj}@alibaba-inc.com
[2] The University of Sydney, Sydney, Australia
[3] Zhejiang University, Hangzhou, China
zczju@zju.edu.cn
[4] Zhejiang Lab, Hangzhou, China
zhouying@zhejianglab.com

Abstract. Online travel platforms (OTPs), *e.g.*, Ctrip.com or Fliggy.com, can effectively provide travel-related products or services to users. In this paper, we focus on the multi-scenario click-through rate (CTR) prediction, *i.e.*, training a unified model to serve all scenarios. Existing multi-scenario based CTR methods struggle in the context of OTP setting due to the ignorance of the cold-start users who have very limited data. To fill this gap, we propose a novel method named Cold-Start based Multi-scenario Network (CSMN). Specifically, it consists of two basic components including: 1) User Interest Projection Network (UIPN), which firstly purifies users' behaviors by eliminating the scenario-irrelevant information in behaviors with respect to the visiting scenario, followed by obtaining users' scenario-specific interests by summarizing the purified behaviors with respect to the target item via an attention mechanism; and 2) User Representation Memory Network (URMN), which benefits cold-start users from users with rich behaviors through a memory read and write mechanism. CSMN seamlessly integrates both components in an end-to-end learning framework. Extensive experiments on real-world offline dataset and online A/B test demonstrate the superiority of CSMN over state-of-the-art methods.

Keywords: Click-Through Rate Prediction · Multi-Scenario · Cold-Start Recommendation · User's Scenario-Specific Interest

1 Introduction

With large-scale travel-related products and services available on the Online Travel Platforms (OTPs), *e.g.*, Ctrip.com or Fliggy.com, Click-Through Rate (CTR) prediction [15,26,29], which aims at predicting the probability of users clicking items, has been playing an increasing role for delivering high-quality recommendation results and boosting the final platform revenues. Nowadays,

X. Wang et al. (Eds.): DASFAA 2023, LNCS 13944, pp. 64–79, 2023.
https://doi.org/10.1007/978-3-031-30672-3_5

the majority of CTR models are mainly built for single-scenario problems, *i.e.*, providing online service exactly for a scenario after trained only with data from the scenario. Here, a scenario refers to a specific spot where items are displayed, *e.g.*, *Guess You Like* module in our app homepage. However, in many industrial applications of travelling recommendation, a user may be engaged with multiple travel scenarios, *e.g.*, *Hot Spring*, *Skiing*, *etc.*, where each scenario has its corresponding relevant candidate items to display, such as *ski spots nearby* could be displayed to users who are visiting the *Skiing* scenarios. In addition, the users in different scenarios also have different travel intentions such as *leisure* for *Hot Spring*, and *adventure* for *Skiing*. Consequently, multi-scenario CTR prediction, which is of great practical significance but has been largely under-explored, deserves more research efforts from academia and industry.

A straightforward strategy is to build an individual model for each business scenario using the single-scenario CTR prediction methods, *i.e.*, DIN [29] or DIEN [28]. However, it has two apparent shortcomings: 1) training models for small-scale scenarios may suffer from severe data sparsity problem, and 2) maintaining different models for all business scenarios will be costly, which motivates us to devise a unified CTR prediction model to serve multiple scenarios simultaneously. Another possible strategy is to employ Multi-Task Learning (MTL) methods [2,14], *i.e.*, one task for the corresponding scenario. However, we argue that MTL methods have significant difference from multi-scenario methods, where MTL methods simultaneously address various types of tasks in the same scenario, such as jointly predicting the CTR and conversion rate (CVR) tasks [23–25], while multi-scenario methods always focus on the same task, *i.e.*, CTR task, across multiple scenarios. What's more, it is very necessary for multi-scenario CTR methods to capture the scenario-shared information of various scenarios and explore scenario-specific characteristics simultaneously, where scenario-shared information means the overlapping users and candidate items among multiple scenarios, and scenario-specific characteristics indicates that users' interests with respect to different scenarios would be significantly different due to the data scale or topic-specific preferences among different scenarios. For facilitating following narration, we regard scenario-shared information and scenario-specific characteristics across various scenarios as the *commonality* and *discrimination* property, respectively. In fact, how to devise a unified and elaborate CTR prediction model for multiple scenarios is very challenging, especially exploiting the *commonality* and *discrimination* properties simultaneously.

To achieve this goal, several representative works towards the multi-scenario CTR prediction have been proposed. For example, STAR [18] trains a single model to serve all scenarios simultaneously by leveraging data from all scenarios, capturing users' interests effectively by employing shared centered parameters and scenario-specific parameters to exploit the *commonality* and *discrimination* property, respectively. SAR-Net [17] learns users' scenario-specific interests by harnessing the abundant data from different scenarios via two specific attention modules, leveraging the scenario features and item features to modulate users' behaviors for exploring the *discrimination* property effectively. Meanwhile, it utilizes the bottom-shared embedding parameters to exploit the *commonality* property. DADNN [8] devises a unified model to serve multiple scenarios, where shared bottom block among all scenarios is employed to exploit the *commonality* property, while scenario-specific head captures the characteristics of every scenario, *i.e.*, exploiting the *discrimination* property. Although aforementioned methods have achieved remarkable performance for multi-scenario CTR prediction task with the consideration of the *commonality* and *discrimination* properties, they all neglect the cold-start issue which are high-frequently encountered in the OTPs setting. In practice, users' behaviors on OTPs are quite sparse or even absent compared with other e-commerce platforms since travel is a low-frequency demand, resulting in the cold-start issue and making it ineffective to learn the cold-start users' personalized preferences. How to tackle the cold-start issue in the context of multiple scenarios, especially considering the *commonality* and *discrimination* properties, has been unexplored and remains challenging.

To fully tackle the cold-start issue while exploiting the *commonality* and *discrimination* properties, in this paper, we propose a novel method named Cold-Start based Multi-scenario Network (CSMN). It consists of two fundamental components including a User Interest Projection Network (UIPN) and a User Representation Memory Network (URMN). Specifically, UIPN firstly purifies users' behavior representations by eliminating the scenario-irrelevant information in behaviors with respect to the visiting scenario, then summarizes the purified behaviors with respect to the target item via an attention mechanism. In other words, even though a user has same and fixed behaviors, each behavior will obtain varied yet purified representation across different scenarios, followed by extracting user scenario-specific interest via attention mechanism, thus exploiting the *discrimination* property. In addition, the resultant representation from UIPN component would be delivered to the URMN component, further being as a supplement cue to infer the interests of cold-start users. Specifically, URMN can make users with sparse behaviors benefit from users with rich behaviors through a memory read and write mechanism, where each slot in the memory can be regarded as a cluster of users who share similar profiles given target item at specific scenario, resulting in cold-start users can absorb well-purified interest representations from users with similar profiles yet rich behaviors, mitigating the cold-start issue effectively. Meanwhile, CSMN utilizes shared bottom block among all scenarios to address the *commonality* property. The contributions of this paper is three-fold:

- We propose a novel method named Cold-Start based Multi-scenario Network (CSMN) for multi-scenario CTR prediction task, which facilitates learning users' distinct interest representations across multiple scenarios.
- We devise two key components, including URMN and UIPN, to jointly address the aforementioned cold-start issue in the multi-scenario setting in an end-to-end manner, where the *commonality* and *discrimination* properties are effectively exploited for multi-scenario modelling.
- We conduct extensive experiments on both real-world offline dataset and online A/B test. The results demonstrate the superiority of the proposed

CSMN over representative methods. CSMN now serves millions of users in our online travel platform, achieving 3.85% CTR improvement.

2 Related Work

CTR Prediction: Recently, academia and industry communities have paid a lot of attention on CTR prediction not only from the perspective of feature interactions, *e.g.*, DCN [22], NCF [9], DeepFM [7], AutoInt [19], but also from the perspective of users' sequential modelling, *e.g.*, DIN [29], DIEN [28], MIMN [15]. Apart from these methods for single-scenario CTR prediction, multi-scenario CTR predictions also have drawn increasing attention, *i.e.*, training a unified model to serve all scenarios simultaneously. For example, STAR [18] employs shared centered parameters and scenario-specific parameters to learn users' interest representation across multiple scenarios. SAR-Net [17] learns users' scenario-specific interests by harnessing the abundant data from different scenarios via specific attention modules. However, both methods ignore the cold-start issue in the OTPs setting, thereby struggling in effectively discovering cold users' real interests across multiple scenarios. By contrast, our proposed CSMN model leverages a User Representation Memory Network to interpret the interest of users with sparse behaviors from those of users with rich behaviors.

Cold Start Recommendation: Cold start issue has been widely recognized in representative recommender systems. Typically, there are three kinds of solutions to address it including: 1) resorting to more generalized auxiliary and contextual information [1,4,13]; 2) cross-domain transfer (CDT) [10,27], *i.e.*, users may have interactions with items in one domain while not in the other relevant domain. The goal of CDT is to effectively infer cold-start users' preferences based on their interactions from one domain to the other relevant domain; and 3) meta-learning approaches [5,12], which argue that users with similar inherent profiles should be recommended with similar items by leveraging users' few behaviors. Despite effective, the SOTA methods for cold-start issue still struggle in the OTPs setting, since they do not address the cold-start issue in the context of multi-scenario CTR prediction. By contrast, our proposed CSMN can benefit users with sparse or even absent behaviors from users with rich behaviors through external memory mechanism, where each slot can be regarded as a cluster of users who share similar profiles, resulting in cold-start users can obtain interest representations from users with similar profiles yet rich behaviors.

Multi-Task Learning: Multi-Task Learning (MTL) [2,14,16,20] has been widely used in recommender systems, which benefits from the multi-objective optimization. For example, MMoE [14] extends the efficient Mixture-of-Experts (MoE) shared-bottom structure to exploit a light-weight gating network to model the relationship of various tasks, which has been demonstrated to handle the task-specific information in a highly efficient manner. Going one step further, to address the seesaw phenomenon, PLE [20] adopts a progressive routing mechanism to gradually extract and separate deeper semantic knowledge. In the context of the multi-scenario prediction task, it makes prediction for multiple scenarios towards the same task, *i.e.*, the CTR task, where the label spaces are same.

Although we can build individual network for corresponding scenario on top of a shared-bottom structure, followed by employing classical MTL approaches for multi-objective optimization. However, the consistency and discrepancy of various scenarios are coupled with each other tightly, resulting in the sophisticated relationships of multiple scenarios are difficult to disentangle. By contrast, we propose a User Interest Projection Network to disentangle scenario-specific interests from users' historical behaviors.

3   The Proposed Approach

In this paper, we propose a novel model named Cold-Start based Multi-scenario Network (CSMN) for multi-scenario CTR prediction. As depicted in Fig. 1, it consists of three basic components including Embedding Layer, User Interest Projection Network (UIPN), and User Representation Memory Network (URMN). We will introduce them in detail.

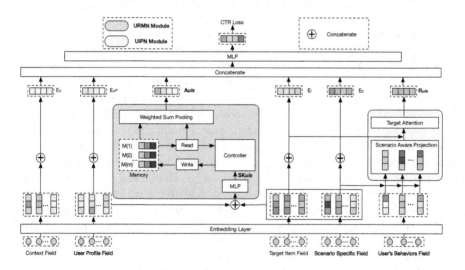

Fig. 1. The overview architecture of the proposed CSMN model, which consists of Embedding Layer, User Representation Memory Network (URMN), and User Interest Projection Network (UIPN). Symbol SK_{uis}, A_{uis}, R_{uis} denotes the comprehensive representation of the users' profiles, target item and current visiting scenario, user's augmented interest for the target item at specific scenario, users' purified scenario-specific interests by leveraging the target items and current visiting scenarios, respectively.

3.1   Problem Definition

In this section, we formally define the problem of multi-scenario CTR prediction task. Let $\mathcal{U} = \{u_1, u_2, ..., u_N\}$, $\mathcal{I} = \{i_1, i_2, ..., i_M\}$, $\mathcal{S} = \{s_1, s_2, ..., s_K\}$, $\mathcal{C} = \{c_1, c_2, ..., c_L\}$ be a set of N users, a set of M items, a set of K scenarios, a set of L contexts, respectively. For facilitating the following narration, we omit the subscript and use symbols u, i, s, c to denote user u_n, item i_m, scenario s_k, context c_l, respectively. And the user-item interaction at specific scenario is typically formulated as a matrix $Y = \{y_{uis}\}_{N \times M \times K}$. Specifically, $y_{uis} = 1$ means user u has clicked item i at the scenario s, otherwise $y_{uis} = 0$. In this paper, we mainly employ five types of input features namely *User Profiles* u^P, *User Behaviors* u^B, *Target item* i, *Context* c and *scenario* s for each sample, where u^P contains *age, sex, purchase power, etc.*, u^B denotes the sequential list of users visiting the set of items, i contains *item ID, item's category ID, etc.*, c contains *weather, time, etc.*, s contains *scenario ID, scenario's accumulated CTR, etc.*, and u can be defined as $\{u^B; u^P\}$. Now, our learning goal is to train a unified CTR model to predict the probability $\hat{y}_{uis}$ of user u clicks the target item i at scenario s given u, i, s, c, formulated as: $\hat{y}_{uis} = \mathcal{F}(u, i, s, c; \theta)$, where $\mathcal{F}$ and θ denote the learning objective and model parameters for the multi-scenario CTR prediction task, respectively.

3.2   Embedding Layer

Most data from industrial recommender systems are presented in a multi-field manner, where the fine-grained feature in each field is normally transformed into high-dimensional sparse one-hot features. For example, the one-hot vector representation of *male* from the *user sex* field can be decoded as $[1, 0]^T$. Without loss of generality, we divide the raw data into five groups: target item i, user' historical behaviors u^B, user-specific profiles u^P, scenario information s and context information c. Assuming the concatenation results of different fields' one-hot vectors from these five groups as X_i, X_{u^B}, X_{u^P}, X_s and X_c, respectively, they can be further transformed into low dimensional dense representations by multiplying corresponding embedding matrices, denoted as E_i, E_{u^B}, E_{u^P}, E_s and E_c, respectively, *e.g.*, $E_{u^B} = [e_1; e_2; ...; e_T]$, where T and e_t represent the length of users' behaviors and the embedding feature of the t-th behavior, respectively. In this paper, we also employ bottom-shared embedding parameters to exploit the *commonality* property. Since the bottom-shared embedding parameters make up the majority of the trainable parameters, they can be learned sufficiently by information sharing among overlapping users and candidate items, thereby avoiding the overfitting issue.

3.3   User Interest Projection Network

Generally, users' interests can be effectively extracted from users' historical behaviors. For example, as an excellent representative method for users' interest extraction, DIN [29] firstly employs the attention mechanism to dynamically compute the weight of users' historical behaviors with respect to different target items, followed by utilizing a weighted-sum pooling operation to adaptively generate users' interest representation. Despite effective, we figure out it is not directly suitable for the multi-scenario CTR task in the OTPs setting, where not only the target item but also the scenario-specific information can affect users' interests. For example, when a user comes into two scenarios with varied topics, *e.g.*, *Hot Springs* and *Skiing* scenarios, user's interest can be significantly different since the user may be concerned about the *leisure* in the *Hot Springs* scenario while preferring *outdoor adventure* in the *Skiing* scenario.

Specifically, we argue that even though given the same target item and fixed user's behaviors, the representations of users' behaviors with respect to different scenarios will be inevitably varied. Therefore, a straightforward strategy towards the multi-scenario CTR prediction is to disentangle the scenario-specific characteristics from users' behaviors with respect to the current visiting scenario, which motivates us to propose the User Interest Projection Network (UIPN) module. Specifically, the representation of each element in E_{u^B} will be projected into the orthogonal space of the scenario embedding E_s to eliminate the scenario-irrelevant information. Formally, without loss of generality, we illustrate the orthogonal mapping process with a randomly selected element e_i from E_{u^B} and the dense feature E_s of scenario s. Firstly, we embed e_i, E_s into the same space by multiplying individual mapping matrix W_o, W_s, respectively, *i.e.*, $f_i = W_o e_i$, $f_s = W_s E_s$. Then, the refined preference representation vector f_i^p with respect to scenario s can be obtained by projecting the vector f_i onto the direction of vector f_s, defined as $f_i^p = project(f_i, f_s)$, where $project(,)$ denotes the scenario aware projection operator, *i.e.*, $project(a, b) = \frac{ab}{|b|}\frac{b}{|b|}$. $|\cdot|$ denotes the norm of a vector. In this manner, the original E_{u^B} can be formulated as $f_{u^B} = \{f_1^p; f_2^p; ...; f_T^p\}$. Inspired by the Multi-Head Self-Attention (MHSA) mechanism [21], which can effectively capture the dependency between any pair within the sequence despite their distance, we use it to further enhance the representation of users' preference. Specifically, given $f_{u^B} = \{f_1^p; f_2^p; ...; f_T^p\}$, we can obtain the enhanced representation $f'_{u^B} = \left\{ f_1'^p; f_2'^p; ...; f_T'^p \right\}$ after applying MHSA on f_{u^B}.

Next, to obtain users' purified interests from f'_{u^B}, we need to calculate the similarity between the target item i and each element of f'_{u^B}, which can be formulated as $\alpha_t = Relu(z^T tanh(W_i E_i + W_f f_t'^p + b))$, where, z, W_i, W_f, and b are all learnable parameters. After normalization, *i.e.*, $\alpha_t = \frac{exp(\alpha_t)}{\sum_{i=1}^T exp(\alpha_i)}$, denoting the weight for the t-th behavior with respect to the target item i. Therefore, the final user' purified interest representation R_u from UIPN can be calculated as $R_{uis} = \sum_{i=1}^T \alpha_i f_i'^p$ via the weighted-sum pooling operation. In this way, UIPN can effectively achieve the interests of users who have rich historical behaviors with respect to current visiting scenario and the target item within it, which can be further exploited as a supplement cue to infer the interests of cold-start users who have similar profiles with rich behavior users.

3.4   User Representation Memory Network

In the OTP settings, users' behaviors are quite sparse compared with other typical e-commerce platforms, *i.e.*, resulting in the cold-start issue and making it difficult to extract users' interests from their behaviors. However, we argue that users' interests not only can be reflected from their behaviors but also from their inherent profiles. In other word, users' behaviors can be regarded as the embodiment of their inherent profiles. For example, when providing online service for a user with *adventure spirit*, we can probably infer the user prefers *Skiing* more than *Hot Springs*, even though the user has no any online behaviors before. Therefore, when users' historical behaviors are sparse or even absent, users' inherent profiles can act as a kind of supplementary cues to discover users' interests. However, how to effectively extract users' interests from their profiles within the unified framework is nontrivial. To this end, we propose a User Representation Memory Network (URMN).

Specifically, URMN customizes an augmented vector A_{uis} for each sample to represent user's augmented interest for the target item at specific scenario, followed by concatenating it with other representation features, together for model training. To obtain A_{uis}, we borrow the idea from Neural Turing Machine [6] which can store information in a fixed size of external memory. Specifically, we firstly generate a specific key SK_{uis}, which can be regarded as the comprehensive representation of the users' profiles, target item, and current visiting scenario. Then, we traverse all the slots of the external memory in URMN and generate each slot's weight with respect to the specific key SK_{uis}. Finally, we achieve the augmented vector A_{uis} by a weighted-sum memory summarization. We will detail them as follows.

First, the specific key SK_{uis} defined as $SK_{uis} = F(\sum_{j=1}^{P} w_j e_{p_j}; E_i; E_s)$, where P denotes the number of user' profiles, $F(.)$ represents three MLP layers with ReLU activation function, w_j is the weight of user's profile e_{p_j} with respect to the target item E_i via attention mechanism. Intuitively, similar keys SK_{uis} cluster together, implying that given specific scenario and target item, users having similar profiles probably share the similar interests, thus cold start users can benefit from users with similar profiles yet rich behaviors. From another perspective, all the learning parameters in $F(.)$ are shared, *e.g.*, shared MLP parameters, which also implies the representation of current key can be affected by the representations of other keys.

Next, we detail the structure of the memory in URMN with its parameters denoted as Mem. It consists of q memory slots $\{Mem_i\}|_{i=1}^{q}$ with each slot containing corresponding key Mem_i^{Key} and value Mem_i^{Value}, i.e., $Mem_i \triangleq \{Mem_i^{Key}, Mem_i^{Value}\}$. Each slot can be regarded as a cluster, where Mem_i^{Key} (resp. Mem_i^{Value}) is updated by itself and SK_{uis} (resp. R_{uis}). Moreover, R_{uis} from the resultant representation of UIPN component denotes users' purified interests, depicted in Fig. 1. Specifically, two basic operations of URMN are *Memory Read* and *Memory Write*, which interact with memory through a controller.

Memory Read: During the *Memory Read* process, the controller generates a read key SK_{uis} as mentioned above to access the memory. Formally, it can be formulated as follows: $w_{uis}^j = \frac{exp(F_{xy}(SK_{uis}, Mem_j^{Key}))}{\sum_{j=1}^{q} exp(F_{xy}(SK_{uis}, Mem_j^{Key}))}, j = 1, ..., q$, where $F_{xy}(x, y) = \frac{x^T y}{\|x\|\|y\|}$, w_{uis}^j is the weight of SK_{uis} with respect to the key of slot j.

Then, we obtain user's augmented interest vector A_{uis} by weighted-sum pooling, defined as: $A_{uis} = \sum_{j=1}^{q} w_{uis}^{j} Mem_{j}^{Value}$.

Memory Write: Using the same weight w_{uis}^{j}, the update process of memory key and value is defined as follows:

$$Mem_{j}^{Key} = \alpha_k w_{uis}^{j} SK_{uis} + (1 - \alpha_k)Mem_{j}^{Key}, \tag{1}$$

$$Mem_{j}^{Value} = \alpha_v w_{uis}^{j} R_{uis} + (1 - \alpha_v)Mem_{j}^{Value}, \tag{2}$$

where α_k and α_v are the hyper-parameters valued in $[0, 1]$, controlling the update rate of each memory slot's key and value, respectively. Mem_{j}^{Key} and Mem_{j}^{Value} are randomly initialized. In this way, URMN can distribute the interest of users with rich behaviors to the users with sparse or absent behaviors in the same cluster, alleviating the cold-start issue effectively.

Finally, all the representation vectors including E_i, E_{uP}, E_s, E_c, A_{uis} and R_{uis} are concatenated, followed by feeding them to multiple MLP layers to generate the final predicted probability $\hat{y}_{uis}$. Now, given the predicted $\hat{y}_{uis}$ and the ground truth $y_{uis} \in \{0, 1\}$, we define the objective function as the negative log-likelihood function, formulated as:

$$Loss = -\frac{1}{Num} \sum (y_{uis} log\hat{y}_{uis} + (1 - y_{uis})log(1 - \hat{y}_{uis})), \tag{3}$$

where, $Loss$ is the total loss and Num denotes the number of training data collected from all the scenarios.

4   Experiments

In this section, we conduct extensive offline and online experiments to comprehensively evaluate the effectiveness of the proposed CSMN, and try to answer following questions:

- **Q1**: How about the overall performance of the proposed CSMN compared with state-of-art methods?
- **Q2**: How about the impact of each component of the proposed CSMN?
- **Q3**: How about the influence of key hyper-parameters, *e.g.*, the number of URMN slots?
- **Q4**: How about the online performance of the proposed CSMN compared with other methods?

4.1   Experiments Settings

Dataset Description. To the extent of our knowledge, there are no public datasets suited for the multi-scenario CTR prediction task in the OTP settings. We make the offline dataset by collecting users' traffic logs from our OTP platform, which contain 29 million users and 0.61 million travel items from 20 scenarios in consecutive 30 days, *i.e.*, from 2022-05-22 to 2022-06-20. They are further divided into the disjoint training set and testing set, where the training set is from 2022-05-22 to 2022-06-19, while the testing set is from the left days. The statistics of this offline dataset are listed in Table 1, where *CSU Ratio* representing cold-start users' ratio. We can find that the data scales and distributions among these travel scenarios are significantly different. In addition, we observe that over 28% (resp. 40%) of users do not have any behaviors in recent 180 (resp. 90) days from the training data, which indeed implies the cold-start issue.

Table 1. The statistics of the offline dataset.

Scenario	#1	#2	#3	#4	#5	#6	#7	#8	#9	#10
#User	2.3M	2.1M	8.2M	0.6M	3.4M	0.4M	0.3M	0.3M	2.5M	0.2M
#Item	53K	37K	122K	23K	37K	51K	33K	18K	86K	63K
CTR	2.57%	1.22%	1.64%	5.54%	8.10%	1.27%	6.75%	2.26%	11.61%	6.51%
CSU Ratio	27.65%	23.92%	11.49%	18.66%	16.49%	31.28%	14.54%	22.93%	28.73%	38.52%
Scenario	#11	#12	#13	#14	#15	#16	#17	#18	#19	#20
#User	1.1M	0.6M	0.6M	1.7M	9.1M	0.17M	0.16M	0.14M	0.13M	0.11M
#Item	135K	41K	57K	27K	11K	5K	66K	23K	31K	16K
CTR	18.70%	9.87%	4.05%	14.38%	5.80%	1.72%	8.02%	4.23%	3.03%	1.62%
CSU Ratio	32.69%	16.77%	28.46%	19.55%	27.44%	26.22%	25.37%	10.45%	30.07%	34.76%

Competitors. To verify the effectiveness of the proposed CSMN, we compare it with following methods:

- **WDL** [3]: It consists of wide linear and deep neural parts, which combines the benefits of memorization and generalization for CTR prediction.
- **DeepFM** [7]: It imposes a factorization machine as a "wide" part in WDL to eliminate feature engineering.
- **DIN** [29]: It extracts users' dynamic interest from their historical behavior via attention mechanism.
- **MMOE** [14]: It models the relationship among different tasks by employing gating networks and multi-task learning framework. We also adapt MMoE for multi-scenario task by assigning each output for corresponding scenario.
- **PLE** [20]: It contains shared components and task-specific components, and adopts a progressive routing mechanism to extract and separate deeper knowledge, enabling the efficiency of representation across multiple tasks.
- **STAR** [18]: It trains a unified model to serve all scenarios simultaneously, containing shared centered parameters and scenario-specific parameters.
- **SAR-Net** [17]: It predicts users' scenario-specific interests from scenario/target item features and adaptively extracts scenario-specific information across multiple scenarios.

Metrics and Implementation Details. To comprehensively evaluate the performance of different methods, we adopt two widely used metrics in recommender systems, *i.e.*, Area Under Curve (AUC) [29] and Relative Improvement (RI)

Table 2. Hyper-parameters of all competitors.

Hyper-parameters	Choice
Loss function	Logistic Loss
Optimizer	Adam
Number of layers in MLP	4
Dimensions of layers in MLP	[512,256,128,32]
Batch size	1024
Learning rate	0.001
Dropout ratio	0.5

[17], where, the larger AUC means better ranking performance, and RI provides an intuitive comparison measure by calculating the relative improvement of a target model over the baseline model. In addition, the proposed CSMN and other competitors are implemented by distributed Tensorflow 1.4, where learning rate, mini-batch, and optimizer, are set as 0.001, 1024, Adam [11], respectively. In addition, there are 4 layers in the MLP. Logistic loss is used as the loss function for all the competitors, as summarized in Table 2.

4.2 Experimental Results (Q1)

In this subsection, we report the AUC results of all the competitors on the offline test set. As illustrated in Table 3, the consistent improvement of the proposed CSMN over other competitors validates its effectiveness. It achieves the best AUC results in each single scenario. Note that compared with WDL, DeepFM, DIN, MMOE and PLE, the multi-scenario CTR methods, e.g., STAR, SAR-Net and the proposed CSMN, consistently achieves better performance, demonstrating that ignoring the scenario difference during the extraction of users' scenario-specific interests will seriously degenerate the performance of multi-scenario CTR prediction models. Nevertheless, SAR-Net still struggles in extracting users' real interests across multiple different scenarios, since it cannot address the cold-start issue in the OTPs setting. By contrast, our CSMN leverages the URMN to specifically mitigate the adverse effect of them, respectively. Consequently, it achieves an improvement of 2.41% RI over SAR-Net. Moreover, STAR also neglects the cold-start problem in the OTPs setting. Therefore, it has worse performance than our CSMN, especially for users with sparse (or even absent) behaviors. For example, for scenario #10 and scenario #20, which have very large portion of cold-start users, i.e., over 38.52% and 34.76% respectively, depicted in Table 1, CSMN achieves a larger AUC improvement of 3.11% and 2.93% over SAR-Net ,respectively. These results demonstrate the effectiveness of the proposed URMN in dealing with the cold-start issue, which clusters users according to the representation of their profiles and obtains the interest of users with sparse behaviors from neighboring users with rich behaviors.

Table 3. The results of all methods on the offline dataset.

scenario	WDL	DeepFM	DIN	MMOE	PLE	STAR	SAR-Net	CSMN	RI
#1	0.6317	0.6342	0.6368	0.6504	0.6506	0.6513	<u>0.6527</u>	**0.6546**	1.24%
#2	0.6695	0.6710	0.6718	0.6897	0.6901	0.6925	<u>0.6933</u>	**0.6959**	1.35%
#3	0.7186	0.7228	0.7231	0.7306	0.7307	<u>0.7328</u>	0.7321	**0.7331**	0.13%
#4	0.6500	0.6511	0.6559	0.6678	0.6681	0.6709	<u>0.6714</u>	**0.6738**	1.40%
#5	0.6822	0.6861	0.6882	0.7018	0.7016	0.7057	<u>0.7063</u>	**0.7079**	0.78%
#6	0.6257	0.6304	0.6322	0.6482	0.6485	<u>0.6501</u>	0.6496	**0.6541**	2.66%
#7	0.6388	0.6419	0.6447	0.6501	0.6522	0.6593	<u>0.6602</u>	**0.6609**	0.44%
#8	0.6535	0.6533	0.6575	0.6624	0.6626	<u>0.6631</u>	0.6628	**0.6654**	1.41%
#9	0.7032	0.7037	0.7058	0.7126	0.7131	0.7138	<u>0.7165</u>	**0.7186**	0.97%
#10	0.6867	0.6921	0.6913	0.7003	0.7001	0.7012	<u>0.7024</u>	**0.7087**	3.11%
#11	0.6701	0.6757	0.6772	0.6795	0.6802	0.6815	<u>0.6822</u>	**0.6864**	2.31%
#12	0.7025	0.7042	0.7016	0.7163	0.7169	0.7176	<u>0.7189</u>	**0.7205**	0.73%
#13	0.7258	0.7291	0.7294	0.7431	0.7438	0.7460	<u>0.7468</u>	**0.7504**	1.46%
#14	0.6662	0.6708	0.6720	0.6789	0.6800	0.6813	<u>0.6862</u>	**0.6884**	1.18%
#15	0.7135	0.7138	0.7131	0.7319	0.7323	0.7342	<u>0.7347</u>	**0.7399**	2.22%
#16	0.6411	0.6532	0.6560	0.6698	0.7005	0.7019	<u>0.7025</u>	**0.7064**	1.93%
#17	0.6739	0.6728	0.6775	0.6947	0.6953	0.6981	<u>0.6998</u>	**0.7027**	1.45%
#18	0.6242	0.6280	0.6300	0.6394	0.6395	0.6412	<u>0.6417</u>	**0.6443**	1.83%
#19	0.6193	0.6236	0.6219	0.6465	0.6462	0.6498	<u>0.6505</u>	**0.6542**	2.46%
#20	0.6287	0.6271	0.6346	0.6437	0.6434	0.6487	<u>0.6503</u>	**0.6547**	2.93%
Overall	0.6729	0.6763	0.6788	0.6864	0.6883	0.6922	<u>0.6954</u>	**0.7001**	2.41%

4.3   Ablation Study (Q2)

To investigate the effectiveness of each component in the proposed CSMN, we conduct several ablation experiments.

The Effectiveness of UIPN. UIPN is devised to extract users' scenario-specific interests from their historical behaviors with respect to the target item and the visiting scenario simultaneously. Here, we devise two variant models including:

- **CSMN w/o UIPN + T attention:** It removes UIPN while employing attention mechanism to extract users' interests from their behaviors only with respect to the Target item (T), ignoring the scenario information.
- **CSMN w/o UIPN + TS attention:** It removes UIPN while leveraging two specific attention modules to re-weigh users' historical behaviors with respect to Target item and the visiting Scenario (TS), respectively, followed by summarizing users' behaviors by weighted-sum to obtain users' interests.

As shown in Table 4, CSMN achieves the best performance compared with the other two variants. For example, compared with CSMN, **CSMN w/o UIPN + T attention** observes a performance drop of 2.51% RI, which demonstrates the

Table 4. The effectiveness of UIPN.

Model	AUC	RI
CSMN	**0.7001**	0.00
CSMN w/o UIPN + T attention	0.6952	−2.51%
CSMN w/o UIPN + TS attention	0.6979	−1.11 %

Table 5. The effectiveness of URMN.

Model	AUC	RI
CSMN	**0.7001**	0.00
CSMN w/o URMN	0.6969	−1.63%

importance of extraction of users' interests with respect to different scenarios. **CSMN w/o UIPN + TS attention** observes a performance drop of 1.11% RI, which demonstrates the effectiveness of eliminating the scenario-irrelevant information in users' behaviors with respect to the visiting scenario to further refine the representations of users' scenario-specific interests.

The Effectiveness of URMN. To demonstrate the effectiveness of URMN, we remove it from the model, resulting in a variant model named **CSMN w/o URMN**. As shown in Table 5, **CSMN w/o URMN** observes a performance drop of 1.63% RI compared with CSMN, which demonstrates CSMN can effectively alleviate the cold-start issue attributing to URMN. Furthermore, we argue that the more sparse users' behaviors are, the greater relative improvement CSMN achieves. As shown in Table 3, CSMN gets the largest RI improvement over other competitors in scenario #10 and scenario #20, which have the largest portion of cold-start users.

4.4 Parameter Sensitivity (Q3)

To further understand the adverse effect of the cold-start issue in OTPs, we investigate the influence of the key hyper-parameters related to the issue, *i.e.*, the number of memory slots q, α_k (resp. α_v) controlling the update rate of each memory slot's key (resp. value). First, as shown in Table 6, CSMN achieves the best performance when q takes the value of 1,000. Intuitively, the smaller q is, the fewer the numbers of formed users' clusters are, and vice versa. Taking two extreme cases as example, one is $q = 1$, where all the users have the same augmented interest vector, resulting in the difficulty of distinguishing different users' interests, especially for those with sparse behaviors. The other is an extremely large q, *i.e.*, $q = 10,000$, where each slot of the memory will be too spare to update sufficiently. Next, we conduct seven groups of experiments, where each experiment setting the value of α_k and α_v as the same. As depicted in Table 7, CSMN obtains the best performance when α_k and α_v take the value of 0.3.

Table 6. The effectiveness of memory size q.

Memory Size	10	100	1,000	10,000
AUC	0.6962	0.6979	**0.7001**	0.6956

Table 7. The effectiveness of update rate α_k and α_v.

Update Rate	0.0	0.1	0.2	0.3	0.5	0.8	1.0
AUC	0.6947	0.6964	0.6982	**0.7001**	0.6971	0.6955	0.6950

Intuitively, when taking 0.0 as the update rate, Mem_j^{Key} and Mem_j^{Value} always follow the initialized values, neglecting the fact that parameters are continuously updating during model training, while taking as the update rate 1.0, Mem_j^{Key} and Mem_j^{Value} always taking the latest representation, abandoning the accumulated representation before. Obviously, both situations could not achieve the best performance, confirming that suitable updating rate is promising.

4.5 Online A/B Test (Q4)

To further demonstrate the effectiveness of the proposed CSMN, we deploy it on our travel platform for A/B test, where the **Base** model is SAR-Net [17] and the evaluation metric is online CTR, *i.e.*, the number of clicks over the number of impression items. To make the online evaluation fair, confident, and comparable, both methods includes same number of users, *e.g.*, millions of users. We find the proposed CSMN achieves consistent improvement over SAR-Net model in consecutive seven days, *e.g.*, achieving an average improvement of 3.85% CTR. To go a step forward, we find a more significant improvement is observed as was expected, *e.g.*, achieving an average improvement of 4.62% CTR, for cold-start users, further demonstrating the effectiveness of CSMN dealing with cold-start issue. In a nutshell, the online A/B test results again demonstrate the effectiveness and practicability of our CSMN model in the industrial setting. Now, CSMN has been deployed on our platform and is serving all the traffic of twenty travel scenarios simultaneously.

5 Conclusions

In this paper, we propose a novel method named CSMN to deliver the unified click-through rate prediction task among the multiple scenarios on the online travel platforms. Specifically, it consists of two basic components including a User Interest Projection Network (UIPN) and a User Representation Memory Network (URMN), which can mitigate the cold-start issue effectively by exploiting scenario-shared and scenario-specific information of various scenarios simultaneously. Extensive experiments on both real-world offline dataset and online A/B test demonstrate the superiority of CSMN over state-of-the-art methods.

How to employ the principle of meta-learning framework to further exploit the *commonality* and *discrimination* properties in multi-scenario CTR prediction task is an interesting topic and deserves more research efforts.

Acknowledgments. This work is supported by National Key Research and Development Program of China under Grant 2020AAA0107400 and National Natural Science Foundation of China (Grant No: 62206248).

References

1. Barjasteh, I., Forsati, R., Ross, D., Esfahanian, A.H., Radha, H.: Cold-start recommendation with provable guarantees: a decoupled approach. IEEE Trans. Knowl. Data Eng. **28**(6), 1462–1474 (2016)
2. Caruana, R.: Multitask learning. Mach. Learn. **28**(1), 41–75 (1997)
3. Cheng, H.T., et al.: Wide & deep learning for recommender systems. In: Proceedings of the 1st Workshop on Deep Learning for Recommender Systems, pp. 7–10 (2016)
4. Chou, S.Y., Yang, Y.H., Jang, J.S.R., Lin, Y.C.: Addressing cold start for next-song recommendation. In: Proceedings of the 10th ACM Conference on Recommender Systems, pp. 115–118 (2016)
5. Dong, M., Yuan, F., Yao, L., Xu, X., Zhu, L.: Mamo: memory-augmented meta-optimization for cold-start recommendation. In: Proceedings of the 26th ACM SIGKDD International Conference on Knowledge Discovery & Data Mining, pp. 688–697 (2020)
6. Graves, A., Wayne, G., Danihelka, I.: Neural turing machines. arXiv preprint arXiv:1410.5401 (2014)
7. Guo, H., Tang, R., Ye, Y., Li, Z., He, X.: Deepfm: a factorization-machine based neural network for ctr prediction. arXiv preprint arXiv:1703.04247 (2017)
8. He, J., Mei, G., Xing, F., Yang, X., Bao, Y., Yan, W.: Dadnn: multi-scene ctr prediction via domain-aware deep neural network. arXiv preprint arXiv:2011.11938 (2020)
9. He, X., Liao, L., Zhang, H., Nie, L., Hu, X., Chua, T.S.: Neural collaborative filtering. In: Proceedings of the 26th International Conference on World Wide Web, pp. 173–182 (2017)
10. Kang, S., Hwang, J., Lee, D., Yu, H.: Semi-supervised learning for cross-domain recommendation to cold-start users. In: Proceedings of the 28th ACM International Conference on Information and Knowledge Management, pp. 1563–1572 (2019)
11. Kingma, D.P., Ba, J.: Adam: a method for stochastic optimization. arXiv preprint arXiv:1412.6980 (2014)
12. Lee, H., Im, J., Jang, S., Cho, H., Chung, S.: Melu: Meta-learned user preference estimator for cold-start recommendation. In: Proceedings of the 25th ACM SIGKDD International Conference on Knowledge Discovery & Data Mining, pp. 1073–1082 (2019)
13. Li, J., Jing, M., Lu, K., Zhu, L., Yang, Y., Huang, Z.: From zero-shot learning to cold-start recommendation. In: Proceedings of the AAAI Conference on Artificial Intelligence, vol. 33, pp. 4189–4196 (2019)
14. Ma, J., Zhao, Z., Yi, X., Chen, J., Hong, L., Chi, E.H.: Modeling task relationships in multi-task learning with multi-gate mixture-of-experts. In: Proceedings of the 24th ACM SIGKDD International Conference on Knowledge Discovery & Data Mining, pp. 1930–1939 (2018)

15. Pi, Q., et al.: Search-based user interest modeling with lifelong sequential behavior data for click-through rate prediction. In: Proceedings of the 29th ACM International Conference on Information & Knowledge Management, pp. 2685–2692 (2020)
16. Ruder, S.: An overview of multi-task learning in deep neural networks. arXiv preprint arXiv:1706.05098 (2017)
17. Shen, Q., Tao, W., Zhang, J., Wen, H., Chen, Z., Lu, Q.: Sar-net: a scenario-aware ranking network for personalized fair recommendation in hundreds of travel scenarios. In: Proceedings of the 30th ACM International Conference on Information & Knowledge Management, pp. 4094–4103 (2021)
18. Sheng, X.R., et al.: One model to serve all: star topology adaptive recommender for multi-domain ctr prediction. In: Proceedings of the 30th ACM International Conference on Information & Knowledge Management, pp. 4104–4113 (2021)
19. Song, W., et al.: Autoint: automatic feature interaction learning via self-attentive neural networks. In: Proceedings of the 28th ACM International Conference on Information and Knowledge Management, pp. 1161–1170 (2019)
20. Tang, H., Liu, J., Zhao, M., Gong, X.: Progressive layered extraction (ple): a novel multi-task learning (mtl) model for personalized recommendations. In: Fourteenth ACM Conference on Recommender Systems, pp. 269–278 (2020)
21. Vaswani, A., etal.: Attention is all you need. Advances in neural information processing systems 30 (2017)
22. Wang, R., Fu, B., Fu, G., Wang, M.: Deep & cross network for ad click predictions. In: Proceedings of the ADKDD'17, pp. 1–7 (2017)
23. Wen, H., Zhang, J., Lin, Q., Yang, K., Huang, P.: Multi-level deep cascade trees for conversion rate prediction in recommendation system. In: Proceedings of the AAAI Conference on Artificial Intelligence, vol. 33, pp. 338–345 (2019)
24. Wen, H., Zhang, J., Lv, F., Bao, W., Wang, T., Chen, Z.: Hierarchically modeling micro and macro behaviors via multi-task learning for conversion rate prediction. arXiv preprint arXiv:2104.09713 (2021)
25. Wen, H., et al.: Entire space multi-task modeling via post-click behavior decomposition for conversion rate prediction. In: Proceedings of the 43rd International ACM SIGIR Conference on Research and Development in Information Retrieval, pp. 2377–2386 (2020)
26. Zhang, J., Tao, D.: Empowering things with intelligence: a survey of the progress, challenges, and opportunities in artificial intelligence of things. IEEE Internet Things J. 8(10), 7789–7817 (2020)
27. Zhao, C., Li, C., Xiao, R., Deng, H., Sun, A.: Catn: cross-domain recommendation for cold-start users via aspect transfer network. In: Proceedings of the 43rd International ACM SIGIR Conference on Research and Development in Information Retrieval, pp. 229–238 (2020)
28. Zhou, G., et al.: Deep interest evolution network for click-through rate prediction. In: Proceedings of the AAAI Conference on Artificial Intelligence (2019)
29. Zhou, G., et al.: Deep interest network for click-through rate prediction. In: Proceedings of the 24th ACM SIGKDD International Conference on Knowledge Discovery & Data Mining, pp. 1059–1068. ACM (2018)

Query2Trip: Dual-Debiased Learning for Neural Trip Recommendation

Peipei Wang[1], Lin Li[1(✉)], Ru Wang[2], and Xiaohui Tao[3]

[1] Wuhan University of Technology, Wuhan 430070, China
{ppwang07,cathylilin}@whut.edu.cn
[2] Shandong Normal University, Jinan 250014, China
[3] University of Southern Queensland, Toowoomba, Australia
Xiaohui.Tao@usq.edu.au

Abstract. Trip recommendation (TripRec) seeks to recommend a trip that consists of an ordered sequence of points-of-interest (POIs) for a tourist through a user-specific query. Recent neural TripRec methods with sequence-to-sequence models have achieved remarkable performance. However, alongside the exposure bias in general recommender systems, the selection bias caused by the lack of explicit feedback (e.g., ratings) from the trip data exacerbates the tendency toward users' unsatisfactory experience in TripRec. To this end, a novel debiased representation learning method for neural TripRec is proposed to fulfill sequence generation from Query to Trip named Query2Trip. It develops dual-debiased learning to mitigate selection bias and exposure bias in TripRec. The former happens as the visit by a user does not necessarily mean the user exhibits a positive preference for the visit. Benefiting from the query provided by a user, Query2Trip designs a debiased adversarial learning module by conditional guidance to alleviate this selection bias from positives (visited). The latter happens as unvisited is not equivalent to negative. Query2Trip devises a debiased contrastive learning module by negative weighting to mitigate this exposure bias from negatives (unvisited). Experiments conducted on eight real-world datasets empirically demonstrate the superior performance of Query2Trip compared to the state-of-the-art baselines.

Keywords: Trip recommendation · Debiased learning · Sequence generation

1 Introduction

The massive check-in data for points-of-interest (POIs) generated by users in location-based social networks (LBSNs) provides unprecedented opportunities to understand human mobility patterns. Various downstream applications such as POI recommendation, next location prediction, and trip recommendation [2,13] have emerged, facilitating better user decision-making and generally improving user experience. These fields have attracted much attention in academia and

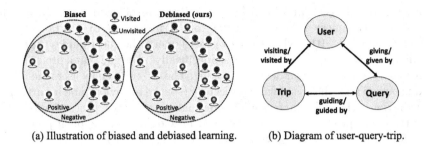

(a) Illustration of biased and debiased learning.   (b) Diagram of user-query-trip.

Fig. 1. Schematic of biased/debiased learning and user-query-trip in TripRec.

industry due to their high value for tourism. In particular, trip recommendation (TripRec) is one of the more critical tasks in tourism-related research. TripRec is also known as tour, travel, or itinerary recommendation/planning across studies [26]. The typical TripRec aims to recommend a trip that consists of an ordered sequence of POIs for a tourist (user) through a user-specific query [5,6,9,12,24]. In general, a query given by the user consists of the origin POI, the destination POI, the number of POIs to be visited, and so forth.

Prior Work and Limitations. Traditional orienteering problem-based solutions for TripRec [7,9,13,17] usually rely on predefined objectives or heuristic algorithms to learn human potential transition preferences. Recent deep learning-based neural TripRec methods take advantage of data-driven methods to achieve remarkable performance. Intuitively, the query given by a tourist is a sequence containing the origin POI and destination POI. The subsequent recommended trip to the tourist is also a sequence composed of multiple POIs that may be of interest. Therefore, recent neural TripRec methods [5,10,12,26,27] utilize sequence-to-sequence (seq2seq) models to learn the sequential information and semantic relationship among POIs. However, most of these studies generally treat the visited trips as positive preferences and the unvisited trips as negative preferences in TripRec modeling. In actuality, there are various biases from both positive and negative samples in TripRec, and current interpretations can lead to biased representation learning in TripRec. To be specific, unvisited (unobserved visit) is not always negative, since users are only exposed to a part of specific items, known as exposure bias [8]. Meanwhile, visited (observed visit) is not always positive since there is no explicit feedback (e.g., ratings) from the observed trip data, known as selection bias in our work.

Motivations. Inspired by existing studies on general recommender systems [14, 15,22], as described in Fig. 1(a), we observed that selection bias occurs as the visited trips by users are not always in their positive preferences, while exposure bias happens as the unvisited trips by users are not always in their negative preferences. Hence, in the process of learning, we mitigate selection bias and exposure bias in TripRec by dual-debiased learning to improve representation ability for a more effective neural TripRec. Generally speaking, a completed trip process includes two basic tourist behaviors: decision-making behavior and spatial behavior [18].

Intuitively, the decision-making behavior and spatial behavior are respectively equivalent to the query and user trajectory of a TripRec task to a certain extent. As illustrated in Fig. 1(b), when the user gives a query, the query guides the corresponding trip, which the user then visits. Specifically, we argue that a preference is expressed by tourists in their decision-making. The query given by users (decision-making behavior) is the basis of user trajectory (spatial behavior), and in turn, spatial behavior is the embodiment of decision-making behavior. Motivated by this, using the query provided by a user, we regard the query as an explicit preference signal for tourists to learn query-based preference information.

Our Method. We propose a novel dual-debiased learning method for neural TripRec, called Query2Trip. The basic idea of our Query2Trip is to fulfill sequence generation from a query given by the user to a trip recommended to the user. A dual-debiased learning method consisting of a debiased adversarial learning (DAL) module and a debiased contrastive learning (DCL) module is developed to mitigate the selection bias and exposure bias in TripRec. Specifically, our DAL module adopts a conditional adversarial manner to guide user preference generation and alleviate the selection bias from positives (visited). The DCL module designs a weight-based negative sampling to punish improper negative samples and lighten the exposure bias from negatives (unvisited).

Contributions. In brief, our key contributions of this work are summarized:

- A novel Query2Trip is proposed for neural TripRec, which combines sequence generation and debiased learning to explore user potential transition patterns for better human mobility behavior understanding.
- A dual-debiased learning method consisting of debiased adversarial learning and debiased contrastive learning is developed to mitigate the selection bias and exposure bias in TripRec. To the best of our knowledge, this is the first attempt to study debiased sequential learning in TripRec.
- A set of experimental results on eight real-world datasets to demonstrate the superiority of our Query2Trip method over comparative results from the state of the arts.

2   Preliminary

To facilitate understanding of our approach, the necessary notations and definitions for the TripRec are presented in this section.

*Definition 1 (**Point-of-interest**).* A POI represents the user's reachable preference location at a certain time and scene. In our TripRec, we mainly consider the latitude lat, longitude lng, and category cat of POIs. We use P to represent the POI set, i.e., $P = \{p|p = (lat, lng, cat)\} \in \mathbb{R}_{|P|}^{dim_p}$, where $|P|$ is the number of POIs and dim_p is the dimension of POI embedding.

*Definition 2 (**Query**).* A query is usually given by the user and represents the user's decision-making behavior in TriRec. A query usually consists of the origin POI, the destination POI, and the number of POIs to be visited (a.k.a length).

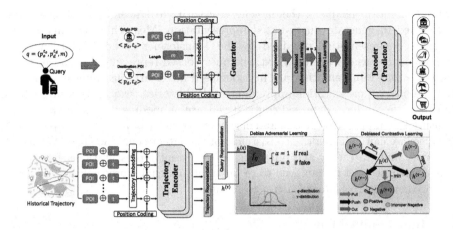

Fig. 2. Overview of the proposed Query2Trip. The selection bias from positives (visited) and exposure bias from negatives (unvisited) can be mitigated by our debiased adversarial learning and debiased contrastive learning, respectively.

In this study, the query set is denoted as $Q = \{q|q = (p_o^{t_o}, p_d^{t_d}, m)\} \in \mathbb{R}_{|Q|}^{dim_q} \times m$, where $|Q|$ is the number of queries, $p_o^{t_o}$ is the origin POI with starting time, $p_d^{t_d}$ is the destination POI with ending time, m is the number of POIs to be visited, and dim_q is the dimension of query embedding.

*Definition 3 (**Trajectory**).* A trajectory (i.e., trip) consists of an ordered sequence of POIs and represents the user's spatial behavior. Let $U = \{u\}$ be the user set and $\Gamma = \{\tau\}$ be the trajectory set. A trajectory is denoted as $\tau = \{< p_1, t_1 > \rightarrow < p_2, t_2 > \rightarrow \ ... \ \rightarrow < p_m, t_m >\} \in \mathbb{R}_{|\Gamma|}^{dim_\tau}$, where $< p_j^{(u)}, t_j >$ is the j-th POI of the trajectory for user u, t_j is the visit timestamp, $|\Gamma|$ is the number of trajectories, and dim_τ is the dimension of trajectory embedding.

*Problem Definition (**Trip Recommendation**).* Given a dataset $D = \{(u, p, \tau)\}$ recording the POI set and trajectory set, and the corresponding user set, when a user provides a query $q^{(u)} = \{p_o^{t_o}, p_d^{t_d}, m\}$, we aim to generate a preferred trip $\tau^{(u)} = \{< p_o^{(u)}, t_o > \rightarrow < p_2^{(u)}, t_2 > \rightarrow \ ... \ \rightarrow < p_d^{(u)}, t_d >\} \in \mathbb{R}_{|\Gamma|}^{dim_{(\tau)}}$ for the user that has not been visited. Formally, $\tilde{\tau} = \arg\max_{\theta} \mathcal{P}(\tau^{(u)}|q^{(u)}, D)$.

3 The Proposed Method

3.1 Query2Trip Overview

Figure 2 shows the overview of our Query2Trip, which as a whole is a sequence-to-sequence learning process. As the core of Query2Trip, our proposed dual-debiased learning consists of debiased adversarial learning and debiased contrastive learning. Firstly, a query given by the user is jointly embedded based on the content of the query, then a generator with the Transformer [20] encoder is designed to generate query-based representation. Meanwhile, the historical trajectory representation obtained by the Transformer encoder is utilized to support query-based learning. Next, the query-based representation and historical trajectory representation are in a debiased adversarial learning module so that the query representation approximates the historical trajectory representation space. Secondly, a debiased contrastive learning module is presented, which punishes the improper negatives and ensures the quality of negative samples. By doing so, we can enhance the consistency of the latent semantic space between the query and trajectory and obtain more consistent representations. Lastly, a Transformer decoder is employed to reconstruct hidden features, which also acts as a predictor to recommend a trip to the user.

3.2   Input Embedding

POI Embedding. Inspired by the prior work [5,9,26], the POI embedding is implemented from three main aspects in our work. First, each POI is initialized randomly by sampling from a truncated Gaussian distribution. Each initialized POI embedding is denoted as $\mathbf{v}_{em}^{(p_j)} \in \mathbb{R}^{dim_p}$. Second, since the geographic coordinate (lat, lng) of an arbitrarily POI p_j is usually near the coordinates of the origin POI p_o or destination POI p_d in a trajectory, the spatial context relationship between POIs is embedded, denoted as $\mathbf{v}_s^{(p_j)} \in \mathbb{R}^{dim_p}$. Third, considering the temporal preference information (visiting duration) of different users for different POIs, we also map the temporal information into a dense embedding, denoted as $\mathbf{v}_t^{(p_j)} \in \mathbb{R}^{dim_p}$. Lastly, the above three embeddings are combined as a unified embedding by the concatenation operation:

$$\mathbf{v}^{(p_j)} = \text{Concat}(\mathbf{v}_{em}^{(p_j)}, \mathbf{v}_s^{(p_j)}, \mathbf{v}_t^{(p_j)}). \tag{1}$$

Query Embedding. For a given query $q = \{p_o^{t_o}, p_d^{t_d}, m\} \in Q$, based on our trained POI embedding, a joint embedding operation is designed to capture the multiple interactions between elements:

$$\begin{aligned}
\mathbf{v}^{(q)} &= f_{(q)}(p_o^{t_o}, p_d^{t_d}, m) \\
&= \text{LeakyReLU}(\{\mathbf{v}^{(p_o)} \parallel t_o\}_1, \parallel \cdots \parallel, \{\mathbf{v}^{(p_d)} \parallel t_d\}_m + b^{(q)}),
\end{aligned} \tag{2}$$

where $\parallel$ is the concatenation operation. $b^{(q)}$ is the bias of $f_{(q)}$. The number of randomly initialized POIs is m-2 excluding the origin and destination. $\mathbf{v}^{(q)} \in \mathbb{R}^{(dim_p+dim_t) \times m}$ is the query embedding. To enhance the deep semantic learning of the query and facilitate our task from query to trip, we design a generative Transformer (Trm) encoder with position coding for the query. The query representation is then obtained as:

$$\mathbf{h}^{(q)} = \mathscr{G}_\delta(\mathbf{v}^{(q)} \parallel \mathbf{v}^{(\text{position})}), \tag{3}$$

where $\mathbf{v}^{(\text{position})}$ is the position coding, $\parallel$ is concatenation operation, δ is the parameter of generator, and $\mathscr{G}_\delta$ is the generator composed of l-layer TrmEncoder.

Trajectory Embedding. For a trajectory τ consisting of m-POIs, all POIs with the vising timestamp in the trajectory are combined by the concatenation operation. Different from query embedding, all POIs in the trajectory are derived from historical check-in preference feature information, which are not randomly initialized. The embedding of each trajectory is computed as:

$$\begin{aligned}
\boldsymbol{v}^{(\tau)} &= f_{(\tau)}(p_1^{t_1}, p_2^{t_2}, p_3^{t_3}, ..., p_{m-1}^{t_{m-1}}, p_m^{t_m}) \\
&= \text{LeakyReLU}(\{\boldsymbol{v}^{(p_1)} \parallel t_1\}_1, \parallel \cdots \parallel, \{\boldsymbol{v}^{(p_m)} \parallel t_m\}_m + b^{(\tau)}),
\end{aligned} \tag{4}$$

where $\mathbf{v}^{(\tau)} \in \mathbb{R}^{(dim_p + dim_t) \times m}$ denotes the trajectory embedding, $\parallel$ is the concatenation operation, and $b^{(\tau)}$ is the bias of $f_{(\tau)}$. For the trajectory representation, we employ the same Transformer layer as the query representation to learn potential transition patterns and obtain the trajectory representation $\boldsymbol{h}^{(\tau)}$.

3.3 Dual-Debiased Learning

Debiased Adversarial Learning. As mentioned before, there is no explicit feedback (e.g., ratings) on what users have visited in TripRec. Thus, we argue that selection bias happens as users may be unsatisfied with what they have visited. To tackle this problem, a debiased adversarial learning (DAL) module is designed to alleviate this selection bias from positive samples. Specifically, we generate the query-based trajectory representation by the generator, and then perform adversarial learning with the historical trajectory representation. Generative adversarial networks (GANs) have been shown to be effective in recommendation task, however, GAN suffers from perfect discriminator and unstable training. Inspired by Wasserstein GANs (WGANs) [1], instead of Jensen-Shannon divergence, we first define Earth-Mover or Wasserstein distance:

$$W(\boldsymbol{h}^{(\tau)}, \boldsymbol{h}^{(q)}) = \inf_{\gamma \sim \sqcap(\boldsymbol{h}^{(\tau)}, \boldsymbol{h}^{(q)})} \mathbb{E}_{(\boldsymbol{h}^{(\tau)}, \boldsymbol{h}^{(q)}) \sim \gamma}[\parallel \boldsymbol{h}^{(\tau)} - \boldsymbol{h}^{(q)} \parallel], \tag{5}$$

where inf is short for infimum. $\boldsymbol{h}^{(\tau)}$ and $\boldsymbol{h}^{(q)}$ are the representation of historical trajectory and query-based trajectory; $\gamma \sim \sqcap(\boldsymbol{h}^{(\tau)}, \boldsymbol{h}^{(q)})$ is a candidate set of all pair of $\boldsymbol{h}^{(\tau)}$ and $\boldsymbol{h}^{(q)}$; and $\parallel \parallel$ is the absolute value of distance between two representations. The generator parameter δ and discriminator parameter η are optimized. The general optimization objective of WGANs is defined as:

$$\min_{\delta} \max_{\eta} \mathbb{E}_{h \sim h^{(\tau)}}[f_\eta(\boldsymbol{h}^{(\tau)})] - \mathbb{E}_{h \sim h^{(q)}}[f_\eta(\mathscr{G}_\delta(\boldsymbol{h}^{(q)})]. \tag{6}$$

Essentially, the query-based trajectory and user historical trajectory have the same conditions: origin POI, destination POI, and the number of POIs. Thus, we adopt the original query ($q = \{p_o^{t_o}, p_d^{t_d}, m\}$) as the condition of WGANs to transport to the generator and discriminator. A conditional Wasserstein generative adversarial network is designed to achieve debiased adversarial learning. The optimization objective can be re-defined as:

$$\min_{\delta} \max_{\eta} \mathbb{E}_{h \sim h^{(\tau)}}[f_\eta(\boldsymbol{h}^{(\tau)}|c_{(query)})] - \mathbb{E}_{h \sim h^{(q)}}[f_\eta(\mathscr{G}_\delta(\boldsymbol{h}^{(q)}|c_{(query)}))], \tag{7}$$

where $c_{(query)}$ denotes the original query condition. By doing so, we can guide the direction of the data generated by the model. The conditional adversarial learning between the query-based trajectory and user historical trajectory is performed, which can not only alleviate the selection bias from positives and guide user real preference generation, but also ensure the semantic consistency of representation spaces.

Debiased Contrastive Learning. In the given trip training data, it is observed that user unvisited trips do not always represent negative user preferences due to the exposure bias problem. Motivated by this observation, we focus on negative samples, hence, a debiased contrastive learning (DCL) module is devised to penalize improper negative samples and alleviate exposure bias from negatives. As a self-supervised learning technique, contrastive learning has been widely used to learn representations by comparing positive and negative pairs. Moreover, it has proven to be an effective way to mitigate the exposure bias in recommender systems [16]. The noise contrastive estimation (NCE) based strategy has become prevalent in various applications. The general contrastive loss [16,23] is defined as:

$$\mathbb{E}_{\substack{\chi \sim \gamma, \chi^+ \sim \gamma^+ \\ \{\chi_k^-\}_{k=1}^K \sim \gamma^-}} \left[-\log \frac{e^{\mathscr{G}_\delta(\chi^{(q)})^{\mathrm{T}} TrmEnc(\chi^{(\tau+)})/\varsigma}}{e^{\mathscr{G}_\delta(\chi^{(q)})^{\mathrm{T}} TrmEnc(\chi^{(\tau+)})/\varsigma} + \sum_{k=1}^K e^{\mathscr{G}_\delta(\chi^{(q)})^{\mathrm{T}} TrmEnc(\chi_k^{(\tau-)})/\varsigma}} \right],$$

$$(8)$$

where χ denotes a batch in $\gamma \sim \sqcap(\boldsymbol{h}^{(\tau)}, \boldsymbol{h}^{(q)})$, $\chi^+ \sim \sqcap(\boldsymbol{h}^{(q)}, \boldsymbol{h}^{(\tau+)})$ is a positive pair, $\chi^- \sim \sqcap(\boldsymbol{h}^{(q)}, \boldsymbol{h}^{(\tau-)})$ is a negative pair, $\sqcap()$ denotes a set, K is the number of negative samples, and ς is the temperature parameter. Note that we match a query to a trajectory to obtain the positive pair, while the rest are the negative pair from the batch. To ensure that the positive pairs involved in contrastive learning are unbiased, we introduce a parameter α as a control condition for performing contrastive learning. Thus, the contrastive loss can be re-written as:

$$\mathbb{E}_{\substack{\chi \sim \gamma, \chi^+ \sim \gamma^+ \\ \{\chi_k^-\}_{k=1}^K \sim \gamma^-}} \left[-\log \frac{e^{\alpha * \mathscr{G}_\delta(\chi^{(q)})^{\mathrm{T}} TrmEnc(\chi^{(\tau+)})/\varsigma}}{e^{\mathscr{G}_\delta(\chi^{(q)})^{\mathrm{T}} TrmEnc(\chi^{(\tau+)})/\varsigma} + \sum_{k=1}^K e^{\mathscr{G}_\delta(\chi^{(q)})^{\mathrm{T}} TrmEnc(\chi_k^{(\tau-)})/\varsigma}} \right],$$

$$(9)$$

$$\text{where} \quad \alpha = \begin{cases} 1, & \text{if real} \\ 0, & \text{if fake} \end{cases}, \tag{10}$$

where α determines whether the sample participates in DCL. When α is 1, contrastive learning is performed, otherwise, it is not performed.

Usually, existing studies produce randomly negative samples from the batch or whole dataset. However, the quality of negative samples produced by random sampling has rarely been considered, resulting in negative sampling bias that leads to suboptimal results. This bias is due to the fact that the similarity between negative samples and positive samples in random or batch is greater than a certain threshold, so the self-supervised learning ability of contrastive learning is disturbed. Inspired by the recent work of [19,28], it is observed that not all negatives are equal. Therefore, we assign different weights to each negative sample, then the improper negative sample with high similarity is punished. Based on negative weighting, our debiased contrastive learning objective (batch-based) is defined as:

$$\mathbb{E}_{\substack{\chi \sim \gamma, \chi^+ \sim \gamma^+ \\ \{\chi_k^-\}_{k=1}^K \sim \gamma^-}} \left[-\log \frac{e^{\alpha * \mathscr{G}_\delta(\chi^{(q)})^\mathrm{T} TrmEnc(\chi^{(\tau+)})/\varsigma}}{e^{\mathscr{G}_\delta(\chi^{(q)})^\mathrm{T} TrmEnc(\chi^{(\tau+)})/\varsigma} + \sum_{k=1}^K e^{\beta * \mathscr{G}_\delta(\chi^{(q)})^\mathrm{T} TrmEnc(\chi_k^{(\tau-)})/\varsigma}} \right],$$
(11)

$$\text{where} \quad \beta = \begin{cases} 1, & sim(\chi^{(q)}, \chi_k^{(\tau-)}) < \Phi \\ 0, & sim(\chi^{(q)}, \chi_k^{(\tau-)}) \geq \Phi \end{cases}, \tag{12}$$

where Φ is a similarity threshold between the query and negative trajectory. When the similarity is greater than or equal to this threshold, β is 0, so then this negative sample is removed from the negative sample set in the batch, otherwise, it is reserved. Following the state-of-the-art methods as discussed in [28], cosine measurement is employed to obtain the similarity between two representations. Other similarity calculation methods can also be further explored in the future.

3.4 Trip Recommendation

Based on our problem definition in Sect. 2, in this section, we perform the decoder transformation on the hidden layer features from Transformer encoder. As a key difference from previous work by [5,26], our decoder is a Transformer decoder trained to reconstruct the trajectory. Additionally, our decoder also acts as a trip recommendation operator. Our decoder transformation is:

$$h^{(dec)} = \text{TrmDecoder}_\varphi(h^{(l)}), \tag{13}$$

where φ is the parameter of TrmDecoder and $h^{(l)}$ is the hidden features from TrmEncoder with l-layers. Based on our obtained $h^{(dec)}$, when generating the recommended trajectory, the prediction probability for each POI in the generated trajectory is calculated, and the POI with the highest probability is then selected. The joint probability of the entire trajectory can be obtained through:

$$\mathcal{P}(p_j|\tau^{(u)}, q^{(u)}) = \frac{\exp(p_j|h^{(dec)})}{\sum_{j=1}^{|P|} \exp(p_j|h^{(dec)})}, \tag{14}$$

$$\mathcal{P}(\tilde{\tau} = (p_o^{t_o}, p_2^{t_2}, ..., p_{m-2}^{t_{m-2}}, p_d^{t_d})) = \prod_{j=1}^m \mathcal{P}(p_j|\tau^{(u)}, q^{(u)}), \tag{15}$$

where $|P|$ is the number of POIs, m is the length of trajectory in the user-specific query, and $\prod$ denotes the continued product operation. The length of the recommended trajectory varies with the number of POIs given by the user-specific query input. In our task, softmax function is applied to compute the probability distribution. Cross entropy is used to measure the loss between the historical trajectory and predicted trajectory.

3.5   Optimization

As introduced previously, $\mathcal{L}_{DAL}$ and $\mathcal{L}_{DCL}$ are defined in Eq. (7) and Eq. (11), respectively. The reconstruction loss $\mathcal{L}_{REC}$ is defined in cross entropy [5,26]. Thus, the total optimization objective of our Query2Trip is formalized as:

$$\mathcal{L}_{total} = \mathcal{L}_{DAL} + \mathcal{L}_{DCL} + \mathcal{L}_{REC}. \tag{16}$$

Specifically, stochastic gradient descent is utilized to update the key parameters of our Query2Trip. A parameter is iteratively optimized, while the remaining parameters are fixed. Regarding our DAL module, the generator parameter δ and the discriminator parameter η are optimized. N is the batch size. We number the sample with i. The updated formulas for $\mathcal{L}_{DAL}$ w.r.t. δ and η can be written respectively as:

$$- \nabla_\delta \frac{1}{N} \sum\nolimits_{i=1}^{N} \big[f_\eta(\mathscr{G}_\delta(\boldsymbol{h}_i^{(q)}|c_{(query)})) \big], \tag{17}$$

$$\nabla_\eta \Big[\frac{1}{N} \sum\nolimits_{i=1}^{N} [f_\eta(\boldsymbol{h}_i^{(\tau)}|c_{(query)})] - \frac{1}{N} \sum\nolimits_{i=1}^{N} [f_\eta(\mathscr{G}_\delta(\boldsymbol{h}_i^{(q)}|c_{(query)}))] \Big]. \tag{18}$$

For our DCL module, the generator parameter δ and Transformer encoder parameter ϕ are optimized. Similarly, the updated formula for $\mathcal{L}_{DCL}$ w.r.t. δ or ϕ can be written as:

$$\nabla_{\delta \ or \ \phi} \Big[(1-\alpha)*\mathscr{G}_\delta(\chi)^{\mathrm{T}} TrmEnc_\phi(\chi^+)) - \sum\nolimits_{k=1}^{K} (\beta * \mathscr{G}_\delta(\chi^{(q)})^{\mathrm{T}} TrmEnc_\phi(\chi_k^{(\tau-)})) \Big]. \tag{19}$$

Moreover, the Transformer encoder parameter ϕ and Transformer decoder parameter φ of reconstruction are updated. Hence, the computational complexity of $\mathcal{L}_{total}$ is $O(n_{epoch}(\frac{n_{total}}{N}(2|\phi| + 2|\delta| + |\varphi| + N|\eta|)))$ in terms of the training process of our Query2Trip, where n_{total} is the number of total samples.

4   Experiments and Analysis

4.1   Experimental Settings

Datasets. In the TripRec task, real-world POI check-in datasets from Flicker and Foursquare are often widely used for experimental evaluations [2,4,5,9,13, 24], thus we utilise these for our experiments. Eight cities are selected from the datasets[1], i.e., Edinburgh, Glasgow, Osaka, Toronto, Budapest, Vienna, Melbourne, and Tokyo. Among them, the first seven are from Flicker, while the last one is from Foursquare. All user-POI visits in each city contain POI longitude and latitude, POI theme, dateTaken and user ID. Each trajectory corresponds to a specific user. The trajectory is composed of several ordered POIs and each query can be produced based on the trajectory. Regarding the necessary data prepossessing, short trajectories with less than three POIs are filtered out. The basic dataset statistics are shown in Table 1, where the name of each city has a corresponding abbreviation.

---

[1] https://sites.google.com/site/limkwanhui/datacode.

Table 1. Basic dataset statistics.

City Name	#User	#POI Visits	#Trajectory
Edinburgh (Edi.)	1,454	33,944	5,028
Glasgow (Gla.)	601	11,434	2,227
Osaka (Osa.)	450	7,747	1,115
Toronto (Tor.)	1,395	39,419	6,057
Budapest (Bud.)	935	18,513	2,361
Vienna (Vie.)	1,155	34,515	3,193
Melbourne (Mel.)	1,000	23,995	5,106
Tokyo (Tok.)	200	8,721	6,414

Evaluation Metrics and Baselines. Generally, the precision, recall, F1 score and pairs-F1 score are commonly used in state-of-the-art TripRec methods [2, 5,9,13,21,24,26,27]. Considering that F1 score is calculated based on precision (Pre, $\frac{|\tau \cap \tilde{\tau}|}{|\tilde{\tau}|}$) and recall ($Rec$, $\frac{|\tau \cap \tilde{\tau}|}{|\tau|}$), therefore the two well-established metrics of F1 score and pairs-F1 score are chosen to evaluate performance in this work. The F1 score measures whether the POI is correctly recommended, while the pairs-F1 score considers both the POI identity and visiting order by calculating the F1 score for each pair of POI [2]. The values of F1 and pairs-F1 score range from 0 to 1. The higher the value, the better the performance.

$$\text{F}_1 = \frac{2 \times Pre \times Rec}{Pre + Rec}; \qquad \text{pairs-F}_1 = \frac{2 \times Pre_{\text{pair}} \times Rec_{\text{pair}}}{Pre_{\text{pair}} + Rec_{\text{pair}}}. \qquad (20)$$

Ten baselines are compared to verify the effectiveness of our Query2Trip: **(i) Traditional TripRec**[2]: RANDOM [2], PersTour [13], RANK+Markov [2], POIRANK [2], and TRAR [7]. **(ii) Neural TripRec**: C-ILP [9], C-ALNS [9], CATHI[3] [27], DeepTrip[4] [5], PT-VPMF [24], and CTLTR [26]. See the related work (Sect. 5) of this paper for a more detailed description. We reproduce all baselines based on the relevant code following the given original hyperparameter settings. For a fair comparison, we take the optimal value of the result given by the baseline and our reproduced result.

Implementation Details. We implement our proposed Query2Trip on TensorFlow with Python and deploy our model on RTX 3090×2 GPU and Intel(R) Xeon(R) CPU 4208. The leave-one-out cross-validation is adopted to train and test models [5,26,27]. The hyperparameter settings of our model are as follows: we set POI/trajectory embedding dimension as 256, batch size as 16, time embedding dimension as 32, hidden size as 128, learning rate as 0.5 (Trm), 0.01 (DAL), and 0.001 (DCL) respectively, the number of encoder layer as 5, and the number of attention heads as 8. We run our model five times to average the results for experimental analysis.

---

[2] https://bitbucket.org/d-chen/tour-cikm16.
[3] https://github.com/Shellyyue/CATHI.
[4] https://github.com/gcooq/DeepTrip.

Table 2. Performance comparison of TripRec in terms of F1 and pairs-F1. The optimal results are in bold and the suboptimal results are with "*".

Method	F1							
	Datasets							
	Edi.	Gla.	Osa.	Tor.	Bud.	Vie.	Mel.	Tok.
RANDOM [2]	0.570	0.632	0.621	0.621	0.289	0.351	0.558	0.581
PersTour [13]	0.656	0.801	0.686	0.720	0.748	0.651	0.483	0.701
POIRANK [2]	0.700	0.768	0.745	0.754	0.784	0.662	0.637	0.728
RANK+Markov [2]	0.659	0.754	0.715	0.723	0.752	0.645	0.613	0.677
C-ILP [9]	0.769	0.853	0.763	0.818	0.832	0.750	0.729	0.835
C-ALNS [9]	0.768	0.852	0.762	0.815	0.829	0.747	0.728	0.833
CATHI [27]	0.772	0.815	0.758	0.807	0.819	0.737	0.723	0.809
TRAR [7]	0.770	0.820	0.770	0.801	0.820	0.728	0.713	0.812
DeepTrip [5]	0.765	0.831	0.834	0.808	0.837	0.730	0.683	0.826
PT-VPMF [24]	0.710	0.817	0.809	0.779	0.823	0.719	0.677	0.820
CTLTR [26]	0.853*	0.874*	0.889*	0.874*	0.842*	0.785*	0.778*	0.860*
Query2Trip (ours)	**0.858**	**0.890**	**0.895**	**0.884**	**0.852**	**0.803**	**0.793**	**0.878**
Method	Pairs-F1							
	Datasets							
	Edi.	Gla.	Osa.	Tor.	Bud.	Vie.	Mel.	Tok.
RANDOM [2]	0.261	0.320	0.304	0.310	0.237	0.268	0.248	0.261
PersTour [13]	0.417	0.643	0.468	0.504	0.534	0.409	0.216	0.268
POIRANK [2]	0.432	0.548	0.511	0.518	0.548	0.510	0.339	0.313
RANK+Markov [2]	0.444	0.545	0.486	0.512	0.542	0.507	0.351	0.399
C-ILP [9]	0.509	0.697	0.683	0.575	0.623	0.593	0.487	0.566
C-ALNS [9]	0.506	0.693	0.680	0.573	0.619	0.589	0.485	0.561
CATHI [27]	0.515	0.523	0.516	0.559	0.605	0.571	0.460	0.526
TRAR [7]	0.548	0.539	0.546	0.556	0.613	0.560	0.557	0.635
DeepTrip [5]	0.660	0.782	0.755	0.748	0.724	0.645	0.546	0.643
PT-VPMF [24]	0.625	0.732	0.713	0.706	0.693	0.624	0.533	0.630
CTLTR [26]	0.729*	0.807*	0.834*	0.850*	0.752*	0.709*	0.648*	0.729*
Query2Trip (ours)	**0.734**	**0.823**	**0.841**	**0.869**	**0.766**	**0.720**	**0.661**	**0.736**

4.2   Performance Comparison

From the results in Table 2, our proposed Query2Trip consistently obtains the best recommendation performance across datasets for both metrics. We can make the following important observations:

(1) **Performance of seq2seq-Based Neural TripRec.** Generally, seq2seq model-based neural TripRec methods (CATHI, DeepTrip, CTLTR, and ours) yield better results than other TripRec methods. This confirms our motivation for using the seq2seq model as the basic framework. Although CATHI, Deep-Trip, and CTLTR exhibit competitive results, our Query2Trip outperforms the

Table 3. Ablation results of different components on eight datasets.

Method	F1							
	Datasets							
	Edi.	Gla.	Osa.	Tor.	Bud.	Vie.	Mel.	Tok.
Variant I (w/o DAL)	0.807	0.830	0.843	0.824	0.836	0.747	0.732	0.845
Variant II (w/o DCL)	0.815	0.853	0.851	0.833	0.840	0.773	0.743	0.859
Variant III (w/o negative weighting)	0.841	0.879	0.867	0.879	0.847	0.786	0.785	0.870
Query2Trip	**0.858**	**0.890**	**0.895**	**0.884**	**0.852**	**0.803**	**0.793**	**0.878**
Method	Pairs-F1							
	Datasets							
	Edi.	Gla.	Osa.	Tor.	Bud.	Vie.	Mel.	Tok.
Variant I (w/o DAL)	0.688	0.787	0.768	0.799	0.733	0.664	0.603	0.685
Variant II (w/o DCL)	0.697	0.796	0.770	0.829	0.746	0.681	0.612	0.694
Variant III (w/o negative weighting)	0.722	0.815	0.797	0.858	0.757	0.699	0.638	0.728
Query2Trip	**0.734**	**0.823**	**0.841**	**0.869**	**0.766**	**0.720**	**0.661**	**0.736**

best baseline CTLTR method by 1.47% and 1.52% on average in terms of F1 and pairs-F1, respectively. There are two possible reasons. The first reason is that we treat the user's given query as an explicit preference signal to learn debiased representations from positives and negatives to mitigate selection bias and exposure bias. Besides, our Transformer with position coding architecture ensures the global coherence of the query sequence and trajectory sequence.

(2) Performance of Non-seq2seq-Based Neural TripRec. Even though C-ILP, C-ALNS, and PT-VPMF are deep learning-based methods, seq2seq-based models are not adopted. Furthermore, they regard the user-specific query as a post-processing constraint and ignore the latent preference information contained in the query. Therefore, compared to ours, they obtain unsatisfactory results. Our method exhibits an advantage over PT-VPMF by an obvious margin, where the improvement achieves 11.66% (F1) and 16.85% (pairs-F1) on average. Also, it can be seen that basic embedding is far from enough for performance improvement. This drives us to design dual-debiased learning to improve representation ability for a more effective TripRec.

(3) Performance of Orienteering Problem-Based TripRec. Recent orienteering problem-based baselines (C-ILP, C-ALNS, and TRAR) perform worse than DeepTrip, CTLTR, and ours. This may be because they are overly dependent on predefined objectives or heuristic algorithms and neglect underlying semantic information of the user-specific query.

4.3 Ablation Study

To better demonstrate the performance gains of the different components of our method, three degraded versions of Query2Trip are discussed. **Variant I** removes our debiased adversarial learning (DAL). A selection bias from positives exists in modeling. The generator for DAL is retained for the purpose of the sequence

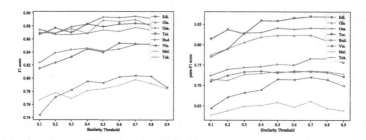

Fig. 3. Impact of varying similarity threshold on performance.

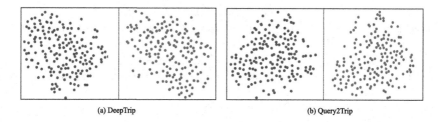

(a) DeepTrip   (b) Query2Trip

Fig. 4. Comparing distributions of learned query and trajectory representations using t-SNE. Blue and red denote query and trajectory representation, respectively. (Color figure online)

generation task. **Variant II** removes our debiased contrastive learning (DCL). This addresses the exposure bias from negatives that exists in modeling. **Variant III** removes our designed weight-based negative sampling strategy. Table 3 summarizes the ablation results on all of the eight datasets.

(1) Impact of DAL. Compared with Variant I, Query2Trip contributes to an additional 6.08% and 7.44% in terms of F1-score and pairs-F1 score on average, separately. The possible reason is the lack of conditional adversarial learning between the historical trajectory and query-based trajectory generated by the generator. Selection bias from positives affects user preference modeling.

(2) Impact of DCL. Compared with Variant II, Query2Trip is improved by 4.40% and 5.65% performance on average in terms of F1-score and pairs-F1 score, separately. This may be because the exposure bias problem from negatives hurts model performance to a certain degree.

(3) Impact of Negative Weighting. Compared with Query2Trip, Variant III show a 1.47% and 2.29% performance drop by averaging the F1 and pairs-F1 results across the eight datasets, respectively. This demonstrates that our negative weighting strategy contributes to performance enhancement. This reason may be because the negative sample quality generated by random sampling is rarely considered, resulting in negative sampling bias.

4.4 Similarity Threshold Analysis

In our DCL module, a similarity threshold Φ is introduced to estimate whether a negative sample is improper. Before the similarity threshold is introduced, the statistical results regarding the similarity distribution between an input query and negative trajectories on all our datasets are analyzed. We can observe that given an input query, about 15%–20% of in-batch negatives have a greater than 0.7 similarity to the input query. Figure 3 shows the impact of similarity threshold on performance by tuning Φ. From the results, the Φ has a crucial effect on the performance of our Query2Trip. Based on the results from all datasets, when Φ is 0.7, our method performs well as a whole. Generally, the results on the eight datasets show that when the value of Φ is too large or too small, the performance of Query2Trip decreases to a certain extent. If Φ is too large, it may not be able to effectively penalize improper negative samples, since only a few negative samples have excessive similarity. If Φ is too small, some profitable negative samples may be misjudged as improper negative samples.

4.5 Representation Performance Analysis

To further illustrate the effectiveness of our Query2Trip learned representations, Fig. 4 illustrates the representation performance comparison of different models on 200 randomly selected samples from the Osa. dataset. As a representative neural TripRec method, DeepTrip with biased learning is compared against our Query2Trip to explore the distribution relevance between the query and trajectory. From the visualization results in Fig. 4, we can observe that: **(i)** Regardless of whether the learned representation is from DeepTrip or Query2Trip, the distribution of the query representation and the trajectory representation are approximate. This indicates that our solution of mapping from query to trip is in favor of sequence learning. **(ii)** The distributions learned by DeepTrip are better but still not similar due to biased learning. Our Query2Trip aligns the distributions of learned query and trajectory representations well, and their distributions are visually similar. This also verifies that our debiased learning can learn more effective representations.

5 Related Work

5.1 Trip Recommendation

Most **traditional** studies solved TripRec problems as a variant of the orienteering problem (OP). PersTour [13] was developed to recommend a personalized trip based on POI popularity and user interest preferences. Furthermore, the transition probability-based POIRANK and RANK+Markov [2] were proposed as feature-driven approaches, which can learn from past behaviors without devising specific processes. Subsequently, Gu et al. [7] presented an enhanced personalized trip recommender with attractive routes method called TRAR. For **neural** TripRec, He et al. [9] presented C-ILP and C-ALNS based on a joint context-aware embedding. Besides, PT-VPMF [24] used convolutional neural networks to capture visual contents in geo-tagged photos, then introduced visual-enhanced probabilistic matrix factorization for personalized TripRec. Due to the outstanding performance of the encoder-decoder model in the TripRec, other methods such as CATHI [27], MDTRP [10], DeepTrip [5], and CTLTR [26] followed a similar line of reasoning by also employing the recurrent neural network (RNN) or long short-term memory network (LSTM) based model to learn underlying semantic relationships between POIs in the trajectory. Also, Jiang et al. [12] researched adversarial neural TripRec with reinforcement learning. Recent CTLTR [26] captures semantic sub-trajectories and corresponding sequential information by using RNN and contrastive learning. Different from these methods, we combine sequence generation with Transformer architecture and debiased learning in TripRec modeling to learn more effective representations.

5.2 Debiased Learning in Recommendation

There are many types of biases in recommender systems, which have been shown to seriously hurt the performance of a recommendation [3]. Gupta et al. [8] proposed three estimators by leveraging known exposure probabilities to correct exposure bias in link recommendation. Zheng et al. [25] developed a CBR method with self-attention to address the context bias caused by complex interactions among multiple items and model user preferences. Liu et al. [14] presented a self-supervised learning framework to solve selection bias in recommendation. Meanwhile, DANCER [11] considered the dynamic scenario in recommendation where selection bias and user preferences change over time. Inspired by these related studies, we take a first step towards a dual-debiased sequential learning by targeting mitigation of selection bias and exposure bias in TripRec.

6 Conclusion and Future Work

Our study explored using debiased sequence learning from query to trip for neural TripRec. From this work, a novel Query2Trip with dual-debiased learning is presented to mitigate both selection and exposure bias in TripRec through our debiased adversarial learning with conditional guidance and debiased contrastive learning with negative weighting. Empirical results from the analysis of eight real-world datasets demonstrate the effectiveness of our method compared to state-of-the-art baseline methods. In future work, we will attempt to draw on the user's diversified social information and extend to group-oriented neural TripRec for comprehensive understanding of human behavior.

Acknowledgements. This work is supported in part by the National Natural Science Foundation of China (62276196) and the Key Research and Development Program of Hubei Province (2021BAA030).

References

1. Arjovsky, M., Chintala, S., Bottou, L.: Wasserstein generative adversarial networks. In: ICML, vol. 70, pp. 214–223 (2017)
2. Chen, D., Ong, C.S., Xie, L.: Learning points and routes to recommend trajectories. In: CIKM, pp. 2227–2232 (2016)
3. Ding, S., Feng, F., He, X., Jin, J., Wang, W., Liao, Y., Zhang, Y.: Interpolative distillation for unifying biased and debiased recommendation. In: SIGIR, pp. 40–49 (2022)
4. Gao, Q., Wang, W., Zhang, K., Yang, X., Miao, C., Li, T.: Self-supervised representation learning for trip recommendation. Knowl. Based Syst. **247**, 108791 (2022)
5. Gao, Q., Zhou, F., Zhang, K., Zhang, F., Trajcevski, G.: Adversarial human trajectory learning for trip recommendation. IEEE Trans. Neural Netw. Learn. Syst., 1–13 (2021)
6. Gao, Y., Gao, X., Li, X., Yao, B., Chen, G.: An embedded GRASP-VNS based two-layer framework for tour recommendation. IEEE Trans. Serv. Comput. **15**(2), 847–859 (2022)
7. Gu, J., Song, C., Jiang, W., Wang, X., Liu, M.: Enhancing personalized trip recommendation with attractive routes. In: AAAI, pp. 662–669 (2020)
8. Gupta, S., Wang, H., Lipton, Z.C., Wang, Y.: Correcting exposure bias for link recommendation. In: ICML, vol. 139, pp. 3953–3963 (2021)
9. He, J., Qi, J., Ramamohanarao, K.: A joint context-aware embedding for trip recommendations. In: ICDE, pp. 292–303 (2019)
10. Huang, F., Xu, J., Weng, J.: Multi-task travel route planning with a flexible deep learning framework. IEEE Trans. Intell. Transp. Syst. **22**(7), 3907–3918 (2021)
11. Huang, J., Oosterhuis, H., de Rijke, M.: It is different when items are older: debiasing recommendations when selection bias and user preferences are dynamic. In: WSDM, pp. 381–389 (2022)
12. Jiang, L., et al.: Adversarial neural trip recommendation. CoRR abs/2109.11731 (2021)
13. Lim, K.H., Chan, J., Leckie, C., Karunasekera, S.: Personalized tour recommendation based on user interests and points of interest visit durations. In: IJCAI, pp. 1778–1784 (2015)
14. Liu, H., et al.: Rating distribution calibration for selection bias mitigation in recommendations. In: WWW, pp. 2048–2057 (2022)
15. Mansoury, M., Abdollahpouri, H., Pechenizkiy, M., Mobasher, B., Burke, R.: A graph-based approach for mitigating multi-sided exposure bias in recommender systems. ACM Trans. Inf. Syst. **40**(2), 32:1–32:31 (2022)
16. Qiu, R., Huang, Z., Yin, H., Wang, Z.: Contrastive learning for representation degeneration problem in sequential recommendation. In: WSDM, pp. 813–823 (2022)
17. Rakesh, V., Jadhav, N., Kotov, A., Reddy, C.K.: Orienteering algorithms for generating travel itineraries. In: WSDM, pp. 180–188 (2018)
18. Smallman, C., Moore, K.: Process studies of tourist' decision-making. Ann. Tour. Res. **37**(2), 397–422 (2010)
19. Suresh, V., Ong, D.C.: Not all negatives are equal: label-aware contrastive loss for fine-grained text classification. In: EMNLP, pp. 4381–4394 (2021)
20. Vaswani, A., et al.: Attention is all you need. In: NeurIPS, pp. 5998–6008 (2017)

21. Wang, J., Wu, N., Zhao, W.X., Peng, F., Lin, X.: Empowering a* search algorithms with neural networks for personalized route recommendation. In: KDD, pp. 539–547 (2019)
22. Wang, P., Li, L., Xie, Q., Wang, R., Xu, G.: Social dual-effect driven group modeling for neural group recommendation. Neurocomputing **481**, 258–269 (2022)
23. Wang, R., Li, L., Tao, X., Wang, P., Liu, P.: Contrastive and attentive graph learning for multi-view clustering. Inf. Process. Manage. **59**(4), 102967 (2022)
24. Zhao, P., et al.: Photo2trip: exploiting visual contents in geo-tagged photos for personalized tour recommendation. IEEE Trans. Knowl. Data Eng. **33**(4), 1708–1721 (2021)
25. Zheng, Z., et al.: CBR: context bias aware recommendation for debiasing user modeling and click prediction. In: WWW, pp. 2268–2276 (2022)
26. Zhou, F., Wang, P., Xu, X., Tai, W., Trajcevski, G.: Contrastive trajectory learning for tour recommendation. ACM Trans. Intell. Syst. Technol. **13**(1), 4:1–4:25 (2022)
27. Zhou, F., Yue, X., Trajcevski, G., Zhong, T., Zhang, K.: Context-aware variational trajectory encoding and human mobility inference. In: WWW, pp. 3469–3475 (2019)
28. Zhou, K., Zhang, B., Zhao, X., Wen, J.: Debiased contrastive learning of unsupervised sentence representations. In: ACL, pp. 6120–6130 (2022)

A New Reconstruction Attack: User Latent Vector Leakage in Federated Recommendation

Zheng Zhang and Wei Song[✉]

School of Computer Science, Wuhan University, Wuhan, China
{zhangzheng,songwei}@whu.edu.cn

Abstract. Federated Recommendation (FR) has received considerable attention in the past few years. For each user in FR, its latent vector and interaction data are kept on its local device and thus are private to others. However, keeping the training data locally can not ensure the user's privacy is compromised. In this paper, we show that the existing FR is vulnerable to a new reconstruction attack in which the attacker leverages the semi-trusted FR server to lunch the reconstruction attack. In this attack, the server rigidly follows the protocol of FR, but the attacker may compromise the system security by analyzing the gradient updates received by the server. Specifically, we design Generative Reconstruction Network (GRN), a model reconstructing attack against FR aiming to generate the target user's (i.e., the victim) latent vector including user's sensitive information. Moreover, a server-side generator is designed to take random vectors as inputs and outputs generated latent vectors. The generator is trained by the distance between the real victim's gradient updates and the generated gradient updates. We explain that the generator will successfully learn the target latent vector distribution to probe into the victim's privacy. The experimental results demonstrate the proposed attack's effectiveness and superiority over the baseline attacks.

Keywords: federated recommendation · reconstruction attack · federated learning · recommender system

1 Introduction

Recommender systems have become one of the major channels for people to obtain information and can directly influence people's perceptions while recommending items. However, traditional recommender systems suffer badly from attacks in many domains. References [7,14,15] present various data poisoning attacks against deep learning, matrix factorization, and graph-based recommender systems, respectively. These attacks assume the adversary has access to the user-item interactions and can control some malicious users to generate fake interactions accordingly, and raise the exposure ratio of target items. It is really not acceptable, especially in some recommender applications involving the personal data.

© The Author(s), under exclusive license to Springer Nature Switzerland AG 2023
X. Wang et al. (Eds.): DASFAA 2023, LNCS 13944, pp. 97–112, 2023.
https://doi.org/10.1007/978-3-031-30672-3_7

Meanwhile, people pay increasing attention to privacy protection issue in recommender system. In the traditional centralized recommender system, every user must upload its interaction records with items to the central server for recommender system training. This service mode makes users lose control of their data. Considering federated learning presents significant advantages in privacy preservation, some studies tried to apply this framework to the recommender system [1,16,22,32], namely federated recommendation.

Federated recommendation (FR) is considered a solution to recommend items under the circumstances of protecting users' privacy. As shown in Fig. 1, every user downloads the shared model (e.g., the recommender system) and trains it based on its private data locally. Then it updates its latent vector locally and sends other shared model parameter gradients back to the server, where all the updates are aggregated and accumulated to the current shared model. Because users' private data (e.g., its latent vector, interaction matrix) never leave their device, it is impractical for an adversary to access users' interaction data, and most existing attacks against traditional recommender systems become invalid.

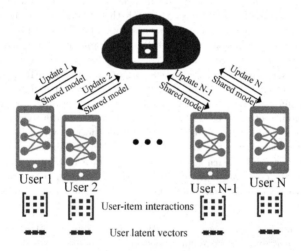

Fig. 1. The framework of federated recommendation. The server sends the shared model (e.g., recommendation system) to each user, who trains the shared model locally by its private data. Then, the gradient updates from all the users are aggregated to the server to improve the shared model collaboratively.

Although introducing federated learning framework is a big step forward in building privacy-preserving recommendation system, it still suffers from some other attacks. In most scenarios, people assume that gradient updates exchanged in federated learning are safe to share and will not expose the private data. However, some recent studies show that gradient updates will reveal the users' training data privacy even the attacker has not accessed to the data. For example, [31] assumed a malicious server, which utilized the gradient updates for training a multi-task GAN [8] and generated the facial image of the target user. Reference [18] designed a general attack based on multiple prediction outputs for neural networks to investigate the privacy leakages in vertical federated learning.

Motivated by these studies in existing attacking techniques, we propose a more generic and practical reconstruction attack in FR on the semi-honest server side. Instead of getting target items recommended to as many users as possible [26, 27, 37], we aim to give a precise reconstruction of the target user's (e.g., the victim) private training data (e.g., the latent vector), which is a more severe threat to its privacy. Inspired by [18], we present a **Generative Reconstruction Network** (GRN) which can obtain the victim's latent vector even if the latent vector is stored on the victim's device and never leaves.

To perform the attack, we first conduct a pair of random latent vectors and label vectors, then input them into a generator and obtain a couple of generated latent vectors and label vectors. Then we derive the generated gradients from the generated data (e.g., generated latent vectors and label vectors) by performing the usual forward and backward, and optimizing the generator's parameters to minimize the distance between the generated and the real gradients. Matching the gradients makes the generated latent vector close to the real victim's latent vector. When the optimization finishes, the private training data (e.g., the latent vector of the victim) will be fully revealed. The main contributions of this work can be summarized as:

- To the best of our knowledge, we are the first to present reconstruction attacks against FR from the perspective of the semi-honest server. Instead of getting target items recommended to more users, we aim to recover the victim's privacy (e.g., its latent vector), which is a more severe threat.
- Correspondingly, we propose a generic attack model – Generative Reconstruction Network (GRN), which will learn the distribution of the victim's private latent vector, achieving an attack on the victim's privacy.
- We implement the proposed attacks and conduct extensive evaluations on different datasets. The results demonstrate the effectiveness of our attacks and superiority over the baseline attacks, e,g., the MSE score of our attack is 10 times smaller than others'.

2 Related Work

2.1 Attacks Against Traditional Recommendation System

Existing studies showed that traditional recommender systems are vulnerable to security attacks, such as data poisoning attacks and profile pollution attacks.

Data poisoning attack is the mainstream attack against the traditional recommendation system. In such attacks, the adversary aims to promote a target item and recommend it to as many users as possible. Roughly speaking, the adversary injects a small number of fake users and generates well-crafted interactions by manipulating the fake users to interact with certain specific items. These fake data will be included in the training dataset of the target recommender system and then poisons the training process. Data poisoning attacks could be divided into two categories, according to whether the attack is focused on a specific type of recommender system: algorithm-agnostic and algorithm-specific [14]. The former attack does not consider the algorithm used by the recommender system. Therefore, it often has limited effectiveness [9,21,24]. The latter attack is optimized to a specific recommender system and has been developed for different recommender systems, such as matrix factorization based recommender systems [6,15], graph-based recommender systems [7], neighborhood-based recommender systems [5], and association rule-based recommender systems [35]. However, these attacks are impractical in FR scenarios because each user's interactions are kept locally, and the adversary has no access to them.

Profile pollution is a niche type of attack on the recommender system, which aims to pollute a user's profile (e.g., historical records) via cross-site request forgery (CSRF) [36]. Reference [34] proposed profile pollution attacks to recommender systems in web services (e.g., Google, YouTube). They showed that all these services are vulnerable to the attack. However, profile pollution attacks can not be applied to recommender systems because the adversary cannot pollute the profile of an item [35].

2.2 Attacks Against Federated Learning

As data privacy becomes a concern for users, Federated Learning (FL) [19,25,30] attracts more attention and has been applied in many private domains [3,10,38]. In every training round of FL, the central server shares the shared model with users. Each of these users trains the shared model locally with the training data kept on its device, then uploads the gradients of model parameters to the server. At the end of the round, the central server updates the shared model by averaging the uploaded gradients.

Recently, more and more studies on how to attack FL have arisen [2,17, 23,28,33]. Reference [29] exploited the differences in the models' outputs on training and non-training inputs to decide whether a given sample belongs to the training data set. Reference [20] demonstrated the membership attack by conducting a batch property classifier to infer the victim's targeted property. Reference [18] designed a general attack based on multiple prediction outputs for neural networks and random forest models to investigate the privacy leakages in vertical FL. Given the victim's updates, [31] assumed the server was malicious. By constructing a multi-tasks GAN, the adversary was capable of making a targeted attack on the victim's privacy image.

2.3 Attacks Against Federated Recommendation

There are a few existing attacks against the federated recommendation. Reference [27] proposed PipAttack, the first framework of model poisoning attack against FR. Its experimental results show that PipAttack is effective in FR. However, the attack will cause significant degradation in recommendation accuracy, which increases the probability of being detected. Reference [37] presented model poisoning attacks to FR named FedRecAttack. However, the attack requires the adversary to have prior knowledge (i.e., the side information that reflects each item's popularity or some public user-item interactions), which is not generic in all FR scenarios. Reference [39] recovered a small training image from the victim's update by designing 'dummy data' and 'dummy label'. Although there is increasing concern about the security issues of FR, the attacks against FR are still under-explored.

3   Preliminaries

3.1   Recommender Systems

Neural Collaborative Filtering (NCF) [13] is one of the most widely used deep learning based recommender models and has state-of-the-art recommendation performance. Without loss of generality, we adopt NCF as our base recommender model.

Respectively, let M and N denote the number of users and items in the recommender system. Also, we use U and I to denote the set of M users and N items. Each $u \in U$ has an embedding vector p_u, which describes its latent feature. Similarly, each item $i \in I$ has an embedding vector q_i. We denote the user-item interaction matrix $Y \in \mathbb{R}^{M \times N}$ from users' feedback as:

$$y_{ui} = \begin{cases} 1; \; if \; interaction \; (u, i) \; is \; observed; \\ 0; \; otherwise. \end{cases} \quad (1)$$

We use $\hat{Y}_{ui}$ to denote the predicted score between user u and item i, which indicates the preference of user u for item i. In NCF, $\hat{Y}_{ui}$ is predicted as follows:

$$\hat{Y}_{ui} = f(p_u, q_i) \quad (2)$$

where f is the user-item interaction function. To supercharge modeling with non-linearities, NFC proposed to leverage a Multi-Layer Perceptron (MLP) to learn the user-item interaction function as follows:

$$f(p_u, q_i) = a_{out}(\Phi(p_u \oplus q_i)) \quad (3)$$

where a_{out}, Φ and $\oplus$ denote the activation function, the MLP function and the vector concatenation, respectively. NCF use the sigmoid function as a_{out} to restrict the model output to be in $(0, 1)$. In NCF with L hidden layers, the multi-layer neural function Φ is as follows:

$$\begin{aligned} \Phi(x) &= \Phi_L(...(\Phi_2(\Phi_1(x))) \\ \Phi_k(x) &= a_k(W_k \, x + b_k) \end{aligned} \quad (4)$$

where a_k, W_k and b_k denote the activation function, the weight matrix and the bias vector in the $k\,th$ layer of the perceptron, respectively. To achieve better performance of recommendation, we use Rectifier (ReLU) as the activation function a_k in all perceptron layers.

3.2   Federated Recommendation

We train our base recommender model distributedly under the framework of federated learning. More specifically, there is a semi-honest server and N users in federated learning scenarios. Each of the users owns its private dataset. Let D_u denote the training dataset of user u, which consists of item-score pairs (i, Y_{ui}). If user u has interacted with item i, $Y_{ui} = 1$ (i.e., a positive instance). Otherwise, $Y_{ui} = 0$ (i.e., a negative instance). Note that, for each user u, since there are many uninteracted items, the negative instances in D_u are sampled with a ratio of $r : 1$ as the number of negative instances against that of positive, like [26].

Let P, Q respectively denote the latent matrices of users and items. p_u is $u\,th$ row of P, which indicates the latent vector of user u, and q_i is $i\,th$ row of Q, which indicates the latent vector of item i. For each user u, p_u, and D_u are stored locally on its device and never sent to others. Meanwhile, all trainable model parameters Θ (e.g., $Q, h, W1, b1, W2, b2 \cdots W_L, b_L$), except for users' latent matrix P, are stored on the central server.

At each training iteration, the parameters Θ of the current recommender system is downloaded from the server to users, and then the model is trained locally on each user. Let p_u^t and Θ^t denote the user u latent vector and model parameters in the $t\,th$ round, respectively. In the $t\,th$ round, the server sends each user a copy of Θ^t. Each user u derives the gradients of p_u^t and Θ_u^t, which are denoted by ∇p_u^t and $\nabla \Theta_u^t$, by computing its loss function L_u. Moreover, we adopt the Binary Cross-Entropy (BCE) loss, like [26], to quantify the difference between the model-predicted scores and the ground-truth scores on the training dataset. Then user u updates its latent vector p_u and sends the model's parameters gradients to the server. The server aggregates the uploaded parameter gradients $\nabla \Theta^t$ and updates the recommender system model:

4   Our Attack

4.1   System Model and Thread Model

Following the training procedure of FR described in Sect. 3.2, we provide a high-level overview of FR in our paper as Fig. 2.

Assuming there are M users and N items in the system, we denote the sets of users and items as U and I, where $|U| = M$ and $|I| = N$. All clients' data is considered non-IID distributed, which is also consistent with the setting of federated learning. Specifically, the data distribution of an arbitrary user D_u is independent of any other one. In addition to D_u, each user $u \in U$ has its latent vector p_u. The latent vector will be defined as p_u^t at iteration t. Especially, we define p_v^t to represent the latent vector of the victim. Meanwhile, the server owns the item latent matrix Q. At the $t\,th$ iteration, each user downloads the shared model's parameter Θ^t (e.g., including the recommender system's parameters and item latent matrix Q) from the server and trains the model based on its latent vector p_u and its user-item interaction matrix D_u. We adopt the Binary Cross-Entropy (BCE) loss as the loss function L_u to quantify the distance between the

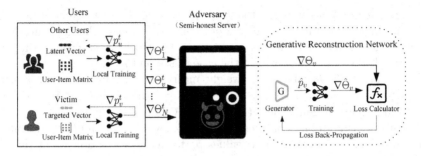

Fig. 2. Illustration of the proposed GRN from a semi-honest server in the federated recommendation.

predicted and ground-truth scores. The L_u could be considered as a function of p_u and Θ as follows:

$$L_u(p_u, \theta) = - \sum_{i, Y_{ui} \in D_u} Y_{ui} log \hat{Y}_{ui} + (1 - Y_{ui}) log(1 - \hat{Y}_{ui}), \tag{5}$$

where $\hat{Y}_{ui}$ is the predicted score between user u and item i. With L_u computed, each of users u derives the gradients of p_u^t and Θ^t, which are denoted by ∇p_u^t and $\nabla \Theta_u^t$. Then each user u updates p_u locally with learning rate η as follows:

$$p_u^{t+1} = p_u^t - \eta \nabla p_u^t. \tag{6}$$

Then u sends $\nabla \Theta_u^t$ back to the server, which updates Θ by aggregating the uploaded gradients as follows:

$$\Theta^{t+1} = \Theta^t - \eta \sum_{u \in U} \nabla \Theta_u^t. \tag{7}$$

In this paper, the FR server is considered to be semi-honest. The semi-honest server will follow the protocol to build the shared model, but it maybe analyze the periodic updates from the users to recover the victim user's privacy data due to the curiosity. It aims to reconstruct latent vector p_v of the victim. Note that the reconstruction attack is launched in the last round of training (e.g., $t = T$), where the victim's latent vector p_v has been trained thoroughly and contains lots of the victim's privacy (e.g., its interests or orientation). In this iteration, the semi-honest server receives the victim's gradient update $\nabla \Theta_v^T$. Then the adversary will analyze this gradient update to reconstruct the victim's latent vector based on the proposed generative reconstruction network (GRN).

4.2 Generative Reconstruction Network

From the perspective of the semi-honest server, the only information he has on the victim is its gradient updates $\nabla \Theta_v$. Moreover, the adversary's objective is to obtain the latent vector of the victim p_v. Intuitively, the basic idea is to figure out the overall correlations between the gradients $\nabla\Theta_v$ and the targeted vector p_v. The main idea behind this is that in a deep learning model, whether centralized or distributed, gradient updates are always computed by back-propagating the loss through the entire network. In the case of sequential fully connected layers $h_l, h_{l+1}(h_{l+1} = W_l \cdot h_l)$, W_l is the weight matrix, the gradient of error E with respect to W_l is computed as $\frac{\partial E}{\partial W_l} = \frac{\partial E}{\partial h_{l+1}} \cdot h_l$. So, the gradient updates can be used to infer features and reflect the users' private training data.

Upon this, the problem of inferring the unknown vector values is equivalent to the problem of generating new values $\hat{p}_v$ to match the gradients of the FR model, where $\hat{p}_v$ target follows a probability distribution determined by the victim's uploaded gradient values $\nabla\Theta_v$. Figure 3 illustrates our attack in detail. To learn such a probability distribution, we build a generator model (in grey color in Fig. 3), which takes a set of random variables (e.g., r_v, r_y) as inputs and produces an estimation of the unknown latent vector values $\hat{p}_v$ and $\hat{y}_v$. Consequently, the generator model is trained by minimizing the loss between the gradients of the generated samples $\nabla\hat{\Theta}_v$ and the ground-truth gradients $\nabla\Theta_v$.

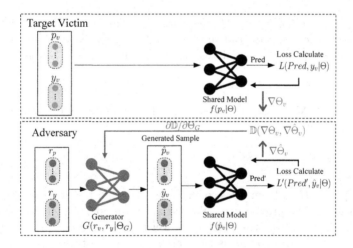

Fig. 3. The overview of the proposed GRN algorithm. While the victim calculates $\nabla\Theta_v$ to update the parameter using its private training data, including its latent vector p_v, the adversary updates its generator to minimize the gradients distance $\mathbb{D}(\nabla\Theta_v, \nabla\hat{\Theta}_v)$. When the optimization finishes, the adversary could reconstruct p_v of the victim.

Note that the random vector r_y is an indispensable component of GRN. The training of generator G depends on the distance of two gradients (e.g., $\mathbb{D}(\nabla\Theta_v, \nabla\hat{\Theta}_v)$). Moreover, the calculation of gradients $\nabla\hat{\Theta}_v$ depends on the real labels of victim y_v, which is as follows:

$$\nabla\hat{\Theta}_v = \frac{\partial L(f(\hat{p}_u, Q|\Theta), y_v)}{\partial\Theta} \tag{8}$$

However, from the perspective of the adversary, he could not know the real labels of the victim without a priori knowledge. Our attack considers this strict case and generates the labels $\hat{y}_v$ by r_v, which is a substitute for real labels.

We assume that the users' latent vector size is k, which are all set by the semi-honest server. Moreover, the adversary has collected the gradient updates of the victim Θ_v. The adversary's objective is to train a generator model, say Θ_G, such that given a random value vector r_p with size k and a random value vector r_y, the generator outputs the corresponding estimation $(\hat{p}_v, \hat{y}_v)$ of the real vector (p_v, y_v). To train the model, the adversary can apply the Adam gradient descent method. The objective function is as follows:

$$\min_{\Theta_G} L_G\left(\frac{\partial L(f(f_G(r_p|\Theta_G)), f_G(r_v|\Theta_G))}{\partial \Theta}, \nabla\Theta_v\right) \tag{9}$$

where Θ and Θ_G are the parameters of the shared model and the generator model, respectively. Moreover, f_G denotes the output of the generator, i.e., $\hat{p}_v$ and $\hat{y}_v$, and f denotes the output of the shared model given the generated sample.

Algorithm 1 presents the training method for GRN. Specifically, in each iteration, the adversary feeds the random vector r_p and v_y into the generator, obtaining $\hat{p}_v$ and $\hat{y}_v$. Then the adversary input the generated latent vector $\hat{p}_v$ and makes a prediction using the shared model. Next, the adversary feeds the generated labels $\hat{y}_v$ and gets "generated gradients" $\hat{\Theta}_v$. As a result, the loss of this generated sample can be calculated by a loss function L_G, e.g., MSE, with the ground-truth gradient Θ_v. After obtaining the loss, the adversary back-propagates the aggregated loss to update the parameters of the generator model Θ_G. Finally, the adversary can obtain the trained generator model after all the training epochs. After obtaining Θ_G, the adversary can use it to infer the unknown latent vector values p_v.

Algorithm 1. Generative reconstruction network training

Input: $\nabla\Theta_v$: gradients of the victim, α: learning rate
Output: Θ_G^*: parameters of the generator model

1: $\Theta_G \leftarrow \mathcal{N}(0,1)$ // Initialize generator model parameters
2: $r_p \leftarrow \mathcal{N}(0,1), r_v \leftarrow \mathcal{N}(0,1)$ // Initialize random latent vector and label vector.
3: **for** $t = 0 \to n$ **do**
4:     $\hat{p}_v, \hat{y}_v \leftarrow f_G(r_p, r_y|\Theta_G)$
5:     $\nabla\hat{\Theta}_v \mathrel{-}= \frac{\partial L(f(\hat{p}_u, Q|\Theta), y_v)}{\partial \Theta}$ //Compute generated gradients
6:     $\mathbb{D} \leftarrow ||\nabla\hat{\Theta}_v - \nabla\Theta_v||^2$
7:     $\Theta_G \leftarrow \Theta_G - \alpha \cdot \nabla_{\Theta_G}\mathbb{D}$ //Update parameters
8: **end for**
9: **return** Θ_G

4.3   Discussion

The GRN attack method is general because it takes the trained shared model as a black-box, as long as that model's objective function is differentiable, so the prediction loss could be back-propagated to the generator model. Moreover, unlike existing reconstruction attacks, our attack method considers the most severe case: it relies on no background information of the attack target's data distribution (such as statistics or marginal feature distribution). In particular, the adversary only needs the victim's gradient updates, which is not beyond his power. Meanwhile, our attack does not affect the federated recommendation. We assume a semi-honest server as the adversary, who analyses the gradient updates from the victim and never modifies any data of the FR. From the victim's perspective, it just executes a training epoch. Therefore, our attack is hard to be detected.

5   Experiments

5.1   Experimental Settings

We implement the proposed attack algorithms in Python and conduct experiments on machines running Windows 10 equipped with AMD Ryzen 9 5900X, 32 GB RAM, and an NVIDIA 3090 GPU card. Specifically, we adopt PyTorch2 for training the FR model.

Dataset. We experiment with three popular accessible datasets in three completely different scenarios (e.g., movie recommendation, music recommendation, and game recommendation) for our experiments: MovieLens-1M (ML) [11], Amazon Digital Music (AZ) [12], and Steam-200K (Steam) [4]. For ML, we use the version containing 1,000,208 ratings involving 6,040 users and 3,706 movies. For AZ, we use the version containing 169,781 reviews involving 16,566 users and 11,797 digital music products. For Steam, we use the version containing 114,713 interactions involving 3,753 users and 5,134 game items. In all datasets, we convert the user-item interactions (i.e., ratings and reviews) into implicit data following [13], and divide each user's interactions into the training set and test set in the ratio of $4 : 1$, like [26]. Note that we intentionally used three datasets of completely different sizes to validate our attacks' robustness.

FR Model and GRN Model. We adopt NCF with different sizes of hidden layers as our base recommender model. The dimensions of hidden layers are different, which are set to $16, 8, 4, 1$. The learning rate η for the user is set to 0.001. Moreover, let k, T, N_u and N_i denote the user latent vector length, the training epoch number, the user number and the item number, respectively.

Our GRN is a double output network consisting of 4 full connect layers. The input layer takes the concatenation of the random target vector r_p and the random label r_v vector as input and the output layer outputs two vector: the generated latent vector $\hat{p}_v$ and the generated label vector $\hat{y}_v$. The tow hidden layers are $600 \rightarrow 300$ and $300 \rightarrow 100$, respectively.

Metrics. We use the mean square error (MSE) to measure the overall accuracy of the generated latent vector. Specifically, the MSE is calculated as:

$$MSE = \frac{1}{k} \sum_{i=1}^{k} (\hat{p}_{v,i} - p_{v,i})^2 \tag{10}$$

where k is the latent vector length, $\hat{p}_v$ and p_v are the generated vector values and the ground-truth, respectively.

Baseline. In our experiments, we compare our GRN with several baseline attack methods. DLG [39] is a typical reconstruction attack in FL, which has been introduced in Sect. 2.3. DLG is always used for generating the victim's training images, whose nature is a multidimensional vector. We fine-tune the output format of DLG so that its products are victim's latent vectors. Meanwhile, We use two other baselines that randomly generate samples from $(-1, 1)$ according to a Uniform distribution $U(-1, 1)$ and a Gaussian distribution $N(0, 0.5)$. We call these two baselines random guesses in the following presentation.

5.2 Comparison of Attack Effectiveness

We compare the effectiveness of baseline attacks and ours on ML, AZ, and Steam. All four above attacks, including three baseline attacks and our attack, have been performed on the same FR settings. The basic settings are shown in Table 1.

Table 1. FR Setting on Three Datasets

Dataset	User Number N_u	Item Number N_i	Interaction	Latent Vector Length k	Epoch T
ML	6,040	3,706	1,000,208	8	20
AZ	16,566	11,797	169,781	16	30
Steam	3,753	5,134	114,713	8	20

To ensure convergence, the base shared model (e.g., recommender system) is trained for 30 epochs on AZ and 20 epochs on both ML and AZ. Moreover, we set the length of the user latent vector and item latent vector to 16 on AZ and 8 on both ML and Steam. And the learning rate for the user is set to 0.001. Figure 4 shows the attack performance MSE scores of baseline attacks and our attack on different datasets under the same FR settings. Our attack has almost perfect performance for all the datasets, e.g., MSE scores on all three datasets are less than 0.01. By contrast, other attacks have higher MSE scores, e.g., the MSE scores of other attacks are 10 times our attack score, which means the generated by baseline attacks have a great distance to the ground truth.

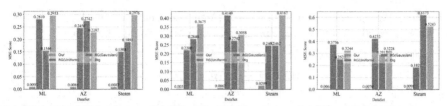

(a) FR with 2 hidden layers (b) FR with 3 hidden layers (c) FR with 4 hidden layers

Fig. 4. Comparison of attack effectiveness with different hidden layers.

To further explore the effectiveness of our attacks under certain circumstances, we evaluate our attacks with different FR settings, including the user latent vector length k, the FR training epoch number T, and the user number N_u.

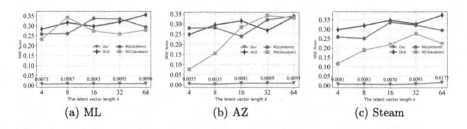

(a) ML   (b) AZ   (c) Steam

Fig. 5. MSE Score *vs.* the user latent vector length k.

5.3   Effect of Different Factors on Attack

Effect of the User Latent Vector Length. We first evaluate the attack performance *w.r.t.* the user latent vector length k. As Fig. 5 shows, our attack achieves reasonably high performance, with all MSE scores being less than 0.01 on ML, AZ, and Steam. For $k = 4, 8, 16, 32, 64$, our attack has lower MSE scores than other baseline attacks, e.g., the user latent vector generated by our attack is closer to the real value than the vector generated by other attacks.

We notice that the performance degrades slightly for our attack as k increases, e.g., our attack's MSE scores increase. The reason is that the longer the target vector, the harder it will be to generate it. Our attack reconstructs the target vector by training a generator, and a more extended vector means more neurons to train, which will cause the attack performance decreases. Meanwhile, we also notice that fixed k, the MSE scores of the three datasets are very similar, which means our attack is robust and versatile for different types of datasets, e.g., movies, digital music, and games.

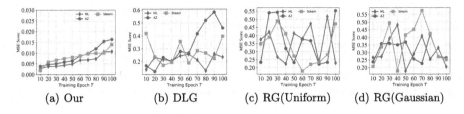

(a) Our   (b) DLG   (c) RG(Uniform)   (d) RG(Gaussian)

Fig. 6. MSE Score *vs.* the FR training epoch number T.

Effect of the FR Training Epoch Number. Then we evaluate the effect of the FR training epoch number T on our attack. Figure 6 shows the performance of four attacks on three datasets with different FR training epochs. In all datasets and $Epoch = 10, 20, 30, 40, 50, 60, 70, 80, 90, 100$, our attack still worked much better than the others, e.g., the MSE scores of other attacks are 10 times more than we ours.

We also observe that the performance decreases slightly for our attack as T increases, e.g., our attack's MSE scores increase. Intuitively, the more training rounds, the smaller the gradient updates. Our attack relies on the distance between two gradients (e.g., real and generated gradients), and the small gap makes the generator challenging to update. Meanwhile, as T increases, the performance of the baselines attack fluctuates dramatically. By contrast, our attacks are stable and do not change dramatically as T increases.

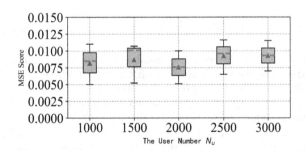

Fig. 7. Our attack's MSE Score *vs.* the user number N_u.

Effect of the User Number. To quantify the effect of the user number N_u on our attack's performance, we trained our attack on the ML, AZ, and Steam dataset with $1000, 1500, 2000, 2500, 3000$ users. Figure 7 shows the attack performance of our attack with different user numbers w.r.t. MSE scores. Our attack still works well in all datasets and different user numbers, e.g., all MSE scores are less than 0.012.

We also notice that in the case of different users, the MSE median and mean of the three datasets are all around 0.008 (e.g., the orange dashed and green triangle), which means the effect of our attack does not depend on the user number, and our attack can be applied to FR with different sizes of users. The reason is that our attack relies on gradient updates from the victim, and the number of others' gradients does not affect our generator's updates.

6   Conclusion and Future Work

This paper investigated a new privacy risk of Federated Recommendation (FR), which is considered a privacy-preserving learning framework working in a recommender system. We propose a generic and practical reconstruction attack named Generative Reconstruction Network, which enables a semi-honest server to reconstruct the target user latent vector although the vector is stored on its local device and never leaves. To our knowledge, this is the first work that investigates the reconstruction attack in FR. Experimental results demonstrate that our attack is more precise than baseline attacks and proves improvements should be made in FR. As part of future work, we will further explore the methods to detect the attacks in the federated recommendation.

Acknowledgement. This work is partially supported by National Natural Science Foundation of China Nos. 62072349, U1811263, Technological Innovation Major Program of Hubei Province No. 2021BEE057, and Technological Innovation Major Program of China Tobacco Corporation No. 110202102031.

References

1. Ammad-ud-din, M., et al.: Federated collaborative filtering for privacy-preserving personalized recommendation system. CoRR abs/1901.09888 (2019)
2. Bhagoji, A.N., Chakraborty, S., Mittal, P., Calo, S.B.: Analyzing federated learning through an adversarial lens. In: International Conference on Machine Learning, ICML 2019. Proceedings of Machine Learning Research, vol. 97, pp. 634–643 (2019)
3. Brisimi, T.S., Chen, R., Mela, T., Olshevsky, A., Paschalidis, I.C., Shi, W.: Federated learning of predictive models from federated electronic health records. Int. J. Med. Inf. **112**, 59–67 (2018)
4. Cerda, G.C., Guzmán, J., Parra, D.: Recommender systems for online video game platforms: the case of STEAM. In: Conference on World Wide Web, WWW 2019, pp. 763–771 (2019)
5. Chen, L., Xu, Y., Xie, F., Huang, M., Zheng, Z.: Data poisoning attacks on neighborhood-based recommender systems. Trans. Emerg. Telecommun. Technol. **32**(6) (2021)
6. Fang, M., Gong, N.Z., Liu, J.: Influence function based data poisoning attacks to top-n recommender systems. In: The Web Conference 2020, WWW 2020, pp. 3019–3025 (2020)
7. Fang, M., Yang, G., Gong, N.Z., Liu, J.: Poisoning attacks to graph-based recommender systems. In: Computer Security Applications Conference, ACSAC 2018, pp. 381–392 (2018)

8. Goodfellow, I.J., et al.: Generative adversarial nets. In: Advances in Neural Information Processing Systems, NIPS 2014, pp. 2672–2680 (2014)
9. Gupta, V., Kapoor, S., Kumar, R.: A review of attacks and its detection attributes on collaborative recommender systems. Int. J. Adv. Res. Comput. Sci. **8** (2017)
10. Hard, A., et al.: Federated learning for mobile keyboard prediction. CoRR abs/1811.03604 (2018)
11. Harper, F.M., Konstan, J.A.: The movielens datasets: history and context. ACM Trans. Interact. Intell. Syst. **5**(4), 19:1-19:19 (2016)
12. He, R., McAuley, J.J.: Ups and downs: modeling the visual evolution of fashion trends with one-class collaborative filtering. In: Conference on World Wide Web, WWW 2016, pp. 507–517 (2016)
13. He, X., Liao, L., Zhang, H., Nie, L., Hu, X., Chua, T.: Neural collaborative filtering. In: Conference on World Wide Web, WWW 2017, pp. 173–182 (2017)
14. Huang, H., Mu, J., Gong, N.Z., Li, Q., Liu, B., Xu, M.: Data poisoning attacks to deep learning based recommender systems. In: Network and Distributed System Security Symposium, NDSS 2021 (2021)
15. Li, B., Wang, Y., Singh, A., Vorobeychik, Y.: Data poisoning attacks on factorization-based collaborative filtering. In: Advances in Neural Information Processing Systems, NIPS 2016, pp. 1885–1893 (2016)
16. Liang, F., Pan, W., Ming, Z.: FedRec++: lossless federated recommendation with explicit feedback. In: 35th AAAI Conference on Artificial Intelligence, AAAI 2021, pp. 4224–4231 (2021)
17. Liu, K., Dolan-Gavitt, B., Garg, S.: Fine-pruning: defending against backdooring attacks on deep neural networks. In: Bailey, M., Holz, T., Stamatogiannakis, M., Ioannidis, S. (eds.) RAID 2018. LNCS, vol. 11050, pp. 273–294. Springer, Cham (2018). https://doi.org/10.1007/978-3-030-00470-5_13
18. Luo, X., Wu, Y., Xiao, X., Ooi, B.C.: Feature inference attack on model predictions in vertical federated learning. In: International Conference on Data Engineering, ICDE 2021, pp. 181–192 (2021)
19. McMahan, B., Moore, E., Ramage, D., Hampson, S., y Arcas, B.A.: Communication-efficient learning of deep networks from decentralized data. In: International Conference on Artificial Intelligence and Statistics, AISTATS 2017. Proceedings of Machine Learning Research, vol. 54, pp. 1273–1282 (2017)
20. Melis, L., Song, C., Cristofaro, E.D., Shmatikov, V.: Exploiting unintended feature leakage in collaborative learning. In: Symposium on Security and Privacy, S&P 2019, pp. 691–706 (2019)
21. Mobasher, B., Burke, R.D., Bhaumik, R., Sandvig, J.J.: Attacks and remedies in collaborative recommendation. IEEE Intell. Syst. **22**(3), 56–63 (2007)
22. Muhammad, K., et al.: FedFast: going beyond average for faster training of federated recommender systems. In: Conference on Knowledge Discovery and Data Mining, 2020, pp. 1234–1242 (2020)
23. Nasr, M., Shokri, R., Houmansadr, A.: Comprehensive privacy analysis of deep learning: passive and active white-box inference attacks against centralized and federated learning. In: Symposium on Security and Privacy, SP 2019, pp. 739–753 (2019)
24. O'Mahony, M.P., Hurley, N.J., Silvestre, G.C.M.: Recommender systems: attack types and strategies. In: Conference on Artificial Intelligence, AAAI 2005, pp. 334–339 (2005)
25. Patarasuk, P., Yuan, X.: Bandwidth optimal all-reduce algorithms for clusters of workstations. J. Parallel Distrib. Comput. **69**(2), 117–124 (2009)

26. Rong, D., He, Q., Chen, J.: Poisoning deep learning based recommender model in federated learning scenario. In: Advances in Neural Information Processing Systems, NIPS 2022 (2022)
27. Rong, D., Ye, S., Zhao, R., Yuen, H.N., Chen, J., He, Q.: FedRecAttack: model poisoning attack to federated recommendation. CoRR abs/2204.01499 (2022)
28. Shafahi, A., et al.: Poison frogs! targeted clean-label poisoning attacks on neural networks. In: Advances in Neural Information Processing Systems, NeurIPS 2018, pp. 6106–6116 (2018)
29. Shokri, R., Stronati, M., Song, C., Shmatikov, V.: Membership inference attacks against machine learning models. In: Symposium on Security and Privacy, S&P, 2017, pp. 3–18 (2017)
30. Tong, Y., et al.: Hu-fu: Efficient and secure spatial queries over data federation. Proc. VLDB Endow. **15**(6), 1159–1172 (2022)
31. Wang, Z., Song, M., Zhang, Z., Song, Y., Wang, Q., Qi, H.: Beyond inferring class representatives: User-level privacy leakage from federated learning. In: Conference on Computer Communications, INFOCOM 2019, pp. 2512–2520 (2019)
32. Wu, J., et al.: Hierarchical personalized federated learning for user modeling. In: The Web Conference 2021, WWW 2021, pp. 957–968 (2021)
33. Xie, C., Huang, K., Chen, P., Li, B.: DBA: distributed backdoor attacks against federated learning. In: International Conference on Learning Representations, ICLR 2020 (2020)
34. Xing, X., et al.: Take this personally: Pollution attacks on personalized services. In: 2013 Proceedings of the 22th USENIX Security Symposium, pp. 671–686 (2013)
35. Yang, G., Gong, N.Z., Cai, Y.: Fake co-visitation injection attacks to recommender systems. In: Network and Distributed System Security Symposium, NDSS 2017 (2017)
36. Zeller, W., Felten, E.W.: Cross-site request forgeries: exploitation and prevention. The New York Times, pp. 1–13 (2009)
37. Zhang, S., Yin, H., Chen, T., Huang, Z., Nguyen, Q.V.H., Cui, L.: PipAttack: poisoning federated recommender systems formanipulating item promotion. CoRR abs/2110.10926 (2021)
38. Zheng, W., Yan, L., Gou, C., Wang, F.: Federated meta-learning for fraudulent credit card detection. In: International Joint Conference on Artificial Intelligence, IJCAI 2020, pp. 4654–4660 (2020)
39. Zhu, L., Liu, Z., Han, S.: Deep leakage from gradients. In: Advances in Neural Information Processing Systems, NIPS 2019, pp. 14747–14756 (2019)

Dual-View Self-supervised Co-training for Knowledge Graph Recommendation

Ruoyi Zhang[1], Huifang Ma[1,2]($\boxtimes$), Qingfeng Li[1], Yike Wang[1],
and Zhixin Li[2]

[1] College of Computer Science and Engineering, Northwest Normal University,
Lanzhou, Gansu, China
mahuifang@yeah.net
[2] Guangxi Key Lab of Multi-source Information Mining and Security,
Guangxi Normal University, Guilin, Guangxi, China

Abstract. Knowledge Graph Recommendation (KGR), which aims to incorporate Knowledge Graphs (KGs) as auxiliary information into recommender systems and effectively improve model performance, has attracted considerable interest. Currently, KGR community has focused on designing Graph Neural Networks (GNNs)-based end-to-end KGR models. Unfortunately, existing GNNs-based KGR models are focused on extracting high-order attributes (knowledge) but suffer from restrictions in several vital aspects, such as 1) neglect of finer-grained feature interaction information via GNNs and 2) lack of adequate supervised signals, leading to undesirable performance. To tackle these gaps, we propose a novel **D**ual-view **S**elf-supervised **C**o-training for **K**nowledge **G**raph Recommendation (DSCKG). We consider two different views, covering user-item collaborative view and KGs structural view. Precisely, for the collaborative view, we first extract high-order collaborative user/item representations with GNNs. Next, we impose a discrepancy regularization term to augment the self-discrimination of the user/item representations. As for the structural view, we initially utilize GNNs to extract high-order features. Next, we utilize novel Dual-core Convolutional Neural Networks to extract bit- and vector-level finer-grained feature interaction signals. DSCKG hence performs a high-quality self-supervised co-training paradigm across dual views, improving the node representation learning capability. Experimental results demonstrate DSCKG achieves remarkable improvements over SOTA methods.

Keywords: Knowledge Graph Recommendation · Convolutional Neural Networks · Graph Neural Networks · Self-supervised Learning

1 Introduction

Nowadays, Recommender systems aim to alleviate information overload, and are pivotal in fields such as social networks [17]. The classical recommendation approach: Collaborative Filtering (CF), which aims at modeling user preferences based on user-item similarities of historical interactions. However, CF-based recommendation models often fall into the dilemma of data sparsity and cold-start issues. To alleviate the above dilemma, researchers have suggested incorporating

X. Wang et al. (Eds.): DASFAA 2023, LNCS 13944, pp. 113–128, 2023.
https://doi.org/10.1007/978-3-031-30672-3_8

side information into recommendation model. Knowledge Graphs (KGs), which consist of real-world objective facts and fruitful entities, are significantly helpful for improving recommendation performance. Hence, Knowledge Graph Recommendation (a.k.a., KGR) has recently attracted considerable attention.

KGR's development lines can be divided into three phases: 1) Embedding-based KGR approaches, which mainly utilize knowledge graph embedding (e.g., TransR) techniques to learn high-quality entity-relationship representations. Unfortunately, the method is unable to extract long-range connections, leading to unsatisfactory recommendation performance. 2) Path-based KGR methods, which mainly exploit meta-path techniques [3] to extract long-range connections [20]. However, this type of methods require expert knowledge with manually defined meta-paths, which is not effective in practice. Meanwhile, the widespread utilization of 3) Graph Neural Networks (GNNs) and Graph Attention networks (GATs) [8] techniques, which can adaptively extract high-order knowledge (attribute information), leads to State-Of-The-Art (SOTA) for downstream recommendation tasks.

Primary Motivation. Despite the promising performance of current GNNs-based KGR methods, the following two aspects are still under-explored:

- **Existing GNNs-based KGR methods fail to extract feature interaction signals**. Current GNNs-based KGR methods for integrating high-order features commonly select mechanistic aggregation strategies (e.g., max-pooling or summation) without feature interaction signals, which is insufficient. On the one hand, selecting conventional and naive high-order feature aggregation strategies could easily lead to over-smoothing and representation degradation issues. On the other hand, potentially valuable finer-grained feature interaction signal is underutilized, which would greatly hinder the representation capability of the model. Intuitively, designing a finer-grained feature interaction mode among high-order features improves model representation capability and recommendation performance.
- **Supervision interaction data scarcity**. Graph-based methods are more susceptible to data sparsity dilemmas. Since established in a supervised mode, existing GNNs-based KGR methods rely on observed user-item inter-actions as supervised signals to perform graph representation learning in KGs. However, practical training often implies a lack of training labels because of the sparsity problem inherent in recommender systems. This issue causes the model failing to learn reliable representations, thus limiting the performance gains of the KGR model.

Considering the limitations of these two lines, we believe that it is essential to properly model finer-grained feature interaction and explore high-quality self-supervised signals in the KGR task.

Towards this end, we a novel **D**ual-view **S**elf-supervised **C**o-training for **K**nowledge **G**raph Recommendation (DSCKG). Specifically, upon the user-item collaborative view, we employ a LightGCN [2] to generate prototype user/item representations. Besides, we impose a novel discrepancy regularization strategy to encourage user/item representations independence from each other. Upon the KGs structural view, we first build the GNNs-based knowledge-aware attention module as the backbone networks to obtain the user/item embeddings. Next, we execute the finer-grained feature interaction model for the item-side. Specifically, we innovatively employ the Dual-core Convolutional Neural Networks (DuCNNs) to extract bit-level (vertical convolution) and vector-level (horizontal convolution) finer-grained feature interaction signals. Immediately after, we rely on two views, which collaboratively utilize different view information to generate extra supervised signals to guide the multi-view self-supervised co-training paradigm. Finally, we jointly employ the cross-entropy primary loss and the rest of the regularization terms to optimize the DSCKG. Overall, we summarize the three-fold contributions of the proposed DSCKG as follows:

- **Comprehensive aspects.** We point out the issues of neglecting finer-grained feature interaction modelling and the lack of supervised signals in KGR tasks, and we advocate the employment of finer-grained feature interactions and self-supervised co-training strategies to mitigate the above issues for better node representation learning.
- **Novel Approach.** We propose a new model, DSCKG, which designs a novel self-supervised co-training mechanism for KGR. Precisely, there are three major novelties: 1) We design a novel discrepancy regularization strategy to encourage the generated user/item representation to be self-identifying on the collaborative view. 2) DSCKG designs a new CNNs-based finer-gained feature interaction pattern for KGs structural view to enhance item-side representations. 3) We propose a novel self-supervised co-training mode to identify high-quality positive/negative samples better.
- **Multi-faceted Experiments.** We perform empirical experiments on three real-world KGs-based recommendation datasets whose results demonstrate the superiority of DSCKG over compelling SOTA baselines.

2 Related Work

Next, we review the most relevant current work related to the proposed approach: 1) GNNs-based KGR methods, and 2) Self-supervised modes in recommendations.

GNNs-Based Recommendation Methods. The GNNs-based KGR approaches are established on the high-order neighbor information aggregation mechanism on graphs, which can iteratively carry out the propagation mechanism to integrate multi-hop neighbor information into node representations. KGAT [14] integrates user-item-entity into a unified heterogeneous Collaborative Knowledge Graph (CKG) and recursively aggregates CKG via **G**raph **A**ttention Ne**T**works (GATs [8]). KGCN [11] and KGNN-LS [10] primarily recursively integrate item-side high-order neighbor information via Graph Convolutional Networks (GCNs) to obtain item-side high-order embeddings. KGIN [15] utilizes intent-aware techniques to disentangle the intent behind user-item interactions, performing recursive aggregation operations on KGs by leveraging relation-aware mechanisms. CKAN [16] employs a heterogeneous propagation mechanism that can enrich user- and item-side representations by capturing CF signals as well as knowledge-aware signals through an attention mechanism. Regrettably, mainstream GNNs-based methods have three significant drawbacks: 1) Failure to model finer-grained feature interactions; 2) Inevitable over-smoothing phenomenon; 3) Serious issue of representation degradation.

Self-supervised Modes in Recommendations. Self-supervised methods aim to build correct positive/negative sample pairs to direct better node/graph representation learning. The recommender systems community has inherited this technological trend and applies self-supervised learning modes to various recommendation tasks [4]. One line of current approaches is dedicated to graph augmentation techniques. To be precise, new views(a.k.a., views) are generated based on perturbations of the original data, e.g., SGL [18] uses three graph augmentation methods (i.e., node/edge dropout, and random walk). Another line of current approaches is aimed at constructing different views, e.g., COTREC [19] constructs item and session views for high-quality self-supervised learning. Cross-CBR [5] constructs complementary views based on two data sources for bundle recommendation. Unfortunately, current work has not fully explored the vast potential of self-supervised learning and KGs-based recommendation.

3   Preliminaries

User-Item Interaction Data. In a typical recommendation case, we have a user set of M users $U = \{u_1, \ldots, u_M\}$ and a item set of N items $V = \{v_1, \ldots, v_N\}$. The interaction matrix $\mathbf{Y} \in \{0,1\}^{|M| \times |N|}$ is determined by user-item (u, v) implicit feedback (i.e., $y_{u,v} = 1$ indicates that an interaction between u and v is observed, otherwise $y_{u,v} = 0$).

Knowledge Graph. Let Knowledge Graph $G = (\mathcal{E}, \mathcal{R})$, which is an un-directed graph composed of knowledge triples (h, r, t). Where $h, t \in \mathcal{E}, r \in \mathcal{R}$ are on behalf of head and tail-entity, as well as relation of a knowledge triple correspondingly. $\mathcal{E}$ and $\mathcal{R}$ denotes the entities sets and relations in KGs. Besides, we define an item-entity alignment set $\mathcal{A} = \{(v, e) \mid v \in V, e \in \mathcal{E}\}$ that is denoted to uncover the alignment operations of items in both the user-item interaction matrix and KGs. Ultimately, we aim to learn a match function $\tilde{y}_{uv} = \mathcal{F}(u, v|\Theta, \mathbf{Y}, G)$, where $\tilde{y}_{uv}$ defines the possibility that a user u would adopt a item v, and Θ is the model parameters set.

Task Description

- Input: Knowledge Graph G, U-I interaction matrix $\mathbf{Y}$, and the model parameters set Θ.
- Output: The probability $\tilde{y}_{uv}$ that the user u interacts with the item v.

4   DSCKG Model Methodology

Overview. The framework of DSCKG is shown in Fig. 1. It consists of three key modules. 1) Collaborative View Encoder: It relies on historical user-item

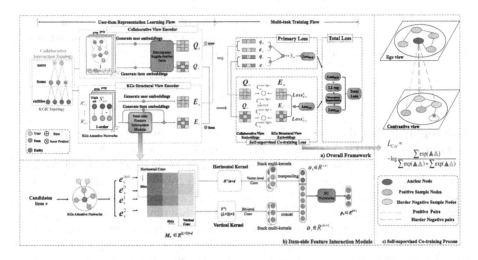

Fig. 1. Illustration of the proposed DSCKG framework. Best viewed in color.

interaction data to build user/item collaborative view prototype embeddings. And imposes a discrepancy regularization term on the generated embeddings. 2) Structural view encoder: It uses knowledge-aware attention networks to generate user/item prototype embeddings. And CNN is used to extract finer-grained item-side feature interaction embeddings. 3) Multi-task training module: Joint primary loss and the rest of optimization terms to guide the training and optimization of the whole model.

4.1 Collaborative View Encoder

We consider utilizing the GNNs-based method by performing iterative propagation on the user-item interaction graph to generate a high-order representation of the user/item from collaborative view. Inspired by the precious Collaborative Filtering (CF) work, we utilize LightGCN to model the long-range connectivity from user-item interactions. LightGCN [2] dispenses with complex message passing as well as feature fusion mechanisms, and eliminates the need for feature transformation and nonlinear activation functions and have high computational efficiency. In the $l'+1$-th layer, the aggregation proceeding can be formulated as follows:

$$q_u^{(l'+1)} = \sum_{v \in \mathcal{N}_u} \frac{1}{\sqrt{|\mathcal{N}_u||\mathcal{N}_v|}} q_v^{(l')}, q_v^{(l'+1)} = \sum_{u \in \mathcal{N}_v} \frac{1}{\sqrt{|\mathcal{N}_u||\mathcal{N}_v|}} q_u^{(l')} \qquad (1)$$

Here $l' \in \{0, 1, 2, \ldots, L'\}$. After L'-order propagation, we adopt the *sum* aggregator as the high-order integration function to incorporate all layer embeddings to yield the final embedding q_u and q_v as such:

$$q_u = \sum_{l'=0}^{L'} q_u^{(l')}, q_v = \sum_{l'=0}^{L'} q_v^{(l')} \tag{2}$$

Discrepancy Regularization Term. The user/item prototype representations generated by LightGCN should contain diversity and preference information. Due to the plague of data sparsity and long-tail issues, the extracted collaborative interaction signals will have some degree of information overlap and redundancy, thus negatively affecting the model. Besides, with the increase of LightGCN depths, node embeddings are extremely vulnerable to local similarity. We employ an embedding optimization strategy to encourage representations of self-discrimination and independence from each other [15]. Precisely, we introduce a discrepancy regularization module to guide the prototype user/item representation learning of diversity.

Inspired by metric learning techniques, we minimize the mutual information between two different embeddings to quantify their difference. Such an idea can be considered an ego-view self-supervised learning technique. More specifically, taking the user-side embedding optimization process as an instance, the discrepancy regularization term modelling is:

$$\mathcal{L}_{Dis}^u = \sum_{u \in U} - \log \frac{e^{s(q_u, q_u)/\tau'}}{\sum_{u' \in U} e^{s(q_u, q_{u'})/\tau'}} \tag{3}$$

Here $s(\cdot)$ is shown as the cosine similarity function between any two embeddings, and τ' is denoted the temperature coefficient. The item-side $\mathcal{L}_{Dis}^v$ and the user-side $\mathcal{L}_{Dis}^u$ have the same construction flow. Therefore, the overall discrepancy regularization loss is:

$$\mathcal{L}_{Dis} = \mathcal{L}_{Dis}^u + \mathcal{L}_{Dis}^v \tag{4}$$

We optimize discrepancy regularization loss to encourage divergence between different user/item embeddings and allow these embeddings to contain richer semantic information for the aim of embedding optimization.

4.2   KGs Structural View Encoder

Backbone Networks. The KGs structural view encoder relies on a knowledge-aware attention mechanism to extract high-order neighbor feature information, which follows the prior KGR approach [9,16]. The 0 and l-order receptive field R are denoted as:

$$\begin{cases} R_u^0 = \{e \mid v \in \{v \mid y_{uv} = 1\} \wedge (v, e) \in \mathcal{A}\} \\ R_v^0 = \{e \mid v^* \in \{v^* \mid \exists u \in U, y_{uv^*} = 1 \wedge y_{uv} = 1\} \wedge (v^*, e) \in \mathcal{A}\} \end{cases} \tag{5}$$

$$R_{u/v}^l = \left\{ t \mid (h, r, t) \in G \wedge h \in R_{u/v}^{l-1} \right\} \tag{6}$$

where $l \in \{1, 2, \dots, L\}$. Next, we elaborately describe the user u and item v 1-order representation learning process. We also need to define the l-th knowledge triples set S. The triplet set size directly affects the size of related high-order KGs entities.

$$S_{u/v}^l = \left\{ (h, r, t) \mid (h, r, t) \in G \wedge h \in R_{u/v}^{l-1} \right\}. \tag{7}$$

We demonstrate the process of computing 1-order embeddings via the knowledge-aware attention mechanism as such:

$$\alpha_{att} = Softmax \left(\text{MLP} \left[e_i^h \| e_i^r \right] \right) \tag{8}$$

Here $i \in \left\{ 1, 2, \dots, \left| S_{u/v}^1 \right| \right\}$, and MLP$(\cdot)$ denotes a multi-layer perceptron (MLP) and $\|$ denotes the vector *concat* operation. Then we can obtain the user and item embedding e_u^1 and e_v^1 after 1-layer propagation as follows:

$$e_{u/v}^1 = \sum_{i=1}^{\left| S_{u/v}^1 \right|} \alpha_{att} \cdot e_i^t \tag{9}$$

Besides, we introduce the user/item 0-order embeddings e_u^0, e_v^0 and the item v itself initial embedding e_v^{init} as follows:

$$e_{u/v}^0 = \frac{1}{\left| R_{u/v}^0 \right|} \sum_{e \in R_{u/v}^0} e, \quad e_v^{init} = \frac{1}{\left| \{e \mid (v, e) \in \mathcal{A}\} \right|} \sum_{\{e \mid (v,e) \in \mathcal{A}\}} e \tag{10}$$

The high-order user/item embeddings can be generalized from the 1-order embedding computation procedure described above to derive the user/item high-order embedding sets naturally: $\left\{ e_u^0, e_u^1, \dots, e_u^L \right\}$, $\left\{ e_v^{init}, e_v^0, e_v^1, \dots, e_v^L \right\}$. And we integrate high-order embeddings by *sum* aggregator. Thus, we acquire user u and unfinished item v integration embeddings: e_u and $\tilde{e}_v$.

DuCNNs. Current GNNs-based KGR approaches acquire the user/item ultimate embedding (e.g. KGAT [14], CKAN [16], and KGIN [15]) only by mechanically aggregating high-order feature representations while ignoring finer-grained feature interactions modelling [22], which is insufficient. Finer-grained feature interaction modelling can further explore the potential feature co-occurrence signals in the KGR task, which can directly improve the recommendation performance. Thus, we firmly believe finer-grained feature interaction are also urgently needed and deficient for the KGR task. We creatively generalize CNNs to the KGR task and utilize them as high-order feature aggregators for item-side 2-D multi-order embedding matrix $M_v \in \mathbb{R}^{(2+L) \times d}$. We innovatively apply two different convolution kernels (i.e., **Horizontal** and **Vertical convolutional kernel**) for capturing two diverse high-order feature interaction signals.

Exactly, we denote vertical convolutional kernel $V^t \in \mathbb{R}^{(2+L) \times 1}$ slides column by column on the matrix M_v to extract dimension-specific finer-grained feature interaction signals, referred the *bit-level feature interaction pattern*. Analogously, we denote the horizontal convolutional kernel $H^t \in \mathbb{R}^{h \times d}$ slides matrix

M_v by rows to extract feature interaction signals between neighboring features, referred the *vector-level feature interaction mode*. In addition, to obtain richer feature interaction information, we stack multiple [8] convolutional kernels for both classes of convolutional patterns. The following is the specific workflow of Dual-core Convolutional Neural Networks (DuCNNs).

Vector-Level Feature Interaction Mode. A horizontal convolutional kernel $H^t \in \mathbb{R}^{h \times d}$ is applied to extract the adjacent-order feature interaction signals, and the detailed process is shown in the subgraph-b upper part of Fig. 1. It is necessary to employ multiple convolutional kernels when we employ the multi-head mechanism, with the aim of capturing richer feature interaction information, as described above. Hence, $h \in \{1, 2, \ldots, (2 + L)\}$ is referred to the horizontal convolution kernel height and $t \in \{1, 2, \ldots, \tilde{n}\}$. The following equation shows the computation of the i-th convolutional value:

$$\tilde{c}_i^t = \phi_c \left(M_v[i : i + h - 1, :] \odot H^t \right) \tag{11}$$

where $\phi_c(\cdot)$ is denoted as the activation function (ReLU) of the convolutional layer, and $\odot$ is denoted as the inner product symbol. The horizontal convolutional output result (vector) $\tilde{c}^t \in \mathbb{R}^{(2+L)-h+1}$ is expressed as:

$$\tilde{c}^t = \left[\tilde{c}_1^t, \ldots, \tilde{c}_i^t, \ldots, \tilde{c}_{((2+L)-h+1)}^t \right] \tag{12}$$

We require extracting the highest significant feature interaction signals in the vector-level feature interaction pattern and filtering out the noise signals unrelated to the feature interaction as much as possible. Hence, we implement the *maxpooling* operation and the final horizontal convolutional vector $o_h \in \mathbb{R}^{\tilde{n}}$ is:

$$o_h = \left[maxpooling \left(\tilde{c}^1 \right), \ldots, maxpooling \left(\tilde{c}^{\tilde{n}} \right) \right] \tag{13}$$

Bit-Level Feature Interaction Mode. The vertical convolutional kernel $V^t \in \mathbb{R}^{(2+L) \times 1}$ is applied to extract dimension-specific significant feature interactions information, and the detailed process is shown in the subgraph-b lower part of Fig. 1. Then the kernel V^t overlays the 2-D embedding matrix M_v and shift along the dimensional direction. Likewise, the following equation shows the computation of the i-th convolutional value:

$$c_i^t = \phi_c \left(M_v[:, i] \odot V^t \right) \tag{14}$$

where $V^t \in \mathbb{R}^{(2+L) \times 1}$. The vertical convolutional output result (vector) $c^t \in \mathbb{R}^d$ is:

$$c^t = \left[c_1^t, \ldots, c_i^t, \ldots, c_d^t \right] \tag{15}$$

Here d is denoted as the fixed embedding dimension. We concatenate n kernels in sequence. The *maxpooling* operation is discarded because the feature information of each dimension needs to be retained to the maximum extent. The final vertical convolutional $o_v \in \mathbb{R}^{dn}$ is given by:

$$o_v = \left[c^1 || c^2 || \ldots || c^n \right] \tag{16}$$

where $\|$ is shown as the vector connection operator.

Fully Connected (FC) Networks. We concatenate the above two convolutional vectors, fed into FC networks to extract global interactive features [7], and output the item feature-enhanced convolutional embedding $\boldsymbol{p}_v$ as such:

$$\boldsymbol{p}_v = \sigma\left(\mathbf{W} \cdot [\boldsymbol{o}_v \| \boldsymbol{o}_h]\right) \tag{17}$$

where $\mathbf{W} \in \mathbb{R}^{d \times (\tilde{n}+dn)}$ is the transformation matrix and the convolutional embedding $\boldsymbol{p}_v \in \mathbb{R}^d$. Then the final item-side representation: $\boldsymbol{e}_v = \tilde{\boldsymbol{e}}_v + \boldsymbol{p}_v$.

4.3   Dual-View Self-supervised Co-training Mode

We present our novel Dual-view self-supervised co-training mode in terms of the quality of the selected harder negative sample [24]. We consider that high-quality negative samples have the following quality: Having a sufficient and significant contribution of self-supervised signal, called **harder negative samples**. Moreover, we firmly believe that an appropriately large positive sample size can also help implement the self-supervised learning strategy. We deploy a novel scheme to search for positive/negative samples of **anchor embedding** $\boldsymbol{e}_v$ as follows:

$$\phi_v^{cov} = Softmax\left(\boldsymbol{value}_v^{cov}\right), \boldsymbol{value}_v^{cov} = \boldsymbol{Q}_v \cdot \boldsymbol{q}_v \tag{18}$$

Here $\boldsymbol{Q}_v \in \mathbb{R}^{N \times d}$ are the matrix of users samples in the collaborative view, $\boldsymbol{q}_v \in \mathbb{R}^{d \times 1}$ is the anchor embedding (target item v) in the collaborative view, $\phi_v^{cov} \in \mathbb{R}^{N \times 1}$ is shown as the probability of the target item v with other items in collaborative view.

With the calculated similarity scores, we select the top-k items with the highest confidence level as a positive sample. Formally, the positive sample set $C_v^{cov^+}$ is selected as follows:

$$C_v^{cov^+} = \mathrm{top}-k\left(\phi_v^{cov}\right) \tag{19}$$

One of the simplest methods is to classify all items other than the top-k items of the similarity as negative samples. However, the contribution of such an approach to self-supervised learning is limited.

Therefore, we randomly select k negative samples from the items ranked in top 10% in ϕ_v^{cov} excluding the positive samples to construct harder negative samples set $C_v^{cov^-}$, these items can be regarded as harder negative samples that can supply sufficient self-supervised signals. Symmetrically, we adopt the same method to choose suitable positive/negative samples for anchor embedding $\boldsymbol{q}_v$ in the KGs structural view, and to construct $C_v^{kgv^-}/C_v^{kgv^+}$. Consequently, for the generated pseudo-labels, we optimize the node representation learning for dual-views by self-supervised learning mode. Since the calculation process $\mathcal{L}_v^{cov}/\mathcal{L}_v^{kgv}$ are symmetric, we only show the calculation process of $\mathcal{L}_v^{kgv}$:

$$\mathcal{L}_v^{kgv} = -\log \frac{\sum_{i \in C_v^{cov^+}} e^{s(\boldsymbol{e}_v, \boldsymbol{q}_i)/\tau}}{\sum_{j \in \left\{C_v^{cov^+} \cup C_v^{cov^-}\right\}} e^{s(\boldsymbol{e}_v, \boldsymbol{q}_j)/\tau}} \tag{20}$$

We follow the InfoNCE [6] paradigm. Where τ and $s(\cdot)$ are consistent with the previous definitions (*c.f.*, Eq. (3)). We perform the same operation on the user-side. Hence, the overall formalization of $\mathcal{L}_{Cot}$ as such:

$$\mathcal{L}_{Cot} = \sum_{v \in V} \left[\mathcal{L}_v^{kgv} + \mathcal{L}_v^{cov} \right] + \sum_{u \in U} \left[\mathcal{L}_u^{kgv} + \mathcal{L}_u^{cov} \right] \tag{21}$$

4.4 Model Prediction and Optimization

We integrate the representations of the two views and calculate the user-item matching scores by the inner product:

$$\tilde{y}_{uv} = \sigma \left(\left[e_u + q_u \right]^\top \left[e_v + q_v \right] \right) \tag{22}$$

where $\sigma(\cdot)$ is the *sigmoid* function. Next, We adopt the Binary Cross-Entropy (BCE) loss as primary loss, and the overall loss function is shown as follows:

$$\mathcal{L}_{total} = \mathcal{L}_{BCE} + \lambda_1 \cdot \mathcal{L}_{Cot} + \lambda_2 \cdot \mathcal{L}_{Dis} + \lambda_3 \cdot \|\Theta\|_2^2 \tag{23}$$

Here λ_1 is the hyper-parameter to determine the self-supervised loss ratio. λ_2 is the hyper-parameter to determine the discrepancy regularization loss ratio. λ_3 is the L2-regularization coefficient for reducing overfitting.

5 Experimental Results

Below, we aim to answer three study questions: (RQ1) How does DSCKG perform compared to other recommendation models? (RQ2) How do hyperparameters affect DSCKG performance? (RQ3) What impact do the various components have on DSCKG?

Table 1. Statistics for the three KGR datasets.

Datasets	Last.FM	Dianping-Food	MovieLens-1M
Users	1872	2298698	6036
Items	3846	1362	2445
Interactions	42346	23416418	753772
Entites	9366	28115	182011
Relations	60	7	12
Triples	15518	160519	1241995

5.1 Dataset Preparation

We evaluate DSCKG on three real-world KGR datasets. (1) **Last.FM**[1] is a online music listening dataset compiled from about 2,000 Last.FM online music systems. (2) **Dianping-Food**[2] is a business dataset from Meituan, consisting

---

[1] grouplens.org/datasets/hetrec-2011/.
[2] https://www.dianping.com/.

of over 10 million different interactions between about 2 million users and 1,000 restaurants. (3) **MovieLens-1M**[3] is a classic movie recommendation dataset that contains about 1 million explicit ratings. The KGs build construction tool uses Microsoft and Meituan, and all data pre-processing is consistent with the baseline methods [9,10]. (See Table 1 for statistics on the three datasets)

5.2   Competitors

To illustrate the validity of our model, we have chosen eight baselines, covering:

- **CKE** [21]: It's an embedding-based KGR approach that incorporates multi-modal knowledge in a single framework.
- **PER** [20]: It's a classical path-based approach that captures long-distance connections between users and items via meta-paths.
- **RippleNet** [9]: It's an embedding- and propagation-based approach that fully simulates the users' preferences in KGs. And treats users' preferences as Ripple that captures high-order attributes on the KGs.
- **KGCN** [11]: It's a GNNs-based method which enriches item embed-dings via iteratively integrating nearby information.
- **KGNN-LS** [10]: It's a GNNs-based method which enriches item em-bedding via label smoothing regularization.
- **KGAT** [14]: It's a GNNs-based method that iteratively integrates neighbors on the CKG via the attention mechanism to acquire the user/item representation.
- **CKAN** [16]: It's a GNNs-based method which independently propagate CF signals and KGs signals.
- **KGIN** [15]: It's a SOTA GNNs-based approach which combines intentions and KG relations with a finer long-range semantics of intentions and relational paths for node representation.

5.3   Experimental Settings and Evaluation Metrics

We strictly follow the data split ratio of 6:2:2 to compare with all models. We implement our DSCKG in Pytorch and carefully tune the core parameters. We use the Adam optimizer and the Xavier algorithm to optimize and initialize the model parameters. The embedding dimension size in the range of $\{8, 16,..., 256\}$. During KGs and collaborative view, the depth of GNNs layer and Light-GCN layer are both adjusted among $\{1, 2, 3, 4\}$. For temperature coefficient τ, we search for the number within $\{0.5, 0.4,..., 0.1\}$, the self-supervised learning weight is tuned amongst $\{1e-8, 1e-7,..., 1e-4\}$. The optimal settings of hyper-parameters in all methods were studied by empirical studies or original papers. We use AUC, F1, and Recall@50,100 as evaluation metrics.

---

[3] https://grouplens.org/datasets/movielens/1m/.

Table 2. Results on two KGR task. *Indicates the significant improvement between our DSCKG and the best performed baseline with $p < 0.05$.

Method	Last.FM				Dianping-Food				MovieLens-1M			
	AUC	F1	R@50	R@100	AUC	F1	R@50	R@100	AUC	F1	R@50	R@100
CKE	0.747	0.674	0.181	0.297	0.812	0.741	0.305	0.439	0.906	0.802	0.375	0.522
PER	0.641	0.603	0.117	0.177	0.766	0.697	0.256	0.355	0.712	0.667	0.171	0.251
RippleNet	0.776	0.702	0.235	0.356	0.863	0.783	0.392	0.545	0.919	0.842	0.317	0.460
KGCN	0.802	0.708	0.298	0.345	0.845	0.774	0.331	0.441	0.909	0.836	0.358	0.512
KGNN-LS	0.805	0.722	0.271	0.349	0.852	0.778	0.342	0.488	0.914	0.841	0.359	0.513
KGAT	0.829	0.742	0.365	0.451	0.846	0.785	0.389	0.561	0.914	0.844	0.385	0.533
CKAN	0.842	0.769	0.342	0.432	0.878	0.802	0.413	0.574	0.915	0.842	0.395	0.540
KGIN	0.848	0.760	0.367	0.462	0.879	0.799	0.415	0.577	0.919	0.844	0.402	0.544
DSCKG	0.868*	0.776*	0.375*	0.472*	0.887*	0.807*	0.421*	0.584*	0.929*	0.858*	0.417*	0.552*
Gain%	2.35%	0.91%	2.17%	2.16%	0.91%	0.87%	1.44%	1.21%	1.08%	1.65%	3.73%	1.47%

5.4 Overall Performance Analysis (RQ1)

To answer the RQ 1, we report the experimental results of DSCKG with eight baselines in Table 2. From the analysis of the experimental results, we have the following observations.

- **Embedding- and path-based methods have unsatisfactory performance.** We first analyze three conventional KGR approaches. PER is typical of meta-path-based methods. CKE utilizes only 1-order knowledge entities as complementary information for recommendations. RippleNet combines path and embedding approaches. We can find that they all achieve unsatisfactory results. We can conclude that 1) path-based methods require human-defined paths, which are often difficult to optimize. 2) Embedding-based methods cannot capture long-range connectivity information. 3) The above methods cannot adaptively capture high-order entity information.
- **GNNs-based KGR methods are dominant.** Analyzing several GNNs-based KGR models, their performance confirms that capturing high-order knowledge and using attention mechanisms effectively enhance recommendation. This inspires us that adaptive mining of long-range information on graphs is essential to boost recommendation effectiveness.
- **Our proposed DSCKG achieves the best performance.** DSCKG performs best in two recommendation tasks, which demonstrates that DSCKG is highly competitive. The main performance improvements are due to: 1) The foundational role of GNNs and attention mechanisms for extracting high-order knowledge; 2) The powerful of feature interaction modelling; 3) The usefulness of self-supervised learning for augmenting node representation learning ability.

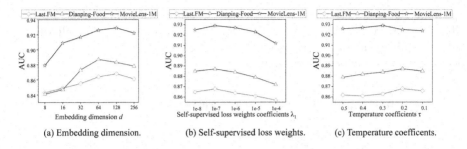

(a) Embedding dimension.   (b) Self-supervised loss weights.   (c) Temperature coefficents.

Fig. 2. Parameter analysis.

5.5 Sensitivity Analysis (RQ2)

Next, we investigate the impact of key hyperparameters on the performance of our DSCKG, including embedding dimension (d), self-supervised loss weights coefficients (λ_1), temperature coefficients (τ), and two view depths (L/L').

Figure 2(a) shows the effect of embedding dimension d. Our DSCKG achieves the best performance with the settings of $d = 128, 64, 128$, indicating that it encodes the user/item and KGs entity information well. We also find that the performance of our model first increases as d grows and then decreases. We analyze the reason may be that too small dimensions do not have enough ability to capture the necessary information, while too large sizes introduce unnecessary noise and reduce the generalization ability of the model.

Figure 2(b) displays the effect of self-supervised loss weights coefficients λ_1. We have the following observation: The model performance reaches the best when λ_1 is taken as uniformly $1e-7$. To analyze the cause, when self-supervised loss weight is set too large, it misleads the direction of the recommendation model optimization and leads to more prominent side effects. Conversely, when the value of λ_1 is taken too small, self-supervised learning fails to achieve significant optimization results and even hinders the correct representation of the model. Therefore, choosing the suitable λ_1 value can augment the node representation learning ability and thus improve the recommendation performance.

Figure 2(c) summarizes the effect of temperature coefficients τ. We consistently conclude that the model performance is best when τ is taken as $[0.2, 0.3]$. After analysis, increasing the value of τ (e.g., 0.5) will lead to poorer model performance, indicating that self-supervised learning cannot correctly distinguish between positive/negative samples. Conversely, when τ is fixed at a too small value (e.g., 0.1), the model performance is compromised, suggesting that a smaller value of τ could misdirect the gradient optimization direction. This is consistent with the findings of previous work [18].

To investigate the effect of dual-view depth on model performance, we vary L and L' in the range $\{1, 2, 3, 4\}$ and show the performance comparison of the three datasets in Table 3. We have the following findings: The model performance is optimal when $L = 2, 1, 2$ and $L' = 2, 1, 2$. This indicates that 1 or 2 orders are sufficient to aggregate enough neighbors or remote knowledge for the purpose of enriching the information in different views, whether it is collaborative view or KGs structural view. Conversely, if the model depth is too large, the well-known oversmoothing and the introduction of irrelevant noise will occur, leading to degradation of the model performance.

Table 3. AUC results of DSCKG with depth in dual-view.

Depth in KGs structural view	1	2	3	4
Last.FM	0.864	**0.868**	0.865	0.863
Dianping-Food	**0.887**	0.883	0.881	0.878
MovieLens-1M	0.926	**0.929**	0.925	0.922
Depth in collaborative view	1	2	3	4
Last.FM	0.863	**0.868**	0.866	0.867
Dianping-Food	**0.887**	0.886	0.883	0.882
MovieLens-1M	0.927	**0.929**	0.924	0.923

5.6 Ablation Studies (RQ3)

Lastly, as shown in Fig. 3, here we examine the contribution of the main components of our model to the final performance by comparing DSCKG with the following four variants.

Variants. 1) DSCKG$_{W/O}$Cot: Remove self-supervised co-training mode. 2) DSCKG$_{W/O}$Dis: Remove discrepancy regularization term. 3) DSCKG$_{W/O}$ CoView: Remove the collaborative view encoder completely. 4) DSCKG-Light: Retain only the backbone networks.

Conclusion. We have the following observations: 1) Removing the self-supervised learning module significantly reduces model performance, indicating that self-supervised co-training mode is critical for node representation learning. 2) Removing the discrepancy regularization term slightly reduces model performance. This reflects the advantage and contribution of the discrepancy regularization to encourage diversity in the representation of collaborative view nodes. 3) Removing the collaborative view modeling means that neither the self-supervised learning nor the discrepancy constraints work properly. We find noticeable performance drop, which justifies the construction of dual views to extract interaction information from different sources and the basis for self-supervised learning. 4) To study in-depth the contribution of our proposed feature interaction (DuCNNs) module, we create a variant called DSCKG-light based on DSCKG$_{W/O}$CoView, i.e., only the backbone network is kept. When DuCNNs are further erased, a significant performance degradation occurs, indicating that mining finer-grained feature interaction signals is indispensable for the KGR model and plays a decisive role in performance improvement.

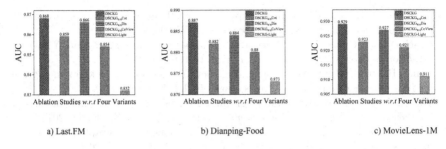

a) Last.FM   b) Dianping-Food   c) MovieLens-1M

Fig. 3. Ablation studies.

6 Conclusion and Future Work

In this paper, we aim to explore the knowledge graph recommendation task. Specifically, to mitigate the neglect of feature interaction modeling and the lack of supervised signal problems of conventional KGR methods, we propose a novel DSCKG framework to deal with the above issues via dual-core convolutional neural networks and a self-supervised co-training mode. Extensive experiments validate the effectiveness of our DSCKG framework. There are two directions for future efforts: 1) Design a more reasonable self-supervised paradigm [1,4] and 2) Transfer our novel self-supervised co-training paradigm to other domains (e.g., knowledge tracking [12,13] and session recommendation [23,24]).

Acknowledgement. This work is supported by the Industrial Support Project of Gansu Colleges (2022CYZC-11), the Gansu Natural Science Foundation Project (21JR7RA114), the National Natural Science Foundation of China (62276073, 61762078) and NWNU Teachers Research Capacity Promotion Plan (NWNU-LKQN2019-2). NWNU Graduate Research Project Funding Program (2021KYZZ02103).

References

1. Gao, Z., Ma, H., Zhang, X., Wu, Z., Li, Z.: Co-contrastive self-supervised learning for drug-disease association prediction. In: Proceedings of PRICAI (2022)
2. He, X., Deng, K., Wang, X., Li, Y., Zhang, Y., Wang, M.: LightGCN: simplifying and powering graph convolution network for recommendation. In: Proceedings of SIGIR (2020)
3. Jiang, Y., Ma, H., Zhang, X., Li, Z., Chang, L.: An effective two-way metapath encoder over heterogeneous information network for recommendation. In: Proceedings of ICMR (2022)
4. Li, Q., Ma, H., Zhang, R., Jin, W., Li, Z.: Co-contrastive learning for multi-behavior recommendation. In: Proceedings of PRICAI (2022)
5. Ma, Y., He, Y., Zhang, A., Wang, X., Chua, T.-S.: CrossCBR: cross-view contrastive learning for bundle recommendation. In: Proceedings of SIGKDD (2022)
6. van den Oord, A., Li, Y., Vinyals, O.: Representation learning with contrastive predictive coding. arXiv:1807.03748 (2018)

7. Tang, J., Wang, K.: Personalized top-n sequential recommendation via convolutional sequence embedding. In: Proceedings of WSDM (2018)
8. Veličković, P., Cucurull, G., Casanova, A., Romero, A., Lio, P., Bengio, Y.: Graph attention networks. arXiv preprint arXiv:1710.10903 (2017)
9. Wang, H., et al.: RippleNet: propagating user preferences on the knowledge graph for recommender systems. In: Proceedings of CIKM (2018)
10. Wang, H., et al.: Knowledge-aware graph neural networks with label smoothness regularization for recommender systems. In: Proceedings of SIGKDD (2019)
11. Wang, H., Zhao, M., Xie, X., Li, W., Guo, M.: Knowledge graph convolutional networks for recommender systems. In: Proceedings of WWW (2019)
12. Wang, W., Ma, H., Zhao, Y., Li, Z., He, X.: Tracking knowledge proficiency of students with calibrated q-matrix. Exp. Syst. Appl. (2022)
13. Wang, W., Ma, H., Zhao, Y., Yang, F., Chang, L.: SEEP: semantic-enhanced question embeddings pre-training for improving knowledge tracing. Inf. Sci. (2022)
14. Wang, X., He, X., Cao, Y., Liu, M., Chua, T.S.: KGAT: Knowledge graph attention network for recommendation. In: Proceedings of WWW (2019)
15. Wang, X., et al.: Learning intents behind interactions with knowledge graph for recommendation. In: Proceedings of the Web Conference (2021)
16. Wang, Z., Lin, G., Tan, H., Chen, Q., Liu, X.: CKAN: collaborative knowledge-aware attentive network for recommender systems. In: Proceedings of SIGIR (2020)
17. Wei, Y., Ma, H., Zhang, R., Li, Z., Chang, L.: Exploring implicit relationships in social network for recommendation systems. In: Proceedings of PAKDD (2021)
18. Wu, J., et al.: Self-supervised graph learning for recommendation. In: Proceedings of SIGIR (2021)
19. Xia, X., Yin, H., Yu, J., Shao, Y., Cui, L.: Self-supervised graph co-training for session-based recommendation. In: Proceedings of CIKM (2021)
20. Yu, X., et al.: Personalized entity recommendation: a heterogeneous information network approach. In: Proceedings of WSDM (2014)
21. Zhang, F., Yuan, N.J., Lian, D., Xie, X., Ma, W.Y.: Collaborative knowledge base embedding for recommender systems. In: Proceedings of SIGKDD (2016)
22. Zhang, R., Ma, H., Li, Q., Wang, Y., Li, Z.:. Fire: knowledge-enhanced recommendation with feature interaction and intent-aware attention networks. Appl. Intell. (2022)
23. Zhang, X., Ma, H., Gao, Z., Li, Z., Chang, L.: Exploiting cross-session information for knowledge-aware session-based recommendation via graph attention networks. Int. J. Intell. Syst. (2022)
24. Zhang, X., Ma, H., Yang, F., Li, Z., Chang, L.: Cross-view contrastive learning for knowledge-aware session-based recommendation. In: Proceedings of PRICAI (2022)

PIDE: Propagating Influence of Dynamic Evolution on Interaction Networks for Recommendation

Chunjing Xiao[1(✉)], Shenkai Lv[1], Wanlin Ji[1], Yuxiang Zhang[1], Haiying Pan[2], and Lingshan Wu[2]

[1] School of Computer Science and Technology, Civil Aviation University of China, Tianjin, China
{cjxiao,2020052053,2019051024,yxzhang}@cauc.edu.cn
[2] Digital Committee of Xiamen Airline, Xiamen, China
{panhaiying,wulingshan}@xiamenair.com

Abstract. Modelling dynamic interactions between users and items is very crucial for many recommendation systems. Although existing methods have achieved great success in modelling dynamic interactions, they ignore propagating influence from interacting users/items to their neighbors along different paths at the same time. In this paper, we focus on how to effectively propagate influence of dynamic evolution on interaction networks. Specifically, we propose a new influence-enhanced deep co-evolutionary method called *PIDE*, which captures not only *mutual influence* between the interacting users and items but also *propagation influence* from them to their neighbors along high-order connectivity. Our model consists of two main components: *interaction mutual influence component* and *influence propagation component*. The former evolves the effective representations for the interacting user and item by considering the direct mutual influence from each other when a new interaction happens. The latter refines the representations of users and items via aggregating the effective interaction influence propagated from multiple interacting nodes along different high-order connectivity at the same time. Finally, the refined representations of users and items are used to predict which item the user is most likely to interact with. To the best of our knowledge, *this is the first attempt to propagate multiple interaction influences at the same time on the dynamic network for better recommendation*. The experimental results illustrate that our method significantly outperforms the state-of-the-art methods on three real-world datasets (Our codes and datasets are available at: https://github.com/pidecode/PIDE.git.).

Keywords: Recommendation system · Interaction network · Influence propagation · Graph neural network

1 Introduction

In daily life, we constantly interact with many items and these interactions form a complex *dynamic interaction network* [1,2], in which nodes correspond to users/ items and edges represent interactions between them, containing timestamps and interaction features. Learning from it has benefited many real-world services and applications, such as rating prediction and item recommendation [3–5].

Representation learning is a powerful tool for recommendation in interaction networks, in which it is crucial how to accurately represent the *properties* of users and items, especially for the *evolution* of these properties. Currently, several deep co-evolutionary methods have been proposed to model the dynamic interaction process and refine representations of users and items involved in the interactions [1,3,6–9]. For example, DeepCoevolve [3] applies Recurrent Neural Network (RNN) to capture complex mutual interactions and the feature evolution over time. JODIE [7] and CoPE [8] not only model the dynamic interaction but also learn the future trajectories of users/items over time.

Although these deep co-evolution methods [3,6–8] have achieved great success in modelling mutual influence between users and items, they ignore the influence propagation from each interacting user/item to its neighbors. A few methods such as PPGN [10], Ripplenet [11] propagate user preference to their neighbors by considering high-order connectivities. However, they are designed for static graphs by completely ignoring the evolving structures of interaction network, while in the real world interaction network is inherently dynamic with new nodes and edges constantly emerging. In fact, the introduction of a new interaction can drive the co-evolution of user preferences and item features, and the changes can further affect the neighbors that are "close" to them because of the changes of network structure and node embeddings. As shown in Fig. 1(b), when a new interaction between user u_1 and item i_1 happens, influence from u_1 will be propagated to i_3, i_4 and i_6, and even higher-order neighbors u_3, u_4 and u_5. In addition, it is reasonable that one node can be influenced by different interacting nodes at the same time since multiple interactions can happen simultaneously. For example, the user u_4 can be simultaneously influenced by two interactions (u_1, i_1, t_c) and (u_2, i_2, t_c) in Fig. 1(b).

In this paper, we focus on how to Propagate Influence of Dynamic Evolution (PIDE) on interaction networks, which faces the challenges of which neighbors should be selected and how much influence will be propagated. We propose *PIDE* to model it, which can capture not only user-item *mutual influence* but also the *propagation influence* from multiple interacting users/items at the same time on the dynamic interaction network. PIDE consists of two major components: interaction mutual influence component and influence propagation component. Specifically, the former aims to evolve effective representations for interacting user and item by respectively adopting two Gate Recurrent Units (GRUs) when a new interaction happens. The latter is constructed to further refine user and item representations via aggregating the influence propagated from multiple interacting nodes (users and items) at the same time. In the influence propagation process, a time-related neighbor selection mechanism is designed for selecting a more effective propagation influence to reduce the computational cost and accelerate the training. Finally, the refined representations are used to compute the matching scores by optimizing the pairwise Bayesian Personalized Ranking (BPR) loss. Our contributions can be summarized as follows:

- To the best of our knowledge, this is the first attempt to propagate interaction influence on a dynamic network and highlight the importance of explicitly exploiting the interaction influence propagated from multiple interacting users and items to their neighbors at the same time.
- We propose PIDE, an influence-enhanced deep co-evolutionary model, to capture not only mutual influence between the interacting users and items but also propagation influence from them to their neighbors along high-order connectivity on the interaction network.
- We design a time-related neighbor selection mechanism to propagate and aggregate more effective interaction influence from multiple interacting nodes into the current node. This can significantly reduce the computational cost and accelerate the training.

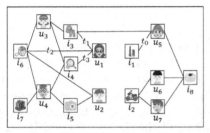 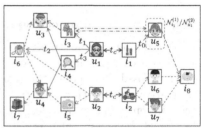

(a) An example of a user-item interaction graph

(b) An example of influence propagations when two interactions happen at time t_c

Fig. 1. (a) An example of a user-item interaction graph of seven users and eight items. Each edge represents an interaction with associated timestamp t. For the sake of brevity, only some edges are labeled with timestamps. (b) An illustration of propagating the influences from four interacting nodes (i.e., u_1, i_1, u_2 and i_2) through l-hop ($l = 1, 2, 3$) neighbors when two new interactions (i.e., (u_1, i_1, t_c) and (u_2, i_2, t_c)) happen at the same time t_c. Each arrow represents a propagation influence on the existing network.

2 Related Work

2.1 Dynamic Network Embedding

Dynamic network embedding aims to find a low-dimensional vector space that can maximally preserve the structural information and properties of dynamic networks. The existing methods can be broadly classified into two categories: embedding snapshot networks [12–14] and modelling temporal evolution [15, 16] The former is to divide the network into a sequence of network snapshots and learn node embedding for each snapshot, such as Dyngraph2vec [13], Dynamic-Triad [17]. While the latter is to capture the evolution of node embeddings and the graph itself, such as CTDNE [18], EvolveGCN [16] and TADGE [19].

2.2 Deep Recurrent Recommendation

Recently, several models employ Recurrent Neural Networks (RNNs) and variants (such as LSTMs and GRUs) to build recommendation systems [3,7,20–22]. RNN-based models such as RRN [20] and DGRec [23] encode user's behaviors into hidden states to predict items. Some methods such as Time-LSTM [21] and LatentCross [22] incorporate features into the embeddings. However, most of them can only generate static embedding and fail to dynamically update the embeddings to capture time-varying properties in the interaction process. To overcome this drawback, several deep co-evolution methods [3,6,7,9] have been proposed to model the dynamic interaction and update user and item embeddings after each interaction. DeepCoevolve [3], KGCR [6], JODIE [7] and DyGNN [9] use two RNNs to learn user and item embeddings and feature evolution over time from user-item interactions.

2.3 Sequential Recommendation

Sequential recommendation attempts to predict a user's next item by leveraging the user's historical behavior sequence. The early works, such as FPMC [24] and its extension Fossil [25], model the transition in user interaction sequences by utilizing markov chains. However, they have no ability to deal with long-term sequences. Recently, many deep learning-based methods have been proposed to capture complex behaviors, such as GRU4Rec [26], SR-GNN [27], TiSASRec [28], TMER [29] and STOSA [30]. The methods mentioned above ignore the dynamic collaborative signals across different user sequences, so some methods model the different user sequences to improve sequential recommendation performance, such as DGSR [31], HyperRec [32] and SURGE [33].

2.4 Message Propagation Recommendation

Graph neural networks pay more attention to model graph structure, especially high-order connectivities (*i.e.*, message propagation among nodes), to guide the embedding learning [1,10]. Motivated by this, some existing methods employ the message propagation mechanism to capture the influences from neighbors of a node. It is worth mentioning that several recent works inspire us to develop PIDE, such as EvolveGCN [16], NGCF [34], LightGCN [35]. Some other methods such as PPGN [10], Ripplenet [11] and DGRec [23] also exploit high-order connectivities and propagate user preference to neighbors.

Although these aforementioned methods have achieved great success in different directions, almost all of these methods ignore propagating the interaction influence from the interacting nodes to their neighbors. Our PIDE overcomes this shortcoming by modelling the influence propagation on the dynamic network.

3 Methodology: PIDE

In this section, we propose a method PIDE to model the influence propagation of dynamic evolution on the interaction network. The proposed method consists of three components: (1) *interaction mutual influence*, in which two GRUs are applied to capture the mutual influence between the user and the item directly involved in each interaction and to evolve separately their embeddings; (2) *influence propagation*, in which the propagation unit is designed to capture the more effective propagation influence from the interacting nodes and to refine the user and item embeddings by recursively propagating and aggregating influences; (3) *user preference prediction*, in which the BPR loss function is used to leverage the refined embeddings to compute the matching scores of user-item pairs to decide which items will be recommended to the user.

3.1   Interaction Mutual Influence Component

As aforementioned analysis, the interactions will further drive the evolution of user interests and item features with the interactions happening. For example, a retro camera (item), initially designed for middle-aged and elderly people, will probably become a fashion item for young people if more and more young people buy it (this indicates the user interests can affect the item features), and other young users who are not interested in the camera may gradually like it and buy it (this indicates the item features can affect the user interests). Inspired by previous methods [3,7], here we propose a framework that employs two GRUs on the interaction-evolving network to automatically capture the user-item mutual influence and update the embeddings of users and items involved in the interactions over time.

For each user u and item i, we use $\mathbf{e}_u(t) \in \mathbb{R}^d$ and $\mathbf{e}_i(t) \in \mathbb{R}^d$ to represent their embeddings at time t. When an interaction (u, i, t) between user u and item i happens at time t, the embeddings $\mathbf{e}_u(t)$ and $\mathbf{e}_i(t)$ will be dynamically updated by respectively adopting two GRUs. Specifically, we use GRU_U and GRU_I to update the embeddings of the interacting user and item, which are represented by the hidden states of user GRU_U and item GRU_I, respectively.

To this end, we model the dynamic evolution of embeddings $\mathbf{e}_u(t)$ and $\mathbf{e}_i(t)$ using two update functions. For the user u, we formulate the embedding $\mathbf{e}_u(t)$ after the interaction (u, i, t) as:

$$\mathbf{e}_u(t) = \sigma\Big(W_1\mathbf{e}_i(t-) + W_2\mathbf{e}_u(t-) + W_3\Delta t_u + W_4\mathbf{f}\Big) \tag{1}$$

where $t-$ is the time point just before time t, Δt_u is the time interval between the user u's last and current interaction, which makes the user u's feature change smoothly over time. $\mathbf{f}$ is the interaction context vector. W_1, W_2, W_3, W_4 are trainable parameters of GRU_U. $\sigma(\cdot)$ is the activation function. The first term $W_1\mathbf{e}_i(t-)$ is used to capture the influence from the interacting item i before time t, *i.e.*, the item embedding influences the update of the user embedding after each interaction. The second term $W_2\mathbf{e}_u(t-)$ explores the intrinsic evolution of the user's features in the sense that the user's current interest is determined by his/her previous interest. The third term $W_3\Delta t_u$ exploits the influence from the time interval between the consecutive interaction of the given user, which is based on the fact that the user's interests smoothly change over time by external influence.

The final term $W_4\mathbf{f}$ captures the influence from the interaction context feature, which is the additional information that happened in the user-item interaction. For example, the context features are the business reviews on Yelp, while on Reddit they are the posts and comments.

Analogously, for the item i, we update $\mathbf{e}_i(t)$ after the interaction (u, i, t) as:

$$\mathbf{e}_i(t) = \sigma\left(V_1\mathbf{e}_u(t-) + V_2\mathbf{e}_i(t-) + V_3\Delta t_i + V_4\mathbf{f}\right) \tag{2}$$

where Δt_i is the time interval between the item i's last and current interaction, and V_1, V_2, V_3, V_4 are the trainable parameters of GRU_I.

3.2 Influence Propagation Component

In the previous subsection, the interaction mutual influence component only considers the evolution of two interacting nodes involved in the new interaction. However, in reality, the newly emerging interaction changes the existing local structure of the graph. Thus, the interaction can influence some nodes which are the direct and high-order neighbors of the interacting nodes. The example in Fig. 1(b) illustrates that the items $\{i_3, i_4, i_6\}$, which are the 1-hop neighbors of the interacting user u_1, are influenced by the interaction (u_1, i_1, t_c).

To capture the interaction influence propagation along the interaction network, we first define the l-hop *neighbor set* for an interacting user/item, and then present the time-related *influence weight* to control the strength of influence propagation on each existing interaction (*i.e.*, edges in Fig. 1(a)), to select more effective propagation to reduce the computational cost and accelerate the training. Finally, the processes of influence embedding *propagation and aggregation* along high-order connectivity are given. Figure 1(b) illustrates the influence propagation and aggregation along high order connectivity when interactions (u_1, i_1, t_c) and (u_2, i_2, t_c) happen at the same time t_c.

l-hop Neighbor Set Definition. Let (u, i, t) be an interaction between u and i at time t. For the given interacting user u, his/her l-hop neighbor set at time t is defined as:

$$\mathcal{N}_u^{(l)}(t) = \begin{cases} \{u\}, & l = 0 \\ \{i_q \in I \mid (u_p, i_q) \in E \wedge u_p \in \mathcal{N}_u^{(l-1)}(t)\}, & l = 1, 3, \ldots \\ \{u_p \in U \mid (i_q, u_p) \in E \wedge i_q \in \mathcal{N}_u^{(l-1)}(t)\}, & l = 2, 4, \ldots \end{cases} \tag{3}$$

Note that the user u's l-hop neighbor set $\mathcal{N}_u^{(l)}(t)$ is defined recursively by his/her $(l-1)$-hop neighbor set $\mathcal{N}_u^{(l-1)}(t)$ at time t. In addition, the l-hop neighbors of interacting user u are items where l is an odd number, while his/her l-hop neighbors are users where l is an even number. For example, for an interacting user u_1 in Fig. 1(b), his/her l-hop neighbor sets are defined as $\mathcal{N}_{u_1}^{(0)}(t) = \{u_1\}$, $\mathcal{N}_{u_1}^{(1)}(t) = \{i_3, i_4, i_6\}$, $\mathcal{N}_{u_1}^{(2)}(t) = \{u_3, u_4, u_5\}$ and $\mathcal{N}_{u_1}^{(3)}(t) = \{i_7, i_8\}$.

Analogously, for the interacting item i, its l-hop neighbor set is defined as:

$$
\mathcal{N}_i^{(l)}(t) = \begin{cases} \{i\}, & l = 0 \\ \{u_p \in U \mid (i_q, u_p) \in E \wedge i_q \in \mathcal{N}_i^{(l-1)}(t)\}, & l = 1, 3, \ldots \\ \{i_q \in I \mid (u_p, i_q) \in E \wedge u_p \in \mathcal{N}_i^{(l-1)}(t)\}, & l = 2, 4, \ldots \end{cases} \tag{4}
$$

In this definition, the l-hop neighbors of interacting item i are users where l is an odd number, while its l-hop neighbors are items where l is an even number. Take the interacting item i_1 as an example in Fig. 1(b), its l-hop neighbor set is defined as $\mathcal{N}_{i_1}^{(0)}(t) = \{i_1\}$, $\mathcal{N}_{i_1}^{(1)}(t) = \{u_5\}$, $\mathcal{N}_{i_1}^{(2)}(t) = \{i_3, i_8\}$ and $\mathcal{N}_{i_1}^{(3)}(t) = \{u_3\}$. Note that in the two aforementioned examples, u_3, u_5, i_3, and i_8 are neighbors of both user u_1 and i_1 with the different hops. This indicates the same node (user or item) can be influenced by the different interacting nodes several times at the same time.

Neighbor Selection and Influence Weight Computation. In the influence propagation process, for a given interacting node, we first need to control the number of user and item nodes, which are influenced by this interacting node. Because with the increase of hops of the given interacting node, the number of its l-hop neighbors rises sharply and the strength of propagation influence will weaken layer by layer. These will lead to a sharp increase in computational cost and even a decrease in recommendation performance.

As similar with the intuition that newer interactions should have more impact on the node evolution, an interaction influence should be propagated to the newer neighbors. Thus, we use a time threshold method to select the directly influenced neighbors for a given node (user or item). Let m be the given node, t be the current interaction time. For any m's direct (1-hop) neighbor $n \in \mathcal{N}_m^{(1)}$, we first compute the time interval between the current and last interaction (m, n, t_{mn}) as $\Delta t = t - t_{mn}$, where t_{mn} is the last timestamp of interaction between m and n. Then we sort m's direct neighbors in descending order according to Δt and remove the top $d\%$ of them.

Thus, it is desired to consider the *influence weight*, which is related to the time interval of the interactions in the propagation process, to decide the strength of the influence propagation. For a given node m and its any influenced neighbor n, we define the influence weight between them at time t as:

$$
\omega_{mn}(t) = \frac{\exp(t_{mn} - t)}{\sum_{i \in \mathcal{N}_m^{(1)}} \exp(t_{mi} - t)} \tag{5}
$$

Note that $\mathcal{N}_m^{(1)}$ excludes m's direct neighbors that have been removed. This formula can enhance the influence propagated to neighbors that are closer to the current node. Each propagation layer performs the neighbor selection process.

Edge Influence Embedding Construction. To capture the influence from the multiple interacting nodes and refine the embeddings of users and items influenced by them at the same time, we first need to construct the influence embedding for each existing edge influenced by the new interactions.

Let $E(t) = \{(u, i, t) \mid u \in U, i \in I\}$ be a set of interactions at a fixed time t, $V(t) = \{u \in U, i \in I \mid (u, i, t) \in E(t)\}$ be the set of interacting nodes at time t. The set of *influenced nodes* at time t, which are all neighbors influenced by the interacting nodes (*i.e.*, belonging to $V(t)$) within the l-top, is defined as:

$$\mathcal{N}_V^{(l)}(t) = \bigcup_{n \in V(t)} \mathcal{N}_n^{(l)}(t) \tag{6}$$

As illustrated in Fig. 1(b), at time t_c, the set of interactions (*i.e.*, interacting edges) is $E(t) = \{(u_1, i_1, t_c), (u_2, i_2, t_c)\}$, the set of interacting nodes is $V(t) = \{u_1, i_1, u_2, i_2\}$, and then the set of nodes influenced by the new interactions within the 1-hop is $\mathcal{N}_V^{(1)}(t) = \{i_3, i_4, i_6, u_5, u_6, u_7\}$, where the neighbor (*i.e.*, $i_5 \in \mathcal{N}_{u_2}^{(1)}$) should be forgone if its influence weight is small.

For an existing interaction/edge $(q^{(l-1)}, p^{(l)})$, in which the two nodes are influenced by one of the interacting nodes at time t, (*i.e.*, $q^{(l-1)} \in \mathcal{N}_V^{(l-1)}(t)$ and $p^{(l)} \in \mathcal{N}_V^{(l)}(t)$), the *edge influence embedding* on it is defined as:

$$\mathbf{m}_{q^{(l-1)} \rightarrow p^{(l)}}(t) = \omega_{q^{(l-1)}p^{(l)}}(t) \cdot \left(W_1^{(l)} \mathbf{e}_{q^{(l-1)}}(t) + W_2^{(l)} \left(\mathbf{e}_{q^{(l-1)}}(t) \odot \mathbf{e}_{p^{(l)}}(t-) \right) \right) \tag{7}$$

where $W_1^{(l)}$ and $W_2^{(l)}$ are trainable parameters, and $\odot$ denotes the element-wise product. ω is a weight to control the strength of influence propagated from node

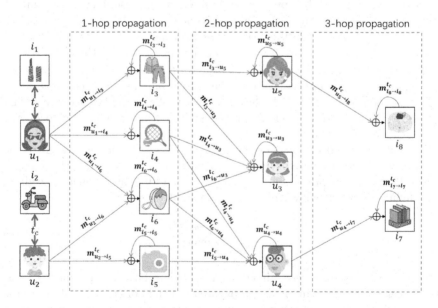

Fig. 2. The propagation of influence embeddings. Some partial embeddings propagated from two interacting users (u_1 and u_2) within 3-hop are shown when two interactions ((u_1, i_1, t_c) and (u_2, i_2, t_c)) happen at time t_c in Fig. 1(b).

$q^{(l-1)}$ to node $p^{(l)}$, and is calculated by Eq. (5). $\mathbf{e}_{q^{(l-1)}}$ is the embedding of node $q^{(l-1)}$, storing the messages which will be conveyed to node $p^{(l)}$. $\mathbf{e}_{q^{(l-1)}}(t) \odot \mathbf{e}_{p^{(l)}}(t-)$ is the similarity between these two nodes, which means passing more influences from the similar nodes.

Influence Embedding Aggregation. For each given user or item influenced by the interacting nodes, the final phase is to aggregate the embeddings from its incoming influenced edges and to refine its embedding. Given an influenced user $u^{(l)} \in \mathcal{N}_V^{(l)}(t)$ at time t, its i-th incoming influenced edge can be denoted by $(q_i^{(l-1)}, u^{(l)}) \in E(t)$, where $q_i^{(l-1)} \in \mathcal{N}_V^{(l-1)}(t)$. The user $u^{(l)}$'s embedding is refined through aggregating all the embeddings of its incoming influenced edges as:

$$\mathbf{e}_{u^{(l)}}(t) = \sigma\left(\mathbf{m}_{u^{(l)}->u^{(l)}}(t) + \sum\nolimits_{q_i^{(l-1)} \in \mathcal{N}_V^{(l-1)}(t)} \mathbf{m}_{q_i^{(l-1)}->u^{(l)}}(t)\right) \tag{8}$$

where $\mathbf{m}_{u^{(l)}->u^{(l)}} = W_1^{(l)}\mathbf{e}_{u^{(l)}}$ is used to retain the original features of user $u^{(l)}$ ($W_1^{(l)}$ is the weight matrix shared with the one used in Eq. (7), *i.e.*, the self-propagation of user u is taken into consideration.) $\mathbf{m}_{q_i^{(l-1)}->u^{(l)}}(t)$ represents the edge influence embedding propagated from $q_i^{(l-1)}$ to $u^{(l)}$. σ is the activation function, here we select Sigmoid function as the our function.

Analogously, for a given influenced item $i^{(l)} \in \mathcal{N}_V^{(l)}(t)$, its embedding is aggregated from the embeddings of its incoming edges (*i.e.*, $(p_u^{(l-1)}, i^{(l)}) \in E(t)$, $p_u^{(l-1)} \in \mathcal{N}_V^{(l-1)}(t)$), and updated as:

$$\mathbf{e}_{i^{(l)}}(t) = \sigma\left(\mathbf{m}_{i^{(l)}->i^{(l)}}(t) + \sum\nolimits_{p_u^{(l-1)} \in \mathcal{N}_V^{(l-1)}(t)} \mathbf{m}_{p_u^{(l-1)}->i^{(l)}}(t)\right) \tag{9}$$

Figure 2 illustrates an example of propagating the influence from two interacting users (u_1 and u_2) within 3-hop when two new interactions ((u_1, i_1, t_c) and (u_2, i_2, t_c)) happen at the same time t_c, shown in Fig. 1(b). For example, u_1 propagates the influence to i_3, i_4 and i_6, then they further influence their neighbors u_3, u_4 and u_5. After the influence propagation and aggregation are finished, we obtain the final representations of users and items.

3.3   Prediction and Optimization

The major task of our model is to predict which item user u will interact with at time t. Given a pair of user and item, we estimate the user's preference towards the target item by conducting the corresponding inner product of $\mathbf{e}_u(t)$ and $\mathbf{e}_i(t)$, where $\mathbf{e}_u(t)$ and $\mathbf{e}_i(t)$ are the latest and refined representations of them. To this end, we rank all items in descending order in terms of the matching scores:

$$\hat{y}_{ui}(t) = \sigma\left(\mathbf{e}_u(t)^\top \mathbf{e}_i(t)\right), \quad u \in U, i \in I \tag{10}$$

$\sigma(\cdot)$ is the exponential function selected in our experiments.

To learn model parameters, we optimize the pairwise Bayesian Personalized Ranking (BPR) loss, which has been widely used in recommender systems. It considers the relative order between observed and unobserved user-item interactions. Specifically, BPR loss assumes that the observed interactions, which are more reflective of a user's preferences, should be assigned higher prediction values than unobserved ones. The objective function is as follows:

$$\mathcal{L} = \sum_{(u,i,j) \in O} - \ln \left(\hat{y}_{ui}(t) - \hat{y}_{uj}(t) \right) + \lambda ||\Theta||_2^2 \tag{11}$$

where $O = \{(u, i, j) \mid (u, i) \in R^+, (u, j) \in R^-\}$ denotes the pairwise training data of the current user u, and R^+ represents the observed interactions, R^- represents the unobserved interactions. Θ represents the trainable parameters. λ is the L_2 regularization strength which helps to prevent overfitting.

4   Experiments

4.1   Dataset Description

We evaluate our method PIDE on three real-world datasets: Reddit, IPTV, and Yelp which are collected by the work [3] and widely used in recommendation systems, and they involve different domains, sizes, and sparsity.

Reddit. This is the public discussion dataset and includes the 14816 posts made by 1,000 users on 1,403 active subreddits (groups) for the month of January 2014. All bot users and their posts are removed from this dataset.

IPTV. This is a publicly accessible dataset and contains around 2M events of 7100 users on 436 TV programs from Jan 1 2012 to Nov 30 2012. It also includes 1420 movie features, which includes 1,073 actors, 312 directors, 22 genres, 8 countries, and 5 years.

Yelp. This dataset includes rich reviews for various businesses from Oct 2004 to Dec 2015, which was available in Yelp dataset challenge Round 7. We select 291,716 reviews of 1,005 users on 47,924 businesses to evaluate our PIDE.

As with existing methods [3], the more interactions each user has, the better recommendation performance the method would obtain. For Reddit, IPTV, and Yelp datasets, their average interactions per user are 14.8, 281.7, and 290.2, respectively. Obviously, Yelp has a higher density than other datasets. However, we note that IPTV only contains 436 TV programs, while Yelp has 47,924 businesses to be chosen. This means that the users in Yelp have more diverse tastes compared with IPTV. Reddit has the lowest average interactions and fewer users and items. Thus, the effectiveness of PIDE can be well evaluated because the three datasets used by us have different sizes and average interactions.

4.2   Experimental Settings

Evaluation Metrics. In the recommendation process, we predict every user's preference scores over all items that the user has not interacted with and rank them in descending order. We adopt Mean Average Rank (MAR) and Hit Ratio (HR@k) to evaluate the results, which are two widely used metrics. MAR is the average rank of actual interacted items of each user in ranking list. So a smaller value indicates better performance. HR@k is the ratio to measure the accuracy of top-k recommended items. The higher HR@k indicates the better performance of the model. We vary the metric k in 5, 10, and 20 in our experiments.

Baselines. To demonstrate the effectiveness and efficiency of our proposed PIDE, we compare it with the following state-of-arts methods:

- CTDNE [14]: This is a random walk based embedding method, which randomly generates the sequence of entities using temporal random walk from continuous-time dynamic networks, and then learns static time-respecting embeddings from the generated sequence.
- DeepCoevolve(DCOE) [3]: This is a deep co-evolutionary method, which uses RNNs to update user and item embeddings over evolving networks. This allows the model to capture complex mutual influence between users and items, and the feature evolution over time.
- JODIE [7]: This is an extension of DeepCoevolve, which not only updates the embeddings of user and item involved in an interaction, but also predicts the future embedding trajectories of them before the incoming interaction.
- PPGN [10]: This is a graph convolution method that proposes the preference propagation GraphNet to address knowledge transfer in cross-domain graphs and captures high-order information propagation for recommendation.
- EvolveGCN(EGCN) [16]: This is a dynamic graph algorithm. It evolves GCN parameters by using RNN to capture the dynamics of graph sequences.
- KGCR [6]: This is also an extension of DeepCoevolve, which extracts different semantic paths between the user-item pair on the dynamic network to update their embeddings when an interaction happens.
- NGCF [34]: This is a collaborative filtering algorithm based on graph neural networks. It updates node representations by recursively propagating the node embedding layer by layer on a static interaction network.

Parameter Settings. For the models, the default Xavier initializer is used to initialize the model parameters. We optimize models with the Adam optimizer. The optimal parameter settings are found through experiments. We fix the embedding size of all models to 128 through experiments. The learning rate is set to 0.001. The first 70% interactions are used for training, in which the 5-fold cross-validation is applied, and the last remaining 30% are used for test. The L_2 regularization strength λ is set to 0.1 for all three datasets. In PIDE, DCOE, PPGN, and KGCR, the batch sizes are set to 100 for Reddit and IPTV, and 64 for Yelp dataset and it is set to 1024 for NGCF. JODIE applies t-batch technique [7] for training and determines the batch size in terms of the time scale metric, which is set to 500. We set the number of batches per training epoch to 100 for Reddit and IPTV, and 50 for Yelp. In NGCF, the number of embedding layers is set to 3. The GCN layers are set to 2 for EvolveGCN. In addition, an early stopping strategy is used in four co-evolutionary methods (including PIDE, DCOE, JODIE, and KGCR), $i.e.$, premature stopping if HR@20 metrics do not increase for 50 epochs.

4.3   Sensitivity Analysis of Parameters

Impacts of Propagation Depth. To investigate how PIDE benefits from interaction influence propagation, we vary the propagation depth in the range of $\{0, 1, 2, 3, 4\}$. Table 1 summarizes the experimental results, in which PIDE-3 indicates the model with three embedding propagation layers and similar notations for others. We observe that increasing propagation depth from 0 to 3, the performance is improved substantially. Obviously, PIDE-0 (no influence propagation) is overall worse than others. We attribute the improvements to the effectiveness of propagating the interaction influence. PIDE-2 and PIDE-3 achieve consistent improvements over PIDE-1, which also proves the effectiveness of propagating the interaction influence to higher-order neighbors. When the propagation depth reaches 4, the recommendation performance begins to decrease. The possible reason is that the influence propagates to some neighbors having less correlation with the interacting node, which may introduce noises to reduce the performance. It suggests that considering appropriate propagation depth is beneficial to capture the influence of propagation to improve performance. Even PIDE-1 outperforms other baselines in most cases by jointly analyzing Table 1 and Table 2. This verifies the superiority of propagating influence again.

Table 1. Impacts of influence propagation depth. It compares the performance of PIDE with different propagation depths, in terms of MAR and HR@k. The best performance in each column is boldfaced and the second best is labeled with '*'.

Method	Reddit				IPTV				Yelp			
	MAR	HR@5	HR@10	HR@20	MAR	HR@5	HR@10	HR@20	MAR	HR@5	HR@10	HR@20
PIDE-0	73.25	0.4252	0.6860	0.8674	56.12	0.1494	0.2757	0.4980	13053.21	0.0045	0.0093	0.0187
PIDE-1	64.23	0.4408	0.6989	0.8701	26.71	0.2451*	0.3386	0.5615	7063.62	0.1047	0.2080	0.3750
PIDE-2	**52.45**	0.4749*	0.7212*	0.8840*	**15.22**	**0.3368**	**0.4688**	**0.6381**	**6341.55**	**0.2158**	**0.2918**	**0.3808**
PIDE-3	52.79*	**0.4751**	**0.7246**	**0.8879**	18.23*	0.2237	0.4421	0.6031	6825.21*	0.1368*	0.2147	0.3774*
PIDE-4	54.70	0.4718	0.7204	0.8632	21.45	0.2189	0.4512*	0.6134*	6469.55	0.1378	0.2192*	0.3746

Impacts of Neighbor Dropout. To empirically investigate the impact of neighbor dropout on recommendation accuracy, we remove some neighbors with different ratios ($d\%$) in $\{0.0, 0.1, 0.2, ..., 0.9\}$ in terms of the time interval in descending order. That is to say, the corresponding ratio of neighbors with the earliest timestamps will be removed. Figure 3 depicts the results on the three datasets. From the results, we observe that as the ratio of neighbor dropout increase, HR@k increases firstly and then decrease gradually. The results demonstrate that the performance of PIDE increases when we remove the outdated neighbors, while its performance decreases if removing too many neighbors (this means the useful neighbors are also removed). This verifies our intuition and confirms previous findings that both propagating an interaction influence and selecting useful neighbors can boost the effectiveness of PIDE.

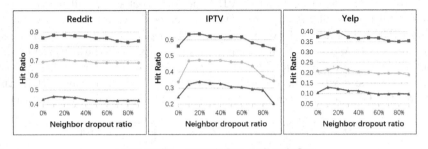

Fig. 3. Impacts of neighbor dropout

Impacts of Embedding Size. We validate the effect of different settings of embedding size on PIDE and other baselines. We vary it from 10 to 128 and the results in term of HR@10 on three datasets are shown in Fig. 4. The results show that PIDE performs the best overall improvements and is less affected by the embedding size, while the performance of other baselines increases with the increase of the embedding size and achieves the worst performance when the dimension size is 10 in most cases. This shows that PIDE can overcome the shortcoming of little coding information when the embedding size is small by propagating interaction influence to neighbors.

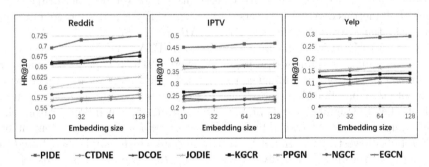

Fig. 4. Impacts of embedding size

4.4   Robustness to the Proportion of Training Data

To validate the robustness of PIDE, we vary the proportion of training data and compare its performance with baselines. The smaller proportion implies users may have fewer interactions used in training. We select interactions with a certain proportion (from 10% to 70%) as the training data and then take the next 30% interactions after the training data as the test data. Thus the robustness analysis is done on the same test data size. Figure 5 illustrates the results of HR@10 on three datasets. PIDE performs better overall performance and is relatively stable as the proportion increases on IPTV and Yelp, which shows PIDE is less affected by the training data size and has good robustness. However, PIDE and almost all baselines sharply vary with the increase of the training data size on Reddit. This shows that their performance relies on the training data size when the dataset is relatively smaller. However, PIDE still achieves the best performance, which shows it can adapt to a smaller dataset.

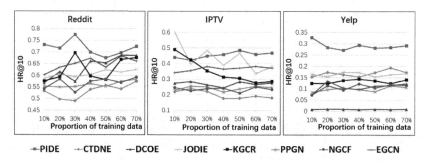

Fig. 5. Impacts of proportion of training data

4.5   Performance Comparison

We compare PIDE with baselines on three datasets. The results are presented in Table 2 and we achieve that PIDE consistently yields the best performance in terms of all metrics on three datasets. In particular, PIDE gets average improvement over the strongest baselines *w.r.t.* MAR by 24.78%, 6.11%, and 1.11%, HR@k by 5.79%, 31.52% and 68.9% on Reddit, IPTV, and Yelp, respectively. By stacking multiple influence propagation layers, PIDE is capable of exploring the impact of propagating influence to high-order neighbors in an explicit way, while all other baselines have no ability to propagate the interaction influence. This verifies the significance of propagating the interaction influence from the interacting nodes to their neighbors. PIDE achieves a highest improvement by HR@k but a lowest improvement by MAR on Yelp, which may be attributed to the fact that Yelp has higher diversity than the other two datasets. This means that the users in Yelp dataset have more diverse tastes compared with others.

Deep co-evolutionary methods (*i.e.*, DCOE, JODIE, KGCR) achieve better performance than other baselines in most cases. This indicates that they can effectively capture the mutual influence between interacting users and items, while others (which are static embedding-based methods) can not make full use of interactions. Except for the mutual influence, KGCR exploits the semantic paths between user-item pairs and JODIE predicts the embedding trajectories to enrich the representations of users/items. Thus, they achieve comparable performance, and are both slightly better than DCOE. EGCN employs GCN to capture the dynamics of graph and evolves GCN parameters by using RNN, while PPGN and NGCF are static graph methods. So EGCN outperforms PPGN and NGCF. Further, PPGN and NGCF get better performance than CTDNE on Reddit and IPTV in most cases. The possible reason is that PPGN and NGCF take preference propagation into account, while CTDNE only introduces the temporal sequence information into user/item representations and forgoes the dynamic interaction information. The performance of CTDNE is better on Yelp. One possible reason is that CTDNE is a random walk based method that can better suit Yelp with higher diversity, while other methods may be affected by this.

Table 2. Performance of all comparison approaches on three datasets across all the evaluation metrics. The best performance in each column is boldfaced; the second best is labeled with '*'; the line '%Improve.' indicates the relative improvements that PIDE achieves *w.r.t.* the best performance of baselines.

Method	Reddit				IPTV				Yelp			
	MAR	HR@5	HR@10	HR@20	MAR	HR@5	HR@10	HR@20	MAR	HR@5	HR@10	HR@20
CTDNE	145.31	0.4340*	0.5740	0.6927	112.47	0.1129	0.2243	0.3025	16733.21	0.0929	0.1727*	0.3134*
DCOE	73.25	0.4252	0.6860*	0.8674	56.12	0.1494	0.2757	0.4980	13053.20	0.0041	0.0094	0.0182
JODIE	91.43	0.3916	0.6256	0.8177	16.21*	0.2138*	0.3789*	0.5574*	6412.71*	0.0934	0.1678	0.2680
KGCR	69.73*	0.4178	0.6764	0.8682*	29.02	0.1554	0.2859	0.5039	13371.45	0.0998*	0.1402	0.2739
PPGN	279.16	0.3910	0.5829	0.7743	54.56	0.1456	0.2455	0.4728	15683.24	0.0554	0.1055	0.1324
NGCF	138.33	0.4027	0.5931	0.7524	77.64	0.1369	0.2331	0.4664	14967.25	0.0633	0.1151	0.2194
EGCN	97.65	0.4133	0.6613	0.8053	48.21	0.2033	0.3714	0.5424	7619.27	0.0804	0.1225	0.2665
PIDE	**52.45**	**0.4751**	**0.7246**	**0.8879**	**15.22**	**0.3350**	**0.4680**	**0.6374**	**6341.55**	**0.2158**	**0.2918**	**0.3808**
%Improve.	24.78%	9.47%	5.63%	2.27%	6.11%	56.69%	23.52%	14.35%	1.11%	116.23%	68.96%	21.51%

4.6   Running Time Comparison

To evaluate the efficiency of PIDE, we compare its running time with other baselines. Algorithmically, DCOE and JODIE mainly train two mutually-recursive RNNs, while PIDE, KGCR, EGCN, PPGN, and CTDNE include more complex operations, such as interaction influence propagation, path sampling and sequence generation based on temporal random walk. Figure 6 shows the running time (in seconds) of one training epoch of Reddit and Yelp (IPTV is similar with Yelp and not shown due to limited space), in which the propagation depth is set to 2 because the performance is better observed from Table 1. We find that on Reddit PIDE is as fast as DCOE and JODIE, and obviously faster than other baselines. This shows that PIDE is able to train influence propagation in equivalent time as the mutually-recursive models when the dataset is small. That is because PIDE can minimize training time on the interaction influence propagation by the neighbor selection mechanism. On Yelp with larger dataset sizes, PIDE is slower than DCOE and JODIE, but still faster than other models. This demonstrates that PIDE takes more time to select neighbors and propagate the interaction influence, so it is slower than mutually-recursive models, but faster than other models.

In addition, Fig. 6 shows the running time of PIDE with the increase of the neighbor dropout ratio and propagation depth on Reddit (here the results of Reddit are shown as an example, and the results of Yelp and IPTV are similar with its). With the fixed ratio of neighbor dropout, the running time increases as the propagation depth increases because it would spend more time propagating the influence to higher-order neighbors. With the fixed propagation depth, the running time decreases with increase of the neighbor dropout ratio. This also verifies and confirms that the neighbor selection can reduce computational cost and accelerate the training.

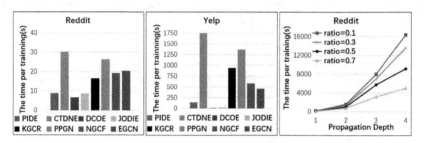

Fig. 6. The running time of PIDE and all baselines and the running time of PIDE with different propagation depths and neighbor dropout proportions.

5   Conclusions

In this paper, we proposed a novel influence-enhanced deep co-evolutionary method PIDE that captures not only the mutual influence between users and items but also the propagation influence from multiple interacting nodes to their neighbors. PIDE overcame the limitations of existing methods by introducing influence propagation into the interaction network evolution, which automatically propagates the interaction influence to l-hop neighbors of interacting users/items to refine the embeddings of them. This led our model to explore the better embeddings of users and items and gave superior prediction performance. We conducted extensive experiments on three real-world datasets, and the experiment results proved the significance of influence propagation and validated the effectiveness and efficiency of PIDE on interaction prediction tasks.

In the future, we plan to consider more interactive features to select more effective neighbors to propagate the influence to. Another is extending our method to integrate more features of users and items into the interaction network to achieve better performance.

Acknowledgment. This work was partially supported by grants from the Fundamental Research Funds for the Central Universities (Grant No. 3122021089) and the Scientific Research Project of Tianjin Educational Committee (Grant No. 2021ZD002).

References

1. Cao, J., et al.: Deep structural point process for learning temporal interaction networks. In: Oliver, N., Pérez-Cruz, F., Kramer, S., Read, J., Lozano, J.A. (eds.) ECML PKDD 2021. LNCS (LNAI), vol. 12975, pp. 305–320. Springer, Cham (2021). https://doi.org/10.1007/978-3-030-86486-6_19
2. Wang, S., Hu, L., Wang, Y., et al.: Graph learning based recommender systems: a review. In: IJCAI, pp. 4644–4652 (2021)
3. Dai, H., Wang, Y., Trivedi, R., et al.: Deep coevolutionary network: embedding user and item features for recommendation. In: SIGKDD (2017)
4. Peng, H., Zhang, R., Dou, Y., et al.: Reinforced neighborhood selection guided multi-relational graph neural networks. ACM Inform. Syst. **40**(4), 1–46 (2021)
5. Deng, Z., Peng, H., Xia, C., et al.: Hierarchical bi-directional self-attention networks for paper review rating recommendation. In: COLING (2020)
6. Xiao, C., Sun, L., Ji, W.: Temporal knowledge graph incremental construction model for recommendation. In: APWEB-WAIM, pp. 352–359 (2020)
7. Kumar, S., Zhang, X., Leskovec, J.: Predicting dynamic embedding trajectory in temporal interaction networks. In: SIGKDD, pp. 1269–1278 (2019)
8. Zhang, Y., Xiong, Y., Li, D., et al.: CoPE: modeling continuous propagation and evolution on interaction graph. In: CIKM, pp. 2627–2636 (2021)
9. Ma, Y., Guo, Z., Ren, Z., et al.: Streaming graph neural networks. In: SIGIR, pp. 719–728 (2020)
10. Zhao, C., Li, C., Fu, C.: Cross-domain recommendation via preference propagation graphnet. In: CIKM, pp. 2165–2168 (2019)
11. Wang, H., Zhang, F., Wang, J., et al.: RippleNet: propagating user preferences on the knowledge graph for recommender systems. In: CIKM, pp. 417–426 (2018)
12. Mao, C., Yao, L., Luo, Y.: MedGCN: medication recommendation and lab test imputation via graph convolutional networks. J. Biomed. Inform. **127** (2022)
13. Goyal, P., Chhetri, S.R., Canedo, A.: dyngraph2vec: capturing network dynamics using dynamic graph representation learning. KBS **187**, 14–22 (2020)
14. Nguyen, G.H., Lee, J.B., Rossi, R.A., et al.: Continuous-time dynamic network embeddings. In: WWW, pp. 969–976 (2018)
15. Wang, Y., Li, P., Bai, C., et al.: TEDIC: neural modeling of behavioral patterns in dynamic social interaction networks. In: WWW, pp. 693–705 (2021)
16. Pareja, A., Domeniconi, G., Chen, J., et al.: EvolveGCN: evolving graph convolutional networks for dynamic graphs. In: AAAI, pp. 5363–5370 (2020)
17. Zhou, L., Yang, Y., Ren, X., et al.: Dynamic network embedding by modeling triadic closure process. In: AAAI, pp. 571–578 (2018)
18. Lu, Y., Wang, X., Shi, C., et al.: Temporal network embedding with micro-and macro-dynamics. In: CIKM, pp. 469–478 (2019)
19. Yang, Y., Yin, H., Cao, J., et al.: Time-aware dynamic graph embedding for asynchronous structural evolution. arXiv preprint arXiv:2207.00594 (2022)
20. Gao, X., Xu, X., Li, D.: Accuracy analysis of triage recommendation based on CNN, RNN and RCNN models. In: IPEC, pp. 1323–1327 (2021)
21. Zhu, Y., Li, H., Liao, Y., et al.: What to do next: Modeling user behaviors by time-LSTM. In: IJCAI, pp. 3602–3608 (2017)
22. Beutel, A., Covington, P., Jain, S., et al.: Latent cross: making use of context in recurrent recommender systems. In: WSDM, pp. 46–54 (2018)
23. Song, W., Charlin, L., Xiao, Z., et al.: Session-based social recommendation via dynamic graph attention networks. In: WSDM, pp. 555–563 (2019)

24. Rendle, S., Freudenthaler, C., Schmidt-Thieme, L.: Factorizing personalized Markov chains for next-basket recommendation. In: WWW, pp. 811–820 (2010)
25. He, R., McAuley, J.: Fusing similarity models with markov chains for sparse sequential recommendation. In: ICDM, pp. 191–200 (2016)
26. Hidasi, B., Karatzoglou, A., Baltrunas, L., Tikk, D.: Session-based recommendations with recurrent neural networks. In: ICLR (2016)
27. Wu, S., Tang, Y., Zhu, Y., et al.: Session-based recommendation with graph neural networks. In: AAAI, pp. 346–353 (2019)
28. Li, J., Wang, Y., McAuley, J.: Time interval aware self-attention for sequential recommendation. In: WSDM, pp. 322–330 (2020)
29. Chen, H., Li, Y., Sun, X., et al.: Temporal meta-path guided explainable recommendation. In: WSDM, pp. 1056–1064 (2021)
30. Fan, Z., Liu, Z., Wang, Y., et al.: Sequential recommendation via stochastic self-attention. In: WWW, pp. 2036–2047 (2022)
31. Zhang, M., Wu, S., Yu, X., Wang, L.: Dynamic graph neural networks for sequential recommendation. TKDE, 1–1 (2022)
32. Wang, J., Ding, K., Hong, L., et al.: Next-item recommendation with sequential hypergraphs. In: SIGIR, pp. 1101–1110 (2020)
33. Chang, J., Gao, C., Zheng, Y., et al.: Sequential recommendation with graph neural networks. In: SIGIR, pp. 378–387 (2021)
34. Wang, X., He, X., Wang, M., et al.: Neural graph collaborative filtering. In: SIGIR, pp. 165–174 (2019)
35. He, X., Deng, K., Wang, X., et al.: LightGCN: simplifying and powering graph convolution network for recommendation. In: SIGIR, pp. 639–648 (2020)

Intra- and Inter-behavior Contrastive Learning for Multi-behavior Recommendation

Qingfeng Li[1], Huifang Ma[1,2(✉)], Ruoyi Zhang[1], Wangyu Jin[1], and Zhixin Li[2]

[1] College of Computer Science and Engineering, Northwest Normal University, Lanzhou, Gansu, China
mahuifang@nwnu.edu.cn
[2] Guangxi Key Lab of Multi-source Information Mining and Security, Guangxi Normal University, Guilin, Guangxi, China
lizx@gxnu.edu.cn

Abstract. Multi-behavior recommendation (MBR) aims to improve the prediction of target behavior by exploiting multi-typed auxiliary behaviors. However, most MBR suffers from data sparsity in real-world scenarios and thus performs mediocrely. In this paper, we propose a novel **I**ntra- and **I**nter-behavior **C**ontrastive **L**earning (IICL) framework to exploit contrastive learning to enhance MBR with two key challenges to be addressed: i). Difficult to learn reliable representations under different behaviors; ii). Sparse supervised signals under target behavior. For the first challenge, we devise an intra-behavior contrastive learning objective, which enhances the representation learning by incorporating potential neighbors into the contrastive learning from the graph topological space and the semantic space, respectively. As for the second challenge, we design an inter-behavior contrastive learning objective, which has the benefit of capturing commonalities between different behaviors and integrating them into the target behavior to alleviate the sparse supervised signal problem. In addition, we also propose an adaptive weight network to customize the integration of all losses efficiently. Extensive experiments on three real-world benchmark datasets show that our proposed method IICL is significantly superior to various state-of-the-art recommendation methods.

Keywords: Multi-Behavior Recommendation · Contrastive Learning · Graph Neural Network

1 Introduction

Personalized recommendations aim to capture user preferences from historical interaction data and recommend appropriate items for users [7]. In real-world scenarios, there are generally multiple types of interactions between users and items. For example, on an e-commerce platform, users can purchase items, share items with friends, add items to their shopping cart and etc. Traditional recommendation methods [4,11] mostly consider only a single type of user-item

© The Author(s), under exclusive license to Springer Nature Switzerland AG 2023
X. Wang et al. (Eds.): DASFAA 2023, LNCS 13944, pp. 147–162, 2023.
https://doi.org/10.1007/978-3-031-30672-3_10

interaction, which is difficult to handle the real scenarios. To cope with this dilemma, Multi-behavior Recommendation (MBR) is proposed, which jointly considers multiple types of behaviors to better learn user preferences and has been widely explored and validated in practice [2,17]. With the success of Graph Neural Networks (GNNs), some studies [1,6,18,22] consider employing GNNs to capture high-order collaborative signals and generate node-level embeddings.

Despite the promising results achieved by the above methods, these MBR methods inevitably suffer from the issue of data sparsity and thus degrading the actual recommendation performance to some extent. Inspired by the success of contrastive learning in alleviating the data sparsity issue, some recent works attempt to adopt contrastive learning, which can be roughly divided into two categories. One type of methods [15,21,23] is to use graph augmentation on the original graph to construct contrastive views, and then maximizing the consistency of the same node under different views and minimizing the consistency between other nodes. The other type of methods [8,19] discards the graph augmentation operation and instead constructs the contrastive space with the information from different aspects of the original graph. In a multi-behavior scenario, multi-typed behaviors can naturally be regard as multiple views for contrastive learning. Recently, some studies [3,13,16] consider the dependency between different behaviors to alleviate the sparse supervision signal problem.

Although existing methods alleviate the data sparsity issue to some extent, they still suffer from two limitations: (1). **Difficult to learn reliable representations under different behaviors.** The feedback we generally collect from users is sparse and implicit, which means that interactions are often accompanied by noise. For example, a user interacts with an item casually, but soon finds out that it is not of interest. In the graph-based neighborhood aggregation paradigm, the representation is extremely susceptible to interaction noise and thus may not learn a reliable representation; (2). **Sparse supervised signals under target behavior.** Most current MBR methods are based on a supervised learning paradigm, which are trained with supervised signals in an end-to-end manner. This means that sufficient target behavior interaction data is needed to learn target user preferences and thus to forecast target user behavior. Unfortunately, in real-world scenarios the target behavior is usually much sparser than the auxiliary behaviors. For example, on an e-commerce platform, a user may view many items, but purchase a tiny fraction of them. Therefore, sparse supervised signals for target behaviors significantly degrade the quality of graph learning and thus degrade recommendation performance.

In light of the aforementioned challenges, we focus on exploiting contrastive learning to enhance the representation and alleviating the issue of sparse supervised signal under target behavior. Towards this end, we develop a novel Intra- and Inter-behavior Contrastive Learning (IICL) framework for MBR. In intra-behavior contrastive learning, we propose two neighbor-enhanced approaches to fully exploit useful potential relations in both structural and semantic spaces. Specifically, we treat the k-th (i.e., even number) layer output of GNNs as a structural neighbor representative embedding of a node, since GNNs naturally aggregate homogeneous structural neighbor information within k-hop. We then develop a structure-aware contrastive learning objective and treat the user and his structural neighbors as positive pairs. To explore the potential relations of users in the semantic space, we recognize users with similar interaction behaviors as semantic neighbors and incorporate these semantic neighbors as positive samples into a classical contrastive learning objective. In inter-behavior contrastive learning, we design a cross-behavior (i.e., target behavior-auxiliary behavior) contrastive learning objective to alleviate the issue of sparse supervised signals under target behavior. Specifically, we treat the same users with different behaviors as positive sample pairs. In addition, unlike simply treating all remaining users as negative samples, we propose to treat neighbors under the user-user semantic graph as difficult negative samples. By capturing the commonalities of target and auxiliary behaviors and integrating them into the target behavior in this way, the supervised signal under target behavior is thus enriched. Moreover, adaptive weights are generated with user embeddings to fuse multiple contrastive learning tasks to capture user personalized behavior preferences. Finally, intra- and inter-behavior contrastive tasks and the multi-behavior Bayesian Personalized Ranking (BPR) [10] are jointly optimized in a multi-task learning (MTL) [18] framework.

In summary, our contributions of this work are as below:

- We propose a novel contrastive learning paradigm IICL for multi-behavior recommendation by emphasizing intra-behavior potential neighbor relations and inter-behavior transferable knowledge.
- In our proposed IICL, we design two levels of contrastive learning tasks. In intra-behavior, we design two neighbor-enhanced schemes to capture potential relations in the structure space and semantic space. In inter-behavior, we design target-auxiliary contrastive learning to extract commonalities and incorporate them into sparse target behavior modeling. In addition, the adaptive weight network customizes the integration of multiple losses to preserve user personalized behavior preferences.
- Extensive experiments are conducted on three benchmark datasets. The results show that IICL consistently outperforms the state-of-the-art baselines, especially for cold-start users.

2   Related Work

Graph Contrastive Learning. Contrastive learning, as a classical self-supervised technique, is considered an antidote to the sparse supervised signals issue [5,12,15]. The core of contrastive learning is to learn high-quality discriminative representations by maximizing the consistency between positive samples and minimizing the consistency between negative samples. For example, HeCo [12] develops a cross-view contrastive learning framework to refine node representations by maximizing the consistency of the two complementary views. In addition, SGL [15] creates two auxiliary views by data augmentation operations and performs contrastive learning on these two views as an auxiliary self-supervised task to enhance the main task. CCL [7] captures both local and high-order information by maximizing the consistency of the two enhanced views. Existing methods generally construct sample sets by simply sampling randomly, without considering the relations between users in recommendation scenario. To fill this gap, we design the intra-behavior contrastive learning to explicitly model user neighbors from structure space and semantic space to capture useful potential relations.

Multi-behavior Recommendation. The purpose of multi-behavior recommendation is to leverage multi-typed behavior data to improve recommendation performance of target behavior [1,2,17,18]. A research line considers generating behavior-specific embeddings and incorporates attention mechanisms to integrate other behaviors to enhance the target behavior. For example, MBGCN [6] builds on the message-passing architecture to obtain discriminative behavior representations and additionally models the similarity between items. Another research line considers exploring the differences between behaviors to learn the target behavior. For example, AMR [14] captures hidden relations in user-item interaction network by constructing multi-relation graphs with different behavior types. CML [13] utilizes contrastive learning to explore differences in different behaviors to augment target behavior prediction. However, most of them ignore the sparse supervised signal problem. Based on this, we design the target-auxiliary contrastive learning to provide additional informative supervision signals.

3   Preliminaries

In this paper, we define the set of users and items as $\mathcal{U}$ and $\mathcal{V}$ respectively. $\mathbf{R}^{(k)}$ denotes the user-item interaction matrix under k^{th} behavior, where $r_{u,v}^k = 1$ indicates that the user u has interacted with the item v under the kth behavior, and otherwise $r_{u,v}^k = 0$. Therefore, user-item multi-behavior interaction data in multi-behavior recommendation scenario is formalized as $\{\mathbf{R}^{(1)}, \ldots, \mathbf{R}^{(k)}, \ldots, \mathbf{R}^{(K)}\}$. Generally, the Kth behavior denotes the *target behavior* (i.e., purchase) and the first $K-1$ behaviors denote the *auxiliary behaviors* (e.g., view, add-to-cart, share). Based on this, we further define the relevant concepts as follows:

User-Item Multi-behavior Interaction Graph G_c. With the awareness of different types of user-item interactions, a multi-behavior interaction graph is defined as: $G_c = \{\mathcal{U}, \mathcal{V}, \mathcal{E}_c\}$, in which the edge set $\mathcal{E}_c$ represents observed interactions with k types of behaviors. In $\mathcal{E}_c$, edge $e_{u,v}^k$ between u and v indicates that $r_{u,v}^k = 1$.

User-User Multi-behavior Semantic Graph G_s. In order to exploit semantically relevant information between users, we define graph $G_s = \{\mathcal{U}, \mathcal{E}_s\}$ to characterize multiplex dependencies across users with the consideration of their high-order relations. In $\mathcal{E}_s$, edge $e_{u,u'}^k$ linked between user u and u' with their co-acting relations under k^{th} behavior. For example, $e_{u,u'}^k = 1$ represents user u and user u' have interacted with the same item under k^{th} behavior, otherwise $e_{u,u'}^k = 0$.

Problem Statement. Given the user-item multi-behavior interaction graph G_c and user-user multi-behavior semantic graph G_s. The task of our model is to predict the interaction probability between user u and item v under the target behavior.

4 Methodology

In this section, we introduce our IICL framework as shown in Fig. 1, which consists of three important parts: intra-behavior neighbor-enhanced contrastive learning, inter-behavior target-auxiliary contrastive learning and adaptive multi-behavior weight network.

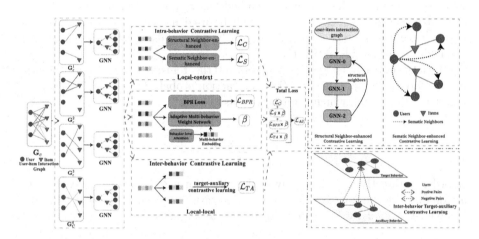

Fig. 1. Illustration of the proposed IICL model. The left subfigure shows model framework of IICL; and the right subfigure presents the details of intra- and inter-behavior contrastive learning mechanism. Best viewed in color.

4.1 Lightweight Behavior-Context Graph Neural Network

In this section, we describe how to learn embeddings that contain the context of user behavior on user-item multi-behavior interaction graph. The high-hop graph structure of user-item heterogeneous interactions contains rich semantic information carrying collaborative signals, so we first develop a lightweight graph-based behavior-aware message passing mechanism. Inspired by the findings of LightGCN [4], we not only eliminate redundant transformations and activations, but also simplify the multiplexed information aggregation, which can be defined as below:

$$\mathbf{z}_u^{k,l} = \sum_{v \in N_u^k} \mathbf{z}_v^{k,l-1}, \tag{1}$$

where $\mathbf{z}_u^{k,l} \in \mathbb{R}^d$ denotes the obtained embedding of the user u at l^{th} propagation layer under k^{th} behavior. N_u^k denotes the set of immediate neighbors of user u under k^{th} behavior. After obtaining user behavior-specific embeddings, we adopt behavior-level attention mechanism to aggregate behaviors to obtain multi-behavior embeddings $\mathbf{z}_u^{K+1,l}$, which preserve behavioral context information at specific layers. As the embedding propagates L layers, we generate L embeddings for each user under each behavior. The embeddings from different layers emphasize information from different perceptual domains, so we need to fuse multiple layers of embeddings, and here our method is concatenation, which is defined as follows:

$$\mathbf{z}_u^k = \mathbf{W}_{cat}^T(\mathbf{W}^0\mathbf{z}_u^{k,1} \ || \ \cdots \ || \ \mathbf{W}^{L-1}\mathbf{z}_u^{k,L}), \tag{2}$$

where $\mathbf{W}_{cat} \in \mathbb{R}^{Ld \times d}$ denotes the concatenating transformation matrix, $\mathbf{W}^{L-1} \in \mathbb{R}^{d \times d}$ represents the transformation matrix of the Lth propagation layer. So far, we obtain the final behavior-specific embedding $\mathbf{z}_u^k$ and the final multi-behavior context-preserving embedding $\mathbf{z}_u^{K+1}$ (will be used in Sect. 4.4) after synthesizing multiple layers, respectively. Analogously, we can also get the corresponding embeddings on the side of the items.

With the final embeddings of the users and the items, we use the inner product to calculate the prediction score (i.e., interaction probability) of user u with item v under k^{th} behavior, which is formalized as follows:

$$\hat{r}_{uv}^k = \sum_{i=1}^d \mathbf{z}_{ui}^k \mathbf{z}_{vi}^k. \tag{3}$$

In order to directly and effectively utilize the information in the observed interaction data, we employ BPR as a supervised task to optimize the model and the behavior-specific BPR loss can be formalized as follows:

$$\mathcal{L}_{BPR}^k = \sum_{(u,v^+,v^-)\in\mathcal{O}_k} -log(sigmoid(\hat{r}_{uv^+}^k - \hat{r}_{uv^-}^k)), \tag{4}$$

where $\mathcal{O}_k = \{(u, v^+, v^-) \mid (u, v^+) \in \mathcal{O}_k^+, (u, v^-) \in \mathcal{O}_k^-\}$ represents the pairwise training data under k^{th} behavior, and $\mathcal{O}_k^+$ denotes observed interactions of user u. $\mathcal{O}_k^- = (\mathcal{U} \times \mathcal{V}) - \mathcal{O}_k^+$ denotes all unobserved interactions. Behavior-specific BPR loss $\mathcal{L}_{BPR}^k$ directly models the user-item interaction under k^{th} behavior.

4.2   Intra-behavior Neighbor-Enhanced Contrastive Learning

Existing multi-behavior contrastive learning frameworks are generally based on GNNs to learn user behavior preferences via observed interaction data, while the potential relations among users are not explicitly captured from these interaction data. To take full advantage of contrastive learning, we propose *two generic neighbor-enhanced schemes* to model potential relations among users and fine-grained modeling of users personalized behavior preferences as shown in Fig. 1, top right.

Structural Neighbor-Enhanced Contrastive Learning. After performing multiple message propagation on the user-item multi-behavior interaction graph G_c, we can obtain multi-layer embeddings of specific behaviors (e.g., $\mathbf{z}_u^{k,l}$). The even-numbered layers of message propagation (e.g., 2, 4, 6) naturally aggregates information about the user homogeneous structural neighbors and outputs homogeneous domain representations, which help us model the relations between the user and his/her structural neighbors.

Inspired by previous work [8], we can use these different layers of user embeddings to explicitly model the potential relations between users and their homogeneous structural neighbors. Specifically, for user u under k^{th} behavior, we consider the original embedding (e.g., $\mathbf{z}_u^{k,0}$) and the embedding from the even-numbered layer as positive pairs, while all remaining users are considered as negative pairs. Following the classical structure of InfoNCE [9], we aim to maximize the consistency between positive pairs while minimizing the consistency between negative pairs, which is formalized as follows:

$$\mathcal{L}_C^k = \sum_{u \in \mathcal{U}} -log \frac{exp(\phi(\mathbf{z}_u^{k,l}, \mathbf{z}_u^{k,0})/\tau)}{\sum_{j \in \mathcal{U}} exp(\phi(\mathbf{z}_u^{k,l}, \mathbf{z}_j^{k,0})/\tau)}, \tag{5}$$

where $\mathbf{z}_u^{k,l}$ denotes the embedding of user u generating from even-numbered layer l under k^{th} the behavior, τ is the temperature coefficient to control the strength of discrimination. $\phi(\cdot)$ is a function that measures the similarity of two embeddings, such as inner product or cosine similarity. Since the structural neighbors are equally important under different behaviors, the total structural neighbor-enhanced contrastive loss is defined as $\mathcal{L}_C = \frac{1}{K} \sum_{k=1}^{K} \mathcal{L}_C^k$.

Sematic Neighbor-Enhanced Contrastive Learning. Structural neighbor-enhanced contrastive learning expands the user neighborhoods on the interaction graph, which is used to capture potential relations between users and homogeneous neighbors. However, treating all homogeneous neighbors equally inevitably introduces noise in the loss of structural contrastive learning. To reduce the impact of this noise, we introduce semantic neighbors to further expand user neighbors. Semantic neighbors are defined as users who have similar behavior preferences to the target user, i.e., neighbors on the user-user multi-behavior semantic graph G_S. The intuition behind this is that the embeddings of users with similar behavior preferences should be close, and the embeddings of other users should be far away. Specifically, for user u under k^{th} behavior, with his/her corresponding neighbors on the graph G_S are defined as positive sample set $\mathcal{P}_u^k$ and all other users as negative sample set, which is formalized as follows:

$$\mathcal{L}_{S,u}^k = -log \frac{\sum_{i \in \mathcal{P}_u^k} exp(\phi(\mathbf{z}_u^k, \mathbf{z}_i^k)/\tau)}{\sum_{j \in \mathcal{U}} exp(\phi(\mathbf{z}_u^k, \mathbf{z}_j^k)/\tau)}. \tag{6}$$

4.3   Inter-behavior Target-Auxiliary Contrastive Learning

In this section, we will introduce our inter-behavior target-auxiliary contrastive learning. The most important step in contrastive learning is the reasonable selection of positive and negative samples, and a straightforward idea is that we consider the embeddings of the same user under different behaviors as the set of positive samples and the other users as the set of negative samples. However, these simple negative samples with large differences commonly contribute little information, so we propose a hard negative sample selection strategy. We narrow down the number of negative samples by selecting only the neighbors of that user on the semantic graph G_S under the auxiliary behavior as the set of negative samples, as shown in Fig. 1, bottom right. Specifically, for the target behavior K and the auxiliary behavior k', the inter-behavior contrastive learning loss is formalized as:

$$\mathcal{L}_{TA,u}^{K,k'} = -log \frac{exp(\phi(\mathbf{z}_u^K, \mathbf{z}_u^{k'})/\tau)}{\sum_{j \in S_u^{k'}} exp(\phi(\mathbf{z}_u^K, \mathbf{z}_j^{k'})/\tau)}, \qquad (7)$$

where $S_u^{k'}$ is the set that holds both the neighbors of user u on the semantic graph G_S under the auxiliary behavior k' and user u. To capture user personalized behavior patterns, we need to customize to integrate this series of contrastive learning losses.

4.4   Adaptive Multi-behavior Weight Network

In a real multi-behavior recommendation scenario, different users have diverse behavior preferences and interact with different items. Personalized behavior patterns from users bring diversity to item interactions, so modeling the diverse dependencies between different user behaviors is critical to enable accurate recommendations. To achieve this, we propose an adaptive multi-behavior weight network to explicitly learn user personalized behavior weights, which are then used to integrate multi-behavior contrastive loss.

Multi-behavior Knowledge Encoder. In our adaptive multi-behavior weight network, we first learn user-specific behavior dependencies. Inspired by previous work on feature extraction [19], we propose two different integration techniques for multi-behavior knowledge encoder as follows (k' denotes the auxiliary behavior):

$$\mathbf{M}_{u,1}^{K,k'} = \mathbf{z}_u^K \,\|\, \mathbf{z}_u^{k'} \,\|\, \mathbf{z}_u^{K+1}, \mathbf{M}_{u,2}^{K,k'} = \mathbf{z}_u^K \odot \mathbf{z}_u^{k'} \odot \mathbf{z}_u^{K+1}, \qquad (8)$$

where the encoded multi-behavior knowledge is represented as $\mathbf{M}_{u,1}^{K,k'}, \mathbf{M}_{u,2}^{K,k'}$, which is designed to preserve multi-behavior interaction contexts and target-auxiliary behavior dependencies. $\|$ and $\odot$ are represented as concatenation and element-wise operations, respectively.

Adaptive Weight Network. After effectively encoding the multi-behavior knowledge of specific user, we design a mapping function $\mathcal{F}(\cdot)$ to transform the learned knowledge into contrastive loss weights. This module enables our IICL to customize the contrastive learning paradigm for users to learn their personalized behavior preferences under different behavior intentions, which can be formalized as follows:

$$\mathcal{F}(\mathbf{M}_u^{K,k'}) = \text{PReLu}(\mathbf{W}_f \mathbf{M}_u^{K,k'} + \mathbf{b}_f), \qquad (9)$$

where $\mathbf{W}_f \in \mathbb{R}^{d \times d}$ and $\mathbf{b}_f \in \mathbb{R}^d$ denote the projection weight matrix and bias, respectively. The activation function is chosen PReLU to endow the non-linearity capability. Then, we can use our proposed adaptive multi-behavior weight network to generate user personalized multi-behavior contrastive loss weights as follows:

$$\beta_u^{K,k'} = \mathcal{F}(\mathbf{M}_{u,1}^{K,k'}) + \mathcal{F}(\mathbf{M}_{u,2}^{K,k'}), \tag{10}$$

where $\beta_u^{K,k'}$ is denoted as an explicit dependency between the target behavior K and the auxiliary behavior k' for specific-user u. Therefore, in this way we can generate customized weight lists for self-supervised learning-based contrastive loss and supervised learning-based BPR recommendation loss. Note that for intra-behavior contrastive loss, the knowledge encoder only requires behavior-specific embeddings $\mathbf{z}_u^k$ and multi-behavior contextual embeddings $\mathbf{z}_u^{K+1}$.

4.5 Joint Optimization

After obtaining the user-specific inter-behavior target-auxiliary contrastive loss and the corresponding weights, we integrate to obtain the final contrastive loss as follows:

$$\mathcal{L}_{TA} = \sum_{u \in \mathcal{U}} \sum_{k'=1}^{K-1} \beta_u^{K,k'} \mathcal{L}_{TA,u}^{K,k'}. \tag{11}$$

Analogously, we can obtain the total intra-behavior semantic contrastive loss $\mathcal{L}_S$ and the total BPR loss $\mathcal{L}_{BPR}$. Then, we simultaneously optimize the above three different but related tasks, which can be formalized as follows:

$$\mathcal{L}_{All} = \mathcal{L}_{BPR} + \gamma \mathcal{L}_C + \eta \mathcal{L}_S + \zeta \mathcal{L}_{TA} + \lambda \| \Theta \|^2, \tag{12}$$

where γ, η, ζ are hyper-parameters to control the proportion of different self-supervised contrastive learning tasks. Θ denotes the learnable parameters in the model and λ is used to control the weights of L_2 regularization.

5 Experiments

To justify the effectiveness of our proposed IICL, we conduct extensive experiments on three real-world datasets to answer the following research questions: **(RQ1)** How does IICL perform compared with the various SOTA baselines in MBR? **(RQ2)** How do the different designed modules contribute to the model performance? **(RQ3)** How does IICL perform w.r.t different interaction sparsity levels as compared to representative competitors? **(RQ4)** How do different hyper-parameter settings affect the final performance?

5.1 Datasets

We evaluate IICL on three publicly available MBR datasets: **Tmall**[1]: It is collected by Tmall, one of the largest e-commerce platforms in China. There are

---

[1] https://tianchi.aliyun.com/dataset/dataDetail?dataId=649.

31,822 users, 31,232 items and 1,451,219 interaction records, which contains four behaviors including page view, favorites, cart and purchases. **IJCAI-Contest**[2]: It is collected from a business-to-customer retail system. There are 17,435 users, 35,920 items, 799,368 interaction records in this dataset. It has the same behavior type as the Tmall dataset. **Retailrocket**[3]: It is collected from Retailrocket, which contains 2,174 users, 31,232 items, 97,381 interaction records and three behaviors including page view, cart and transaction.

5.2   Baselines

To demonstrate the effectiveness of our IICL model, we compare it with several state-of-the-art methods. The baselines are classified into two categories: Single-behavior recommendation Methods and Multi-behavior recommendation Methods. We completely follow the tuning strategy of the original papers to set parameters.

Single-Behavior Recommendation Methods Include:

- **BPR** [10]: It is a widely used pairwise learning method for item recommendation.
- **PinSage** [20]: It is a method that defines surrounding important neighboring nodes to perform graph convolution.
- **NGCF** [11]: It is a popular GNN-based method which uses convolutional messaging mechanism to enhance collaborative filtering.
- **LightGCN** [4]: It is a state-of-the-art method which is better suited to recommendation scenarios by simplifying redundant operations in graph convolutional networks.
- **SGL** [15]: It obtains augmented views through data augmentation (e.g., node and edge dropout) and maximizes consistency between them.

Multi-behavior Recommendation Methods Include:

- **NMTR** [2]: It models multiple types of user interaction behavior dependencies and optimizes them jointly with multi-task learning strategies.
- **MATN** [17]: It uses the self-attention mechanism to examine the effects between different behaviors.
- **MBGCN** [6]: It is a representative GCN-based method which captures user behavior patterns by introducing high-order connectivity on the user-item interaction graph.
- **KHGT** [1]: It uses the attention mechanism to study inter-behavior effects and integrates temporal information into modeling.
- **EHCF** [18]: It models the differences between multiple behaviors and enhances target behavior predictions through knowledge transfer.
- **CML** [13]: It preserves behavioral contextual information through meta-encoders and captures inter-behavioral transferable knowledge to enhance target behavior modelling.

---

[2] https://tianchi.aliyun.com/dataset/dataDetail?dataId=47.
[3] https://www.kaggle.com/retailrocket/ecommerce-dataset.

To validate the effectiveness of IICL in modelling heterogeneous behavior relations, we additionally compare it with two state-of-the-art heterogeneous neural network models.

- **HGT** [5]: It models heterogeneous relations using an attention mechanism based on heterogeneous graphs. We use the heterogeneous message-passing mechanism to model multiplex behavior.
- **HeCo** [12]: It employs network schema and meta-path as two views to capture different factors and performs the contrastive learning across them. We use semantic views to generate the required meta-path relations.

5.3   Hyper-parameters and Metrics

The embedding size is set to 16. The embedding initialization is performed using Xavier. The model uses the AdamW optimizer and the learning rate is adjusted by searching in $\{0.6e^{-4}, 1e^{-3}, 2e^{-3}, 5e^{-3}\}$. The graph encoder depth is adjusted from $\{1, 2, 3, 4\}$. The weights of L2 regularization are adjusted from $\{1e^{-3}, 5e^{-3}, 1e^{-2}\}$. To mitigate the overfitting problem, we set the dropout rate to 0.8. After tuning, the weights of supervised loss $\mathcal{L}_{BPR}$ and the three contrastive learning loss (i.e., $\mathcal{L}_C, \mathcal{L}_S, \mathcal{L}_{TA}$) are set to 1.0, 0.2, 0.05, 0.05, respectively. For all baselines, we conduct a grid search for parameter selections. We adopt the leave-one-out strategy and use the last-interacted items under the target behavior (i.e., purchase/transaction) as the test set. Performance is analyzed using two representative ranking-based metrics: HR (Hit Ratio) and NDCG (Normalized Discounted Cumulative Gain).

5.4   Performance Comparison (RQ1)

Table 1 reports the experimental results of the overall performance, where the results of IICL and the best-performing baseline are highlighted in bold and underlined, respectively. From the results, the following observations can be made:

(1). IICL consistently outperforms all types of recommendation methods on all metrics across the three datasets. Significance level as $p < 0.05$, which indicates a statistically significant improvement between our method and baselines. This is attributed to two aspects: 1). In intra-behavior, IICL obtains rich user-personalized preference information by neighbor-enriched contrastive learning objectives. 2). In inter-behavior, we provide auxiliary self-supervision signals by exploring commonalities between different behaviors and provide reliable information gradients for graph collaborative filtering.

(2). Multi-behavior recommendation methods generally perform better than single-behavior recommendation methods. It verifies the effectiveness of considering multi-behavior information. Moreover, different from the self-supervised method SGL, which contrasts the original graph-derived representation with the augmented graph and ignores other potential relations (e.g., user similarity) in the recommender system. We consider both intra-behavior structural and semantic information and inter-behavior transferable knowledge to fit the multi-behavioral recommendation.

(3). IICL performs better than the heterogeneous graph neural network methods. This demonstrates the ability of our model to efficiently encode relational heterogeneity by relying on behavior association contrastive learning and adaptive weight network.

Table 1. Performance comparison of all compared methods on different datasets in terms of NDCG@10 and HR@10. * indicates the statistical significance for $p < 0.05$ compared to the best baseline.

Dataset	Metric	BPR	PinSage	NGCF	LightGCN	SGL	HGT	HeCO	NMTR	MBGCN	MATA	KHGT	EHCF	CML	IICL	Imprv.
Tmall	HR	0.243	0.274	0.322	0.342	0.350	0.357	0.358	0.362	0.381	0.406	0.391	0.433	0.543	**0.558***	2.8%
	NDCG	0.143	0.151	0.184	0.205	0.210	0.210	0.199	0.215	0.213	0.225	0.323	0.262	0.327	**0.339***	3.7%
IJCAL-Context	HR	0.163	0.176	0.256	0.257	0.249	0.250	0.262	0.294	0.304	0.369	0.317	0.409	0.477	**0.487***	2.1%
	NDCG	0.085	0.091	0.124	0.122	0.123	0.119	0.121	0.161	0.160	0.209	0.182	0.237	0.283	**0.297***	4.9%
RetailRocket	HR	0.235	0.247	0.260	0.261	0.263	0.305	0.297	0.314	0.308	0.301	0.324	0.321	0.356	**0.368***	3.4%
	NDCG	0.146	0.139	0.140	0.152	0.165	0.176	0.178	0.201	0.181	0.181	0.202	0.207	0.222	**0.231***	4.1%

5.5   Ablation and Effectiveness Analyses (RQ2)

In this section, we further design ablation experiments to elucidate improvements in performance, demonstrating that IICL can address the two challenges mentioned in the introduction. Analysis details are summarized as:

Table 2. Ablation study on key components of IICL.

Variants	Tmall		IJCAL-Context		RetailRocket	
Metrics	HR	NDCG	HR	NDCG	HR	NDCG
w/o-MCL	0.4654	0.2748	0.3626	0.1971	0.3026	0.1854
w/o-ACL	0.5183	0.2916	0.4473	0.2572	0.3488	0.2099
w/o-SCL	0.5311	0.3128	0.4602	0.2713	0.3569	0.2154
w/o-AWN	0.5498	0.3302	0.4783	0.2913	0.3606	0.2219
IICL	**0.5582**	**0.3393**	**0.4873**	**0.2972**	**0.3684**	**0.2312**

Effect of Intra- and Inter-behavior Contrastive Learning. We first have to answer whether our proposed contrastive learning framework is effective? To this end, we generate model variant IICL(w/o)-MCL by removing all contrastive learning tasks including intra-behavior and inter-behavior. Instead, we rely solely on lightweight graph neural networks to learn behavioral relations. Table 2 presents the experimental results, we can find that IICL consistently outperforms IICL(w/o)-MCL, which is attributed to the fine-grained learning of preferences within behaviors and the additional informative signals generated by capturing commonalities between different behaviors. Furthermore, we also generate variants of IICL(w/o)-ACL by disabling intra-behavior contrastive learning. The joint analysis of IICL(w/o)-ACL and IICL shows that IICL performs better, which validates that fine-grained modeling of preferences for different behaviors helps to learn reliable and high-quality representations. These representations are more accurate and discriminating, which is beneficial for inter-behavior contrastive learning to extract behavioral commonalities to combat the data sparsity problem.

Effect of Semantic Neighbor-Enhanced Contrastive Learning. To investigate whether semantic neighbor contrastive learning improves structural neighbor contrastive learning, we propose another variant IICL(w/o)-SCL, which learns behavior-specific user preferences based only on structural neighbors. Obviously, IICL gets better performance over IICL(w/o)-ACL, which proves that semantic neighbor contrastive can filter out noisy information in structural neighbors and provide more accurate information for learning user preferences.

Effect of Adaptive Multi-behavior Weight Network. To verify the impact of adaptive weight network on our model, we generate variants of IICL(w/o)-AWN to perform ablation studies by removing the adaptive weight network. Instead, we employ a gating mechanism to uniformly integrate the contrastive losses across different behaviors. The performance degradation demonstrates that the adaptive weight network can discriminate the effects between different behaviors and retain the user personalized behavior preferences.

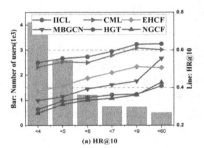 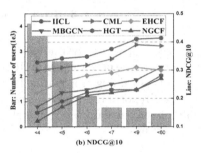

Fig. 2. Performance comparison w.r.t different interaction sparsity degrees on Tmall data.

5.6   Performance v.s. Sparsity Degrees (RQ3)

In this section, we aim to demonstrate the effectiveness of the model in alleviating the data sparsity problem. Specifically, we first divide the users in the Tmall dataset into six groups based on the number of interactions, where the total number of users in each group is shown on the left side of the y-axis. We use HR and NDCG to measure the model performance, which is obtained from the average of users in each group and shown on the right side of the y-axis. Due to space limitation, we selected a few representative baselines for comparison, and the results are shown in Fig. 2. We can know that: i). The performance of all methods improves as the number of interactions increases. It is because that sufficient interaction data brings rich supervised signals to learn more accurate preferences. ii). Multi-behavior methods achieve better performance compared to single-behavior methods, which demonstrates that considering behavioral diversity can alleviate data sparsity to some extent. iii). IICL consistently outperforms other methods under different number of interactions. This proves that IICL can better alleviate the data sparsity problem by mining more user behavior preferences and maintaining behavior heterogeneity through intra- and inter-behavior contrastive learning paradigms.

5.7   Hyper-parameter Analysis on IICL (RQ4)

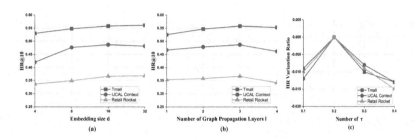

Fig. 3. Hyper-parameter analysis of IICL.

In this section, we study the effect of different settings of several key hyper-parameters on the performance in IICL, including the embedding dimension d, the number of embedding propagation layers l, and the temperature hyper-parameter τ. Figure 3 reports the experimental results. Each time we study the influence of only one hyper-parameter, and the other hyper-parameters are kept at their default settings.

Embedding Dimension d. In Fig. 3(a), it can be seen that IICL can achieve good performance with embedding dimension $16 \leq d \leq 32$. This demonstrates that IICL can accurately characterize users and items in a small hidden state dimensionality, which is attributed to the employ of contrastive learning to fully and effectively enhance multi-typed user-item interactions.

Embedding Propagation Layer l. We can observe that the model performance improves as l increases when $l \leq 3$. This validates the reasonableness of considering the collaborative signals of high-order neighbors. However, as l continues to increase, this may introduce additional noise into the user representation and cause over-smoothing problems [18].

The Temperature Hyper-parameter τ. τ is the parameter used to control the effect of contrastive learning discrimination. From Fig. 3(c), we can find that IICL achieves the best performance at $\tau = 0.2$. Inappropriate τ will reduce the effectiveness of the contrastive learning tasks and lead to poor performance.

6 Conclusion

In this work, we employ contrastive learning to enhance the MBR. Specifically, our IICL explores user structural neighbors and semantic neighbors within behaviors to obtain reliable embeddings. To alleviate the sparsity problem of target behavior, we capture transferable user-item relations between behaviors to incorporate auxiliary self-supervised signals into user modeling. In addition, the adaptive weight network customizes the integration of multiple losses to preserve the user personalized behavior patterns. Extensive experimental results validate the effectiveness of IICL and its superiority on the cold-start problem.

Acknowledgment. This study is supported by the Industrial Support Project of Gansu Colleges (No. 2022CYZC-11), National Natural Science Foundation of China (62276073, 6176028), Northwest Normal University Young Teachers Research Capacity Promotion plan (NWNU-LKQN2019-2), Natural Science Foundation of Gansu Province (21JR7RA114).

References

1. Chen, C., Zhang, M., Zhang, Y., Ma, W., Liu, Y., Ma, S.: Efficient heterogeneous collaborative filtering without negative sampling for recommendation. In: Proceedings of the AAAI Conference on Artificial Intelligence, vol. 34, pp. 19–26 (2020)
2. Gao, C., et al.: Neural multi-task recommendation from multi-behavior data. In: Proceedings of the 35th International Conference on Data Engineering (ICDE), pp. 1554–1557. IEEE (2019)
3. Gu, S., Wang, X., Shi, C., Xiao, D.: Self-supervised graph neural networks for multi-behavior recommendation. In: International Joint Conference on Artificial Intelligence (IJCAI) (2022)
4. He, X., Deng, K., Wang, X., Li, Y., Zhang, Y., Wang, M.: LightGCN: simplifying and powering graph convolution network for recommendation. In: Proceedings of the 43rd International ACM SIGIR Conference on Research and Development in Information Retrieval (SIGIR), pp. 639–648 (2020)
5. Hu, Z., Dong, Y., Wang, K., Sun, Y.: Heterogeneous graph transformer. In: Proceedings of The Web Conference 2020, pp. 2704–2710 (2020)
6. Jin, B., Gao, C., He, X., Jin, D., Li, Y.: Multi-behavior recommendation with graph convolutional networks. In: Proceedings of the 43rd International ACM SIGIR Conference on Research and Development in Information Retrieval (SIGIR), pp. 659–668 (2020)
7. Li, Q., Ma, H., Zhang, R., Jin, W., Li, Z.: Co-contrastive learning for multi-behavior recommendation. In: Proceedings of the PRICAI 2022: Trends in Artificial Intelligence, pp. 32–45 (2022)
8. Lin, Z., Tian, C., Hou, Y., Zhao, W.X.: Improving graph collaborative filtering with neighborhood-enriched contrastive learning. In: Proceedings of the ACM Web Conference 2022, pp. 2320–2329 (2022)
9. Oord, A.V.D., Li, Y., Vinyals, O.: Representation learning with contrastive predictive coding. arXiv preprint arXiv:1807.03748 (2018)
10. Rendle, S., Freudenthaler, C., Gantner, Z., Schmidt-Thieme, L.: BPR: Bayesian personalized ranking from implicit feedback. In: Proceedings of the Conference on Uncertainty in Artificial Intelligence (UAI), pp. 452–461 (2009)

11. Wang, X., He, X., Wang, M., Feng, F., Chua, T.S.: Neural graph collaborative filtering. In: Proceedings of the 42nd international ACM SIGIR Conference on Research and Development in Information Retrieval (SIGIR), pp. 165–174 (2019)
12. Wang, X., Liu, N., Han, H., Shi, C.: Self-supervised heterogeneous graph neural network with co-contrastive learning. In: Proceedings of the 27th ACM SIGKDD Conference on Knowledge Discovery & Data Mining, pp. 1726–1736 (2021)
13. Wei, W., Huang, C., Xia, L., Xu, Y., Zhao, J., Yin, D.: Contrastive meta learning with behavior multiplicity for recommendation. In: Proceedings of the Fifteenth ACM International Conference on Web Search and Data Mining, pp. 1120–1128 (2022)
14. Wei, Y., Ma, H., Wang, Y., Li, Z., Chang, L.: Multi-behavior recommendation with two-level graph attentional networks. In: Proceedings of the 27th International Conference on Database Systems for Advanced Applications (DASFAA), pp. 248–255 (2022)
15. Wu, J., et al.: Self-supervised graph learning for recommendation. In: Proceedings of the 44th International ACM SIGIR Conference on Research and Development in Information Retrieval (SIGIR), pp. 726–735 (2021)
16. Wu, Y., et al.: Multi-view multi-behavior contrastive learning in recommendation. In: International Conference on Database Systems for Advanced Applications, vol. 13246, pp. 166–182. Springer, Cham (2022). https://doi.org/10.1007/978-3-031-00126-0_11
17. Xia, L., Huang, C., Xu, Y., Dai, P., Zhang, B., Bo, L.: Multiplex behavioral relation learning for recommendation via memory augmented transformer network. In: Proceedings of the 43rd International ACM SIGIR Conference on Research and Development in Information Retrieval (SIGIR), pp. 2397–2406 (2020)
18. Xia, L., et al.: Knowledge-enhanced hierarchical graph transformer network for multi-behavior recommendation. In: Proceedings of the AAAI Conference on Artificial Intelligence, vol. 35, pp. 4486–4493 (2021)
19. Xia, X., Yin, H., Yu, J., Shao, Y., Cui, L.: Self-supervised graph co-training for session-based recommendation. In: Proceedings of the 30th ACM International Conference on Information & Knowledge Management, pp. 2180–2190 (2021)
20. Ying, R., He, R., Chen, K., Eksombatchai, P., Hamilton, W.L., Leskovec, J.: Graph convolutional neural networks for web-scale recommender systems. In: Proceedings of the 24th ACM SIGKDD International Conference on Knowledge Discovery & Data Mining, pp. 974–983 (2018)
21. Yu, J., Yin, H., Xia, X., Chen, T., Cui, L., Nguyen, Q.V.H.: Are graph augmentations necessary? Simple graph contrastive learning for recommendation. In: Proceedings of the 45th International ACM SIGIR Conference on Research and Development in Information Retrieval, pp. 1294–1303 (2022)
22. Zhang, R., Ma, H., Li, Q., Wang, Y., Li, Z.: Fire: knowledge-enhanced recommendation with feature interaction and intent-aware attention networks. Appl. Intell. 1–21 (2022). https://doi.org/10.1007/s10489-022-04300-x
23. Zhang, Y., Zhu, H., Song, Z., Koniusz, P., King, I.: COSTA: covariance-preserving feature augmentation for graph contrastive learning. In: Proceedings of the 28th ACM SIGKDD Conference on Knowledge Discovery and Data Mining, pp. 2524–2534 (2022)

Disentangled Contrastive Learning for Cross-Domain Recommendation

Ruohan Zhang, Tianzi Zang, Yanmin Zhu$^{(\boxtimes)}$, Chunyang Wang, Ke Wang, and Jiadi Yu

Department of Computer Science and Engineering,
Shanghai Jiao Tong University, Shanghai, China
{Zhangruohan,zangtianzi,yzhu,wangchy,onecall,jiadiyu}@sjtu.edu.cn

Abstract. Cross-Domain Recommendation (CDR) has been proved helpful in dealing with two bottlenecks in recommendation scenarios: data sparsity and cold start. Recent research reveals that identifying domain-invariant and domain-specific features behind interactions aids in generating comprehensive user and item representations. However, we argue that existing methods fail to separate domain-invariant and domain-specific representations from each other, which may contain noise and redundancy when treating domain-invariant representations as shared information across domains and harm recommendation performance. In this paper, we propose a novel **D**isentangled **C**ontrastive Learning for **C**ross-**D**omain **R**ecommendation framework (DCCDR) to disentangle domain-invariant and domain-specific representations to make them more informative. Specifically, we propose a *separate representation generation* component to generate separate domain-invariant and domain-specific representations for each domain. Next, We enrich the representations through multi-order collaborative information with GNNs. Moreover, we design a mutual-information-based contrastive learning objective to produce additional supervision signals for disentanglement and enhance the informativeness of disentangled representations by reducing noise and redundancy. Extensive experiments on two real-world datasets show that our proposed DCCDR model outperforms state-of-the-art single-domain and cross-domain recommendation approaches.

Keywords: Cross-domain Recommendation · Contrastive Learning · Disentangled Representation Learning · Graph Convolutional Networks

1 Introduction

Recommender systems have become key components for e-commerce and social media platforms, assisting users in accurately finding items they are potentially interested in amid overloaded information [36]. With the number of new items and users increasing, data sparsity and cold-start become two important issues hurting the efficacy of traditional recommender systems. Cross-domain recommendation (CDR) appears to be a solution to these two bottlenecks [35,42].

X. Wang et al. (Eds.): DASFAA 2023, LNCS 13944, pp. 163–178, 2023.
https://doi.org/10.1007/978-3-031-30672-3_11

Existing CDR approaches generally leverage rich information (i.e., ratings, feedback, tags, reviews) from source domains to improve recommendation performance in a target domain. Collective matrix factorization (CMF)-based methods [17,23,31] generate shared representations of users and items. Embedding and mapping-based methods [20] utilizing embedding mapping and dual knowledge transfer-based methods [9,14,15,40] have demonstrated efficacy in sharing knowledge across domains. Graph neural networks (GNNs)-based methods [4,12,32,37,41] in CDR utilize high-order collaborative information to improve performance. However, these methods consider that users have consistent interests across domains, so they only directly share common features when generating user representations, which ignores users may have different interests across domains. Recent approaches [3,10,16,33,38] consider domain-invariant and domain-specific features that are shared and distinct across domains [25]. They initialize two separate embeddings to represent domain-invariant and domain-specific features and transfer shared knowledge across domains to learn domain-invariant and domain-specific representations. Nonetheless, they fail to generate informative representations containing diverse semantics as it is difficult to distinguish between these two kinds of highly entangled features.

It is critical to separate domain-invariant and domain-specific representations from each other to enable them to contain more diverse semantic information. However, for existing CDR methods that identify domain-invariant and domain-specific features, their generated representations may have two main limitations as shown in Fig. 1, First, the domain-invariant representations may involve domain-specific features, which introduces noise caused by domain-specific features when treating domain-invariant representations as shared information to transfer, leading to the negative transfer problem [35]. Second, the domain-invariant and domain-specific representations may have some redundancy representing identical features, which decreases their expression ability greatly and results in sub-optimal results. Therefore, our method wants to separate domain-invariant and domain-specific representations to obtain more informative representations without noise and redundancy.

To reduce noise and redundancy contained in the generated domain-invariant and domain-specific representations and improve their informativeness, we are faced with the following challenges. The first challenge is *how to distinguish between domain-invariant and domain-specific features from user-item interactions*. Generally, user-item interactions are explicit ratings or implicit feedback in the datasets, so domain-invariant and domain-specific features behind user-item interactions are highly entangled, which makes it difficult to disentangle them. The second challenge is *how to enhance the informativeness of domain-invariant and domain-specific representations*. If domain-invariant and domain-specific representations are obtained, they may involve noise caused by irrelevant features, which hinders the effectiveness of sharing knowledge across domains and hurts recommendation performance. The generated representations may have some redundant information, which reduces the informativeness of representations.

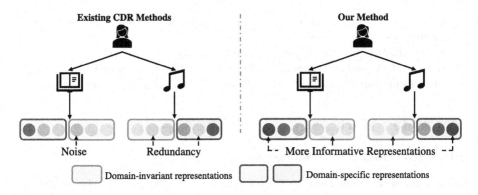

Fig. 1. Existing CDR models generate domain-invariant and domain-specific representations which contain noise and redundancy, whereas our methods generate more informative representations.

Therefore, it is difficult to remove noise and redundancy from the generated representations to make them more informative.

In this paper, we develop a new model called **D**isentangled **C**ontrastive Learning for **C**ross-**D**omain **R**ecommendation (DCCDR) to disentangle domain-invariant and domain-specific representations. To tackle the aforementioned challenges, we propose a *disentangled contrastive learning module* to generate more informative domain-invariant and domain-specific representations. To address the first challenge, we develop a *separate representation generation* component to generate separate domain-invariant and domain-specific representations. To enrich representations, we perform a GNNs-based approach to utilize high-order collaborative information in the user-item interaction graph. To deal with the second challenge, we develop a *representation informativeness enhancement* component to supervise the disentanglement and enhance the informativeness of representations by reducing noise and redundancy. Specifically, the contrastive learning objective maximizes the mutual information between domain-invariant representations of users across domains to reduce noise. To make disentangled representations contain more diverse semantics, the objective minimizes the mutual information between domain-invariant and domain-specific representations within domains and that between domain-specific representations across domains. Finally, we concatenate the domain-invariant and domain-specific representations to generate the final ones and predict the probability of given user-item pairs. In summary, the main contributions of this paper are as follows:

- We emphasize the importance of distinguishing domain-invariant and domain-specific features in cross-domain recommendation. A novel model DCCDR is proposed to disentangle domain-invariant and domain-specific representations to enable them to be more informative and contain diverse semantics.
- We develop the *representation informativeness enhancement* to supervise the disentanglement and enhance the informativeness of disentangled representations by reducing noise and redundancy. A mutual-information-based

contrastive learning objective is designed to add supervision signals for model training and representation enhancement.
- We conduct extensive experiments on real-world Amazon and Douban datasets. Comprehensive results demonstrate that our model significantly outperforms the state-of-the-art methods of cross-domain recommendation.

2   Related Work

2.1   Cross-Domain Recommendation

Cross-domain recommendation (CDR), which takes advantage of the abundant knowledge in source domains, often improves recommendation performance in a sparse target domain. Traditional approaches factorize rating matrices jointly and capture common user preferences [17,23]. As deep learning techniques gain popularity, methods use mapping functions [20], domain adaption skills [5,34], dual knowledge transfer mechanism [9,14,40], and graph neural networks (GNNs) [4,12,32,37,41] to transfer or leverage shared knowledge across domains. Recent approaches consider domain-invariant and domain-specific features when sharing knowledge across domains [3,10,16,33,38]. For instance, Zhao et al. propose MSDCR [38] to learn domain-specific and domain-invariant user preferences at the aspect level by transferring the user's complementary aspect preferences across domains. However, existing CDR approaches fail to separate domain-invariant and domain-specific representations from each other, which introduces noise and redundancy. Unlike prior research, Our DCCDR aims to disentangle domain-invariant and domain-specific representations to enhance their informativeness.

2.2   Disentangled Learning in Recommendation

Disentangled Learning [1] is proposed to learn distinct representations from multiple latent factors that influence the data, which is well aligned with recommendation tasks. Disentangled learning has proved effective in single-domain recommendation [18], sequential recommendation [19,39], and social recommendation [13] Researchers have shown that GNNs are effective to learn disentangled representations from graphs for recommendation. GNN-based disentangled methods leverage the interaction graph [26], the heterogeneous graph [27], and the knowledge graph [29]. Recently, learning disentangled representations has been introduced in CDR tasks [2]. Yet, the aforementioned works fail to separate domain-invariant and domain-specific representations from each other and enhance their informativeness, which leads to sub-optimal recommendation performance in CDR. These short-comings can be addressed by our proposed DCCDR model, which disentangles domain-invariant and domain-specific representations with generated supervised signals to make them contain more diverse semantics.

3   Notations and Problem Definition

Let D^A and D^B denote two distinct domains which share the same set of users denoted as $\mathcal{U}$. $\mathcal{I}^A$ and $\mathcal{I}^B$ denote non-overlapped sets of items in domain D^A

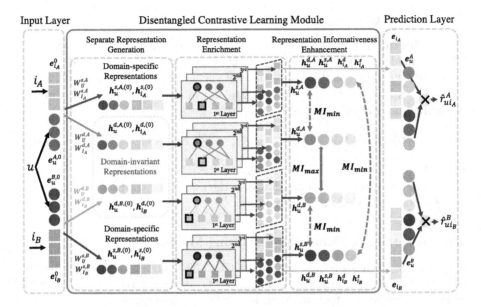

Fig. 2. An architecture overview of our model **DCCDR**. The core module of DCCDR is the *Disentangled Contrastive Learning Module*, which contains three key components: (1) the *Separate Representation Generation*, (2) the *Representation Enrichment*, and (3) the *Representation Informativeness Enhancement*.

and D^B, among which no item is in common. $|\mathcal{U}|$, $|\mathcal{I}^A|$, and $|\mathcal{I}^B|$ are the number of shared users and items in each domain. $\mathbf{G}^A = (\mathcal{U},\ \mathcal{I}^A,\ \mathbf{R}^A)$ and $\mathbf{G}^B = (\mathcal{U},\ \mathcal{I}^B,\ \mathbf{R}^B)$ denote the interaction graph in D^A and D^B separately, where $\mathbf{R}^A \in \mathbb{R}^{|\mathcal{U}| \times |\mathcal{I}^A|}$ and $\mathbf{R}^B \in \mathbb{R}^{|\mathcal{U}| \times |\mathcal{I}^B|}$ are user-item interaction matrices. $\mathbf{R}_{ui} = 1$ indicates an observed interaction between the user u and item i, otherwise 0.

With two domains D^A and D^B and a set of common users $\mathcal{U}$, two sets of non-overlapped items $\mathcal{I}^A$, $\mathcal{I}^B$, and corresponding interaction matrices $\mathbf{R}^A$ and $\mathbf{R}^B$ given, We consider Top-N recommendation with implicit feedback for all users in each domain. That is to say, we recommend a set of items $\mathcal{I}^a \subset \mathcal{I}^A$, $\mathcal{I}^b \subset \mathcal{I}^B$ that users have not interacted with but are most likely to be interested in to improve the recommendation performance in both domains simultaneously.

4   DCCDR

The main structure of our **D**isentangled **C**ontrastive learning networks for **C**ross-**D**omain **R**ecommendation (DCCDR) is shown in Fig. 2, which contains the *input layer*, the *disentangled contrastive learning module* and the *prediction layer*. In the following, we will introduce it in detail.

4.1   Input Layer

Considering that two domains share the same user set, the inputs of our proposed DCCDR model in the two domains are the user-item pairs (u, i_A) and (u, i_B). We denote both users and items by one-hot encodings, i.e., $\mathbf{x}_u \in \{0,1\}^{|\mathcal{U}|}$, $\mathbf{x}_{i_A} \in \{0,1\}^{|\mathcal{I}^A|}$, $\mathbf{x}_{i_B} \in \{0,1\}^{|\mathcal{I}^B|}$. We generate the initialized embeddings through embedding matrices $\mathbf{E}_{UA}, \mathbf{E}_{UB}, \mathbf{E}_{IA}, \mathbf{E}_{IB}$:

$$\mathbf{e}_u^{A,0} = \mathbf{E}_{UA}^T \mathbf{x}_u, \quad \mathbf{e}_u^{B,0} = \mathbf{E}_{UB}^T \mathbf{x}_u, \quad \mathbf{e}_{i_A}^0 = \mathbf{E}_{IA}^T \mathbf{x}_{i_A}, \quad \mathbf{e}_{i_B}^0 = \mathbf{E}_{IB}^T \mathbf{x}_{i_B}, \qquad (1)$$

where $\mathbf{E}_{UA} \in \mathbb{R}^{|\mathcal{U}| \times d}, \mathbf{E}_{UB} \in \mathbb{R}^{|\mathcal{U}| \times d}, \mathbf{E}_{IA} \in \mathbb{R}^{|\mathcal{I}^A| \times d}, \mathbf{E}_{IB} \in \mathbb{R}^{|\mathcal{I}^B| \times d}, \mathbf{e}_u^0$ and $\mathbf{e}_i^0$ denote the generated initial embeddings of the user u and the item i respectively, and d is the dimension of all embeddings.

4.2   Disentangled Contrastive Learning Module

This module is the core of our proposed DCCDR model, which aims to disentangle domain-invariant and domain-specific representations to make them more separate and informative. It is crucial to disentangle to reduce noise caused by irrelevant features when sharing domain-invariant representations across domains. It will also help to reduce the redundancy of these two representations to obtain informative representations with more diverse semantics. There are three problems faced with the disentanglement. The first one is how to extract domain-invariant and domain-specific features and generate two separate representations. The second one is how to leverage collaborative information to enrich representations. The third one is how to enhance the informativeness of disentangled representations. We propose a *separate representation generation* component, a *representation enrichment* component and a *representation informativeness enhancement* component to tackle these problems.

Separate Representation Generation. We employ latent space projection to generate separate domain-invariant and domain-specific representations. Specifically, for the user u, different from other recommendation methods [9, 23, 40] which capture a uniform user interest and generate a holistic representation, we extract domain-invariant and domain-specific features as two independent parts of user interest. Formally, we project the original user embedding $\mathbf{e}_u^0$ into different latent spaces:

$$\mathbf{h}_u^{d,(0)} = \sigma(\mathbf{W}_U^d \mathbf{e}_u^0), \qquad \mathbf{h}_u^{s,(0)} = \sigma(\mathbf{W}_U^s \mathbf{e}_u^0), \qquad (2)$$

where d and s denote the domain-invariant and domain-specific latent space respectively, W_U is the projection matrix and $\sigma(\cdot)$ is the activation function. The representations of each u would be composed of two parts, i.e., $\mathbf{e}_u^{(0)} = [\mathbf{h}_u^{d,(0)}, \mathbf{h}_u^{s,(0)}]$, where $\mathbf{h}_u^{d,(0)}$ and $\mathbf{h}_u^{s,(0)}$ denote the domain-invariant and domain-specific representation of the user u. Analogously, we can generate separate representations $\mathbf{h}_i^{d,(0)}, \mathbf{h}_i^{s,(0)}$ for item i.

Representation Enrichment. To enrich the representations, we leverage high-order collaborative information in the interaction graphs due to the effectiveness of GNNs [6,11,24] proved in recent studies. Research has shown that applying the embedding propagation mechanism on graph structure can extract useful information by aggregating information from neighbors and updating original nodes. The basic idea of Graph Convolutional Networks (GCNs) [11] is to learn node representations by smoothing features over the graph. It performs the following neighborhood aggregation iteratively to achieve a new representation of a target user node u with K convolutional layers:

$$e_u^{(k+1)} = AGG(e_u^{(k)}, e_i^{(k)} : i \in \mathcal{N}_u), \tag{3}$$

where k indicates the current convolutional layer, $e_u^{(k)}$ denotes the user embedding in the kth layer, $\mathcal{N}_u$ is the set of neighbors of u in the interaction graph and AGG symbolizes the chosen aggregation strategy.

We adopt LightGCN [7] to enrich domain-invariant and domain-specific representations in each domain due to its low number of parameters. Take the domain-invariant representations in domain D^A as an example and the aggregation for a user u and an item i_A can be summarized as follows:

$$\mathbf{h}_u^{d,A,(k+1)} = \sum_{i \in \mathcal{N}_u^A} \frac{1}{\sqrt{|\mathcal{N}_u^A||\mathcal{N}_i|}} \mathbf{h}_i^{d,(k)}, \quad \mathbf{h}_{i_A}^{d,(k+1)} = \sum_{u \in \mathcal{N}_{i_A}} \frac{1}{\sqrt{|\mathcal{N}_u^A||\mathcal{N}_{i_A}|}} \mathbf{h}_u^{d,A,(k)}. \tag{4}$$

To achieve more comprehensive representations from multi-order neighbors, we combine the embeddings learned in each layer since different embedding layers capture different semantics as follows:

$$\mathbf{h}_u^{d,A} = \mathbf{h}_u^{d,A,(0)} \ || \ ... \ || \ \mathbf{h}_u^{d,A,(K)}, \quad \mathbf{h}_{i_A}^{d} = \mathbf{h}_{i_A}^{d,(0)} \ || \ ... \ || \ \mathbf{h}_{i_A}^{d,(K)}, \tag{5}$$

where $||$ denotes the concatenation operation. The enrichment of domain-specific representations can be performed similarly. This enrichment process is the same for both D^A and D^B.

Representation Informativeness Enhancement. To supervise the disentanglement and enhance the informativeness of disentangled representations, we utilize self-supervised learning (SSL) [22,28,30] to generate extra supervision signals for model training. Specifically, this component adopts the mutual information (MI) maximization mechanism [21] to construct a contrastive learning objective. The core idea of mutual information maximization is to select *positive pairs* whose MI should be maximized (i.e., MI_{max} in Fig. 2) and *negative pairs* whose MI should be minimized (i.e., MI_{min} in Fig. 2).

To reduce noise caused by irrelevant features when treating domain-invariant user representations as knowledge to be shared across domains, we treat the two domain-invariant user representations of the same user as a *positive pair* and minimize the difference between them because intuitively users' domain-invariant features are shared and consistent. To reduce redundancy of the disentangled representations and enhance their informativeness, we propose two negative pairs for our designed contrastive learning objective. First, if domain-invariant and domain-specific representations contain distinguishable semantics in each domain, all the shared knowledge will be squeezed into domain-invariant representations. Hence, the domain-specific and domain-invariant representations of the same user in each domain are treated as a *negative pair*. Second, based on the assumption that all the common knowledge is supposed to be represented in domain-invariant representations, it is convincing that the domain-specific representations of users across domains should have little mutual information and can be considered as another *negative pair*. Therefore, the contrastive learning objective is determined as follows:

$$\mathcal{L}_{CL} = -log \frac{exp(f(\mathbf{h}_u^{d,A}, \ \mathbf{h}_u^{d,B}))/\tau)}{\substack{exp(f(\mathbf{h}_u^{d,A}, \ \mathbf{h}_u^{d,B}))/\tau) + exp(f(\mathbf{h}_u^{d,A}, \ \mathbf{h}_u^{s,A}))/\tau) + \\ exp(f(\mathbf{h}_u^{d,B}, \ \mathbf{h}_u^{s,B}))/\tau) + exp(f(\mathbf{h}_u^{s,A}, \ \mathbf{h}_u^{s,B}))/\tau)}}, \quad (6)$$

where $f(\cdot)$ is a function measuring the mutual information contained in representation and τ is a hyper parameter for softmax temperature. Here we utilize the cosine similarity and other functions can be adopted.

$$f(\mathbf{h}_u^d, \ \mathbf{h}_u^s) = cos(\mathbf{h}_u^d, \ \mathbf{h}_u^s) = \frac{{\mathbf{h}_u^d}^T \mathbf{h}_u^s}{||\mathbf{h}_u^d|| \ ||\mathbf{h}_u^s||}. \quad (7)$$

4.3   Prediction Layer

Through the *disentangled contrastive learning module*, we can obtain disentangled domain-invariant and domain-specific representations for users and items. To make representations more comprehensive, we adopt a concatenation operation to fuse the disentangled representations as follows:

$$\mathbf{e}_u^A = \mathbf{h}_u^{d,A} \ || \ \mathbf{h}_u^{s,A}, \quad \mathbf{e}_{i_A} = \mathbf{h}_{i_A}^d \ || \ \mathbf{h}_{i_A}^s, \quad (8)$$

where $\mathbf{e}_u^A$ and $\mathbf{e}_{i_A}$ are the final representation for user u and item i_A in D^A respectively. Finally, we utilize the dot product to calculate the probability of the interaction between u and i_A in D^A:

$$\hat{r}_{ui_A}^A = \hat{y}^A(u, \ i_A) = \sigma({\mathbf{e}_u^A}^T \mathbf{e}_{i_A}), \quad (9)$$

where σ is the sigmoid function to map real multiplication results to probability of interactions. Note that $\hat{r}_{ui_B}^B$ can be achieved through a similar process.

4.4   Model Training

We employ the *Bayesian Personalized Ranking* (BPR) loss, which is a typical pairwise loss encouraging a higher predicted probability of an observed interaction compared with unobserved ones.

$$\mathcal{L}_{BPR}' = - \sum_{(u,i)\in \mathbf{R}^+, (u,j)\in \mathbf{R}^-} ln \ \sigma(\hat{r}_{ui} - \hat{r}_{uj}), \quad (10)$$

where $'$ denotes the chosen domain D^A or D^B to estimate corresponding BPR loss, i.e., $\mathcal{L}^A_{BPR}$ and $\mathcal{L}^B_{BPR}$. $\mathbf{R}^+$ is the set of observed interactions between users and items while $\mathbf{R}^-$ is the sampled set of unobserved interactions. Considering the determined disentangled contrastive learning objective in Formula (6), the total joint loss function is defined as follows:

$$\mathcal{L} = \mathcal{L}^A_{BPR} + \mathcal{L}^B_{BPR} + \beta\mathcal{L}_{CL} + \lambda||\Theta||^2_2, \tag{11}$$

where β is the weight of $\mathcal{L}_{CL}$ and λ controls the L_2 regularization on the parameter set Θ to prevent overfitting.

5 Experiments

In this section, we discuss the experimental setup. We adopt two real-world datasets to conduct experiments and demonstrate the effectiveness of the proposed model. We expect to find answers to the following research questions.

- RQ1: Can our proposed model outperform other state-of-the-art approaches?
- RQ2: Do our designs aid in enhancing the performance of our model?
- RQ3: How do various hyper-parameter values affect our performance?
- RQ4: Are the representations we learned really disentangled?

5.1 Experimental Settings

Datasets We use two real-world datasets in our experiments to assess the performance of our proposed approach. The first dataset, **Amazon**[1], is the most often utilized for CDR. We select four domains for evaluation: "Movies and TV," "Digital Music," "Cell Phones and Accessories," and "Electronics" (abbreviated "Amazon-Movie," "Amazon-Music," "Amazon-Cell," and "Amazon-Elec"). The second dataset is the **Douban** dataset, which is crawled from the Douban website[2], a prominent online social network. There are three domains denoted as "Douban-Movie", "Douban-Music" and "Douban-Book", respectively. For each dataset, we treat the ratings of 4–5 as positive samples and others as negative ones, where each interaction is marked as 1 otherwise 0. For each task, we select the common users across both domains who have more than 3 interactions in each domain and limit each domain to less than or equal to 10000 items. Table 1 summarizes the detailed statistics of the datasets.

Evaluation Metrics. We adopt the leave-one-out strategy to evaluate our approach and baselines [40]. We randomly select one interaction as the test item for each user and determine hyper parameters by randomly sampling another interaction as the validated item. We randomly choose 999 negative items that are not interacted with by the user and rank the test item among the combined

---

[1] http://jmcauley.ucsd.edu/data/amazon/.
[2] https://www.douban.com.

Table 1. Experimental datasets and tasks.

Dataset	Domain	#Users	#Items	#Interaction	sparsity
Amazon	Movie	6995	10000	215299	99.69%
	Music	6995	10000	162779	99.77%
	Cell	7988	8455	86784	99.87%
	Elec	7988	9513	51398	99.93%
Douban	Movie	4494	10000	2038134	95.46%
	Book	4494	10000	303329	99.33%
	Music	6529	10000	704858	98.92%
	Book	6529	10000	467335	99.28%

1000 items because we are interested in the top-N recommendation tasks. This process is repeated five times, and the average ranking results are displayed. The recommendation performance is evaluated by two metrics: *Hit Ratio* (HR) and *Normalized Discounted Cumulative Gain* (NDCG). HR examines whether the test item is in the top-N ranking list, and NDCG measures the ranking quality by assigning higher scores to hits at top ranks [16].

Comparison Methods. We compare our proposed DCCDR model with both single-domain and cross-domain recommendation methods.

- **Single-domain recommendation. NeuMF** [8] combines matrix factorization (MF) and Deep Neural Networks (DNNs) to model user-item latent interactions. **LightGCN** [7] proposes a simplified neighborhood aggregation through normalized sum to generate representations.
- **Cross-domain recommendation. CMF** [23] jointly factorizes matrices and shares the latent factors of overlapped users. **DeepAPF** [33] captures both cross-site common and site-specific interests with weights learned by the attentional network. **CoNet** [9] introduces cross-connection units to conduct a dual knowledge transfer. **PPGN** [37] constructs a cross-domain preference matrix to maintain the cross-domain interactions and captures high-order connections. **BiTGCF** [16] conducts a bi-direction transfer learning through graph collaborative filtering. **MSDCR** [38] enhances domain-specific aspect preferences through adversarial training to form comprehensive preferences.

Experiment Setup. We use PyTorch to develop DCCDR[3], and all experiments are run on an NVIDIA TITAN Xp GPU. The dimension d of both the user and item embeddings is set to 64. The number of graph convolutional layers is set to 3. We set the disentangled contrastive learning objective's weight β to 0.001. With a learning rate of 0.001, we use the Adam optimizer in a mini-batch mode to update parameters. The batch size is set to 1024. Furthermore, to prevent

---

[3] https://github.com/wangshanyw/DCCDR.

Table 2. Performance comparison between **DCCDR** and different methods. Best baselines are underlined. $\star$ indicates the statistical significance for $p \leq 0.01$ compared with the best baseline method based on the paired t-test.

Metric	H@2	H@5	N@5	H@2	H@5	N@5	H@2	H@5	N@5	H@2	H@5	N@5
Datasets	Amazon-Movie			Amazon-Music			Amazon-Cell			Amazon-Elec		
NeuMF	0.045	0.090	0.057	0.064	0.117	0.076	0.051	0.080	0.056	0.047	0.079	0.055
LightGCN	0.120	0.232	0.141	0.150	0.229	0.171	0.141	0.260	0.170	0.155	0.247	0.180
CMF	0.085	0.150	0.101	0.106	0.176	0.124	0.127	0.192	0.141	0.109	0.159	0.119
DeepAPF	0.067	0.119	0.080	0.095	0.151	0.106	0.072	0.112	0.080	0.058	0.095	0.067
CoNet	0.153	0.237	0.170	0.151	0.235	0.169	0.119	0.176	0.132	0.122	0.179	0.135
PPGN	0.178	0.268	0.200	0.241	0.356	0.268	0.307	0.436	0.336	0.202	0.306	0.226
BiTGCF	0.179	0.283	0.204	0.230	0.351	0.257	0.250	0.359	0.272	0.233	0.338	0.258
MSDCR	<u>0.215</u>	<u>0.328</u>	<u>0.245</u>	<u>0.298</u>	<u>0.420</u>	<u>0.311</u>	<u>0.410</u>	<u>0.548</u>	<u>0.422</u>	<u>0.282</u>	<u>0.415</u>	<u>0.310</u>
DCCDR	0.259*	0.376*	0.285*	0.330*	0.471*	0.359*	0.476*	0.616*	0.493*	0.315*	0.465*	0.347*
Improv.	**21%**	**15%**	**16%**	**11%**	**12%**	**16%**	**16%**	**12%**	**17%**	**12%**	**12%**	**12%**
Datasets	Douban-Movie			Douban-Book			Douban-Music			Douban-Book		
NeuMF	0.088	0.160	0.105	0.092	0.173	0.112	0.096	0.175	0.115	0.097	0.178	0.118
LightGCN	0.093	0.166	0.112	0.135	0.227	0.155	0.122	0.119	0.143	0.119	0.160	0.158
CMF	0.060	0.106	0.070	0.086	0.152	0.101	0.079	0.142	0.096	0.086	0.152	0.103
DeepAPF	0.086	0.160	0.107	0.132	0.225	0.154	0.110	0.197	0.132	0.119	0.203	0.140
CoNet	0.237	0.357	0.265	0.243	<u>0.357</u>	0.264	0.134	<u>0.228</u>	0.170	0.151	0.221	<u>0.186</u>
PPGN	0.097	0.176	0.116	0.149	0.244	0.168	0.122	0.210	0.146	0.128	0.218	0.150
BiTGCF	0.084	0.160	0.105	0.184	0.277	0.203	0.146	0.224	0.167	<u>0.154</u>	<u>0.227</u>	0.176
MSDCR	<u>0.260</u>	<u>0.360</u>	<u>0.285</u>	<u>0.250</u>	0.354	<u>0.275</u>	<u>0.180</u>	0.217	<u>0.190</u>	0.148	0.201	0.163
DCCDR	0.309*	0.386*	0.314*	0.288*	0.375*	0.307*	0.198*	0.241*	0.211*	0.175*	0.238*	0.208*
Improv.	**19%**	**7%**	**10%**	**15%**	**6%**	**12%**	**10%**	**6%**	**11%**	**13%**	**5%**	**12%**

overfitting in graph convolution, we implement the message dropout mechanism during propagation with a dropout of 0.4 in training and disable it during testing.

5.2 Performance Comparison (RQ1)

Table 2 summarizes the results of our experiments on four tasks using HR@2 (H@2), HR@5 (H@5), and NDCG@5 (N@5). We can see that CDR methods (such as CMF, CoNet, and MSDCR) outperform single-domain recommendation approaches in general (i.e. NeuMF). This demonstrates how useful data across domains help improve recommendation performance. Moreover, in most circumstances, GNN-based recommender systems surpass non-graph recommendation methods (e.g., LightGCN vs NeuMF, PPGN, vs CoNet, etc.). This demonstrates the efficacy of using the graph to model high-order relationships. Additionally, among previous methods identifying domain-invariant and domain-specific features, MSDCR outperforms other approaches (i.e. DeepAPF, BiTGCF) but is still weaker than our model, which disentangles domain-invariant and domain-specific representations from each other to make them more informative. Our proposed DCCDR model performs optimally on all tasks. Over four pairs of

Table 3. Results of ablation study.

Metric	HR@2	HR@5	NDCG@5	HR@2	HR@5	NDCG@5
Variants	Amazon-Movie			Amazon-Music		
w/o.g	0.1459	0.2011	0.1620	0.1495	0.2106	0.1704
w/o.cl	0.1609	0.2213	0.1838	0.1740	0.2353	0.2165
w/o.dt	0.2357	0.3410	0.2630	0.3150	0.4647	0.3471
DCCDR	**0.2588**	**0.3763**	**0.2854**	**0.3297**	**0.4712**	**0.3590**

tasks, the average performance improvement is 15.2%, 13.5%, 11.5%, and 9.5%, demonstrating the effectiveness of our design.

5.3 Ablation Study (RQ2)

We conduct an ablation study to compare DCCDR with three variants to evaluate the effectiveness of each designed module in DCCDR. *w/o.srg* replaces the *separate representation generation* component with two initialized embeddings for domain-invariant and domain-specific representations for each domain. *w/o.g* replaces LightGCN in the *representation enrichment* component with matrix factorization (MF). *w/o.cl* removes the contrastive learning objective from the joint loss (i.e. $\beta = 0$). The outcomes of the experiments are shown in Table 3.

We can see that without GNNs leveraging multi-hop connections in graphs, *w/o.g* performs the poorest, with an average drop of 49.32%. This demonstrates the importance of modeling high-order relationships to enrich user and item representations. The recommendation performance of *w/o.cl* decreases by 41.93% on average, which demonstrates contrastive learning helps share information across domains and learn more informative disentangled representations. The drop of *w/o.dt* (i.e. 1.38–9.38%) demonstrates the importance of separating domain-invariant and domain-specific representations. Our DCCDR greatly outperforms all variants in terms of HR@2, HR@5, and NDCG@5, indicating each designed component truly contributes to performance improvement.

5.4 Impact of Hyper-parameter Settings (RQ3)

Impact of the weight β **of** $\mathcal{L}_{CL}$. We first investigate the impact of the weight β of contrastive learning objective $\mathcal{L}_{CL}$. Taking "Amazon-Cell↔Amazon-Elec" and "Douban-Music↔Douban-Book" as examples, we select {0.0001, 0.0005, 0.001, 0.002, 0.005, 0.01} as β respectively, and show experimental results in Fig 3. The recommendation performance improves as β increases and peaks at 0.001. Then the recommendation performance decreases as β becomes larger. We think this is because when β reaches 0.001, it keeps a good balance between

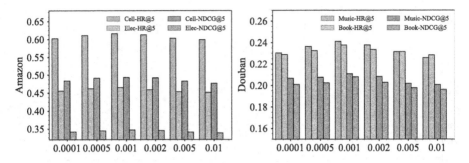

Fig. 3. Impact of the weight β of $\mathcal{L}_{CL}$.

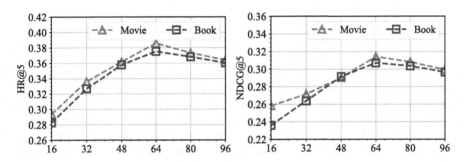

Fig. 4. Impact of the embedding size d.

BPR loss and contrastive learning objective, which helps learn more informative representations. Therefore, we set β as 0.001.

Impact of the Embedding Size d. We then investigate the impact of the dimension of user/item representations. Taking the "Douban-Movie$\leftrightarrow$ Douban-Music" task as an example, we range the embedding size within $\{16, 32, 48, 64, 80, 96\}$ and plot the results in Fig. 4. In the beginning, as the size of the embeddings increases, both HR@5 and NDCG@5 increase. This may be because a relatively large size of embeddings helps represent more semantics. The recommendation performance peaks when the dimension reaches 64 and declines when the size of embeddings is larger than 64. This may be caused by the overfitting of the model. Therefore, in our experiments, we set the embedding size to 64.

5.5   Visualization (RQ4)

We visualize the disentangled user representations to see if domain-invariant and domain-specific representations are separate from each other and contain diverse semantics. In the "Douban-Music$\leftrightarrow$Douban-Book" task, we randomly select 20 groups of disentangled user representations in both domains $\mathbf{h}_u^{d,A}$, $\mathbf{h}_u^{s,A}$, $\mathbf{h}_u^{d,B}$, $\mathbf{h}_u^{s,B}$. Using the t-SNE technique, we project the high-dimensional representations into 2D space. The visualization results are displayed in Fig. 5, where red

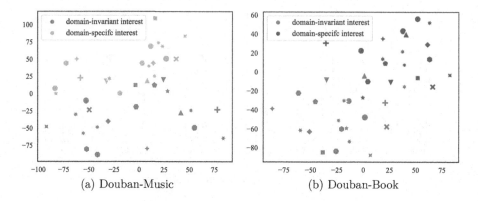

Fig. 5. Visualization of disentangled user representations.

represents domain-invariant interest and blue represents domain-specific one. Distinct shapes (i.e., squares and circles) represent different groups of disentangled user representations. We can see that the clustering centers of domain-invariant and domain-specific interests are separate, suggesting that interactions have distinguishable domain-invariant and domain-specific properties. Moreover, the representations of the same user (i.e., the same shape) are far apart, indicating that our model can disentangle user representations.

6   Conclusion and Future Work

In this paper, we propose DCCDR to disentangle domain-invariant and domain-specific representations to make them more informative. A *separate representation generation* component is designed to generate separate domain-invariant and domain-specific representations. We design a mutual-information based contrastive learning objective to generate supervised signals and enhance representation informativeness by reducing noise and redundancy. Extensive experiments demonstrate the effectiveness of our proposed model. Currently, we only disentangle representations based on the interaction graph and the disentanglement can be applied to other graphs (e.g. social graph). In the future, we will research more efficient ways to generate disentangled representations based on various types of information and further improve recommendation performance.

Acknowledgement. This research is supported in part by National Science Foundation of China (No. 62072304, No. 61772341, No. 61832013, No. 62172277, No. 62272320), Shanghai Municipal Science and Technology Commission (No. 19510760500, No. 21511104700, No. 19511120300), and Zhejiang Aoxin Co. Ltd.

References

1. Bengio, Y., Courville, A., Vincent, P.: Representation learning: a review and new perspectives. IEEE Trans. Pattern Anal. Mach. Intell. **35**(8), 1798–1828 (2013)
2. Cao, J., Lin, X., Cong, X., Ya, J., Liu, T., Wang, B.: DisenCDR: learning disentangled representations for cross-domain recommendation. In: SIGIR, pp. 267–277 (2022)
3. Chen, C., et al.: An efficient adaptive transfer neural network for social-aware recommendation. In: SIGIR, pp. 225–234 (2019)
4. Cui, Q., Wei, T., Zhang, Y., Zhang, Q.: HeroGRAPH: a heterogeneous graph framework for multi-target cross-domain recommendation. In: ORSUM@ RecSys (2020)
5. Duan, L., Tsang, I.W., Xu, D., Chua, T.S.: Domain adaptation from multiple sources via auxiliary classifiers. In: ICML, pp. 289–296 (2009)
6. Hamilton, W., Ying, Z., Leskovec, J.: Inductive representation learning on large graphs. In: Advances in Neural Information Processing Systems, vol. 30 (2017)
7. He, X., Deng, K., Wang, X., Li, Y., Zhang, Y., Wang, M.: LightGCN: simplifying and powering graph convolution network for recommendation. In: SIGIR, pp. 639–648 (2020)
8. He, X., Liao, L., Zhang, H., Nie, L., Hu, X., Chua, T.: Neural collaborative filtering. In: WWW, pp. 173–182 (2017)
9. Hu, G., Zhang, Y., Yang, Q.: CoNet: collaborative cross networks for cross-domain recommendation. In: CIKM, pp. 667–676 (2018)
10. Huang, L., Zhao, Z.L., Wang, C.D., Huang, D., Chao, H.Y.: LSCD: low-rank and sparse cross-domain recommendation. Neurocomputing **366**, 86–96 (2019)
11. Kipf, T.N., Welling, M.: Semi-supervised classification with graph convolutional networks. arXiv preprint arXiv:1609.02907 (2016)
12. Li, J., Peng, Z., Wang, S., Xu, X., Yu, P.S., Hao, Z.: Heterogeneous graph embedding for cross-domain recommendation through adversarial learning. In: Nah, Y., Cui, B., Lee, S.-W., Yu, J.X., Moon, Y.-S., Whang, S.E. (eds.) DASFAA 2020. LNCS, vol. 12114, pp. 507–522. Springer, Cham (2020). https://doi.org/10.1007/978-3-030-59419-0_31
13. Li, N., Gao, C., Jin, D., Liao, Q.: Disentangled modeling of social homophily and influence for social recommendation. In: TKDE (2022)
14. Li, P., Tuzhilin, A.: DDTCDR: deep dual transfer cross domain recommendation. In: WSDM, pp. 331–339 (2020)
15. Liu, J., et al.: Exploiting aesthetic preference in deep cross networks for cross-domain recommendation. In: Proceedings of The Web Conference 2020, pp. 2768–2774 (2020)
16. Liu, M., Li, J., Li, G., Pan, P.: Cross domain recommendation via bi-directional transfer graph collaborative filtering networks. In: CIKM, pp. 885–894 (2020)
17. Ma, H., Yang, H., Lyu, M.R., King, I.: SoRec: social recommendation using probabilistic matrix factorization. In: CIKM, pp. 931–940 (2008)
18. Ma, J., Zhou, C., Cui, P., Yang, H., Zhu, W.: Learning disentangled representations for recommendation. In: NeuIPS, vol. 32 (2019)
19. Ma, J., Zhou, C., Yang, H., Cui, P., Wang, X., Zhu, W.: Disentangled self-supervision in sequential recommenders. In: KDD, pp. 483–491 (2020)
20. Man, T., Shen, H., Jin, X., Cheng, X.: Cross-domain recommendation: an embedding and mapping approach. In: IJCAI, vol. 17, pp. 2464–2470 (2017)

21. Oord, A.V.D., Li, Y., Vinyals, O.: Representation learning with contrastive predictive coding. arXiv preprint arXiv:1807.03748 (2018)
22. Qiu, R., Huang, Z., Yin, H.: Memory augmented multi-instance contrastive predictive coding for sequential recommendation. In: ICDM, pp. 519–528 (2021)
23. Singh, A.P., Gordon, G.J.: Relational learning via collective matrix factorization. In: KDD, pp. 650–658 (2008)
24. Velickovic, P., Cucurull, G., Casanova, A., Romero, A., Lio, P., Bengio, Y.: Graph attention networks. Stat **1050**, 20 (2017)
25. Wang, K., Zhu, Y., Liu, H., Zang, T., Wang, C., Liu, K.: Inter-and intra-domain relation-aware heterogeneous graph convolutional networks for cross-domain recommendation. In: DASFAA, pp. 53–68 (2022)
26. Wang, X., Jin, H., Zhang, A., He, X., Xu, T., Chua, T.S.: Disentangled graph collaborative filtering. In: SIGIR, pp. 1001–1010 (2020)
27. Wang, Y., Tang, S., Lei, Y., Song, W., Wang, S., Zhang, M.: DisenHAN: disentangled heterogeneous graph attention network for recommendation. In: CIKM, pp. 1605–1614 (2020)
28. Wu, J., et al.: Self-supervised graph learning for recommendation. In: SIGIR, pp. 726–735 (2021)
29. Wu, J., et al.: DisenKGAT: knowledge graph embedding with disentangled graph attention network. In: CIKM, pp. 2140–2149 (2021)
30. Xia, X., Yin, H., Yu, J., Wang, Q., Cui, L., Zhang, X.: Self-supervised hypergraph convolutional networks for session-based recommendation. In: AAAI, vol. 35, pp. 4503–4511 (2021)
31. Xin, X., Liu, Z., Lin, C.Y., Huang, H., Wei, X., Guo, P.: Cross-domain collaborative filtering with review text. In: IJCAI (2015)
32. Xu, K., Xie, Y., Chen, L., Zheng, Z.: Expanding relationship for cross domain recommendation. In: CIKM, pp. 2251–2260 (2021)
33. Yan, H., Chen, X., Gao, C., Li, Y., Jin, D.: DeepAPF: deep attentive probabilistic factorization for multi-site video recommendation. TC **2**(130), 17–883 (2019)
34. Yu, W., Lin, X., Ge, J., Ou, W., Qin, Z.: Semi-supervised collaborative filtering by text-enhanced domain adaptation. In: KDD, pp. 2136–2144 (2020)
35. Zang, T., Zhu, Y., Liu, H., Zhang, R., Yu, J.: A survey on cross-domain recommendation: taxonomies, methods, and future directions. ACM Trans. Inf. Syst. **41**(2), 1–39 (2022)
36. Zang, T., Zhu, Y., Zhu, J., Xu, Y., Liu, H.: MPAN: multi-parallel attention network for session-based recommendation. Neurocomputing **471**, 230–241 (2022)
37. Zhao, C., Li, C., Fu, C.: Cross-domain recommendation via preference propagation GraphNet. In: CIKM, pp. 2165–2168 (2019)
38. Zhao, X., Yang, N., Yu, P.S.: Multi-sparse-domain collaborative recommendation via enhanced comprehensive aspect preference learning. In: WSDM, pp. 1452–1460 (2022)
39. Zheng, Y., et al.: Disentangling long and short-term interests for recommendation. In: Proceedings of the ACM Web Conference 2022, pp. 2256–2267 (2022)
40. Zhu, F., Chen, C., Wang, Y., Liu, G., Zheng, X.: DTCDR: a framework for dual-target cross-domain recommendation. In: CIKM, pp. 1533–1542 (2019)
41. Zhu, F., Wang, Y., Chen, C., Liu, G., Zheng, X.: A graphical and attentional framework for dual-target cross-domain recommendation. In: IJCAI, pp. 3001–3008 (2020)
42. Zhu, F., Wang, Y., Chen, C., Zhou, J., Li, L., Liu, G.: Cross-domain recommendation: challenges, progress, and prospects. arXiv preprint arXiv:2103.01696 (2021)

Intention-Aware User Modeling for Personalized News Recommendation

Rongyao Wang[1], Shoujin Wang[2], Wenpeng Lu[1(✉)], Xueping Peng[3], Weiyu Zhang[1], Chaoqun Zheng[1], and Xinxiao Qiao[1]

[1] School of Computer, Qilu University of Technology
(Shandong Academy of Sciences), Jinan, China
{wenpeng.lu,zwy,qxxyn}@qlu.edu.cn

[2] Data Science Institute, University of Technology Sydney, Sydney, Australia
shoujin.wang@uts.edu.au

[3] Australian Artificial Intelligence Institute, University of Technology Sydney, Sydney, Australia
xueping.peng@uts.edu.au

Abstract. Although tremendous efforts have been made in the field of personalized news recommendations, how to accurately model users' reading preferences to recommend satisfied news remains a critical challenge. In fact, users' reading preferences are often driven by his/her high-level goal-oriented *intentions*. For example, in order to satisfy the intention of traveling, a user may prefer to read news about national parks or hiking activities. However, existing methods for news recommendations often focus on capturing users' low-level preferences towards specific news only, neglecting to model their intrinsic reading intentions, leading to insufficient modeling of users and thus suboptimal recommendation performance. To address this problem, in this paper, we propose a novel intention-aware personalized news recommendation model (IPNR), to accurately model both a user's reading intentions and his/her preference for personalized next-news recommendations. In addition to modeling users' reading preferences, our proposed model IPNR can also capture users' reading intentions and the transitions over intentions for better predicting the next piece of news which may interest the user. Extensive experimental results on real-world datasets demonstrate that IPNR outperforms the state-of-the-art news recommendation methods in terms of recommendation accuracy (The source code is available at: https://github.com/whonor/IPNR).

Keywords: News recommendation · Intention-aware user modeling · User preference · Graph convolutional network

The work is partly supported by Key R&D Program of Shandong under Grant No. 2020CXGC010901, Nature Science Foundation of Shandong under Grant No. ZR2022MF243, National Nature Science Foundation of China under Grant No. 61502259, Program of Science and Technology of Qilu University of Technology under Grant No. 2021JC02010.

X. Wang et al. (Eds.): DASFAA 2023, LNCS 13944, pp. 179–194, 2023.
https://doi.org/10.1007/978-3-031-30672-3_12

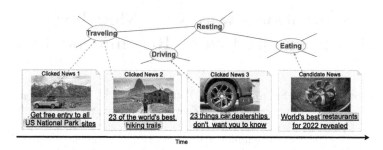

Fig. 1. An example of user reading behaviours with three clicked news and a candidate one. Each news reading behaviour is driven by a specific intention which is denoted by the word in the corresponding circle.

1 Introduction

With the prevalence of online news platforms, such as Apple News and Google News, users are overwhelmed with a large amount of online news covering various topics every day. This makes it difficult for users to quickly find out interesting news. To alleviate such an information overload problem, it is essential to recommend a small set of news that interests users according to their preferences for saving their time and improving their reading experience. Therefore, news recommender systems (NRS) have become a critical component of online news reading platforms, and they have attracted much attention from both industry and academia in recent years [1–3].

Many efforts have been devoted to news recommender systems and thus different research directions have been formed. Earlier methods strive to utilize topic models and collaborative filtering to represent news and users to recommend suitable news [4,5]. In recent years, deep learning has gained exceptional success in news recommendations, which focuses on modeling user reading preferences from different perspectives for recommendations. For example, some works focus on generating accurate news representations with convolutional neural networks (CNN) and attention mechanisms [6,7]. However, they only consider users' single and static preferences hidden in the user's reading history. To learn users' dynamic preferences over time, some works attempt to model users' long-term and short-term preferences with recurrent neural networks (RNN) [8] and graph neural networks (GNN) [9] respectively. However, they neglect to consider the diversity of users' preferences. More recently, to solve such a problem, some works have been done to model the potential multiple preferences with a parallel network [10] or poly attention mechanism [2]. Although these existing works have achieved better performance, they merely focus on modelling users' preferences, while neglecting to model users' intrinsic intentions which essentially drive users' reading behaviours and affect users' reading preferences. In practice, for a given user, his/her reading behaviours are often jointly determined by both his/her intentions and preferences [11]. Users' intentions are intrinsic and a high-level signal to indicate users' behaviour direction (e.g., to read political news or read economic news) while preferences are specific and a low-level signal to decide which specific news piece may be of the user's interest [12]. In most cases, users often first have a goal to read some topic/event-related news, such as economic news, which indicates their reading intentions. Then, for the same intention (e.g., to read economic news), different users may have different preferences towards specific news and thus may read different pieces of economic news. For instance, some users may like to read economic news in Wall Street Journal while others may read economic news in The Economist. Although both intentions and preferences are important for determining users' reading behaviours, most existing works fail to capture users' reading intentions, which inevitably results in insufficient modeling of users and thus sub-optimal news recommendation performance [13,14]. Actually, how to accurately model users' reading intentions is a very critical problem for NRSs.

Some pioneering works have attempted to capture user intentions in next-basket recommendations [15,16]. Moreover, some other related works strive to construct knowledge graphs to model interest transitions in session-based recommendations [17,18]. Although these works have achieved great success, they are devised for product recommendations only in the e-commerce domain. They often use a relatively simple model to handle product IDs without modelling any semantic and descriptive information [19]. This greatly differentiates them from news recommendations. Different from products, in addition to news IDs, a piece of news also contains an article with rich content and semantic information, which provide essential clues for us to capture users' reading intentions and preferences. Such an informative textual article should be well-modelled with more complicated and powerful models. This prevents the aforementioned works for product recommendations from being applied to news recommendations directly. Therefore, it is an urgent demand to devise novel news recommender systems with the capability of good modelling of both users' reading intentions and preferences.

Based on our observations, users' reading behaviours often show the following two unique characteristics. First, a user Tom may click multiple preferred news articles which are fit his/her high-level reading intentions. For example, as shown in Fig. 1, in the beginning, a given user has the intention of *traveling* and thus he preferred to click travel-related news, namely News 1 and 2, corresponding to national parks and hiking trails respectively. Apparently, the preferences towards news are driven by high-level intentions. Second, a user's reading intentions keep changing over time. For example, at the current time, Tom's intentions may be *traveling* and *driving*. However, the next time, his intention may transit to *resting* or *eating* as shown in Fig. 1. As a result, some news articles about restaurants may be preferred by Tom for his next click. Apparently, modelling the transitions of intentions is critical for accurate news recommendations.

To address the unique challenges triggered by the aforementioned unique characteristics in news recommendations, we propose a novel news recommendation framework called Intention-aware Personalized News Recommendation (IPNR). Thanks to the careful and unique design, IPNR is able to accurately and effectively model each user's reading intentions and his/her reading preferences for accurate next-news recommendations. In IPNR, first, a novel *reading intention module* based on graph neural networks is devised to generate the representation of the user's current reading intention by comprehensively modelling his/her intention transitions over time. At the same time, a novel *reading preference module* is devised to learn the representation of the user's reading preference from the historical news which has been read by the user. Then, a *gate network* is carefully designed to smartly aggregate the learned reading intention representation and reading preference representation together to form the informative user's representation, called intention-aware user representation in this work. Finally, a *prediction module* is designed to predict the click probability of each candidate news in the user's next click action. In this module, a special *candidate-aware attention* network was designed to more accurately select the useful information for prediction via taking the candidate news information as a guidance signal.

We summarize the main contributions of this work below:

- We propose modelling users' reading intentions for accurate next-news recommendations. As far as we know, this is the first work to comprehensively model users' reading intentions and their transitions for news recommendations.
- We devise a novel intention-aware personalized news recommendation model called IPNR, to simultaneously and effectively model users' reading intentions and reading preferences. IPNR not only models reading intention transitions over time but also detects reading preferences from users' reading history.
- A novel *reading intention module* is particularly designed to first detect possible reading intentions of a given user and then to model complex transitions of reading intentions over time to infer his/her next reading intention(s).

2   Related Work

2.1   Personalized News Recommender Systems

Personalized news recommendation is critical to improving the reading experience of users [20]. Many researchers have made a great effort to enhance the performance of news recommendations with various deep learning methods, including convolutional neural networks (CNNs), attention mechanisms, recurrent neural networks (RNNs) and graph neural networks (GNNs). For example, CNNs and 3D convolutions were utilized to encode fine-grained user representations and capture interactions between users and candidate news [21,22]. The self-attention mechanism was applied to select important information from reading history to model user preferences [13,23,24]. Multi-head self-attention was employed to detect potential multiple interests in parallel so as to model the diversity of user preferences [7,10]. RNNs were used to capture users' long- and short-term preferences from their recently browsed news, to learn the transitions of user preferences over time [8]. GNNs were utilized to model semantic interactions of news content and cluster-structural representation of users' reading history for further modeling the complex transitions of users' reading preferences [9,25]. Although these existing works have achieved better performances, they merely focus on modeling user preferences, while neglecting to model the intrinsical high-level intentions that drive users to click preferred news articles. This actually inspires us to explore users' reading intentions to recommend suitable news more accurately.

2.2   Intention-Aware Recommender Systems

Recently, how to capture users' potential intentions has received much attention in next-basket/session-based recommendations [26]. Some earlier works utilized RNNs together with specially-devised intention recognizers to capture user intentions in next-basket recommendations [15,16]. Recently, GNN-based methods are popular in recommendation communities. The LP-MRGNN model constructed multi-relational-item graphs over all sessions and employed GNNs to model interest transitions of users in session-based recommendations [17]. The ISRec model extracted user intentions from historical sequences and constructed an intention graph to model intention transitions [18]. The Satori model first constructed a heterogeneous graph with users, items, and categories as the user intention graph, then leveraged a graph attention network to model user intentions and preferences respectively [27]. Although the existing works have achieved great success by exploring the role of user intentions, they are carefully devised for product recommendations, whose inputs are IDs of products without any semantic information. Apparently, there is abundant text in news articles, which conveys enough semantic information to represent users' reading intentions. If the intentions are captured, they are beneficial to improve the performance of personalized news recommender systems. However, the existing works ignore the importance of users' reading intentions. To the best of our knowledge, there is no existing work that leverages both users' reading intentions and preferences simultaneously to recommend news articles.

3   Problem Formulation

A user-news interaction dataset consists of the interaction sequence of each user, which records users' reading or clicking behaviours with news articles. Let $\mathcal{D} = \{\mathcal{S}_1, \cdots, \mathcal{S}_u, \cdots, \mathcal{S}_{|\mathcal{U}|}\}$ denotes a user-news interaction dataset, where $\mathcal{U}$ refers to the set of all users, $\mathcal{S}_u$ means the interaction sequence of the user u. $\mathcal{S}_u = \{v_1, \cdots, v_t\}$ consists of t pieces of news which are sequentially interacted by the user u, where $v \in \mathcal{A}$, $\mathcal{A}$ refers to the set of all news articles.

For each user u, given the $(t-1)$ pieces of clicked news, denoted as a reading history $C_u = \{v_1, \cdots, v_{t-1}\}$, the goal of our proposed model $\mathcal{M}$ (i.e., IPNR) is to learn users' reading intentions and preferences from C_u and predict the click probability of each candidate news.

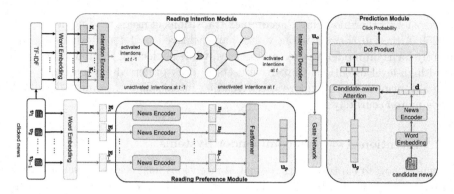

Fig. 2. Framework of IPNR, which mainly consists of four modules: a *reading intention module*, a *reading preference module*, a *gate network* and a *prediction module*. In the *reading intention module*, the nodes with/without colours indicate activated/inactivated intentions respectively.

4 The IPNR Model

4.1 Framework of IPNR

As illustrated in Fig. 2, our proposed IPNR model mainly contains four modules: (1) a *reading intention module*, which leverages GNNs to model the transitions of user intentions to generate user intention representations; (2) a *reading preference module*, which leverages CNNs and transformer networks to capture user preferences to generate user preference representations; (3) a *gate network*, which aggregates the two former representations to generate the intention-aware user representation; (4) a *prediction module*, which leverages a special *candidate-aware attention* network to incorporate candidate news features into the intention-aware user representation and predict the click probability of each candidate news.

IPNR is fed with a user's reading history $C_u = \{v_1, \cdots, v_i, \cdots, v_{t-1}\}$. We first map the content of each news v_i consisting of the news title, abstract and category, into the news embedding $\mathbf{E}_i$, which is initialized with the pre-trained Glove embeddings [28]. In addition, we utilize a TF-IDF component to extract keywords from the user's reading history, and then map the keywords of each news v_i into the keyword embedding $\mathbf{K}_i$. Afterwards, we feed the news embedding $\mathbf{E}_i$ and the keyword embedding $\mathbf{K}_i$ into the *reading preference module* and the *reading intention module*, respectively.

Users' intention graphs are critical for IPNR to leverage GNNs to model users' intentions. We denote a user's intention graph as $\mathcal{G} = (\mathcal{V}, \mathcal{E})$, where $\mathcal{V}$ denotes the set of nodes consisting of all available keywords and $\mathcal{E}$ denotes the set of edges including all directed edges. The keywords are extracted by TF-IDF from news content. Once a keyword can be matched with a concept in ConceptNet[1], it will

---

[1] http://conceptnet.io/.

be taken as a node in $\mathcal{G}$. The edges are connected and are set weights according to the semantic relations and weights in ConceptNet. We refer to these keyword nodes in the user's intention graph as possible intentions.

4.2   Reading Intention Module

To model a user's potential intentions from the browsed news sequences, we devise a *reading intention module* to infer the possible reading intentions at the current time $(t - 1)$ and then predict reading intentions at the next time (t) by modelling the intention transitions over the intention graph. This module contains an *intention encoder*, a *graph neural network* and an *intention decoder*.

Intention Encoder. The *intention encoder* aims to infer the possible intentions from a user's reading history. Taking the keyword embeddings of the user's clicked news $\{\mathbf{K}_1, \cdots, \mathbf{K}_i, \cdots, \mathbf{K}_{t-1}\}$ as the input, we first employ a CNN to encode these embeddings, described as:

$$\mathbf{c}_i = \mathrm{ReLU}\left(\mathbf{W}' * \mathbf{K}_{(i-f):(i+f)} + \mathbf{b}'\right), \tag{1}$$

where $\mathbf{K}_{(i-f):(i+f)}$ is the concatenation of the keyword embeddings from the position $(i - f)$ to $(i + f)$, $\mathbf{W}'$ is the learnable parameter of CNN filters, $\mathbf{b}'$ is the bias, $*$ indicates the convolutional operator, ReLU denotes a non-linear activation function. The output $\mathbf{c}_i$ is the convolutional keyword vector.

To infer the possible intentions from the keyword vectors of clicked news articles, we introduce a keyword embedding matrix $\mathbf{H}$ which is initialized by all node embeddings in the user's intention graph $\mathcal{G}$. Since only a part of keywords extracted from the news content belongs to the intention graph, we need to infer the possible keywords by capturing the relevance between the convolutional keyword vectors $\mathbf{C}_i = [\mathbf{c}_1, \mathbf{c}_2, \ldots, \mathbf{c}_{t-1}]$ and the keyword embedding matrix $\mathbf{H}$ to generate the possible keyword embedding matrix. Inspired by the work of He et al. [29], we first filter the possible keywords through a transformation matrix $\mathbf{M}$ which can learn the relations between the convolutional keyword vectors $\mathbf{C}_i$ and the keyword embedding matrix $\mathbf{H}$. Then we are able to obtain the possible keywords embedding matrix by calculating the distribution of the possible keywords in the keyword embedding matrix. The operations are specified as follows:

$$\mathbf{W}_p = \mathrm{softmax}\left(\mathbf{HMC}_i\right), \quad \mathbf{C}_p = \mathbf{W}_p\mathbf{H}, \tag{2}$$

where $\mathbf{M}$ is the learnable parameter, softmax is the normalized operator, $\mathbf{W}_p$ is the learnable weight matrix, $\mathbf{C}_p$ indicates the possible keyword matrix.

Graph Neural Network. Once the possible keyword matrix is ready, we feed it into a graph neural network. To learn the transitions of user intentions on his/her intention graph, we employ a graph convolutional network (GCN) inspired by Li et al. [18]. Specifically, the operation of the lth GCN layer is specified as follows:

$$\mathbf{H}^{l+1} = \mathrm{ReLU}\left(\widetilde{\mathbf{D}}^{-\frac{1}{2}}\widetilde{\mathbf{A}}\widetilde{\mathbf{D}}^{-\frac{1}{2}}\mathbf{H}^l\mathbf{W}^l\right), \tag{3}$$

where $\mathbf{H}^l$ is the node representation of the lth GCN layer, $\mathbf{W}^l$ is a learnable matrix in the lth layer, $\tilde{\mathbf{D}}$ is a diagonal degree matrix, $\tilde{\mathbf{A}} = \mathbf{A} + \mathbf{I}$, $\mathbf{I}$ is the identity matrix. To be specific, the input of the first layer is the keyword matrix $\mathbf{C}_p$, i.e., $\mathbf{H}^0 = \mathbf{C}_p$. Through n GCN layers, the user intention at the time (t) can be represented with $\mathbf{C}_t = \mathbf{H}^n$.

Intention Decoder. After $\mathbf{C}_t$ is built by GCN, we employ the self-attention mechanism to devise the decoder to generate the representation of the user's reading intentions, as follows:

$$\alpha_j = \frac{\exp\left(\varphi\left(\mathbf{c}_j\right)\right)}{\sum\limits_{l=1}^{R} \exp\left(\varphi\left(\mathbf{c}_l\right)\right)}, \quad \mathbf{u}_o = \sum_{j=1}^{R} \alpha_j \mathbf{c}_j, \tag{4}$$

where $\mathbf{c}_j$ indicates the jth row of $\mathbf{C}_t$, R is the number of keywords. The output $\mathbf{u}_o$ is the representation of the user's reading intentions.

4.3 Reading Preference Module

Since the intention is the high-level representation which is not able to model the specific user preferences, we devise the *reading preference module* to learn fine-grained user preferences that are also an important part of user representations. This module consists of a *news encoder* and a *Fastformer*. The former learns contextual news representations for each piece of clicked news, and the latter learns the representation of the user's reading preferences from a sequence of clicked news representations.

News Encoder. As shown in Fig. 3, once the ith news embedding $\mathbf{E}_i$ is obtained from the word embedding layer, we take it as the input of the *news encoder*. A news embedding contains various meta-news information, e.g., title, abstract, category and subcategory.

For the news title, we denote its embedding as $\mathbf{E}^t = \left[\mathbf{e}_1^t, \mathbf{e}_2^t, \ldots, \mathbf{e}_{|\mathbf{E}^t|}^t\right]$. To reserve position information in the sentence, we apply the positional embedding [30], e.g., $\mathbf{P}^w = \left[\mathbf{p}_1^w, \mathbf{p}_2^w, \ldots, \mathbf{p}_{|\mathbf{P}^w|}^w\right]$. Specifically, we firstly concatenate the title embedding and the positional embedding for each position x, e.g., $\mathbf{h}_x^w = \mathbf{e}_x^w + \mathbf{p}_x^w$, then feed its embedding into CNN aiming to capture important words:

$$\mathbf{t}_x = \text{ReLU}\left(\mathbf{W}_c * \mathbf{h}_{(x-f):(x+f)}^w + \mathbf{b}_c\right), \tag{5}$$

where $*$ indicates a convolutional operator, $\mathbf{W}_c$ is a learnable parameter, $\mathbf{h}_{(x-f):(x+f)}^w$ is the aggregation of word embeddings from the position $(x - f)$ to $(x + f)$.

Afterwards, an attention network is utilized to learn the final title representation $\mathbf{h}^t$ from the convolutional title representation $\mathbf{t}_x$ as follows:

$$\alpha_x^t = \text{softmax}\left(\mathbf{v}_t^\top \tanh\left(\mathbf{W}_t \mathbf{t}_x + \mathbf{b}_t\right)\right), \quad \mathbf{h}^t = \sum_{x=1}^{T} \alpha_x^t \mathbf{t}_x, \tag{6}$$

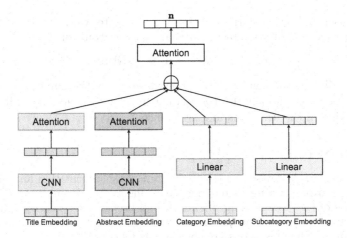

Fig. 3. Architecture of news encoder.

where $\mathbf{v}_t$, $\mathbf{W}_t$, $\mathbf{b}_t$ are learnable parameters, T is the length of a title. For an abstract, we can adopt the similar process mentioned above to generate the abstract representation $\mathbf{h}^a$. Inspired by the work of Wu et al. [31], a simple linear layer is applied to learn the category representation $\mathbf{h}^c$ and subcategory representation $\mathbf{h}^{sc}$. Finally, we aggregate all information and employ the attention mechanism to obtain the final news representation $\mathbf{n}$ as follows:

$$\mathbf{h} = \left[\mathbf{h}^t; \mathbf{h}^a; \mathbf{h}^c, \mathbf{h}^{sc}\right], \quad \alpha_x^h = \mathrm{softmax}\left(\mathbf{v}_h^\top \tanh\left(\mathbf{W}_h \mathbf{h}_x + \mathbf{b}_h\right)\right), \quad \mathbf{n} = \sum_{x=1}^{L} \alpha_x^h \mathbf{h}_x, \quad (7)$$

where $\mathbf{v}_h$, $\mathbf{W}_h$, $\mathbf{b}_h$ are learnable parameters, L is the length of meta news information, i.e., 4, $\mathbf{h}_x$ indicates one word embedding in a news embedding $\mathbf{h}$.

Fastformer. Aiming to model the informative behaviour interactions from a long news document, we utilize a state-of-the-art transformer network called *Fastformer* [32]. To be specific, we take the operation of an arbitrary attention head in *Fastformer* as example [33]. The *Fastformer* first aggregates global contexts into a query embedding $\mathbf{q}$. Next, it transforms the embedding of each token according to their relatedness with global contexts. Specifically,

$$\mathbf{q}_i = \mathbf{W}_q \mathbf{e}_i^w, \quad \mathbf{q} = \mathrm{Att}\left(\mathbf{q}_1, \mathbf{q}_2, \ldots, \mathbf{q}_N\right), \quad (8)$$

$$\mathbf{k}_i = \mathbf{W}_k \mathbf{e}_i^w, \quad \mathbf{k} = \mathrm{Att}\left(\mathbf{q} \odot \mathbf{k}_1, \mathbf{q} \odot \mathbf{k}_2, \ldots, \mathbf{q} \odot \mathbf{k}_N\right), \quad (9)$$

$$\mathbf{v}_i = \mathbf{W}_v \mathbf{e}_i^w, \quad \hat{\mathbf{e}}_i = \mathbf{W}_o\left(\mathbf{k} \odot \mathbf{v}_i\right), \quad (10)$$

where $\mathbf{W}_q$, $\mathbf{W}_k$, $\mathbf{W}_v$, $\mathbf{W}_o$ are learnable parameters, $\mathbf{e}_i^w$ indicate the ith token embedding, Att indicates the attention pooling network, $\odot$ indicates the element-wise product. $\hat{\mathbf{e}}_i$ indicates the output of ith token embedding in the sequence, which is generated by the current attention head. Afterwards, we build the reading preference representation by concatenating the outputs of all attention heads:

$$\mathbf{d}_k = [\hat{\mathbf{e}}_1^k; \hat{\mathbf{e}}_2^k; \cdots ; \hat{\mathbf{e}}_M^k], \quad \mathbf{u}_p = [\mathbf{d}_1; \mathbf{d}_2; \cdots ; \mathbf{d}_N], \tag{11}$$

where $[;]$ indicates the concatenation operation, M is the number of attention heads, N indicates the length of a reading history sequence, $\mathbf{u}_p$ indicates the representation of the user's reading preferences.

4.4   Gate Network

The *gate network* is devised to select the important information and aggregate the representations of a user's reading intentions and preferences. Once the representations of the user reading intentions and preferences (i.e., $\mathbf{u}_o$ and $\mathbf{u}_p$) are ready, we feed them into the *gate network* to generate the intention-aware user representation. The operations are specified as follows:

$$\mathbf{g} = \text{ReLU}\left(\mathbf{W}_g\left[\mathbf{u}_o; \mathbf{u}_p\right] + \mathbf{b}_g\right), \tag{12}$$

$$\mathbf{u}_g = \mathbf{g} \odot \tanh\left(\mathbf{V}\mathbf{u}_o + \mathbf{v}\right) + (1 - \mathbf{g}) \odot \mathbf{u}_p, \tag{13}$$

where $\mathbf{g}$ is a gate embedding, $\mathbf{u}_g$ is the intention-aware user representation.

4.5   Prediction Module

Before predicting the next news, we devise a *candidate-aware attention* to incorporate candidate news features into the intention-aware user representation. And then, take the candidate news representation $\mathbf{d}$ and the final user representation $\mathbf{u}$ as inputs, and we employ the dot product to predict the next piece of news. The operation is specified as follows:

$$\hat{\alpha} = \text{Att}\left(\mathbf{W}_Q\mathbf{d}, \mathbf{W}_K\mathbf{u}_g\right), \quad \mathbf{u} = \sum_{i=1}^{U} \hat{\alpha}_i \mathbf{u}_{g,i}, \quad \hat{y} = \mathbf{u}^\top \cdot \mathbf{d}, \tag{14}$$

where $\mathbf{W}_Q$, $\mathbf{W}_K$ are learnable parameters, $\mathbf{d}$ indicates the representation of the candidate news generated by the *news encoder* shown in Fig. 3, U is the length of a user's reading history, the output $\mathbf{u}$ is the final user representation, $\hat{y}$ indicates the click probability of the candidate news.

4.6   Model Training

We utilize negative sampling strategy to train our model and employ the log-likelihood function as a loss function:

$$\mathcal{L} = -\sum_{j=1}^{S} \log \frac{\exp\left(\hat{y}_j^+\right)}{\exp\left(\hat{y}_j^+\right) + \sum_{i=1}^{\mathcal{N}} \exp\left(\hat{y}_{j,i}^-\right)}, \tag{15}$$

where $\hat{y}_j^+$ indicates the probability of the jth positive sample, $\hat{y}_{j,i}^-$ indicates the probability of the ith negative sample w.r.t the jth positive sample, S is the number of the training positive samples, $\mathcal{N}$ is the number of negative samples.

5   Experiment and Evaluation

5.1   Dataset and Experimental Settings

Table 1. Performance Comparison with Baselines.

Model	MIND-small				MIND-200k			
	AUC	MRR	nDCG@5	nDCG@10	AUC	MRR	nDCG@5	nDCG@10
libFM	0.6001	0.2764	0.2992	0.3595	0.6116	0.2788	0.3006	0.3644
DKN	0.6394	0.2999	0.3246	0.3941	0.6543	0.3030	0.3316	0.3993
LSTUR	0.6611	0.3100	0.3419	0.4066	0.6752	0.3238	0.3610	0.4242
NRMS	0.6682	0.3184	0.3517	0.4158	0.6701	0.3185	0.3534	0.4175
TANR	0.6455	0.3107	0.3367	0.4017	0.6611	0.3148	0.3467	0.4114
NAML	0.6588	0.3092	0.3411	0.4058	0.6765	<u>0.3269</u>	0.3623	<u>0.4270</u>
NPA	0.6613	0.3174	0.3510	0.4140	0.6734	0.3259	0.3598	0.4228
NNR	<u>0.6771</u>	<u>0.3239</u>	<u>0.3592</u>	<u>0.4222</u>	<u>0.6828</u>	0.3252	<u>0.3634</u>	0.4266
IPNR	**0.6825**	**0.3247**	**0.3615**	**0.4236**	**0.6995**	**0.3401**	**0.3785**	**0.4414**
Improv.[a] (%)	0.80%	0.25%	0.64%	0.33%	2.45%	4.04%	4.16%	3.37%

[a] The improvement over the best-performing baselines which is underlined.

The real-world news recommendation dataset MIND[2] is utilized to conduct our experiments, which contains two versions: MIND-small and MIND-large. Due to MIND-large being quite large-scale and hard to process, following the previous works [25], we randomly sample 200,000 users' behaviour logs to build a new version named MIND-200k. Besides, as the limitation of licences, we can not obtain the labels of samples in the test set of MIND-large. Therefore, we randomly split half of the original validation set into a new validation set and a new test set respectively. For experimental settings, we apply Adam optimizer to optimize the process of training. The learning rate and dropout rate are set to $2e-5$, 0.2 respectively. The ratio of negative sampling S is set to 4 and the batch size is 64. The number of GCN layers is set to 5. We evaluate the performance of our model in terms of ROC curve (AUC), mean reciprocal rank (MRR) and normalized discounted cumulative gain (NDCG).

5.2   Baselines

We select eight state-of-the-art methods to compare with our model:[3] libFM [34], a classical matrix factorization model for news recommendations. DKN [35], a deep news recommender system, which enriches news content with external entities in a knowledge graph and employs a knowledge-aware CNN to generate news representations. TANR [36], a deep news recommender system with topic-aware news representations, which employs CNNs and attention networks to learn news representations, jointly optimized with an auxiliary topic classification task.

---

[2] https://msnews.github.io.

[3] We adopt the official code to re-implement all baselines on the datasets.

NAML [31], an attentive multi-view recommendation model, which learns news representation from multiple kinds of news information with CNNs and attention mechanisms. NRMS [7], a deep news recommendation model, which utilizes multi-head self-attention to model news representations from news titles, and employs multi-head self-attention to capture the relatedness of browsed news to generate user representations. LSTUR [8], a neural news recommendation model based on short- and long-term user interests, which employs gated recurrent network (GRU) and user IDs to generate the representations of user's short- and long-term interests. NPA [6], a neural news recommendation model based on personalized attention, which devises a personalized attention network to recognize the important words in news content according to user preferences. NNR [25], a deep recommender system based on collaborative news encoding (CNE) and structural user encoding (SUE), which employs biLSTM and cross-attention to realize CNE and utilizes GCNs to implement SUE.

5.3 Performance Comparison with Baselines

The recommendation performance of our proposed IPNR and those of eight baselines are reported in Table 1. We have the following observations.

First, the traditional method based on matrix factorization (i.e., libFM) perform worse significantly than the other deep neural methods. This demonstrates the superiority of deep models in handling the news and users data, which can capture more sophisticated potential features in news articles and reading history.

Second, among the deep models, DKN perform worst. Although DKN introduces external entities to enrich news representation, it employs a news-level attention network which can not capture the important word-level information within a news article, and thus it performs badly.

Third, out of the last six baselines, LSTUR employs GRUs and user IDs to model users' short- and long-term preferences. It merely utilizes IDs to embed long-term preferences, which is hard to capture enough user information. NRMS utilizes multi-head self-attention to capture word and news interactions to generate news and user representations. However, it merely utilizes news title information, which inevitably misses the important features contained in news abstract and content. TANR, NAML and NPA employ various CNNs and attention mechanisms to capture important information to generate news and user representations. However, they rely on CNNs, which are not good at modelling sequential features contained in reading history and thus limit their performance. NNR utilizes biLSTM and cross-attention to realize collaborative news encoding and employs GCNs to implement structural user encoding. With the support of GCNs, NNR can capture more structural user features and can show more powerful performance than the other baselines. Interestingly, NAML performs better than NNR in terms of MRR and nDCG@10 on the MIND-200k dataset. This may be because multi-view news information adopted by NAML is efficient to model accurate news representations over a large number of news articles.

Finally, our proposed IPNR achieves the best performance on both datasets. This verifies the superiority of IPNR, which not only effectively models users' reading preferences and efficiently learns users' reading intentions to perform personalized news recommendations. Specifically, IPNR significantly outperforms the best-performing news recommendation methods (e.g., NNR) with an average of 1.98% in terms of all metrics on two datasets.

5.4 Ablation Study

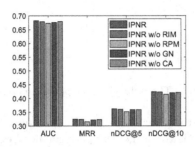

Fig. 4. Performance comparison with variants.

We design the variants of our proposed IPNR to analyze the effectiveness of each key module. According to Fig. 2, we remove the reading intention module (RIM), reading preference module (RPM) and gate network (GN) respectively to obtain three variants, i.e., *IPNR w/o RIM, IPNR w/o RPM* and *IPNR w/o GN*. Further, we remove the component of the *candidate-aware attention* in the prediction module to obtain *IPNR w/o CA*.

Because training on the MIND-200k dataset requires too expensive GPU cost, we only utilize the MIND-small dataset to conduct the ablation study. As shown in Fig. 4, we can see that removing each module leads to suboptimal performance. This demonstrates that each module is critical and effective in IPNR. Comparing *IPNR w/o RIM* and *IPNR w/o RPM*, the former is better than the latter. This demonstrates that modelling preferences are more important than modelling intentions for IPNR. This is reasonable because that RPM captures more abundant news information, including titles, abstracts, etc., while RIM merely depends on the keywords extracted from news content. RIM can be applied as the complement of RPM, but can not replace RPM. Besides, the performance decrease of *IPNR w/o GN* and *IPNR w/o CA* means that the aggregation method of different representations also plays a key role in IPNR.

5.5 Hyperparameter Analysis

Influence of Number of Keywords. Due to the keywords are taken as the input of *intention encoder* in IPNR, we evaluate the influence of different numbers of keywords as shown in the upper-left panel in Fig. 5. According to these results, when the number is set to 5, our model can achieve the best performance. When the number of keywords is too small, the performance is worse. This is because a few keywords can not provide enough key semantic information to the

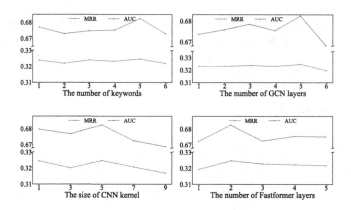

Fig. 5. The influence of hyperparameters.

intention encoder, and thus hurt the modelling of potential intentions. When the number of keywords is too large, the performance begins to decline. This is because too many keywords are easy to induce noise intention information.

Influence of Number of GCN Layers. In order to explore the influence of GCN layers, we conduct several experiments as shown in the upper-right panel in Fig. 5. When the number of layers is set to 5, our model can achieve the best performance. This is probably because GCN can effectively aggregate neighbour information through 5 layers, which benefits the transitions of user intentions. When the number of layers is too small or too large, the performance will decrease. This is probably because GCN fails to efficiently capture the transitions of user intentions under the inappropriate settings on layers.

Influence of CNN Kernel Size. Because CNN is utilized in the *news encoder* in IPNR, we explore the influence of different kernel sizes in convolutional networks. As shown in the lower-left panel in Fig. 5, when the kernel size is set to 5, our model can achieve the best performance. When the kernel size is too small, our model performs worse. This is because useful information can not be fully captured, and thus it is difficult to learn accurate news representations. However, when the kernel size is set to too large, the performance of our model consistently declines. This may be because some noisy information hurts news representations.

Influence of Number of Fastformer Layers. As Fastformer is a key component in IPNR, we analyze the influence of the number of Fastformer layers. As shown in the lower-right panel in Fig. 5, when the number of Fastformer layers is set to 2, our model achieves the best performance. When the number of Fastformer layers becomes too large, the performance of our model begins to decline. The reason may be that too many layers cause the over-smoothing issue. In the case of the model with one Fastformer layer, the performance is worse. This is because one Fastformer layer can not support IPNR to accurately learn the representation of a user's reading preferences.

6 Conclusion

In this work, we pay attention to a novel and important research problem: how to effectively model a user's reading intentions for next-news recommendations. In order to solve this problem, we have proposed an intention-aware personalized news recommendation (IPNR) model to accurately model both a user's reading intentions and his/her preferences. Extensive experimental results on real-world datasets demonstrate IPNR outperforms the state-of-the-art news recommendation methods in terms of recommendation accuracy. In future, we will investigate more effective methods (e.g., pre-trained language model) to further enhance user representation learning.

References

1. Qiu, Z., Hu, Y., Wu, X.: Graph neural news recommendation with user existing and potential interest modeling. ACM Trans. Knowl. Discov. Data **16**(5), 1–17 (2022)
2. Li, J., Zhu, J., Bi, Q., et al.: MINER: multi-interest matching network for news recommendation. In: ACL, pp. 343–352 (2022)
3. Wang, S., Pasi, G., Hu, L., Cao, L.: The era of intelligent recommendation: editorial on intelligent recommendation with advanced AI and learning. IEEE Intell. Syst. **35**(5), 3–6 (2020)
4. Bansal, T., Das, M., Bhattacharyya, C.: Content driven user profiling for comment-worthy recommendations of news and blog articles. In: RecSys, pp. 195–202 (2015)
5. Ge, S., Wu, C., et al.: Graph enhanced representation learning for news recommendation. In: WWW, pp. 2863–2869 (2020)
6. Wu, C., Wu, F., An, M., Huang, J., Huang, Y., Xie, X.: NPA: neural news recommendation with personalized attention. In: KDD, pp. 2576–2584 (2019)
7. Wu, C., Wu, F., Ge, S., Qi, T., Huang, Y., Xie, X.: Neural news recommendation with multi-head self-attention. In: EMNLP, pp. 6390–6395 (2019)
8. An, M., Wu, F., Wu, C., Zhang, K., Liu, Z., Xie, X.: Neural news recommendation with long-and short-term user representations. In: ACL, pp. 336–345 (2019)
9. Hu, L., Li, C., et al.: Graph neural news recommendation with long-term and short-term interest modeling. Inf. Process. Manage. **57**(2), 102142 (2020)
10. Wang, R., Wang, S., Lu, W., Peng, X.: News recommendation via multi-interest news sequence modelling. In: ICASSP, pp. 7942–7946 (2022)
11. Alley, S.J., Schoeppe, S., Rebar, A.L., et al.: Age differences in physical activity intentions and implementation intention preferences. J. Behav. Med. **41**(3), 406–415 (2018)
12. Albarracin, D., Wyer Jr., R.S.: The cognitive impact of past behavior: influences on beliefs, attitudes, and future behavioral decisions. J. Pers. Soc. Psychol. **79**(1), 5 (2000)
13. Qi, T., Wu, F., Wu, C., et al.: HieRec: hierarchical user interest modeling for personalized news recommendation. In: ACL, pp. 5446–5456 (2021)
14. Wang, J., Chen, Y., Wang, Z., Zhao, W.: Popularity-enhanced news recommendation with multi-view interest representation. In: CIKM, pp. 1949–1958 (2021)
15. Wang, S., Hu, L., et al.: Intention nets: psychology-inspired user choice behavior modeling for next-basket prediction. In: AAAI, pp. 6259–6266 (2020)

16. Wang, S., Hu, L., Wang, Y., et al.: Intention2Basket: a neural intention-driven approach for dynamic next-basket planning. In: IJCAI, pp. 2333–2339 (2021)
17. Wang, W., Zhang, W., Liu, S., et al.: Incorporating link prediction into multi-relational item graph modeling for session-based recommendation. IEEE Trans. Knowl. Data Eng. **35**(3), 2683–2696 (2021)
18. Li, H., Wang, X., Zhang, Z., Ma, J., Cui, P., Zhu, W.: Intention-aware sequential recommendation with structured intent transition. IEEE Trans. Knowl. Data Eng. **34**(11), 5403–5414 (2022)
19. Wang, S., Cao, L., Hu, L., et al.: Hierarchical attentive transaction embedding with intra-and inter-transaction dependencies for next-item recommendation. IEEE Intell. Syst. **36**(4), 56–64 (2020)
20. Wang, S., Xu, X., Zhang, X., et al.: Veracity-aware and event-driven personalized news recommendation for fake news mitigation. In: WWW, pp. 3673–3684 (2022)
21. Wang, H., Wu, F., Liu, Z., Xie, X.: Fine-grained interest matching for neural news recommendation. In: ACL, pp. 836–845 (2020)
22. Lu, W., Wang, R., Wang, S., et al.: Aspect-driven user preference and news representation learning for news recommendation. IEEE Trans. Intell. Transp. Syst. **23**(12), 25297–25307 (2022)
23. Qi, T., Wu, F., Wu, C., Huang, Y.: News recommendation with candidate-aware user modeling. In: SIGIR, pp. 1917–1921 (2022)
24. Lu, W., et al.: Chinese sentence semantic matching based on multi-level relevance extraction and aggregation for intelligent human-robot interaction. Appl. Soft Comput. **131**, 109795 (2022)
25. Mao, Z., Zeng, X., Wong, K.F.: Neural news recommendation with collaborative news encoding and structural user encoding. In: EMNLP, pp. 46–55 (2021)
26. Wang, S., Zhang, X., Wang, Y., et al.: Trustworthy recommender systems. arXiv preprint arXiv:2208.06265 **1**(1), 1–16 (2022)
27. Chen, J., Cao, Y., Zhang, F., Sun, P., Wei, K.: Sequential intention-aware recommender based on user interaction graph. In: ICMR, pp. 118–126 (2022)
28. Pennington, J., Socher, R., Manning, C.D.: Glove: global vectors for word representation. In: EMNLP, pp. 1532–1543 (2014)
29. He, R., Lee, W.S., Ng, H.T., Dahlmeier, D.: An unsupervised neural attention model for aspect extraction. In: ACL, pp. 388–397 (2017)
30. Vaswani, A., et al.: Attention is all you need. In: NeurIPS, pp. 6000–6010 (2017)
31. Wu, C., Wu, F., An, M., Huang, J., Huang, Y.: Neural news recommendation with attentive multi-view learning. In: IJCAI, pp. 3863–3869 (2019)
32. Wu, C., Wu, F., Qi, T., Huang, Y., Xie, X.: FastFormer: additive attention can be all you need. arXiv preprint arXiv:2108.09084 **1**(1), 1–11 (2021)
33. Qi, T., Wu, F., Wu, C., Huang, Y.: FUM: fine-grained and fast user modeling for news recommendation. In: SIGIR, pp. 1974–1978 (2022)
34. Koren, Y., Bell, R., Volinsky, C.: Matrix factorization techniques for recommender systems. Computer **42**(8), 30–37 (2009)
35. Wang, H., Zhang, F., Xie, X., Guo, M.: DKN: deep knowledge-aware network for news recommendation. In: WWW, pp. 1835–1844 (2018)
36. Wu, C., Wu, F., An, M., Huang, Y., Xie, X.: Neural news recommendation with topic-aware news representation. In: ACL, pp. 1154–1159 (2019)

Deep User and Item Inter-matching Network for CTR Prediction

Zhiyang Yuan[1,2], Yingyuan Xiao[1,2]($\boxtimes$), Peilin Yang[1,2], Qingbo Hao[1,2], and Hongya Wang[3]

[1] Engineering Research Center of Learning-Based Intelligent System, Ministry of Education, Tianjin University of Technology, Tianjin 300384, China
yyxiao@tjut.edu.cn
[2] Tianjin Key Laboratory of Intelligence Computing and Novel Software Technology, Tianjin University of Technology, Tianjin 300384, China
[3] College of Computer Science and Engineering, Donghua University, Shanghai, China
hywang@dhu.edu.cn

Abstract. CTR prediction plays an important role in increasing company revenue and user experience, and many efforts start with historical behavior to uncover user interest. There are two main problems with previous works: (1) When most previous works mined interests from users' historical behaviors, they only focus on implicit or explicit interests. (2) When most previous works mined user interests through the relationship between target users and similar users, the representation of target users was not rich and accurate enough, resulting in less accurate item recommendations. Therefore the Deep User and Item Inter-Matching Network (DUIIN) is proposed in this paper to solve the above problems. First, we design Item-to-Item Network (IIN), using two different sub-networks Evolving Interest Network (EIN) and Feature Interaction Network (FIN) to mine the hidden interests and explicit interests shown by users' historical behaviors, respectively. Then the User-to-User Network (UUN) is designed to mine user interests through the relationship between target users and similar users after representing the target users more accurately and richly. The experimental results show that the DUIIN model proposed in this paper performs better than other state-of-the-art models.

Keywords: CTR Prediction · Recommendation · Deep learning · Behavior sequence

1 Introduction

In recent years, more and more work has been devoted to CTR prediction. FM [11] and FFM [6] learned the weights between different feature pairs. WDL [2], Deep & Cross Network [13], and Autoint [12] used two sub-networks to mine both low-order and high-order feature interactions. CAN [1] used a method similar to the MLP structure to mine the relationship between different feature

X. Wang et al. (Eds.): DASFAA 2023, LNCS 13944, pp. 195–204, 2023.
https://doi.org/10.1007/978-3-031-30672-3_13

pairs. DIN [20] and DIEN [21] used attention mechanisms and GRU to tap into the interest shown by user behavior. DSIN [3] used Bi-LSTM to mine the relationship between different sessions. DMIN [19] used multi-head self-attention to uncover the diversity of interests contained in users at the same time as the latent dominant interest hidden in user behavior. These methods mentioned above mine user interests from the perspective of feature interactions and user behavior sequences, respectively, but they are not good at mining the hidden information in item behavior.

Subsequently, a number of works began to focus on the item behavior. DIB [4] learned the dynamic vector representation of an item. TIEN [7] used GRU to mine users who are similar to the target users. These methods mainly mine the relationship between target users and similar users to find the interests of target users, but the mining of users' historical behaviors is not comprehensive enough.

Based on the above observations, we propose the Deep User and Item Inter-Matching Network (DUIIN) to solve the above proposed problem.

The main contributions of this paper are as follows:

- We design a model called DUIIN, using two sub-networks to mine user's historical behavior and item historical behavior respectively.
- In IIN, first EIN uses multi-head self-attention and new PAGRU to mine users' hidden interests over time then FIN uses a structure similar to MLP to mine users' explicit interests.
- In UUN, we richly represent the target user and find the association between similar users and him, so as to infer the interest preferences of the target user.
- We evaluate our proposed model on four public datasets in terms of CTR prediction. The results show Deep User and Item Inter-Matching Network (DUIIN) outperforms the state-of-the-art solution.

2   Related Work

2.1   User History Behavior

PNN [10] introduced a product layer. ONN [16] assigned different feature vectors to different features. DREAM [17] used RNN structure to mine the sequential behavior of a user's historical purchase records. DHAN [14] found that users' interests are transformed from broad directions to small directions. DMR [9] used two networks to process historical user behavior and target items in two different ways. These aforementioned methods mine the explicit interests and the hidden interests. However, the information hidden between similar users in the item behavior sequence is ignored.

2.2 Item History Behavior

DUMN [5] used a User Representation Layer to find correlations between the target and similar users. DRINK [18] used a series of interactions, using interactive users and timestamps to represent candidate items, while focusing on the multi-representational characteristics of the item. The above methods mine the information in the item behavior sequence, but the information in the user behavior is not sufficiently and comprehensively mined.

Based on these observations, inspired by DIEN [21], CAN [1], DRINK [18], we design the Deep User and Item Inter-Matching Network(DUIIN) model, which can be a better solution to the problems above.

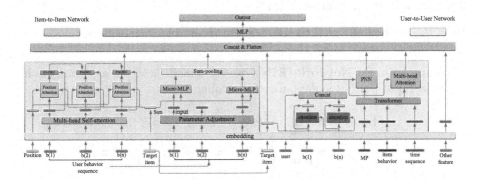

Fig. 1. The architecture of the proposed DUIIN

3 The Proposed Method

In this section, we introduce the Deep User and Item Inter-Matching Network (DUIIN) model in detail.

3.1 Embedding Layer

In IIN, the item features can be denoted as $I = [i_1, i_2, \ldots i_l, \ldots, i_I]$, the user history behavior can be denoted as $x_b = [i_1^u, i_2^u, \ldots, i_T^u] \in \mathbb{R}^{T \times d_i}$, where T is the number of user history behaviors, d_i is the dimension of item embedding, and we use p_t to denote the position code of t-th item position information.

In UUN, user information can be represented as $U = [u_1, u_2, \ldots, u_U]$, item history behavior as $x_i = [u_1^i, u_2^i, \ldots, u_N^i] \in \mathbb{R}^{N \times d_u}$, where N is the number of similar users, and d_u represents the embedding dimension of the user. In addition, item time can be represented as $x_t = [y_1, y_2, \ldots, y_M] \in \mathbb{R}^{M \times d_t}$, M is the total length of time, and d_t represents the dimension of item features after embedding. For user characteristics, we denote it as x_p.

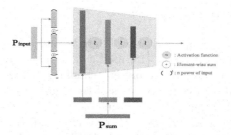

Fig. 2. The architecture of the micro-MLP

3.2   Item-to-Item Network (IIN)

From the left half of Fig. 1, the IIN is divided into two main sub-modules.

Evolving Interest Network (EIN). we use a multi-head self-attention mechanism [8,15] to purify the data after embedding,

$$
\begin{aligned}
\text{head}_h &= \text{Attention}\left(x_b W_h^{Q_1}, x_b W_h^{K_1}, x_b W_h^{V_1}\right) \\
&= \text{Softmax}\left(\frac{x_b W_h^Q \cdot \left(x_b W_h^K\right)^T}{\sqrt{d_h}} \cdot x_b W_h^V\right)
\end{aligned}
\tag{1}
$$

where $W_h^{Q_1}, W_h^{K_1}, W_h^{V_1} \in \mathbb{R}^{d_i \times d_h}$ are projection matrices of the h-th head for query, key, and value separately. The results of the previous step are concatenated to obtain the new item matrix,

$$
Z(P) = \text{concat}\left(\text{head}_1, \text{head}_2, \ldots, \text{head}_{h_1}\right) W^1
\tag{2}
$$

where $W^1 \in \mathbb{R}^{d_i \times d_i}$. In addition to this, we use an auxiliary loss function,

$$
L_{aux} = -\frac{1}{N}\left(\sum_{i=1}^{N}\sum_{t}\log\sigma\left(\langle z(p)_t^i, e_{t+1}^i\rangle\right) + \log\left(1 - \sigma\left(\langle z(p)_t^i, \hat{e}_{t+1}^i\rangle\right)\right)\right)
\tag{3}
$$

where $\sigma(.)$ is the sigmoid activation function, $\langle,\rangle$ is the inner product, $\hat{e}_{t+1}^i$ is the original embedding of the negative sample, N represents the number of training samples. We then pass these information into the attention network,

$$
x_p = [i_1 + p_1, \ldots, i_i + p_i, \ldots, i_N + p_N]
\tag{4}
$$

$$
pa_t = \frac{\exp\left(Z(P)_t W x_{p_i}\right)}{\sum_{j=1}^{T}\exp\left(Z(P)_j W x_{p_i}\right)}
\tag{5}
$$

where $Z(P)_t$ denotes the t-th user behavior after multi-head self-attention, x_{p_i} is the i-th vector of x_p. The obtained pa_t and $Z(P)_i$ are then passed into GRU with position attention (PAGRU),

$$u_t = \sigma \left(W^u i_t + U^u h_{t-1} + b^u \right) \tag{6}$$

$$\tilde{u}_t = pa_t * u_t \tag{7}$$

$$r_t = \sigma \left(W^r i_t + U^r h_{t-1} + b^r \right) \tag{8}$$

$$\tilde{h}_t = \tanh \left(W^h i_t + r_t \circ U^h h_{t-1} + b^h \right) \tag{9}$$

$$h_t = (1 - \tilde{u}_t) \circ h_{t-1} + \tilde{u}_t \circ \tilde{h}_t \tag{10}$$

where $W^h \in R^{n_H \times d_i}$, $U^u, U^r, U^h \in R^{n_H \times n_H}$, n_H is hidden size, σ is the sigmoid activation function, $\circ$ is the element-wise product, $i_t = Z(P)_t$ denotes the user t-th behavior.

Feature Interaction Network (FIN). We select F_{input} and F_{sum} from the user's historical behavior x_b and the item I, respectively. For a specific behavior in the user's history $i^u \in F_{input}$, we obtain $P_{sum} \in \mathbb{R}^{D'}$ after Parameter Adjustment. For item feature $i_l \in F_{sum}$ after the same processing to obtain $P_{input} \in \mathbb{R}^D$. The structure of micro-MLP is shown in Fig. 2, next we divide p_{sum} into the weights of each layer in the micro-MLP,

$$\|_{i=0}^{L-1} (s_i) = P_{sum} \tag{11}$$

$$\sum_{i=0}^{L-1} (|s_i|) = |P_{sum}| = D' \tag{12}$$

where $\|$ indicates connection operation, L denotes the number of layers in the micro-MLP, s_i is the weight of the i-th layer, and $| \cdot |$ is used to calculate the number of variables. p_{input} is then passed into the micro-MLP and sum-pooling layer,

$$h_0 = P_{input} \tag{13}$$

$$h_i = \sigma \left(s_{i-1} \otimes h_{i-1} \right), \quad i = 1, 2, \ldots, L \tag{14}$$

$$F \left(u_{o'}, m_o \right) = H \left(P_{sum}, P_{input} \right) = \|_{i=1}^{L} h_i \tag{15}$$

$$H \left(P_{sum}, P_{seq} \right) = H \left(P_{sum}, \sum_{t=1}^{T} P_{b(t)} \right) \tag{16}$$

where σ is the activation function, $\otimes$ denotes matrix multiplication, H denotes the operation performed in micro-MLP. Next, we use an sum-pooling layer to get a better result.

3.3 User-to-User Network(UUN)

User-to-User Network (UUN) is designed to mine the information in the item behavior, as shown in the right half of Fig. 1. We first connect the historical behavior x_i of the item after embedding, the temporal embedding x_t and the multiple representation of the item,

$$R^u = \text{concat} \left(MP_1, \ldots, MP_n, x_i + x_t \right) \tag{17}$$

where MP_1 denotes the first multi-representation of the item and n denotes the number of perspectives. Next we pass R^u into the multi-head self-attention,

$$\text{head}_i = \text{Attention}\left(R^u W_i^{Q_2}, R^u W_i^{K_2}, R^u W_i^{V_2}\right) \tag{18}$$

$$\text{Attention}(Q, K, V) = \text{softmax}\left(\frac{QK^T}{\sqrt{d_k}} V\right) \tag{19}$$

$$Z(R^u) = \text{concat}(\text{head}_1, \dots, \text{head}_{h2}) W^2 \tag{20}$$

where $W_i^{Q_2}, W_i^{K_2}, W_i^{V_2} \in R^{d \times d_k}$, $W^2 \in R^{d \times d}$ are training parameters, h_2 is the number of the head. We represent the results obtained in the previous section in terms of B^u, and pass B^u and the target user U into the multi-head self-attention mechanism as follows,

$$\text{head}_i = \text{Attention}\left(M^U W_i^{Q_3}, B^u W_i^{K_3}, B^u W_i^{V_3}\right) \tag{21}$$

$$Z(M^U, B^u) = \text{concat}(\text{head}_1, \dots, \text{head}_{h3}) W^3 \tag{22}$$

In order to better represent users, we integrate users' initial interest preferences and user information,

$$\beta_k = W_a^T \sigma\left(W_{c_1} i_n^u + W_{c_2} i_l + b_a\right) \tag{23}$$

$$ua_t = \frac{\exp(\beta_k)}{\sum_{j=1}^T \exp(\beta_j)} \tag{24}$$

$$u' = \sum_{n=1}^T ua_t i_n^u \tag{25}$$

$$M^U = \text{concat}(u', u) \tag{26}$$

where σ is sigmoid function, $W_a^T, W_{c_1}, W_{c_2}, b_a$ are learning parameters, ua_t is the attention weight. We use Negative log-likelihood as loss function,

$$L_{\text{target}} = -\frac{1}{N} \sum_{(x,y) \in \mathbb{D}} (y \log(p(x)) + (1 - y) \log(1 - p(x))) \tag{27}$$

where $\mathcal{D}$ is the training set of size N. $y \in \{0, 1\}$ represents whether the user will click the target item. The output of our network is $p(x)$. In the IIN, we use an auxiliary function, so the global auxiliary function can be expressed as,

$$L = L_{\text{target}} + \eta * L_{\text{aux}} \tag{28}$$

where η is the hyper-parameter in order to balance the two sub tasks.

4   Experiments

To verify the superiority of the proposed model in this paper, we conduct experiments on four real-world datasets to answer the following questions:

- (RQ1) How our model DUIIN compare to other state-of-the-art methods in terms of CTR prediction?
- (RQ2) Do several of the components proposed in this paper necessarily have a positive impact on the model results?
- (RQ3) What are the hyperparameters that have an impact on the prediction results of the models mentioned in the article?

4.1 Experimental Setup

Datasets. To better compare the validity of the models proposed in this paper, we follow [5,21] and use Amazon's e-commerce recommendation dataset[1] to validate whether DUIIN is really valid. In this paper, we choose four datasets. The statistics of the selected datasets are summarized in Table 1. Same as DUMN [5], 85% of data is used as training and validation and 15% data is used as the testing set.

Table 1. The statistics of four datasets

Dataset	Users	Items	Categories	Reviews	Samples
Grocery	14681	8713	129	151254	273146
Beauty	22363	12101	221	198502	352278
Phones	27879	10429	51	194439	333120
Digital	5541	3568	134	64706	118330

Evaluation Metrics AUC: The Area Under ROC (Receiver Operating Characteristic) Curve evaluation indicators are used widely in classification problems.

Compared Models. DUIIN is compared with some mainstream CTR prediction methods: WDL [2], PNN [10], DIN [20], DIEN [21], DHAN [14], DMIN [19], CAN [1], DUMN [5].

Implementation Details. In this experiment, all models are implemented in TensorFlow. For all comparison models, the parameters are set according to those in the original paper. The embedding_dim is set to 18, hidden_size and attention_size are set to 36. And the batch_size is set to 128.

Table 2. The AUC performance on real-world datasets

Model	Grocery	Beauty	Phones	Digital
WDL	0.7021	0.8007	0.7581	0.8182
PNN	0.6958	0.8195	0.7662	0.8274
DIN	0.6897	0.8130	0.7674	0.8288
DIEN	0.6875	0.8194	0.7672	0.8328
DIEN_neg	0.7794	0.8546	0.8758	0.8401
DHAN	0.7925	0.8690	0.8776	0.8390
DMIN	0.7856	0.8803	0.8769	0.8263
CAN	0.7790	0.8723	0.8683	0.8512
DUMN	0.8015	0.8503	0.8295	0.8527
DUIIN	**0.8473**	**0.9015**	**0.9136**	**0.8726**

---

[1] http://jmcauley.ucsd.edu/data/amazon/.

4.2   Performance Comparison(RQ1)

From Table 2, we can find WDL with manually designed features performs not well. PNN automatically learns the interaction between features with improved results. DIN and DIEN use the attention mechanism to allow for further improvement in results. And DIEN_neg introduces an auxiliary loss, and thus the effect is substantially improved. DHAN, DMIN, and DUMN improved on the previous work by using different methods to explore users' interest preferences from different perspectives, and all achieved better experimental results. CAN mines the information between user and item through multiple MLPs. The effectiveness of DUIIN can be seen from the fact that our proposed DUIIN gives the best results on all data sets.

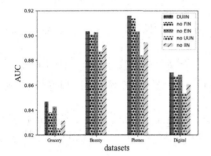

Fig. 3. The impact of the main structure

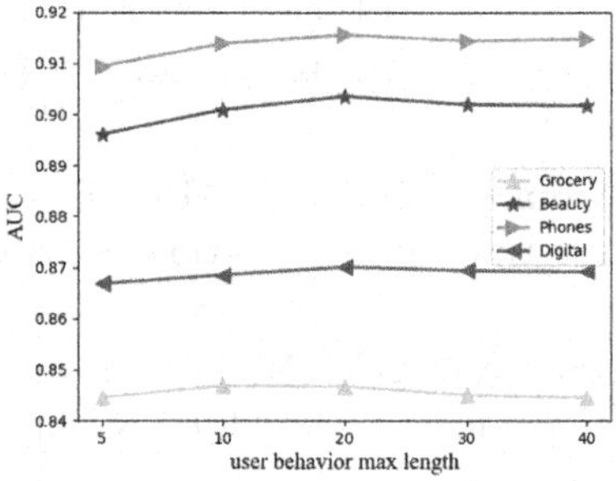

Fig. 4. The impact of the maximum length of the user historical behavior

4.3   Ablation Experiment (RQ2)

As can be seen from Fig. 3, the experimental effects are reduced to different degrees after removing several major modules mentioned in the paper. It can be seen that all the modules proposed in this paper have their own roles.

4.4   Hyper-parameter Tuning (RQ3)

This experiment mines the information in the user and product behavior sequences separately as a way to infer the user's interest preferences. Therefore, we speculate that the length of the sequences affects the experiment. From Fig. 4 and Fig. 5, we can see that as the length of user and product behavior sequences increase, they show an increasing, then decreasing and continuous rising trend, respectively.

In FIN, we use a structure similar to MLP, and we can see from Fig. 6 that the AUC tends to rise as the number of MLP layers increases on most of the datasets.

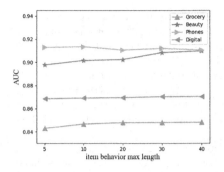

Fig. 5. The impact of the maximum length of the item historical behavior

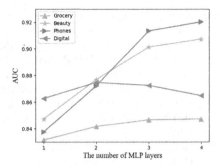

Fig. 6. The impact of the MLP layers

5 Conclusion

In this paper, we design the DUIIN to mine user interest preferences from user and item behavior sequences, respectively. For user behavior sequences, EIN and FIN are specifically used to mine explicit and implicit interests. For item behavior sequences, we mine the relationship between target user and similar users. However, the mining for Item-to-User in this paper is not sufficient, and we will continue to explore this direction in future work.

References

1. Bian, W., Wu, K., Ren, L., et al.: CAN: feature co-action network for click-through rate prediction. In: WSDM 2022: The Fifteenth ACM International Conference on Web Search and Data Mining, Virtual Event/Tempe, AZ, USA, 21–25 February 2022 (2022)
2. Cheng, H.T., Koc, L., Harmsen, J., Shaked, T., et al.: Wide & deep learning for recommender systems. In: Proceedings of the 1st Workshop on Deep Learning for Recommender Systems (2016)
3. Feng, Y., Lv, F., Shen, W., Wang, M., Sun, F., et al.: Deep session interest network for click-through rate prediction. In: Proceedings of the Twenty-Eighth International Joint Conference on Artificial Intelligence, IJCAI 2019, Macao, China, 10–16 August 2019, pp. 2301–2307 (2019)
4. Guo, G., Ouyang, S., He, X., et al.: Dynamic item block and prediction enhancing block for sequential recommendation. In: Proceedings of the Twenty-Eighth International Joint Conference on Artificial Intelligence, IJCAI 2019, Macao, China, 10–16 August 2019, pp. 1373–1379. ijcai.org (2019)
5. Huang, Z., Tao, M., Zhang, B.: Deep user match network for click-through rate prediction. In: SIGIR 2021: The 44th International ACM SIGIR Conference on Research and Development in Information Retrieval, Virtual Event, Canada, 11–15 July 2021, pp. 1890–1894. ACM (2021)
6. Juan, Y., Zhuang, Y., Chin, W., et al.: Field-aware factorization machines for CTR prediction. In: Proceedings of the 10th ACM Conference on Recommender Systems, Boston, MA, USA, 15–19 September 2016, pp. 43–50. ACM (2016)

7. Li, X., Wang, C., Tong, B., et al.: Deep time-aware item evolution network for click-through rate prediction. In: CIKM 2020: The 29th ACM International Conference on Information and Knowledge Management, Virtual Event, Ireland, 19–23 October 2020, pp. 785–794. ACM (2020)

8. Long, L., Yin, Y., Huang, F.: Hierarchical attention factorization machine for CTR prediction. In: Database Systems for Advanced Applications - 27th International Conference, DASFAA 2022, Virtual Event, 11–14 April 2022, Proceedings, Part II. LNCS, vol. 13246. Springer, Cham (2022). https://doi.org/10.1007/978-3-031-00126-0_27

9. Lyu, Z., Dong, Y., Huo, C., et al.: Deep match to rank model for personalized click-through rate prediction. In: The Thirty-Fourth AAAI Conference on Artificial Intelligence, AAAI 2020, New York, NY, USA, 7–12 February 2020

10. Qu, Y., Han, C., Kan, R., et al.: Product-based neural networks for user response prediction. In: 2016 IEEE 16th International Conference on Data Mining (ICDM) (2016)

11. Rendle, S.: Factorization machines. In: ICDM 2010, The 10th IEEE International Conference on Data Mining, Sydney, Australia, 14–17 December 2010, pp. 995–1000. IEEE Computer Society (2010)

12. Song, W., Shi, C., Xiao, Z., et al.: AutoInt: automatic feature interaction learning via self-attentive neural networks. In: the 28th ACM International Conference (2019)

13. Wang, R., Fu, B., Fu, G., Wang, M.: Deep & cross network for ad click predictions. In: ADKDD 2017 (2017)

14. Xu, W., He, H., Tan, M., et al.: Deep interest with hierarchical attention network for click-through rate prediction. In: Proceedings of the 43rd International ACM SIGIR Conference on Research and development in Information Retrieval, SIGIR 2020, Virtual Event, China, 25–30 July 2020, pp. 1905–1908. ACM (2020)

15. Yang, P., Xiao, Y., Zheng, W., et al.: MAN: main-auxiliary network with attentive interactions for review-based recommendation, pp. 1–16. Springer, Cham (2022)

16. Yi, Y., Xu, B., Shen, S., Shen, F., Zhao, J.: Operation-aware neural networks for user response prediction. Neural Networks **121**, 161–168 (2020)

17. Yu, F., Liu, Q., Wu, S., et al.: A dynamic recurrent model for next basket recommendation. In: Proceedings of the 39th International ACM SIGIR conference on Research and Development in Information Retrieval, SIGIR 2016, Pisa, Italy, 17–21 July 2016 (2016)

18. Zhang, J., Lin, F., Yang, C., et al.: Deep multi-representational item network for CTR prediction. In: SIGIR 2022: The 45th International ACM SIGIR Conference on Research and Development in Information Retrieval, Madrid, Spain, 11–15 July 2022, pp. 2277–2281. ACM (2022)

19. Xiao, Z., Yang, L., Jiang, W., et al.: Deep multi-interest network for click-through rate prediction. In: CIKM 2020: The 29th ACM International Conference on Information and Knowledge Management, Virtual Event, Ireland, 11–15 October 2020, pp. 2265–2268. ACM (2020)

20. Zhou, G., Gai, K., Zhu, X., et al.: Deep interest network for click-through rate prediction, pp. 1059–1068 (2018)

21. Zhou, G., Mou, N., Fan, Y., et al.: Deep interest evolution network for click-through rate prediction. Proc. AAAI Conf. Artif. Intell. **33**, 5941–5948 (2019)

Towards Lightweight Cross-Domain Sequential Recommendation via External Attention-Enhanced Graph Convolution Network

Jinyu Zhang, Huichuan Duan, Lei Guo, Liancheng Xu, and Xinhua Wang[✉]

School of Information Science and Engineering,
Shandong Normal University, Jinan 250358, China
{hcduan,lcxu,wangxinhua}@sdnu.edu.cn

Abstract. Cross-domain Sequential Recommendation (CSR) is an emerging yet challenging task that depicts the evolution of behavior patterns for overlapped users by modeling their interactions from multiple domains. Existing studies on CSR mainly focus on using composite or in-depth structures that achieve significant improvement in accuracy but bring a huge burden to the model training. Moreover, to learn the user-specific sequence representations, existing works usually adopt the global relevance weighting strategy (e.g., self-attention mechanism), which has quadratic computational complexity. In this work, we introduce a lightweight external attention-enhanced GCN-based framework to solve the above challenges, namely LEA-GCN. Specifically, by only keeping the neighborhood aggregation component and using the Single-Layer Aggregating Protocol (SLAP), our lightweight GCN encoder performs more efficiently to capture the collaborative filtering signals of the items from both domains. To further alleviate the framework structure and aggregate the user-specific sequential pattern, we devise a novel dual-channel External Attention (EA) component, which calculates the correlation among all items via a lightweight linear structure. Extensive experiments are conducted on two real-world datasets, demonstrating that LEA-GCN requires a smaller volume and less training time without affecting the accuracy compared with several state-of-the-art methods.

Keywords: recommendation systems · cross-domain sequential recommendation · attention mechanism · graph neural network · collaborative filtering

1 Introduction

Sequential Recommendation (SR) aims to capture the user's dynamic behavioral pattern from the interaction sequence that has attracted immense research attention and has wide applications in many domains, such as electronic commerce, online retrieval, and mobile services [1,6,17,21,36]. Though some SR methods have achieved great success in many popular tasks, they meet the challenge while characterizing user preferences from the sparse data or cold-start scenarios. Therefore, Cross-domain Sequential Recommendation (CSR) is gaining increasing research attention that mitigates the above problem by leveraging the side information from other domains [37]. The key idea of CSR is to recommend the next-item to the overlapped user whose historical interactions can observe in multiple domains during the same period.

Early Cross-domain Sequential Recommender systems (CSRs) incorporate a Recurrent Neural Network (RNN)-based structure [26,31] to capture the sequential dependencies from the hybrid sequence but fail to model the associations among cross-domain entities. Then, the attention-based methods, e.g., Chen et al. [2] and Li et al. [20], adopt dual-attention structures that attentively transfer the users' sequential preference between domains but have difficulty excavating structural patterns inside the sequential transitions. Recently proposed Graph Convolution Network (GCN)-based methods [10,12,37] for CSR tasks bridge two domains by the Cross-domain Sequential (CDS) graph and transfer the fine-grained domain knowledge by considering structural information. However, the volume and computational complexity of such graph-based methods are enormous, resulting in the low running efficiency of models, which hinders their deployment on generic devices with limited memory (e.g., GPUs).

To realize the memory-efficient lightweight recommender system, recent studies [4,23,28] predominantly focus on compressing the original collaborative filtering matrix to improve the recommendation efficiency. However, the dependency between the origin representations might be disturbed in the process of embedding transformation, and the user's interest migration in a period also can not be sufficiently modeled. Then, to explore the evolution and migration of users' preferences from a lightweight perspective, some researchers [21,27] shift their focus to transformer-based models. Nevertheless, the self-attention mechanism within the Transformer [32] has quadratic complexity that brings a heavy parameter scale. Similar problems also exist in CSRs, even worse, because they need to share information across two domains, which leads to a doubling of the parameter size. Hence, developing the lightweight CSRs is an ongoing trend, but accompanied by two significant challenges: 1) As mentioned earlier, the current CSRs begin to use complex and in-depth models such as graph convolution networks [10,37] to learn the primary preference for users, which brings a lot of burden for model training but contributes little to the node representation learning. 2) The global relevance weighting strategy (e.g., the Transformer-based methods) [12,21,27] may not be the optimal scheme to capture the sequential behavioral pattern, as it has enormous computational complexity and ignore the positional relations of the item from the dual-domain hybrid sequence.

In this work, we propose a novel Lightweight External Attention-enhanced Graph Convolution Network (LEA-GCN) to address the above challenges. Concretely, we first extract the positional information of each item in the original hybrid sequence to better capture the inter-domain behavior evolution of overlapped users. Then, we construct the Cross-domain Sequential (CDS) graph and model the complicated inner-domain associations among users and items, such as the user-item interactions and the sequential orders of items in each domain. After that, we take two steps to simplify the GCN encoder: 1) By removing the feature transformation matrices and the non-linear activation function (i.e., like the LightGCN [15] does on NGCF [33]), we only keep the simple weighted sum aggregator in GCN to capture the collaborative filtering signals from both domains. 2) Then, we adopt the Single-Layer Aggregating Protocol (SLAP), which reduces the complexity of layer propagation and simultaneously avoids the interference caused by high-order connectivity in the CDS graph. To address the second challenge, we adopt a newly proposed technology, named External Attention (EA), which has attracted extensive research attention in the field of Computer Vision (CV) [7,13]. EA uses two external memory units to optimize the computational complexity of the traditional self-attention mechanism, which surprisingly matches our lightweight purpose. However, the external storage units are independent of the input features, which leads to the deficiency in modeling items' collaborative filtering signal. To avoid that, we devise a dual-channel EA-based sequence encoder to learn the user-specific sequential pattern. It simultaneously calculates the correlation between items and the external memory units by a multi-head structure and the relation score of each item in the sequence via a Multi-Layer Perceptron (MLP). The main contributions of this work can be summarized as follows:

- After pointing out the defects of existing CSR methods in parameter scale and training efficiency, we propose a lightweight GCN-based scheme, namely LEA-GCN, for the memory-efficient cross-domain sequential recommendation.
- We improve the GCN by simplifying the network structure and using the single-layer aggregation protocols. Then, we devise a dual-channel external attention to model the user's sequential preference in a lightweight perspective.
- Extensive experiments on two real-world datasets demonstrate that LEA-GCN performs better and requires fewer parameters than several state-of-the-art baselines.

2   Related Work

2.1   Sequential Recommendation

As Sequential Recommender Systems (SRs) propose to model the user-item interaction sequence [21], it has been proven effective in capturing the evolution of user behavioral patterns [17]. Existing studies on SR can categorize into traditional methods and deep-learning-based methods. Early traditional SR methods usually incorporate Markov chain assumption to capture high-order sequential patterns [5,14]. With the development of deep neural networks, researchers have applied the RNN-based [30,34], Graph Neural Network (GNN)-based [6,36,38], transformer-based [1,3,18], and self-supervised [29,35] methods to the SR task. These methods have the powerful capability of representation learning but have difficulty addressing the challenges caused by data sparsity or cold-start.

2.2   Cross-Domain Sequential Recommendation

By treating the information from other domains as a supplement, the Cross-domain Sequential Recommendation (CSR) approaches can alleviate the data sparsity and the cold-start problems for SR [25]. In early explorations, π-net [26] and PSJNet [31] are two RNN-based solutions for CSR that parallel share the information between domains and simultaneously learn the sequence representations for both of them. Then, Zheng et al. [37] and Guo et al. [10,12] address CSR from the graph-based perspective, which first builds the CDS graph and attentively learns the user-specific representations in both local and global aspects. Another research direction is attention-based methods, such as Chen et al. [2] and Li et al. [20], which provide cross-domain recommendations by matching the user's sequential preference with candidate items through a dual-attention learning mechanism. With the increasing depth of the neural networks and the complexity of the model structure, these CSR methods are gradually becoming uncontrollable on the size of parameters or the memory overhead, exceeding the load of most conventional devices.

2.3   Lightweight Recommendation

To simplify the structure of the recommender system yet make it easier to be implemented on various devices, the concept of lightweight recommendation has attracted a lot of attention [28]. In traditional methods, recent studies have focused on lightening the structure of DeepFM [24], DNN [22,23], GCN [28], and transformer [27] to improve the memory efficiency of recommenders. Li et al. [21] are the first ones that introduce a lightweight solution for Sequential Recommendation (SR) via twin-attention networks to simultaneously address the challenges of lightening the parameter and discovering the temporal signals from all interacted items. As CSR usually requires auxiliary structures to bridge multiple domains, they often need more training time and a larger parameter scale, but the relevant lightweight solutions are mostly unexplored.

3   Method

3.1   Preliminary

CSR task tends to recommend the next item for an overlapped user by modeling her/his historical interactions from the hybrid sequences [37]. Suppose that $U = \{U_1, U_2, \ldots, U_k, \ldots, U_p\}$ is the set of overlapped users whose historical behaviors are available in two domains, where $U_k \in \mathcal{U}$ ($1 \leq k \leq p$) denotes an independent user in $\mathcal{U}$. Let S_H be the original hybrid sequence of an overlapped user, we further split the S_H into $S_A = \{A_1, A_2, \ldots, A_i, \ldots, A_m\}$ and $S_B = \{B_1, B_2, \ldots, B_j, \ldots, B_n\}$, which denote the interaction sequences in domain A and B respectively, where $A_i \in \mathcal{A}$ ($1 \leq i \leq m$) represents the items in domain A and $B_j \in \mathcal{B}$ ($1 \leq j \leq n$) represents the items in domain B. Then we let $P_A = \{P_{A_1}, P_{A_2}, \ldots, P_{A_i}, \ldots, P_{A_m}\}$ and $S_B = \{P_{B_1}, P_{B_2}, \ldots, P_{B_j}, \ldots, P_{B_n}\}$

be the positional information for the sequence S_A and S_B respectively, which accurately record the position of items from the original hybrid sequence S_H.

The probabilities of being recommended for all candidate items in both domains can be denoted as:

$$P(A_{i+1}|S_A, S_B) \sim f_A(S_A, S_B), \tag{1}$$

$$P(B_{j+1}|S_B, S_A) \sim f_B(S_B, S_A), \tag{2}$$

where $P(A_{i+1}|S_A, S_B)$ is the probability of recommending A_{i+1} as the next consumed item in domain A based on S_A and S_B. And $f_A(S_A, S_B)$ denotes the learning function utilized to estimate the probability. And the similar definition for domain B can be denoted as $P(B_{j+1}|S_B, S_A)$ and $f_B(S_B, S_A)$.

3.2 Overview

The key idea of LEA-GCN is to develop a lightweight graph-based solution for CSR without affecting prediction accuracy. In LEA-GCN, we optimize the structure of GCN by adopting the Single-Layer Aggregating Protocol (SLAP) and simplify the sequence encoder by taking advantage of the External Attention (EA) mechanism. As shown in Fig. 1, we first record the position orders of all the items in the hybrid sequences to retain the information on users' inter-domain behavioral patterns. Secondly, by selecting items of each domain with the inner-domain sequential orders fixed from the hybrid sequences, we result in the subsequences S_A and S_B for domains A and B, respectively. Then, we follow the same composition rules as DA-GCN [10] to construct the CDS graph by considering the sophisticated associations among users and items (i.e., inter-domain user-item interactive relations and inner-domain item-item order relations). After constructing the CDS graph, we adopt the Light-GCN [15] graph encoder to linearly propagate embedding on the CDS graph and adopt the SLAP to learn the node representations and optimize the parameter scale synchronously. Subsequently, to further capture the sequential patterns from users' interactions in a lower calculation complexity, we devise a dual-channel External Attention (EA)-based sequence encoder. That calculates the correlation between all the items and considers the positional information (i.e., the V_A and V_B) extracted from the hybrid sequence. Then, the resulting sequence-level representations can be denoted as H_{S_A} and H_{S_B} for both domains. We finally feed the concatenation of H_{S_A} and H_{S_B} to the prediction layer.

3.3 Lightweight Node Representation Learning

Graph Construction. Inspired by Guo et al. [12], we construct the CDS graph to link two domains by considering two types of associations: 1) user-item interactions between both domains; 2) item-item sequential transitions within both domains, where users and items in each domain are nodes and their associations are edges. However, such a composition method only retains the inner-domain sequential transferring characteristics, but discards the order dependency of items from the original hybrid sequences. To fix the above defects,

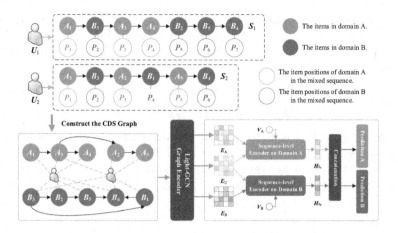

Fig. 1. An overview of our proposed LEA-GCN, where S_1 and S_2 denote the hybrid sequence of two different overlapped users U_1 and U_2, respectively.

we additionally record the positional information of items in the hybrid sequence before constructing the CDS graph, so as to use them in the sequence-level representation learning.

Then, the CDS graph $\mathcal{G} \in \mathbb{R}^{(m+p+n)\times(m+p+n)}$ can be described in a matrix-form as:

$$\mathcal{G} = \begin{bmatrix} A & 0 \\ 0 & B, \end{bmatrix}; \qquad A = \begin{bmatrix} 0 & R_{AU} \\ R_{AU}^T & 0, \end{bmatrix}; \qquad B = \begin{bmatrix} 0 & R_{BU} \\ R_{BU}^T & 0, \end{bmatrix}, \qquad (3)$$

where $A \in \mathbb{R}^{(m+p)\times(m+p)}$ and $B \in \mathbb{R}^{(p+n)\times(p+n)}$ respectively denote the compressed Laplace matrices of both domains, $R_{AU} \in \mathbb{R}^{m\times p}$ and $R_{BU} \in \mathbb{R}^{p\times n}$ represent the user-item interaction matrix of domain A and B, respectively, $R_{AU}^T \in \mathbb{R}^{m\times p}$ and $R_{BU}^T \in \mathbb{R}^{p\times n}$ are the transpose matrices.

Single-Layer Aggregation Protocol (SLAP). It has been proven effective in early proposed state-of-the-art methods NGCF [33] and LightGCN [15], which adopt a Multi-Layer Aggregating Protocol (MLAP) to propagate embeddings from nodes layer by layer. The core idea of the MLAP is to use the high-order connectivity on user-item bipartite graph to obtain the potential association between them, thereby enhancing the performance of node representation learning. The node representation learning on the l-th layer can be detailed as:

$$E^{(l)} = H(E^{(l-1)}, \mathcal{G}), \qquad (4)$$

where $\mathcal{G}$ is the user-item bipartite graph and $\boldsymbol{E}^{(l)}$ denotes the node representations at the l-th layer, $\boldsymbol{E}^{(l-1)}$ is that of the previous layer. $H(\cdot)$ represents the function for neighbor aggregation.

However, the MLAP may not suitable for the CDS graph. For example, when learning the item representation of domain A by considering the high-order connectivity of the items in domain B. It will also bring high-order domain-specific structural information from domain B, which might be the noise message for domain A, interfering with the interest expression of overlapped users. Hence, we develop a single-layer propagation rule (a.k.a., the Single-Layer Aggregating Protocol (SLAP)) to consider first-order neighbor aggregation on the CDS graph as:

$$E_U = \sum_{k \in U} (\sum_{i \in S_A} \frac{1}{\sqrt{|N_U|}\sqrt{|N_A|}} e_{i \rightarrow k} + \sum_{j \in S_B} \frac{1}{\sqrt{|N_U|}\sqrt{|N_B|}} e_{j \rightarrow k}); \qquad (5)$$

$$E_A = \sum_{i \in S_A} \sum_{k \in U} \frac{1}{\sqrt{|N_U|}\sqrt{|N_A|}} e_{k \rightarrow i}; \qquad (6)$$

$$E_B = \sum_{j \in S_B} \sum_{k \in U} \frac{1}{\sqrt{|N_U|}\sqrt{|N_B|}} e_{k \rightarrow j}, \qquad (7)$$

where $e_{i \rightarrow k}$ and $e_{j \rightarrow k}$ denote the passing message from the items to the overlapped user, $e_{k \rightarrow i}$ and $e_{k \rightarrow j}$ respectively denote the message transferred from users to items of domain A and domain B, N_U is the set of overlapped users that interact with item A_i or item B_j, N_A and N_B are the set of items that are interacted by user u. By adopting the SLAP, the knowledge between domains could be transferred with less impact from the domain-specific information and simultaneously reduces the parameter scale.

To support our view, we conduct a series of ablation experiments to investigate the impact of the layer depth. Due to the space limitation, we only report the experimental results of DOUBAN (i.e., a real-world dataset which will be detailed in Sect. 4.1) on Table 1.

As shown in Table 1, we search the performance of LEA-GCN at different layers compared with two state-of-the-art GCN-based methods (i.e., LightGCN [15] and DA-GCN [10]). To make the LightGCN comparable to other two CSR methods, we simultaneously report its performance on both domains. Then, we have the following observations: 1) Increasing the number of layers can improve the performance of LightGCN, demonstrating the effectiveness of considering the high-order connectivity while propagating embedding on the user-item graph within a single domain. 2) DA-GCN and LEA-GCN with SLAP perform better than them with MLAP, which supports our hypothesis that simply aggregates message from the high-order connectivity on the CDS graph may bring more noise to the node representation learning and proves the significance of adopting SLAP for the CSR task.

Table 1. Performance (%) comparison between LightGCN, DA-GCN and our proposed GCN-based solution at different layers.

Domain		A			B		
Layer Numbers	Method	RC10	MRR10	NDCG10	RC10	MRR10	NDCG10
@ 1 Layer	LightGCN	78.03	75.06	48.24	68.28	52.03	37.52
	DA-GCN	83.55	80.84	51.53	71.91	58.31	40.67
	LEA-GCN	**83.83**	**81.22**	**52.06**	**76.14**	**66.15**	**45.01**
@ 2 Layers	LightGCN	78.33	75.25	48.89	68.19	52.00	37.66
	DA-GCN	83.12	80.55	50.92	71.42	57.45	40.16
	LEA-GCN	83.53	81.08	51.35	74.63	63.67	42.91
@ 3 Layers	LightGCN	78.52	75.35	48.93	68.32	52.11	37.79
	DA-GCN	81.99	80.08	49.35	71.08	57.17	39.85
	LEA-GCN	83.40	81.02	51.05	74.22	63.29	42.11

3.4 External Attention (EA)-Based Sequence Encoder

Self-attention uses the combination of self values to refine the input sequence representations, which only considers the relation between items within a sequence but ignores implied relationships between items in different sequences [13]. And the high computational complexity of $O(N^2)$ presents another significant drawback to use self-attention. To accurately measure different items' contributions to the sequence yet with lower computational costs, we replace the self-attention-based algorithms [12,27] by External Attention (EA) mechanism which have achieved great success on the task of image classification, object detection, and semantic segmentation [13].

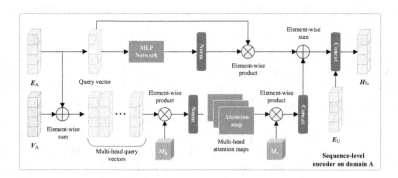

Fig. 2. The workflow of our proposed dual-channel external attention-based sequence encoder (take domain A as an example), where M_k and M_v are two different external memory units.

Different from the self-attention mechanism, External Attention (EA) uses external storage units to reserve the global sharing weights. As shown in Fig. 2, We devise a dual-channels composite linear structure to implement the EA (take domain A as an example). Specifically, in the first channel, we attach the positional information V_A from the hybrid sequence to the item representation E_A. Then, we calculate the external attention between items and two external memory units (i.e., M_k and M_v), which respectively act as the key and value matrices. The score calculations can be formulated as:

$$T_{S_A}^{(1)} = Norm[(E_A + \alpha V_A)M_k^T]; \tag{8}$$

$$H_{S_A}^{(1)} = T_{S_A}^{(1)} M_v^T, \tag{9}$$

where α is a hyper-parameter that controls the participation of the positional information, $T_{S_A}^{(1)}$ is the external attention map, $H_{S_A}^{(1)}$ represents the final output embedding of sequence S_A from the first channel. For the i-th item A_i in S_A, the calculation can be further detailed as:

$$a_{i,j}^{(1)} = Norm[(e_{A_i} \oplus (\alpha \cdot v_{A_i}))m_j^k]; \tag{10}$$

$$H_{S_A}^{(1)} = \sum_{i=1}^{|S_A|} a_{i,j}^{(1)} m_i^v, \tag{11}$$

where $a_{i,j}^{(1)}$ is the pair-wise affinity between i-th items e_{A_i} and the j-th row m_j^k of matrix M_k. v_{A_i} denotes the positional information of A_i and m_i^v denotes the i-th row of the second memory unit.

Inspired by the Transformer [32], we improve the capacity of EA by adopting a multi-head manner [13] to better capture different relations between items as:

$$Z_h^{(1)} = ExternalAttention(E_A, V_A, M_k, M_v); \tag{12}$$

$$H_{S_A}^{(1)} = Concat(Z_h^{(1)}, \ldots, Z_\beta^{(1)})W_1, \tag{13}$$

where W_1 is the linear transformation matrix to align the dimensions of input and output, Z_h is the h-th head, and β controls the number of heads. In experiments, we search β in [1, 2, 4, 8, 16] and report the experimental results in Sect. 4.5.

Although external attention can well solve the problem of computational complexity, its final output representation is generated by M_v. As the external memory units are independent of the input features (i.e., the E_A), it will lead to the deficiency in modeling items' collaborative filtering signals. To address above questions, in the second channel, an MLP network with a smooth normalization layer (Inspired by Guo et al. [12]) is used to extract the collaborative filtering signals of E_A. We measure the correlation between e_i and e_j as:

$$a_{i,j}^{(2)} = Norm(f(e_{A_i}, e_{A_j})) = Norm(W_3^T ReLU(W_2[e_{A_i} \oplus e_{A_j}] + b)), \tag{14}$$

where $f(\cdot)$ is a score function implemented by an MLP network, $\boldsymbol{W}_2$ and $\boldsymbol{W}_3$ are two weight matrices, and $\boldsymbol{b}$ is the bias vector. Then we can get the sequence representation $\boldsymbol{H}_{S_A}^{(2)}$ from the second channel:

$$\boldsymbol{H}_{S_A}^{(2)} = \sum_{i=1}^{|S_A|} a_{i,j}^{(2)} e_{A_i}. \tag{15}$$

Hence, the resulting user-specific sequence-level representation $\boldsymbol{H}_{S_A}$ for domain A can be denoted as:

$$\boldsymbol{H}_{S_A} = Concat((\boldsymbol{H}_{S_A}^{(1)} + \boldsymbol{H}_{S_A}^{(2)}), \boldsymbol{E_U}). \tag{16}$$

3.5   Prediction Layers

After the sequence representation learning, LEA-GCN gets the sequence embedding $\boldsymbol{H}_{S_A}$ and $\boldsymbol{H}_{S_B}$ for domain A and B, respectively. Then, for leveraging the information in both domains, we feed the concatenation of them to the prediction layer:

$$P(A_{i+1}|S_A, S_B) = softmax(\boldsymbol{W}_A \cdot [\boldsymbol{H}_{S_A}, \boldsymbol{H}_{S_B}]^{\mathrm{T}} + \boldsymbol{b}_A); \tag{17}$$

$$P(B_{j+1}|S_B, S_A) = softmax(\boldsymbol{W}_B \cdot [\boldsymbol{H}_{S_B}, \boldsymbol{H}_{S_A}]^{\mathrm{T}} + \boldsymbol{b}_B), \tag{18}$$

where $\boldsymbol{W}_A$ and $\boldsymbol{W}_B$ are the weight matrix of all items in domain A and B, respectively; $\boldsymbol{b}_A$ and $\boldsymbol{b}_A$ are the bias term for both domains. Then, to avoid the seesaw phenomenons in π-net [26] and DA-GCN [10], we adopt the cross-entropy loss and optimize them independently on both domains:

$$\mathcal{L}_A = -\frac{1}{|\mathcal{S}|} \sum_{S_A, S_B \in \mathcal{S}} \sum_{A_i \in S_A} \log P(A_{i+1}|S_A, S_B), \tag{19}$$

$$\mathcal{L}_B = -\frac{1}{|\mathcal{S}|} \sum_{S_B, S_A \in \mathcal{S}} \sum_{B_j \in S_B} \log P(B_{j+1}|S_B, S_A), \tag{20}$$

where $\mathcal{S}$ denotes the training sequences in both domains.

4   Experiment

We conduct extensive experiments on two real-world datasets to validate the effectiveness of LEA-GCN. In this section, we aim to answer the following Research Questions (RQ):

RQ1: Does LEA-GCN work on lightening the model's weights? How is the training efficiency of the LEA-GCN?

RQ2: How does the LEA-GCN perform compared with other state-of-the-art baselines? Does our lightweight strategy lead to the deterioration of recommendation performance?

RQ3: Is it helpful to consider the items' positional relationship in the hybrid input sequences? Does it work by using external attention to learn sequence representation for both domains?

RQ4: How do the hyper-parameters affect the performance of LEA-GCN?

4.1   Experimental Setup

Datasets and Evaluation Protocols.   We evaluate LEA-GCN on two real-world datasets (i.e., DOUBAN [39] and AMAZON [8]). DOUBAN contains historical interactions of overlapped users on domain A and domain B (i.e., douban movies and douban books), which are collected from the well-known Chinese social media platform Douban[1] [39]. AMAZON is a product review dataset collected by Fu et al. [8]. It contains overlapped users' review behaviors on two different amazon[2] platforms, i.e., amazon-book (domain A) and amazon-movie (domain B). As shown in Table 2, We randomly choose 80% of all the hybrid sequences of both datasets as the training sets, and the rest 20% as the testing sets. Moreover, for the pretreatments on both datasets, we filter out the cold users with less than ten historical interactions and those cold items which only noticed less than five times [17].

For evaluation, we first treat the last two observed items in each hybrid sequence as the ground truth items for both domains. Secondly, we employ three frequently used metrics (i.e., RC@10, MRR@10, and NDCG@10) [10,11] to evaluate each instance on the testing sets and report their average values.

Table 2. Statistics of the datasets, where A and B represent different domains for both datasets.

Dataset	DOUBAN		AMAZON	
Domain	A	B	A	B
Items	14,636	2,940	126,526	61,362
Interactions	607,523	360,798	1,678,006	978,226
Users	6,582		9,204	
Sequences (Train)	42,062		90,574	
Sequences (Test)	10,431		14,463	

Baselines.   To validate the performance of LEA-GCN, we compared our proposed method with the following baselines: 1) Traditional recommendations: NCF [16], NGCF [33], and LightGCN [15]. We adapt the traditional methods with sequential inputs and report their experimental results in each domain. 2) Sequential recommendations: GRU4REC [17] and HRNN [30]. We report their performance in each domain. 3) Cross-domain Sequential recommendations: π-Net [26], PSJNet [31], DA-GCN [10], and TiDA-GCN [12].

---

[1] http://www.douban.com/.

[2] http://jmcauley.ucsd.edu/data/amazon/.

Implementation Details.   We implement LEA-GCN[3] by TensorFlow and accelerate the model training by NVIDIA Tesla K80M GPU. For parameters, we employ Xavier [9] for initialization and optimize them by Adam [19]. To train the model, we set the batch-size as 256, the dropout ratio as 0.1, and the learning rate as 0.002 for domain A and 0.004 for domain B, respectively. For LEA-GCN, we set the embedding size as 16 and the regularization ratio as 1e−7. The hyper-parameter α is searched in [0–1] with a step size of 0.1 to adjust the participation of the positional information and the number of attention head β is explored within [1, 2, 4, 8, 16] to reach the best performance for LEA-GCN. We detail the experimental results for α and β in Sect. 4.4. Moreover, we uniformly set the embedding-size to 16 for all the reference baselines to make their results comparable. As for other hyper-parameters, we refer to the best settings of their papers and fine-tune them on both datasets.

4.2   Parameter Scale and Training Efficiency (RQ1)

In this section, we first conduct a series of experiments by changing the ratio of input data in [0.2–1.0] on DOUBAN and AMAZON to measure the time consumption of the model training. Second, we analyze the performance of LEA-GCN in lightweight modeling ability by measuring the scale of its parameters compared with two most competitive baselines (i.e., PSJNet and TiDA-GCN). Then we have the following observations: 1) From Fig. 3 (a) and (b), we notice that LEA-GCN costs lower training time than TiDA-GCN and PSJNet, which demonstrates that LEA-GCN has a better training efficiency and is scalable to the large-scale datasets. 2) From Fig. 3 (c) and (d), we observe that LEA-GCN and LEA-All need far fewer parameters than PSJNet and TiDA-GCN, providing a positive answer to RQ1. Note that, the LEA-All is a variant method that only keeps the GCN encoder in LEA-GCN.

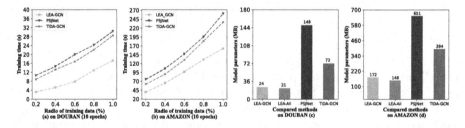

Fig. 3. Time consumption and the parameter scale of LEA-GCN compared with PSJNet and TiDA-GCN.

---

[3] https://github.com/JinyuZ1996/LEA-GCN.

4.3 Performance on Recommendation Accuracy (RQ2)

Table 3 shows the experimental results of LEA-GCN compared with other state-of-the-art methods on both datasets. The observations are summarized as follows: 1) LEA-GCN achieves the best performance on both domains of AMAZON and outperforms other state-of-the-art baselines in most evaluation metrics on DOUBAN, demonstrating that our lightweight strategy in LEA-GCN has little impact on the prediction accuracy, even improves the model's performance. 2) The CSR solutions outperform other state-of-the-art methods (i.e., traditional recommenders and sequential recommenders), demonstrating the significance of simultaneously modeling users' sequential preference and cross-domain characteristics. 3) LEA-GCN outperforms all the CSR baselines, indicating the effectiveness of modeling user-specific preference from the CDS graph by a lightweight structure, and displays the superiority of external attention in capturing users' sequential patterns.

Table 3. Experiment results (%) of compared methods on DOUBAN and AMAZON. Note that, the bold value denotes the best result in terms of the corresponding metric. Significant improvements are marked with† (paired samples t-test, $p < .05$).

Dataset	DOUBAN						AMAZON					
Domain	A			B			A			B		
Metric (@10)	RC	MRR	NDCG	RC	MRR	NDCG	RC	MRR	NDCG	RC	MRR	NDCG
NCF	69.75	58.05	40.26	35.24	23.29	18.28	15.59	11.30	6.12	17.38	13.20	5.29
NGCF	79.21	77.82	49.12	67.37	54.41	37.26	21.52	18.74	12.08	26.55	25.37	17.45
LightGCN	78.52	75.35	48.93	68.32	52.11	37.79	21.67	17.41	10.13	26.49	24.23	15.09
GRU4REC	80.13	75.41	47.81	66.66	54.10	38.03	21.51	17.11	9.66	24.51	22.13	12.94
HRNN	81.25	77.90	48.88	68.32	54.99	38.93	21.92	17.30	9.87	25.10	22.48	13.44
π-net	83.22	80.71	51.22	69.54	55.72	39.18	24.33	20.52	11.80	27.66	25.03	16.20
PSJNet	83.54	80.96	51.72	71.59	58.36	40.71	25.03	21.09	13.54	31.24	28.35	18.93
DA-GCN	83.55	80.84	51.53	71.91	58.31	40.67	24.62	20.91	13.18	31.12	28.21	18.85
TiDA-GCN	83.68	**81.27**	52.02	72.56	60.27	41.38	25.05	21.23	14.68	32.84	29.65	19.12
LEA-GCN	**83.83**†	81.22	**52.06**	**76.14**†	**66.15**†	**45.01**†	**25.47**†	**21.57**†	**14.93**†	**33.97**†	**30.65**†	**20.46**†

4.4 Ablation Study (RQ3)

In this section, we conduct a series of ablation studies on DOUBAN and AMAZON to explore the impact of different components on LEA-GCN. Due to space limitations, we only report the results on DOUBAN. As shown in Table 4, the LEA-Pos is a variant that disables the participation of the positional information from the hybrid sequence. LEA-EA is another variant model that removes the external attention-based sequence encoder. LEA-All is a variant that disables both the positional information and the EA. The observations of Table 4 are summarized as follows: 1) LEA-GCN outperforms LEA-All and LEA-Pos, demonstrating the importance of the positional information of items from the original hybrid sequence for learning overlapped users' sequential characteristics.

Table 4. The experimental results (%) of ablation studies on the DOUBAN dataset.

Domain	A			B		
Metric (@10)	RC	MRR	NDCG	RC	MRR	NDCG
LEA-Pos	83.67	81.16	52.00	75.10	64.35	44.38
LEA-EA	82.62	79.14	50.43	72.82	61.92	42.41
LEA-All	82.12	78.10	49.64	71.28	59.43	41.28
LEA-GCN	**83.83**	**81.22**	**52.06**	**76.14**	**66.15**	**45.01**

2) LEA-GCN performs better than LEA-EA, demonstrating the contribution of the external attention component for sequence-level representation learning.

4.5 Hyper-parameters Analysis (RQ4)

The hyper-parameter α controls the participation of the positional information attached to the item representations for both domains. Figure 4 (a) to (f) show the performance of LEA-GCN with different $\alpha \in [0\text{--}1]$. The experimental results prove the significance of leveraging the positional information from the original hybrid sequences. However, it is not advisable to regard it as equally important as the sequence representation.

The hyper-parameter β controls the head number of the multi-head external attention component. The experimental results in Fig. 4 (g) to (l) demonstrate that only with an appropriate number of heads, does the EA mechanism benefits the sequence representation learning process.

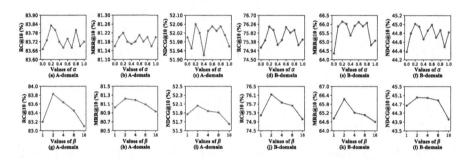

Fig. 4. Impact of hyper-parameters α and β on DOUBAN.

5   Conclusions

In this work, we propose a lightweight GCN-based solution for CSR, which simultaneously simplifies the structure of GCNs and optimizes the calculation complexity of the sequence encoder. Specifically, we only keep the neighborhood aggregation to reduce the parameter scale of the GCN encoder and propose the Single-Layer Aggregating Protocol (SLAP) to propagate embedding on the CDS graph. Then, we devise a dual-channel External Attention (EA)-based sequence encoder to calculate the correlation among all items via a lighter linear structure. The experimental results on two real-world datasets demonstrate the superiority of our lightweight solution.

References

1. Ai, Z., Wang, S., Jia, S., Guo, S.: Core interests focused self-attention for sequential recommendation. In: DASFAA, pp. 306–314 (2022)
2. Chen, C., Guo, J., Song, B.: Dual attention transfer in session-based recommendation with multi-dimensional integration. In: SIGIR, pp. 869–878 (2021)
3. Chen, H., et al.: Denoising self-attentive sequential recommendation. In: RecSys, pp. 92–101 (2022)
4. Chen, T., Yin, H., Zheng, Y., Huang, Z., Wang, Y., Wang, M.: Learning elastic embeddings for customizing on-device recommenders. In: KDD, pp. 138–147 (2021)
5. Cheng, C., Yang, H., Lyu, M.R., King, I.: Where you like to go next: successive point-of-interest recommendation. In: IJCAI, pp. 2605–2611 (2013)
6. Dong, X., Jin, B., Zhuo, W., Li, B., Xue, T.: Sirius: sequential recommendation with feature augmented graph neural networks. In: Jensen, C.S., et al. (eds.) DASFAA 2021. LNCS, vol. 12683, pp. 315–320. Springer, Cham (2021). https://doi.org/10.1007/978-3-030-73200-4_21
7. Fang, J., Yang, C., Shi, Y., Wang, N., Zhao, Y.: External attention based TransUNet and label expansion strategy for crack detection. In: TITS, pp. 19054–19063 (2022)
8. Fu, W., Peng, Z., Wang, S., Xu, Y., Li, J.: Deeply fusing reviews and contents for cold start users in cross-domain recommendation systems. In: AAAI, pp. 94–101 (2019)
9. Glorot, X., Bengio, Y.: Understanding the difficulty of training deep feedforward neural networks. In: AISTATS, pp. 249–256. JMLR (2010)
10. Guo, L., Tang, L., Chen, T., Zhu, L., Nguyen, Q.V.H., Yin, H.: DA-GCN: a domain-aware attentive graph convolution network for shared-account cross-domain sequential recommendation. In: IJCAI, pp. 2483–2489 (2021)
11. Guo, L., Zhang, J., Chen, T., Wang, X., Yin, H.: Reinforcement learning-enhanced shared-account cross-domain sequential recommendation. In: TKDE (2022)
12. Guo, L., Zhang, J., Tang, L., Chen, T., Zhu, L., Yin, H.: Time interval-enhanced graph neural network for shared-account cross-domain sequential recommendation In: TNNLs (2022)
13. Guo, M.H., Liu, Z.N., Mu, T.J., Hu, S.M.: Beyond self-attention: External attention using two linear layers for visual tasks. In: TPAMI (2022)
14. He, R., Kang, W., McAuley, J.J.: Translation-based recommendation. In: RecSys, pp. 161–169 (2017)
15. He, X., Deng, K., Wang, X., Li, Y., Zhang, Y., Wang, M.: LightGCN: simplifying and powering graph convolution network for recommendation. In: SIGIR, pp. 639–648 (2020)
16. He, X., Liao, L., Zhang, H., Nie, L., Hu, X., Chua, T.S.: Neural collaborative filtering. In: WWW, pp. 173–182 (2017)

17. Hidasi, B., Karatzoglou, A., Baltrunas, L., Tikk, D.: Session-based recommendations with recurrent neural networks. In: ICLR (2016)
18. Kang, W., McAuley, J.J.: Self-attentive sequential recommendation. In: ICDM, pp. 197–206 (2018)
19. Kingma, D.P., Ba, J.: Adam: a method for stochastic optimization. In: ICLR (2015)
20. Li, P., Jiang, Z., Que, M., Hu, Y., Tuzhilin, A.: Dual attentive sequential learning for cross-domain click-through rate prediction. In: KDD, pp. 3172–3180 (2021)
21. Li, Y., Chen, T., Zhang, P., Yin, H.: Lightweight self-attentive sequential recommendation. In: CIKM, pp. 967–977 (2021)
22. Lian, D., Wang, H., Liu, Z., Lian, J., Chen, E., Xie, X.: LightREC: a memory and search-efficient recommender system. In: WWW, pp. 695–705 (2020)
23. Liu, H., Zhao, X., Wang, C., Liu, X., Tang, J.: Automated embedding size search in deep recommender systems. In: SIGIR, pp. 2307–2316 (2020)
24. Liu, S., Gao, C., Chen, Y., Jin, D., Li, Y.: Learnable embedding sizes for recommender systems. In: ICLR (2021)
25. Ma, M., et al.: Mixed information flow for cross-domain sequential recommendations. In: TKDD pp. 64:1–64:32 (2022)
26. Ma, M., Ren, P., Lin, Y., Chen, Z., Ma, J., Rijke, M.D.: π-Net: a parallel information-sharing network for shared-account cross-domain sequential recommendations. In: SIGIR, pp. 685–694 (2019)
27. Mei, M.J., Zuber, C., Khazaeni, Y.: A lightweight transformer for next-item product recommendation. In: RecSys, pp. 546–549 (2022)
28. Miao, H., Li, A., Yang, B.: Meta-path enhanced lightweight graph neural network for social recommendation. In: DASFAA, pp. 134–149 (2022)
29. Qiu, R., Huang, Z., Yin, H., Wang, Z.: Contrastive learning for representation degeneration problem in sequential recommendation. In: WSDM, pp. 813–823 (2022)
30. Quadrana, M., Karatzoglou, A., Hidasi, B., Cremonesi, P.: Personalizing session-based recommendations with hierarchical recurrent neural networks. In: RecSys, pp. 130–137 (2017)
31. Sun, W., et al.: Parallel split-join networks for shared account cross-domain sequential recommendations. In: TKDE (2021)
32. Vaswani, A., et al.: Attention is all you need. In: NIPS, pp. 5998–6008 (2017)
33. Wang, X., He, X., Wang, M., Feng, F., Chua, T.: Neural graph collaborative filtering. In: SIGIR, pp. 165–174 (2019)
34. Wu, C., Ahmed, A., Beutel, A., Smola, A.J., Jing, H.: Recurrent recommender networks. In: WSDM, pp. 495–503 (2017)
35. Xie, X., et al.: Contrastive learning for sequential recommendation. In: ICDE, pp. 1259–1273 (2022)
36. Zang, Y., et al.: GISDCN: a graph-based interpolation sequential recommender with deformable convolutional network. In: DASFAA. LNCS, vol. 13246, pp. 289–297. Springer, Cham (2022). https://doi.org/10.1007/978-3-031-00126-0_21
37. Zheng, X., Su, J., Liu, W., Chen, C.: DDGHM: dual dynamic graph with hybrid metric training for cross-domain sequential recommendation. In: MM, pp. 471–481 (2022)
38. Zheng, Y., Liu, S., Li, Z., Wu, S.: DGTN: dual-channel graph transition network for session-based recommendation. In: ICDM, pp. 236–242 (2020)
39. Zhuang, F., et al.: Sequential recommendation via cross-domain novelty seeking trait mining. In: JCST, pp. 305–319 (2020)

ALGCN: Accelerated Light Graph Convolution Network for Recommendation

Ronghai Xu[1,2], Haijun Zhao[1,2], Zhi-Yuan Li[1,2], and Chang-Dong Wang[1,2(✉)]

[1] School of Computer Science and Engineering, Sun Yat-sen University,
Guangzhou, Guangdong, China
{xurh6,zhaohj23,lizhiy225}@mail2.sysu.edu.cn, changdongwang@hotmail.com
[2] Key Laboratory of Machine Intelligence and Advanced Computing,
Ministry of Education, Beijing, China

Abstract. Recently, Graph Convolutional Network (GCN) has been widely applied in the field of collaborative filtering (CF) with tremendous success, since its message-passing mechanism can efficiently aggregate neighborhood information between users and items. However, most of the existing GCN-based CF models suffer from low convergence rates during training, mainly because they follow the design of standard GCN using a simple uniform average to aggregate the neighborhood information. We also find that the scale of embedding across different layers oscillates. We argue that these issues can be alleviated by our proposed graph convolution framework, namely Accelerated Light Graph Convolutional Network (ALGCN). ALGCN mainly contains two components: influence-aware graph convolution operation and augmentation-free in-batch contrastive loss on the unit sphere. Empirical evaluations on three large and public datasets demonstrate that the proposed method achieves remarkable training speedups over LightGCN and substantially outperforms the state-of-the-art GCN-based CF models. Our method also shows a great improvement in long-tail recommendation.

Keywords: Collaborative filtering · Recommendation systems · Graph neural networks

1 Introduction

Recommendation systems are considered effective techniques for personalized information filtering to deliver content. The task of recommendation systems is to predict whether a user will interact with an item. The interactions can be in diverse ways, *e.g.*, click, rate, and purchase. A standard solution for the task of recommendation systems is collaborative filtering (CF) [4], which exploits the historical user-item interactions to predict the current potential user-item interactions. Most of the CF methods share the same paradigm of learning a unique embedding for each user/item and performing prediction based on the

X. Wang et al. (Eds.): DASFAA 2023, LNCS 13944, pp. 221–236, 2023.
https://doi.org/10.1007/978-3-031-30672-3_15

embedding. Matrix factorization [9], a fundamental method of CF, directly learns the embeddings by factorizing the user-item interaction matrix.

Later, various methods are devised to better learn the embedding from the history of user-item interactions. Among them, considering the user-item interaction matrix as the adjacency matrix of the user-item interaction graph makes it possible to utilize the development of deep graph learning [2,22]. For example, Wang et al. propose NGCF [16], which makes use of the standard GCN [8] to propagate the features on the user-item interaction graph. Multiple orders of neighbor features are aggregated on multiple propagation layers. Due to GCN's capability of high order representation, NGCF achieves promising results over most non-GCN CF methods. Later progress [13,17] shows that some of the modules in standard GCN are unnecessary in graph representation. LightGCN [6] simplifies the operations in graph CF methods and achieves the state-of-the-art performance.

Moreover, due to the strong learning capability of contrastive learning (CL) [1], many efforts have been made in applying CL to recommendation, which has shown a considerable performance gain [11,18,21]. Most CL methods first conduct augmentation to generate different views of positive samples, and then operate on a single type of node embedding (*e.g. only user or only item*) to maximize the consistency of representation in different views. Recently, Yu *et al.* [21] experimentally show that graph augmentation is redundant for graph based CF, and CL loss term improves the model performance by regulating the uniformity of representation space.

Though current graph collaborative filtering methods show effectiveness in their performance, their relatively high training costs are tremendous. Training costs come from two aspects. First of all, propagation on graphs takes more time than conventional MF methods. Besides, the graph embeddings are obtained by a non-parameterized uniform average aggregation in the forward process. Regarding the backward update, the gradients are also evenly divided to update the embedding of each neighbor node, which makes the learning process stable but slow to converge. Thereby, an aggregation module exploiting the property of graph may help faster convergence. Moreover, CL methods [20] are also proved to help accelerate.

In this work, we propose a new approach called Accelerated Light Graph Convolution Network (ALGCN) for collaborative filtering. ALGCN contains two components: influence-aware graph convolution operation and augmentation-free in-batch contrastive loss on the unit hypersphere. By scaling the representation with the node influence, graph neighborhood information is propagated faster on graph convolution operation. Therefore the training of the graph CF model is accelerated to converge. While the current CL methods take different views of embedding of the same node as a positive pair and contrast between the single type of embeddings (*e.g.* user-user or item-item), noting that both user and item share the same representation space, it is intuitive to directly contrast between user embedding and item embedding. In the proposed contrastive loss, positive pairs are formed by users and their interacted items, and negative pairs are formed by users and randomly sampled items. Besides, the proposed CL loss term working on unit hypersphere disentangles the magnitude and direction of the representation and keeps the most representation capability while regulating the direction uniformity of representation.

The major contributions of this paper are summarized as follows:

- We experimentally discover two defects of the conventional LightGCN, namely *low convergence rate* and *scale ocsillation*. We propose a new graph collaborating filtering method called Accerlerated Light Graph Convolution Network (ALGCN), to ameliorate LightGCN for better performance and faster convergence rate by designing delicate modules to tackle these issues.
- We propose a new form of contrastive loss for graph collaborative filtering. Different from the current CL methods, the proposed contrastive loss forms positive pairs and negative pairs between user representation and item representation. Contrasting on the unit hypersphere successfully regulates the uniformity of representation while maintaining the representation capability of each embedding.
- Combining the above two components, the proposed ALGCN method achieves the new state-of-the-art performance on multiple real-world datasets with far fewer training epochs. We conduct extensive experiments on ablation study to validate the effectiveness of each component.

2 Preliminaries

2.1 GCN-Based CF Paradigm

We consider Top-K item recommendations with M users $\mathcal{U}$ and N items $\mathcal{I}$. In details, we use $\mathcal{U} = \{u_1, u_2, ...u_M\}$ and $\mathcal{I} = \{i_1, i_2, ...i_N\}$ to represent the set of users and items. Let an interaction matrix $\mathbf{R} \in \mathbb{R}^{M \times N}$ be the implicit feedback of users. $R_{u,i} = 1$ if user u has interacted with item i, and $R_{u,i} = 0$ otherwise. We define a bipartite graph as $\mathcal{G} = (\mathcal{U} \cup \mathcal{I}, \mathcal{E})$, where $\mathcal{E} = \{R_{u,i} = 1 | u \in \mathcal{U}, i \in \mathcal{I}\}$. User u (item i) which is considered as a node on the bipartite graph is described with an embedding vector $\mathbf{e}_u^{(0)} \in \mathbb{R}^d (\mathbf{e}_i^{(0)} \in \mathbb{R}^d)$ at the 0-th layer. Based on the matrix $\mathbf{R}$, we obtain the adjacency matrix $\mathbf{A} \in \mathbb{R}^{(M+N) \times (M+N)}$ of the user-item bipartite graph as:

$$\mathbf{A} = \begin{pmatrix} 0 & \mathbf{R} \\ \mathbf{R}^T & 0 \end{pmatrix}. \tag{1}$$

The neighborhood aggregation of the k-th graph convolutional layer can be formulated as follows:

$$\mathbf{E}^{(k+1)} = \sigma(\hat{\mathbf{A}} \mathbf{E}^{(k)} \mathbf{W}^{(k+1)}). \tag{2}$$

where $\hat{\mathbf{A}} = \mathbf{D}^{-1/2} \mathbf{A} \mathbf{D}^{-1/2}$ is a symmetrically normalized Laplacian transformation matrix, and $\mathbf{D}$ is a $(M+N) \times (M+N)$ diagonal node degree matrix. $\mathbf{E}^{(k)}$ is the k-th layer representation of users and items, and the initial state is the

0-th layer embedding matrix $\mathbf{E}^{(0)}$. The final embeddings of users and items are generated by a pooling function:

$$\mathbf{E}' = \text{pooling}(\mathbf{E}^{(0)}, ..., \mathbf{E}^{(K)}). \tag{3}$$

In the end, the prediction is usually defined as the inner product of the final embeddings:

$$\hat{y}_{u,i} = \mathbf{e}'^T_u \mathbf{e}'_i. \tag{4}$$

2.2   LightGCN Brief

To achieve better performance and effectiveness, LightGCN [6] removes the feature transformation and non-linear activation components and simply utilizes the Laplacian norm to aggregate neighbors. The graph convolution operation in LightGCN is as follows:

$$
\begin{aligned}
\mathbf{e}_u^{(k+1)} &= \sum_{i \in \mathcal{N}_u} \frac{1}{\sqrt{|\mathcal{N}_u|}\sqrt{|\mathcal{N}_i|}} \mathbf{e}_i^{(k)}, \\
\mathbf{e}_i^{(k+1)} &= \sum_{u \in \mathcal{N}_i} \frac{1}{\sqrt{|\mathcal{N}_i|}\sqrt{|\mathcal{N}_u|}} \mathbf{e}_u^{(k)},
\end{aligned}
\tag{5}
$$

where $\mathbf{e}_u^{(k)}$ and $\mathbf{e}_i^{(k)}$ indicate the k-th layer hidden embeddings of user u and item i, respectively. $\mathcal{N}_u$ denotes the set of user u's first-hop neighbors, and $|\mathcal{N}_u|$ denotes the number of user u's first-hop neighbors. After K-layer message propagation, LightGCN generates the final representations of user u and item i by weighted combination:

$$
\begin{aligned}
\mathbf{e}'_u &= \sum_{k=0}^{K} \frac{1}{k+1} \mathbf{e}_i^{(k)}, \\
\mathbf{e}'_i &= \sum_{k=0}^{K} \frac{1}{k+1} \mathbf{e}_u^{(k)}.
\end{aligned}
\tag{6}
$$

Compared with the conventional MF methods, LightGCN benefits from the user-item interaction graph. First, the final representation is generated via the graph encoder, and thereby contains higher-order interaction information on the user-item interactions graph. So the representation provides more fine-grained information than the initial layer of embedding and is more informative to obtain a more accurate prediction. However, this graph convolution operation is quite time-consuming, which causes the slow convergence rate problem to be more unacceptable than non-GCN based methods.

2.3   Empirical Study on LightGCN

To investigate the convergence speed of LightGCN, we first record the epochs needed during training of LightGCN, MF, and the proposed ALGCN on

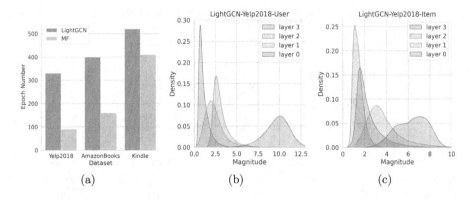

Fig. 1. Empirical study on LightGCN.

Yelp2018, Amazon-Book, and Kindle. To further study of properties of Light-GCN, we plot the distribution of user/item embedding and hidden layer representation's magnitudes of a pretrained 3-layer LightGCN on Yelp2018, and the results are shown in Fig. 1 (b) and (c). In brief, we have the following findings:

- **Low Convergence Rate.** In Fig. 1(a), we can observe that LightGCN exhibits the lowest speed during training in all methods and is much slower than traditional MF, which may be due to the symmetrically normalized Laplacian convolution operation in LightGCN.
- **The Scale Oscillation Issue.** As shown in Fig. 1(b) and (c), Layer 0 to 3 mean different layer representations of users or items. We observe that user/item 0-th layer embedding has the largest magnitude overall. As the number of layers grows larger, the distribution of magnitude seems to converge more to the uniform distribution. We define the *scale oscillation* issue as: *The scale of the representation vectors is neither stable in a specific range nor shares the same trend of increasing or decreasing across different layers. Instead, it oscillates.* As an example, in Fig. 1(b), the ascending order of the mean magnitude of different layers is layer 3, layer 1, layer 2, and layer 0 instead of layer 3, layer 2, layer 1, and layer 0 which is oscillatory and unstable.

3   Methodology

In the former section, we have discussed several issues raised with the current methods of graph collaborative filtering. To alleviate these issues, we propose the ALGCN model, which is illustrated in Fig. 2. There are three components in ALGCN: (1) graph convolution operation with random-walk normalized Laplacian; (2) influence-aware graph convolutional operation that enhances message aggregated by the head user (or item) for accelerating convergence rate; (3) simplified and augmentation-free contrastive loss to regulate the uniformity of representation effectively. Furthermore, we analyze the time complexity of ALGCN and other existing methods.

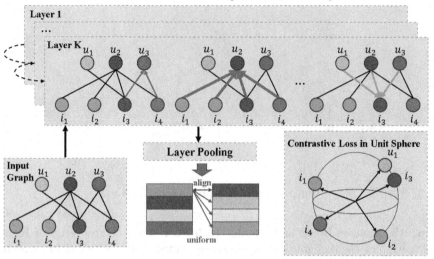

Fig. 2. Overview of our proposed ALGCN model. The dark blue or orange nodes represent the head users or items. And the thickness of the arrow in the convolution layer indicates the influence of target nodes. (Color figure online)

3.1 Random-Walk Normalized Laplacian

Unlike the symmetric Laplacian normalization term, we have applied the target node's in-degree normalization term to balance the effect across different layers of embeddings, also called the random-walk normalized Laplacian. The graph convolution operation is defined as:

$$
\begin{aligned}
e_u^{(k+1)} &= \sum_{i \in \mathcal{N}_u} \frac{1}{|\mathcal{N}_u|} e_i^{(k)}, \\
e_i^{(k+1)} &= \sum_{u \in \mathcal{N}_i} \frac{1}{|\mathcal{N}_i|} e_u^{(k)}.
\end{aligned}
\tag{7}
$$

The normalization term $\frac{1}{|\mathcal{N}_i|}$ or $\frac{1}{|\mathcal{N}_u|}$ only consider the degree of the target nodes. Then we can provide the matrix equivalent as:

$$
\mathbf{E}^{(k+1)} = \mathbf{D}^{-1} \mathbf{A} \mathbf{E}^{(k)}.
\tag{8}
$$

3.2 Influence-Aware Graph Convolution Module

In real-world, the distribution of node's degree in many graphs is long-tailed, and bipartite graphs in the recommendation are no exception. On the one hand, the number of item interactions follows a power-law distribution where most items only have a few interactions. On the other hand, a few items with high degrees have a significant influence on the whole graph. However, the existing GCN-based methods like LightGCN ignore this perspective and suffer from inefficient graph convolution operation.

Based on the discussion above, we assume that the message aggregated by high-degree nodes from neighbors is more important and influential in the whole graph. On the contrary, low-degree nodes that lack supervision signals may bring more noise during propagation. To enhance the high-degree node's message during the neighbor aggregation process, we devise a novel influence-aware graph convolution network intuitively as follows:

$$
\begin{aligned}
e_u^{(k+1)} &= f(|\mathcal{N}_u|) \sum_{i \in \mathcal{N}_u} \frac{1}{|\mathcal{N}_u|} e_i^{(k)}, \\
e_i^{(k+1)} &= f(|\mathcal{N}_i|) \sum_{i \in \mathcal{N}_i} \frac{1}{|\mathcal{N}_i|} e_u^{(k)}.
\end{aligned}
\tag{9}
$$

where $f(\cdot)$ denotes an influence function to scale the weight of the aggregation. And the matrix equivalent is as follows:

$$
\mathbf{E}^{(k+1)} = F(\mathbf{D})\mathbf{D}^{-1}\mathbf{A}\mathbf{E}^{(k)}.
\tag{10}
$$

where $F(\cdot)$ is an element-wise function, which maps each node to its influence based on its degree.

3.3   Influence Functions

We hope that the influence function applied to the embedding enhancement should have the following properties:

- **Non-Negative.** Since the interaction involved in the graph is considered as positive feedback, influence function should not reverse.
- **Monotone-Increasing.** Nodes with higher degree should be of more influence.
- **Passing Through Specific Points.** We define that nodes with zero degree have no influence, and nodes with one degree have one unit influence. So the function should pass through two specific points.

Several elementary functions are available options for the above properties. Here are two examples:

$$
\begin{aligned}
f_{power}(|\mathcal{N}_i|) &= |\mathcal{N}_i|^{\alpha}, \\
f_{log}(|\mathcal{N}_i|) &= \log_2(|\mathcal{N}_i| + 1).
\end{aligned}
\tag{11}
$$

In our proposed ALGCN, the power function is selected. The power α of the power function can be a tunable hyperparameter and provides the power function more flexibility across different datasets (Fig. 3).

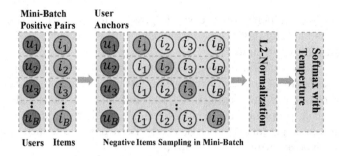

Fig. 3. Flowchart of contrastive loss with negative sampling in mini-batch.

3.4 Simplified Contrastive Loss

Different from the current CL methods [11,18] generating multiple views for the same use (item) with auxiliary data augmentation, we propose a novel simplified contrastive loss with efficient in-batch negative sampling. Recent studies [20, 21] have re-considered the necessity of graph augmentation, which may bring limited performance gains and need more running times. Besides, CL methods' outstanding performance is due to intensify uniformity of the feature distribution [15]. We intuitively treat the interacted user-item pair (u_j, i_j) as the positive pair and the user-item pair (u_j, i_k) as the negative pair when $j \neq k$ in the same mini-batch. Firstly, we directly matching user (item) on the unit hypersphere to easily preserve uniformity [15,21]:

$$\tilde{\mathbf{e}}' = \frac{\mathbf{e}'}{||\mathbf{e}'||_2}. \tag{12}$$

Following the design of InfoNCE [12], we get the objective function as:

$$L_{contrastive} = -\sum_{j=1}^{B} \log \frac{\exp(\frac{\langle \tilde{\mathbf{e}}'_{u_j}, \tilde{\mathbf{e}}'_{i_j} \rangle}{\tau})}{\exp(\frac{\langle \tilde{\mathbf{e}}'_{u_j}, \tilde{\mathbf{e}}'_{i_j} \rangle}{\tau}) + \sum_{k=1, k \neq j}^{B} \exp(\frac{\langle \tilde{\mathbf{e}}'_{u_j}, \tilde{\mathbf{e}}'_{i_k} \rangle}{\tau})}, \tag{13}$$

where B indicates the batch size and j is the index of user (item) in the same mini-batch.

3.5 Optimization

BPR Loss. We utilize the *Bayesian Personalized Ranking* (BPR) loss [14] to encourage the observed entries have higher prediction, which is widely used in Top-K recommendation. And the loss function is defined as follows:

$$L_{bpr} = \sum_{(u,i^+,i^-) \in \mathcal{O}} -ln\sigma(\hat{y}_{ui+} - \hat{y}_{ui-}), \tag{14}$$

Table 1. The comparison of time complexity.

Model	LightGCN	SGL	SimGCL	XSimGCL	ALGCN
Adjacency matrix	$\mathcal{O}(|\mathbf{A}|)$	$\mathcal{O}(|\mathbf{A}| + 2\rho|\mathbf{A}|)$	$\mathcal{O}(|\mathbf{A}|)$	$\mathcal{O}(|\mathbf{A}|)$	$\mathcal{O}(|\mathbf{A}|)$
Graph encoding	$\mathcal{O}(|\mathbf{A}|Kd)$	$\mathcal{O}((1+2\rho)|\mathbf{A}|Kd)$	$\mathcal{O}(3|\mathbf{A}|Kd)$	$\mathcal{O}(|\mathbf{A}|Kd)$	$\mathcal{O}(|\mathbf{A}|Kd)$
Contrastive loss	–	$\mathcal{O}(B^2d)$	$\mathcal{O}(B^2d)$	$\mathcal{O}(B^2d)$	$\mathcal{O}(B^2d)$

Table 2. The statistics of experimental datasets.

Dataset	# Users	# Items	# Interactions	Density
Yelp2018	31668	38048	1561406	0.00130
Amazon-Book	52643	91599	2984108	0.00062
Kindle	138333	81614	1607813	0.00007

where $\mathcal{O}$ is the observed training set, in which i^+ and i^- indicate a positive item and a negative item of user u respectively. The overall loss function is defined as:

$$L = L_{bpr} + \lambda L_{contrastive} + \lambda_{reg}\left\|\mathbf{E}^{(0)}\right\|, \tag{15}$$

where λ_{reg} controls the strength of L_2 regularization.

3.6 Complexity Analysis

To analyze the theoretical time complexity of ALGCN, we compare it with LightGCN and its graph-augmentation-based CL methods. Let $|\mathbf{A}|/2$ be the edge number in the bipartite graph, B be the batch size, d be the embedding size, and ρ be the edge keep rate in SGL [18]. The results are listed in Table 1.

SGL and SimGCL need the highest costs in these models with additional augmentation operations. Moreover, we can see that ALGCN is as theoretically time-efficient as XSimGCL [20], a lightweight counterpart of SimGCL [21], due to the simplified contrastive loss (Table 2).

4 Experiments

4.1 Experimental Settings

Datasets. We use three publicly available benchmark datasets: Yelp2018 [6], Amazon-Book (Amazon for short) [18], and Kindle [18] to evaluate the performance of ALGCN. To keep the comparison fair and reliable, we use the same experimental settings (including train/test splits) as those of [6,20,21]. The evaluation metrics are Recall@20 and Normalized Discounted Cumulative Gain (NDCG@20) [5] for Top-K recommendation.

Hyper-parameter Settings. For a fair comparison, the embedding size is fixed to 64, and the batch size is 2048 for all the models. The L_2 regularization λ_{reg} is $1e^{-4}$ and we use the Xavier method [3] to initialize all the embeddings. Besides, we optimize all the baselines with Adam and set a default learning rate of 0.001. In ALGCN, we set the temperature $\tau = 0.1$, which is a general setting.

Compared Baselines

- **MF (2012)** [9] Matrix factorization optimized by the Bayesian personalized ranking (BPR) loss is a way to learn users' and items' latent features by directly exploiting the explicit user-item interactions.
- **LightGCN (2020)** [6] is an effective and widely used GCN-based CF which removes the feature transformation and non-linear activation.
- **MixGCF (2021)** [7] Different from sampling raw negatives from data, MixGCF designs the hop mixing technique to synthesize hard negatives for improving GNN-based recommender systems.
- **BUIR (2021)** [10] adopts two distinct encoder networks that learn from each other, which does not require negative sampling.
- **DNN+SSL (2021)** [19] is a model architecture agnostic self-supervised learning framework that conducts feature masking for CL. Furthermore, here we build it based on a two-tower DNN to follow the original paper.
- **NCL (2022)** [11] is a very recent contrastive model which designs a proto-typical contrastive approach optimized with EM algorithm to learn the correlations between a user(or an item) and its neighbors from graph structure and semantic space.
- **SGL-ED (2021)** [18] SGL is a self-supervised learning framework that provides three types of graph augmentation operation. And SGL-ED is the best version on paper.
- **SimGCL (2022)** [21] is a recent simple CL method without graph augmentation, which generates contrastive views by adding uniform noises.
- **XSimGCL (2022)** [20] makes a step further to design an extremely simple graph contrastive learning method for recommendation with lower computational complexity (Table 3).

4.2 Performance Comparison with LightGCN

As one of the core ideas of this work is that symmetrically normalized Laplacian transformation matrix suffers from many limitations, in this section, we perform a detailed comparison with LightGCN, which is a typical and widely used GCN-based CF. We show the performance and relative improvement at different layers and record the running epochs during training. Besides, we further study the benefits of ALGCN from the long-tail perspective and the *scale oscillation* issue. We have the following observations:

- **Performance.** ALGCN outperforms LightGCN by a large margin on the three datasets. Especially on Amazon-Book, we can see that the proposed

Table 3. Performance comparison with LightGCN.

Dataset	Layer#	1 Layer		2 Layers		3 Layers	
	Method	LightGCN	ALGCN	LightGCN	ALGCN	LightGCN	ALGCN
Yelp2018	Recall↑	0.0566	0.0682	0.0593	0.0723	0.0621	0.0728
		–	+20.5%	–	+21.9%	–	+17.2%
	NDCG↑	0.0462	0.0565	0.0485	0.0599	0.0508	0.0600
		–	+22.3%	–	+23.5%	–	+18.1%
	Epoch	110	10	250	8	330	8
		–	11x	–	31.3x	–	41.3x
Amazon	Recall↑	0.0355	0.0549	0.0383	0.0564	0.0404	0.0581
		–	+54.6%	–	+47.3%	–	+43.8%
	NDCG↑	0.0279	0.0434	0.0297	0.0451	0.0316	0.0466
		–	+55.6%	–	+51.9%	–	+47.5%
	Epoch	100	18	250	19	400	14
		–	5.6x	–	13.2x	–	28.6x
Kindle	Recall↑	0.1720	0.2374	0.1922	0.2379	0.1909	0.2415
		–	+38.0%	–	+23.8%	–	+26.5%
	NDCG↑	0.1038	0.1596	0.1186	0.1599	0.1170	0.1655
		–	+53.8%	–	+34.8%	–	+41.5%
	Epoch	390	33	420	22	520	17
		–	11.8x	–	19.1x	–	30.6x

model is around 50% relative higher than LightGCN at all three layers. On quite sparse Amazon-Book and Kindle datasets, ALGCN outperforms Light-GCN with fewer layers, which shows that our model can leverage neighborhood information more effectively.

- **Training Efficiency.** Different from the slow convergence of LightGCN, ALGCN achieves significant speedups over it. For example, under the three layers setting, on Yelp2018, ALGCN only needs eight epochs during training, while LightGCN requires 330 epochs which is 41.3 times higher than ALGCN. Another interesting finding is that LightGCN needs more epochs, but ALGCN needs fewer epochs when the number of layers increases.

- **Long Tail Recommendation.** Following the [18] settings, we split the items into ten groups, and the smaller the group id is, the smaller degrees the items have. Due to the space limit, we only report the results on Yelp2018 and Amazon-Book with the Recall@20 metric. Figure 4 shows that LightGCN is inclined to expose popular items, especially on Amazon, due to the sparse interaction signals. In contrast, our ALGCN only performs worse in the 10-th group but significantly better in others, which admits that ALGCN alleviates the long-tail issue.

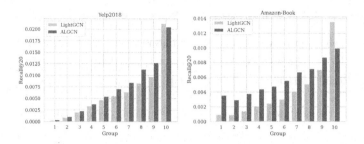

Fig. 4. Performance Comparison of Long-tail Recommendation.

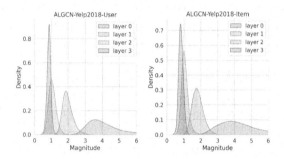

Fig. 5. The distribution of representation's Magnitude in different layers.

- **The Scale Oscillation Issue.** Similar to the experiment setting as before, we also plot the distribution of different layer representation's magnitudes of ALGCN on Yelp2018. Figure 5 shows that ALGCN effectively deals with the scale oscillation issue on both the user and item sides, which indicates that ALGCN can generate more stable scales between different layers.

4.3 Performance Comparison with the State-of-the-Arts

To confirm the ALGCN's outstanding performance, we further compare it with the recent contrastive learning-based state-of-the-art methods. Table 4 shows the overall performance comparison in terms of Recall@20 and NDCG@20. In summary, we have the following observations:

- ALGCN achieves the overall best performance on all three datasets. In detail, the proposed model reaches an improvement over the strongest baseline in terms of Recall@20 by 0.69%, 14.4%, 12.4% on Yelp2018, Amazon-Book, and Kindle, and only slightly worse than XSimGCL in terms of NDCG@20 by 0.66% on Yelp2018, probably because ALGCN is inclined to learn the rare interaction signals on sparse datasets.
- The GCN-based CF is better than others in general, showing the benefits of leveraging neighbor messages for representation learning. On the other hand,

Table 4. Overall performance comparison.

Method	Yelp2018		Amazon-Book		Kindle	
	Recall@20	NDCG@20	Recall@20	NDCG@20	Recall@20	NDCG@20
MF	0.0489	0.0394	0.0305	0.0236	0.1694	0.1021
BUIR	0.0487	0.0404	0.0260	0.0209	0.0922	0.0528
DNN+SSL	0.0483	0.0382	0.0438	0.0337	0.1520	0.0989
LightGCN	0.0621	0.0508	0.0404	0.0316	0.1909	0.1170
MixGCF	0.0713	0.0589	0.0485	0.0378	0.2098	0.1355
NCL	0.0670	0.0562	0.0429	0.0335	0.2090	0.1348
SGL-ED	0.0675	0.0555	0.0478	0.0379	0.2090	0.1352
SimGCL	0.0721	0.0601	0.0515	0.0414	0.2104	0.1374
XSimGCL	0.0723	**0.0604**	0.0508	0.0405	0.2147	0.1415
ALGCN	**0.0728**	0.0600	**0.0581**	**0.0466**	**0.2415**	**0.1655**
Improv	0.69%	−0.66%	14.4%	12.6%	12.4%	16.9%

non-GCN-based methods like BUIR and DNN+SSL, which need auxiliary user/item features are not powerful, because abundant features are unavailable in our experiments.

- Among the baselines, XSimGCL and SimGCL exhibit the strongest performance, which is consistent with the claim of the literature [20, 21]. This verifies the effectiveness of self-supervised learning in GCN-based CF.

4.4 Ablation Study of ALGCN

In this section, to investigate the effect of influence-aware graph convolution operation and the simplified contrastive loss in ALGCN, we design the following variants of ALGCN: i) w/o cl: the variant of ALGCN without the contrastive loss. ii) w/o inf&cl: we further remove the influence-aware graph convolution operation based on ALGCN$_{w/o\ cl}$ variant. From the results in Fig. 6(a) (b), two variants have a significantly varying decline in terms of Recall@20 and NDCG@20 in all cases. To study the impact on convergence speed, we record the number of epochs needed during training which is summarized in Fig. 6(c). We find that ALGCN$_{w/o\ cl}$ with the influence-aware graph operation accelerates the convergence speed tremendously compared to ALGCN$_{w/o\ inf\&cl}$. Furthermore, ALGCN reaches the least epochs, which indicates that our proposed simplified contrastive loss for regulating uniformity can benefit the convergence rate.

4.5 Parameter Sensitivity Analysis

We present the results of our hyper-parameters of α and λ for controlling the strength of high degree node's influence and the weight of contrastive loss.

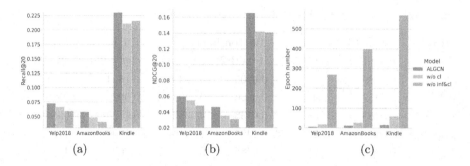

Fig. 6. Effect of ablation study.

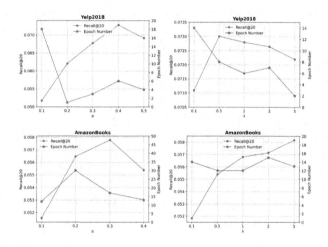

Fig. 7. Impact of α and λ on Yelp2018 and Amazon-Book.

Due to the space limit, we only report the results on Yelp2018 and Amazon-Book with two metrics and needed epochs. In detail, we search α and λ from the range of (0.1, 0.2, 0.3, 0.4, 0.5) and (0.1, 0.5, 1, 2, 5). On Amazon-Book, setting $\alpha = 0.5$ gets Nan during training because of excessive message enhancement, so we do not report it. From the results in Fig. 7, we find that the best performance is achieved by $\alpha = 0.4$ and $\lambda = 0.5$ on Yelp2018, $\alpha = 0.3$ and $\lambda = 5$ on Amazon-Book, which shows that lower α may limit the influence of popular users/items and higher α has the excessive message problem. Moreover, the small value of λ may reduce the uniformity of embeddings, and large value of λ can disturb the downstream task of recommendation.

5   Conclusion

In this paper, we first analyze the limitations of LightGCN, which is a representative work of GCN-based CF. Based on empirical studies, we reveal that

LightGCN is time-consuming in training and suffers from the scale oscillation issue. Then, we propose a novel Accelerated Light Graph Convolution Network (ALGCN) for collaborative filtering. Specifically, influence-aware graph convolution operation scales the representation with the node influence to enhance the messages aggregated by head nodes. To preserve better uniformity of the node's representation, we design a new augmentation-free in-batch contrastive loss on the unit hypersphere. The extensive experiments on three large datasets demonstrate the advantages of our method regarding training convergence, the scale oscillation issue, and long-tail recommendation.

Acknowledgments. This work was supported by NSFC (62276277 and U1911401), and Guangdong Basic and Applied Basic Research Foundation (2022B1515120059).

References

1. Chen, T., Kornblith, S., Norouzi, M., Hinton, G.E.: A simple framework for contrastive learning of visual representations. In: Proceedings of the 37th International Conference on Machine Learning, ICML 2020, 13–18 July 2020, Virtual Event. Proceedings of Machine Learning Research, vol. 119, pp. 1597–1607. PMLR (2020)
2. Chen, Y.H., Huang, L., Wang, C.D., Lai, J.H.: Hybrid-order gated graph neural network for session-based recommendation. IEEE Trans. Ind. Inf. **18**(3), 1458–1467 (2022)
3. Glorot, X., Bengio, Y.: Understanding the difficulty of training deep feedforward neural networks. In: Proceedings of the Thirteenth International Conference on Artificial Intelligence and Statistics, AISTATS 2010, Chia Laguna Resort, Sardinia, Italy, 13–15 May 2010. JMLR Proceedings, vol. 9, pp. 249–256. JMLR.org (2010)
4. Goldberg, D., Nichols, D.A., Oki, B.M., Terry, D.B.: Using collaborative filtering to weave an information tapestry. Commun. ACM **35**(12), 61–70 (1992)
5. He, X., Chen, T., Kan, M.Y., Chen, X.: TriRank: review-aware explainable recommendation by modeling aspects. In: Proceedings of the 24th ACM International Conference on Information and Knowledge Management, CIKM 2015, Melbourne, VIC, Australia, 19–23 October 2015, pp. 1661–1670. ACM (2015)
6. He, X., Deng, K., Wang, X., Li, Y., Zhang, Y.D., Wang, M.: LightGCN: simplifying and powering graph convolution network for recommendation. In: Proceedings of the 43rd International ACM SIGIR Conference on Research and Development in Information Retrieval, SIGIR 2020, Virtual Event, China, 25–30 July 2020, pp. 639–648. ACM (2020)
7. Huang, T., et al.: MixGCF: an improved training method for graph neural network-based recommender systems. In: KDD 2021: The 27th ACM SIGKDD Conference on Knowledge Discovery and Data Mining, Virtual Event, Singapore, 14–18 August 2021, pp. 665–674. ACM (2021)
8. Kipf, T.N., Welling, M.: Semi-supervised classification with graph convolutional networks. CoRR abs/1609.02907 (2016)
9. Koren, Y., Bell, R.M., Volinsky, C.: Matrix factorization techniques for recommender systems. Computer **42**(8), 30–37 (2009)
10. Lee, D., Kang, S., Ju, H., Park, C., Yu, H.: Bootstrapping user and item representations for one-class collaborative filtering. In: SIGIR 2021: The 44th International ACM SIGIR Conference on Research and Development in Information Retrieval, Virtual Event, Canada, 11–15 July 2021, pp. 1513–1522. ACM (2021)

11. Lin, Z., Tian, C., Hou, Y., Zhao, W.X.: Improving graph collaborative filtering with neighborhood-enriched contrastive learning. In: WWW 2022: The ACM Web Conference 2022, Virtual Event, Lyon, France, 25–29 April 2022, pp. 2320–2329. ACM (2022)

12. Liu, Y., et al.: Contrastive predictive coding with transformer for video representation learning. Neurocomputing **482**, 154–162 (2022)

13. Miao, H., Li, A., Yang, B.: Meta-path enhanced lightweight graph neural network for social recommendation. In: Database Systems for Advanced Applications - 27th International Conference, DASFAA 2022, Virtual Event, 11–14 April 2022, Proceedings, Part II. vol. 13246, pp. 134–149. Springer, Cham (2022). https://doi.org/10.1007/978-3-031-00126-0_9

14. Rendle, S., Freudenthaler, C., Gantner, Z., Schmidt-Thieme, L.: BPR: Bayesian personalized ranking from implicit feedback. CoRR abs/1205.2618 (2012)

15. Wang, T., Isola, P.: Understanding contrastive representation learning through alignment and uniformity on the hypersphere. In: Proceedings of the 37th International Conference on Machine Learning, ICML 2020, 13–18 July 2020, Virtual Event. Proceedings of Machine Learning Research, vol. 119, pp. 9929–9939. PMLR (2020)

16. Wang, X., He, X., Wang, M., Feng, F., Chua, T.S.: Neural graph collaborative filtering. In: Proceedings of the 42nd International ACM SIGIR Conference on Research and Development in Information Retrieval, SIGIR 2019, Paris, France, 21–25 July 2019, pp. 165–174. ACM (2019)

17. Wu, F., Zhang, T., de Souza, A.H. Jr., Fifty, C., Yu, T., Weinberger, K.Q.: Simplifying graph convolutional networks. CoRR abs/1902.07153 (2019)

18. Wu, J., et al.: Self-supervised graph learning for recommendation. In: SIGIR 2021: The 44th International ACM SIGIR Conference on Research and Development in Information Retrieval, Virtual Event, Canada, 11–15 July 2021, pp. 726–735. ACM (2021)

19. Yao, T., et al.: Self-supervised learning for large-scale item recommendations. In: CIKM 2021: The 30th ACM International Conference on Information and Knowledge Management, Virtual Event, Queensland, Australia, 1–5 November 2021, pp. 4321–4330. ACM (2021)

20. Yu, J., Xia, X., Chen, T., Cui, L., Hung, N.Q.V., Yin, H.: XSimGCL: towards extremely simple graph contrastive learning for recommendation. CoRR abs/2209.02544 (2022)

21. Yu, J., Yin, H., Xia, X., Chen, T., Cui, L., Nguyen, Q.V.H.: Are graph augmentations necessary?: Simple graph contrastive learning for recommendation. In: SIGIR 2022: The 45th International ACM SIGIR Conference on Research and Development in Information Retrieval, Madrid, Spain, 11–15 July 2022, pp. 1294–1303. ACM (2022)

22. Zhang, X., Sha, C.: Fully utilizing neighbors for session-based recommendation with graph neural networks. In: Database Systems for Advanced Applications - 27th International Conference, DASFAA 2022, Virtual Event, April 11–14, 2022, Proceedings, Part II, vol. 13246, pp. 36–52 (2022). https://doi.org/10.1007/978-3-031-00126-0_3

Multi-view Spatial-Temporal Enhanced Hypergraph Network for Next POI Recommendation

Yantong Lai[1,2], Yijun Su[3,4]($\boxtimes$), Lingwei Wei[1,2], Gaode Chen[1,2], Tianci Wang[1,2], and Daren Zha[1]

[1] Institute of Information Engineering, Chinese Academy of Sciences, Beijing, China
{laiyantong,weilingwei,chengaode,wangtianci,zhadaren}@iie.ac.cn
[2] School of Cyber Security, University of Chinese Academy of Sciences, Beijing, China
[3] JD iCity, JD Technology, Beijing, China
suyijun.ucas@gmail.com
[4] JD Intelligent Cities Research, Beijing, China

Abstract. Next point-of-interest (POI) recommendation has been a prominent and trending task to provide next suitable POI suggestions for users. Current state-of-the-art studies have achieved considerable performances by modeling user-POI interactions or transition patterns via graph- and sequential-based methods. However, most of them still could not well address two major challenges: 1) Ignoring important spatial-temporal correlations during aggregation within user-POI interactions; 2) Insufficiently uncovering complex high-order collaborative signals across users to overcome sparsity issue. To tackle these challenges, we propose a novel method Multi-View Spatial-Temporal Enhanced Hypergraph Network (MSTHN) for next POI recommendation, which jointly learns representations from local and global views. In the local view, we design a spatial-temporal enhanced graph neural network based on user-POI interactions, to aggregate and propagate spatial-temporal correlations in an asymmetric way. In the global view, we propose a stable interactive hypergraph neural network with two-step propagation scheme to capture complex high-order collaborative signals. Furthermore, a user temporal preference augmentation strategy is employed to enhance the representations from both views. Extensive experiments on three real-world datasets validate the superiority of our proposal over the state-of-the-arts. To facilitate future research, we release the codes at https://github.com/icmpnorequest/DASFAA2023_MSTHN.

Keywords: Next POI recommendation · Spatial-temporal graph · Hypergraph neural network

1 Introduction

Location-based social networks (LBSNs) have provided open platforms for users to share their experience at different point-of-interests (POIs), such as restaurants and shopping malls. Therefore, POI recommender systems have been

X. Wang et al. (Eds.): DASFAA 2023, LNCS 13944, pp. 237–252, 2023.
https://doi.org/10.1007/978-3-031-30672-3_16

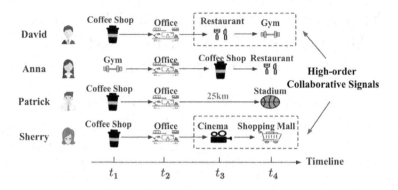

Fig. 1. A motivating example of our proposed framework

widely utilized to help users and service providers for exploration and targeted advertising, respectively. Among various POI recommendation tasks, next POI recommendation is arguably a prominent and trending one. Different from conventional POI recommendation focusing on user's general long-term preference, next POI recommendation considers user's recent spatial-temporal contexts [21] and long- and short-term preferences for next suitable location suggestions [10,16,17,31].

Prior next POI recommendation approaches are mainly based on sequential methods, ranging from Markov chain [2] to recurrent neural networks (RNNs) [3]. These methods treat it as a general sequence prediction task, and ignore important spatial-temporal information. Subsequently, researchers extend various of RNNs [17,31] by incorporating geographical distance, time intervals or spatio-temporal gates. However, RNN-based methods are limited to short-term contiguous visits. Inspired by the great success of self-attention mechanism [18] in natural language processing field, researchers [12,14] employ it to capture long-term dependencies and correlations between non-consecutive POIs. Nevertheless, they only focus on intra-sequence learning but fail to explore beyond sequence information. Recently, graph-based methods [4,8,10,13,15,20] leverage graph neural networks (GNNs) to refine latent representations of POIs from a global view. Despite their success in next POI recommendation, there still exist some limitations to be better explored.

1) First, ignoring spatial-temporal correlations during aggregation within a user-POI interaction graph. In next POI recommendation, previous GNN-based studies [4,8,10,13,15,20] mainly utilize GNNs to enrich representations from a global view. However, they either ignore or could not directly model spatial-temporal correlations during aggregation and propagation with GNNs. Take David and Anna in Fig. 1 for example, they visit the same POIs (coffee shop, office, restaurant and gym) but in different sequential order. If only aggregating their interacted POIs, the embeddings of them would be the same. However, latent representations of David and Anna should be different in fact, due to different sequential order. Additionally, since each user has her/his acceptance on distance, spatial influences should also be taken into account, as illustrated in Patrick's trajectory. Thus, how to mine and fuse spatial-temporal correlations within a user-POI interaction graph is well deserved to be explored in next POI recommendation.

2) Second, insufficient to uncover high-order collaborative signals. Some researchers [10] in next POI recommendation try to capture collaborative signals [19] by sampling one-hop POI neighbors randomly. Unfortunately, they overlook the high-order connectivity among POIs. As shown in Fig. 1, restaurant and gym are high-order neighbors of coffee shop in the trajectory of David. While in Sherry's, its high-order neighbors are cinema and shopping mall. Thus, restaurant, gym, cinema and shopping mall are potentially related, and there might exist implicit high-order collaborative signals among them. If uncovering such signals, it would help alleviate the data sparsity issue.

To this end, we propose a novel framework **Multi-View Spatial-Temporal Enhanced Hypergraph Network (MSTHN)** for next POI recommendation. To capture spatial-temporal correlations within user-POI interactions, we first design a local spatial-temporal enhanced graph neural network, which aggregates and propagates in an asymmetric way. Then, we construct a global interactive hypergraph to sufficiently uncover high-order collaborative signals with a designed two-step propagation scheme. Subsequently, in contrast to simple concatenation, we utilize a user temporal preference augmentation strategy to enhance the representations from both local and global views. Empirical results show that our MSTHN consistently outperforms state-of-the-art methods, e.g., average relative improvement of 36.20% over LightGCN, 20.36% over SGRec, 17.10% over STAN and 11.09% over DHCN in terms of Recall@10.

We summarize our main contributions as follows:

- To the best of our knowledge, this is the first attempt at multi-view spatial-temporal hypergraph network in next POI recommendation, which captures spatial-temporal correlations and high-order collaborative signals from local and global views.
- We propose a novel local spatial-temporal enhanced graph neural network to jointly model complex user-POI interactions, POI-POI sequential relations and non-adjacent POI-POI geographical relations, which aggregates and propagates spatial-temporal correlations in an asymmetric way.
- We design an interactive hypergraph to depict global interaction dependencies, which empowers to distill high-order collaborative signals effectively.
- Extensive experiments on three public available datasets validate the effectiveness of our proposed MSTHN over various state-of-the-art methods for next POI recommendation.

2 Related Work

2.1 Next POI Recommendation

Next POI recommendation aims to suggest next suitable location for users based on their recent spatial-temporal context and visiting behaviours. Early studies in next POI recommendation are mainly based on sequential methods, ranging from Markov chain [2] to recent RNN and its variants [3,17,31]. Limited to short-term contiguous visits in these methods, more recent studies [12,14] solve by utilizing self-attention mechanism [18] to model spatial-temporal information in an explicit or implicit way. However, the above studies only rely on each user's trajectory and overlook the potential collaborative signals among users. Since graph structure is naturally suitable to represent data in LBSN, some researchers have started to leverage graph-based techniques for next POI recommendation, ranging from graph [23] and hypergraph embeddings [24,25] to more recent GNNs [4,8,10,13,15,20]. However, most GNNs-based works [4,8,10,13,20] do not consider important spatial-temporal information within user-POI interactions. Rao et al. [15] noticed the importance of spatial-temporal and chronological information for next POI recommendation, but they still could not directly model such information in GNNs during aggregation and propagation. To tackle the challenge, we propose a local spatial-temporal enhanced graph neural network to capture spatial-temporal correlations during aggregation and propagation within a user-POI interaction graph.

2.2 Hypergraph Neural Network-Based Recommendation

Due to the extension structure and the ability in modeling complex high-order dependencies, hypergraph neural network [1,5] has been recently developed in various recommendation tasks, such as session recommendation [11,22], social recommendation [6,29] and group recommendation [30]. Inspired by these works, we design a learnable interactive hypergraph neural network to uncover global high-order collaborative signals across users.

3   Problem Formulation

Let $\mathcal{U} = \{u_1, u_2, ..., u_{|\mathcal{U}|}\}$ and $\mathcal{L} = \{l_1, l_2, ..., l_{|\mathcal{L}|}\}$ be a set of users and POIs, respectively. Each POI $l \in \mathcal{L}$ has unique geographical coordinates (*longitude, latitude*) tuple, i.e., (*lon, lat*). For each user $u \in \mathcal{U}$, we split her/his trajectory sequence into several sessions by specific time interval (i.e., 1 day) and obtain a trajectory sequence $S^u = \{S_1^u, S_2^u, ..., S_n^u\}$, where n denotes the number of sessions. Each session is denoted as $S_i^u = \{(l_j^u, t_{l_j^u}) | j = 1, 2, ...\}$, where each tuple $(l_j^u, t_{l_j^u})$ indicates user u visited POI l_j^u at timestamp $t_{l_j^u}$.

Given a target user u and her/his trajectory sequence S^u, the goal of next POI recommendation is to recommend top-K POIs that u may visit in the next timestamp.

4   Methodology

In this section, we present our proposed framework **M**ulti-View **S**patial-**T**emporal Enhanced **H**ypergraph **N**etwork (MSTHN) in detail. As illustrated

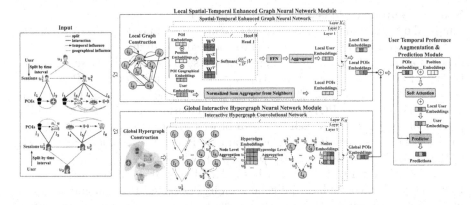

Fig. 2. The framework of our proposed MSTHN. It mainly contains three modules: 1) *Local spatial-temporal enhanced graph neural network module* to capture spatial-temporal correlations; 2) *Global interactive hypergraph neural network module* to uncover high-order collaborative signals; 3) *User temporal preference augmentation module* to augment user preference for prediction.

in Fig. 2, our MSTHN mainly consists of: 1) Local spatial-temporal enhanced graph neural network module captures spatial-temporal correlations within a user-POI interaction graph in the local view; 2) Global interactive hypergraph neural network module uncovers high-order collaborative signals with a two-step propagation scheme in the global view; 3) User temporal preference augmentation module fuses POIs latent representations from both local and global views and augments long- and short-term user temporal preference; 4) Prediction and optimization module predicts visiting probability from the learned POIs and users latent representations.

4.1 Local Spatial-Temporal Enhanced Graph Neural Network Module

The local spatial-temporal enhanced graph neural network module aims to capture spatial-temporal correlations during aggregation and propagation within user-POI interactions in the local view.

Local Spatial-Temporal Enhanced Graph Construction. To represent user-POI interactions and spatial-temporal correlations among interacted POIs, we firstly construct the local spatial-temporal enhanced graph $\mathcal{G}_L = (\mathcal{V}_L, \mathcal{E}_L)$ (Fig. 2). In the local graph $\mathcal{G}_L$, nodes $\mathcal{V}_L$ are users and POIs, and edges $\mathcal{E}_L$ consist of user-POI interactions, POI-POI sequential relations and non-adjacent POI-POI geographical relations.

Spatial-Temporal Message Embedding. To leverage important spatial-temporal information within local spatial-temporal enhanced graph $\mathcal{G}_L$, we firstly sort the interacted POIs of user u chronologically and the sorted POIs set is denoted as $T_u = \{l_1^u, l_2^u, ..., l_m^u\}$, where m is the sequence length of user u. Then, through look-up table, we obtain the initial embeddings for each POI in the sorted set $\mathbf{E}^u = \{\mathbf{e}_1^u, \mathbf{e}_2^u, ..., \mathbf{e}_m^u\}$, where $\mathbf{e}_i^u \in \mathbb{R}^d$ and the embedding dimension is d. Since the interacted POIs are in temporal sequential dependencies, we employ positional encoding [18], which has been proved effective in sequence modeling, to represent the sequential relationship among POIs. The position embeddings of the sorted POI set is $\mathbf{P}^u = \{\mathbf{p}_1^u, \mathbf{p}_2^u, ..., \mathbf{p}_m^u\}$, where $\mathbf{p}_i^u \in \mathbb{R}^d$.

As described in Fig. 1, each user has different spatial acceptance on choosing POIs. Thus, the geographical influence among interacted POIs should be taken into account. To achieve this goal, we construct a geographical adjacent matrix $\mathbf{A}_{geo} \in \mathbb{R}^{m \times m}$ to reflect the edge constraints among interacted POIs. For each (l_i^u, l_j^u) pair in the sorted set T_u, the geographical influence a_{ij} is defined as:

$$a_{ij} = \exp(-dist(d_i, d_j)^2) \tag{1}$$

here we choose Haversine distance as $dist(\cdot, \cdot)$ and d_i denotes the geographical coordinates of POI l_i^u. Additionally, we use Δ_d as the distance threshold, if $dist(d_i, d_j) > \Delta_d$, we set $a_{ij} = 0$. For simplicity, we modify Gaussian kernel function to represent the geographical influence between two POIs, which depicts the inverse correlation between geographical influence and distance, and controls the constraint ranging from 0 to 1. To capture the non-linear geographical influence among interacted POIs, we employ the graph convolutional network [9] as follows:

$$\mathbf{V}^u = \mathbf{A}_{geo}\mathbf{E}^u\mathbf{W}_{geo} + \mathbf{b}_{geo} \tag{2}$$

where $\mathbf{W}_{geo} \in \mathbb{R}^{d \times d}$ represents a transition matrix and $\mathbf{b}_{geo} \in \mathbb{R}^d$ is a bias vector.

Subsequently, we obtain the spatial-temporal message embeddings $\mathbf{Z}^u = \mathbf{E}^u + \mathbf{P}^u + \mathbf{V}^u$ by performing element-wise addition on initial embeddings, position embeddings and geographical embeddings, where $\mathbf{Z}^u \in \mathbb{R}^{m \times d}$.

Spatial-Temporal Graph Aggregation Layer. To aggregate important spatial-temporal message collected from interacted POI neighbors in graph $\mathcal{G}_L$, we design a novel spatial-temporal graph aggregation layer that models temporal dependency and non-linear geographical influence among POIs in the local view.

We utilize self-attention [18], an effective mechanism in sequence modeling, to capture sequential dependency and assign different weights to each POI within the interactions. Given the spatial-temporal message embeddings $\mathbf{Z}^u \in \mathbb{R}^{m \times d}$, the spatial-temporal graph aggregation layer firstly performs multi-head scaled dot-product attention operation to get spatial-temporal aware representations $\mathbf{h}^{T_u} \in \mathbb{R}^{m \times d}$ as follows:

$$\mathbf{h}_i^{T_u} = \text{softmax}\left(\frac{(\mathbf{Z}^u\mathbf{W}_Q)(\mathbf{Z}^u\mathbf{W}_K)^T}{\sqrt{D'}}\right)(\mathbf{Z}^u\mathbf{W}_V) \tag{3}$$

$$\mathbf{h}^{T_u} = \text{FFN}([\mathbf{h}_1^{T_u}; \mathbf{h}_2^{T_u}; ...; \mathbf{h}_H^{T_u}]) \tag{4}$$

where H denotes the number of heads in multi-head attention, $[\cdot; \cdot]$ represents the concatenation operation and $D' = \sqrt{d/H}$. Here, $\mathbf{W}_Q$, $\mathbf{W}_K$ and $\mathbf{W}_V \in \mathbb{R}^{d \times D'}$ are shared weight transformations. Additionally, feed-forward network could be represented as $\mathrm{FFN}(\mathbf{x}) = \mathbf{x}\mathbf{W}_0 + \mathbf{b}_0$, where $\mathbf{W}_0 \in \mathbb{R}^{D' \times d}$ and $\mathbf{b}_0 \in \mathbb{R}^d$ are trainable parameters.

Then, we apply mean pooling to obtain local central user representation $\mathbf{x}_L^u = \frac{1}{m+1} \sum_{i=1}^m \mathbf{h}_i^{T_u}$, where $\mathbf{x}_L^u \in \mathbb{R}^d$. It aggregates spatial-temporal information from one-hop neighbors in the local view and updates corresponding local node embeddings.

The user-item interaction matrix is $\mathbf{R} \in \mathbb{R}^{|\mathcal{U}| \times |\mathcal{L}|}$ and we define the adjacency matrix $\mathbf{A} \in \mathbb{R}^{(|\mathcal{U}|+|\mathcal{L}|) \times (|\mathcal{U}|+|\mathcal{L}|)}$ of local spatial-temporal enhanced graph $\mathcal{G}_L$ as:

$$\mathbf{A} = \begin{pmatrix} \mathbf{0} & \mathbf{R} \\ \mathbf{R}^T & \mathbf{0} \end{pmatrix} \tag{5}$$

Inspired by LightGCN [7], we also omit non-linear transformation and stack several spatial-temporal graph aggregation layers for propagation to update nodes embeddings:

$$\mathbf{X}_L^{(k+1)} = (\mathbf{D}_L^{-\frac{1}{2}} \mathbf{A} \mathbf{D}_L^{-\frac{1}{2}}) \mathbf{X}_L^{(k)} \tag{6}$$

where $\mathbf{D}_L \in \mathbb{R}^{(|\mathcal{U}|+|\mathcal{L}|) \times (|\mathcal{U}|+|\mathcal{L}|)}$ is a diagonal matrix and the 0-layer embedding matrix $\mathbf{X}_L^{(0)} \in \mathbb{R}^{(|\mathcal{U}|+|\mathcal{L}|) \times d}$ contains initial users embeddings and POIs embeddings. After propagating K_L layers, the final local nodes representations $\mathbf{X}_L \in \mathbb{R}^{(|\mathcal{U}|+|\mathcal{L}|) \times d}$ are generated by aggregator (i.e., mean-pooling or sum-pooling).

Different from STAM [27], our proposed spatial-temporal graph aggregation layer models both temporal sequential dependency and non-linear geographical influence among POIs jointly in the local view. Since a user may visit the same POI several times, if taking the chronologically interacted users into account, it would lead to a sub-optimal performance. That is, we only perform spatial-temporal graph aggregation operation in an asymmetric way for local central user node. Detailed empirical analysis would be introduced in Sect. 5.5.

4.2   Global Interactive Hypergraph Neural Network Module

The global interactive hypergraph neural network module aims to uncover high-order collaborative signals effectively with a two-step propagation scheme in the global view.

Global Interactive Hypergraph Construction. Motivated by the strength of hypergraph for unifying nodes beyond pairwise relations, we construct an interactive hypergraph $\mathcal{G}_H = (\mathcal{V}_H, \mathcal{E}_H)$ to uncover high-order collaborative signals across sessions. In the hypergraph $\mathcal{G}_H$, we represent each user's session in her/his trajectory sequence as an hyperedge and the interacted POIs within the session consist of nodes in the hyperedge (Fig. 2). Incidence matrix $\mathbf{H} \in \mathbb{R}^{|\mathcal{L}| \times |S|}$ is introduced to describe the topology structure of hypergraph, with entries defined as:

$$h(v, e) = \begin{cases} 1, & \text{if } e \text{ connects } v, \\ 0, & \text{otherwise} \end{cases} \tag{7}$$

For each node $v \in \mathcal{V}_H$, its degree is defined as $d(v) = \sum_{e \in \mathcal{E}_H} W_e h(v, e)$, calculating the occurrence of node v in all hyperedges. W_e is an assigned positive weight and all the weights formulate a diagonal matrix $\mathbf{W} \in \mathbb{R}^{|S| \times |S|}$. For each hyperedge $e \in \mathcal{E}_H$, its degree is $d(e) = \sum_{v \in \mathcal{V}_H} h(v, e)$. All the node degree and hyperedge degree form diagonal node degree matrix $\mathbf{D}_H$ and diagonal hyperedge degree matrix $\mathbf{B}$ respectively.

Hypergraph Convolutional Network. After the construction of hypergraph $\mathcal{G}_H$, we develop a hypergraph convolutional network with two-step information propagation scheme to capture high-order POI-level relations iteratively. In the node-hyperedge-node propagation scheme, hyperedges serve as mediums for nodes aggregation within the hyperedge and propagation across hyperedges (Fig. 2). Particularly, we design our hypergraph convolutional network as follows:

$$\mathbf{X}_H^{(k+1)} = \mathbf{D}_H^{-\frac{1}{2}} \mathbf{H} \mathbf{W} \mathbf{B}^{-1} \mathbf{H}^{\mathrm{T}} \mathbf{D}_H^{-\frac{1}{2}} \mathbf{X}_H^{(k)} \tag{8}$$

where $\mathbf{X}_H^{(k)} \in \mathbb{R}^{|\mathcal{L}| \times d}$ represents the embeddings of POIs, encoded from the $k-$th hypergraph convolutional network layer. In the first node to hyperedge propagation stage, we use multiplication $\mathbf{H}^T \mathbf{X}_H^{(k)}$ to denote the aggregation process, for $\mathbf{H}^T$ reflects the hyperedge-node relation. After aggregating nodes representations within each hyperedge, we then premultiply $\mathbf{H}$ to aggregate information from hyperedges to nodes. Since incidence matrix $\mathbf{H}$ represents the node-hyperedge relation, the second hyperedge to node propagation stage aims to leverage global information beyond current hyperedge to enrich nodes representations.

Distinct to spectral hypergraph convolutional HGNN [5], we omit nonlinear activation function for simplification. Unlike the simplified row normalization in DHCN [22], we keep the same row normalization as HGNN since it is more stable in propagation than the simplified one $\mathbf{D}^{-1} \mathbf{H} \mathbf{W} \mathbf{B}^{-1} \mathbf{H}^{\mathrm{T}}$ in DHCN. According to [5], the symmetric hypergraph Laplacian matrix $\mathbf{I} - \mathbf{D}^{-\frac{1}{2}} \mathbf{H} \mathbf{W} \mathbf{B}^{-1} \mathbf{H}^{\mathrm{T}} \mathbf{D}^{-\frac{1}{2}}$ is a positive semi-definite matrix, where $\mathbf{I} \in \mathbb{R}^{|\mathcal{L}| \times |\mathcal{L}|}$. Therefore, the eigenvalue of $\mathbf{D}^{-\frac{1}{2}} \mathbf{H} \mathbf{W} \mathbf{B}^{-1} \mathbf{H}^{\mathrm{T}} \mathbf{D}^{-\frac{1}{2}}$ is no larger than 1, solving the instability problem in propagation.

After propagating K_H hypergraph convolutional layers, we average the POIs representations obtained at each layer and output the final global POIs representations $\mathbf{X}_H \in \mathbb{R}^{|\mathcal{L}| \times d}$.

4.3   User Temporal Preference Augmentation Module

The user temporal preference augmentation module aims to fuse the learned representations from both local and global views and augment temporal-aware user preference.

In next POI recommendation, the final decision heavily depends on user's recent preference. Instead of simply aggregating or concatenating the interacted

POIs representations, inspired by [22], we integrate the reversed position embeddings for user temporal preference augmentation. After learning nodes representations from both views, we could obtain the embeddings of all POIs by element-wise addition, e.g., $\mathbf{X}^L = \mathbf{X}_L^L + \mathbf{X}_H$, where $\mathbf{X}_L^L \in \mathbb{R}^{|\mathcal{L}| \times d}$ and $\mathbf{X}_H \in \mathbb{R}^{|\mathcal{L}| \times d}$ denote POIs embeddings in the local and global view, respectively. The i-th POI temporal augmented embedding $\mathbf{x}_i^{u^*}$ in user u's sorted sequence T^u is defined as following:

$$\mathbf{x}_i^{u^*} = \tanh(\mathbf{W}_1[\mathbf{x}_i^u; \mathbf{p}_{m+1-i}] + \mathbf{b}_1) \tag{9}$$

where $\mathbf{W}_1 \in \mathbb{R}^{d \times 2d}$ and $\mathbf{b}_1 \in \mathbb{R}^d$ are trainable parameters. $\mathbf{x}_i^u \in \mathbb{R}^d$ could be indexed from POIs embeddings $\mathbf{X}^L$. Moreover, $\mathbf{p}_{m+1-i} \in \mathbb{R}^d$ denotes reverse position embedding.

Thus, with soft-attention mechanism, we could get temporal preference augmented embedding $\mathbf{x}_T^u \in \mathbb{R}^d$ of user u by assigning different attention weights:

$$\mathbf{x}_T^u = \sum_{i=1}^{m} \alpha_i \mathbf{x}_i^{u^*} \tag{10}$$

$$\alpha_i = \mathbf{q}^T \sigma(\mathbf{W}_2 \mathbf{x}^{u^*} + \mathbf{W}_3 \mathbf{x}_i^{u^*} + \mathbf{b}_2) \tag{11}$$

where $\mathbf{q} \in \mathbb{R}^d$, $\mathbf{W}_2, \mathbf{W}_3 \in \mathbb{R}^{d \times d}$ and $\mathbf{b}_2 \in \mathbb{R}^d$ are trainable attention parameters. $\mathbf{x}^{u^*} \in \mathbb{R}^d$ is aggregated by performing mean-pooling on all the interacted POIs embeddings of user u. σ denotes sigmoid activation function here.

4.4   Prediction and Optimization Module

Having obtained user u's local representation $\mathbf{x}_L^u \in \mathbb{R}^d$ and local-global aware temporal-augmented user representation $\mathbf{x}_T^u \in \mathbb{R}^d$, we apply element-wise addition to get the final user representation as $\mathbf{u} = \mathbf{x}_L^u + \mathbf{x}_T^u, \mathbf{u} \in \mathbb{R}^d$. After that, we compute the score by doing inner product between the final user representation $\mathbf{u}$ and target POI representation $\mathbf{x}^l \in \mathbb{R}^d$:

$$\hat{\mathbf{y}}_l^u = \text{softmax}(\mathbf{u}^T \mathbf{x}^l) \tag{12}$$

We formulate the learning objective as a cross-entropy loss function, which has been largely used in next POI recommendation:

$$\mathcal{J} = -\sum_{u \in \mathcal{U}} \sum_{i \in T^u} \sum_{j=1}^{|\mathcal{L}|} \mathbf{y}_{i,j}^u \log(\hat{\mathbf{y}}_{i,j}^u) + \lambda \|\Theta\|_2 \tag{13}$$

where $\mathbf{y}_{i,j}^u$ is an indicator that is equal to 1 if l_j is the ground truth and 0 otherwise. $\|\Theta\|_2$ represents the $L2$ regularization of all parameters for preventing over-fitting under the control of λ.

5   Experiments

In this section, we present our empirical results to evaluate the effectiveness of our MSTHN.

Table 1. Dataset statistics

	#Users	#POIs	#Check-ins	#Sessions	Sparsity
NYC	834	3,835	44,686	8,841	98.61%
TKY	2,173	7,038	308,566	41,307	97.82%
Gowalla	5,802	40,868	301,080	75,733	99.87%

5.1 Experimental Setting

Datasets. We conduct experiments on three public LBSN datasets: Foursquare-NYC (NYC for abbreviation), Foursquare-TKY (TKY) [26] and Gowalla [28]. NYC and TKY were collected from Apr. 2012 to Feb. 2013 in New York City and Tokyo, respectively, while Gowalla contains check-ins from Feb. 2009 to Oct. 2010. Following [17], we first eliminate unpopular POIs that are visited by less than 10 users and 5 users for Gowalla and Foursquare, respectively. Then, we split each user's complete check-ins into sessions within 1 day and remove those which includes fewer than 3 records. Furthermore, inactive users with less than 5 sessions for Gowalla and 3 sessions for Foursquare are filtered out. According to [17], the first 80% sessions of each user are used for training and the rest for testing. The statistics of pre-processed datasets are shown in Table 1.

Evaluation Metrics. Following previous works in next POI recommendation, we adopt two widely used evaluation metrics: Recall@K and Normalized Discounted Cumulative Gain (NDCG@K). Specifically, Recall@K measures the rate of the label within top-K recommendations and NDCG@K reflects the quality of ranking lists. In this paper, we repeat experiments on each metric for 10 times and report the averaged Recall@K and NDCG@K with the popular $K \in \{5, 10\}$.

Baselines. We compare our MSTHN with following representative methods for next POI recommendation, including 1) statistical-based method UserPop; 2) RNN-based methods GRU, STGN and LSTPM; 3) self-attention-based method STAN; 4) GNN-based methods LightGCN and SGRec and 5) hypergrpah neural network-based method DHCN:

– **UserPop**: It ranks the most popular POIs according to each user's visiting frequency.
– **GRU** [3]: A popular variant of RNN, which controls the information flow with two gates.
– **STGN** [31]: A state-of-the-art LSTM-based model, which introduces spatial and temporal gates for users' long- and short-term preferences.
– **LSTPM** [17]: A state-of-the-art LSTM-based model, which captures long- and short-term preferences with a non-local network and geo-dilated LSTM.
– **STAN** [14]: A state-of-the-art method based on self-attention mechanism, which explicitly models spatial-temporal influences within a user's check-in sequence.

- **LightGCN** [7]: A state-of-the-art simplified GNN-based collaborative filtering framework, which omits the non-linear activation and feature transformation during propagation.
- **SGRec** [10]: A state-of-the-art GNN-based method, which proposes Seq2Graph augmentation and captures collaborative signals among one-hop neighbors. For fairness comparison, we remove the POI categorical information that other methods do not use.
- **DHCN** [22]: A state-of-the-art hypergraph neural network-based method for session recommendation, which could be applied for next POI recommendation.

Parameter Settings. Our experiments are conducted with PyTorch 1.9.1 on a 32 GB Tesla V100 GPU. For baselines, we firstly preserve the settings as provided in original papers and fine-tune each model's hyperparameters on three datasets. For our MSTHN, we adopt Adam as optimizer with a learning rate of $1e-3$, weight decay of $1e-5$ and dropout rate of 0.3. We apply the same dimension size $d = 128$ for user and POI embeddings and set batch size as 100. In each batch, we pad sessions which do not meet the maximum session length in batch. Furthermore, we empirically choose 2.5 km (for NYC and TKY) and 100km (for Gowalla) as distance threshold and use 1 layer spatial-temporal graph aggregation layer in all datasets. The number of stable hypergraph convolutional layer and head of self-attention is chosen from $\{1, 2, 3, 4\}$ and $\{1, 2, 4, 8, 16\}$, respectively.

5.2 Performance Comparison

The results of all the methods are reported in Table 2. For the results, we have the following observations.

Our Proposed MSTHN Achieves the Best Results on All Datasets. On NYC dataset, our MSTHN improves the performance over the best baseline by 8.84%–15.58%. Additionally, MSTHN outperforms the best results by 5.12%–13.69% on TKY dataset and 7.12%–8.98% on Gowalla dataset. We contribute the

Table 2. Performances comparison on three datasets. The best and the second best performances are bolded and underlined, respectively. The improvements are calculated between the best and the second best scores.

Method	NYC				TKY				Gowalla			
	Rec@5	Rec@10	NDCG@5	NDCG@10	Rec@5	Rec@10	NDCG@5	NDCG@10	Rec@5	Rec@10	NDCG@5	NDCG@10
UserPop	0.2866	0.3297	0.2283	0.2423	0.2229	0.2668	0.1718	0.1861	0.0982	0.1489	0.0907	0.1336
GRU	0.236	0.2471	0.2252	0.2279	0.1549	0.1734	0.1371	0.1436	0.1282	0.1606	0.1102	0.1225
STGN	0.2371	0.2594	0.2261	0.2307	0.2112	0.2587	0.1482	0.1589	0.1600	0.2041	0.1191	0.1333
LSTPM	0.2495	0.2668	0.2425	0.2483	0.2203	0.2703	0.1556	0.1734	0.2021	0.2510	0.1523	0.1681
STAN	0.3523	0.3827	0.3025	0.3137	0.2621	0.3317	0.2074	0.2189	0.2449	0.2878	0.1837	0.1942
LightGCN	0.3221	0.3488	0.2958	0.3042	0.2213	0.2594	0.1977	0.2098	0.2356	0.2590	0.1801	0.1915
SGRec	0.3451	0.3723	0.3052	0.3178	0.2537	0.3213	0.2221	0.2447	0.2395	0.2813	0.1862	0.2002
DHCN	0.3745	0.3966	0.3126	0.3203	0.3172	0.3454	0.2442	0.2543	0.2653	0.3124	0.2038	0.2191
MSTHN	**0.4076**	**0.4398**	**0.3612**	**0.3702**	**0.3378**	**0.3927**	**0.2567**	**0.2721**	**0.2842**	**0.3396**	**0.2221**	**0.2365**
%Improv.	8.84%	10.89%	15.55%	15.58%	6.49%	13.69%	5.12%	7.00%	7.12%	8.71%	8.98%	7.94%

improvements to the following aspects: 1) Capturing important spatial-temporal correlations within user-POI interactions in the local view. 2) Uncovering complex high-order collaborative signals in the global view.

Capturing Spatial-temporal Correlations is Important for Next POI Recommendation. Methods which leverage spatial-temporal information explicitly or implicitly perform better than that do not use. For example, our MSTHN reaches up to 15.58% on NDCG@10 on NYC dataset than DHCN. On sparser Gowalla dataset, our MSTHN still outperforms DHCN on both metrics. Leveraging temporal information, SGRec surpasses LightGCN by 23.86% in terms of Recall@10 on TKY dataset. Beneficial from well-designed spatial-temporal gates, LSTPM and STGN also outperforms GRU on three datasets, especially on more sparser Gowalla dataset.

Uncovering High-Order Collaborative Signals is Effective and Significant to Improve Quality of Recommendation. From Table 2, hypergraph neural network-based methods (our MSTHN and DHCN) perform better than other baselines. For example, on TKY dataset, our MSTHN improves Recall@10 by 22.22% and 51.39% against SGRec and LighGCN. The major reason for Light-GCN is over-smoothing effect that makes nodes representation indistinguishable with deeper layer. Since SGRec performs one-hop neighbors sampling randomly, it would cause information losses. STAN performs better than SGRec on Recall metrics on three datasets, but worse on NDCG. Thus, uncovering high-order collaborative signals contributes more on exploring potential POIs but lacks of exploiting sequential dependency.

5.3 Ablation Study

Next we investigate the underlining mechanism of our MSTHN with three ablated models: 1) $\text{MSTHN}_{w/o\ local}$ that removes local spatial-temporal enhanced graph neural network module; 2) $\text{MSTHN}_{w/o\ global}$ that removes global interactive hypergraph neural network module; 3) $\text{MSTHN}_{w/o\ temporal}$ that removes user temporal preference augmentation module; 4) $\text{Local}_{w/o\ spatial}$ that removes spatial information in the local view; 5) $\text{Local}_{w/o\ temporal}$ that removes temporal information in the local view. From Table 3, we have the following observations:

Table 3. Ablation study on MSTHN w.r.t. Recall@10 and NDCG@10.

Method	NYC		TKY		Gowalla	
	Recall@10	NDCG@10	Recall@10	NDCG@10	Recall@10	NDCG@10
$\text{MSTHN}_{w/o\ local}$	0.4275	0.3620	0.3734	0.2606	0.3125	0.2207
$\text{MSTHN}_{w/o\ global}$	0.3812	0.3033	0.3105	0.2388	0.2832	0.1917
$\text{MSTHN}_{w/o\ temporal}$	0.4013	0.3105	0.3562	0.2541	0.3098	0.2126
$\text{Local}_{w/o\ spatial}$	0.4333	0.3651	0.3791	0.2595	0.3174	0.2187
$\text{Local}_{w/o\ temporal}$	0.4387	0.3698	0.3874	0.2632	0.3259	0.2241
MSTHN	0.4398	0.3702	0.3927	0.2721	0.3396	0.2365

First, when removing the local view module, MSTHN$_{w/o\ local}$ decreases slightly compared with the other two variants due to the losses of spatial-temporal correlations. The local view affects the correctness of recommendation more than the quality. Specifically, the average decline rates on three datasets are 5.57% on Recall@10 and 4.61% on NDCG@10. Second, when removing the global view module, MSTHN$_{w/o\ global}$ drops clearly. It strongly indicates the importance of global hypergraph network, for it could represent beyond pairwise relations and model distant POIs. Additionally, it proves the significance of high-order collaborative signals for next POI recommendation. Third, when removing the user temporal preference augmentation module, the variant MSTHN$_{w/o\ temporal}$ is less competitive than the complete MSTHN. It implies the effectiveness of user temporal preference augmentation for next POI recommendation. Fourth, spatial-temporal information is essential in the local view and spatial information contributes more to performances.

5.4   Hyperparameter Analysis

We further qualitatively analyze the impacts of layer number and head number in MSTHN.

Impact of Layer Number. To investigate the impact of stacking hypergraph convolutional layers, we conduct experiments with number of layer in $\{1, 2, 3, 4\}$. As illustrated in Fig. 3, our MSTHN achieves the best performances by stacking 3 layers on NYC dataset, 4 layers on TKY dataset, and 2 layers on Gowalla dataset. The results prove that our MSTHN could uncover and distill high-order collaborative signals effectively, especially on denser dataset (i.e., TKY). The possible cause of dropping would be the over-smoothing issue.

Impact of Heads Number. To explore the impact of choosing number of heads in spatial-temporal aggregation layer, we search from set $\{1, 2, 4, 8, 16\}$. From Fig. 4, our MSTHN is insensitive to the number of heads on both Recall and NDCG metrics, and obtains the best performances with 8 heads on three datasets. With number of heads increasing from 8 to 16, Recall@10 and NDCG@10 on three datasets drop. The possible cause would be the over-fitting

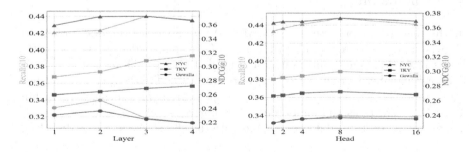

Fig. 3. Impact of Layer Number   **Fig. 4.** Impact of Head Number

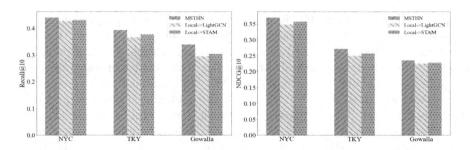

Fig. 5. Effect of Local View on Recall@10

Fig. 6. Effect of Local View on NDCG@10

in capturing spatial-temporal correlations within user-POI interactions in the local view.

5.5 Further Study

To explore the effect of our proposed local view, we maintain other parts of MSTHN and replace local view with STAM [27] and LightGCN [7]. From Fig. 5–6, on both Recall@10 and NDCG@10, our MSTHN outperforms these variants and the variant with STAM performs better than that with LightGCN. It proves the effectiveness of capturing spatial-temporal correlations within user-POI interactions. STAM utilizes users sequential dependency to update representations of POIs (e.g., if a POI has been visited by user $u_1 \rightarrow u_2 \rightarrow u_1$, STAM would take $u_1 \rightarrow u_2$ as input), which ignores repeated visiting patterns and another existing sequential dependency (i.e., $u_2 \rightarrow u_1$), and leads to a suboptimal performance. Thus, our proposed local spatial-temporal enhanced graph neural network could well address the limitation by learning in an asymmetric way. Moreover, the results against variant with STAM also indicate the significance of non-adjacent POI-POI geographical relations for next POI recommendation.

6   Conclusion

In this paper, we propose a novel **M**ulti-**V**iew **S**patial-**T**emporal Enhanced **H**yper-graph **N**etwork (MSTHN) for next POI recommendation, which jointly learns representations from both local and global views. Through the spatial-temporal enhanced graph neural network and interactive hypergraph neural network, MSTHN could capture important spatial-temporal correlations within user-POI interactions and high-order collaborative signals across users. Experimental results on three datasets demonstrate the effectiveness of our MSTHN.

References

1. Bai, S., Zhang, F., Torr, P.H.: Hypergraph convolution and hypergraph attention. Pattern Recogn. **110**, 107637 (2021)
2. Cheng, C., Yang, H., Lyu, M.R., King, I.: Where you like to go next: successive point-of-interest recommendation. In: Twenty-Third International Joint Conference on Artificial Intelligence (2013)
3. Cho, K., et al.: Learning phrase representations using RNN encoder-decoder for statistical machine translation. In: EMNLP, pp. 1724–1734. ACL (2014)
4. Dang, W., et al.: Predicting human mobility via graph convolutional dual-attentive networks. In: Proceedings of the Fifteenth ACM International Conference on Web Search and Data Mining, pp. 192–200 (2022)
5. Feng, Y., You, H., Zhang, Z., Ji, R., Gao, Y.: Hypergraph neural networks. In: Proceedings of the AAAI Conference on Artificial Intelligence, vol. 33, pp. 3558–3565 (2019)
6. Han, J., Tao, Q., Tang, Y., Xia, Y.: DH-HGCN: dual homogeneity hypergraph convolutional network for multiple social recommendations. In: Proceedings of the 45th International ACM SIGIR Conference on Research and Development in Information Retrieval, pp. 2190–2194 (2022)
7. He, X., Deng, K., Wang, X., Li, Y., Zhang, Y., Wang, M.: LightGCN: simplifying and powering graph convolution network for recommendation. In: Proceedings of the 43rd International ACM SIGIR Conference on Research and Development in Information Retrieval, pp. 639–648 (2020)
8. Huang, Z., Ma, J., Dong, Y., Foutz, N.Z., Li, J.: Empowering next poi recommendation with multi-relational modeling. In: Proceedings of the 45th International ACM SIGIR Conference on Research and Development in Information Retrieval, pp. 2034–2038 (2022)
9. Kipf, T.N., Welling, M.: Semi-supervised classification with graph convolutional networks. In: International Conference on Learning Representations (2017)
10. Li, Y., Chen, T., Luo, Y., Yin, H., Huang, Z.: Discovering collaborative signals for next poi recommendation with iterative Seq2Graph augmentation. In: Proceedings of the 30th IJCAI, pp. 1491–1497 (2021)
11. Li, Y., Gao, C., Luo, H., Jin, D., Li, Y.: Enhancing hypergraph neural networks with intent disentanglement for session-based recommendation. In: Proceedings of the 45th International ACM SIGIR Conference on Research and Development in Information Retrieval, pp. 1997–2002 (2022)
12. Lian, D., Wu, Y., Ge, Y., Xie, X., Chen, E.: Geography-aware sequential location recommendation. In: Proceedings of the 26th ACM SIGKDD International Conference on Knowledge Discovery & Data Mining, pp. 2009–2019 (2020)
13. Lim, N., Hooi, B., Ng, S.K., Goh, Y.L., Weng, R., Tan, R.: Hierarchical multi-task graph recurrent network for next poi recommendation. In: Proceedings of the 45th International ACM SIGIR Conference on Research and Development in Information Retrieval (2022)
14. Luo, Y., Liu, Q., Liu, Z.: STAN: spatio-temporal attention network for next location recommendation. In: Proceedings of the Web Conference 2021, pp. 2177–2185 (2021)
15. Rao, X., Chen, L., Liu, Y., Shang, S., Yao, B., Han, P.: Graph-flashback network for next location recommendation. In: Proceedings of the 28th ACM SIGKDD Conference on Knowledge Discovery and Data Mining, pp. 1463–1471 (2022)

16. Su, Y., Li, X., Tang, W., Xiang, J., He, Y.: Next check-in location prediction via footprints and friendship on location-based social networks. In: 2018 19th IEEE International Conference on Mobile Data Management (MDM), pp. 251–256. IEEE (2018)

17. Sun, K., Qian, T., Chen, T., Liang, Y., Nguyen, Q.V.H., Yin, H.: Where to go next: modeling long-and short-term user preferences for point-of-interest recommendation. In: Proceedings of the AAAI Conference on Artificial Intelligence, vol. 34, pp. 214–221 (2020)

18. Vaswani, A., et al.: Attention is all you need. In: Advances in Neural Information Processing Systems, vol. 30 (2017)

19. Wang, X., He, X., Wang, M., Feng, F., Chua, T.S.: Neural graph collaborative filtering. In: Proceedings of the 42nd International ACM SIGIR Conference on Research and Development in Information Retrieval, pp. 165–174 (2019)

20. Wang, Z., Zhu, Y., Liu, H., Wang, C.: Learning graph-based disentangled representations for next poi recommendation. In: Proceedings of the 45th International ACM SIGIR Conference on Research and Development in Information Retrieval, pp. 1154–1163 (2022)

21. Wang, Z., Zhu, Y., Zhang, Q., Liu, H., Wang, C., Liu, T.: Graph-enhanced spatial-temporal network for next poi recommendation. ACM Trans. Knowl. Discovery From Data (TKDD) **16**(6), 1–21 (2022)

22. Xia, X., Yin, H., Yu, J., Wang, Q., Cui, L., Zhang, X.: Self-supervised hypergraph convolutional networks for session-based recommendation. In: Proceedings of the AAAI Conference on Artificial Intelligence, vol. 35, pp. 4503–4511 (2021)

23. Xie, M., Yin, H., Wang, H., Xu, F., Chen, W., Wang, S.: Learning graph-based POI embedding for location-based recommendation. In: Proceedings of the 25th ACM International on Conference on Information and Knowledge Management, pp. 15–24 (2016)

24. Yang, D., Qu, B., Yang, J., Cudre-Mauroux, P.: Revisiting user mobility and social relationships in LBSNs: a hypergraph embedding approach. In: The World Wide Web Conference, pp. 2147–2157 (2019)

25. Yang, D., Qu, B., Yang, J., Cudré-Mauroux, P.: LBSN2Vec++: heterogeneous hypergraph embedding for location-based social networks. IEEE Trans. Knowl. Data Eng. (2020)

26. Yang, D., Zhang, D., Zheng, V.W., Yu, Z.: Modeling user activity preference by leveraging user spatial temporal characteristics in LBSNs. IEEE Trans. Syst. Man Cybern. Syst. **45**(1), 129–142 (2014)

27. Yang, Z., Ding, M., Xu, B., Yang, H., Tang, J.: STAM: a spatiotemporal aggregation method for graph neural network-based recommendation. In: Proceedings of the ACM Web Conference 2022, pp. 3217–3228 (2022)

28. Yin, H., Cui, B., Chen, L., Hu, Z., Zhang, C.: Modeling location-based user rating profiles for personalized recommendation. ACM Trans. Knowl. Discovery From Data (TKDD) **9**(3), 1–41 (2015)

29. Yu, J., Yin, H., Li, J., Wang, Q., Hung, N.Q.V., Zhang, X.: Self-supervised multi-channel hypergraph convolutional network for social recommendation. In: Proceedings of the Web Conference 2021, pp. 413–424 (2021)

30. Zhang, J., Gao, M., Yu, J., Guo, L., Li, J., Yin, H.: Double-scale self-supervised hypergraph learning for group recommendation. In: Proceedings of the 30th ACM International Conference on Information & Knowledge Management, pp. 2557–2567 (2021)

31. Zhao, P., et al.: Where to go next: a spatio-temporal gated network for next poi recommendation. IEEE Trans. Knowl. Data Eng. **34**, 2512–2524 (2020)

ML-KGCL: Multi-level Knowledge Graph Contrastive Learning for Recommendation

Gong Chen and Xiaoyuan Xie[(⊠)]

School of Computer Science, Wuhan University, Wuhan, China
{chengongcg,xxie}@whu.edu.cn

Abstract. The knowledge graph-based (KG-based) recommender systems have achieved excellent results in the recommendation domain. However, the long-tail issue hinders the model from mining the real interests of users. Existing research has shown that Contrastive Learning (CL) can alleviate the long-tail issue, but the existing graph contrastive learning methods are not completely compatible with KG-based recommendation. To fill this gap, we propose a *Multi-Level Knowledge Graph Contrastive Learning framework (ML-KGCL)* to introduce CL into the KG-based recommendation. ML-KGCL makes the CL task more compatible with the recommendation task while mitigating the long-tail issue by performing fine-grained node representation learning. Firstly, we generate positive samples via graph augmentation strategy. Then, we divide the KG-based recommendation into three levels: user-level, entity-level and user-item-level, and perform fine-grained multi-level CL to optimize the node representations. Next, we obtain the final node representations through the signal integration strategy. Finally, the model is trained by the joint learning paradigm. The experimental results on three public datasets are better than the baseline models.

Keywords: Recommender system · Knowledge graph · Contrastive learning · Graph neural network

1 Introduction

Recommender systems have been successfully and widely applied in web applications. In previous work Matrix Factorization maps ID of each user or item to an embedding vector space [23]. Collaborative Filtering makes use of the historical interactions to learn improved vector representations and predicts interests of users [6].

Recently, graph-based recommendation methods utilize graph structure to model user-item interactions and incorporate multi-hop neighbors into representation learning of graph nodes to improve recommendation accuracy. There are two directions for improving graph-based recommendation methods [4]. One direction is the self-supervised learning methods based on the user-item graphs, which use the unlabeled data space to extract additional supervised signals from the data itself, and achieve higher quality representations via CL. The other improved approach is to introduce the KG as a useful external source of data into the recommender systems to build KG-based recommender models. Such methods build a hybrid structure of KG and user-item bipartite graph, called Collaborative knowledge Graph (CKG), which relates items to their attributes, and builds high-order relationships between items.

Although the current KG-based recommender models have certain effectiveness, there are also some limitations [23]. First, the observed interactions are extremely sparse compared with the whole interaction space, which is not sufficient for learning high-quality representations of graph nodes. Second, both entity and item data show long tail distribution. Due to the skewed distribution of the data, the models are unable to capture true interest of users. This results in exposure bias, whereby highly exposed items are recommended to the user in lieu of items that the user is actually interested in. Lastly, current graph-based CL methods are not directly transferable to the more complex KG-based recommendation. Figure 1 shows the phenomenon of the long-tail distribution in two datasets, where the exposure counts are represented on the X-axis and the corresponding numbers are represented on the Y-axis.

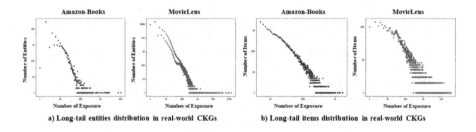

a) Long-tail entities distribution in real-world CKGs   b) Long-tail items distribution in real-world CKGs

Fig. 1. Long-tail distributions in real-world CKGs

In this paper, we propose a *Multi-level Knowledge Graph Contrastive Learning framework (ML-KGCL)* to address above issues. ML-KGCL performs various levels CL on CKG. Specifically, at three levels, namely the user-level, entity-level, and user-item-level, the fine-grained CL method is carried out, which makes the CL more compatible with the KG-based recommendation.

ML-KGCL can optimize recommendation at three levels: the discrimination of user interest, the embedding representation distribution of entities and items, and correlation of items and user interest. Combined with fine-grained CL at three levels, we can obtain more accurate recommendation results. It is worth mentioning that our ML-KGCL is model-agnostic. In the experiment of this paper, we choose the KGAT model as the graph encoder to implement the ML-KGCL. Experimental results on three public datasets show that ML-KGCL significantly improves the recommendation accuracy, and the performance improvement comes from the accurate recommendation of long-tail entities, which alleviates the exposure bias of the recommender systems. The contributions can be summarized as follows:

1. We propose ML-KGCL, through fine-grained CL at different levels, the CL is more compatible with the KG-based recommendation. It provides a new research perspective for the knowledge-aware recommendation.
2. ML-KGCL is a further exploration of the KG-based CL. It can improve the accuracy of the recommendation models and alleviate the long-tail issue in the real world datasets.
3. We conduct experiments on three public datasets, and the experimental results demonstrate that the ML-KGCL outperforms the baseline models. Further ablation experiments verify the effectiveness and rationality of CL at different levels. The source code and datasets are available on Github[1].

2   Preliminaries

We denote user and item as: $u \in \mathcal{U}$, $i \in \mathcal{I}$, where $\mathcal{U}$, $\mathcal{I}$ denote the user set and item set, respectively. And the *Interact* $\in \mathcal{Y}^{|\mathcal{U}| \times |\mathcal{I}|}$ is used to represent the interaction between user and item, where *Interact* $\in \{0, 1\}$. *Interact* $= 1$ indicates the existence of interaction, otherwise *Interact* $= 0$. Based on the interaction relationships present in the interaction matrix $\mathcal{Y}$, the established user-item bipartite graph is expressed as $\mathcal{G}_{ui} = \{(u, Interact, i)\}$.

KG is a network consists of triplets, denoted by [head entity: h, relationship: r, tail entity: t]. CKG is obtained by associating KG with $\mathcal{G}_{ui}$ based on the alignment of items in $\mathcal{G}_{ui}$ with entities in KG. CKG is represented as $\mathcal{G} = \{(h, r, t) | h, t \in \mathcal{E} \cup \mathcal{U}, r \in \mathcal{R}\}$, where $\mathcal{E}$ denotes the set of entities, $\mathcal{R}$ denotes the set of relations and *Interact* $= 1$ is regarded as one type of relations.

Taking $\mathcal{G}$ as input data, we predict the probability $\hat{y}_{ui}$ of recommending item i to user u, and the items with the *topK* probability rankings are recommended to user u.

3   Methodology

In this section, we introduce *Multi-level Knowledge Graph Contrastive Learning framework (ML-KGCL)*. ML-KGCL follows the joint learning paradigm to train the model, consisting of a recommendation task and CL tasks at three levels. We first show the general framework of ML-KGCL in Fig. 2. Then we present the four components of ML-KGCL. They are graph augmentation layer, representation learning layer, contrastive learning layer and joint learning layer.

3.1   Graph Augmentation

Dropout is often used in deep neural networks to prevent overfitting. We use it as a graph augmentation method. We randomly drop elements in the CKG with a certain probability. By forwarding the input CKG twice through the graph encoder, two different views can be generated. They are regarded as a positive sample pair for CL. Specifically, we adopt two dropout methods: node dropout and edge dropout.

---

[1] https://github.com/mlkgcl/mlkgcl.

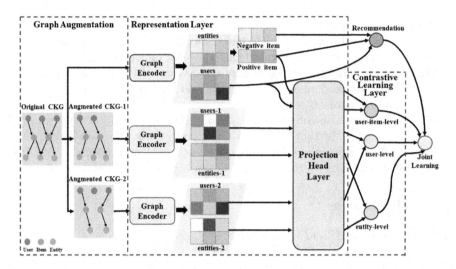

Fig. 2. The framework of ML-KGCL

Node Dropout. Each node of the graph and its connecting edges are dropped with probability ρ. This augmentation method helps identify more influential nodes within the graph.

Edge Dropout. Edges in the graph that are dropped with probability ρ. This augmentation method helps capture more influential substructures within the graph.

3.2   Representation Learning Layer

In the Representation Learning layer, the graph encoder is used to learn the node representations. The output is the embedding vector of each node. Representation learning layer has three components: attentive embedding propagation layers, high-order propagation, and supervised signal integration.

Attentive Embedding Propagation Layers. Attentive embedding propagation layer uses first-order connection information to associate users, items and knowledge entity nodes. It has the following three components.

Information Propagation is achieved by computing a linear combination of first-order connected graph of triplets containing same entities. For an entity h, it is represented as:

$$v_{\mathcal{H}} = \sum_{(h,r,t)\in\mathcal{H}} \xi\left(h,r,t\right) v_t \tag{1}$$

where $\mathcal{H}$ denotes the set of triplets containing head entity h.

Knowledge-aware Attention adopts the attention-aware mechanism to implement $\xi\left(h,r,t\right)$, indicating how much information is conditioned on r from t propagates to h. Its expression is as $\xi\left(h,r,t\right) = \left(W_r v_t\right)^T \tanh\left(\left(W_r v_h + v_r\right)\right)$.

Adopt the softmax function to normalize the coefficients of all triplets concatenated with h:

$$\xi(h, r, t) = \frac{\exp(\xi(h, r, t))}{\sum_{(h, r', t') \in \mathcal{H}} \exp(\xi(h, r', t'))} \tag{2}$$

Information Aggregation stage aggregates the entity representations v_h and its first-order connected graph representations $v_{\mathcal{H}}$ into a new representation of h using LeakyRelu function. The calculation is as follows:

$$v_h^{(1)} = f(v_h, v_{\mathcal{H}}) = LeakyRelu\left(W'(v_h + v_{\mathcal{H}})\right) + LeakyRelu\left(W''(v_h \odot v_{\mathcal{H}})\right) \tag{3}$$

High-Order Propagation. Further stacking more attentive embedding propagation layers can collect information propagated from higher-hop neighbors.

$$v_h^{(l)} = f\left(v_h^{(l-1)}, v_{\mathcal{H}}^{(l-1)}\right) \tag{4}$$

The representation of each step is concatenated into a single vector:

$$v_u^* = v_u^{(0)} \| \cdots \| v_u^{(L)}, \ v_i^* = v_i^{(0)} \| \cdots \| v_i^{(L)}, \ v_e^* = v_e^{(0)} \| \cdots \| v_e^{(L)} \tag{5}$$

Supervised Signal Integration. Since the learned embedding representations are often dense, it is difficult to keep dividing them more uniformly in the dense space for CL. It is necessary to map the learned embedding representations into a sparse high-dimensional space. For this reason, the projection head layer [19] is introduced into the representation layer to further integrate the supervised signal. The implementation of projection head layer is a two layer Multi Layer Perceptron (MLP). Here is the formal representation:

$$v = MLP(v^*) = W_2\sigma(W_1 v^*) \tag{6}$$

3.3 Multi-level Contrastive Learning

The CL loss is composed of three levels: user-level, entity-level and user-item-level. The CL loss is labeled as L_{cl} has the following representation:

$$L_{cl} = L_{user-level} + L_{entity-level} + L_{user-item-level} \tag{7}$$

The CL loss encourages the consistency of pairs of positive samples while minimizing the consistency of pairs of negative samples. InfoNCE loss is used as the CL loss in each level. Formally defined as:

$$L_{cl} = \sum_{i \in B} -\log \frac{\exp\left(z_i'^T z_i'' / \tau\right)}{\sum_{j \in B} \exp\left(z_i'^T z_j'' / \tau\right)} \tag{8}$$

where i, j are the sampling in batch B. (z_i', z_i'') denote the positive sample pair, (z_i', z_j'') denote the negative sample pair, and $\tau > 0$ (e.g., 0.1) is the temperature. At user-level, z is specifically denoted as z_u. At entity-level, z is specifically denoted as z_e. In user-item-level, z is specifically denoted as z_{ui}.

User-Level Contrastive Learning can shorten the distance between users with similar interests and widen the distance between users with different interests in the embedding space. Through user-level CL, the degree of discrimination between users with similar interests can be increased, so that the representations of user nodes are more independent. Thereby, more detailed interests of users can be mined to increase the accuracy of recommendation.

Entity-Level Contrastive Learning can increase the degree of discrimination between different entities, the distribution of entity node representations in the embedding space becomes more uniform, alleviating the long-tail issue of entity nodes.

User-Item-Level Contrastive Learning is to make the CL task more compatible with the recommendation task. In the embedding space, the distance between users and their interaction items is shortened, which maximizes the similarity between the interacting items and the item that the user is really interested in. In this way, we can mine the items that users are really interested in, instead of recommending highly exposure items. In both the recommendation task and the CL task at user-item-level, the representations of user nodes and item nodes are optimized to be more compatible.

3.4   Joint Learning Paradigm

In this paper, we use joint learning paradigm to optimize recommendation task and multi-level CL tasks. And the recommendation loss is composed of the KG loss and the BPR loss. Formally defined as:

$$L_{joint} = L_{kg} + L_{bpr} + \lambda_1 L_{cl} + \lambda_2 \|\theta\|_2^2 \tag{9}$$

where L_{kg} is the KG loss. L_{bpr} is the BPR loss. L_{cl} is the multi-level CL loss. L_2 is the regularization loss. λ_1 and λ_2 are hyper-parameters, and θ is variance.

KG models the triplets by embedding entities and relations as vectors. For a specific triplet (h_0, r_0, t_0), the rationality score is expressed as $Score(h_0, r_0, t_0) = \|W_r v_{h_0} + v_{r_0} - W_r v_{t_0}\|_2^2$. We use pairwise loss to train the triplets with sigmod function $\sigma(\cdot)$ as the activation function, which is calculated as follows:

$$L_{kg} = \sum_{(h,r,t)\in\mathcal{G},(h,r,t')\notin\mathcal{G}} -\ln\sigma\left(Score(h,r,t') - Score(h,r,t)\right) \tag{10}$$

The BPR loss is calculated as follows [9]:

$$L_{bpr} = \sum_{(u,i)\in R^+,(u,j)\in R^-} -\ln\sigma\left(\hat{y}(u,i) - \hat{y}(u,j)\right) \tag{11}$$

where R^+ denotes the set of positive sample pairs, R^- denotes the set of negative sample pairs.

3.5 Model Prediction

Compute the inner product of the user and item representations to predict their match scores:

$$\hat{y}(u, i) = v_u^{*T} v_i^* \tag{12}$$

4 Experiments

In this experiment, we mainly research the three questions:

– **RQ1: How does ML-KGCL perform compared with SOTA methods?**
– **RQ2: How do different CL levels and components affect ML-KGCL?**
– **RQ3: Can our model alleviate the long-tail issue of knowledge-aware recommendation?**

4.1 Model Implementation

The Recbole library has implemented many well-known recommendation models. It has good support and integration for the development of recommendation models so that researchers can easily develop their own recommendation models. So we implemented ML-KGCL based on the Recbole library [24].

4.2 Datasets Description

To assess the efficacy of ML-KGCL, we use three widely used public datasets: MovieLens-1m, MovieLens-10m and Amazon-Books. The datasets are different in domain, size and sparsity.

To prevent the datasets being too sparse, we filter the datasets using a 10-core setting. Only users who have at least 10 interactions with the items are retained. The statistical results of the three datasets are shown in Table 1.

Table 1. Datasets statistics.

		MovieLens-1m	MovieLens-10m	Amazon-Books
User-Item Graph	Users	6034	69585	135110
	Items	3124	9176	115173
	Interaction	834449	8233567	4042382
Avg. Actions		138.31	118.33	29.92
Knowledge Graph	Entities	79368	531254	284820
	Relations	51	51	22
	Triplets	385923	3123868	522475

For each dataset, we split the training dataset, validation dataset, and testing dataset using the {8:1:1} ratio. We treat the observed interaction as positive example pair. And we use negative sampling strategy to select an item that has not interacted to compose negative sample pair.

4.3   Experiment Settings

Evaluation Metrics. In order to assess of $topK$ recommended items and interest ranking, two commonly used evaluation metrics $Recall@K$ and $NDCG@K$ were adopted. Recall represents the ratio of the number of related results retrieved from the topK results to all of the related results in the database. NDCG is used to measure the ranking quality. By default, we set $K = 10$.

Baseline. We use the implementation of Recbole library with optimal parameters for the baseline models compared in this experiment. We selected representative models as follows:

- **BPR** [9]. It is a classic conventional CF recommendation model.
- **KGAT** [14]. This model implements an attentive propagation strategy based on CKGs to explore high-order connection information.
- **SGL** [17]. SGL achieves SOTA performance through self-supervised learning based on graph structure.
- **KGCN** [13]. KGCN encodes high-order dependencies based on semantic information in KGs for entity representation.
- **LightGCN** [5]. This model achieves SOTA GCN-based recommendation methods.
- **CFKG** [1]. CFKG represents the recommendation task as a confidence prediction that there is an interaction between users and items.
- **KGNN-LS** [12]. KGNN-LS generates a user-specific weighted graph using KG and then generates user-specific item representations.

Parameter Settings. For all models, the embedding dimension is set to 64. We use the Adam optimizer to optimize all models. We set the batch size to 2048. The Xavier [3] initializer is used to initialize parameters. We set the training to stop after 500 epochs. When Recall does not increase for 30 continuous epochs in the validation set, the early-stop strategy is performed.

The hyper-parameter λ_1 is the learning rate for CL. The adjusting experimental results are shown in Table 2. And the optimal setting obtained from experiment is $\lambda_1 = 0.01$.

Table 2. Impact of learning rate λ_1 on MovieLens-1m dataset.

λ_1	0.005	0.01	0.05	0.1	0.2
Total Epoch	74	82	113	125	149
Recall	0.1842	**0.1901**	0.1864	0.1831	0.1788
NDCG	0.2489	**0.2553**	0.2491	0.2464	0.2425

4.4 RQ1: How Does ML-KGCL Perform Compared with SOTA Methods?

Table 3 shows the experimental results for all the models. Analyzing the performance of each model in the table, the following observations are made:

In our experiments, the SGL model has the best overall performance among all the baseline models and the KGAT model achieves the best results among all the KG-based models on three datasets. Therefore, we choose the optimal KGAT model in the KG-based models as the graph encoder.

ML-KGCL consistently achieves the best performance on three datasets. On three datasets, compared with the KGAT model with the highest performance, Recall@10 of ML-KGCL is 7.95%, 9.62% and 29.28% higher, and NDCG@10 is 7.68%, 9.91% and 36.73% higher, respectively.

Combining Table 1 and Table 3 for analysis, it can be seen that the improvement effect of ML-KGCL is more obvious on the Amazon-Books dataset. It indicates that the KGAT model can learn a better embedding representations for the dataset with intensive interaction, but the embedding representations learned by the KGAT model for the dataset with sparse interaction are not good. Through multi-level CL, ML-KGCL can optimize the node embedding representations to obtain a more uniform embedding distribution by learning the relative relationship between the embedding representations, thereby alleviating the sparse interaction problem. Therefore, the ML-KGCL can learn better embedding representations on datasets with sparse interactions, which makes the improvement of recommendation effect more obvious.

Table 3. Overall performance of models.

	MovieLens-1m		MovieLens-10m		Amazon-Books	
Model	Recall	NDCG	Recall	NDCG	Recall	NDCG
BPR	0.1561	0.2165	0.2028	0.2219	0.0865	0.0471
LightGCN	0.1789	0.2346	0.2233	0.2537	0.1259	0.0652
SGL	0.1807	0.2359	0.2285	0.2556	0.1271	0.0694
KGCN	0.154	0.2152	0.2024	0.2215	0.0861	0.0468
CFKG	0.1648	0.224	0.1989	0.2266	0.0972	0.0521
KGAT	0.1761	0.2371	0.2138	0.2433	0.1045	0.0588
KGNN-LS	0.1581	0.2171	0.2034	0.2223	0.087	0.0474
ML-KGCL	**0.1901**	**0.2553**	**0.2361**	**0.2674**	**0.1351**	**0.0804**

4.5 RQ2: How Do Different CL Levels and Components Affect ML-KGCL?

We conduct ablation experiments on the projection head layer, CL at different levels and model depth.

Effect of Projection Head Layer. We first performed ablation experiments on the projection head layer, a component that integrates supervised signals. Analyzing the experimental results in Table 4, the following conclusions can be drawn:

Using the projection head layer to integrate supervised signals can significantly improve the recommendation effect, which verifies the effectiveness of the projection head layer.

We can also observe that when the projection head layer is not used, adding multi-level CL tasks will reduce the recommendation effect of the original model. It has an even lower recommendation effect than the KGAT model. This result shows that it is difficult to represent the data distribution uniformly in dense space. And due to the data distribution being too dense, it will reduce the recommendation effect. It also illustrates the importance of mapping embeddings into a high-dimensional sparse space for CL.

Table 4. Ablation experiment results.

	MovieLens-1m		MovieLens-10m		Amazon-Books	
Model	Recall	NDCG	Recall	NDCG	Recall	NDCG
ML-KGCL without PH	0.1491	0.208	0.1964	0.2091	0.0962	0.0567
ML-KGCL-UL	0.179	0.241	0.2233	0.2515	0.1163	0.0676
ML-KGCL-EL	0.1786	0.2397	0.2219	0.2513	0.1144	0.0654
ML-KGCL-UIL	0.1801	0.2429	0.2325	0.2622	0.1259	0.0742
ML-KGCL-UL&UIL	0.1809	0.2451	0.2335	0.2629	0.1289	0.0762
ML-KGCL-UL&EL	0.1799	0.242	0.2235	0.2522	0.1198	0.0696
ML-KGCL-EL&UIL	0.1845	0.2495	0.2349	0.2642	0.1298	0.0769
KGAT	0.1761	0.2371	0.2138	0.2433	0.1045	0.0588
ML-KGCL	**0.1901**	**0.2553**	**0.2361**	**0.2674**	**0.1351**	**0.0804**

Effect of CL at Different Levels. We also conduct ablation experiments for CL at each level, and the experimental results are given in Table 4. To analyze the effect of combination of CL at different levels more intuitively, we have calculated the improvement of the combination of different levels compared with the KGAT model. Analyzing the results in Fig. 3, the following observations can be drawn:

The combination of CL at all three levels achieves the best recommendation performance on all three datasets. The recommendation effect of using CL at each level alone has improved, which shows that using CL at different levels can achieve certain benefits. Using CL at the user-item-level achieves the best results, because the CL task at this level is more directly related to the recommendation task. The CL task at user-item-level and the recommendation task jointly optimize the embedding representations of positive items, which are the samples

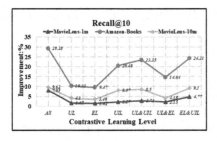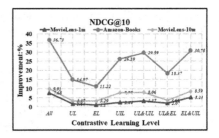

Fig. 3. Performance improvement of ML-KGCL with combination of CL at different levels

that users are really interested in. High-quality representations of positive items can significantly improve the recommendation effect of the model. Therefore, the CL task at user-item-level has a more obvious impact on the recommendation effect, and it also proves that the CL task that is more compatible with the recommendation task has a greater improvement in the recommendation effect.

The recommendation performance of combination of CL at two levels is better than using CL at a single level. It proves that in the combination of CL at two levels, more benefits can be obtained by optimizing different types of nodes at the same time. The recommendation effect of the combination of CL at entity-level and user-item-level is the best of the two-level CL combined. The second is the combination of CL at user-level and user-item-level. And the last is the combination of CL at user-level and entity-level. It can be seen that the combination with user-item-level can achieve better recommendation effect. It is again proved that the CL task that is more compatible with the recommendation task has a greater improvement in the recommendation effect. By comparing the results of CL at a single level and combination of CL at two levels, it can be found that the effect of CL at user-item-level alone is better than the combination of CL at user-level and entity-level. It is shown that simply accumulating CL tasks can definitely achieve better recommendation effects, and a CL task that is more relevant to the recommendation task will improve the effect even more.

Effect of Model Depth. The depth of the ML-KGCL is chosen within the scope of $\{1, 2, 3, 4\}$. An analysis of the results in the Table 5 yields the following observations:

Increasing the depth of the graph encoder can significantly improve the performance. Both ML-KGCL-2 and ML-KGCL-3 achieve a certain improvement over ML-KGCL-1. We attribute this improvement to the fact that deeper models can more effectively model high-order relationships between users, items and entities.

But the improvement reaches marginal in M-KGCL-4. It suggests that ML-KGCL with three layers is sufficient to capture high-order relations between entities, which is consistent with the findings of [15]. Thus, the best setting for the depth of the ML-KGCL is 3-layer.

Analyzing Table 5 and Table 3 jointly, ML-KGCL performs better than KGAT for all depths. It again verifies the effectiveness of this fine-grained CL method.

Table 5. Performance of ML-KGCL with different model depths.

Model	MovieLens-1m		MovieLens-10m		Amazon-Books	
	Recall	NDCG	Recall	NDCG	Recall	NDCG
ML-KGCL-1	0.186	0.2503	0.2331	0.2629	0.1307	0.0771
ML-KGCL-2	0.1873	0.2532	0.2343	0.2646	0.1332	0.0783
ML-KGCL-3	**0.1901**	**0.2553**	**0.2361**	**0.2674**	**0.1351**	**0.0804**
ML-KGCL-4	0.1846	0.248	0.2307	0.2603	0.1345	0.0789

4.6 RQ3: Can Our Model Alleviate the Long-Tail Issue of Knowledge-Aware Recommendation?

To verify whether ML-KGCL can alleviate the long-tail issue in knowledge-aware recommendation, we successively divide items into 10 groups according to their popularity from small to large. And each group has the same total number of interactions. Then, the contribution of each group to Recall is calculated separately, with the following formula:

$$Recall = \frac{1}{\mathcal{N}} \sum_{u=1}^{\mathcal{N}} \frac{\sum_{\mathcal{M}=1}^{10} \left|(l_{rec}^u)^{\mathcal{M}} \cap l_{interact}^u\right|}{|l_{interact}^u|} = \sum_{\mathcal{M}=1}^{10} Recall^{\mathcal{M}}$$

where l_{rec}^u and $l_{interact}^u$ respectively denote items recommended to user u and all items which user u interacts with. $\mathcal{N}$ is the number of users. $Recall^{\mathcal{M}}$ represents the contribution of the m-th group.

It can be seen from the calculation results in Fig. 4 that KGAT tends to recommend items with high exposure. On the MovieLens-10m and Amazon-Books datasets, although group 10 only contains 0.81% and 0.94% of the item space, the proportion of the total Recall score is 27.86% and 32.75%, respectively. However, the results of ML-KGCL show that the contribution of group 10 decreases to 19.62% and 28.42%, respectively. This suggests that it is difficult for KGAT to learn high-quality representations of long-tail items because of sparse interaction signals. ML-KGCL brings the ability to alleviate the long-tail issue, and its performance improvement comes from the accurate recommendation of sparse interaction items. In other words, ML-KGCL can recommend unpopular items with less interaction instead of only recommending popular items, which alleviates the exposure bias.

To further understand why ML-KGCL significantly outperforms the KGAT, we visualize the trained embedding representations of the two models at user-level, entity-level, and user-item-level. Specifically, we first randomly sample the

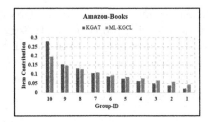 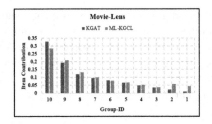

Fig. 4. Performance comparison over different groups between KGAT and ML-KGCL

embedding representations of nodes. Then, t-SNE [8] data dimensionality reduction method is used to reduce the dimension of the representations and draw them on a 2D plane. The visualization results are given in Fig. 5.

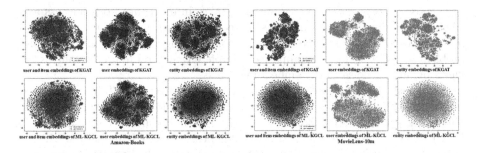

Fig. 5. Distribution of representations in the embedding space

From the figure, we can see that the ML-KGCL has a more uniform distribution of embedding representations at each level compared to the KGAT model. And users and items have similar and uniform embedding representations, so that unpopular items will have a greater chance to be recommended to users. Multi-level CL can learn high-quality node representations, which can significantly improve the recommendation effect.

5 Related Work

5.1 GNN-Based Recommendation

Graph Neural Networks (GNNs) [2] have become widely recognized as a powerful framework for modeling recommendation tasks. GCN [7], as the most common GNN variant, uses the information of neighbor nodes in the user-item graph to optimize the embedding of the target node to achieve graph reasoning [18]. Among these methods, the LightGCN [5] model with simple structure and excellent performance is the most popular. In recent years, KG-based recommendation has become increasingly important to incorporate KGs [10] into recommender systems to improve the performance and interpretability of recommender systems. KGAT [14] implements an attentive propagation strategy based on CKGs to explore high-order connection information. RippleNet [11] proposed an embedding-guided preference propagation approach for KG-based recommendation. KGNN-LS [12] generates a user-specific weighted graph using KG and then generates user-specific item representations.

5.2   CL-Based Recommendation

Motivated by the successful use of CL in other domains, some studies have applied CL in recommender systems as a method to alleviate data sparsity [18,23]. S^3-Rec [25] maximizes interaction information by creating contrastive views of sequences using the dropout-based approach. CLCRec [16] uses information-theoretical methods to maximize the relevance between items content and collaboration signals to mitigate data sparsity. CLRec [21] uses CL to propose a deep matching method that takes into account both fairness and efficiency to alleviate exposure bias in recommendation. SGL [17] proposes a self-supervised model for the limitations of the long-tail issue and data noise problem in GCN. QRec [22] proposes a simple CL method to create contrastive views by adding noise to the embedding space.

The most recent research is KGCL [20] which introduces CL into KG-based recommendation, and uses CL cross-view mode to denoise the knowledge graph, alleviating the data noise problem in KGs. But it is quite different from ML-KGCL. Starting from the different types of nodes, ML-KGCL divides the CL task into different levels and carries out more fine-grained multi-level CL. Optimizing node representations at different levels makes the CL task more closely related to the recommendation task, thereby achieving benefits. In particular, CL at user-level increases the discrimination between users with similar interests. And Cl at entity-level makes the representation distribution of entity nodes more uniform in the embedding space. Both the recommendation task and the CL task at user-item-level are optimizing the node representations of users/items to make the two tasks more compatible. It provides a new perspective for KG-based recommendation.

6   Conclusion and Future Work

In this paper, we propose ML-KGCL to alleviate the long-tail issue by designing fine-grain CL that is compatible with the recommendation task. ML-KGCL generates contrastive views of the inputs through the dropout-based graph augmentation strategy, improving the node representations at three levels. And through the projection head layer to integrate the supervised signal to improve the effect of model training. A joint learning paradigm is used to simultaneously optimize the recommendation task and the three-level CL tasks. Finally, the evaluation indicators Recall@10 on three public datasets MovieLens-1m, MovieLens-10m and Amazon-Books increased by 7.95%, 9.62% and 29.28% and

NDCG@10 increased by 7.68%, 9.91% and 36.73%, respectively. The significant improvement in performance metrics demonstrates the efficacy of ML-LGCL. This work is an exploration of CL in knowledge-aware recommendation domain, and provides a new perspective for subsequent CL research.

The graph augmentation strategies adopted in this paper are relatively simple, and more effective graph augmentation strategies can significantly improve the effect of CL. Future work should discuss specific graph augmentation strategies at different levels, especially mining hard negative examples to explore more influential data to improve the CL task.

Acknowledgements. This work was partially supported by the National Natural Science Foundation of China under the grant numbers 62250610224, 61972289 and 61832009. And the numerical calculations in this work have been partially done on the supercomputing system in the upercomputing Center of Wuhan University. Xiaoyuan Xie is the corresponding author.

References

1. Ai, Q., Azizi, V., Chen, X., Zhang, Y.: Learning heterogeneous knowledge base embeddings for explainable recommendation. Algorithms **11**(9), 137 (2018)
2. Gao, C., et al.: Graph neural networks for recommender systems: challenges, methods, and directions. arXiv preprint arXiv:2109.12843 (2021)
3. Glorot, X., Bengio, Y.: Understanding the difficulty of training deep feedforward neural networks. In: Proceedings of the Thirteenth International Conference on Artificial Intelligence and Statistics, pp. 249–256. JMLR Workshop and Conference Proceedings (2010)
4. Guo, Q., et al.: A survey on knowledge graph-based recommender systems. IEEE Trans. Knowl. Data Eng. **34**(8), 3549–3568 (2020)
5. He, X., Deng, K., Wang, X., Li, Y., Zhang, Y., Wang, M.: LightGCN: simplifying and powering graph convolution network for recommendation. In: Proceedings of the 43rd International ACM SIGIR Conference on Research and Development in Information Retrieval, pp. 639–648 (2020)
6. He, X., Liao, L., Zhang, H., Nie, L., Hu, X., Chua, T.S.: Neural collaborative filtering. In: Proceedings of the 26th International Conference on World Wide Web, pp. 173–182 (2017)
7. Kipf, T.N., Welling, M.: Semi-supervised classification with graph convolutional networks. arXiv preprint arXiv:1609.02907 (2016)
8. Van der Maaten, L., Hinton, G.: Visualizing data using t-SNE. J. Mach. Learn. Res. **9**(11), 2579–2605 (2008)
9. Rendle, S., Freudenthaler, C., Gantner, Z., Schmidt-Thieme, L.: BPR: Bayesian personalized ranking from implicit feedback. arXiv preprint arXiv:1205.2618 (2012)
10. Tang, X., Sun, T., Zhu, R., Wang, S.: CKG: dynamic representation based on context and knowledge graph. In: 2020 25th International Conference on Pattern Recognition (ICPR), pp. 2889–2895. IEEE (2021)
11. Wang, H., et al.: RippleNet: propagating user preferences on the knowledge graph for recommender systems. In: Proceedings of the 27th ACM International Conference on Information and Knowledge Management, pp. 417–426 (2018)

12. Wang, H., et al.: Knowledge-aware graph neural networks with label smoothness regularization for recommender systems. In: Proceedings of the 25th ACM SIGKDD International Conference on Knowledge Discovery & Data Mining, pp. 968–977 (2019)
13. Wang, H., Zhao, M., Xie, X., Li, W., Guo, M.: Knowledge graph convolutional networks for recommender systems. In: The World Wide Web Conference, pp. 3307–3313 (2019)
14. Wang, X., He, X., Cao, Y., Liu, M., Chua, T.S.: KGAT: knowledge graph attention network for recommendation. In: Proceedings of the 25th ACM SIGKDD International Conference on Knowledge Discovery & Data Mining, pp. 950–958 (2019)
15. Wang, X., Wang, D., Xu, C., He, X., Cao, Y., Chua, T.S.: Explainable reasoning over knowledge graphs for recommendation. In: Proceedings of the AAAI Conference on Artificial Intelligence, vol. 33, pp. 5329–5336 (2019)
16. Wei, Y., et al.: Contrastive learning for cold-start recommendation. In: Proceedings of the 29th ACM International Conference on Multimedia, pp. 5382–5390 (2021)
17. Wu, J., et al.: Self-supervised graph learning for recommendation. In: Proceedings of the 44th International ACM SIGIR Conference on Research and Development in Information Retrieval, pp. 726–735 (2021)
18. Wu, S., Sun, F., Zhang, W., Xie, X., Cui, B.: Graph neural networks in recommender systems: a survey. ACM Comput. Surv. 55(5), 1–37 (2022)
19. Yan, Y., Li, R., Wang, S., Zhang, F., Wu, W., Xu, W.: ConSERT: a contrastive framework for self-supervised sentence representation transfer. arXiv preprint arXiv:2105.11741 (2021)
20. Yang, Y., Huang, C., Xia, L., Li, C.: Knowledge graph contrastive learning for recommendation. In: Proceedings of the 45th International ACM SIGIR Conference on Research and Development in Information Retrieval, pp. 1434–1443 (2022)
21. Yu, J., Gao, M., Li, J., Yin, H., Liu, H.: Adaptive implicit friends identification over heterogeneous network for social recommendation. In: Proceedings of the 27th ACM International Conference on Information and Knowledge Management, pp. 357–366 (2018)
22. Yu, J., Yin, H., Xia, X., Chen, T., Cui, L., Nguyen, Q.V.H.: Are graph augmentations necessary? simple graph contrastive learning for recommendation. In: Proceedings of the 45th International ACM SIGIR Conference on Research and Development in Information Retrieval, pp. 1294–1303 (2022)
23. Yu, J., Yin, H., Xia, X., Chen, T., Li, J., Huang, Z.: Self-supervised learning for recommender systems: a survey. arXiv preprint arXiv:2203.15876 (2022)
24. Zhao, W.X., et al.: RecBole: towards a unified, comprehensive and efficient framework for recommendation algorithms. In: Proceedings of the 30th ACM International Conference on Information & Knowledge Management, pp. 4653–4664 (2021)
25. Zhou, K., et al.: S3-Rec: self-supervised learning for sequential recommendation with mutual information maximization. In: Proceedings of the 29th ACM International Conference on Information & Knowledge Management, pp. 1893–1902 (2020)

Temporal-Aware Multi-behavior Contrastive Recommendation

Hongrui Xuan[1] and Bohan Li[1,2,3(✉)]

[1] College of Computer Science and Technology, Nanjing University of Aeronautics and Astronautics, Nanjing, China
`{1692595335,bhli}@nuaa.edu.cn`
[2] Key Laboratory of Safety-Critical Software, Ministry of Industry and Information Technology, Nanjing, China
[3] National Engineering Laboratory for Integrated Aero-Space-GroundOcean Big Data Application Technology, Nanjing, China

Abstract. Modeling various types of users' interactions and jointly considering individual preferences from multiple perspectives, multi-behavior recommendation has attracted increasing attention recently. However, most existing multi-behavior recommendations only focus on the behavioral interaction itself, attempting to extract user preferences merely by modeling behaviors, while ignoring the properties of the interaction (e.g., the temporal information). Meanwhile, the introduction of multi-behavior information is bound to generate additional noise signals, which further affects the accuracy of recommendation. Thus, to tackle the above challenges, two aspects need to be explored: (1) How to fuse interaction attributes to construct unified cross-behavior dependencies while distinguishing subtle differences between individuals' behaviors. (2) How to mitigate the effects of noise to a certain extent while preserving multi-behavioral information. In this work, we propose a Temporal-Aware Multi-Behavior Contrastive Learning (TMCL) framework to explore the patterns of multiple behaviors of individuals through temporal information, jointly capture the correlation of users' preference evolution, and alleviate the impact of multi-behavior noise. Extensive experiments and ablation tests on the three datasets indicate our TMCL outperforms various state-of-the-art methods and verify the effectiveness of our model.

Keywords: Multi-Behavior Recommendation · Temporal Effects · Contrastive Learning

1 Introduction

Recommender systems learn individual preferences from historical interaction data, modeling feature representations of users and items [1], which has become the critical component for various platforms to alleviate information overload and provide personalized services [3]. Most traditional methods are based on collaborative filtering techniques to construct individual similarity representations from observed data [2]. Then, with the boosting of deep learning, researchers

X. Wang et al. (Eds.): DASFAA 2023, LNCS 13944, pp. 269–285, 2023.
https://doi.org/10.1007/978-3-031-30672-3_18

have focused on utilizing neural networks for recommendation. For instance, in 2017, M. Quadrana et al. proposed an information encoding paradigm based on RNN [4]. In 2018, by designing a convolution-based kernel function, J. Tang et al. achieved item sequential pattern modeling with slot information [5]. Furthermore, another promising research line is to introduce graph networks into recommendation, generating feature representations through information propagation on user-item interaction graph (e.g., LightGCN [6] and MA-GNN [7]).

Nevertheless, most existing technologies merely focus on a singular type of interaction (e.g., purchase) between users and items, while ignoring the multi-behavior information that generally exists in real-world scenarios (e.g., click, tag-as-favorite), which causes models unable to uniformly consider users' preferences from multiple perspectives to provide sufficient recommendations. Therefore, a challenging but meaningful work is to design a paradigm to capture cross-behavior dependencies while discriminating underlying differences between behaviors. Incoming various works are devoted to the study of multi-behavior recommendation. Based on global graph perspective, MBGCN [8] develops a relation-aware embedding layer to aggregate multi-behavioral information from higher-order neighbors. MMCLR [9] stacks multi-view on the basis of multi-behavior, combining the coarse-grained commonalities and the fine-grained differences between multi-behavior of users through contrastive learning tasks.

Despite the above methods have shown promising results, there are still two critical technical deficiencies in research: (1) The properties of the interaction are not fully utilized: Most multi-behavior recommendations merely consider cross-behavior dependencies and focus on how to model behavioral patterns, while ignoring the attribute characteristics of the interaction (e.g., the temporal information). Actually, individual preferences tend to change over time depending on different contexts [10], while a person's preferences in a period may not last for a long time, so certain historical preferences and behavioral habits contribute to a lower degree of recommendation than current preferences. Therefore, ignoring temporal information and treating all interaction behaviors as unified contributions will damage the feature construction ability of recommendations. (2) Influence of the noise by the introduction of multi-behavior information: There exists a lot of noise in multi-behavior information, which is not considered by the existing recommendation technologies, but it actually exists in real scenarios. For instance, in online shopping platforms, there are many unconscious and purposeless click operations of users, which will affect the model's judgment of individual preferences. So how to eliminate outliers in multi-behavior information to profile accurate individual portraits is crucial.

In view of the aforementioned deficiencies, we propose a Temporal-Aware Multi-Behavior Contrastive Learning (TMCL) framework for recommendation. Specifically, to handle the temporal relationships in multi-behavioral information, we design a time-aware feature aggregation mechanism to capture behavioral context-dependent temporal embeddings of nodes. Then, a subgraph extraction algorithm based on temporal stable nodes is proposed for the user-item interaction graph, which combines multi-behavior information denoising with behavior generality. We propose a contrastive learning (CL) paradigm based on denoised subgraphs, leveraging stable multi-behavior information to guide data augmentation of multi-view self-supervised signals. Through the injection of behaviors information and the perception of temporal relationships, TMCL dynamically captures individual preferences from different dimensions. Furthermore, combined with the CL paradigm, TMCL achieves the modeling of coarse-grained commonalities of behaviors and fine-grained differences of individuals while utilizing self-supervised data for signal enhancement.

In summary, the contributions of our work are highlighted as follows:

- We propose a new recommendation framework TMCL, which dynamically models feature embeddings from the perspective of preference evolution by combining multi-behavior information with temporal signals, emphasizing the criticality of temporal-aware cross-type behavior representation for recommendation.
- In TMCL framework, we propose a noise reduction technique based on temporal stable nodes to alleviate the interference effect of multi-behavior noise signals on the preference modeling of the system. Besides, with designed CL task, TMCL further refines the coarse-grained commonalities between behaviors and fine-grained differences between individuals to improve model performance.
- We conduct diverse experiments on three public datasets to demonstrate that our framework outperforms compared to various state-of-the-art recommendation methods. Furthermore, the ablation analysis is performed to better understand the design of TMCL and justifies the effectiveness of key components.

2 Preliminaries

In this research task, we consider a typical recommendation scenario with setting $\mathcal{U}$ and $\mathcal{I}$ to represent the set of users and items: $\mathcal{U} = \{u_1, ..., u_m, ..., u_M\}$ and $\mathcal{I} = \{i_1, ..., i_n, ..., i_N\}$, where M and N are the number of users and items, respectively. The detailed definitions and key notions are given as follows:

User-Item Multi-Behavior Interaction Graph. Based on various types of behavioral relationships, a user-item multi-behavior interaction graph is defined as: $G = (\mathcal{U}, \mathcal{I}, \mathcal{R})$, where $\mathcal{R} = \{r^1, ..., r^k, ..., r^K\}$ denotes the set of edges with multiple behaviors indexed by k. Note that $r^k_{m,n} = 1$ if user u_m interacts with item i_n under the k-th behavior, and $r^k_{m,n} = 0$ otherwise.

User-Item Interaction Temporal Graph. Given a certain behavior type k, we extract the temporal information of all interactions under this behavior, constructing the following graph: $G_t = (\Theta, \mathcal{T})$, where Θ is each interaction behavior message (i.e., user and item), and $\mathcal{T}$ is the temporal information of the interaction.

Task Formulation. Based on the above definitions, we formulate the temporal-aware multi-behavior contrastive recommendation as:

Input: the user-item multi-behavior interaction graph $G = (\mathcal{U}, \mathcal{I}, \mathcal{R})$ and the user-item interaction temporal graph $G_t = (\Theta, \mathcal{T})$.

Output: a prediction function which estimates the interaction probability $y^k_{m,n}$ of user u_m and item i_n under the k-th behavior.

3 Methodology

We present the TMCL model in this section, the specific framework of which is shown in Fig. 1. Key components and technical details are elaborated in subsequent sub-sections.

3.1 Dynamic Preference Context Modeling

In this process, the multi-behavior interaction graph and temporal graph are fed into the feature encoder as input. Through the designed context transformer network, TMCL constructs user-item representations that contain behavioral temporal information. In particular, we propose a novel temporal attention vector to distinguish projection embeddings under different temporal shifts.

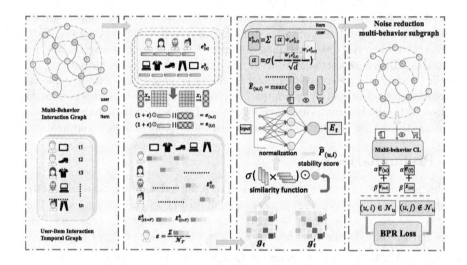

Fig. 1. The model architecture of TMCL framework.

Graph Information Encoding. Graph-based recommendation propensity extracts node features from interaction data to obtain latent representations that contain raw individual preferences and item attributes. To leverage behavioral messages while jointly injecting temporal signals into multiplex relationships, we develop a general contextual information encoder to learn initial representations for nodes:

$$e_{(u)}^k, e_{(i)}^k, E_{(t)}^k = Encoder^k(G, G_t, k), \tag{1}$$

where $e_{(u)}^k$ and $e_{(i)}^k$ indicate the representations of user u and item i under the behavior k. $E_{(t)}^k$ is the set of all interaction temporal embeddings with behavior k. Inspired by the lightweight architecture of LightGCN [6] and the positional embedding technology in Transformer [11], TMCL adopts following operations as $Encoder^k(\sim)$ to calculate the above representation:

$$e_{2i}^k = sin(\frac{\phi(t_{m,n}^k)}{10000^{\frac{2i}{d}}}); \qquad e_{2i+1}^k = cos(\frac{\phi(t_{m,n}^k)}{10000^{\frac{2i}{d}}}); \qquad E_{(t)}^k = e^k \cdot W_t; \qquad (2)$$

$$e_{(u)}^{k,(l+1)} = \sum_{i \in \mathcal{N}_u^k} \frac{e_{(i)}^{k,l}}{\sqrt{\mathcal{N}_u^k}\sqrt{\mathcal{N}_i^k}}; \qquad e_{(i)}^{k,(l+1)} = \sum_{u \in \mathcal{N}_i^k} \frac{e_{(u)}^{k,l}}{\sqrt{\mathcal{N}_i^k}\sqrt{\mathcal{N}_u^k}}. \qquad (3)$$

Here $\mathcal{N}_u^k$, $\mathcal{N}_i^k$ represent the neighbors of user u and item i under k-th behavior. l denotes the graph propagation layer to perform the operation. Based on the work of Xia et al. [12], we define $\phi(\sim)$ as a time period function that maps interaction timestamps $t_{m,n}^k$ indexed by $2i$ and $2i + 1$ (representing odd and even position indices) to initial time slots. Besides, $W_t \in \mathcal{R}^{2d \times d}$ is a trainable temporal transformation parameter with d as the latent dimension.

Temporal Attention Vector. Previous methods exploiting time-series signals, e.g., TGSRec [13] and KHGT [12], naively concatenate node embeddings with temporal representations and propagate them through linear layers. However, following the research by Srijan Kumar et al. [14], the above operations unable to achieve efficient modeling of input concatenated features. To tackle this challenge, we propose a novel temporal attention vector to distinguish individual projections based on historical interactions to efficiently construct users' preference embeddings. The specific formulas are as follows:

$$e_{(u,t)}^k = (1 + \varepsilon) \odot e_{(u)}^k; \qquad \varepsilon = \frac{\sum\limits_{t \in \mathcal{N}_t} E_{(t)}^k}{\mathcal{N}_t}. \qquad (4)$$

$\mathcal{N}_t$ is the set of user $u's$ interactions. Due to the different evolution trends of preferences, ε varies from person to person, which reflects user's attention to behavioral temporal signals. Motivated by JODIE [14], we utilize $(1 + \varepsilon)$ to scale the original node embeddings $e_{(u)}^k$. When $\varepsilon = 0$, $e_{(u,t)}^k$ degenerates into node representation without temporal information. So, we can manually set time intercepts to select time segments which we pay more attention to change the value of ε, thereby affecting the individual's local attention preference. We rewrite ε to $\varepsilon = \frac{\sum_{\mathcal{T}} E_{(t>\mathcal{T})}^k}{\mathcal{N}_{\mathcal{T}}}$, where $\mathcal{T}$ is the time intercept. Through various settings of $\mathcal{T}$, TMCL distinguishes the attitudes of different behaviors towards temporal signals. For example, in "click" behavior, we consider that the long-ago interaction has little contribution to the current preference recommendation, and even has a negative impact. Therefore, TMCL has a low tolerance for these interactions tending to select closer time intercept. Conversely, the model can tolerate the existence of more previous interactions for "tag-as-favorite" operations.

Analogous to edge features in the field of image processing, there exist some feature nodes in interaction temporal graphs, which reflect the step changes of individual behavior and contain special preference information. Previous work has been devoted to dealing with edge features in temporal graphs. By optimizing the spectral-based GCN, Simonovsky et al. [15] achieved the merging of edge features. In 2018, Battaglia [16] et al. proposed a novel GNN framework that can process edge features. It is encouraging that TGAT [17] achieves the processing of edge features in temporal graphs with a natural way through message propagation. So inspired by TGAT, we rewrite Eq. (4) as follows:

$$e_{(u,t)}^k = (1 + \varepsilon) \odot e_{(u)}^k \| \mathcal{X}_u; \qquad \varepsilon = \frac{\sum_T E_{(t>T)}^k}{N_T}, \qquad (5)$$

where $\mathcal{X}_u$ represents the feature vector of user's interaction behavior matrix. Through the above steps, TMCL will propagate edge features during message aggregation to the hidden representation of the target node and the next layer of neural network. Similar operations also apply to item-side information.

Graph Information Aggregation. To capture the effect of neighbor temporal embeddings on target nodes, while jointly considering behavior heterogeneity, we design a general information propagation paradigm to unify temporal collaborative signals and multi-behavioral information. Specifically, for each behavior k, TMCL adopts the following formulas (for the convenience of writing, the symbol l representing the layers of the neural network is omitted below):

$$E_{(u,t)}^k = \sum_{i \in \mathcal{N}_u} \alpha_{(i,t)}^u W_u e_{(i,t)}^k;$$

$$\alpha_{(i,t)}^u = \sigma\left(\frac{(W_1 e_{(i,t)}^k)^T (W_2 e_{(u,t)}^k)}{\sqrt{d}}\right), \qquad (6)$$

where both $W_u \in \mathcal{R}^{2d}$ and $W_1, W_2 \in \mathcal{R}^d$ are trainable parameters, and $\alpha_{(i,t)}^u$ denotes the importance weight of item i interacted at time t to user u. $\sigma(\sim)$ is the softmax function. Note that we utilize $\alpha_{(i,t)}^u$ to represent the co-attention signal of historical interaction (u, i, t) to user embedding $e_{(u,t)}^k$ based on temporal inference. As proposed in Work [13], this collaborative attention jointly considers adjacent interactions and temporal inference information, being a better mechanism than self-attention to capture temporal collaborative signals.

Since the advantage of multi-behavior recommendation is to make up for the data sparsity issue caused by single-behavior representations, we propose to perform cross-type behavioral information aggregation to construct unified individual embeddings to reflect the multi-dimensional preferences of users:

$$E_{(u,t)} = FFN(mean(E_{(u,t)}^1 \oplus, ..., \oplus E_{(u,t)}^k \oplus, ..., \oplus E_{(u,t)}^K))); \qquad (7)$$

$FFN(\sim)$ is a two-layer feed-forward neural network with ReLU as activation function [11]. Through the above operations, TMCL realizes dynamic modeling of contextual individual preferences for behaviors utilizing temporal signals, with obtaining users' representations which contain evolutionary information. Similar steps are adopted to acquire the corresponding item-side representations.

3.2 Temporal-Aware Global Learning

Different from conventional recommendation algorithms, multi-behavior recommendation utilizes additional behaviors as auxiliary information to support target behavior recommendation, which greatly improves the efficiency and accuracy of models. However, the introduction of additional messages is bound to bring more serious noise effects. Therefore, in this section, we propose a temporal-aware global subgraph extraction algorithm to alleviate the noise effect caused by multi-behavior information to gain more stable feature representations.

Stability Calculation on Temporal Graph. Compared with target behavior information, auxiliary multi-behavior data contain more noise information. For example, in "click" behavior, there exists a large of meaningless click items, which interfere with the model's inference of individual preferences. Motivated by work [18,19], if a node has strong noise immunity with better stability, the representation of the node from different subgraphs should be similar. Therefore, TMCL proposes an intuitive method to calculate the stability of temporal nodes:

$$p = Sim(g_t(E_{(t)}), g'_t(E_{(t)})); \qquad g_t, g'_t = Sub(G_t), \tag{8}$$

where g_t, g'_t are the subgraphs selected by the sampling function $Sub(\sim)$ with setting random seeds. p is the stability score of temporal interaction obtained by the similarity calculation function $Sim(\sim)$. Noted that the higher the stability score p, the better the anti-noise ability of the corresponding temporal node, which will further guide model to optimize the multi-behavior information.

TMCL attempts to guide the denoising of multi-behavior interaction graphs with stable temporal nodes. We deem that interactive information with structural stability better reflects the real preferences of individuals, which should have a greater probability of being selected. But for comprehensive consideration, we intend to strengthen the influence of individual-specific preferences on interaction choices from user-level. In reality, there exist certain users who have consumption impulses with changeable hobbies, which causes their historical interaction nodes with low stability. Therefore, we intend to utilize global information reflecting overall profile of individual to optimize the selection probability:

$$p_{u,i} = Norm(\sigma(e_{(i)}^T e_{(u)}) \odot p), \tag{9}$$

$$\hat{p_{u,i}} = (1 - p_{u,i})a + p_{u,i}b, \tag{10}$$

where $p_{u,i}$ represents the initial sampling probability, which is based on individual behavior profiles and node stability score, obtained by the normalization function $Norm(\sim)$. After that, TMCL maps $p_{u,i}$ to the interval [a,b] to obtain the final sampling probability for filtering edges of interaction graph. In particular, the range of the interval represents the tolerance of different behaviors k to low-stability information. Through the above operations, TMCL samples different behavior interaction graphs according to respective probabilities, finally obtaining the noise reduction multi-behavior subgraph.

Global Behavior Contrastive Learning. Recent work has demonstrated that, by exploiting multiple views to construct different features as self-supervised auxiliary signals, CL alleviates the data sparsity issue to a certain extent [19,20]. Furthermore, with well-designed pretext tasks, CL minimizes the difference between homogeneous nodes while maximizing the difference between heterogeneous nodes [9]. The crucial point of CL is to construct multiple perspectives. In multi-behavior recommendation, we naturally deem that various types of behaviors reflect user different preference dimensions, which is suitable for CL.

Based on the sampling probabilities mentioned above, we sample the user-item interaction graphs under different behaviors to obtain multi-behavior subgraphs after noise reduction. TMCL treat them as multi-view of individual behavioral preferences, from which model extracts valid positive-negative sample pairs. Following the previous work [9,20], for user embedding e_x^k, we consider setting positive and negative sample pairs as: $\{e_x^k, e_x^{k'}\}, \{e_x^k, e_y^{k'}\}$, where $x, y \in \mathcal{U}$ and $x \neq y$. Furthermore, TMCL defines the multi-behavior contrastive loss based on InfoNCE loss [21] as follows:

$$\mathcal{L}_{CL} = \sum_{k'=1}^{K} \sum_{x \in \mathcal{U}} -log \frac{exp(s(e_x^k, e_x^{k'})/\tau)}{\sum_{y \in \mathcal{U}} exp(s(e_x^k, e_y^{k'})/\tau)}, \tag{11}$$

where $s(\sim)$ is the function to estimate the similarity of sample pairs, with τ as the temperature hyperparameter to control curve smoothness. Through the above noise reduction operations and CL task, TMCL learns more robust node embeddings from low-noise information, models latent behavior patterns from multiple dimensions, and captures more discriminative preference representations.

3.3 Process of Model Optimization

So far, TMCL realizes the construction of four kinds of node embeddings, namely: $e_{(u)}, e_{(i)}, E_{(u,t)}, E_{(i,t)}$. Specifically, $e_{(u)}$ and $e_{(i)}$ are derived from the optimized structure of multi-behavior contrastive loss. Please note that although we leverage the temporal graph to guide the noise reduction of multi-behavior graph, the generation and propagation of $e_{(u)}$ and $e_{(i)}$ do not contain any temporal information, so they can be treated as static global preferences. Whereas $E_{(u,t)}$ and $E_{(i,t)}$ reflect the embedded expression of interactions evolving over time, in this case we regard them as dynamic local preferences. Therefore, TMCL gives the following settings:

$$e_u = \alpha e_{(u)} + \beta E_{(u,t)}, \qquad e_i = \alpha e_{(i)} + \beta E_{(i,t)}, \tag{12}$$

where α and β are utilized to dynamically adjust the ratio of global preference to local preference. In the optimization stage, due to the generality and simplicity of Bayesian Personalized Ranking [22], TMCL adopts BPR as the main task loss function to calculate the corresponding loss:

$$\mathcal{L}_{Main} = - \sum_{(u,i) \in \mathcal{N}_u} \sum_{(u,j) \notin \mathcal{N}_u} ln\sigma(e_u^T e_i - e_u^T e_j). \tag{13}$$

Here i denotes the observed interaction item while j represents the non-interacted item. To sum up, we integrate contrastive loss with main loss to give the overall loss definition:

$$\mathcal{L} = \mathcal{L}_{Main} + \lambda 1 \mathcal{L}_{cl} + \lambda 2 ||\Omega||_2^2, \tag{14}$$

where Ω denotes the set of learnable model hyperparameters, with $\lambda 1$ and $\lambda 2$ determining the ratio of self-supervised signals and the weight of regularization, respectively.

4 Experiments

In this section, we conduct validation experiments of TMCL on various datasets, aiming to answer the following questions: **(RQ1)** How does TMCL perform when compared with other SOTA methods? **(RQ2)** What are the effects of key components in TMCL on recommended performance? **(RQ3)** How effective is the strategy designed in TMCL in mitigating the effects of multi-behavior noise? **(RQ4)** How different hyperparameter settings affect TMCL performance?

Table 1. Statistics of datasets

Dataset	Users	Items	Interactions	Interactive Behavior Type
IJCAI	17,435	35,920	799,368	{Page View, Favorite, Cart, Purchase}
Tmall	31,882	31,232	1.451,219	{Page View, Favorite, Cart, Purchase}
Retail	147,894	99,037	1,584,238	{Favorite, Cart, Purchase}

Datasets for Experiments. To verify the effectiveness of the model, we conduct extensive experiments on three real recommendation datasets. **IJCAl:** The dataset is released by the IJCAI competition with the online activities of users in e-commerce platforms. According to the work [20], user's behaviors include *page view, tag-as-favorite, add-to-cart* and *purchase*. **Tmall:** The dataset is collected on one of the largest online consumer platforms in China. It shares the same user's behavior settings with the IJCAI dataset. **Retail:** This dataset contains multiple types of interaction behaviors in online retail scenarios. TMCL applies the following behaviors, including *tag-as-favorite, add-to-cart* and *purchase*, as training data for the model. Detailed statistics for datasets are further summarized in Table 1.

Methods for Comparison. We compare TMCL with recommendation methods from various research lines: (To save space, we select several comparison techniques for a brief introduction. All comparison baselines are shown in Table 2.)

Collaborative Filtering-Based Method:

- **LightGCN** [6]: It utilizes a simplified convolution operation for information propagation with applying collaborative filtering to graph neural networks.
- **NGCF** [1]: It considers the higher-order connectivity of nodes and performs collaborative filtering message passing based on graph neural networks.

Temporal-Aware Recommendation Method:

- **NARM** [27]: It is a recommendation method that combines recurrent neural network and attention mechanism to capture individual interaction patterns based on temporal sequential information.
- **SR-GNN** [28]: It constructs global sequence into session subsequences, attempting to mine node representations that contain transition features of subsequences through gated mechanism.

Multi-Behavior Recommendation Method:

- **MB-GMN** [31]: It is an approach that leverages meta-learning to model the underlying dependencies behind individual behavioral heterogeneities.
- **CML** [20]: It proposes meta-contrastive coding, which attempts to adopt contrast task to assist the model to learn behaviors preference features.

Parameter Settings. We implement TMCL model with Pytorch, construct the initialization embedding through the Xavier method, and adopt Adma optimizer with learning rate of $3e^{-4}$. In particular, we set the embedding dimension of nodes to 32. We leverage L2 regularization with a decaying weight configuration $\lambda2$, with the range of $\lambda2$ being $\{5e^{-2}, 1e^{-2}, 5e^{-3}, 1e^{-3}, 5e^{-4}, 1e^{-4}\}$. The number of neural network layers applied for message propagation in TMCL is set from the range $\{1, 2, 3, 4\}$. We specify that the sum of the embedded combination weights α and β is 1. In the experiment, the value range of α is selected from $\{0.6, 0.7, 0.8, 0.9\}$, with the corresponding value of β changing. For the compared baseline methods, we set the corresponding implementation parameters according to the original papers.

4.1 Performance Validation (RQ1)

We present the performance of all baseline methods and TMCL on HR@10 and NDCG@10 metrics in Table 2, and summarize the following conclusions:

TMCL consistently outperforms other baselines on each dataset, which confirms the effectiveness of the model. Meanwhile, TMCL has excellent generality in view of the large differences in data size and sparsity of validation datasets. According to the experimental analysis, we attribute the superiority of model performance to three aspects: (1) Due to the introduction of behavioral temporal information, TMCL dynamically captures the evolution of individual preference patterns, according to historical interactions adjusting the time attention interval to emphasize the modeling of local preferences. (2) Multi-behavior information makes up for the issue of data sparsity in traditional recommendation, enabling TMCL to profile individual portraits from multiple perspectives with personalized information. In addition, the CL task further facilitates the model to explore high-dimensional features of nodes and distinguish information gradients between instances. (3) Benefiting from the designed subgraph extraction algorithm, TMCL drops out the unstable nodes in the multi-behavior interaction graph and alleviates the noise effect caused by extra information.

It can be observed from Table 2 that temporal-aware and multi-behavior methods consistently outperform collaborative filtering-based methods on all indicators, which indicates that auxiliary signals indeed promote context-dependent modeling and provide effective personalized information for recommendation. Furthermore, CML consistently outperforms various baselines, demonstrating the superiority and rationality of CL combined with multi-behavior recommendation. As a powerful tool to improve the performance of the main task, the contrasting auxiliary task enables the model to finely distinguish coarse-grained commonalities between behaviors and fine-grained differences between individuals through various semantic guidance.

Table 2. Results on different datasets in terms of HR@10 and NDCG@10

model	dataset					
	IJCAI		Tmall		Retail	
	HR	NDCG	HR	NDCG	HR	NDCG
BPR	0.1626	0.0839	0.2443	0.1501	0.2608	0.1653
DMF	0.2316	0.1093	0.3027	0.1746	0.2984	0.1806
LightGCN	0.2558	0.1214	0.3418	0.2048	0.3065	0.1847
NGCF	0.2539	0.1226	0.3288	0.1964	0.3027	0.1844
NARM	0.2843	0.1476	0.3421	0.2013	0.3009	0.1816
MA-GNN	0.2921	0.1471	0.3547	0.2062	0.3088	0.1887
SR-GNN	0.3006	0.1583	0.3603	0.2114	0.3109	0.1901
NMTR	0.3161	0.1627	0.3623	0.2097	0.3112	0.1869
MATN	0.3681	0.2053	0.4302	0.2437	0.3490	0.2183
KHGT	0.3319	0.1962	0.4032	0.2531	0.4199	0.2471
MB-GMN	0.3917	0.2281	0.4502	0.2705	0.3974	0.2406
CML	0.4775	0.2851	0.5185	0.3055	0.4256	0.2503
TMCL	**0.4889**	**0.2984**	**0.5574**	**0.3453**	**0.4521**	**0.2714**

4.2   Model Ablation Study (RQ2)

To verify the effect of the designed key components on the performance improvement, we conducted additional ablation experiments to analyze the detailed experimental results of the following three TMCL variants.

- **w/o-Mul:** We design a variant w/o-Mul to explore the impact of multi-behavior information on performance. The variant discards auxiliary behaviors message with learning from single behavior of individual. In particular, due to the discarding of multi-behavioral information, the variant disables the multi-behavior contrastive task, merely relying on the target behavior and temporal message to assist the main task for recommendation. As shown in Table 3, compared with the complete TMCL, the performance of w/o-Mul variant on both metrics drop significantly. Such results are reasonable and easy to understand. Since various behaviors reflect preferences from multiple dimensions. These intertwined behavioral patterns collectively profile the individual's intrinsic characteristics. Meanwhile, the CL task strengthens the model's ability to understand of the coarse-grained commonality of behaviors and the fine-grained differences of individuals, making it easier to distinguish between similar instances and different instances, thereby improving the accuracy of recommendation.

- **w/o-Ti:** To demonstrate that the temporal information can make model capture the evolution law of individual preferences, we propose a variant w/o-Ti, which disables all temporal components such as temporal attention vector, aggregation mechanism based on temporal information, etc. Comparing TMCL with w/o-Ti, it shows that with temporal information, TMCL achieves better performance by mining behavior temporal features to focus on individual local preferences. The designed temporal attention vector efficiently constructs node embeddings based on specific historical interactions while preserving edge features in the temporal graph. While the aggregation mechanism is able to propagate the temporal features of neighbors according to context dependencies.

- **w/o-Sc:** TMCL considers that multi-behavior information will introduce additional noise, which in turn reduces the recommendation performance. Therefore, we propose a stability calculation strategy to obtain sampling probabilities to extract nodes with strong anti-interference ability from multi-behavior interaction graphs. The experimental results show that the designed subgraph extraction algorithm based on temporal stable nodes is effective. The noise-reduced multi-behavior subgraph drops out the original interfering information that would affect model learning, making TMCL accurately capture key preference paradigms from effective interactive behaviors to realize more robust individual modeling.

Table 3. Results of ablation experiments

model	dataset					
	IJCAI		Tmall		Retail	
	HR	NDCG	HR	NDCG	HR	NDCG
w/o-Mul	0.3149	0.1612	0.4439	0.2671	0.3246	0.1927
w/o-Ti	0.4506	0.2638	0.4965	0.2843	0.3928	0.2397
w/o-Sc	0.4719	0.2795	0.5211	0.3092	0.4304	0.2558
TMCL	**0.4889**	**0.2984**	**0.5574**	**0.3453**	**0.4521**	**0.2714**

4.3   Performance on Alleviating Noise Effect(RQ3)

In this subsection, we conduct additional verification experiments to explore the performance of TMCL in handling noisy signals. Specifically, we inject noise information into the user-item multi-behavior interaction graph, and conduct validation tests on various superior methods to demonstrate the effectiveness of our model in mitigating the effects of noise. The specific experimental results are shown in Fig. 2, where "X_N" represents the performance of the "X" method processing data with about 10% noise signal.

We conclude that: (1) Under all conditions, TMCL consistently outperforms CML with less affected by noise, proving the superiority of the designed multi-behavior denoised algorithm of contrastive learning. (2) Compared with non-contrastive learning methods MATN and KHGT, and the latter crudely adds temporal signals, we argue that the performance advantage of TMCL suggests that utilizing temporal information to capture individual mainstream preference evolution and discarding the influence of outdated preferences through attention vectors, which is also a way to mitigate noise interference. (3) Experimental results show that when compared with other advanced baseline methods, TMCL consistently achieves the best performance in both recommendation accuracy and noise immunity. This verifies the effectiveness and rationality of various components designed in our model to mitigate noise interference.

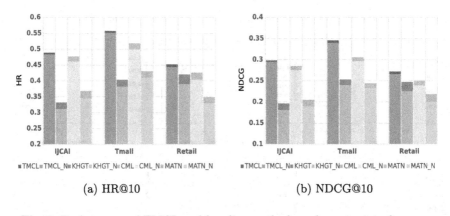

(a) HR@10   (b) NDCG@10

Fig. 2. Performance of TMCL and baseline methods under noise interference.

4.4 Hyperparameter Analyses (RQ4)

Hyperparameter experiments are performed to explore the effect of different parameter settings on model performance, where the research parameters include hidden state dimensionality, graph propagation layers, and the combination weights. The experimental results are shown in Fig. 3. Note that the x-axis represents the various parameter settings, while the y-axis denotes the degree of performance degradation compared to the best performance.

- **Hidden State Dimensionality:** We set the value range of the embedding dimension to $\{8, 16, 32, 64\}$ with keeping other optimal hyperparameters unchanged. Figure 3 shows that when the embedding dimension is increased from 8 to 32, the model performance improves accordingly. However, larger embedding dimension is not always beneficial for more robust feature representation, which will lead to overfitting.

- **Graph Propagation Layers:** The stacking of graph propagation layers enables the model to capture latent contextual dependencies from higher-order neighbors, leveraging the cooperative relationship between higher-order nodes to facilitate accurate representation construction, thereby improving recommendation performance. It can be seen from the results that the model achieves the best results when the number of layers is 2 or 3. Too few or too many propagation layers will lead to a decline in the representational ability of the model.
- **Combination weights:** In order to explore the reasonable contribution weight of multi-behavior information and temporal information, we designed experiments to select appropriate values of α and β. According to the previous settings, the x-axis in Fig. 3 shows the value range of α. We find that the model achieves optimal performance when α is in the interval $\{0.7, 0.8\}$. It proves that TMCL regards multi-behavioral information as the main recommendation signal, with the temporal message being a pivotal auxiliary role in improving performance.

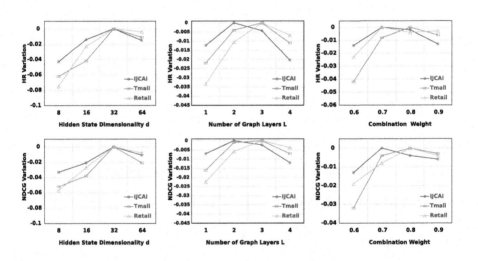

Fig. 3. Hyperparameter analyses of TMCL.

5   Related Work

Multi-behavior Recommendation Models. To reflect individual preferences and alleviate the data sparsity issue faced by single-behavior recommendation, multi-behavior recommendation has received increasing attention in recently [8,29]. By introducing additional item-side auxiliary information, KHGT [12] learns individual behavior paradigms from knowledge-aware multi-behavior networks for recommendation. MATN [30] designs an information aggregation pattern which jointly considers the relationship between weighted attention and behaviors. MB-GMN [31] proposes to leverage meta-learning to distill the latent dependencies behind individual's multi-behavior pattern. While our TMCL mines the temporal information of interactions, with dynamically modeling node features, and considers alleviating the influence of noise in multi-behavior information.

Temporal Graph-Based Recommendation. Temporal graph-based recommendation is usually formulated as temporal-based prediction [14], which closely intersects and integrates with sequential recommendation [23]. Collaborative filtering is first applied to modeling temporal information, where TimeSVD++ [24] treats the bias as a function of time-varying. Then, researchers have considered utilizing neural network techniques to capture temporal representations [26]. Attention-based methods, such as NARM [27] and JODIE [14], enable the model to distinguish the importance of different temporal information and preserve the long-term dependencies of features. In addition, work such as SR-GNN [28] leverages links between nodes to learn embedded representations by GNN, which has achieved certain results.

Recommendation with Contrastive Learning. As the crucial component to improve model performance, CL has been demonstrated to be effective in the field of recommendation [19]. CL attempts to mine comprehensive individual characteristics through multiple views of instances. Long et al. [32] devise a joint training paradigm based on maximizing mutual information between local and global features, propagating over heterogeneous graphs via self-supervised signals. Work [9] proposes to introduce multiple CL tasks in recommendation to model multi-dimensional commonalities in various views. Our TMCL adopts CL to explore the expression paradigm behind multiple behaviors of individuals, mitigating the data sparsity issue of target behaviors while modeling coarse-grained commonalities between behaviors and fine-grained differences of individuals.

6   Conclusion

In this work, we propose TMCL, a novel temporal-aware recommendation framework for multi-behavior contrastive learning. Specifically, we consider introducing the temporal attributes into multi-behavior recommendation to dynamically model behavior patterns, constructing individual representations that conform to the preference paradigm. Besides, through the designed subgraph extraction algorithm and CL task, TMCL learns users' features from multiple perspectives while mitigating the influence of multi-behavior noise, which strengthens model's anti-interference and discriminative instance ability, improving the accuracy of recommendation. Extensive experiments on various datasets validate the effectiveness of TMCL compared to other state-of-the-art methods.

Acknowledgement. This work is supported by National Natural Science Foundation of China (62003379), the "14th Five-Year Plan" Civil Aerospace Pre-Research Project of China (D020101).

References

1. Wang, X., He, X., Wang, M., et al.: Neural graph collaborative filtering. In: Proceedings of SIGIR (2019)
2. He, X., Liao, L., et al.: Neural collaborative filtering. In: Proceedings of WWW (2017)
3. Liu, Y., Li, B., et al.: A Knowledge-aware recommender with attention-enhanced dynamic convolutional network. In: Proceedings of CIKM (2021)
4. Quadrana, M., Karatzoglou, A., et al.: Personalizing session-based recommendations with hierarchical recurrent neural networks. In: Proceedings of Recsys (2017)
5. Tang, J., Wang. K.: Personalized top-n sequential recommendation via convolutional sequence embedding. In: Proceedings of WSDM (2018)
6. He, X., Deng, K., Wang, X., et al.: Lightgcn: Simplifying and powering graph convolution network for recommendation. In: Proceedings of SIGIR (2020)
7. Ma, C,, Ma, L,, Zhang, Y., et al.: Memory augmented graph neural networks for sequential recommendation. In: Proceedings of AAAI (2020)
8. Jin, B., Gao, C., He, X., et al.: Multi-behavior recommendation with graph convolutional networks. In: Proceedings of SIGIR (2020)
9. Wu, Y., Xie, R., Zhu, Y., et al.: Multi-view Multi-behavior Contrastive Learning in Recommendation. In: Proceedings of DASFAA (2022)
10. Xia, L., Huang, C., Xu. Y., et al.: Multi-Behavior Sequential Recommendation with Temporal Graph Transformer. In: Proceedings of TKDE (2022)
11. Vaswani, A., et al.: Attention is all you need. In: Proceedings of NIPS (2017)
12. Xia, L., Huang, C., Xu, Y., et al.: Knowledge-enhanced hierarchical graph transformer network for multi-behavior recommendation. In: Proceedings of AAAI (2021)
13. Fan, Z., Liu, Z., Zhang, J., et al.: Continuous-time sequential recommendation with temporal graph collaborative transformer. In: Proceedings of CIKM (2021)
14. Kumar, S., Zhang, X., Leskovec, J.: Predicting dynamic embedding trajectory in temporal interaction networks. In: Proceedings of SIGKDD (2019)
15. Simonovsky, M., Komodakis, N.: Dynamic edge-conditioned filters in convolutional neural networks on graphs. In: Proceedings of CVPR (2017)
16. Battaglia, P.W., Hamrick, J.B., Bapst, V., et al.: Relational inductive biases, deep learning, and graph networks. arXiv preprint arXiv:1806.01261 (2018)
17. Xu, D., Ruan, C., Korpeoglu, E., et al.: Inductive representation learning on temporal graphs. arXiv preprint arXiv:2002.07962 (2020)
18. Jin W, Zhao Z, Zhang P, et al.: Hierarchical cross-modal graph consistency learning for video-text retrieval. In: Proceedings of SIGIR (2021)
19. Yuhao, Y., Huang, C., Xia, L., et al.: Knowledge graph contrastive learning for recommendation. In: Proceedings of SIGIR (2022)
20. Wei, W., Huang, C., Xia, L., et al.: Contrastive meta learning with behavior multiplicity for recommendation. In: Proceedings of WSDM (2022)
21. Oord, A., Li, Y., Vinyals, O.: Representation learning with contrastive predictive coding. arXiv preprint arXiv:1807.03748 (2018)
22. Rendle, S., Freudenthaler, C., Gantner, Z., et al.: BPR: Bayesian personalized ranking from implicit feedback. arXiv preprint arXiv:1205.2618 (2012)
23. Wang, S., Hu, L., Wang, Y., et al.: Sequential recommender systems: challenges, progress and prospects. In: Proceedings of IJCAl (2019)
24. Koren, Y.: Collaborative filtering with temporal dynamics. In: Proceedings of SIGKDD (2009)

25. Nguyen, G.H., Lee, J.B., Rossi, R.A., et al.: Continuous-time dynamic network embeddings. In: Proceedings of Companion Proceedings of the Web Conference (2018)
26. Fang, H., Zhang, D., Shu, Y., et al.: Deep learning for sequential recommendation: Algorithms, influential factors, and evaluations. In: Proceedings of TOIS (2020)
27. Li. J., Ren, P., Chen, Z., et al.: Neural attentive session-based recommendation. In: Proceedings of CIKM (2017)
28. Wu, S., Tang, Y., Zhu, Y., et al.: Session-based recommendation with graph neural networks. In: Proceedings of AAAI (2019)
29. Gao, C., He, X., Gan, D., et al.: Neural multi-task recommendation from multi-behavior data. In: Proceedings of ICDE (2019)
30. Xia, L., Huang, C., et al.: Multiplex behavioral relation learning for recommendation via memory augmented transformer network. In: Proceedings of SIGIR (2020)
31. Xia, L., Xu, Y., Huang, C, et al.: Graph meta network for multi-behavior recommendation. In: Proceedings of SIGIR (2021)
32. Long, X., Huang, C., Xu, Y., et al.: Social recommendation with self-supervised metagraph informax network. In: Proceedings of CIKM (2021)

GUESR: A Global Unsupervised Data-Enhancement with Bucket-Cluster Sampling for Sequential Recommendation

Yongqiang Han[1], Likang Wu[1], Hao Wang[1(✉)], Guifeng Wang[2],
Mengdi Zhang[3], Zhi Li[1], Defu Lian[1], and Enhong Chen[1]

[1] Anhui Province Key Laboratory of Big Data Analysis and Application,
University of Science and Technology of China, Hefei, China
{harley,wulk,wanghao3,zhili03}@mail.ustc.edu.cn,
{liandefu,cheneh}@ustc.edu.cn
[2] Huawei Technologies Co., Ltd., Hangzhou, China
wangguifeng4@huawei.com
[3] Meituan-Dianping Group, Beijing, China
zhangmengdi02@meituan.com

Abstract. Sequential Recommendation is a widely studied paradigm for learning users' dynamic interests from historical interactions for predicting the next potential item. Although lots of research work has achieved remarkable progress, they are still plagued by the common issues: data sparsity of limited supervised signals and data noise of accidentally clicking. To this end, several works have attempted to address these issues, which ignored the complex association of items across several sequences. Along this line, with the aim of learning representative item embedding to alleviate this dilemma, we propose GUESR, from the view of graph contrastive learning. Specifically, we first construct the Global Item Relationship Graph (GIRG) from all interaction sequences and present the Bucket-Cluster Sampling (BCS) method to conduct the sub-graphs. Then, graph contrastive learning on this reduced graph is developed to enhance item representations with complex associations from the global view. We subsequently extend the CapsNet module with the elaborately introduced target-attention mechanism to derive users' dynamic preferences. Extensive experimental results have demonstrated our proposed GUESR could not only achieve significant improvements but also could be regarded as a general enhancement strategy to improve the performance in combination with other sequential recommendation methods.

Keywords: Sequential Recommendation · Graph Neural Network · Contrastive Learning

1 Introduction

With the rapid development of the Internet, recommendation systems have been widely employed on online platforms. Among these, sequential recommendation (SR), predicting the next item for users by regarding historical interactions as temporally-ordered sequences, has attracted various attention from both academia and industry. In the recent literature, a large number of works have been proposed and achieved remarkable progress. The initial approach is often based on neural networks [7,12]. Considering the different importance of inter-acted items on the next prediction, an attention mechanism is further introduced to quantify the weights of items in the sequence, such as SASRec [8] and BERT4Rec [11]. Some studies explore the adaptation of graph neural networks in SR and capture the complex patterns of items hidden in interaction sequences, such as SR-GNN [1] and GC-SAN [19]. However, most of these works still face the problem of data sparsity and noise, which is prone to fail on limited training signals and complex associations between items. Meanwhile, contrastive learning techniques have made a great breakthrough in representation learning. Inspired by its success, some methods apply contrastive learning to improve sequential recommendation, such as S^3Rec [22] and CL4SRec [18].

Despite these methods usually achieving remarkable success, there still exist some deficiencies that can be improved. Firstly, most sequential methods exploit the local context of items in each sequence individually, where co-occurrence information between items is sensitive to noise, and the associations that cross several sequences are not well captured. Secondly, the popularity of items approximates a long-tail distribution, and the interactions of many items are very sparse. The representation of these items in existing models may introduce selection bias in some cases. Thirdly, although some recent studies have applied contrastive learning to alleviate the sparsity of interaction data, they usually construct the data augmentation randomly and lack consideration on how to design suitable contrastive learning strategies for the characteristics of sequential recommendation tasks.

To this end, in this paper, we introduce graph contrastive learning to the sequential recommendation to learn informative representations of items and provide a solid foundation for the portrayal of users' interests accurately. Specifically, we propose a novel framework GUESR, a **G**lobal **U**nsupervised Data-**E**nhancement method for **S**equential **R**ecommendation. To be specific, we first construct a Global Item Relationship Graph (GIRG) from all interaction sequences. We quantified different order adjacent information of items as edge weights to obtain complex associations between items and set a threshold to filter out the noise. In this setting, more edges will be removed for items with high popularity, thus reducing item popularity bias. Subsequently, we adopt graph contrastive learning to learn this global association information for learning enhanced item representations. Besides, we present a Bucket-Cluster Sampling (BCS) method to alleviate the sampling bias of improper negatives and uniform the representation space, which takes into account both efficiency and effectiveness. Additionally, we further extend CapsNet [10] with a target-attention mechanism to derive the users' preferences on multiple interests, which is formulated as the prediction function. Finally, we jointly optimize this paradigm loss and our proposed auxiliary contrastive learning task. To summarize, the contributions of this paper are as follows,

- We propose a global contrastive data-enhancement framework for the sequential recommendation, termed GUESR, where the graph contrastive learning is adopted on a constructed global graph to capture the complex associations between items across sequences to alleviate the problem of data sparsity and noise.

- To alleviate the influence of improper negatives, we present the Bucket Cluster Sampling (BCS) method with consideration of attribute knowledge, which could benefit from both worlds of efficiency and effectiveness.
- Extensive experiments on publicly available datasets demonstrate that GUESR outperforms state-of-the-art methods. In addition, some analyses further validate that GUESR is a generic module that can improve the performance of other sequentially recommended methods.

2   Related Work

2.1   Sequential Recommendation

Sequential recommendation (SR) aims to predict the next item based on historical interaction sequences. With the development of deep learning, several models based on neural networks have been proposed [7,12]. Furthermore, the attention mechanism is a powerful tool applied in sequential recommendation, such as SAS-Rec [8] and BERT4Rec [11]. In recent years, graph neural networks have achieved state-of-the-art performance in processing graph structure data [13,14,16,17]. Since the powerful GNNs can capture complex item transition patterns hidden in user sequences, there are some studies applying GNNs to SR, such as SR-GNN [1] and GC-SAN [19]. However, most of these methods above are trained by the prediction loss that optimizes the representation of the entire sequence to improve recommendation performance, while ignoring the valuable unsupervised signal.

2.2   Contrastive Learning for Recommendation

Contrastive learning is an emerging unsupervised learning paradigm that has been successfully applied to computer vision and natural language processing. Meanwhile, some models are applying contrastive learning techniques in sequential recommendation scenarios [4,15,18,22], S^3Rec [22] devises four auxiliary self-supervised objectives for data representation learning by using mutual information maximization. CL4SRec [18] applies three data augmentation to generate positive pairs, and contrasts positive pairs for learning robust sequential transition patterns. Despite the achievement, the contrastive learning-based methods for SR mainly focus on learning the self-supervised signals from each sequence. Due to the limited information in a separate sequence, the obtained self-supervised signal is too weak to encode informative embedding.

3   Preliminary

Considering a set of users $U(|U| = M)$ and items $V(|V| = N)$, each user $u \in U$ has a sequence of interacted items sorted in chronological order $I_u = [I_1^u, \ldots, I_t^u, \ldots, I_m^u]$ where m is predefined maximum capacity of interacted items

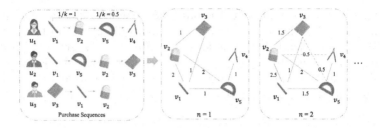

Fig. 1. An example shows the GIRG construction procedure from item sequences.

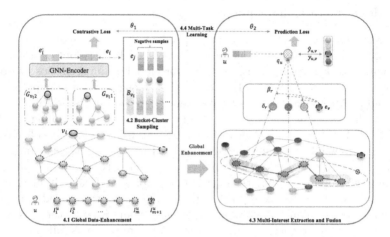

Fig. 2. The network architecture of our proposed GUESR.

and I_t^u is the item interacted at step t. Given an observed sequence I_u, the typical task of sequential recommendation is to predict the next item I_{m+1}^u that the user u is most likely to be interacted with.

Unlike existing methods [7,8,18,22] that individually exploit the local context of items in each sequence, we first generate a Global Item Relationship Graph (GIRG) G from all interaction sequences, where each item is connected by an undirected edge. It is worth mentioning that we do not condense repeated items in a sequence. The weight of the edge depends on the adjacency of each item in sequences. Specifically, we define a n-GIRG where the weight of two edges is the sum of the number of k_{th} ($k = [1, 2, \cdots, n]$) direct connections in all sequences and k represents the adjacent interval of items. This empirical setting is inspired by the success of previous work [6]. For (v_i, v_j), we calculate w_{ij} and then normalize the edge weights to obtain w'_{ij}. The above process is as follows:

$$w_{ij} = \sum_{u \in U} \sum_{k=1}^{n} \frac{\delta(|loc(v_i) - loc(v_j)| = k \mid v_i, v_j \in I_u)}{k}, w'_{ij} = Norm(w_{ij}). \quad (1)$$

Among them, $loc(\cdot)$ represents the position of the item $v_i \in S_u$, δ represents the number of times the positions of the two items differ by k in the sequence, and $deg(\cdot)$ denotes the degree of nodes in G. After that, we set a hyperparameter threshold ε to delete some edges [2] for reducing noise and get final edge weight $\hat{w}_{ij}$. That is, it mitigates the items' popularity bias since items with high popularity will have more edges removed.

4   Methodology

4.1   Global Data-Enhancement

As we know that GCL method is highly dependent on the choice of augmentation scheme [9]. Based on the exploration of previous work, we adopt a probability-based data augmentation method suitable for global graph scenarios. Based on the previous steps, we get the GIRG G for all sequences with the hyperparameters ε and n. After that, for a particular item v_i, we sample its neighbor v_j in G by probability p_{ij}, which is the ratio of edge weight $\hat{w}_{ij}$ and sampling depth D. Please note that the sampling process is iterative, when a node is collected, its adjacent nodes will be put into the sampling pool. In this way, the semantic information of the target node can be preserved. Through two sampling processes, we obtain two augmented views $G_{v_i 1}$ and $G_{v_i 2}$ for a particular item v_i. Then, the LightGCN [6] method is applied as the encoder with the shared parameters in different views. Taking the obtained view $G_{v_i 1}$ as an example, the information propagation and aggregation at the l-th layer of item v_i are as follows:

$$e_i^{(l+1)} = \sum_{v_j \in N_i} \frac{e_j^{(l)}}{\sqrt{deg(v_i)}\sqrt{deg(v_j)}}, \qquad e_i = \sum_{l=0}^{L} \alpha_l e_i^{(l)}, \qquad (2)$$

where $e_i^{(l)}$ denotes the representation of v_i in the l-th layer. N_i denotes the set of v_i's neighbors in $G_{v_i 1}$ and α_l represents the weight in l-th layer for the final embedding, which is trained as a model parameter. After L layers of information propagation on $G_{v_i 1}$, we denote the embedding of v_i as e_i. Similarly, the embedding of v_i in another view $G_{v_i 2}$ is denoted as e_i'.

4.2   Bucket-Cluster Sampling

Previous works [18,22] mostly adopt in-batch negatives or samples from training data at random. Such a way may cause a sampling bias, which will hurt the uniformity of the representation space. To address it, we present a Bucket-Cluster Sampling (BCS) method to alleviate the influence of these improper negatives.

In general, we divide buckets according to the initial attributes and current embeddings iteratively. Firstly, we use coarse-grained attribute information provided in the datasets as the basis for bucket splitting. It decides which bucket the item v_i was originally in. According to the K-means algorithm, the centers of K buckets $(\mu_1, \mu_2, \ldots, \mu_K)$ are calculated as the mean value of their item embeddings. Then we calculate the distance between each item embedding e_i and the cluster center μ_j. Finally, we assign item v_i to the bucket B_{v_i} and uniformly select N_{neg} negative samples in other buckets according to the bucket size. Even though the algorithm is relatively simple, the experiment proves that the sampling method is effective because of the introduction of prior knowledge. The whole process can be formulated as:

$$B_{v_i} = \arg\max_b \left[(1 - \lambda) * \mathbf{I}(v_i \in B_{v_i}^{orig}) + \lambda * \frac{\|e_i - \mu_b\|^2}{\sum_{k=1}^{K} \|e_i - \mu_k\|^2} \right], \qquad (3)$$

here $\mathbf{I}(\cdot)$ indicate whether v_i was originally in bucket $B_{v_i}^{orig}$ or not, the hyperparameter λ is designed to control the weight of the prior knowledge.

4.3 Multi-interest Extraction and Fusion

In this part, we utilize CapsNet [10] to generate the users multiple interests and then conduct the target-attention mechanism to derive the users preferences. Specifically, In order to utilize critical temporal information for the sequential recommendation, we utilize a Transformer to encode the interaction sequence and obtain the sequential patterns $Z \in \mathbb{R}^{m \times d}$ by additionally introducing a residual operation over S after linear projection with parameter $W^z \in \mathbb{R}^{d \times d}$.

Then we define $\mathbf{g} = [g_1, g_2, \cdots, g_m]$ as the agreement score, which indicates the relevance of each item towards capsules. Assume that we have R interest capsules, for the r-th capsule, its hidden representation $h_r \in \mathbb{R}^d$ is summarized over each sequential pattern $z_i \in Z$ by the agreement score with a softmax function. Then, the output of r-th capsule denoted as $o_r \in \mathbb{R}^d$ is derived from a nonlinear squashing function. Immediately after, the agreement score g_i is updated based on the output and the sequential pattern embedding. The above process could be formulated as follows:

$$h_r = \sum_{i=1}^{m} softmax(g_i) z_i, \qquad o_r = \frac{\|h_r\|^2}{1 + \|h_r\|^2} \frac{h_r}{\|h_r\|}, \qquad g_i = g_i + o_r^\top z_i. \quad (4)$$

For each interest capsule, we execute the above process for T iterations. The output in the final iteration is fed into a fully-connected layer and a ReLU activation function to derive the interest representation:

$$\tilde{o}_r = ReLU \left(o_r W_r^o \right). \qquad (5)$$

The weight of interests is affected by the target item [21]. For example, a sports enthusiast may click on a recommended bike, even after he has clicked several books. Through the above process, we get the interest representations $[\tilde{o}_1, \tilde{o}_2 \cdots \tilde{o}_R]$, R is the number of interest capsules. Given a target item with embedding e_v, we utilize the target-attention mechanism to derive the user preference, the process is as follows:

$$\beta_r = \frac{\exp\left(\tilde{o}_r^\top e_v\right)}{\sum_{j=1}^{R} \exp\left(\tilde{o}_j^\top e_v\right)}, \qquad q_u = \sum_{r=1}^{R} \beta_r \tilde{o}_r + u. \qquad (6)$$

Among them, β_j is the attention weight and q_u is the representation of the integrated interest. The user vector u is added to maintain the uniqueness of the users and the recommendation score is calculated by inner product $\hat{y}_{u,v} = q_u e_v$.

4.4   Multi-task Learning

With the main prediction task and contrastive learning task, we jointly optimize them in this section. Concretely, we use the InfoNCE Loss [3] to distinguish the augmented representations of the same item from others. In addition, the Binary Cross Entropy (BCE) loss is implemented for the prediction task.

$$\mathcal{L}_{CL} = -\sum_{v_i \in V} \left[\log \frac{\exp(\mathbf{sim}(e_i, e'_i)/\tau)}{\sum_{j=1}^{N_{neg}} \exp(\mathbf{sim}(e_i, e_j)/\tau)} \right], \tag{7}$$

$$\mathcal{L}_{Pred} = -\sum_{u,v} \left[y_{u,v} \ln(\hat{y}_{u,v}) + (1 - y_{u,v}) \ln(1 - \hat{y}_{u,v}) \right]. \tag{8}$$

Therefore, the final objective function of GUESR is:

$$\mathcal{L}_{Total} = \theta_1 \mathcal{L}_{Pred} + \theta_2 \mathcal{L}_{CL} + \theta_3 \|\Theta\|_2, \tag{9}$$

where θ_3 is the $L2$ regularization parameter to prevent over-fitting.

5   Experiment

5.1   Experimental Settings

Datasets. We conduct experiments on four publicly available datasets in different domains [5]. The detailed statistic of datasets is illustrated in Table 1.

Table 1. The statistics of datasets.

Dataset	# Users	# Items	# interactions	Sparsity
ML-1M	6,040	3,618	836,434	96.18%
Sports	35,598	18,357	296,337	99.95%
Yelp	45,478	30,709	1,777,765	99.87%
Books	58,145	58,052	2,517,437	99.93%

Implementation Details. To start up the experiments, for each user, we select the first 80% of the interaction sequence as training data, the next 10% as validation data, and the remained 10% as the testing data. All Baselines are described in the related work. For a fair comparison, we keep the same experimental environment and search the hyper-parameters carefully. To avoid the sampling bias issues of the candidate selection [20], we adopt the full-ranking strategy. To compare performance with state-of-the-art comparison baselines, we adopt two widely used evaluation metrics Recall@K and NDCG@K, and set the top K to 10 and 20.

5.2   Performance Comparison

In this section, we report the overall recommendation performance by ranking both Recall and NDCG metrics on four public datasets, as shown in Table 2, and conclude the following observations.

Table 2. The performance of different models. The best results are in boldface, and the second best results are tagged with the symbol '*' in this paper.

Dataset	Metric	BPRMF	GRU4Rec	Caser	SASRec	LightGCN	SR-GNN	GC-SAN	S^3Rec	CL4Rec	GUESR
ML-1M	R@10	0.1812	0.1843	0.1756	0.1932	0.1836	0.1885	0.1940	0.1995	0.2034*	**0.2157**
	N@10	0.2455	0.2465	0.2387	0.2605	0.2463	0.2507	0.2572	0.2638	0.2701*	**0.2879**
	R@20	0.2712	0.2734	0.2643	0.2895	0.2729	0.2762	0.2850	0.2940	0.3001*	**0.3147**
	N@20	0.2558	0.2567	0.2434	0.2730	0.2565	0.2625	0.2707	0.2790	0.2823*	**0.2951**
Sports	R@10	0.1032	0.1043	0.1021	0.1145	0.1040	0.1132	0.1176	0.1221	0.1231*	**0.1342**
	N@10	0.0854	0.0893	0.0833	0.0993	0.0885	0.0935	0.0988	0.1041	0.1054*	**0.1118**
	R@20	0.1284	0.1302	0.1139	0.1402	0.1298	0.1335	0.1385	0.1438	0.1492*	**0.1579**
	N@20	0.0987	0.0992	0.0921	0.1093	0.0991	0.1015	0.1069	0.1123	0.1143*	**0.1239**
Yelp	R@10	0.0635	0.0643	0.0634	0.0798	0.0647	0.0732	0.0808	0.0884	0.0892*	**0.0981**
	N@10	0.0452	0.0487	0.0435	0.0601	0.0480	0.0539	0.0610	0.0682*	0.0678	**0.0753**
	R@20	0.1038	0.1046	0.1042	0.1203	0.1044	0.1236	0.1291	0.1346	0.1362*	**0.1457**
	N@20	0.0572	0.0598	0.0578	0.0710	0.0602	0.0643	0.0728	0.0798	0.0812*	**0.0862**
Books	R@10	0.0621	0.0683	0.0624	0.0829	0.0725	0.0763	0.0847	0.0931*	0.0921	**0.1032**
	N@10	0.0431	0.0462	0.0445	0.0583	0.0501	0.0506	0.0587	0.0674*	0.0672	**0.0735**
	R@20	0.0972	0.1032	0.1001	0.1249	0.1056	0.1116	0.1205	0.1295	0.1307*	**0.1401**
	N@20	0.0529	0.0542	0.0546	0.0700	0.0569	0.0609	0.0694	0.0780	0.0795*	**0.0849**

Table 3. The performance achieved by the different modules of GUESR.

Dataset	Metric	GUESR	GUESR-GCL	GUESR-W	GUESR-BCS	CL4Rec
ML-1M	R@20	**0.3147**	0.2802 (−11.0%)	0.3087 (−1.9%)	0.2954 (−6.1%)	0.3001*
	N@20	**0.2951**	0.2625 (−11.0%)	0.2886 (−2.2%)	0.2799 (−5.1%)	0.2823*
Books	R@20	**0.1401**	0.1149 (−18.0%)	0.1332 (−5.0%)	0.1282 (−8.5%)	0.1307*
	N@20	**0.0849**	0.0602 (-29.1%)	0.0821 (−3.3%)	0.0723(−14.8%)	0.0795*

First, we can conclude that SR-GNN and GC-SAN commonly perform better than BPRMF, GRU4Rec, and Caser, which further demonstrates the conclusion of previous work that they can represent more high-order information of users and items. Second, we notice that SASRec, which introduced a self-attention mechanism, has achieved better performance than GRU4Rec and Caser, which indicates that self-attention architecture can be suitable for sequence modeling and better capture the long-term dependencies of items in the sequence. Third, for the self-supervised learning methods, we find that CL4Rec and S^3Rec consistently perform better than other baselines with single paradigm loss, which demonstrates the effectiveness of introducing self-supervised tasks into sequential recommendation problems. Last but not least, we observe that our proposed GUESR consistently performs better than all baselines on both evaluation metrics among all datasets, which indicated the advantage of our proposed global-enriched graph contrastive learning.

5.3 Ablation Study

In this section, we will perform the ablation study to validate and quantify the effectiveness of each component in our proposed GUESR. To be specific, we formulate the following corresponding comparison setting: 1). GUESR-GCL: indicates it removes the graph contrastive learning loss; 2). GUESR-W: represents we create the global item graph without consideration of edge weights and set $n=1$ (n is the max adjacent interval of items.) ; 3). GUESR-BCS: illustrates we use random negative sampling instead of Bucket-Cluster Sampling. The comparison results are presented in Table 3.

From the results shown in Table 3, we can observe the following conclusions: Firstly, the performance of GUESR-GCL decreases dramatically on both evaluation metrics in comparison with the original, which proves the effectiveness of employing contrastive learning to mitigate the problems of data sparsity and noise in the sequential recommendation. Secondly, through comparison with GUESR-W, we can prove that our constructed Global Item Relationship Graph (GIRG) plays an important role in capturing the complex associations of items. Thirdly, by comparing with GUESR-BCS, we can conclude that Bucket Cluster Sampling (BCS) method by introducing attribute knowledge could not only alleviate the influence of improper negatives but also can promote efficiency and further improve the effectiveness of our proposed GUESR.

5.4 Impacts of Enhancement Module

In this part, we will investigate whether our proposed enhancement module in GUESR could be a general framework and flexibly integrate with other sequential recommendation paradigms. Here, we consider GRU4Rec, Caser, and SASRec, and supplement the enhancement module to themselves, which are jointly optimized by both contrastive and prediction loss. From the table, we can observe that all of the enhanced models consistently perform better than the corresponding backbones, which demonstrates the global graph contrastive learning enhancement strategy proposed in GUESR can be a general module directly applied to lots of existing sequential recommendation paradigms.

Table 4. The performance comparison with the proposed enhancement module.

Dataset	Metric	GRU4Rec	EGRU4Rec	Caser	ECaser	SASRec	ESASRec
ML-1M	R@20	0.2734	0.2809 (+2.7%)	0.2643	0.2760 (+4.4%)	0.2895	0.2941 (+1.6%)
	N@20	0.2567	0.2635 (+2.7%)	0.2434	0.2602 (+7.0%)	0.2730	0.2798 (+2.5%)
Sports	R@20	0.1302	0.1354 (+4.0%)	0.1139	0.1332 (+17.0%)	0.1402	0.1435 (+2.4%)
	N@20	0.0992	0.1055 (+6.4%)	0.0921	0.1021 (+10.9%)	0.1093	0.1117 (+2.2%)

6 Conclusion

In this work, we proposed a novel graph contrastive learning paradigm for the sequential recommendation problem, termed GUESR, to explicitly capture potential relevance within both local and global contexts of items. Extensive experiments on four public datasets demonstrate have demonstrated our proposed GUESR could not only achieve significant improvements but also could be regarded as a general enhancement strategy to improve the performance gains in combination with other sequential recommendation methods.

Acknowledgement. This research was supported by grants from the National Natural Science Foundation of China (No. 62202443). This research was also supported by Meituan Group.

References

1. Bera, A., Wharton, Z., Liu, Y., Bessis, N., Behera, A.: Sr-gnn: Spatial relation-aware graph neural network for fine-grained image categorization. IEEE Trans. Image Process. **31**, 6017–6031 (2022)
2. Chang, J., et al.: Sequential recommendation with graph neural networks. In: Proceedings of the 44th International ACM SIGIR Conference on Research and Development in Information Retrieval, pp. 378–387 (2021)
3. Chen, T., Kornblith, S., Norouzi, M., Hinton, G.: A simple framework for contrastive learning of visual representations. In: International Conference on Machine Learning, pp. 1597–1607. PMLR (2020)
4. Chen, Y., Liu, Z., Li, J., McAuley, J., Xiong, C.: Intent contrastive learning for sequential recommendation. In: The ACM Web Conference, pp. 2172–2182 (2022)
5. Haque, T.U., Saber, N.N., Shah, F.M.: Sentiment analysis on large scale amazon product reviews. In: 2018 IEEE International Conference on Innovative Research and Development (ICIRD), pp. 1–6. IEEE (2018)
6. He, X., Deng, K., Wang, X., Li, Y., Zhang, Y., Wang, M.: Lightgcn: Simplifying and powering graph convolution network for recommendation. In: Proceedings of the 43rd International ACM SIGIR Conference on Research and Development in Information Retrieval, pp. 639–648 (2020)
7. Hidasi, B., Karatzoglou, A., Baltrunas, L., Tikk, D.: Session-based recommendations with recurrent neural networks. arXiv preprint arXiv:1511.06939 (2015)
8. Kang, W.C., McAuley, J.: Self-attentive sequential recommendation. In: 2018 IEEE International Conference on Data Mining (ICDM), pp. 197–206. IEEE (2018)
9. Lee, N., Lee, J., Park, C.: Augmentation-free self-supervised learning on graphs. In: AAAI Conference on Artificial Intelligence, vol. 7, pp. 7372–7380 (2022)
10. Sabour, S., Frosst, N., Hinton, G.E.: Dynamic routing between capsules. In: Advances in Neural Information Processing Systems 30 (2017)
11. Sun, F., et al.: Bert4rec: Sequential recommendation with bidirectional encoder representations from transformer. In: Proceedings of the 28th ACM International Conference on Information and Knowledge Management, pp. 1441–1450 (2019)
12. Tang, J., Wang, K.: Personalized top-n sequential recommendation via convolutional sequence embedding. In: Proceedings of the Eleventh ACM International Conference on Web Search and Data Mining, pp. 565–573 (2018)

13. Wang, H., Lian, D., Tong, H., Liu, Q., Huang, Z., Chen, E.: Decoupled representation learning for attributed networks. IEEE Trans. Knowl. Data Eng. (2021)
14. Wang, H., et al.: Mcne: An end-to-end framework for learning multiple conditional network representations of social network. In: Proceedings of the 25th ACM SIGKDD International Conference on Knowledge Discovery & Data Mining, pp. 1064–1072 (2019)
15. Wei, Y., et al.: Contrastive learning for cold-start recommendation. In: Proceedings of the 29th ACM International Conference on Multimedia, pp. 5382–5390 (2021)
16. Wu, L., Li, Z., Zhao, H., Pan, Z., Liu, Q., Chen, E.: Estimating early fundraising performance of innovations via graph-based market environment model. In: Proceedings of the AAAI Conference on Artificial Intelligence, vol. 34, pp. 6396–6403 (2020)
17. Wu, L., Wang, H., Chen, E., Li, Z., Zhao, H., Ma, J.: Preference enhanced social influence modeling for network-aware cascade prediction. In: Proceedings of the 45th International ACM SIGIR Conference on Research and Development in Information Retrieval, pp. 2704–2708 (2022)
18. Xie, X., et al.: Contrastive learning for sequential recommendation. In: 2022 IEEE 38th International Conference on Data Engineering (ICDE), pp. 1259–1273. IEEE (2022)
19. Xu, C., et al.: Graph contextualized self-attention network for session-based recommendation. In: IJCAI, vol. 19, pp. 3940–3946 (2019)
20. Zhao, W.X., Chen, J., Wang, P., Gu, Q., Wen, J.R.: Revisiting alternative experimental settings for evaluating top-n item recommendation algorithms. In: Proceedings of the 29th ACM International Conference on Information & Knowledge Management, pp. 2329–2332 (2020)
21. Zheng, Y., et al.: Disentangling long and short-term interests for recommendation. In: Proceedings of the ACM Web Conference 2022, pp. 2256–2267 (2022)
22. Zhou, K., et al.: S3-rec: Self-supervised learning for sequential recommendation with mutual information maximization. In: Proceedings of the 29th ACM International Conference on Information & Knowledge Management, pp. 1893–1902 (2020)

A Three-Layer Attentional Framework Based on Similar Users for Dual-Target Cross-Domain Recommendation

Jinhu Lu[1], Guohao Sun[1(✉)], Xiu Fang[1], Jian Yang[2], and Wei He[1]

[1] Donghua University, Shanghai, China
{1209126,2212483}@mail.dhu.edu.cn, {ghsun,xiu.fang}@dhu.edu.cn
[2] School of Computing, Macquarie University, Macquarie Park, Australia
jian.yang@mq.edu.au

Abstract. To address the long-standing data sparsity problem in recommender systems (RSs), cross-domain recommendation (CDR) has been proposed. Traditional CDR leverages the relatively richer information from a richer domain to improve recommendation performance in a sparser domain, which is also called single-target CDR. In recent years, dual-target CDR has been proposed to improve recommendation performance in both domains simultaneously. The existing dual-target CDR methods are based on common users between domains, where they extract the embeddings of common users and then transfer the embeddings to the two target domains to improve recommendation performance. However, in real life, the proportion of common users between domains is usually very small, which makes it hard to generate representative and high-quality user embeddings, and thus, limits the performance of the existing methods in real applications. To address this problem, in this paper, we propose a **T**hree-Layer **A**ttentional Framework based on **S**imilar **U**sers, called TASU. In addition to common users, TASU leverages information from similar users to improve the quality of user embeddings. By a three-layer attentional framework, TASU can generate more representative and high-quality user embeddings to improve recommendation performance in both domains. Extensive experiments conducted on three real-world datasets demonstrate that TASU significantly outperforms the state-of-the-art approaches.

Keywords: Cross-domain recommendation · Dual-target recommendation · Attention mechanism · Knowledge transfer

1 Introduction

Recommender systems (RSs) have been proposed for many years and have a wide range of applications [7,16,17,31], while data sparsity is a long-standing problem in RSs. To deal with this problem, Cross-Domain Recommendation (CDR) [2] is proposed to improve recommendation performance when facing sparse data. Traditional CDR is also known as single-target CDR [33], because

X. Wang et al. (Eds.): DASFAA 2023, LNCS 13944, pp. 297–313, 2023.
https://doi.org/10.1007/978-3-031-30672-3_20

it only improves the performance in the sparse domain by using the knowledge in the richer domain with relatively rich information. For example, in Fig. 1, there are two domains, i.e., a movie domain with relatively richer information (more ratings, reviews and tags information) and a sparser music domain. Supposed that Alice is an active common user in these two domains, the traditional CDR method is to transfer Alice's user embeddings learned from the movie domain to the music domain to generate personalized music recommendations. However, in the richer movie domain, there is no improvement in recommendation performance.

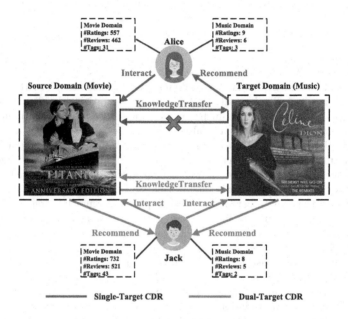

Fig. 1. An example of single-target CDR and dual-target CDR

In the past two years, dual-target CDR [30] has been proposed and become a research hotspot. Its main purpose is to combine the embeddings of common users learned from two domains and transfer them to two domains respectively to improve recommendation performance in each domain.

As shown in Fig. 1, Jack is observed as a common user in both the movie domain and the music domain. The dual-target CDR can transfer Jack's user embeddings from the movie domain to the music domain, and can also transfer his user embeddings from the music domain to the movie domain. Such that, the recommendation performance in each domain can be improved.

The existing single-target CDR and dual-target CDR methods leverage common users as a bridge to achieve knowledge transfer between two domains. Therefore, the proportion of common users between domains will greatly affect the effectiveness of their methods. However, in real life, by the analysis of some scholars [12], the average proportion of the common users to total users is usually small. For example, in the Amazon dataset[1], the proportion of common users is less than 5%. Thus, the performance of the existing methods is limited in real applications.

According to the idea of collaborative filtering [9,23,27] that similar users usually have the same preferences, we realize that to achieve better knowledge transfer, in addition to common users, similar users can also play an important role in the process of knowledge transfer. However, in the process of knowledge transfer of similar users, we need to transfer much more information compared with common users because the number of similar users of one user is usually large. What's more, the transfer of some useless information will even reduce the recommendation performance. Therefore, how to extract key information from similar users and generate representative and high-quality user embeddings is one challenging problem. Google DeepMind [24] proposed an attention mechanism based on human attention behavior patterns, which can capture relatively key elements of user and item features, and extract effective user preferences. In recent years, attention mechanisms-based neural network models have been widely used in computer vision, machine translation [1], and recommendation systems [4,25].

Inspired by the existing work, in this paper, we propose TASU, a three-layer attentional framework based on similar users for dual-target CDR, to effectively leverage common users and similar users between domains to share knowledge, and further improve recommendation performance. The characteristics and contributions of our work are summarized as follows:

- We propose a method of finding similar users between domains, such that the embeddings of both common users and similar users can be transferred between domains. By this method, we obtain transferable high-quality information from similar users.
- We propose a three-layer attentional framework to combine user embeddings at different attention layers. With our new framework, we can generate representative and high-quality user embeddings, such that improve recommendation performance.
- We conduct extensive experiments on three real-world datasets and the results demonstrate that our TASU significantly outperforms the best-performing baselines.

2   Related Work

In this section, We briefly review some representative methods of Single-target CDR and Dual-target CDR.

Single-Target CDR: Single-target CDR is the traditional cross-domain recommendation. Feature-based transfer improves the performance in the sparse domain by using the knowledge in the domain with relatively rich information.

---

[1] http://snap.stanford.edu/data/amazon/productGraph/categoryFiles.

Table 1. Important notations.

Symbol	Definition
$\mathcal{U} = \{u_1, \ldots, u_m\}$	the set of users
U	the embedding matrix of users
UC	the text embedding matrix of users
$\mathcal{V} = \{v_1, \ldots, v_n\}$	the set of items
V	the embedding matrix of items
VC	the text embedding matrix of itmes
$\mathcal{SU}$	the set of similar users across domains
m	the number of users
n	the number of items
$R \in \mathbb{R}^{m \times n}$	the rating matrix
$y_{ij} \in Y$	the interaction of user u_i on item v_j
$Y \in \mathbb{R}^{m \times n}$	the user-item interaction matrix
$*^a$ and $*^b$	the notations for domains a and b
$\hat{*}$	the predicted notations, e.g., $\hat{y}_{ij}$ represents the predicted interactions of u_i on item v_j

One of the classic Single-target CDR methods is EMCDR [18], an embedding and mapping method, which builds the mapping function via common users' representations. SSCDR [12] is a semi-supervised learning framework, which is designed to solve the cold-start problem. More specifically, SSCDR models the users and items in metric spaces, and the similarities among uses and items are represented as their distances. By a distance-based loss function, SSCDR trains a cross-domain mapping function to improve recommendation performance. A new aspect transfer network framework is proposed in CATN [29] for solving the cold-start problem as well. Compared with SSCDR, CATN models user preference transfer at aspect-level derived from reviews, and transfer user preference derived from the source domain to the target domain for the effective and explainable recommendation. TMCDR [34] leverages meta-learning with good generalization ability to novel tasks and learns a task-oriented meta-network to implicitly transform the user embedding in the source domain to the target feature space. PTUPCDR [35] learns a meta-network fed with users' characteristic embeddings to generate personalized bridge functions to achieve personalized transfer of preferences for each user. All the above-mentioned single-target CDR methods only can improve recommendation performance in the sparse domain.

Dual-Target CDR: To improve recommendation performance in both sparse and rich domains, Dual-target CDR has been proposed and DTCDR [30] is the first dual-target CDR method. DTCDR applies fixed combination strategies, e.g., average-pooling, max-pooling, and concatenation, to combine the embeddings of common users learned from both domains. Based on DTCDR, DDTCDR [15] has been proposed and it applies a latent orthogonal mapping to extract user preferences over multiple domains while preserving relations between users across different latent spaces. To optimize the embedding layer and the feature combination layer in DTCDR, GA-DTCDR [32] has been proposed, where

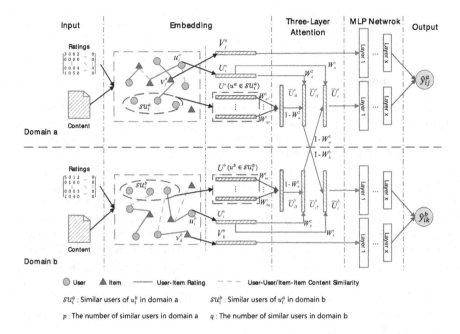

Fig. 2. The architecture of the proposed TASU framework

the embedding layer in GA-DTCDR applies the graph embedding method to generate the embeddings of users and items, and the feature combination layer leverages the element-level attention mechanism to improve the combination of the embeddings of common users. DA-CDR [28] is a deep dual adversarial network, which can maximally match the latent feature spaces of both users and items between the source and the target domains to enhance positive knowledge transfer. In addition, some dual-target works also consider shared-account cross-domain sequential recommendation [6,7], recommendations across different social networks [3]. Although these dual-target CDR methods can improve recommendation performance in both domains, none of these methods can be applied to explore the embeddings of similar users in a cross-domain process. In other words, they only leverage the information of common users that account for a small proportion of two domains in real applications.

3   Methodology

Figure 2 presents the architecture of the proposed TASU. TASU has five main layers: (1) Input Layer (2) Embedding Layer (3) Three-Layer Attention Layer (4) MLP Network Layer (5) Output Layer. The Embedding Layer encodes inputs into latent embeddings. The Three-Layer Attention Layer effectively combines user embeddings and transfers knowledge across domains. The MLP Network Layer learns the nonlinear relationship between user embeddings and item embeddings. We list the important notations used in this section in Table 1 and the detail of each layer is described below.

3.1 Input Layer

The inputs of TASU can be divided into two categories: Ratings and Content. The rating information includes the users' rating matrix for items. The content information includes users' profiles, users' reviews, titles of items, descriptions of items, and categories of items.

For content, we adopt the document embedding technique Doc2vec proposed in [14] to generate text vectors. The details are as follows: in the training set, there are m users and n items, (1) first, for each user u, we leverage user's profiles and the user's reviews to describe u, and collect these information into document d, which can generate $D = \{d_1, \cdots, d_m\}$ for m users; (2) then, for each item v, we leverage the title, description, and categories of the item to describe v, and collect these information into the same document d, which generate $D = \{d_1, \cdots, d_m, \cdots d_{m+n}\}$ for m users and n items; (3) finally, we adopt a widely-used natural language processing tool StandfordCoreNLP proposed in [19] to segment document D, and adopt Doc2vec to generate text embedding matrix UC for users and VC for items.

3.2 Embedding Layer

The work proposed in [32] achieves dual-target CDR based on common users and element-level attention mechanism. Inspired by [32] and collaborative filtering, we further consider that combing the available high-quality similar users in the framework can effectively improve recommendation performance. In particular, non-common users can also obtain transferable knowledge from another domain. The work of the Embedding Layer consists of encoding inputs into latent embeddings and finding similar users.

Encoding and Embedding. We adopt a method similar to the work [32] to construct a heterogeneous graph of user-item interactions for domains a and b, respectively. As shown in *Embedding Layer* in Fig. 2, there are two types of nodes, i.e., user nodes and item nodes. We link user nodes and item nodes via their interaction relationships. The weights of edges are the normalized ratings, i.e., $R/max(R)$. In addition to the interaction relationship between users and items, we also generate the edge between two user nodes or two item nodes based on text embeddings. The weight of the edge between two users or two items is the cosine similarity of text embeddings, and generation probability $P(i, r)$ of the edge between users u_i and u_r is as follows:

$$P(i, r) = \alpha \cdot sim(UC_i, UC_r), \tag{1}$$

where α is a hyper-parameter that controls the generation probability and $sim(UC_i, UC_r)$ is the cosine similarity. Similarly, the generation probability of the edge between two items can be obtained.

By this way, for domains a and b, we construct heterogeneous graphs that contain user-item interaction relationships, user-user similarity relationships, and item-item similarity relationships, respectively. Then, the graph embedding technique, i.e., Node2vec [5], is applied to generate user embedding matrix U and item embedding matrix V. However, for a non-common user u_i^a in domain a (u_i^a is in domain a but not in domain b), we still need a more useful method to obtain transferable knowledge from domain b for u_i^a.

Similar Users between Domains. As we mentioned in the Introduction section, in addition to common users, similar users between domains can also play an important role in the process of knowledge transfer. In this layer, after getting the user and item embeddings, we propose an effective method to find similar users between domains, such that to better transfer knowledge.

Inspired by some natural language processing technologies [20, 22], We leverage content, i.e., text information, to find similar users between domains. The details are as follows: (1) we first apply user's profiles and user's reviews on all items to semantically describe user's characteristics in one domain; (2) then, we apply Doc2vec to generate the user's text embedding in the text semantic space of this domain; (3) finally, we leverage the cosine similarity of users' text embeddings to represent the text semantic similarity of users in text semantic space. For two users in different domains, we put them into text semantic space to compute the similarity of users. For users u_i^a and u_g^b in domains a and b, the users' text embeddings are UC_i^a and UC_g^b respectively, and the similarity between users u_i^a and u_g^b can be represented as:

$$S_{cross}(u_i^a, u_g^b) = sim(UC_i^a, UC_g^b), \tag{2}$$

where $sim(UC_i^a, UC_g^b)$ is the cosine similarity of the users' text embeddings between domains. According to the similarity of users between domains, the transfer probability $P_{transfer}(u_i^a \leftarrow u_g^b)$ of the knowledge of u_g^b in domain b transferred to u_i^a in domain a as:

$$P_{transfer}(u_i^a \leftarrow u_g^b) = \beta \cdot S_{cross}(u_i^a, u_g^b), \tag{3}$$

where β is a hyper-parameter that controls the transfer probability. By this way, we obtain l transferable similar users $\mathcal{SU}_i^b = [u_1^b, u_2^b, \cdots, u_l^b]$ from domain b for u_i^a. Similarly, we can obtain transferable similar users $\mathcal{SU}_g^a$ from domain a for u_g^b.

3.3 Three-Layer Attention Layer

Based on transferable similar users between domains, we further propose a three-layer attention framework to ensure that the embeddings of similar users can be effectively utilized and more representative and high-quality user embeddings will be generated in our framework.

First Layer of Attention. We leverage the first layer of attention to combine embeddings between similar users and set weights to the embeddings of similar users according to the normalized S_{cross}. By this way, the users who are more similar will get higher weights, such that their embeddings will be applied more abundantly. For u_i^a, we combine the embeddings of transferable similar users $\mathcal{SU}_i^b$, which can be represented as:

$$[W_{b1}^1, \cdots, W_{bl}^1] = softmax([S_{cross}(UC_i^a, UC_1^b), \cdots, S_{cross}(UC_i^a, UC_l^b)]), \quad (4)$$

$$\tilde{U}_{iS}^b = W_{b1}^1 \odot U_1^b + W_{b2}^1 \odot U_2^b + \cdots + W_{bl}^1 \odot U_l^b, \quad (5)$$

where $\odot$ is the element-wise multiplication, $U_1^b, \cdots, U_l^b$ are the embeddings of transferable similar users of u_i for domain a. Similarly, we can obtain the combined similar user embedding $\tilde{U}_{gS}^a$ of u_g for domain b.

Second Layer of Attention. After getting the combined embedding $\tilde{U}_{iS}^b$ of similar users by the first layer of attention, the second layer of attention is applied in this stage to better combine it with the embeddings of u_i in domain b. By this way, we generate a combined user embedding $\tilde{U}_{iT}^b$ that can be transferred from domain b to domain a, which can be represented as:

$$\tilde{U}_{iT}^b = W_b^2 \odot U_i^b + (1 - W_b^2) \odot \tilde{U}_{iS}^b + b_S, \quad (6)$$

where $\odot$ is the element-wise multiplication, $W_b^2 \in \mathbb{R}^{m^b \times k}$ is the weight matrix for the attention network in domain b, and b_S is the bias. If u_i is a non-common user, then there is no feature in U_i^b, which means $\tilde{U}_{iT}^b$ only contains the embeddings of similar users. Similarly, we can obtain $\tilde{U}_{iT}^a$ for domain a.

Third Layer of Attention. After getting the combined user embedding $\tilde{U}_{iT}^b$ by the second layer of attention, the third layer of attention is applied in this stage to better combine it with the embeddings of u_i in domain a. By this way, we generate final user embedding $\tilde{U}_i^a$ for domain a that contains knowledge transferred from domain b, which can be represented as:

$$\tilde{U}_i^a = W_a^3 \odot U_i^a + (1 - W_a^3) \odot \tilde{U}_{iT}^b + b_T, \quad (7)$$

where $\odot$ is the element-wise multiplication, $W_a^3 \in \mathbb{R}^{m^a \times k}$ is the weight matrix for the attention network in domain a, and b_T is the bias. Similarly, we can obtain $\tilde{U}_i^b$ for domain b.

Our three-layer attention effectively leverages the embeddings of similar users in the framework. With our new framework, we can generate representative and high-quality user embeddings.

3.4   MLP Network Training

For training, we use normalized cross-entropy as the loss function for domains a and b:

$$\mathcal{L} = - \sum_{y \in Y + \cup Y -} (\frac{y}{max(R)} log\hat{y} + (1 - \frac{y}{max(R)}) log(1 - \hat{y})) + \lambda \|\Theta\|_F^2, \quad (8)$$

where y is the observed interaction and $\hat{y}$ is its corresponding predicted interaction, $max(R)$ is the maximum rating in a domain, $\Theta = \{\tilde{U}, V\}$ is the regularizer and λ is the regularization coefficient.

TASU employs a fully connected network, i.e., MLP, to learn a nonlinear relationship between users and items. The input of MLP Network is the combined user embedding $\tilde{U}$ by Three-Layer Attention, and the item embedding V, respectively. Therefore, the embedding of user u_i and the embedding of item v_j in the output layer of MLP Network can be represented as:

$$\hat{U}_i = \sigma(\ldots\sigma(\sigma(\tilde{U}_i \cdot W_{U_1}) \cdot W_{U_2})),$$
$$\hat{V}_j = \sigma(\ldots\sigma(\sigma(V_j \cdot W_{V_1}) \cdot W_{V_2})),$$

(9)

where σ is the *ReLU* activation function, $W_{U_1}, W_{U_2} \ldots$ and $W_{V_1}, W_{V_2} \ldots$ are the weights of multi-layer networks in different layers, respectively.

Finally, we use the cosine similarity between the embeddings of user u_i and item v_j as the normalized predicted interaction:

$$\hat{y}_{ij} = cosine(\hat{U}_i, \hat{V}_j) = \frac{\hat{U}_i \cdot \hat{V}_j}{\left\|\hat{U}_i\right\| \left\|\hat{V}_j\right\|},$$

(10)

4 Experiments

In this section, we conduct extensive experiments on three real-world datasets to evaluate our TASU framework. First, we give a brief introduction to the datasets, parameter setting, evaluation protocol, and baselines. Then, we design detailed ablation experiments to demonstrate the rationality and effectiveness of the design choices of TASU. Finally, we conduct multiple experiments to study the impact of different parameter settings on the performance of TASU.

Table 2. Details of Datasets and Tasks

	AmazonBook	AmazonCD	AmazonMovie
#Users	2,357	1,918	2,816
#items	9,035	7,179	33,187
#Interactions	22,331	14,631	253,450
Density	0.105%	0.106%	0.271%
	Sparser	Richer	#Overlap
Task 1	AmazonBook	AmazonMovie	117
Task 2	AmazonCD	AmazonMovie	95

4.1 Experimental Setup

Datasets and Tasks. We use three real-world datasets, including Amazon-Movie, AmazonBook, and AmazonCD, to conduct our experiments. Since the complete Amazon datasets are too large, to save experimental time, these three datasets are a portion of interactions randomly sliced from complete Amazon datasets. Table 2 shows the statistics of the used datasets. To make the distribution of the experimental datasets more in line with the realistic application scenarios, compared with the rich AmazonMovie dataset, the proportion of common users in the sparser two datasets (AmazonBook and AmazonCD) accounts for about 5%. Based on these datasets, we set two recommendation tasks (see Table 2) to validate recommendation performance in the CDR scenario.

Parameter Setting. Generally, we optimize the parameters of our TASU and those of the baselines. For *Embedding Layer* in Fig. 2, the parameter α of generation probability of edges and the parameter β of transfer probability are set to 0.05. In *MLP Network Layer*, the structure of the network layers is "$4k \rightarrow 8k \rightarrow k$", and we adopt Gaussian distribution with 0 mean and $1e^{-4}$ standard deviation to initialize the parameters of the MLP network. For training our TASU, we randomly select 7 negative instances into Y^- for each positive instance, which is the same as other works, e.g., DeepMF, DTCDR, and GA-DTCDR. The Adam [13] is applied to train the MLP network, and the maximum number of epochs is set to 50. The learning rate is $1e^{-4}$, the regularization coefficient λ is $1e^{-3}$, and the batch size is 8,192. The dimension k of the embedding varies in $\{8, 16, 32, 64, 128\}$.

Table 3. The comparison of the baselines and our method

Model		Training Data	Embedding	Transfer Strategy
Baselines	NCF	Rating	MLP	–
	EMCDR	Rating	MF	MLP (Mapping)
	PTUPCDR	Rating	DNN	MLP (Meta-Network)
	DDTCDR	Rating	MLP	Dual Transfer Learning
	GA-DTCDR	Rating & Content	Graph Embedding	Combination (Element-wise Attention)
Our Method	TASU	Rating & Content	Graph Embedding	Combination (**Three-Layer Attention**)

Evaluation Protocol. We adopt a ranking-based evaluation strategy, i.e., *Leave-one-out evaluation* [21], as the evaluation strategy for our TASU and all baselines. For each test user, we first select the latest interaction with a test item as the test interaction, and randomly sample 99 unobserved interactions for this test user, constituting a total of 100 interactions. Then, we rank these 100 items and finally perform top-10 recommendations on these 100 test items. HR@10 (*Hit Ratio*) and NDCG@10 (*Normalized Discounted Cumulative Gain*) are adopted as the final evaluation metrics, which are widely used in the literature [8,10,11,26].

Baselines. To evaluate more comprehensively, we compare TASU with various types of competing models, including one Single-Domain Recommendation (SDR) method (NCF), two single-target CDR methods (EMCDR and

PTUPCDR), and two dual-target CDR methods (DDTCDR and GA-DTCDR), all of which are representative or state-of-the-art methods:

- **NCF** [9]: A traditional SDR method for modeling latent features of users and items using collaborative filtering based on MLP neural networks.
- **EMCDR** [18]: An embedding and mapping method that applies matrix factorization to learn the embeddings of users and items, and applies a linear or nonlinear(MLP network) function based on common users to bridge the user embeddings from the auxiliary domain to the target domain.
- **PTUPCDR** [35]: A state-of-the-art single-target CDR method. Differing from EMCDR, PTUPCDR learns a meta-network fed with users' characteristic embeddings to generate personalized bridge functions to achieve personalized transfer of preferences for each user.
- **DDTCDR** [15]: A deep dual transfer CDR method. DDTCDR applies an autoencoder approach to extract the latent essence of feature information and proposes a latent orthogonal mapping to extract users' preferences over different domains while preserving the relations between users across different latent spaces.
- **GA-DTCDR** [32]: A graphical and attentional dual-target CDR method. GA-DTCDR applies graph embedding method to generate latent embeddings of users and items, and applies an element-level attention mechanism to combine the embeddings of common users learned from both domains.

In addition, for ablation studies, we perform two simplified versions on our TASU to combine similar users, i.e., TASU_Average (replacing similarity-based normalization weights with average pooling) and TASU_Max (replacing similarity-based normalization weights with max pooling). For all baselines and TASU, we list their training data types, embedding strategies, and transfer strategies in Table 3.

Table 4. The experimental results (HR@10 & NDCG@10) for Task 1. The best performance among all is in bold while the second best one is underlined.

Model	Domain	$k = 8$		$k = 16$		$k = 32$		$k = 64$		$k = 128$	
		HR	NDCG	HR	NDCG	HR	NDCG	HR	NDCG	HR	NDCG
NCF	Book	.1120	.0496	.1319	.0627	.1913	.0990	.2287	.1292	.2677	.1604
	Movie	.2248	.1133	.2837	.1518	.3455	.1811	.3775	.1972	.3988	.2126
EMCDR	Book	.1027	.0472	.1031	.0442	.1018	.0452	.1146	.0527	.1001	.0442
	Movie	–	–	–	–	–	–	–	–	–	–
PTUPCDR	Book	.1150	.0518	.1112	.0513	.1154	.0537	.1141	.0521	.1154	.0546
	Movie	–	–	–	–	–	–	–	–	–	–
DDTCDR	Book	.1549	.0811	.2028	.1069	.2635	.1545	.3114	.1828	.3250	.1982
	Movie	.3097	.1672	.3192	.1675	.3303	.1737	.3612	.1911	.4183	.2370
GA-DTCDR	Book	.1090	.0524	.1977	.1011	.3360	.1962	.3284	.1942	.3458	.1976
	Movie	.3590	.1910	.3505	.1835	.3594	.1929	.3949	.2064	.4386	.2440
TASU	Book	**.1969**	**.1097**	**.2461**	**.1365**	**.3466**	**.2071**	**.3377**	**.1962**	**.3466**	**.2102**
	Movie	**.3803**	**.2041**	**.3668**	**.1985**	**.4382**	**.2476**	**.4489**	**.2544**	**.4627**	**.2705**

4.2   Performance Comparison and Analysis

Tables 4–5 reports the experimental results of HR@10 and NDCG@10 with different embedding dimensions k for Tasks 1 and 2, respectively. Based on the results, we have the following observations:

Overall, as shown in Tables 4–5, our TASU consistently yields the best performance across all three datasets for Tasks 1 and 2, outperforming all baselines by an average improvement of 18.96%. In particular, our TASU hugely improves the best-performing baseline results by an average of 13.02% for Task 1, and an average of 24.89% for Task 2, respectively. The results of significance testing indicate that our improvements over the current strongest CDR baselines are statistically significant. We attribute the good performance of TASU to the following reasons: (1) Compared with other methods, our TASU not only considers common users in the process of cross-domain knowledge transfer, but also adds the embeddings of transferable similar users to generate more representative embeddings; (2) By the three-layer attention, the user embeddings are combined at different attention layers, and the knowledge transfer between domains is effectively achieved. These advantages together contribute to the superiority of TASU.

Compared with the single-target CDR methods, we observe that the performance of single-target CDR methods EMCDR and PTUPCDR are the worst, even compared with the SDR method NCF. That's because EMCDR and PTUPCDR are completely based on common users to bridge information from the auxiliary domain to the target domain. In the dataset with less common users, their performance will be greatly affected. This observation proves that using the embeddings of similar users to improve recommendation performance has great significance.

Compared with the dual-target CDR methods, our TASU also significantly outperforms two competing dual-target CDR methods, DDTCDR and GA-DTCDR. DDTCDR simply changes the existing single-target transfer learning

Table 5. The experimental results (HR@10 & NDCG@10) for Task 2

Model	Domain	$k = 8$		$k = 16$		$k = 32$		$k = 64$		$k = 128$	
		HR	NDCG	HR	NDCG	HR	NDCG	HR	NDCG	HR	NDCG
NCF	CD	.1064	.0479	.1137	.0497	.1241	.0565	.1230	.0586	.1533	.0764
	Movie	.2248	.1133	.2837	.1518	.3455	.1811	.3775	.1972	.3988	.2126
EMCDR	CD	.0938	.0423	.0912	.0404	.0923	.0444	.0959	.0434	.0886	.0408
	Movie	–	–	–	–	–	–	–	–	–	–
PTUPCDR	CD	.1199	.0561	.1194	.0537	.1189	.0546	.1152	.0531	.1131	.0525
	Movie	–	–	–	–	–	–	–	–	–	–
DDTCDR	CD	.1116	.0508	.1210	.0612	.1048	.0491	.1403	.0728	.2112	.1213
	Movie	.2926	.1597	.3111	.1642	.3420	.1778	.3629	.1948	.4038	.2226
GA-DTCDR	CD	.1246	.0541	.1053	.0479	.1590	.0698	.1627	.0762	.2445	.1356
	Movie	.3636	.1920	.3438	.1824	.3750	.1990	.4105	.2234	.4474	.2518
TASU	CD	.1533	.0767	.1924	.0954	.2195	.1168	.2523	.1391	.2482	.1386
	Movie	.3707	.1956	.3679	.2018	.4158	.2304	.4403	.2479	.4599	.2701

to dual transfer learning, and GA-DTCDR combines the embeddings of common users learned from both domains and transfers them into each domain. In addition to common users, GA-DTCDR only applies non-common users on their respective domains to train the upper-layer neural network. In general, the performance of GA-DTCDR is better than that of DDTCDR. Our proposed TASU takes into account the fact that there are less common users in real-world scenarios and adds the embeddings of similar users between domains on the basis of the common users, thus further improving recommendation performance.

4.3 Ablation Study

Table 6 reports the experimental results of two variants of our TASU (i.e., TASU_Average and TASU_Max). Due to space limitations, we fixed the embedding dimension $k = 32$ (the experimental results with different k are similar). TASU_Average replaces similarity-based normalization weights with average pooling to combine the embeddings of similar users, that is, giving the weight equally, i.e., $\frac{1}{l}$ (l is the number of similar users), to the embedding of each similar user. TASU_Max replaces similarity-based normalization weights with max pooling to generate the combined embeddings of similar users, that is, the largest feature among all similar users is the final output. From Table 6, we have the following two observations: (1) For Tasks 1 and 2, TASU outperforms TASU_Average by an average improvement of 5.10% and 14.82%, respectively; and TASU outperforms TASU_Max by an average of 5.71% and 10.28%, respectively. This demonstrates that similarity-based normalization weights given to the embeddings of similar users play an important role in our TASU. (2) We also observe that for the same dimension k, TASU_Average and TASU_Max still outperform the best-performing baselines, which again demonstrates that our TASU effectively improves recommendation performance with similar users.

Table 6. The experimental results (HR@10 & NDCG@10) for our variants

Tasks	Domain	TASU_Average		TASU_Max		Imp vs. _Average		Imp vs. _Max	
		HR	NDCG	HR	NDCG	HR	NDCG	HR	NDCG
Task 1 $k = 32$	Book	.3398	.2048	.3415	.2023	2.00%	1.12%	1.49%	2.37%
	Movie	.4102	.2242	.4023	.2250	6.83%	10.44%	8.92%	10.04%
Task 2 $k = 32$	CD	.1877	.0930	.1997	.1069	16.94%	25.59%	9.91%	9.26%
	Movie	.3896	.2094	.3810	.2042	6.72%	10.03%	9.13%	12.83%

4.4 Parameter Analysis

For the impact of the embedding dimension k, the results observed in Tables 4 - 5 are consistent with those discussed in the work[32], and thus, will not be repeated

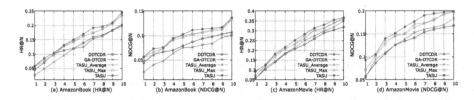

Fig. 3. Results of Top-N recommendation for Task 1 ($k = 16$)

in this paper. We mainly discuss the impact of the Top-N recommendation and the impact of the proportion of common users between domains.

Impact of Top-N. We compare the recommendation performance of top-N with HR@N and NDCG@N, where N ranges from 1 to 10. Due to space limitations, the Top-N recommendation results of DDTCDR, GA-DTCDR, TASU_Average, and TASU_Max are only reported for Task 1 ($k = 16$) in Fig. 3. Similar results are observed for Task 2 and different embedding dimensions k. As shown in Fig. 3, in AmazonBook and AmazonMovie domains, the performance of our TASU, as well as the variants method TASU_Average and TASU_Max consistently outperform the competing methods DDTCDT and GA-DTCDR for different Top-N recommendations. In AmazonBook domain, TASU has an average improvement of 4.14% and 3.43% in terms of HR@N and NDCG@N compared with the best-performing baseline. Also, in AmazonMovie domain, TASU has an average improvement of 6.16% and 7.05% in terms of HR@N and NDCG@N compared with the best-performing baseline.

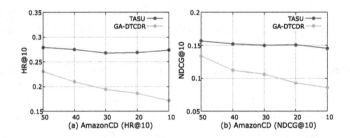

Fig. 4. Impact of proportion of common users

Impact of Proportion of Common Users. As we mentioned in the Introduction section, the proportion of common users will greatly affect the recommendation performance. In this part, we discuss the impact of the proportion of common users. We report the impact of the proportion of common users on the recommendation performance with HR@10 and NDCG@10 compared with the competitive method GA-DTCDR. We randomly sliced the real Amazon datasets to construct 5 groups of AmazonMovie and AmazonCD datasets.

The proportion of common users in the sparser domain (AmazonCD) are set as $10\%, 20\%, 30\%, 40\%$, and 50% for 5 groups of datasets, and the result is reported for AmazonCD ($k = 32$) in Fig. 4. As shown in Fig. 4, our TASU consistently outperforms GA-DTCDR for different proportions of common users and has an average improvement of 38.81% and 45.27% in items of HR@10 and NDCG@10, respectively. In addition, with the decrease in the proportion of common users, the recommendation performance of GA-DTCDR declines, while the results of our TASU keep stable at different proportions of common users. This phenomenon demonstrates that the performance of GA-DTCDR is greatly affected by the proportion of common users, which is a limitation of many cross-domain recommendation methods based on common users. In contrast, our TASU adds the embeddings of similar users on basis of common users to improve recommendation performance and can maintain a stable and effective result even in the case of less common users.

5 Conclusion

In this paper, we propose a three-layer attentional framework based on similar users for the dual-target cross-domain recommendation, called TASU. TASU adds similar users on the basis of common users and employs a three-layers attention mechanism to combine user embeddings at different attention layers, thus generating more representative and high-quality user embeddings to improve recommendation performance in both domains. Also, we have conducted extensive experiments to demonstrate the effectiveness of our TASU. In the future, we plan to extend our method to the multi-domain recommendation to enhance the representation of user embeddings on the target domain by employing common users and similar users in the multi-source domain, so as to improve the recommendation of the target domain.

Acknowledgements. This work was supported by Shanghai Science and Technology Commission (No. 22YF1401100), Fundamental Research Funds for the Central Universities (No. 22D111210, 22D111207), and National Science Fund for Young Scholars (No. 62202095).

References

1. Bahdanau, D., Cho, K., Bengio, Y.: Neural machine translation by jointly learning to align and translate. arXiv preprint arXiv:1409.0473 (2014)
2. Berkovsky, S., Kuflik, T., Ricci, F.: Cross-domain mediation in collaborative filtering. In: Conati, C., McCoy, K., Paliouras, G. (eds.) UM 2007. LNCS (LNAI), vol. 4511, pp. 355–359. Springer, Heidelberg (2007). https://doi.org/10.1007/978-3-540-73078-1_44
3. Chen, H., Yin, H., Sun, X., Chen, T., Gabrys, B., Musial, K.: Multi-level graph convolutional networks for cross-platform anchor link prediction. In: SIGKDD, pp. 1503–1511 (2020)

4. Chen, J., Zhang, H., He, X., Nie, L., Liu, W., Chua, T.S.: Attentive collaborative filtering: Multimedia recommendation with item-and component-level attention. In: SIGIR, pp. 335–344. ACM (2017)
5. Grover, A., Leskovec, J.: node2vec: Scalable feature learning for networks. In: SIGKDD, pp. 855–864. ACM (2016)
6. Guo, L., Tang, L., Chen, T., Zhu, L., Nguyen, Q.V.H., Yin, H.: Da-gcn: A domain-aware attentive graph convolution network for shared-account cross-domain sequential recommendation. arXiv preprint arXiv:2105.03300 (2021)
7. Guo, L., Zhang, J., Chen, T., Wang, X., Yin, H.: Reinforcement learning-enhanced shared-account cross-domain sequential recommendation. In: TKDE (2022)
8. He, X., Chen, T., Kan, M.Y., Chen, X.: Trirank: Review-aware explainable recommendation by modeling aspects. In: CIKM, pp. 1661–1670. ACM (2015)
9. He, X., Liao, L., Zhang, H., Nie, L., Hu, X., Chua, T.S.: Neural collaborative filtering. In: WWW, pp. 173–182 (2017)
10. He, X., Zhang, H., Kan, M.Y., Chua, T.S.: Fast matrix factorization for online recommendation with implicit feedback. In: SIGIR, pp. 549–558. ACM (2016)
11. Hu, G., Zhang, Y., Yang, Q.: Conet: Collaborative cross networks for cross-domain recommendation. In: CIKM, pp. 667–676. ACM (2018)
12. Kang, S., Hwang, J., Lee, D., Yu, H.: Semi-supervised learning for cross-domain recommendation to cold-start users. In: CIKM, pp. 1563–1572. ACM (2019)
13. Kingma, D.P., Ba, J.: Adam: A method for stochastic optimization. arXiv preprint arXiv:1412.6980 (2014)
14. Le, Q., Mikolov, T.: Distributed representations of sentences and documents. In: ICML, pp. 1188–1196. PMLR (2014)
15. Li, P., Tuzhilin, A.: Ddtcdr: Deep dual transfer cross domain recommendation. In: WSDM, pp. 331–339 (2020)
16. Liu, G., Wang, Y., Orgun, M.A.: Trust transitivity in complex social networks. In: AAAI (2011)
17. Liu, G., Wang, Y., Orgun, M.A., Lim, E.P.: Finding the optimal social trust path for the selection of trustworthy service providers in complex social networks. IEEE T Serv Comput. 6(2), 152–167 (2011)
18. Man, T., Shen, H., Jin, X., Cheng, X.: Cross-domain recommendation: An embedding and mapping approach. In: IJCAI, vol. 17, pp. 2464–2470 (2017)
19. Manning, C.D., Surdeanu, M., Bauer, J., Finkel, J.R., Bethard, S., McClosky, D.: The stanford corenlp natural language processing toolkit. In: ACL System Demonstrations, pp. 55–60 (2014)
20. Mihalcea, R., Corley, C., Strapparava, C., et al.: Corpus-based and knowledge-based measures of text semantic similarity. In: AAAI, vol. 6, pp. 775–780 (2006)
21. Rendle, S., Freudenthaler, C., Gantner, Z., Schmidt-Thieme, L.: Bpr: Bayesian personalized ranking from implicit feedback. arXiv preprint arXiv:1205.2618 (2012)
22. Resnik, P.: Using information content to evaluate semantic similarity in a taxonomy. arXiv preprint cmp-lg/9511007 (1995)
23. Sarwar, B., Karypis, G., Konstan, J., Riedl, J.: Item-based collaborative filtering recommendation algorithms. In: WWW, pp. 285–295 (2001)
24. Vaswani, A., et al.: Attention is all you need. In: Advances in Neural Information Processing Systems 30 (2017)
25. Wen, P., Yuan, W., Qin, Q., Sang, S., Zhang, Z.: Neural attention model for recommendation based on factorization machines. Appl. Intell. 51(4), 1829–1844 (2021)
26. Xue, H.J., Dai, X., Zhang, J., Huang, S., Chen, J.: Deep matrix factorization models for recommender systems. In: IJCAI, vol. 17, pp. 3203–3209 (2017)

27. Zhang, H., Shen, F., Liu, W., He, X., Luan, H., Chua, T.S.: Discrete collaborative filtering. In: SIGIR, pp. 325–334. ACM (2016)
28. Zhang, Q., Liao, W., Zhang, G., Yuan, B., Lu, J.: A deep dual adversarial network for cross-domain recommendation. In: TKDE (2021)
29. Zhao, C., Li, C., Xiao, R., Deng, H., Sun, A.: Catn: Cross-domain recommendation for cold-start users via aspect transfer network. In: SIGIR, pp. 229–238. ACM (2020)
30. Zhu, F., Chen, C., Wang, Y., Liu, G., Zheng, X.: Dtcdr: A framework for dual-target cross-domain recommendation. In: CIKM, pp. 1533–1542. ACM (2019)
31. Zhu, F., Wang, Y., Chen, C., Liu, G., Orgun, M., Wu, J.: A deep framework for cross-domain and cross-system recommendations. arXiv preprint arXiv:2009.06215 (2020)
32. Zhu, F., Wang, Y., Chen, C., Liu, G., Zheng, X.: A graphical and attentional framework for dual-target cross-domain recommendation. In: IJCAI, pp. 3001–3008 (2020)
33. Zhu, F., Wang, Y., Chen, C., Zhou, J., Li, L., Liu, G.: Cross-domain recommendation: challenges, progress, and prospects. arXiv preprint arXiv:2103.01696 (2021)
34. Zhu, Y., et al.: Transfer-meta framework for cross-domain recommendation to cold-start users. In: SIGIR, pp. 1813–1817. ACM (2021)
35. Zhu, Y., et al.: Personalized transfer of user preferences for cross-domain recommendation. In: WSDM, pp. 1507–1515 (2022)

CHSR: Cross-view Learning from Heterogeneous Graph for Session-Based Recommendation

Junchen Wang[1], Lei Duan[1(✉)], Runze Ma[1], Yidan Zhang[1], and Zhaohang Luo[2]

[1] School of Computer Science, Sichuan University, Chengdu, China
{wangjunchen,runze_ma,zhangyidan}@stu.scu.edu.cn, leiduan@scu.edu.cn
[2] Nuclear Power Institute of China, Chengdu, China

Abstract. Session-based recommendation (SBR) aims to predict the next item based on short behavior sequences for anonymous users. Most of the current SBR methods consider the scenario that a session just consists of a series of items. However, the multiple item attributes can also reflect user behaviors and provide information for recommendation. In other words, a session in the real world should consist of items and multiple item attributes, which means that the session is heterogeneous. In this paper, we propose a novel method for the anonymous recommendation with heterogeneous item attributes, named CHSR. Firstly, we construct homogeneous session graph and heterogeneous global graph for heterogeneous sessions to map the relationships among different item attributes. Secondly, homo-view and hetero-view of these two kinds of graph encoders are proposed to capture both intra and inter patterns of heterogeneous sessions. Thirdly, a cross-view fusion strategy with consistency loss is introduced to integrate the heterogeneous attribute information by fusing the representations from the two-type views. Finally, the interest preference of anonymous users is represented from the above steps. Extensive experiments conducted on three large-scale real-world datasets demonstrate the superior performance of CHSR over the state-of-the-art methods.

Keywords: Session-based recommendation · Heterogeneous session · Cross-view learning

1 Introduction

Recommendation system, as an effective tool to address information overload problem, plays an essential role in many applications such as e-commerce, social media and news portals [4,15,22]. Most traditional recommendation systems [12,13] predict the next item based on user profiles and long-term historical interactions. However, these kinds of information are not always available or only

This work was supported in part by the National Natural Science Foundation of China (61972268), and the Joint Innovation Foundation of Sichuan University and Nuclear Power Institute of China.

X. Wang et al. (Eds.): DASFAA 2023, LNCS 13944, pp. 314–329, 2023.
https://doi.org/10.1007/978-3-031-30672-3_21

partially available, which limits the performance of recommendation systems. Therefore, session-based recommendation (SBR) is proposed to predict the next item for an anonymous user based on short-term interactions [25].

Existing SBR methods focus on modeling the transition patterns of items, usually based on recurrent neural networks [2,17], attention networks [6,8] and graph neural networks[24,25]. Although achieving impressive performance, the above methods ignore the importance of the item's attributes, such as category, brand and price. And some studies [3,16] indicate that considering the item attribute information in an SBR task can improve the performance of the recommendation systems. It inspires us to consider a general SBR scenario, that is a session consisting of items and multiple item attributes, and we call that a *heterogeneous session*.

In practice, designing an SBR model with heterogeneous information is non-trivial. We need to carefully consider the characteristics of heterogeneous sessions. This requires us to address the following fundamental challenges:

- **(CH1) How to model the heterogeneous information in a session.** In an online shopping scenario, there are many attributes for an item, such as category and brand. These heterogeneous attribute information can reflect the user's preference from different perspectives. For instance, fitness equipment is more likely to attract people keen on sports. The attributes of items provide abundant information to learn higher-level transition patterns, which is of great importance to model the user intent [29]. However, most existing methods perform poorly since they only consider a single-type item attribute (e.g., item ID), which is information loss.
- **(CH2) How to establish relationships among different sessions.** Since each session only contains a limited number of items, it is not enough to capture the user intent using a single session alone, which easily ignores the potential relevance among different sessions. In other words, we need to explore more useful information from other sessions and establish relationships among different sessions. Overall, capturing the heterogeneous information from different sessions is crucial for SBR.
- **(CH3) How to integrate the heterogeneous information in a session.** Recently, researchers have pointed out that information from different views has some consistencies and complementarity [20,26]. The heterogeneous information under different views is desired for SBR, fusing the information from multiple views in a suitable way will help us obtain item representations in high quality. Under this circumstance, it is necessary to find a proper strategy to fuse multi-view information into one comprehensive representation, so that we can accurately infer the user preference in SBR.

To address the above challenges, we propose a novel method for SBR, named CHSR (short for cross-view learning from heterogenous graph for session-based recommendation). **To address CH1**, we identify the concept of heterogeneous session to model the heterogeneity of items and item attributes, which utilizes multiple attribute information to enrich SBR. In detail, we regard transition relationships between each type of attributes as a sequence like item session, since they can reflect higher-level user intent. To explore the transitions among the contexts, we construct a homogeneous session graph for each type of session. Inspired by [25], we adopt a gated graph neural network to update the node information to obtain homo-view node representations. **To address CH2**, we construct a heterogeneous global graph to capture the potential relevance among different sessions. Specifically, it takes into account both the connectivity of entities (i.e., items and item attributes) in a session and among different sessions. After that, we employ a graph attention neural network on the heterogeneous global graph to learn hetero-view node representations. **To address CH3**, we design an adaptive strategy to fuse the representations under homo-view and hetero-view to integrate the heterogeneous information. And we introduce the consistency loss as an auxiliary task to improve the quality of representations. The main contributions of this work are summarized as follows:

- We present the concept of the heterogeneous session to model the heterogeneous short-time interactions of anonymous users.
- We propose a method named CHSR, which obtains the homo-view and hetero-view node representations by propagating information from the homogeneous session graph and heterogeneous global graph, respectively.
- We propose a cross-view fusion strategy with consistency loss to make full use of the heterogeneous information from different views.
- We evaluate the performance of CHSR on three real-world datasets. Experimental results demonstrate that CHSR outperforms the state-of-the-art methods on all metrics.

2   Related Work

2.1   Session-Based Recommendation

Traditional Methods. Most early SBR methods focused on the co-occurrence information of items. Sarwar *et al.* [13] proposed Item-KNN aiming to make recommendations based on item similarity calculated on the co-occurrence. To extract the sequential behavior in session, Rendle *et al.* [12] utilized Markov Chains to model the sequential relations between adjacent items. Shani *et al.* [14] devised a recommendation model to capture item transition based on Markov Decision Processes. However, these methods only modeled sequential behavior between adjacent actions, ignoring the information converted by the session context.

Recurrent Neural Networks Based Methods. Due to the power of recurrent neural networks (RNNs) in modeling sequential data [1,5,7], RNNs are also adopted for SBR tasks. Hidasi *et al.* [2] proposed GRU4Rec to capture the item interaction sequences by GRU layers. Tan *et al.* [18] proposed a data augmentation technique and a method to account for the shifts of data distribution to enhance the performance of the RNNs for SBR. Nevertheless, a session may contain multiple user choices and even noise, which is not sufficient to generate all correct dependencies. To solve this problem, attention mechanism [19] was introduced to distinguish the importance of each item. Li *et al.* [6] proposed NARM to model the user's sequential behavior with a hybrid encoder. Liu *et al.* [8] proposed STMAP that considers both long-term interest and short-term interest to capture the user's comprehensive preference.

Graph Neural Networks Based Methods. Recently, graph neural networks (GNNs) are introduced to SBR tasks because of their ability to capture complex transition patterns. Wu *et al.* [25] first introduced GNNs in SBR, and used gated graph neural network to aggregate neighbor information. Xu *et al.* [28] dynamically constructed a graph structure for session sequences to capture local dependencies. Qiu *et al.* [10] proposed FGNN that uses multi-layered weighted graph attention networks to model the session graph. GCE-GNN [24] employed GNNs to learn item-level and global-level item representations from two graph structures, respectively. DHCN [27] took each session as a hyperedge, and capture beyond-pairwise item relations by a hypergraph convolutional network. Wang *et al.* [21] employed a hypergraph attention mechanism to model the high-order relations among items.

2.2   Heterogeneous Graph Learning

Heterogeneous graph (HG), consisting of multiple types of nodes and relations, has the power to model rich semantics and structural information in real-world data. Thus, many researchers have used it in recommendation systems. For example, Zhao *et al.* [29] introduced the concept of meta-graph to HG-based recommendation to formulate complicated semantics between users and items. HAN [23] used hierarchical attention to capture node-level and semantic-level structures. Pang *et al.* [9] proposed HG-GNN that combines user-item interactions and global co-occurrence items to capture user preference for a personalized recommendation. The items in the real world have multiple attributes which can be represented by a heterogeneous graph, whereas most of the existing works only utilize the single-type item information.

3   Preliminaries

Problem Statement. Let $V = \{v_1, v_2, ..., v_m\}$ denote the set of unique items, where m is the number of items. An item $v_k \in V$ contains some attribute information, like item ID, category and brand. Each session is represented as a sequence $s = [v_{s,1}, v_{s,2}, ..., v_{s,n}]$ in chronological order, where $v_{s,k} \in V$ denotes item v_k clicked by an anonymous user within session s, and the length of s is n. The goal of SBR is to predict the next item $v_{s,n+1}$ for any given session s.

Definition 1 (Heterogeneous Session). *Heterogeneous session is defined as a sequence* $\mathcal{H} = (\mathcal{V}, \mathcal{A}, \phi)$, *where* $\mathcal{V}$ *denotes a set of ordered nodes, and it is associated with a node type mapping function* $\phi : \mathcal{V} \to \mathcal{A}$, *where* $\mathcal{A}$ *denotes a set of node types and* $|\mathcal{A}| > 1$.

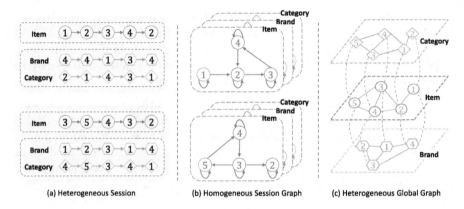

(a) Heterogeneous Session   (b) Homogeneous Session Graph   (c) Heterogeneous Global Graph

Fig. 1. A toy example of heterogeneous session and session graph.

3.1   Homogeneous Session Graph Construction

To model the sequential patterns of pairwise adjacent nodes in the session-based recommendation, we construct a homogeneous session graph $\mathcal{G}_o = (\mathcal{V}_o, \mathcal{E}_o)$ for each heterogeneous session $\mathcal{H} = [v_{s,1}, v_{s,2}, ..., v_{s,n}]$, named HOG. In this HOG, $\mathcal{V}_o$ denotes the set of nodes deriving from $\mathcal{H}$, and $\mathcal{E}_o$ denotes the edge set, in which each edge indicates two adjacent nodes $(v_{s,i}, v_{s,j})$ in $\mathcal{H}$. Following the work [11], we add a self-loop for each node. By propagating information from HOG via a graph neural network, we can learn the intra-session feature of the nodes. Figure 1 (a) illustrates two heterogeneous sessions and Fig. 1 (b) shows their corresponding homogeneous session graphs.

3.2   Heterogeneous Global Graph Construction

In the SBR scenario, each session only contains a limited number of items, the data sparsity problem may limit the benefits of HOG modeling. It is insufficient to use only the current session to capture the user intent, which easily ignores the potential relevance among different sessions. To address the problem, we construct a heterogeneous global graph (HEG) to establish relationships among different sessions. Formally, HEG is an undirected graph generated from a set of given heterogeneous sessions $\mathcal{H}s$, denoted as $\mathcal{G}_e = (\mathcal{V}_e, \mathcal{E}_e, \mathcal{A}, \phi)$, where $\mathcal{V}_e$ denotes the set of nodes deriving from $\mathcal{H}s$ and it is associated with a node type mapping function $\phi : \mathcal{V}_e \rightarrow \mathcal{A}$, where $\mathcal{A}$ denotes a set of node types and $|\mathcal{A}| > 1$. Each edge $e_{ij} \in \mathcal{E}_e$ contains two nodes v_i and v_j, and we use its frequency over all the sessions as its weight w_{ij}. A toy example is illustrated in Fig. 1 (c), which is generated from two heterogeneous sessions shown in Fig. 1 (a).

It is worth noting that some neighbors may be connected due to several unexpected click events. To ensure the relevance of nodes, for each node, we select the n_s most associated adjacent nodes as the final neighbors based on the weight of the edges in the n_g-hop global neighbor.

4   The Design of CHSR

In this section, we introduce the details of the proposed CHSR. As shown in Fig. 2, the framework of CHSR comprises three main modules. (i) *Graph Learning Module*. It learns the homo-view and hetero-view representations via graph neural networks from the constructed homogeneous session graphs and heterogeneous global graph. (ii) *Cross-view Fusion Module*. It employs a cross-view fusion strategy with consistency loss to integrate the heterogeneous information from two-type views. (iii) *Prediction Module*. It outputs the probability of candidate items. In this paper, we use bold capital letters to denote matrices and bold lowercase letters to denote vectors.

4.1   Homo-view Representation Learning

By generating the homogeneous session graph, we obtain pairwise homogeneous transitions within the current session. Inspired by [25], we adopt a gated graph neural network (GGNN) to model the HOG for obtaining the **homo-view** node representations.

We first embed each type of nodes into the same space and let $\mathbf{h}_i^0 \in \mathbb{R}^d$ denote the initial representation of node i with dimension d. Based on the basic principle of iterative propagation in GNN, we use the hidden state in the h-th ($1 \leq h \leq H$) step as the final representations. For a node $v \in \mathcal{G}_e$, the update functions are given as follows:

$$\mathbf{a}_v^h = \left(\mathbf{A}_{v:}^+ + \mathbf{A}_{v:}^-\right)\left[\mathbf{h}_1^{h-1}, \mathbf{h}_2^{h-1}, \ldots, \mathbf{h}_n^{h-1}\right]^\top + \mathbf{b} \tag{1}$$

$$\mathbf{z}_v^h = \sigma\left(\mathbf{W}_{az}\mathbf{a}_v^h + \mathbf{W}_{hz}\mathbf{h}_v^{h-1}\right) \tag{2}$$

$$\mathbf{r}_v^h = \sigma\left(\mathbf{W}_{ar}\mathbf{a}_v^h + \mathbf{W}_{hr}\mathbf{h}_v^{h-1}\right) \tag{3}$$

$$\mathbf{c}_v^h = \tanh\left(\mathbf{W}_{ac}\mathbf{a}_v^h + \mathbf{W}_{hc}\left(\mathbf{r}_v^h \odot \mathbf{h}_v^{h-1}\right)\right) \tag{4}$$

$$\mathbf{h}_v^h = \left(1 - \mathbf{z}_v^h\right) \odot \mathbf{h}_v^{h-1} + \mathbf{z}_v^h \odot \mathbf{c}_v^h \tag{5}$$

where all bold lowercases are the d-dimensional representations and all $\mathbf{W}$s $\in \mathbb{R}^{d \times d}$ are the trainable parameters. $\mathbf{A}^+$ and $\mathbf{A}^-$ denote incoming and outgoing adjacency matrix of $\mathcal{G}_e$, respectively. $\left[\mathbf{h}_1^{h-1}, \mathbf{h}_2^{h-1}, \ldots, \mathbf{h}_n^{h-1}\right]$ is the list of node representations in a session. σ is Sigmoid function and $\odot$ is element-wise multiplication operation. $\mathbf{r}_v^h$ and $\mathbf{z}_v^h$ are reset gate and update gate respectively.

After H steps, we obtain the updated representations of node v. Similarly, we can also calculate other nodes' representations from all HOGs. We regard these as the **homo-view** representations, which are denoted as:

$$\mathbf{H}^o = [\mathbf{h}_1^o, \mathbf{h}_2^o, \ldots, \mathbf{h}_m^o]^\top = \left[\mathbf{h}_1^H, \mathbf{h}_2^H, \ldots, \mathbf{h}_m^H\right]^\top \tag{6}$$

4.2   Hetero-view Representation Learning

The heterogeneous global graph (HEG) contains connection relationships across sessions and multiple types of nodes. In order to explore the complicated heterogeneous transition patterns, we conduct information propagating on HEG.

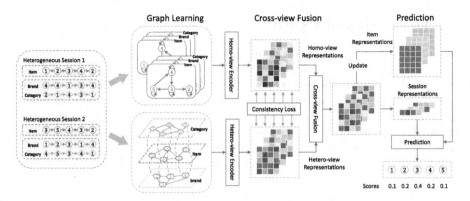

Fig. 2. The framework of the proposed CHSR.

Specifically, we adopt a heterogeneous graph attention network to learn the **_hetero-view_** node representations.

In HEG, a node may be contained in multiple sessions and connected to multiple types of nodes. To propagate more useful information and ensure the relevance of nodes, we utilize the attention mechanism to distinguish the importance of adjacent nodes. For any two connected nodes $v_i, v_j \in \mathcal{G}_e$, the importance weight is defined as:

$$\pi(v_i, v_j) = \mathbf{q}_1^\top \text{LeakyReLU}(\mathbf{W}_1[(\mathbf{s} \odot \mathbf{h}_{v_j}) \parallel w_{ij}]) \tag{7}$$

$$\mathbf{s} = \frac{1}{|S|} \sum_{v_i \in S} \mathbf{h}_{v_i} \tag{8}$$

where $\mathbf{h}_{v_i}$ is the d-dimensional vector of v_i and $\mathbf{s}$ is the vector of current session s by computing the average of each node representation. $w_{ij} \in \mathbb{R}^1$ is the weight of edge (v_i, v_j). $\odot$ is element-wise multiplication operation and $\parallel$ indicates concatenation operation. $\mathbf{q}_1 \in \mathbb{R}^{d+1}$ and $\mathbf{W}_1 \in \mathbb{R}^{(d+1)\times(d+1)}$ are the trainable parameters. Next, a softmax function is used to normalize the attention scores:

$$\pi(v_i, v_j) = \frac{\exp(\pi(v_i, v_j))}{\sum_{v_k \in \mathcal{N}_{v_i}} \exp(\pi(v_i, v_k))} \tag{9}$$

where $\mathcal{N}_{v_i}$ is the neighborhood set of node v_i, which consists of the nodes that interact with node v_i in all sessions. After aggregating information from neighboring nodes, we obtain the updated representations in HEG.

$$\mathbf{h}_{v_i}^e = \sum_{v_j \in \mathcal{N}_{v_i}} \pi(v_i, v_j)\mathbf{h}_{v_j} \tag{10}$$

We regard these as the **_hetero-view_** representations, which are denoted as:

$$\mathbf{H}^e = [\mathbf{h}_1^e, \mathbf{h}_2^e, \ldots, \mathbf{h}_m^e]^\top \tag{11}$$

4.3   Cross-View Representation Learning

After the above stages, we obtain node representations under the homo-view and hetero-view. It is worth noting that the representations of a node under different views have some consistencies and complementarities. To get a joint representation from the two views, we propose an auxiliary loss function to better fuse representations between the two views by measuring the consistency.

Consistency Loss Across Two Views. The representations under the homo-view and hetero-view describe the same node respectively. Consequently, we consider the two representations of the same node have higher consistency compared to either of the node representations with the other node representations.

$\mathbf{H}^o$ and $\mathbf{H}^e$ are learned by different encoders, which means that they are distributed in different feature spaces. However, it is not realistic to compute their consistency directly. To address this problem, we adopt a linear projection function to convert $\mathbf{H}^o$ and $\mathbf{H}^e$ into a common latent vector space. Specially, for a node v_i under the homo-view, the transformation can be defined as:

$$\mathbf{z}_{v_i}^o = \sigma(\mathbf{W}_2^\top \cdot \mathbf{h}_{v_i}^o + \mathbf{b}_1) \tag{12}$$

where $\mathbf{z}_{v_i}^o$ is the projected representation. $\mathbf{W}_2 \in \mathbb{R}^{d \times d}$, $\mathbf{b}_1 \in \mathbb{R}^d$ are the trainable parameters, and σ is the Sigmoid activation function. Similarly, we can also obtain the representation $\mathbf{z}_{v_i}^e$ under the hetero-view after the same transformation.

Taking the projected homo-view representation $\mathbf{z}_{v_i}^o$ as an example, its corresponding positive sample is the projected hetero-view representation $\mathbf{z}_{v_i}^e$, and the consistency loss under the homo-view can be defined as:

$$\mathcal{L}_i^o = -\log \frac{\exp\left(\mathrm{Sim}(\mathbf{z}_{v_i}^o, \mathbf{z}_{v_i}^e)/\tau\right)}{\sum_{j=1}^N \exp\left(\mathrm{Sim}(\mathbf{z}_{v_i}^o, \mathbf{z}_{v_j}^e)/\tau\right)}, \tag{13}$$

where τ is a temperature coefficient. The operation $\mathrm{Sim}(\cdot)$ represents the cosine similarity between two representations. Similarly, we can also calculate the consistency loss under the hetero-view as follows:

$$\mathcal{L}_i^e = -\log \frac{\exp\left(\mathrm{Sim}(\mathbf{z}_{v_i}^e, \mathbf{z}_{v_i}^o)/\tau\right)}{\sum_{j=1}^N \exp\left(\mathrm{Sim}(\mathbf{z}_{v_i}^e, \mathbf{z}_{v_j}^o)/\tau\right)}, \tag{14}$$

Combining the consistency loss in Eq. 13 and Eq. 14 under the two views, we can get the overall consistency loss, which is defined as:

$$\mathcal{L}_{aux} = \frac{1}{2m} \sum_{i=1}^m (\mathcal{L}_i^o + \mathcal{L}_i^e) \tag{15}$$

Cross-View Representation Fusion. To integrate the information from different views, we fuse the homo-view and hetero-view representations by a multi-layer perceptron. For a node v_i, the fusion process can be defined as:

$$\mathbf{z}_{v_i} = \sigma(\mathbf{W}_3^\top (\mathbf{z}_{v_i}^o + \mathbf{z}_{v_i}^e) + \mathbf{b}_2) \tag{16}$$

where $\mathbf{z}_{v_i}$ is the fused representation of node v_i. $\mathbf{W}_3 \in \mathbb{R}^{d \times d}$, $\mathbf{b}_2 \in \mathbb{R}^d$ are trainable parameters and σ is the Sigmoid activation function.

4.4   Interest Preference Generation

Based on the fused representations, we further explore how to obtain the session representation to reflect the anonymous user's interest preference.

The session sequences occur in chronological order, to capture the dynamic process, we adopt the position embeddings to model the sequential relationship. Specifically, we concatenate the reversed position embeddings with the node representations by a trainable matrix $\mathbf{P} = [\mathbf{p_1}, \mathbf{p_2}, ..., \mathbf{p_n}]$, where n is the length of the current session and $\mathbf{p_i} \in \mathbb{R}^d$ represents the vector of position i.

$$\mathbf{z}'_{v_i} = \tanh(\mathbf{W}_4[\mathbf{z}_{v_i} \parallel \mathbf{p}_{n-i+1}] + \mathbf{b}_3) \tag{17}$$

where $\mathbf{z}_{v_i}$ is the representation of i-th node in the session s. $\mathbf{W}_4 \in \mathbb{R}^{d \times d}$ and $\mathbf{b}_3 \in \mathbb{R}^d$ are trainable parameters. Next, we adopt the soft-attention mechanism to obtain the user's interest preference representation as follows:

$$\alpha_i = \mathbf{q}_2^\top (\mathbf{W}_5 \mathbf{z}'_{v_i} + \mathbf{W}_6 \bar{\mathbf{z}} + \mathbf{b}_4) \tag{18}$$

$$\theta_s = \sum_{i=1}^{n} \alpha_i \mathbf{z}'_{v_i} \tag{19}$$

where $\bar{\mathbf{z}} = \frac{1}{n} \sum_{i=1}^{n} \mathbf{z}'_{v_i}$ indicates the average representation of session s and θ_s is the user's interest preference representation. $\mathbf{W}_5, \mathbf{W}_6 \in \mathbb{R}^{d \times d}$ and $\mathbf{b}_4, \mathbf{q}_2 \in \mathbb{R}^d$ are trainable parameters.

4.5   Model Optimization

Given a session s, we compute the recommendation probability $\hat{\mathbf{y}}$ of all candidate items by doing the inner product between the final node representation $\mathbf{z}_{v_i}$ and user's interest preference representation θ_s.

$$\hat{\mathbf{y}}_i = \text{Softmax}(\theta_s^\top \mathbf{z}_{v_i}) \tag{20}$$

where $\mathbf{y}_i \in \hat{\mathbf{y}}$ denotes the probability that the anonymous user will click on next item $v_i \in V$ in the session s. Next, we use cross entropy as the optimization objective, which can be defined as follows:

$$\mathcal{L}_{rec}(\hat{\mathbf{y}}) = - \sum_{i=1}^{m} \mathbf{y}_i \log(\hat{\mathbf{y}}_i) + (1 - \mathbf{y}_i) \log(1 - \hat{\mathbf{y}}_i) \tag{21}$$

where $\mathbf{y}_i$ is the one-hot encoding vector of the ground truth item. Hence, the overall loss $\mathcal{L}$ of our model can be defined as:

$$\mathcal{L} = \mathcal{L}_{rec} + \beta \mathcal{L}_{aux} \tag{22}$$

where β is a weight coefficient to balance the influence of auxiliary consistency loss. Finally, we use the Back-Propagation Through Time (BPTT) algorithm to optimize the proposed CHSR model.

Table 1. Statistics of the datasets.

Dataset	# clicks	# train sessions	# test sessions	# items	# attributes	Avg. length
Tmall	818479	351268	25898	40728	Category: 713 Brand: 4228 Seller: 4415	6.69
Nowplaying	1367963	825304	89824	60417	Category: 11461	7.42
Diginetica	982961	60858	43097	40728	Category: 995 Price: 12	5.12

5 Experiments

To answer the following four key research questions, we conduct experiments on three datasets to evaluate the performance of CHSR.

- **RQ1:** How does the proposed model CHSR perform compared with state-of-the-art baselines? (Sect. 5.1)
- **RQ2:** Does the hetero-view encoder and homo-view encoder with multiple attributes improve the performance of CHSR? (Sect. 5.2)
- **RQ3:** What is the effect of the consistency fusion strategy? (Sect. 5.3)
- **RQ4:** How do the key hyperparameters affect the accuracy of CHSR? (Sect. 5.4)

Datasets and Preprocessing. We evaluate CHSR on three real-world datasets: Tmall[1], Nowplaying[1] and Diginetica[3], which are widely used in session-based recommendation research. The statistics of datasets are summarized in Table 1.

Following the previous works [6,25], we filter out sessions of length 1 and items appearing less than 5 times over all datasets. We generate sequences and corresponding labels by splitting the input sequence. For a session, the last item is viewed as the label and the remaining sequence is used to model user preferences. To be specific, for an input session $S = [v_1, v_2, ..., v_m]$, we generate a series of sequences and labels $\{[v_1], v_2\}$, $\{[v_1, v_2], v_3\}, ...\{[v_1, v_2, ..., v_{m-1}], v_m\}$, where $[v_1, v_2, ..., v_{m-1}]$ is the generated sequences and v_m denotes the label of the sequence. Besides, to integrate heterogeneous attribute information into the model, we establish a relationship between items and multiple attributes.

Baselines. To demonstrate the effectiveness of CHSR, the following methods are chosen for comparison:

- **Item-KNN** [13] recommends items similar to the previously clicked item in the session. We use cosine similarity in our implementation.

---

[1] https://tianchi.aliyun.com/dataset/dataDetail?dataId=42.
[1] http://dbis-nowplaying.uibk.ac.at/#nowplaying.
[3] https://competitions.codalab.org/competitions/11161.

- **FPMC** [12] is a sequential method based on Markov Chain. Following previous work, we ignore user representation when computing scores.
- **GRU4Rec** [2] utilizes Gated Recurrent Unit (GRU) to capture sequential transition between items.
- **NARM** [6] is a RNN-based model with attention mechanism to capture the user's main purpose and sequential behavior.
- **STAMP** [8] employs a bidirectional self-attention architecture to combine user's general interest and current interest of the last item.
- **SR-GNN** [25] utilizes a gated graph neural network to obtain item representations and employs attention mechanism to obtain session representations.
- **GCE-GNN** [24] aggregates the global context and the item sequence in the current sessions to capture the global and local information.
- S^2-**DHCN** [27] regards each session as a hyperedge and construct all sessions as a hypergraph to capture high-order item relations.

Evaluation Metrics. As in [8,25,27], we use P@k (Precision) and M@k (Mean Reciprocal Rank) to evaluate the performance of all methods. We report the results with k = 10, 20.

Implementation Details. We implement the proposed model based on Pytorch framework. Following previous methods [6], the dimension of the latent vectors is fixed to 100, the batch size for mini-batch is set to 100 for all datasets and the L_2 regularization is 10^{-5}. All model parameters are initialized by a Gaussian distribution with a mean of 0 and a standard deviation of 0.1. Besides, we employ the Adam optimizer with the initial learning rate 0.01, which will decay by 0.1 after every 3 epochs. To avoid overfitting, we leverage the early-stop training method with patience 5. We select other hyper-parameters on a validation set which is a random 10% subset of the training set. Furthermore, all experiment results are calculated by a grid search strategy with learning rate ranging from 0.001 to 0.01. The implementation of our method is publicly available on github[4].

5.1   Overall Performance (RQ1)

To demonstrate the overall performance of CHSR, we compare it with other state-of-the-art methods and highlight the best results in boldface. Analyzing the results in Table 2, we can draw the following conclusions:

- Our proposed model CHSR achieves the best performance in terms of all evaluation metrics on all datasets. Specially, CHSR outperforms the best model by 21.4% on Tmall, 7.7% on Nowplaying and 2.8% on Diginetica on average. We believe that the heterogeneous attribute information can enhance the performance of the model. The strength of the model lies in its ability to capture

---

[4] https://github.com/junchen-wang/Rec-CHSR-2023.

Table 2. Performance comparisons of CHSR with baselines over three datasets.

Method	Tmall				Nowplaying				Diginetica			
	P@10	**M@10**	**P@20**	**M@20**	**P@10**	**M@10**	**P@20**	**M@20**	**P@10**	**M@10**	**P@20**	**M@20**
Item-KNN [13]	6.65	3.11	9.15	3.31	10.96	4.55	15.94	4.91	25.07	10.77	35.75	11.57
FPMC [12]	13.10	7.12	16.06	7.32	5.28	2.68	7.36	2.82	15.43	6.20	26.53	6.95
GRU4Rec [2]	9.47	5.78	10.93	5.89	6.74	4.40	7.92	4.48	17.93	7.33	29.45	8.33
NARM [6]	19.17	10.42	23.30	10.70	13.60	6.62	18.59	6.93	35.44	15.13	49.70	16.17
STAMP [8]	22.63	13.12	26.47	13.36	13.22	6.57	17.66	6.88	33.98	14.26	45.64	14.32
SR-GNN [25]	23.41	13.45	27.57	13.72	14.17	7.15	18.87	7.47	38.42	16.89	51.26	17.78
GCE-GNN [24]	28.01	15.08	33.42	15.42	16.94	8.03	22.37	8.4	41.16	18.15	54.22	19.04
S^2-DHCN [27]	26.22	14.60	31.42	15.05	17.35	7.87	23.50	8.18	40.21	17.59	53.66	18.51
$CHSR_{add}$	27.32	14.64	33.05	15.04	16.94	7.54	22.92	7.96	38.63	16.46	52.08	17.40
$CHSR_{mul}$	29.65	15.50	35.37	15.90	16.91	7.29	22.80	7.69	38.98	17.56	51.22	18.40
$CHSR_{concat}$	21.71	11.91	26.55	12.24	16.32	8.27	22.33	8.64	36.43	14.75	50.12	15.69
CHSR	**33.64**	**18.42**	**40.35**	**18.89**	**18.21**	**8.95**	**24.21**	**9.35**	**42.32**	**18.65**	**55.65**	**19.58**
Improv.	20.10%	22.15%	20.74%	22.50%	4.96%	11.5%	3.02%	11.31%	2.82%	2.75%	2.64%	2.84%

the transition relationships of cross-view heterogeneous information. In addition, by considering the consistency of representation under different views and fusing homo-view and hetero-view information, CHSR can obtain a more comprehensive representation to better capture user's interest preference.

- Compared to the traditional methods (e.g., Item-KNN, FPMC), the neural network based methods achieve great performance, which proves the neural network based methods can better model the complex sequential patterns in the session. In particular, among RNN-based methods, NRAM and STAMP achieve more significant performance than GRU4Rec. This is because that GRU4Rec only considers the sequence information of the session and has difficult in handling the user's dynamic interest preference. In contrast, NARM and STAMP adopt an attention mechanism to distinguish the importance of items in the session.

- GNN-based methods are better than RNN-based methods in all metrics, demonstrating the power of graph neural networks in session-based recommendation. Among them, S^2-DHCN outperforms SR-GNN by using hypergraph to model the beyond pairwise relations. Furthermore, GCE-GNN performs best among the baseline in most evaluation metrics, which proves that capturing different levels of information (intra-session and inter-session information) helps accurately predict the user's intent.

5.2 Ablation Study (RQ2)

In order to measure the contribution of different encoders, we derive two variants of CHSR, including CHSR w/o A_k that removes the homogeneous session graph encoder for the k-th attribute, and CHSR w/o Hetero that removes the heterogeneous global graph encoder.

Table 3. The effects of different encoders.

Dataset	Model	P@10	M@10	P@20	M@20
Tmall	CHSR	**33.64**	**18.42**	**40.35**	**18.89**
	w/o A_1	30.51	16.36	37.03	16.82
	w/o A_2	31.67	17.12	38.01	17.56
	w/o A_3	30.87	16.51	37.69	16.98
	w/o A_1A_2	28.15	15.58	34.52	16.03
	w/o A_1A_3	29.72	15.79	36.14	16.22
	w/o A_2A_3	26.13	14.09	32.07	14.48
	w/o $A_1A_2A_3$	23.40	12.63	27.89	12.94
	w/o Hetero	33.01	17.65	39.31	18.09
Nowplaying	CHSR	**18.21**	**8.95**	24.21	**9.35**
	w/o A_1	17.27	8.41	22.90	8.75
	w/o Hetero	18.08	7.25	**24.41**	7.68
Diginetica	CHSR	**42.32**	**18.65**	**55.65**	**19.58**
	w/o A_1	38.64	16.31	52.07	16.70
	w/o A_2	35.19	14.21	48.53	15.13
	w/o A_1A_2	38.51	16.79	51.29	17.68
	w/o Hetero	41.86	18.52	55.07	19.43

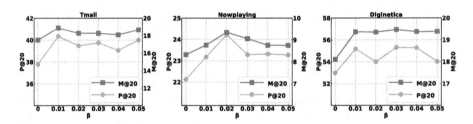

Fig. 3. The impacts of the weight of consistency loss.

Table 3 shows the comparison between different contrast models. From Table 3, we can find that CHSR achieves better performance than CHSR w/o A_k on all metrics and outperforms CHSR w/o Hetero on most metrics. This can demonstrate that integrating both homogeneous and heterogeneous information can help to improve prediction accuracy. It can also be observed that the performance of CHSR decreases on Tmall and Nowplaying as the number of attribute encoders reduces, indicating that considering more auxiliary information can make the model more complete.

5.3 The Effect of Cross-View Learning (RQ3)

We introduce a weight coefficient β to control the weight of consistency loss $\mathcal{L}_{aux}$. To investigate the influence of consistency loss, we report the performance of CHSR with a set of alternative β values in $\{0, 0.01, 0.02, 0.03, 0.04, 0.05\}$.

As illustrated in Fig. 3, we can observe that CHSR achieves the best performance when jointly trained with consistency loss. The most suitable value of β

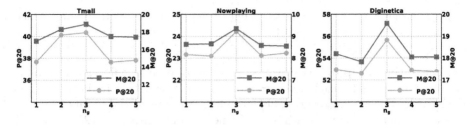

Fig. 4. The impacts of global neighbors n_g.

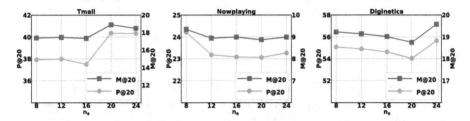

Fig. 5. The impacts of sampling size n_s.

is 0.01 on Tmall, 0.02 on Nowplaying and 0.03 on Diginetica. However, when β continues to increase, the performance of the model decreases markedly. This is because as the weights increase during the training process, the consistency loss becomes a larger proportion of the overall loss. This affects the training of the model on the main task and brings more noise to the model.

Moreover, to demonstrate the effectiveness of our proposed fusion strategy, we design three variant models $CHSR_{add}$, $CHSR_{mul}$ and $CHSR_{concat}$, which fuse the representations of two views by addition, multiplication and concatenation operations respectively. As is shown in Table 2, we can see that our proposed fusion strategy achieves the best results. Besides, $CHSR_{concat}$ performs the worst, which implies that simply concatenating representations from two different views increases the dimension of representation space, from which it may be difficult for the model to learn useful information.

5.4 Hyperparameters Analysis (RQ4)

In this subsection, we conduct experiments to explore how the key hyperparameters, such as the number of global neighbors n_g and sampling size n_s, influence the performance of CHSR.

Effect of the Global Neighbors. To investigate the impact of the number of global neighbors, we report the performance with a set of representative n_g values in $\{1, 2, 3, 4, 5\}$ on all datasets. According to the results in Fig. 4, we can find that CHSR perform best when $n_g = 3$ on all datasets. However, when the neighbor number increases beyond 3, the performance decreases obviously.

This is because a larger neighbor number brings more unrelated nodes, which weakens the learning ability of the model.

Effect of the Sampling Size. Considering the computing efficiency, we sample n_s adjacent nodes for each node according to the weight of edges in the process of building the heterogeneous global graph. According to the results in Fig. 5, we can find that CHSR reaches the highest value when n_s is 20 on Tmall, 8 on Nowplaying and 24 on Diginetica, respectively. It demonstrates that a suitable sampling size can balance between effective information and irrelevant information, so as to achieve the best performance.

6   Conclusion

In this paper, we propose a novel method CHSR for SBR. Specifically, we first put forward the concept of heterogeneous session to integrate multi-type item attribute information. In addition, we obtain the homo-view and hetero-view item representations by learning two session graphs, respectively. Then we propose a cross-view fusion strategy based on consistency loss to obtain more comprehensive item representations. Finally, extensive experiments conducted on three real-world datasets demonstrate that CHSR achieves better performance than existing state-of-the-art models.

References

1. Cui, Q., Wu, S., Liu, Q., Zhong, W., Wang, L.: Mv-rnn: A multi-view recurrent neural network for sequential recommendation. TKDE **32**(2), 317–331 (2018)
2. Hidasi, B., Karatzoglou, A.: Recurrent neural networks with top-k gains for session-based recommendations. In: CIKM, pp. 843–852 (2018)
3. Hu, B., Shi, C., Zhao, W.X., Yu, P.S.: Leveraging meta-path based context for top- N recommendation with a neural co-attention model. In: KDD, pp. 1531–1540 (2018)
4. Huang, C., et al.: Knowledge-aware coupled graph neural network for social recommendation. In: AAAI, pp. 4115–4122 (2021)
5. Jannach, D., Ludewig, M.: When recurrent neural networks meet the neighborhood for session-based recommendation. In: RecSys, pp. 306–310 (2017)
6. Li, J., Ren, P., Chen, Z., Ren, Z., Lian, T., Ma, J.: Neural attentive session-based recommendation. In: CIKM, pp. 1419–1428 (2017)
7. Liu, Q., Wu, S., Wang, D., Li, Z., Wang, L.: Context-aware sequential recommendation. In: ICDM, pp. 1053–1058 (2016)
8. Liu, Q., Zeng, Y., Mokhosi, R., Zhang, H.: STAMP: short-term attention/memory priority model for session-based recommendation. In: KDD, pp. 1831–1839 (2018)
9. Pang, Y., et al.: Heterogeneous global graph neural networks for personalized session-based recommendation. In: WSDM, pp. 775–783 (2022)
10. Qiu, R., Li, J., Huang, Z., Yin, H.: Rethinking the item order in session-based recommendation with graph neural networks. In: CIKM, pp. 579–588 (2019)

11. Qiu, R., Li, J., Huang, Z., Yin, H.: Rethinking the item order in session-based recommendation with graph neural networks. In: CIKM (2019)
12. Rendle, S., Freudenthaler, C., Schmidt-Thieme, L.: Factorizing personalized markov chains for next-basket recommendation. In: WWW, pp. 811–820 (2010)
13. Sarwar, B.M., Karypis, G., Konstan, J.A., Riedl, J.: Item-based collaborative filtering recommendation algorithms. In: WWW, pp. 285–295 (2001)
14. Shani, G., Heckerman, D., Brafman, R.I.: An MDP-based recommender system. J. Mach. Learn. Res. **6**, 1265–1295 (2005)
15. Sheu, H.S., Chu, Z., Qi, D., Li, S.: Knowledge-guided article embedding refinement for session-based news recommendation. In: TNNLS (2021)
16. Shi, C., Hu, B., Zhao, W.X., Yu, P.S.: Heterogeneous information network embedding for recommendation. TKDE **31**(2), 357–370 (2019)
17. Sun, F., et al.: Bert4rec: Sequential recommendation with bidirectional encoder representations from transformer. In: CIKM, pp. 1441–1450 (2019)
18. Tan, Y.K., Xu, X., Liu, Y.: Improved recurrent neural networks for session-based recommendations. In: RecSys, pp. 17–22 (2016)
19. Vaswani, A., et al.: Attention is all you need. In: NIPS, pp. 5998–6008 (2017)
20. Velickovic, P., Fedus, W., Hamilton, W.L., Liò, P., Bengio, Y., Hjelm, R.D.: Deep graph infomax. In: ICLR (2019)
21. Wang, J., Ding, K., Zhu, Z., Caverlee, J.: Session-based recommendation with hypergraph attention networks. In: SDM, pp. 82–90 (2021)
22. Wang, S., Cao, L., Wang, Y., Sheng, Q.Z., Orgun, M.A., Lian, D.: A survey on session-based recommender systems. ACM Comput. Surv. **54**(7), 1–38 (2021)
23. Wang, X., et al.: Heterogeneous graph attention network. In: WWW, pp. 2022–2032 (2019)
24. Wang, Z., Wei, W., Cong, G., Li, X.L., Mao, X.L., Qiu, M.: Global context enhanced graph neural networks for session-based recommendation. In: SIGIR, pp. 169–178 (2020)
25. Wu, S., Tang, Y., Zhu, Y., Wang, L., Xie, X., Tan, T.: Session-based recommendation with graph neural networks. In: AAAI, pp. 346–353 (2019)
26. Xia, X., Yin, H., Yu, J., Shao, Y., Cui, L.: Self-supervised graph co-training for session-based recommendation. In: CIKM, pp. 2180–2190 (2021)
27. Xia, X., Yin, H., Yu, J., Wang, Q., Cui, L., Zhang, X.: Self-supervised hypergraph convolutional networks for session-based recommendation. In: AAAI, pp. 4503–4511 (2021)
28. Xu, C., et al.: Graph contextualized self-attention network for session-based recommendation. In: IJCAI, pp. 3940–3946 (2019)
29. Zhao, H., Yao, Q., Li, J., Song, Y., Lee, D.L.: Meta-graph based recommendation fusion over heterogeneous information networks. In: KDD, pp. 635–644 (2017)

MOEF: Modeling Occasion Evolution in Frequency Domain for Promotion-Aware Click-Through Rate Prediction

Xiaofeng Pan[1(✉)], Yibin Shen[1], Jing Zhang[2], Xu He[1], Yang Huang[1], Hong Wen[1], Chengjun Mao[1], and Bo Cao[1]

[1] Alibaba Group, Hangzhou, China
pxfvintage@163.com, {shilou.syb,hx234688,hy234680,
qinggan.wh,chengjun.mcj,zhizhao.cb}@alibaba-inc.com
[2] The University of Sydney, Sydney, Australia
jing.zhang1@sydney.edu.au

Abstract. Promotions are becoming more important and prevalent in e-commerce to attract customers and boost sales, leading to frequent changes of occasions, which drives users to behave differently. In such situations, most existing Click-Through Rate (CTR) models can't generalize well to online serving due to distribution uncertainty of the upcoming occasion. In this paper, we propose a novel CTR model named MOEF for recommendations under frequent changes of occasions. Firstly, we design a time series that consists of occasion signals generated from the online business scenario. Since occasion signals are more discriminative in the frequency domain, we apply Fourier Transformation to sliding time windows upon the time series, obtaining a sequence of frequency spectrums which is then processed by Occasion Evolution Layer (OEL). In this way, a high-order occasion representation can be learned to handle the online distribution uncertainty. Moreover, we adopt multiple experts to learn feature representations from multiple aspects, which are guided by the occasion representation via an attention mechanism. Accordingly, a mixture of feature representations is obtained adaptively for different occasions to predict the final CTR. Experimental results on real-world datasets validate the superiority of MOEF and online A/B tests also show MOEF outperforms representative CTR models significantly.

Keywords: Click-Through Rate Prediction · E-commerce Promotions · Occasion Evolution · Frequency Domain

1 Introduction

Click-Through Rate (CTR) prediction has been one of the most important tasks in recommender systems [11,19] since it is directly related to user satisfaction, efficiency, and revenue. With the rapid progress of deep neural models, most of CTR models use high-order interactions of features to improve their representation ability [1,5,22], and leverage sequential user behaviors to model users in a dynamic manner [9,10,17,20,21].

X. Wang et al. (Eds.): DASFAA 2023, LNCS 13944, pp. 330–340, 2023.
https://doi.org/10.1007/978-3-031-30672-3_22

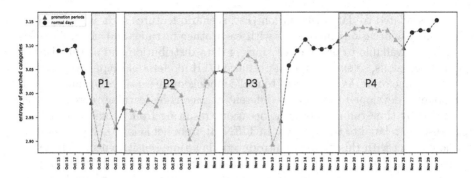

Fig. 1. The entropy of searched categories is calculated as $H = -\sum p_i \, log \, p_i$, where p_i denotes the global search ratio of the i-th category. H intuitively shows the diversity of user interests and its change reflects the change of user data distribution.

Most previous works assume that users' preferences are coherent and change smoothly over time. However, users' behaviors can be significantly influenced by different occasions [16] which are fine-grained time periods related to particular times or events. Especially with the intensification of e-commerce competition, online promotions are becoming more frequent and diverse, which differ from each other in many factors, *e.g.*, scale, duration, density, and regularity in time, resulting in frequent changes of occasions. From Fig. 1, we can observe that the entropy of searched categories, an indicator of user data distribution, fluctuated dramatically as four promotions occurred. The fluctuations during different promotions differed greatly and shared no regularity, illustrating that there exists considerable distribution uncertainty under frequent changes of occasions, although the schedule of promotions is available beforehand. Besides, with different occasions included, training samples can be non-identically distributed, imposing extra difficulty on model training. To achieve promotion-aware CTR prediction, we need to tackle these two issues simultaneously.

An intuitive idea to deal with recommendations for e-commerce promotions is leveraging years of plenty of historical data to train models with better generalization performance. However, directly introducing data with a large time span results in discrepant data distributions between training and serving, which degrades the generalization performance [8]. Besides, promotions driven by emerging consumption trends are inaccessible in historical data. As far as we know, the most widely adopted way to handle recommendations for promotions is temporarily customizing or adding extra models, which highly relies on expert experience to decide when to switch models for online serving because it's uncertain when data distribution will change.

Recently, several studies, *e.g.*, STAR [13], TREEMS [7], and SAR-Net [12], have been carried out to handle Multi-Scenario CTR prediction, which seems applicable to our case. In general, these methods leverage scenario features to learn data distributions of different scenarios to facilitate the Mixture-of-Experts [3] so they can serve all scenarios with one model. Each recommendation scenario locates spatial differently and can be treated as stable over time, with explicit features related to data distribution (*i.e.*, scenario features such as scenario id). In contrast, occasions intertwine with each other in time, and no explicit features are available to provide certainty of data distribution and distinguish different occasions, so existing Multi-Scenario CTR models don't apply to our case. Another line of works [4,16] tries to make models more sensitive to time. In [16], a model is developed to adapt to different occasions by learning representations indexed by timestamps. However, occasions resulting from promotions are not necessarily related to timestamps. In TIEN [4], item behaviors, *i.e.*, a set of users who interact with this item, are introduced via a time-sensitive neural structure, which models items in a dynamic manner to strengthen the ability to predict users' emerging interests. However, local changes captured from item behaviors are not equivalent to global changes of data distribution.

Based on these observations, we propose a novel CTR model named MOEF which models occasion evolution to perceive the changes of occasions and modulate the learning of user-item feature representations. Specifically, with inspiration from time series prediction [18], we generate occasion signals (detailed in Sect. 2.1) from our online business scenario with a proper sampling interval, constructing a time series which is then processed by an elaborated Occasion Representation Network (ORN). In the ORN, we obtain a sequence of frequency spectrums by applying Fast Fourier Transformation (FFT) [14] to time windows that slide on the time series with proper window size and stride, since occasion signals are more discriminative in the frequency domain. Then we process the sequence of frequency spectrums by Occasion Evolution Layer (OEL), where deep neural structures for sequential modeling such as LSTM [2] and Transformer-Encoder [15] can be used to model occasion evolution. In this way, a high-order occasion representation can be learned to help tackle the online distribution uncertainty. Meanwhile, multiple experts are adopted to learn feature representations, mitigating the mutual interference between non-identically distributed training data. Under modulation of the occasion representation, multiple experts can learn feature representations from different aspects, which are mixed via an attention mechanism and used for the final prediction. Our main contributions are summarized as follows:

- To the best of our knowledge, this is the first study of CTR prediction concerning e-commerce promotions. We introduce occasion signals from a perspective of time series prediction and propose a novel MOEF model to achieve promotion-aware CTR prediction.
- We model occasion evolution in the frequency domain to facilitate learning a good occasion representation, which is further used to modulate the learning of multiple experts via an attention mechanism. In this way, both the distribution uncertainty and training difficulty resulting from non-identically distributed training data are handled.
- Experiments on real-world datasets and online tests demonstrate the superiority of MOEF over representative methods. We also perform extensive analyses for better interpretability. The code is publicly available[1].

---

[1] https://github.com/AaronPanXiaoFeng/MOEF.

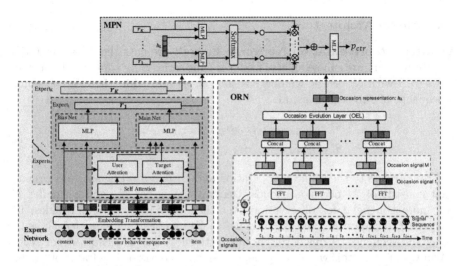

Fig. 2. Framework of the proposed MOEF model, which consists of the Occasion Representation Network (ORN), Experts Network and Mixture Prediction Network (MPN).

2   The Proposed Method

2.1   Model Input

In CTR prediction, the model takes input as $(\boldsymbol{x}, y) \sim (X, Y)$, where $\boldsymbol{x}$ is the feature and $y \in \{0, 1\}$ is the click label. Specifically, the features in this work consist of five parts: 1) user behavior sequence $\boldsymbol{x}^{seq}$; 2) user features $\boldsymbol{x}^u$ including user profile and user statistic features; 3) item features $\boldsymbol{x}^i$ such as item id, category, brand, and related statistic features; 4) context features $\boldsymbol{x}^c$ such as position and time information; 5) occasion time series $\boldsymbol{S}$, which is first proposed in this work and will be detailed below.

Now we detail the construction of the occasion time series $\boldsymbol{S}$. Similar but slightly different to global occasions discussed in [16], occasions in this work are more fine-grained time periods when users' behaviors change significantly. Thus, occasions derived from promotions are not explicit. To observe the changes of occasions, we calculate a set of statistics of the online business scenario (*e.g.*, active user amount, gross merchandise volume, and amount of add-to-cart behavior) with a proper sampling interval T, generating M occasion signals $\boldsymbol{x}^o = \{\boldsymbol{x}_1^o, ..., \boldsymbol{x}_i^o, ..., \boldsymbol{x}_M^o\}$. For each $\boldsymbol{x}_i^o$, we record it as a time-domain sequence and keep its recent N values, *i.e.*, $\boldsymbol{x}_i^o = [x_{i1}^o, x_{i2}^o, ..., x_{iN}^o]$. Combining all occasion signals in the form of sequences over time, we obtain the occasion time series $\boldsymbol{S} = [\boldsymbol{S}_1, ..., \boldsymbol{S}_t, ..., \boldsymbol{S}_N] \in \mathbb{R}^{M \times N}$, where $\boldsymbol{S}_t = [x_{1t}^o, x_{2t}^o, ..., x_{Mt}^o]$ denotes the value vector of occasion signals at the t-th time step. It's noteworthy that N and T should be set according to the characteristics of the specific business. For example, a large N and a small T are needed for the financial quantification business, which however is not necessary for relatively stable businesses such as e-commerce. In this work, we set T to 5 min and N to 96 so that occasion signals cover the time frame of last 8 h.

2.2 Model Architecture

Occasion Representation Network. Figure 2 presents the details of ORN. Firstly, we generate sub-sequences from the time series S by applying sliding windows to S with N_w as the window size and N_s as the stride, therefore obtaining a sequence of sub-sequences, *i.e.*,

$$S^s = [S_1^s, ..., S_t^s, ..., S_L^s], \ L = \lceil N/N_s \rceil,$$
$$S_t^s = [S_{t \times N_s}, ..., S_{t \times N_s + N_w - 1}] \in \mathbb{R}^{M \times N_w}, \tag{1}$$

where $\lceil \cdot \rceil$ refers to the round up operation and S_t^s denotes the t-th sub-sequence that consists of M occasion signals. Generally, a larger N_w and a smaller N_s are more beneficial for observation over the occasion evolution. Besides, it's suggested to choose N_w and N_s carefully so S_L^s can keep the same length as the other sub-sequences. Considering both effectiveness and efficiency, we set N_w to 24 and N_s to 6 in this work, so each sub-sequence covers a time frame of 2 h and the time gap between adjacent sub-sequences is half an hour.

Intuitively, with occasions changing, the frequency components of different sub-sequences are much more discriminative. Motivated by this, we perform N_f-point FFT on each occasion signal in each sub-sequence:

$$F = [F_1, ..., F_t, ..., F_L],$$
$$F_t = |FFT(S_t^s)| \in \mathbb{R}^{M \times N_f}, \tag{2}$$

where $|\cdot|$ denotes calculating the modulus of complex numbers. Then we flatten F_t so frequency spectrums of all occasion signals are merged into a 1-d vector:

$$\widetilde{F_t} = Flatten(F_t) \in \mathbb{R}^d, \ d = M \times N_f, \tag{3}$$

In this way, we obtain $\widetilde{F} = [\widetilde{F_1}, ..., \widetilde{F_t}, ..., \widetilde{F_L}] \in \mathbb{R}^{d \times L}$, a sequence of frequency spectrums of which each position can be treated as a raw representation of the corresponding sliding window.

At last, in the Occasion Evolution Layer (OEL), we apply LSTM to $\widetilde{F}$ to capture the global temporal dependency in the short-term sequence. The LSTM encodes the short-term sequence of frequency spectrums into a hidden output vector h_t at time t. With information from h_{t-1} flowing between cells, we can model the occasion evolution more comprehensively in a dynamic manner. The h_L is passed to subsequent neural structures as the occasion representation.

Experts Network. In this section, we adopt K experts to model the target item and user from multiple aspects, generating a set of feature representations $\{r_1, ..., r_K\}$ for the final prediction. The input of experts network consists of x^u, x^i, x^c and x^{seq}. All these features are processed by a Shared Embedding Layer to obtain their embeddings. In each expert, we perform three kinds of attention calculation on x^{seq}, as shown in Fig. 2. The processed x^{seq} is then concatenated with embeddings of x^u, x^i and x^c, and fed into the MainNet, *i.e.*,

Table 1. Statistics of the established datasets.

#Dataset	#Users	#Items	#Exposures	#Clicks	#Purchases
$\mathcal{D}$	5.85M	0.82M	1.32B	65.82M	819K
$\mathcal{D}_v$	1.87M	0.69M	0.18B	9.57M	91k

a Multi-Layer Perception (MLP). Meanwhile, we feed embeddings of $\boldsymbol{x}^u$ and $\boldsymbol{x}^c$ into another MLP (*i.e.*, the BiasNet) to model the bias that different users in different contexts usually behave differently even to similar items. Finally, the outputs of MainNet and BiasNet are concatenated as the output of an expert.

Mixture Prediction Network. In this section, we integrate the occasion representation and experts via an attention mechanism [15], *i.e.*, treating the occasion representation as Query and the output of experts as Key and Value, therefore obtaining the final occasion-adaptive feature representations to achieve promotion-aware CTR prediction.

As shown in Fig. 2, the final CTR prediction is formulated as:

$$\hat{y} = f_\theta(\boldsymbol{x}) = \mathcal{F}\left(\sum_{i=1}^{K} \alpha_i \cdot \boldsymbol{r}_i\right), \quad \alpha_i = \frac{exp(f_g(\boldsymbol{h}_L, \boldsymbol{r}_i))}{\sum_{j=1}^{K} exp(f_g(\boldsymbol{h}_L, \boldsymbol{r}_j))}, \quad (4)$$

where α_i, calculated via an attention mechanism, is the weight value for the i-th expert, and f_g is a function that projects the input into a scalar. $\mathcal{F}$ is a 3-layer MLP with a hidden unit size [144, 64, 1], of which the last layer uses Sigmoid as activation function while the other layers use ReLU. We adopt the widely-used logloss to train our MOEF model.

3 Experiments

3.1 Experimental Setup

Datasets. We establish the datasets by collecting the users' interaction logs[2] from our e-commerce platform, where promotions are highly frequent and have a considerable impact on the recommender system. Logs are sampled from 2020/10/01 to 2020/12/31, including Double 11, Black Friday, Double 12, and three member day promotions (21st of every month). The entire dataset is split into non-overlapped training set $\mathcal{D}$ (2020/10/01-2020/12/15) and validation set $\mathcal{D}_v$ (2020/12/16-2020/12/31). Both $\mathcal{D}$ and $\mathcal{D}_v$ cover normal days and promotion periods. Table 1 summarizes their statistics.

Evaluation Metrics. Area under ROC curve (AUC) is used as the offline evaluation metric. For online A/B testing, we choose CTR and average number of user clicks (IPV), which are widely adopted in industrial recommender systems.

---

[2] To the extent of our knowledge, there are no public datasets suited for this promotion-aware CTR prediction task.

Table 2. Offline and online comparison results. Bold: best. Underline: runner-up.

Model	Promotion			Normal		
	AUC	CTR Gain	IPV Gain	AUC	CTR Gain	IPV Gain
DeepFM	0.7140	–	–	0.7076	–	–
AutoFIS	0.7151	–	–	0.7098	–	–
AIM	0.7162	–	–	0.7129	–	–
DIEN	0.7174	−1.51%	−3.32%	0.7145	−0.87%	−1.72%
TIEN	0.7143	–	–	0.7140	–	–
HPMN	0.7289	−0.08%	−0.13%	<u>0.7265</u>	+1.27%	+2.64%
MOEF-1E	<u>0.7293</u>	0%	0%	0.7215	0%	0%
MOEF	**0.7457**	**+4.23%**	**+6.47%**	**0.7376**	**+4.61%**	**+6.96%**

Competitors. We compare our MOEF model with two classes of the previous methods, *i.e.*, methods that aim to capture high-order feature interactions, and methods based on sequential modeling of user behaviors. Specifically, **DeepFM** [1] and **DIEN** [20] are representatives in CTR prediction. **AutoFIS** [5] and **AIM** [22] combine the idea of feature interaction and AutoML, proposing to find useful feature interactions and decide how the interaction should be modeled. **HPMN** [10] adopts a hierarchical and periodical updating mechanism to capture multi-scale sequential patterns of user interests, and **TIEN** [4] models items in a dynamic manner by using item behaviors to strengthen the ability to predict users' emerging interests. Besides, **MOEF-1E**, a variant of MOEF that adopts one expert in the Experts Network, is also used for comparison.

Implementation Details. Input features for all the above competitors are the same except that TIEN additionally adopts item behaviors and DeepFM, AutoFIS, and AIM discard the user behavior sequence. In our experiments, AutoFIS uses DeepFM as base model and AIM is implemented without searching embedding size. Except for MOEF and its variants, other models treat occasion signals as context features and process them via log1p transformation [23].

We implement these deep learning models in distributed Tensorflow 1.4. During training, we use 3 parameter servers and 6 NVIDIA Tesla V100 16GB GPU workers. The number of points for FFT is 32 and the hidden unit size of LSTM is 96. Item ID, category ID and brand ID have an embedding size of 32 while 8 for the other categorical features. We use 8-head attention with a hidden unit size of 128 in each expert. Both MainNet and BiasNet are 3-layer MLPs, of which the hidden unit size is [480, 256, 128] and [96, 32, 16] respectively. Adagrad optimizer with a learning rate of 0.01 and a mini-batch size of 256 is used for training. The number of experts K varies from 2 to 5 and is set to 2 by default. We report the results of each method under its empirically optimal hyper-parameter settings.

3.2   Experimental Results

Offline and Online Comparison. To comprehensively evaluate model performance in both normal days and promotion periods, we divide $\mathcal{D}_v$ into two parts accordingly. Each offline evaluation is repeated 3 times and the average results are reported. Based on the offline evaluation, we chose several models and conducted online A/B testing lasting two weeks. Note that the time frame for online A/B testing covered promotion periods and normal days and MOEF-1E was used as the baseline model. The experimental results are presented in Table 2 and the major observations are summarized as follows. **1)** MOEF-1E outperforms DeepFM, AutoFIS, AIM and DIEN impressively, validating the importance of modeling occasion evolution since the main difference is that MOEF-1E process occasion signals with OEL while the other models treat occasion signals as regular context features. **2)** With periodic patterns of user interests captured, HPMN becomes the runner-up method in normal days. However, during promotions when changes of occasions are more frequent, it's not comparable to MOEF-1E and MOEF since it pays little attention to the occasion evolving process and fails to adapt to different occasions. TIEN delivers unsatisfactory performance because it's hard to train the huge user embedding parameters introduced by item behaviors. **3)** For both offline and online, MOEF yields the best performance with ORN guiding multiple experts to learn feature representations from different aspects. Compared to the baseline model (*i.e.*, MOEF-1E), MOEF improves 4.23% on CTR and 6.47% on IPV during promotion periods while 4.61% and 6.96% in normal days, which helps to recommend more attractive items and improves the amount of purchases.

Ablation Study. In this section, we conduct ablation experiments to study how each component in the ORN contributes to the final performance: 1) **MOEF_w/o_FFT** refers to MOEF that processes occasion signals in the time domain via log1p transformation [23] and then feeds them into a MLP to generate the occasion representation; 2) **MOEF_w/o_LSTM** refers to MOEF that uses a MLP instead of LSTM; 3) **MOEF-TE** refers to MOEF that uses Transformer-Encoder [15] instead of LSTM; and 4) **MOEF**. Each model is evaluated on $\mathcal{D}_v$ for 3 times and the AUCs are **0.7148±0.00147, 0.7304±0.00126, 0.7394±0.00118** and **0.7413±0.00122**, respectively. MOEF_w/o_LSTM significantly outperforms MOEF_w/o_FFT, validating the effectiveness of handling occasion signals in the frequency domain. Compared with MOEF_w/o_LSTM, MOEF further improves the performance by modeling the occasion evolution via LSTM, which helps to learn a better occasion representation in a dynamic manner. Moreover, we compare MOEF-TE with MOEF and observe a slight performance degradation when LSTM is replaced by Transformer-Encoder, although the latter is more efficient in computation. Empirically, we choose to use LSTM in our case since the sequence processed by the OEL is short; otherwise, we suggest using Transformer-Encoder instead.

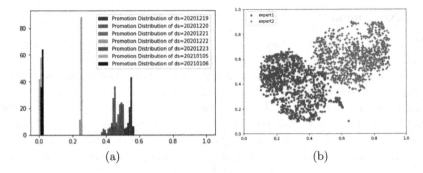

Fig. 3. (a) Distributions of weights in different days. (b) Visualization of feature representations from two experts.

Visualization of the ORN and Experts Network. In this section, we conduct a visual analysis on the ORN and Experts Network to study how they contribute to the final performance. Intuitively, we attempt to observe the differences of the ORN between different occasions to validate the effectiveness of ORN, which is not easy since occasions are not explicit time periods. However, we can alternatively observe the differences of the ORN between different kinds of days of which the occasions vary greatly. Specifically, we feed samples from different days into MOEF so that each sample generates a weight value α for one of the experts ($1 - \alpha$ for another). Then, distributions of α in different days are plotted with different colors as shown in Fig. 3(a), of which the x-axis refers to the value of α and the y-axis refers to the amount of samples. It can be observed that for different kinds of days, e.g., promotion periods (20201219-20201223) and normal days (20210105-20210106), the weights distinguish from each other obviously. While for the same kind of days, e.g., normal days, the weights are very close. Moreover, we find that days right before promotion (20201219-20201220), middle of promotion (20201222), and the other days of promotion (20201221 and 20201223) can be further distinguished. These observations show that the ORN can perceive fine-grained occasions rather than simply distinguish promotion periods from normal days. Therefore, MOEF can adapt to different occasions and handle the distribution uncertainty.

To validate the effectiveness of the Experts Network, we randomly sample hundreds of samples and feed them into MOEF. The output of the two experts is visualized using t-SNE [6]. As shown in Fig. 3(b), representations of the two experts clearly distinguish from each other, which implies that the Experts Network can learn feature representations from multiple aspects with multiple experts under the guidance of the ORN.

Influence of the Number of Experts. In this section, we explore the influence of the number of experts by training MOEF models with the number of experts varying from 1 to 5 and evaluating each model on $\mathcal{D}_v$ for 3 times. The AUCs are **0.7263±0.00097, 0.7413±0.00122, 0.7381±0.00141,**

0.7386±0.00157 and **0.7273±0.00184**, respectively. Intuitively, MOEF with two experts yields better performance than that with a single expert because multiple experts can help handle the mutual interference between non-identically distributed data and learn feature representations from multiple aspects. However, with the number of experts varying from 2 to 5, we observe a decreasing trend of AUC. We suspect that the performance degradation is caused by the increased model complexity introduced by more experts, which have more learnable parameters and may lead to over-fitting.

4 Conclusion

In this paper, we investigate the difficulties of CTR prediction in the context of frequent e-commerce promotions and propose a novel MOEF model. Our MOEF model mainly consists of the ORN and Experts Network, where the former learns the occasion representation and guides the latter to learn feature representations from multiple aspects using multiple experts. In this way, our MOEF model achieves promotion-aware CTR prediction and outperforms representative CTR methods on both real-world offline dataset as well as online A/B testing. The ablation study validates the effectiveness of modeling the occasion evolution in the frequency domain. Further, the visualization of weights distribution and feature representation confirms the effectiveness of our model design.

References

1. Guo, H., Tang, R., Ye, Y., Li, Z., He, X.: Deepfm: a factorization-machine based neural network for ctr prediction. In: IJCAI (2017)
2. Hochreiter, S., Schmidhuber, J.: Long short-term memory. Neural Comput. **9**(8), 1735–1780 (1997)
3. Jacobs, R.A., Jordan, M.I., Nowlan, S.J., Hinton, G.E.: Adaptive mixtures of local experts. Neural Comput. **3**(1), 79–87 (1991)
4. Li, X., Wang, C., Tong, B., Tan, J., Zeng, X., Zhuang, T.: Deep time-aware item evolution network for click-through rate prediction. In: CIKM, pp. 785–794 (2020)
5. Liu, B., et al.: Autofis: automatic feature interaction selection in factorization models for click-through rate prediction. In: KDD, pp. 2636–2645 (2020)
6. Van der Maaten, L., Hinton, G.: Visualizing data using t-sne. J. Mach. Learn. Res. **9**(11) (2008)
7. Niu, X., Li, B., Li, C., Tan, J., Xiao, R., Deng, H.: Heterogeneous graph augmented multi-scenario sharing recommendation with tree-guided expert networks. In: WSDM, pp. 1038–1046 (2021)
8. Pan, X., et al.: Metacvr: conversion rate prediction via meta learning in small-scale recommendation scenarios. In: SIGIR, pp. 2110–2114 (2022)
9. Pi, Q., Bian, W., Zhou, G., Zhu, X., Gai, K.: Practice on long sequential user behavior modeling for click-through rate prediction. In: KDD, pp. 2671–2679 (2019)
10. Ren, K., et al.: Lifelong sequential modeling with personalized memorization for user response prediction. In: SIGIR, pp. 565–574 (2019)
11. Sarwar, B., Karypis, G., Konstan, J., Riedl, J.: Item-based collaborative filtering recommendation algorithms. In: WWW, pp. 285–295 (2001)

12. Shen, Q., Tao, W., Zhang, J., Wen, H., Chen, Z., Lu, Q.: Sar-net: a scenario-aware ranking network for personalized fair recommendation in hundreds of travel scenarios. In: CIKM, pp. 4094–4103 (2021)
13. Sheng, X.R., et al.: One model to serve all: star topology adaptive recommender for multi-domain ctr prediction. In: CIKM, pp. 4104–4113 (2021)
14. Soliman, S.S., Srinath, M.D.: Continuous and discrete signals and systems. Englewood Cliffs (1990)
15. Vaswani, A., et al.: Attention is all you need. In: NeurIPS, pp. 5998–6008 (2017)
16. Wang, J., Louca, R., Hu, D., Cellier, C., Caverlee, J., Hong, L.: Time to shop for valentine's day: shopping occasions and sequential recommendation in e-commerce. In: WSDM, pp. 645–653 (2020)
17. Xiao, Z., Yang, L., Jiang, W., Wei, Y., Hu, Y., Wang, H.: Deep multi-interest network for click-through rate prediction. In: CIKM, pp. 2265–2268 (2020)
18. Yue, Z., et al.: Ts2vec: towards universal representation of time series. In: AAAI (2022)
19. Zhang, J., Tao, D.: Empowering things with intelligence: a survey of the progress, challenges, and opportunities in artificial intelligence of things. IEEE Internet Things J. 8(10), 7789–7817 (2020)
20. Zhou, G., et al.: Deep interest evolution network for click-through rate prediction. In: AAAI (2019)
21. Zhou, G., et al.: Deep interest network for click-through rate prediction. In: KDD (2018)
22. Zhu, C., et al.: Aim: automatic interaction machine for click-through rate prediction. IEEE Trans. Knowl. Data Eng. (2021)
23. Zhuang, H., Wang, X., Bendersky, M., Najork, M.: Feature transformation for neural ranking models. In: SIGIR, pp. 1649–1652 (2020)

Towards Efficient and Effective Transformers for Sequential Recommendation

Wenqi Sun[1,2], Zheng Liu[3], Xinyan Fan[1,2], Ji-Rong Wen[1,2],
and Wayne Xin Zhao[1,2(✉)]

[1] Gaoling School of Artificial Intelligence, Renmin University of China,
Beijing, China
{wenqisun,xinyan.fan,jrwen}@ruc.edu.cn, batmanfly@gmail.com
[2] Beijing Key Laboratory of Big Data Management and Analysis Methods,
Beijing, China
[3] Huawei Technologies Ltd., Co., Shenzhen, China
liuzheng107@huawei.com

Abstract. Transformer and its variants have been intensively applied for sequential recommender systems nowadays as they take advantage of the self-attention mechanism, feed-forward network (FFN) and parallel computing capability to generate the high-quality sequence representation. Recently, a wide range of fast, efficient Transformers have been proposed to facilitate sequence modeling, however, the lack of a well-established benchmark might lead to the non-reproducible and even inconsistent results across different works, making it hard to gain rigorous assessments. In this paper, We provide a benchmark for reproducibility and present a comprehensive empirical study on various Transformer-based recommendation approaches, and key techniques or components in Transformers. Based on this study, we propose a hybrid effective and Efficient Transformer variant for sequential Recommendation (ETRec), which incorporates the scalable long- and short-term preference learning, blocks of items aggregating as interests, and parameter-efficient cross-layer sharing FFN. Extensive experiments on six public benchmark datasets demonstrate the advanced efficacy of the proposed approach.

Keywords: Sequential Recommendation · Efficient Transformers

1 Introduction

Nowadays, sequential recommender systems have been widely deployed in many online services (*e.g.*, e-commerce, music and movies) for capturing dynamic user preference. For sequential recommendation, various approaches based on Markov Chains (MC), Recurrent Neural Networks (RNN) and Convolutional Neural Networks (CNN) have been proposed [10,15,20,23], and in recent years, Transformer [27] and its variants have become state-of-the-art approaches due to the powerful capacity of sequence modeling.

Compared with previous MC/RNN/CNN-based methods, Transformer-based recommendation methods (*e.g.*, SASRec [11] and BERT4Rec [22]) have three

X. Wang et al. (Eds.): DASFAA 2023, LNCS 13944, pp. 341–356, 2023.
https://doi.org/10.1007/978-3-031-30672-3_23

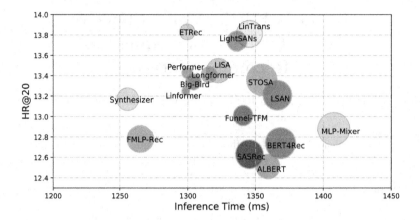

Fig. 1. Efficiency comparisons w.r.t the trade-off between performance (y axis), inference time (x axis) and memory cost (size of circles) on "Books".

major advantages at least. First, the receptive fields in the self-attention mechanism are global, and the representation of user behavior sequence can draw the context from all the user interactions in the past, which makes it more effective on obtaining long-term user preference than CNN-based methods. Second, the feed-forward network, which is often a multi-layer perceptron (MLP), endows the model with non-linearity, and the representation can consider interactions between different dimensions. Third, the representation can be computed in parallel, which makes it more efficient on the time-sensitive online services than RNN-based methods. Considering these merits of Transformers, there is an increasing interest in adapting various Transformers mainly originating from the Natural Language Processing (NLP) field to Recommender Systems (RS). For example, Transformers4Rec [21] provides an open-source library wrapping popular Transformer models (*e.g.,* BERT [5] and XLNet [30]) from HuggingFace[1] on the recommendation tasks.

However, as the recommendation models are deployed in many online services, their efficiency are more emphasized in the field of RS than NLP, therefore, not all models from NLP obtain perfect trade-off between effectiveness and efficiency for recommendations. Recently, a wide spectrum of improved Transformers have been proposed to enhance the efficiency of long sequence modeling. Specifically, several Transformer-based approaches [6,7,17,29] are proposed to advance both effectiveness and efficiency for modeling the dynamics of user behaviors. However, there is still a lack of comprehensive empirical study in term of both effectiveness and efficiency of various Transformer-based approaches, and key techniques in Transformers for recommendations.

---

[1] https://huggingface.co/.

Considering this issue, in this paper, we provide a high-level taxonomy, present a systematic empirical study w.r.t. the effect of various Transformer-based approaches and key techniques or components in Transformers for sequential recommendation, and then build a benchmark, which provides a comprehensive coverage for reproducible research. Furthermore, based on this study, we propose a hybrid effective and efficient Transformer variant ETREC, which incorporates the scalable long- and short-term preference learning, blocks of items aggregating as interests, and parameter-efficient cross-layer sharing FFN.

The major findings are summarized as follows: (1) The representative RNN/CNN-based recommendation methods are still competitive, however, the efficiency of RNN-based methods is limited to the parallel computing capability, and the effectiveness of CNN-based methods is limited to the receptive fields; (2) Low-rank decomposed or kernelized modifications can often achieve better performance with faster inference when adopting the self-attention architecture for recommendations; (3) It is usually effective to model both long- and short-term preference for improving the recommendation performance; (4) Cross-layer sharing FFN parameters can share the learning of item-item transition patterns across different layers with faster convergence on the recommendation tasks. In addition, we have some observations from experiments that a considerable number of models have far less performance improvement than expected and sometimes the reproducing experimental results are even inconsistent with the results reported in the literatures (similar to the finding in recent studies [3,34]).

Our main contributions are summarized as follows: (1) We conduct a comprehensive empirical study on various Transformer-based methods and key techniques in Transformers for recommendations; (2) We propose a hybrid effective and efficient Transformer variant ETREC. Extensive experiments conducted on six public benchmark datasets demonstrate the superior efficacy of ETREC over baselines; (3) We provide a benchmark of efficient Transformers for recommendations, and release our code[2] (both algorithm and evaluation) as additional resources for reproducibility, which will help improve the research in this field.

2   Related Work

Sequential Recommendation. Early works (*e.g.*, FPMC [20]) on sequential recommendation are primarily based on the Markov Chain assumption and focus on learning the item-item transitions to predict the next item. With the development of neural networks, RNN/CNN-based recommendation methods (*e.g.*, GRU4Rec [10], NARM [15], Caser [23], and NextItNet [31]) were introduced to capture sequential patterns. In recent years, Transformer-based methods have become state-of-the-art approaches for sequential recommendation due to powerful sequential modeling capacity [11,17,22]. Despite the success of Transformer-based recommendation methods, we find that these models can be further improved on both effectiveness and efficiency.

---

[2] https://github.com/RUCAIBox/ETRec.

Efficient Transformers. Transformers [27] are ubiquitously state-of-the-art across various tasks from NLP to RS [21,25] nowadays. However, they do not scale very well to long sequences due to the quadratic computational complexity of self-attention. Recently, a wide spectrum of fast, efficient Transformer-based models have been proposed to tackle this problem [2,6,24,28,32], such as Linformer (low-rank approximation) [28], Big-Bird (sparse patterns with local self-attention and global tokens) [32] and LISA (linear self-attention with codeword histogram) [29]. Most of these Transformers are originally proposed for NLP tasks, and there is an increasing interest to adapt them to sequential recommendation considering the similar task formulations in essence (*i.e.*, token or identifier sequence). As a representative work, Transformers4Rec [21] provides an open-source library wrapping popular Transformers for recommendations. However, there are almost no systematic analysis and evaluation on the effect of different Transformer-based approaches, and key techniques or components in Transformers for sequential recommendation, which often lacks a comprehensive consideration on both effectiveness and efficiency.

3 Task Formulation

Sequential recommendation aims to predict the next item that a user is likely to interact with, which is based on the chronological sequence of the user's historical interactions. Formally, given a set of users $\mathcal{U}$, a set of items $\mathcal{V}$, and their associated interactions, we sort the interacted items of each user $u \in \mathcal{U}$ chronologically to form a n-length sequence as $S^u = [v_1, v_2, \ldots, v_n]$, where $v \in \mathcal{V}$ denotes the i-th interacted item in the sequence. Following previous works [11,22], the goal of sequential recommendation is to recommend a top-K ranking list as the potential next items by predicting $p(v_{n+1}|v_1, v_2, \ldots, v_n)$.

In this work, we only consider the most widely adopted input setting, which are sequences of interacted item identifiers (IDs), since existing Transformer-based sequential recommendation methods are mainly designed based on this setting. Therefore, the usage and discussion of other input (*e.g.*, side information) are beyond the scope of this work.

4 Taxonomy of Transformers for Recommendation

In this section, we first analyze the features of Transformer-based methods for sequential recommendation, divide all methods into seven different classes, and provide a high-level taxonomy. Furthermore, we analyze their computational complexity (*i.e.*, total computational cost of self-attention and feed-forward network per layer). The summary w.r.t. features and computational complexity is shown in Table 1. Note that there are similar task formulations on sequence modeling in the field of NLP and RS [8,21], therefore, our taxonomy and benchmark include several sequence modeling methods from NLP. We will analyze various classes of methods in detail next.

Table 1. Summary w.r.t. features and computational complexity of various Transformer-based methods for sequential recommendation. Features in order are: full self-attention (**F**), low-rank approximation (**L**), kernelization (**K**), local self-attention (**C**), scaling layers (**S**), all-MLP (**M**), and additional techniques (**T**). The "✔" symbol indicates the presence of features.

Year	Model	F	L	K	C	S	M	T	Complexity per Layer	Class
2018	SASRec [11]	✔							$\mathcal{O}(N^2D + ND^2)$	VT
2019	BERT4Rec [22]	✔							$\mathcal{O}(N^2D + ND^2)$	VT
2019	ALBERT [14]	✔						✔	$\mathcal{O}(T^2D + TD^2)$	VT
2020	Linformer [28]		✔						$\mathcal{O}(NKD + ND^2)$	LR
2020	Synthesizer [24]		✔						$\mathcal{O}(NKD + ND^2)$	LR
2020	LinTrans [12]			✔					$\mathcal{O}(N^2R + ND^2)$	KN
2021	Performer [2]			✔					$\mathcal{O}(N^2R + ND^2)$	KN
2021	Longformer [1]				✔				$\mathcal{O}(N(W+G)D + ND^2)$	LW
2021	Big-Bird [32]				✔				$\mathcal{O}(N(W+G+M)D + ND^2)$	LW
2020	Funnel-TFM [4]	✔				✔			$\mathcal{O}((\frac{N}{C^\ell})^2D + (\frac{N}{C^\ell})D^2)$	SL
2021	MLP-Mixer [26]						✔		$\mathcal{O}(N^2D + ND^2)$	AM
2022	FMLP-Rec [36]						✔	✔	$\mathcal{O}(ND\log N + ND^2)$	AM
2021	LightSANs [6]		✔					✔	$\mathcal{O}(NKD + ND^2)$	HT
2021	LSAN [17]	✔						✔	$\mathcal{O}(N^2D + ND^2)$	HT
2021	LISA [29]	✔						✔	$\mathcal{O}(NBWD + ND^2)$	HT
2022	STOSA [7]	✔						✔	$\mathcal{O}(ND + N^2D + 4N^2 + \frac{ND^2}{2})$	HT
2023	ETRec (Ours)		✔		✔	✔		✔	$\mathcal{O}(2^{L-\ell}(\frac{N}{2^{L-\ell}})K_\ell D + ND^2)$	HT

Vanilla Transformer (VT). The vanilla Transformer based methods are the fundamental baseline methods leveraging the multi-head self-attention, feed-forward network and position encoding, which can model the dynamics from user's chronological behaviors. SASRec [11] and BERT4Rec [22] utilize the directional and bidirectional full self-attention respectively to explore the implicit user preference presentation, and each item in the sequence attends to the attention weight of all other items to generate the high-capability context. Therefore, the computational complexities of their self-attention and feed-forward network are $\mathcal{O}(N^2D)$ and $\mathcal{O}(ND^2)$ respectively, where N is the item sequence length, and D is the embedding dimension. ALBERT [14] leverages the factorized embedding parameterization with the N to T projection on the item embedding.

However, such a computation of full self-attention is confronted with the quadratic computational complexity of self-attention [1,32] and vulnerability to over-parameterization on the recommendation tasks [6]. To alleviate the above issues, the previous Transformer-based works have carried out various explorations including the low-rank decomposition, kernelization, local-window self-attention, scaling-layer self-attention, all-MLP, and hybrid Transformer.

Low-Rank Decomposed Self-attention (LR). The low-rank decomposed self-attention based methods leverage the low-rank approximation to compress the size of the attention matrix to facilitate the efficiency. These methods [24, 28] generally utilize N to K mapping to reduce the computational complexity ($K \ll N$ in general). Specifically, Linformer [28] reduces the dimension of "key" and "value" in Transformer from N to K via $N \times K$ linear mappings. Synthesizer [24] leverages the synthetic attention weights, which are factorizations of two randomly initialized low-rank matrices.

Kernelized Self-attention (KN). The kernelized self-attention based methods leverage the kernelization to compress the scale of the attention matrix to advance the efficiency, which can be viewed as a form of low-rank approximation [25]. Specifically, Performer [2] and Linear Transformer (LinTrans) [12] reduce the cardinality of hidden dimension from D to R via kernelization. Intuitively, the number of a user' latent interests is less than or equal to the number of interactions, therefore, the majority of user's interactions can be categorized into several latent interests via low-rank decomposed or kernelized self-attention.

Local-Window Self-attention (LW). The local-window self-attention based methods model each local interval of sequence by sliding windows with the fixed stride size of W. Longformer [1] and Big-Bird [32] combine local self-attention and global tokens with the size of G, acting as a form of memory that learns to gather from local interval of sequence via sliding windows, in which Big-Bird uses the random attention (with the size of M) additionally. Intuitively, the local-window self-attention focus more on the local interval thereby to learn more high-quality local patterns of the user's historical interactions.

Scaling-Layer Self-attention (SL). The scaling-layer self-attention based methods achieves the scalable receptive fields of self-attention by pooling (*e.g.*, Funnel-TFM [4]), selecting the most-informative item representation (*e.g.*, Pyramidion [19]) to the next layer, or adjusting the range of the receptive fields according to the different layers. These methods are effective to obtain good scalability. In particular, Funnel-TFM can gradually compress the sequence of hidden states by pooling to a shorter one with the compression ratio of C between layers.

All-MLP (AM). The feed-forward network in Transformers, which is often a multi-layer perceptron (MLP), endows the model with non-linearity and models interactions in different latent dimensions. All-MLP based methods (*e.g.*, MLP-Mixer [26], FMLP-Rec [36] and MLP4Rec [16]) attempt to leverage MLPs only without self-attention to advance the effectiveness and efficiency, which can be viewed as a form of Transformer's variant. Specifically, MLP-Mixer aim to devise effective MLP-based token-mixing and channel-mixing architectures to capture the interactions of the sequence, in which the function of token-mixing is similar to the one of self-attention to some extent. FMLP-Rec leverages MLPs with learnable filters for sequential recommendation.

Hybrid Transformer (HT). The hybrid Transformer based methods generally introduce the extra techniques or components (*e.g.,* the codeword histogram [29], the twin-attention [17] and the Wasserstein self-attention [7]) to adapt to different sequence data. In particular, LightSANs [6] leverages the low-rank decomposed self-attention with the decoupled position encoding. LSAN [17] utilizes the combined networks of Transformer and CNN with a temporal context-aware embedding. LISA [29] obtains the differentiable histograms of codeword distributions on items to scale self-attention linearly. However, these methods mostly designed for different data distributions may be confronted with significant performance fluctuations across different datasets.

5   Methodology

In this section, we introduce a hybrid Transformer variant ETREC, which enhances the effectiveness and efficiency of Transformer-based recommendations.

5.1   Overview

With the theoretical analysis and empirical study w.r.t. existing Transformers for recommendations, we find that three key points are important to consider for sequential recommendation. First, there are different data distributions for different datasets, which requires a scalable architecture to adapt to different user behavior sequences. Second, learning both user's long- and short-term interests can mostly obtain better performance in recommendations. Third, it is important to discover latent interests from a large number of interacted items for user preference modeling. Therefore, we enhance the above points with following improvements: (1) A scalable long- and short-term layer, in which we divide the item sequence on average into blocks with a dynamic block size; (2) A block aggregation as interest layer, which can aggregate a block of items as interests by low-rank decomposed self-attention; (3) Parameter-efficient cross-layer sharing FFN, which can share the learning of item-item transition patterns across different Transformer layers with faster convergence on the recommendation tasks. The overall framework of our proposed ETREC is depicted in Fig. 2.

5.2   ETREC: A Hybrid Transformer Variant

Embedding Layer. In the embedding layer, we maintain an item embedding matrix $M_\mathcal{V} \in \mathbb{R}^{|\mathcal{V}| \times D}$ to project the high dimensional one-hot representation of each item to the low dimensional dense representation. Given a user behavior sequence including N items, we employ look-up operation from $M_\mathcal{V}$ to form the item sequence embedding matrix $E \in \mathbb{R}^{N \times D}$. Besides, we incorporate a learnable position encoding matrix $P \in \mathbb{R}^{N \times D}$ to facilitate the presentation of item sequence. The final sequence representation E_N can be obtained by summing the two embedding matrices with the layer-norm and dropout:

$$E_N = \text{Dropout}(\text{LayerNorm}(E + P)) \tag{1}$$

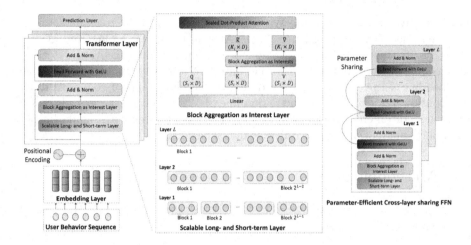

Fig. 2. An overview of the proposed ETREC.

Scalable Long- and Short-term Preference Layer. We divide the sequence of length N on average into blocks in each Transformer layer. In the ℓ-th layer of total L layers, the number of blocks is $2^{L-\ell}$ with the block size of $\frac{N}{2^{L-\ell}}$. If it is not of integer division, the remaining items in the last will be merged into the last block. Therefore, the block size will gradually increase with growing ℓ, until the block size is equal to N in the last layer. In this way, both long- and short-term user preference learning with a scalable architecture.

Block Aggregation as Interests. We first convert the input embedding sequence in this block $\boldsymbol{E}_{S_i}$ into three matrices $\boldsymbol{Q}_{S_i}, \boldsymbol{K}_{S_i}, \boldsymbol{V}_{S_i} \in \mathbb{R}^{S_i \times D}$ through linear projections $\boldsymbol{W}_Q, \boldsymbol{W}_K, \boldsymbol{W}_V \in \mathbb{R}^{D \times D}$, and feed them into the next module, where S_i is the number of items in the i-th block. Assuming that the majority of user's S_i items in the i-th block can be categorized with no more than K_i latent interests ($K_i \leq S_i$), We then leverage the block-aggregation-as-interests operation, which is a learnable projection function $f : \mathbb{R}^{S_i \times D} \to \mathbb{R}^{K_i \times D}$, to aggregate items in the i-th block to latent interests (D is cardinality of hidden dimension):

$$\widetilde{\boldsymbol{H}} = f(\boldsymbol{H}) = (\text{softmax}(\boldsymbol{H} \cdot \boldsymbol{\Theta}^T))^T \cdot \boldsymbol{H}, \qquad (2)$$

where $\boldsymbol{\Theta} \in \mathbb{R}^{K_i \times D}$ is a learnable parameter, and $\boldsymbol{H}$ is $\boldsymbol{K}_{S_i}$ or $\boldsymbol{V}_{S_i}$. Next we project the original $\boldsymbol{K}_{S_i}$ and $\boldsymbol{V}_{S_i}$ to $\widetilde{\boldsymbol{K}}_{S_i}$ and $\widetilde{\boldsymbol{V}}_{S_i}$ via block-aggregation-as-interests and then adapt the scaled dot-product attention to them:

$$\widetilde{\boldsymbol{R}}_{S_i} = \text{softmax}(\frac{\boldsymbol{Q}_{S_i} \cdot \widetilde{\boldsymbol{K}}_{S_i}^\top}{\sqrt{d/h}})\widetilde{\boldsymbol{V}}_{S_i} = \text{softmax}(\frac{\boldsymbol{Q}_{S_i} \cdot f(\boldsymbol{K}_{S_i})^\top}{\sqrt{d/h}})f(\boldsymbol{V}_{S_i}). \qquad (3)$$

Then all representations $\widetilde{\boldsymbol{R}}_{S_i}$, where $i \in \{1, \ldots, \frac{N}{2^{L-\ell}}\}$, will be concatenated as the final output $\widetilde{\boldsymbol{R}}^\ell$ in the ℓ-th layer.

Parameter-Efficient Cross-Layer Sharing FFN. Cross-layer parameter sharing on Transformer-based architectures (*e.g.*, ALBERT [14]) are popular means to obtain the parameter-efficient training with faster convergence. There are multiple sharing combinations, such as sharing FFN parameters across different layers, sharing self-attention parameters and sharing both self-attention and FFN parameters. Intuitively, sharing FFN across different layers can transfer the learning of item-item transition patterns on the recommendation tasks. Specifically, the structure of FFN is denoted as:

$$\widetilde{\boldsymbol{F}}^{\ell} = \text{FFN}(\widetilde{\boldsymbol{R}}^{\ell}) = (\text{GeLU}(\boldsymbol{W}_1\widetilde{\boldsymbol{R}}^{\ell} + \boldsymbol{b}_1))\boldsymbol{W}_2 + \boldsymbol{b}_2, \tag{4}$$

where $\widetilde{\boldsymbol{R}}^{\ell}$ means the output of low-rank decomposed self-attention after the residual connection and layer normalization in the ℓ-th layer, and the parameters in FFN (*i.e.*, $\boldsymbol{W}_1$, $\boldsymbol{b}_1$, $\boldsymbol{W}_2$ and $\boldsymbol{b}_2$) share across all layers.

5.3   Model Training

Prediction Layer. We utilize a fully-connected feed-forward network to $\widetilde{\boldsymbol{R}}^{\ell}$ and obtain the result $\widetilde{\boldsymbol{F}}^{\ell}$. With the first t items encoded, the next item is predicted based on the last layer's output $\widetilde{\boldsymbol{F}}_t^{L}$ of the t-th item, where L is the number of layers. We use inner-product to measure user preference on an arbitrary item i:

$$\text{P}(i|\{i_1^u, ..., i_t^u\}) = \langle\widetilde{\boldsymbol{F}}_t^{L}, \boldsymbol{E}_i\rangle. \tag{5}$$

Learning Objective. We adopt the cross-entropy loss to train the model:

$$\mathcal{L} = -\log\frac{\exp{(\langle\widetilde{\boldsymbol{F}}_t^{L}, \boldsymbol{E}_g\rangle)}}{\Sigma_{i=1}^{|\mathcal{V}|}\exp{(\langle\widetilde{\boldsymbol{F}}_t^{L}, \boldsymbol{E}_i\rangle)}} \tag{6}$$

where item g is the ground truth item, and $|\mathcal{V}|$ is the number of all items.

Time and Space Complexity. The computational complexity of ETREC is dominated by the scalable long- and short-term preference layer, block aggregation as interest layer, and feed-forward networks. In the ℓ-th layer of total L layers, a block of items aggregate as K_{ℓ} interests, and the overall time complexity is $\mathcal{O}(2^{L-\ell}(\frac{N}{2^{L-\ell}})K_{\ell}D + ND^2)$. The learnable parameters in ETREC are from item embedding, self-attention, FFN and layer normalization, and the overall space complexity is $\mathcal{O}(|\mathcal{V}|D + 2^{L-\ell}((\frac{N}{2^{L-\ell}})K_{\ell} + K_{\ell}D) + D^2)$.

Discussion. In this part, we compare ETREC with existing Transformer-based recommendation methods. Compared with vanilla Transformers, ETREC learns both long- and short-term preference with lower computational complexity. Compared with low-rank decomposed or kernelized self-attention, ETREC can learn the short-term preference and transfer the learning of item-item transition patterns by sharing FFN parameters across different layers. Compared with all-MLP, ETREC can model the long- and short-term preference more scalable with the self-attention mechanism. Compared with existing Hybrid Transformers, ETREC achieve a scalable architecture adapting to more different data distributions at low computational cost.

Table 2. Statistics of the datasets after preprocessing. "Avg. n" denotes the average length of user behavior sequences.

Dataset	#Users	#Items	#Interactions	Avg. n	Sparsity
ML-1M	$6,040$	$3,629$	$836,478$	138.51	96.18%
ML-25M	$35,475$	$5,907$	$6,370,892$	179.59	96.96%
Yelp	$56,590$	$75,159$	$2,290,516$	40.48	99.95%
Books	$19,214$	$60,707$	$1,733,934$	90.24	99.85%
Beauty	$10,082$	$64,172$	$177,282$	17.59	99.97%
Tmall	$28,805$	$22,886$	$4,746,961$	164.80	99.28%

6   Experiments

In this section, we validate the effectiveness and efficiency of the existing methods and our proposed ETREC on the recommendation tasks.

6.1   Experimental Setup

Datasets and Evaluation Metrics. We conduct experiments on six public benchmark datasets (*i.e.*, ML-1M, ML-25M [9], Yelp, Amazon-Books, Amazon-Beauty [18], and Tmall). The statistical details of the preprocessed datasets are shown in Table 2. For each dataset, we group interactions by user IDs, and then generate a chronological item sequence for each user. Following previous works [11,22], we use the leave-one-out strategy and employ hit ratio (HR) and normalized discounted cumulative gain (NDCG) on top-K ranked items to evaluate the performance. As suggested in [13], we rank all items in test sets instead of sampling evaluation. In addition, we utilize floating point operations (FLOPs) to evaluate efficiency from the perspective of computational cost.

Compared Approaches and Implementation Details. We compare the representative MC/RNN/CNN-based methods and the selective Transformer-based methods in various classes as shown in Sect. 4. All models are implemented on an open-source recommendation framework RecBole [33,35] in PyTorch. With regard to fair comparisons, we set item embedding with the dimension D of 64, the attention heads with the number H of 2, and the layers with the number L of 2. All trainable parameters in these models are optimized by leveraging Adam optimiser with the batch size of 1024, learning rate of 0.001, drop rate of 0.1, and L2 regularisation strength λ of $1e-5$.

6.2   Overall Performance

Table 3 lists the evaluation results of different methods including the representative MC/RNN/CNN-based, Transformer-based methods, and our proposed ETREC on six public benchmark datasets. We have following observations:

Table 3. Overall performance (%) of different recommendation methods. The best and second best methods are denoted in bold and underlined fonts, respectively. "*" denotes that the improvements are significant over baselines (t-test with $p < 0.05$).

Model	ML-1M		Yelp		Books	
	HR@10	NDCG@10	HR@10	NDCG@10	HR@10	NDCG@10
FPMC [20]	13.11	6.05	2.64	1.51	8.03	4.16
NARM [15]	21.73	11.29	3.80	1.93	8.23	4.69
SASRec [11]	21.18	10.55	5.20	2.84	8.38	4.25
BERT4Rec [22]	21.12	10.39	4.89	2.62	8.41	4.37
Linformer [28]	25.35	13.46	4.96	2.70	9.12	5.02
Performer [2]	24.13	12.76	5.04	2.69	9.32	5.28
LinTrans [12]	24.53	12.97	4.92	2.67	9.38	5.33
Big-Bird [32]	24.67	12.69	4.94	2.63	9.29	5.29
Funnel-TFM [4]	26.89	15.01	4.39	2.31	8.97	4.63
MLP-Mixer [26]	23.77	12.40	5.36	2.82	8.27	4.17
LightSANs [6]	25.22	13.42	5.25	2.84	9.52	5.31
LSAN [17]	25.63	13.89	4.95	2.69	8.92	4.89
ETRec	**27.52***	**15.42***	**5.46**	**2.94***	**9.71***	**5.43***
Improv	+2.3%	+2.7%	+1.9%	+3.5%	+2.0%	+2.4%
Model	ML-25M		Beauty		Tmall	
	HR@10	NDCG@10	HR@10	NDCG@10	HR@10	NDCG@10
FPMC [20]	7.36	3.79	3.73	2.31	28.14	22.73
NARM [15]	11.79	6.02	3.81	2.37	36.87	30.18
SASRec [11]	10.48	5.21	4.63	2.49	37.82	30.48
BERT4Rec [22]	10.21	5.19	4.67	2.57	35.98	30.04
Linformer [28]	13.74	7.06	4.82	2.70	37.91	30.29
Performer [2]	13.11	6.57	4.31	2.43	38.94	31.04
LinTrans [12]	13.11	6.89	4.41	2.46	36.79	30.14
Big-Bird [32]	12.86	6.44	4.53	2.62	36.65	30.19
Funnel-TFM [4]	13.61	6.89	4.34	2.42	38.24	31.09
MLP-Mixer [26]	12.95	6.84	3.27	1.79	37.92	30.71
LightSANs [6]	13.67	6.98	4.59	2.58	36.74	30.09
LSAN [17]	12.94	6.79	4.55	2.54	37.94	30.59
ETRec	**13.98**	**7.29***	**5.01***	**2.92***	**42.74***	**34.23***
Improv.	+1.7%	+3.3%	+3.9%	+8.1%	+9.8%	+10.3%

(1) ETREC achieves the best overall results and displays its superiority to the other methods. The relative improvements range from 1.7% to 10.3%, demonstrating the superiority of our proposed method. We attribute the improvements to several factors: first, the low-rank decomposed self-attention aggregates a

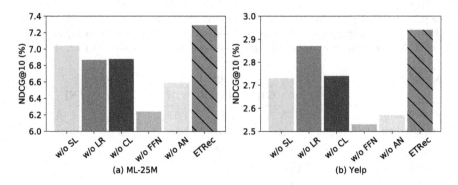

Fig. 3. Ablation study of ETREC's variants on "ML-25M" and "Yelp".

block of interacted items to a small number of latent interests because the number of latent interests is less than or equal to the number of interactions in a block intuitively. Second, the scalable long- and short-term layer achieves the flexible long- and short-term preference learning, thus adapting to more different data distributions. Third, cross-layer sharing parameters of FFN can transfer the learning of item-item transition patterns across different layers.

(2) The overall comparisons w.r.t. various classes of methods are as follows. First, the representative RNN/CNN-based methods are still competitive, however, the efficiency of RNN-based methods is limited to the parallel computing capability, and the effectiveness of CNN-based methods is limited to the receptive fields. Second, vanilla Transformers are the baselines of next discussion on efficient Transformers. Low-rank decomposed or kernelized self-attention based methods outperform vanilla Transformer based methods across almost all evaluation metrics on ML-1M, ML-25M and Tmall, demonstrating that low-rank decomposed or kernelized modifications boost the performance on the relatively dense user behavior datas. Scaling-layer self-attention based methods obtain good performance on relatively dense user behavior datas. Hybrid Transformers are competitive, however, each of them performs well on a few datasets, indicating that these methods mostly designed for different data distributions are confronted with significant performance fluctuations across different datasets. On the whole, low-rank approximation or kernelization can often obtain better performance with faster inference, and it is usually effective to model both long- and short-term preference for improving the recommendation performance.

6.3 Further Analysis

Ablation Study. In this part, we aim to evaluate how each of the proposed techniques or components affects the final performance. We design five variants of the proposed ETREC model by removing or replacing certain components for comparisons, including: (1) w/o SL: without scalable long- and short-term layer. (2) w/o LR: without block aggregation as interests by low-rank approximation. (3) w/o CL: without cross-layer sharing parameters of FFN (*i.e.,* keeping

Table 4. Performance, #Parameters and #FLOPs comparisons w.r.t. Transformer's variants on "ML-1M". "↑" and "↓" denote higher and lower performance respectively compared with SASRec.

Variant	HR@10	NDCG@10	# Parameters	# FLOPs
SASRec	21.18	10.55	0.34M	28.91G (1.00x)
Intermediate Width (×1)	20.27↓	9.67↓	0.29M	22.64G (0.78x)
Intermediate Width (×2)	20.61↓	10.21↓	0.30M	24.78G (0.86x)
Intermediate Width (×8)	21.45	10.52↓	0.37M	37.30G (1.29x)
Cross-layer Self-attention	20.37↓	10.12↓	0.31M	28.91G (1.00x)
Cross-layer FFN	22.42	11.31	0.30M	28.91G (1.00x)
Cross-layer Both	20.48↓	10.39↓	0.28M	28.90G (1.00x)
Removing Position	19.13↓	9.37↓	0.33M	28.90G (1.00x)
Relative Position	21.09↓	10.57	0.34M	28.91G (1.00x)
Decoupled Position	24.15	12.81	0.35M	34.43G (1.19x)

FFN independently per layer). (4) w/o FFN: without feed-forward network. (5) w/o AN: without skip-connection and normalization.

The experimental results of ETREC and its variants are reported in Fig. 3. We can observe that all the proposed techniques or components are useful to improve the recommendation performance. The performance of the variant w/o FFN and w/o AN are poor because the skip-connection, layer normalization and feed-forward network are the necessary techniques of Transformer-based architectures. The performance degradations of w/o SL, w/o LR and w/o CL indicate that scalable long- and short-term layer, block aggregation as interests layer and cross-layer sharing FFN are effective on the recommendation tasks.

Key Technique Analysis in Transformers. In this part, we aim to analyze the key techniques or components in Transformers. We construct three types of Transformer's variants, compare them with the vanilla Transformer (*i.e.*, SAS-Rec with $L = 2$, "intermediate width" of "hidden dimension"×4 in FFN, independent FFN per layer, and absolute position encoding).

From Table 4, We have following observations: first, the intermediate width in FFN varies from D (hidden dimension) ×4 to ×1, ×2 and ×8. In the settings of $D = 1$ and $D = 2$, performance drops significantly with decreased FLOPs. In the setting of $D = 8$, there is little performance gain with greatly increased FLOPs. Hence, the setting of $D = 4$ is appropriate for most of recommendations. Second, cross-layer sharing parameters of self-attention, FFN and both (self-attention and FFN) indicate that cross-layer sharing FFN is effective and parameter-efficient to boost performance with faster convergence for recommendations. Third, the results on removing, relative, decoupled position encoding instead of absolute position encoding demonstrate that there is significant performance degradation without position encoding, therefore, position encoding

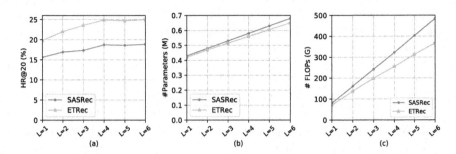

Fig. 4. Performance, #Parameters and #FLOPs comparisons as varying the number of layers (L) from 1 to 6 on "ML-25M".

Table 5. Efficiency comparisons w.r.t. training and inference speed on "Yelp".

Model	Time/Epoch	#Epoch	Training Time	Inference Time
SASRec	324.21 s	51	275.58 min	3.80 s
Linformer	308.84 s	49	252.22 min	3.54 s
Funnel-TFM	287.97 s	47	225.58 min	3.56 s
MLP-Mixer	373.67 s	42	261.57 min	3.89 s
LightSANs	320.82 s	54	288.74 min	3.57 s
ETREC	303.79 s	39	197.46 min	3.48 s

is necessary in Transformers. The performance of decoupled position encoding outperforms the other position encoding with increasing computational cost.

Efficiency Analysis. We analyze efficiency of different Transformer-based methods w.r.t. trade-off between performance, inference time, and memory cost on "Books" as shown in Fig. 1, and training and inference speed of selected methods on "Yelp" as listed in Table 5. Furthermore, we analyze performance and efficiency including SASRec and ETREC as the number of layers vary from $L = 1$ to 6 on "ML-25M" (shown in Fig. 4).

We have following observations: first, ETREC's training and inference speed (*i.e.*, Time/Epoch, Training Time and Inference Time) are close to Linformer, and it obtains fast inference and just needs 39 epochs to converge, which is much less than SASRec, leading to only 197.46 min for total training (around 1.4x and 1.5x speedup compared with SASRec and LightSANs respectively), demonstrating the advanced efficiency of ETREC. The empirical results are generally in line with the theoretical computational complexity as highlighted in Sect. 4. Second, with the number of layers (L) varying from 1 to 6, the performance gain is very small after more than 2 layers, however, #FLOPs increase almost linearly. Hence, $L = 2$ as the number of layers is efficient for most of Transformer-based recommendation methods. Third, the number of parameters mainly depends on the item embedding for sequential recommendation, leading to small change of #parameters with varying of layers.

7 Conclusion and Future Work

In this paper, We conduct a comprehensive empirical study of various Transformer-based recommendation methods and key techniques in Transformers, and present several important findings to improve Transformer-based recommendation methods. Based on this study, we propose a hybrid effective and efficient Transformer variant ETREC. Extensive experiments demonstrate the superior efficacy of ETREC over baselines. Furthermore, we provide a benchmark of efficient Transformers for sequential recommendation, in which we implement all models under a unified framework, and release our codes as an additional resource for reproducibility, which will help improve the research in this field. As future work, we will explore both effectiveness and efficiency of Transformer-based recommendation methods with other input (*e.g.,* item descriptions).

Acknowledgements. This work was partially supported by National Natural Science Foundation of China under Grant No. 62222215, Beijing Natural Science Foundation under Grant No. 4222027, and Beijing Outstanding Young Scientist Program under Grant No. BJJWZYJH012019100020098. Xin Zhao is the corresponding author.

References

1. Beltagy, I., Peters, M.E., Cohan, A.: Longformer: the long-document transformer. CoRR abs/2004.05150 (2020)
2. Choromanski, K.M., Likhosherstov, V., Dohan, D., Song, X., Gane, A., et al.: Rethinking attention with performers. In: ICLR (2021)
3. Dacrema, M.F., Cremonesi, P., Jannach, D.: Are we really making much progress? A worrying analysis of recent neural recommendation approaches. In: RecSys (2019)
4. Dai, Z., Lai, G., Yang, Y., Le, Q.: Funnel-transformer: filtering out sequential redundancy for efficient language processing. In: NeurIPS (2020)
5. Devlin, J., Chang, M., Lee, K., Toutanova, K.: BERT: pre-training of deep bidirectional transformers for language understanding. In: NAACL (2019)
6. Fan, X., Liu, Z., Lian, J., Zhao, W.X., et al.: Lighter and better: low-rank decomposed self-attention networks for next-item recommendation. In: SIGIR (2021)
7. Fan, Z., e al.: Sequential recommendation via stochastic self-attention. In: TheWebConf (2022)
8. Geng, S., Liu, S., Fu, Z., et al.: Recommendation as language processing (RLP): a unified pretrain, personalized prompt & predict paradigm (P5). In: RecSys (2022)
9. Harper, F.M., Konstan, J.A.: The movielens datasets: history and context. ACM Trans. Interact. Intell. Syst. (2016)
10. Hidasi, B., Karatzoglou, A., Baltrunas, L., Tikk, D.: Session-based recommendations with recurrent neural networks. In: ICLR (2016)
11. Kang, W.C., et al.: Self-attentive sequential recommendation. In: ICDM (2018)
12. Katharopoulos, A., Vyas, A., Pappas, N., Fleuret, F.: Transformers are RNNs: fast autoregressive transformers with linear attention. In: ICML (2020)
13. Krichene, W., Rendle, S.: On sampled metrics for item recommendation. In: SIGKDD (2020)

14. Lan, Z., Chen, M., Goodman, S., Gimpel, K., Sharma, P., Soricut, R.: Albert: a lite bert for self-supervised learning of language representations. In: ICLR (2020)
15. Li, J., Ren, P., Chen, Z., Ren, Z., Lian, T., Ma, J.: Neural attentive session-based recommendation. In: CIKM (2017)
16. Li, M., Zhao, X., Lyu, C., Zhao, M., Wu, R., Guo, R.: MLP4Rec: a pure MLP architecture for sequential recommendations. In: IJCAI (2022)
17. Li, Y., Chen, T., Zhang, P.F., Yin, H.: Lightweight self-attentive sequential recommendation. In: CIKM (2021)
18. McAuley, J., Targett, C., Shi, Q., Van Den Hengel, A.: Image-based recommendations on styles and substitutes. In: SIGIR (2015)
19. Pietruszka, M., Borchmann, L., Gralinski, F.: Sparsifying transformer models with trainable representation pooling. In: ACL (2022)
20. Rendle, S., Freudenthaler, C., Schmidt-Thieme, L.: Factorizing personalized markov chains for next-basket recommendation. In: WWW (2010)
21. de Souza Pereira Moreira, G., Rabhi, S., et al.: Transformers4Rec: bridging the gap between nlp and sequential/session-based recommendation. In: RecSys (2021)
22. Sun, F., Liu, J., Wu, J., Pei, C., Lin, X., et al.: BERT4Rec: sequential recommendation with bidirectional encoder representations from transformer. In: CIKM (2019)
23. Tang, J., Wang, K.: Personalized top-n sequential recommendation via convolutional sequence embedding. In: WSDM (2018)
24. Tay, Y., Bahri, D., Metzler, D., Juan, D.C., Zhao, Z., Zheng, C.: Synthesizer: rethinking self-attention for transformer models. In: ICML (2021)
25. Tay, Y., et al.: Long range arena: a benchmark for efficient transformers. In: ICLR (2020)
26. Tolstikhin, I., et al.: MLP-Mixer: an all-MLP architecture for vision. In: NeurIPS (2021)
27. Vaswani, A., et al.: Attention is all you need. In: NIPS (2017)
28. Wang, S., Li, B.Z., Khabsa, M., Fang, H., Ma, H.: Linformer: self-attention with linear complexity. arXiv preprint arXiv:2006.04768 (2020)
29. Wu, Y., Lian, D., Gong, N.Z., Yin, L., Yin, M., et al.: Linear-time self attention with codeword histogram for efficient recommendation. In: TheWebConf (2021)
30. Yang, Z., Dai, Z., Yang, Y., Carbonell, J.G., et al.: XLNet: generalized autoregressive pretraining for language understanding. In: NeurIPS (2019)
31. Yuan, F., Karatzoglou, A., Arapakis, I., Jose, J.M., He, X.: A simple convolutional generative network for next item recommendation. In: WSDM (2019)
32. Zaheer, M., et al.: Big bird: transformers for longer sequences. In: NeurIPS (2020)
33. Zhao, W.X., et al.: RecBole 2.0: towards a more up-to-date recommendation library. In: CIKM (2022)
34. Zhao, W.X., Lin, Z., Feng, Z., Wang, P., Wen, J.R.: A revisiting study of appropriate offline evaluation for top-n recommendation algorithms. ACM Trans. Inf. Syst. 41(2) (2022)
35. Zhao, W.X., Mu, S., Hou, Y., Lin, Z., et al.: RecBole: towards a unified, comprehensive and efficient framework for recommendation algorithms. In: CIKM (2021)
36. Zhou, K., Yu, H., Zhao, W.X., Wen, J.: Filter-enhanced MLP is all you need for sequential recommendation. In: TheWebConf (2022)

Adversarial Learning Enhanced Social Interest Diffusion Model for Recommendation

Jiaxi Wang, Haochen Li, Tong Mo, and Weiping Li[✉]

School of Software and Microelectronics, Peking University, Beijing, China
{wangjxi,haochenli}@pku.edu.cn, {motong,wpli}@ss.pku.edu.cn

Abstract. With the rapid development of recommendation system, various side information has been utilized to remedy data sparsity and cold start problem. Social recommendation performs by modeling social information which brings high-order information beyond user-item interaction. However, existing works relay on GNN based social network embedding which may lead to over-smoothing problem. The process of graph diffusion encodes high-order feature also takes much noise into the model. We argue that the latent influence of social relations cannot be well captured which had not be well addressed in previous work. In this work, we propose a new recommendation framework named adversarial learning enhanced social influence graph neural network (SI-GAN) that can inherently fuses the adversarial learning enhanced social network feature and graph interaction feature. Specifically, we propose an interest-wise influence diffusion network which modeling the user-item interaction and learning the embedding of users and items through influence diffusion. We adopt the embedding of user by both interaction information and adversarial learning enhanced social network which are efficiently fused by feature fusion model. We utilize the structure of adversarial network to address the problem of over-smoothing and digging the latent feature representation. Comprehensive experiments on three real-world datasets demonstrate the superiority of our proposed model.

Keywords: Social recommendation · Recommender systems · Graph convolutional network

1 Introduction

Recommender system has played an important role in various scenarios especially in social network related applications. The Collaborative Filtering (CF) utilizes the interaction to learn the latent embedding between users and items [11]. However, data sparsity and cold start problem still exist among CF-based methods [16,21]. Traditional recommender systems hold the assumption of independent and identically distribution which ignore the asymmetric distribution of users and complexity of social connections.

To address these problems, researchers have attempted to improve CF by including auxiliary information outside users' activities called side information.

Various side information have been utilized to bring more independent sources to describing both users, items and interactions between them such as reviews of items [8,13], commodities metadata [24], location of users [32] and items or social relations [19].

Graph Neural Networks (GNNs), which are cutting-edge techniques for graph representation learning, have seen a surge in popularity in recent years [28]. Data in recommender system may be naturally represented as graphs with nodes representing user and item and edges representing interactions. GNN-based recommender systems learn user and item embeddings by stacking layers of graph convolution operations and iteratively propagating messages across graph nodes. However, most users have few behavior data, recommender system models that depend only on user-item history interactions also suffer from data sparsity. Fortunately, with the fast growth of social networks, an increasing number of works deeply investigate social relationships and model the social connections as critical side information to enhance the embeddings of users and items. Recently, social recommendation has emerged as a promising research direction as well as being well backed by the theory of social influence [2]. The idea of social homophily [18] reveals the truth of the interest similarity among specific groups of people. Based on this intuitive finding, social recommendation aims to utilize the social information by extracting the high-order relationship and interactions.

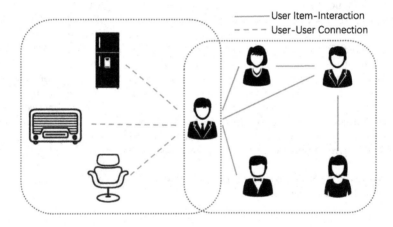

Fig. 1. User social network and user-item interaction network

User-user relationships and user-item relationships are two different types of relationships in the social recommendation. We can model these relationships using graphs. Figure 1 illustrates how users with friend connections can be viewed as nodes and connected to create a user-user social graph. It is possible to accurately model the social relationships between users and the interactions between users and items by using GNNs, which partly simulates real social relation connectivity. The fundamental concept behind GNNs is how to transmit node information throughout the graph by repeatedly aggregating feature information from small graph neighbors using neural networks to represent node information. GNNs can therefore be used for social recommendation.

However, several studies propose that node embeddings are made up of low-frequency features and high-frequency noise from the graph signal processing [12,25]. They demonstrated that combining neighborhood node embeddings iteratively in GNN-based models is a special form of low-pass filtering on graph signals. As a result, GNN models employ low-pass filtering to reduce noise and learn richer graph representations and enhance performance. Such operations on the user-item interaction network make sense since the data to be predicted and past interactions are formed from the same distribution. The social network differs depending on the situation. If the different users are been modeling in the same distribution, it may set limit to model's efficiency and reduce its performance. The diffusion process includes more neighbor to obtain better embeddings but also introduces irrelevant nodes and edges. This also may lead to over-smoothing and bring much noise into the model.

To tackle the issues above, we propose an adversarial learning enhanced social influence GNN-based model called SI-GAN that can inherently fuses the adversarial learning enhanced social network feature and graph interaction feature. We first adopt the embedding of user by both interaction information and adversarial learning enhanced social network. We utilize the structure of adversarial network to remedy the problem of over-smoothing and noise from diffusion process. Our contributions of this paper are summarized as follows:

- We propose an interest-wise influence diffusion network that modeling the user-item interaction and learning the embedding of users and items through influence diffusion.
- We propose an adversarial learning enhanced social network to improve the social recommendation predictive performance. To the best of our knowledge, we are the first to inherently fuse the adversarial learning enhanced social network feature and graph interaction feature.
- We perform extensive experiments on real-world datasets to demonstrate the superiority of our proposed model.

2   Related Work

In this section, we mainly review social recommendation, GNN-based recommendation and adversarial learning in GNN-based recommender system.

2.1   Social Recommendation

Before the era of deep learning, social recommendation has been studied since 1997 [21] and mainly based on collaborative filtering. SocialMF [9] and Social Regularization [17] are typical regularization method which utilizes matrix factorization method to regulate the preference of a user. STE [15] and mTrust [20] are categorized as ensemble method which predict the missing rating by the linear combination of ratings from the neighbor users. Co-factorization methods are widely adopt by non-deep learning social recommendation methods. SoRec [16] uses the probability matrix to perform co-factorization on both user-item rating matrix and user-user social relation matrix. Although collaborative filtering have gained great success since the efficiency and performance, it still suffer from data sparsity issues and cold start problem. Deep learning-based methods can modeling the social embedding of users and items, especially using GNNs. GraphRec [4] is a GNN-based model for social recommendation rating prediction that aggregates representations for items and users from their linked neighbors. Diffnet [26]creates a layer-by-layer influence propagation system to model how consumers' preferences develop as social influence spreads recursively. LightGCN [6] linearly propagate the user and item embeddings with weighted sum. HOSR [14] propose to iteratively "propagate" embeddings along the social network, effectively introducing the influence of high-order neighbors into user representation, to overcome the issue that high-order neighbors rise rapidly with the order size. MHCN [30] utilizes hierarchical mutual information maximization to retrieve connection information by incorporating self-supervised learning into the training of the hypergraph convolutional network.

2.2   Adversarial Learning in GNN-Based RS

The adversarial learning has achieved huge success in many areas. GAN-based recommendation have been deeply investigated in many aspect [1,3,27,33]. An extension of Adversarial Learning for graph structure called GraphGAN [23] is employed to adopt representations of latent neighbors in an adversarial way. A GraphGAN model has two parts, much like an adversarial learning model: a generator and a discriminator. False vertex representations are created using a certain connectivity distribution. The connectivity pattern is used to distinguish between fake and real vertex representations. Throughout the training, they play a minimax game and steadily improve both of their performances. As the successful usages of generative adversarial networks in many areas. Some adversarial learning based models are proposed for social recommendation. DASO [5] adopted two adversarial learning components to model interaction between two domains. RSGAN [29] present a Generative Adversarial Nets-based end-to-end social recommendation framework (GAN). The framework is made up of two parts: a generator that generates friends that might potentially improve the social recommendation model, and a discriminator that evaluates these created friends and ranks the things based on both the current user's and her friends' preferences. APR [7] is a adversarial learning based model. By using adversarial training, the pairwise ranking technique Bayesian Personalized Ranking (BPR) is improved. It may be regarded as a minimax game in which the minimisation of the BPR objective function defends an opponent, which introduces adversarial perturbations on model parameters to maximise the BPR objective function.

3   Methodology

In this section, we propose a novel Social Interest diffusion model by utilizing Generative Adversarial Network for social recommendation, named SI-GAN as Fig. 2. In general, SI-GAN consists of three modules. (1) Interest-wise influence diffusion module: modeling the user-item interaction and learning the embedding of users and items through influence diffusion; (2) Social relation adversarial network: instead of using influence-based GNN architecture, utilizing adversarial network can modeling a more accurate embedding without the trend of over-smoothing in node representation. (3) Information fusion and prediction module: we design attention model to find the crucial features from two modules above and make the prediction by employing MLP.

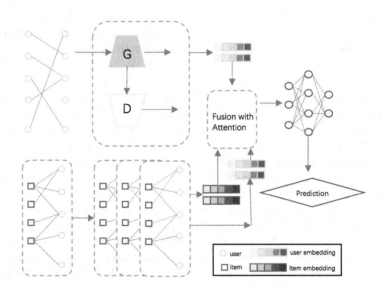

Fig. 2. The Overview of SI-GAN model, including three modules, Interest-wise influence diffusion module calculate the interest of each user to preform a better diffusion process. Social relation adversarial network utilizes GAN to modeling social network which demonstrate superior embedding efficiency and can avoid the over-fitting problem. Information fusion and prediction module adopts attention to fuse the two part embedding feature.

3.1   Definitions and Problem Formulation

The context of social recommendation basically contains two types of entities. Let $U = \{u_1, u_2, \ldots, u_n\}$ and $V = \{v_1, v_2, \ldots, v_m\}$ be the set of users and items respectively where n and m are the numbers of users and items. The interaction graphs are formed from U and V. User-item interaction graph can be denoted as

G_{ui}, we can use the adjacency matrix to represent the interactions in the graph. $M_{inter} \in \mathbb{R}^{n \times m}$ is the interaction matrix between users and items. If there is an interaction between the user u and the item i, y_{ui} is 1, otherwise it is 0. User-user social network can be denoted as G_{uu}, $M_{social} \in \mathbb{R}^{n \times n}$ is the interaction matrix between users. We formulate the studied recommendation task in this paper as:

Input: The User set U and the item set V, $M_{inter} \in \mathbb{R}^{n \times m}$ the adjacency matrix representation of the user-item interaction graph G_{ui}, $M_{social} \in \mathbb{R}^{n \times n}$ the adjacency matrix representation of the user-user social relation network G_{uu}.

Output: A predictive function $\text{Predict}(z_{ij}; \theta_P)$ that effectively forecasts the future user-item interaction.

3.2   Interest-Wise Influence Diffusion

The ratings for items given by users can show the potential preference of a user. At the initial start of learning process, a free embedding and the feature vector are fused as u_i^0 for the following part. Similarly, for an item j, a free embedding and the feature vector are fused as v_j^0. We utilize multi-layer diffusion to model the users' potential interest and personal preference. For a user i, u_i^k is the k-th layer embedding, Unobserved interpersonal connections are hardly to be fully investigated. We use link prediction algorithms to model unseen interpersonal connections in order to target unobserved influence among social network. It should be highlighted that in a diffusion based graph neural network, the performance of nodes embedding learning and link prediction are highly interaction and mutual influence. As a result, in our GCN learning, the K-th layer predicted links are contributed to the subsequent training process as observed links, allowing the learning of nodes embedding and link prediction to mutually benefit one another. Each item's influence is estimated as an interest value, which is updated throughout each GCN learning iteration and combined with item embedding. A smoothed ratio of linked users to linked users for all products is used to measure the interest of an item. We define item interest as the following:

$$interest_i = (|R_i| + 1) / \left(\sum_{j \in V} |R_j| + |V| \right) \tag{1}$$

where R_i is the set of the users linked to the certain item and V is the set of all items. After each layer's influence iteration, the cosine similarity between the embedding of each item and each user is calculated in order to calculate the interest of each item to each user. As a result, $\alpha_{ij}^{k+1} = interest_j \times \cos\left(u_i^k, v_j^k\right)$ is the interest sensitive weight of a user on$(k + 1)$-th layer. The user embedding in $(k + 1)$-th layer can be formalized as:

$$u_i^{k+1} = \sum_{j \in R_j} \alpha_{ij}^{k+1} v_j^k \tag{2}$$

where u_i^{k+1} is the $(k + 1)$-th layer embedding vector of user i. R_j is the set of items have rated user i. For each item i, given its k-th layer embedding v_i^k, its update embedding for the $(k + 1)$-th layer can be modeled as:

$$v_j^{k+1} = \sum_{i \in R_i} \eta_{ij}^{k+1} u_i^k \tag{3}$$

where v_j^{k+1} is the $(k + 1)$-th layer embedding vector of item j, η_{ij}^{k+1} represents the predefined aggregated weights where R_i is the set of users have rated item i.

3.3   Social Relation Adversarial Network

In order to capture social relation features to improve the performance of recommendations, user embeddings which contain the social relationship information need to be fully investigated. Methods such as GNN and GCN, are suitable for the graph embedding task. However, they are unable to address the social network's data sparsity issue. In order to reduce the problem of data sparsity and data noise, in this work we develop a social relation GAN model to find latent representations of each user. A graphGAN-based network is proposed and made up of two parts: a generator to generate latent friends of a given user by fitting the connectivity pattern distribution in the social relation network and a discriminator to play a minimax game during the training to improve their capability step by step. Inspired by [22] and [31], we use a sigmoid function of the inner product to implement discriminator as follow:

$$D(u; \boldsymbol{\theta}_D) = \frac{1}{1 + \exp\left(-\phi_D(\hat{u}_k; \boldsymbol{\theta}_D)^\top \phi_D(u_i; \boldsymbol{\theta}_D)\right)} \tag{4}$$

where $\phi_D(\cdot; \boldsymbol{\theta}_D)$ denotes the embedding function in discriminator, we utilize graph softmax to implement it as follows:

$$G(u \mid u_i; \theta_G) = \left(\prod_{k=1}^{L_p} p_r\left(u_k^\rho \mid u_{k-1}^\rho\right)\right) \times \left(p_r\left(u_{L_p-1}^\rho \mid u_{L_p}^\rho\right)\right) \tag{5}$$

where L_p is the path length from u_i to $\hat{u}_k$, u^ρ is the user in the path. $p_r(u_j \mid u_i)$ is the conditional probability computed as follows:

$$p_r(u_j \mid u_i) = \frac{\exp\left(\phi_G(u_j; \boldsymbol{\theta}_G)^\top \phi_G(u_i; \boldsymbol{\theta}_G)\right)}{\sum_{u_k \in \mathcal{N}(u_i)} \exp\left(\phi_G(u_k; \boldsymbol{\theta}_G)^\top \phi_G(u_i; \boldsymbol{\theta}_G)\right)}, \tag{6}$$

where the $\phi_G(\cdot; \boldsymbol{\theta}_G)$ denotes the embedding function in generator.

According to the mechanism of adversarial learning, the generator and discriminator play the following two-player minimax game. We minimize the generator (see Eq. 5) and maximize the discriminator (see Eq. 4):

$$\arg \min_{\theta_G} \max_{\theta_D} \mathcal{L} (\theta_G, \theta_D) = \sum_{i=1}^{|U|} \left(\mathbb{E}_{u \sim p_{true}(\cdot | u_i)} [\log D (u, u_i; \theta_D)] \right. \tag{7}$$
$$\left. + \mathbb{E}_{u \sim p_G(\cdot | u_i)} [\log (1 - D (u, u_i; \theta_D))] \right),$$

where the $p_{true} (\cdot \mid u_i)$ is the ground truth value of users. In the iteration of adversarial learning, discriminator is feed with true value from $p_{true} (\cdot \mid u_i)$ and the fake sample from generator. The generator is updated with the regulation from discriminator. The minimax game will continue until the sample from generator is indistinguishable from the ground truth value.

3.4   Feature Fusion and Prediction

The node features in two graphs are combined by an attention mechanism-based multi-graph information fusion model to improve the learning of the vertex representation after being learned from the user-item bipartite graph G_{ui} and the social relation network G_{uu}. An user-item attention network is used to learn a set of weights for all the items associated to each user vector in the user-item graph. A user-user attention network is used to learn a set of weights for each user's neighbors in the social relation network. These two categories of features are combined using a concatenation operation after the weighted features have been created as follow:

$$\mathbf{e}_{ij} = \sigma \left(W_{UI} \times \left((W_R \times \left(\sum_{i,j=1}^{|N(u_i,v_j)|} \omega_{ij} \times e (u_i, v_j) \right) + \beta_R \right) + \beta_{UI} \right) \tag{8}$$

W_{UI} and β_{UI} represent the parameters of user-item aggregation neural network. W_R and β_R represent the parameters of item-rating representation neural network. $e (u_i, v_j))$ denotes the rating representation between u_i and v_j. $omega_{ij}$ denotes the attention weight of neural network. After the training of neural network with attention mechanism we can obtain the representation of user-item pair(u_i, v_j).

Similarly to the user-item attention network, a user-user attention network is to learn a set of weights for all the neighbors for each user in social relation network G_{uu}. This network can be formulated as:

$$\mathbf{e}_i = \sigma \left(W_{UU} \times \left(\sum_{k=1}^{|N(u_i)|} \omega_{ik} \times e (u_i, u_k) \right) + \beta_{UU} \right) \tag{9}$$

where ω_{ik} denotes the corresponding weight of u_i. The attention weight can be produced by the following network:

$$\hat{\omega}_{ik} = W_2 \times \sigma (W_1 \times \sigma (e (u_i, v_j) + \beta_R) + \beta_1) + \beta_2, \tag{10}$$

$$\omega_{ik} = \frac{\exp\left(\hat{\omega}_{ik}\right)}{\sum_{k=1}^{|\mathcal{N}(u_i)|} \exp\left(\hat{\omega}_{ik}\right)} \tag{11}$$

To improve the representation learning, an information fusion model is used to fuse the representation e_{ij} from user-item bipartite graph $\boldsymbol{G_{ui}}$ and the representation e_i from social relation network $\boldsymbol{G_i}$ together. This model is implemented by an l-layers deep neural network model $z_{ij} = \mathrm{Dnn}\,(e_{ij}, e_i; \theta_{Dnn})$. Specifically, the first layer is a concatenation operator that concatenates e_{ij} and e_i, and the remaining several layers construct a multiple-layer Perceptron. The structure of this DNN can be formulated as

$$\begin{aligned} \boldsymbol{a}_{ij}^{(1)} &= \mathbf{e}_{ij} \oplus \mathbf{e}_i, \\ a_{ij}^{(2)} &= \sigma\left(\boldsymbol{W}_2 \times a_{ij}^{(1)} + \beta_2\right), \\ &\cdots \\ \boldsymbol{z}_{ij} &= \sigma\left(W_l \times \boldsymbol{a}_{ij}^{(l)} + \beta_l\right), \end{aligned} \tag{12}$$

where $a^{(j)}$ is the output of j th layer, $j \in [1, \ldots, l]$. The parameter vector is $\theta_{Dnn} = \left(\{\boldsymbol{W}_j\}|_{j=2}^{l}, \{\boldsymbol{\beta}_j\}|_{j=2}^{l}\right)$. Then the fused representations are fed into a linear prediction layer to generate the predicted recommendation as follows:

$$r'_{ij} = \mathrm{Predict}\,(z_{ij}; \theta_P) = W \times z_{ij}. \tag{13}$$

Then the loss function of the model can be formulated as:

$$\mathcal{L}_{\mathrm{p}} = \frac{1}{2|O|} \sum_{i,j \in O} \left(r'_{ij} - r_{ij}\right)^2 \tag{14}$$

where $|O|$ is the number of user-item pair and r_{ij} is the ground truth value of rating. So the overall loss function can be formulated as:

$$\mathcal{L} = \mathcal{L}_{\mathrm{p}} + \mathcal{L}(\boldsymbol{\theta}_G, \boldsymbol{\theta}_D) \tag{15}$$

4   Experiment

In this section, we evaluate the performance of SI-GAN on a series of real-world datasets. Specifically, we choose Epinions, Yelp and Ciao dataset for experiment, to answer the following research questions:

RQ1 Can SI-GAN perform better than other baseline methods?

RQ2 Does our proposed interest-wise influence diffusion network improve recommendation performance? Is adversarial learning enhanced social network producing a more accurate user embedding and improve recommendation performance?

4.1   Experimental Setting

Dataset. In order to validate the performance of the SI-GAN, we conduct experiments on three public datasets, Epinions, Yelp and Ciao. The three datasets are described in detail as follows (Table 1).

Table 1. The statistics of the three datasets

Dataset	Epinions	Yelp	Ciao
#User	22164	17235	7375
#Item	296277	37378	105114
#Ratings	911442	207945	169730
#Relations	574202	169150	57544
#Rating density	0.014%	0.032%	0.035%
#Relation density	0.116%	0.087%	0.313%

- **Epinions:** The dataset comes from a general consumer review site Epinions.com which contains trust relationships interactions and review ratings information.
- **Yelp:** This dataset is a location-based review dataset which also contains social information and ratings.
- **Ciao:** This dataset contains rating information of users given to items, and also contain item category information.

Evaluation Metrics. We evaluate our methods by using two classical metrics Mean Absolute Error (MAE) and Root Mean Squared Error (RMSE) on all datasets. MAE and RMSE metrics are widely used in recommendation system to evaluate the closeness of predicted ratings to the true ratings.

$$MAE = \frac{\sum_{i=1}^{n} |\hat{y}_i - y_i|}{n} \tag{16}$$

$$RMSE = \sqrt{\frac{\sum_{i=1}^{n} (\hat{y}_i - y_i)^2}{n}} \tag{17}$$

Baselines. We compare the proposed SI-GAN with the following baseline models:

- **SoRec** [16]: By employing both user social network data and rating records, the approach perform a factor analysis strategy based on probabilistic matrix factorization to address the issues of data sparsity and low prediction accuracy.
- **SocialMF** [9]: This approach adopts matrix factorization techniques and incorporate the mechanism of trust propagation into the model to reduce cold start problem and make more accurate prediction.

- **GraphRec** [4]: This approach proposes a new GNN-based model which integrates user-user social graph with user-item interaction graph. Also it utilizes attention network to filter out the high-influence users and items.
- **Diffnet** [26]: The approach employs a layer-wise GNN diffusion architecture to model the users' preference.
- **LightGCN** [6]: The user and item embeddings are obtained by LightGCN by linearly propagating them on the user-item interaction graph, and the final embedding is the weighted sum of the embeddings learned at all layers. It is significantly simpler to implement and train this tidy, linear, and simple model.
- **HOSR** [14]: This approach propose to iteratively "propagate" embeddings along the social network, effectively introducing the influence of high-order neighbors into user representation, to overcome the issue that high-order neighbors rise rapidly with the order size.
- **MHCN** [30]: This approach utilizes hierarchical mutual information maximization to retrieve connection information by incorporating self-supervised learning into the training of the hypergraph convolutional network.

Parameter Settings. We implement our proposed model by Pytorch. We utilize Adam [10] as the optimize for the gradient descent processes. For the best result of all the baseline model, we search the learning rate in the range of [0.001, 0.005, 0.1, 0.2, 0.5]. In order to avoid over-fitting, dropout has been adopted with the rate of 0.5. The embedding size is [32, 64, 128, 256]. For the diffusion model, we set the layer number with 2 and 3 respectively. In our evaluations, we adopt early stopping for training termination when the performance degrades for 5 continuous epochs on the validation dataset.

4.2   Overall Comparison (RQ1)

We demonstrate the overall performance by the MAE and RMSE with different dimensional embedding in three datasets. Our model SI-GAN gets better on all data sets which shows the effectiveness of denoising and accurate embedding. As the results, the non-GNN based method such as SoRec and SocialMF has comparatively weak feature extraction ability. Due to the modeling of high-level non-linear feature, neural network-based models typically outperform traditional approaches. Among the different baseline approaches, GNN-based models demonstrate the ability of high-order relations embedding among users items network. Unlike existing GNN methods, our framework incorporates social and social relationships from the social network using an adversarial learning approach.

As Table 2 and Table 3 indicate, the matrix factorization has a significant shortcoming that the side information usually cannot be fully utilized. The diffusion and propagating based method main utilize GNN as the base model. The difference of numbers in diffusion layers vary in methods. GraphRec fuses the first-order interactions on social networks. Diffnet adopts convolutional operations to diffusion in social network to obtain high-order feature. LightGCN

Table 2. Overall Comparison with d = 32

Dataset	Epinions		Yelp		Ciao	
	MAE	RMSE	MAE	RMSE	MAE	RMSE
SoRec	0.9058	1.1960	0.9401	1.2070	0.8510	1.0304
SocialMF	0.8805	1.1509	0.9477	1.2012	0.8203	1.0282
GraphRec	0.8160	1.0633	0.8509	1.1240	0.7380	1.0191
Diffnet	0.7911	1.0094	0.8010	1.0474	0.7051	0.9545
LightGCN	0.7848	0.9944	**0.7931**	0.9804	0.7503	0.9874
HOSR	0.8039	1.0356	0.8031	1.0408	0.7314	1.0096
MHCN	**0.7704**	1.0168	0.8289	0.9980	0.6928	0.9542
SI-GAN	0.7800	**0.9927**	0.8151	**0.9794**	**0.6850**	**0.9498**

Table 3. Overall Comparison with d = 64

Dataset	Epinions		Yelp		Ciao	
	MAE	RMSE	MAE	RMSE	MAE	RMSE
SoRec	0.8806	1.1589	0.9307	1.2068	0.8391	1.0377
SocialMF	0.8756	1.1331	0.9386	1.2065	0.8231	1.0370
GraphRec	0.7701	1.0328	0.8477	1.0570	0.7080	0.9630
Diffnet	0.7523	1.0206	0.8310	1.0390	0.6955	0.9471
LightGCN	0.7806	0.9837	0.8101	0.9730	0.7413	0.9764
HOSR	0.8001	0.9956	0.8219	1.0297	0.7251	1.0138
MHCN	0.7837	0.9968	0.8271	0.9741	0.6900	0.9577
SI-GAN	**0.7709**	**0.9827**	**0.8003**	**0.9694**	**0.6810**	**0.9507**

adopt linear propagating to capture interaction between users and items. HOSR and MHCN perform best among baseline by adopting iterative propagating and mutual information maximization respectively. As the dimension changed from 32 to 64, most of the baseline method have gained the performance due to the improvement of embedding dimension. Then we discuss the impact of embedding size on model performance in next part.

4.3   Ablation Experiments (RQ2)

In this section, we examine the influence of embedding dimension and verify the effect of interest-wise diffusion and adversarial learning enhanced social network. We design three sets of experiments. First we conduct embedding dimension study, we vary the embedding dimension in [32, 64, 128, 256] to obtain the best value. Then we investigate the efficiency of interest-wise diffusion by comparing the interest-wise diffusion model SI-GAN and normal diffusion model SI-GAN$_d$. At last we compare the SI-GAN model with single diffusion model SI-GAN$_s$ to examine the feature exaction result for the adversarial learning enhanced model.

Table 4. Comparisons for Interest-wise Diffusion

Dataset	Epinions		Yelp		Ciao	
	MAE	RMSE	MAE	RMSE	MAE	RMSE
SI-GAN$_d$	0.7831	0.9901	0.8307	0.9971	0.7031	0.9601
SI-GAN	**0.7709**	**0.9827**	**0.8003**	**0.9694**	**0.6810**	**0.9507**

Table 5. Comparisons for Adversarial Learning Enhanced Network

Dataset	Epinions		Yelp		Ciao	
	MAE	RMSE	MAE	RMSE	MAE	RMSE
SI-GAN$_s$	0.8274	1.0490	0.8573	1.0485	0.7503	1.0397
SI-GAN	**0.7709**	**0.9827**	**0.8003**	**0.9694**	**0.6810**	**0.9507**

As Table 4 shows that our SI-GAN is significantly better than SI-GAN$_d$ which implies the positive effect of interest modeling. Moreover, for further investigation, interest modeling can utilize the item-aspect feature and enhance the feature extraction for user. As Table 5 implies that adversarial learning can significantly improve the performance of diffusion based method. As the layer of diffusion increases, the users and items representation has the trend to be the same which has absolutely negative effect on feature extraction. In order to tackle this problem, adversarial learning can be apply to obtain more vivid representation.

Table 6. Comparisons for Embedding Dimension

SI-GAN	Epinions		Yelp		Ciao	
	MAE	RMSE	MAE	RMSE	MAE	RMSE
DIM-32	0.7800	0.9927	0.8151	0.9794	0.6850	**0.9498**
DIM-64	**0.7709**	**0.9827**	**0.8003**	**0.9694**	**0.6810**	0.9507
DIM-128	0.7812	0.9903	0.8031	0.9840	0.6948	0.9637
DIM-256	0.8020	1.0387	0.8274	0.9970	0.7139	0.9746

The result of impact of embedding dimension are shown in Table 6. The sweet point of the value of dimension is around 64. The performance will get worse at 32 and be better at 64, then get worse again at 128 and 256. The increase of embedding size may lead to sparse and the low efficient in neural network back propagation.

5   Conclusion

In this work, we propose a new recommendation framework called SI-GAN to remedy the over-smoothing and noise from diffusion process problem. In order to avoid these problems, we designed an adversarial learning enhanced social embedding to gain a more accurate feature representation. Then we use interest-wise diffusion model to obtain the interactive embeddings for both users and items. Comprehensive experiments on three real-world datasets demonstrate the superiority of our proposed model.

Our model shows promising performance in diffusion-based social recommendation. In the future, we consider improving our method to handle more complex data such as heterogeneous information network and hypergraph.

Acknowledgment. This work was supported by the National Key R&D Program of China [2022YFF0902703].

References

1. Chen, H., Li, J.: Adversarial tensor factorization for context-aware recommendation. In: Proceedings of the 13th ACM Conference on Recommender Systems, pp. 363–367 (2019)
2. Cialdini, R.B., Goldstein, N.J.: Social influence: compliance and conformity. Annu. Rev. Psychol. **55**(1), 591–621 (2004)
3. Ding, J., Quan, Y., He, X., Li, Y., Jin, D.: Reinforced negative sampling for recommendation with exposure data. In: IJCAI, Macao, pp. 2230–2236 (2019)
4. Fan, W., et al.: Graph neural networks for social recommendation. In: The World Wide Web Conference, pp. 417–426 (2019)
5. Fan, W., Ma, Y., Yin, D., Wang, J., Tang, J., Li, Q.: Deep social collaborative filtering. In: Proceedings of the 13th ACM Conference on Recommender Systems, pp. 305–313 (2019)
6. He, X., Deng, K., Wang, X., Li, Y., Zhang, Y., Wang, M.: LightGCN: simplifying and powering graph convolution network for recommendation. In: Proceedings of the 43rd International ACM SIGIR Conference on Research and Development in Information Retrieval, pp. 639–648 (2020)
7. He, X., He, Z., Du, X., Chua, T.S.: Adversarial personalized ranking for recommendation. In: The 41st International ACM SIGIR Conference on Research & Development in Information Retrieval, pp. 355–364 (2018)
8. Huang, C., Jiang, W., Wu, J., Wang, G.: Personalized review recommendation based on users' aspect sentiment. ACM Trans. Internet Technol. **20**(4) (2020)
9. Jamali, M., Ester, M.: A matrix factorization technique with trust propagation for recommendation in social networks. In: Proceedings of the 4th ACM Conference on Recommender Systems, pp. 135–142 (2010)
10. Kingma, D.P., Ba, J.: Adam: a method for stochastic optimization. arXiv preprint arXiv:1412.6980 (2014)
11. Konstan, J.A., Miller, B.N., Maltz, D., Herlocker, J.L., Gordon, L.R., Riedl, J.: GroupLens: applying collaborative filtering to Usenet news. Commun. ACM **40**(3), 77–87 (1997)

12. Li, Q., Wu, X.M., Liu, H., Zhang, X., Guan, Z.: Label efficient semi-supervised learning via graph filtering. In: Proceedings of the IEEE/CVF Conference on Computer Vision and Pattern Recognition, pp. 9582–9591 (2019)
13. Ling, G., Lyu, M.R., King, I.: Ratings meet reviews, a combined approach to recommend. In: Proceedings of the 8th ACM Conference on Recommender Systems, RecSys 2014, pp. 105–112. Association for Computing Machinery, New York, NY, USA (2014)
14. Liu, Y., Liang, C., He, X., Peng, J., Zheng, Z., Tang, J.: Modelling high-order social relations for item recommendation. IEEE Trans. Knowl. Data Eng. (2020)
15. Ma, H., King, I., Lyu, M.R.: Learning to recommend with social trust ensemble. In: Proceedings of the 32nd International ACM SIGIR Conference on Research and Development in Information Retrieval, pp. 203–210 (2009)
16. Ma, H., Yang, H., Lyu, M.R., King, I.: SoRec: social recommendation using probabilistic matrix factorization. In: Proceedings of the 17th ACM Conference on Information and Knowledge Management, pp. 931–940 (2008)
17. Ma, H., Zhou, D., Liu, C., Lyu, M.R., King, I.: Recommender systems with social regularization. In: Proceedings of the 4th ACM International Conference on Web Search and Data Mining, pp. 287–296 (2011)
18. McPherson, M., Smith-Lovin, L., Cook, J.M.: Birds of a feather: homophily in social networks. Ann. Revi. Sociol., 415–444 (2001)
19. Song, W., Xiao, Z., Wang, Y., Charlin, L., Zhang, M., Tang, J.: Session-based social recommendation via dynamic graph attention networks. In: Proceedings of the 12th ACM International Conference on Web Search and Data Mining, pp. 555–563 (2019)
20. Tang, J., Gao, H., Liu, H.: mTrust: discerning multi-faceted trust in a connected world. In: Proceedings of the 5th ACM International Conference on Web Search and Data Mining, pp. 93–102 (2012)
21. Tang, J., Hu, X., Liu, H.: Social recommendation: a review. Soc. Netw. Anal. Min. **3**(4), 1113–1133 (2013). https://doi.org/10.1007/s13278-013-0141-9
22. Wang, H., et al.: GraphGAN: graph representation learning with generative adversarial nets. In: Proceedings of the 32nd AAAI Conference on Artificial Intelligence and 30th Innovative Applications of Artificial Intelligence Conference and 8th AAAI Symposium on Educational Advances in Artificial Intelligence, AAAI'18/IAAI'18/EAAI'18. AAAI Press (2018)
23. Wang, H., et al.: Learning graph representation with generative adversarial nets. IEEE Trans. Knowl. Data Eng. **33**(8), 3090–3103 (2019)
24. Wang, J., Huang, P., Zhao, H., Zhang, Z., Zhao, B., Lee, D.L.: Billion-scale commodity embedding for e-commerce recommendation in alibaba. In: Proceedings of the 24th ACM SIGKDD International Conference on Knowledge Discovery Data Mining, KDD '18, pp. 839–848. Association for Computing Machinery, New York, NY, USA (2018)
25. Wu, F., Souza, A., Zhang, T., Fifty, C., Yu, T., Weinberger, K.: Simplifying graph convolutional networks. In: International Conference on Machine Learning, pp. 6861–6871. PMLR (2019)
26. Wu, L., Sun, P., Fu, Y., Hong, R., Wang, X., Wang, M.: A neural influence diffusion model for social recommendation. In: Proceedings of the 42nd International ACM SIGIR Conference on Research and Development in Information Retrieval, pp. 235–244 (2019)
27. Wu, Q., Liu, Y., Miao, C., Zhao, B., Zhao, Y., Guan, L.: PD-GAN: adversarial learning for personalized diversity-promoting recommendation. In: IJCAI, vol. 19, pp. 3870–3876 (2019)

28. Wu, Z., Pan, S., Chen, F., Long, G., Zhang, C., Philip, S.Y.: A comprehensive survey on graph neural networks. IEEE Trans. Neural Netw. Learn. Syst. **32**(1), 4–24 (2020)
29. Yu, J., Gao, M., Yin, H., Li, J., Gao, C., Wang, Q.: Generating reliable friends via adversarial training to improve social recommendation. In: 2019 IEEE International Conference on Data Mining (ICDM), pp. 768–777. IEEE (2019)
30. Yu, J., Yin, H., Li, J., Wang, Q., Hung, N.Q.V., Zhang, X.: Self-supervised multi-channel hypergraph convolutional network for social recommendation. In: Proceedings of the Web Conference 2021, pp. 413–424 (2021)
31. Zhang, C., Wang, Y., Zhu, L., Song, J., Yin, H.: Multi-graph heterogeneous interaction fusion for social recommendation. ACM Trans. Inf. Syst. (TOIS) **40**(2), 1–26 (2021)
32. Zheng, V.W., Zheng, Y., Xie, X., Yang, Q.: Towards mobile intelligence: learning from GPS history data for collaborative recommendation. Artif. Intell. **184**, 17–37 (2012)
33. Zhou, F., Yin, R., Zhang, K., Trajcevski, G., Zhong, T., Wu, J.: Adversarial point-of-interest recommendation. In: The World Wide Web Conference, pp. 3462–34618 (2019)

Adversarial Learning Data Augmentation for Graph Contrastive Learning in Recommendation

Junjie Huang[1,2], Qi Cao[1], Ruobing Xie[3], Shaoliang Zhang[3], Feng Xia[3],
Huawei Shen[1,2(✉)], and Xueqi Cheng[1,4]

[1] Data Intelligence System Research Center, Institute of Computing Technology,
Chinese Academy of Sciences, Beijing, China
{huangjunjie17s,caoqi,shenhuawei,cxq}@ict.ac.cn
[2] University of Chinese Academy of Sciences, Beijing, China
[3] WeChat, Tencent, Beijing, China
{ruobingxie,modriczhang,xiafengxia}tencent.com
[4] CAS Key Laboratory of Network Data Science and Technology,
Institute of Computing Technology, Chinese Academy of Sciences, Beijing, China

Abstract. Recently, Graph Neural Networks (GNNs) achieve remarkable success in Recommendation. To reduce the influence of data sparsity, Graph Contrastive Learning (GCL) is adopted in GNN-based CF methods for enhancing performance. Most GCL methods consist of data augmentation and contrastive loss (e.g., InfoNCE). GCL methods construct the contrastive pairs by hand-crafted graph augmentations and maximize the agreement between different views of the same node compared to that of other nodes, which is known as the **InfoMax** principle. However, improper data augmentation will hinder the performance of GCL. **InfoMin** principle, that the good set of views shares minimal information and gives guidelines to design better data augmentation. In this paper, we first propose a new data augmentation (i.e., edge-operating including edge-adding and edge-dropping). Then, guided by **InfoMin** principle, we propose a novel theoretical guiding contrastive learning framework, named <u>L</u>earnable <u>D</u>ata <u>A</u>ugmentation for <u>G</u>raph <u>C</u>ontrastive <u>L</u>earning (LDA-GCL). Our methods include data augmentation learning and graph contrastive learning, which follow the **InfoMin** and **InfoMax** principles, respectively. In implementation, our methods optimize the adversarial loss function to learn data augmentation and effective representations of users and items. Extensive experiments on four public benchmark datasets demonstrate the effectiveness of LDA-GCL.

Keywords: Self-supervised Learning · Graph Contrastive Learning · Learning Data Augmentation · Graph Collaborative Filtering

1 Introduction

Collaborative Filtering (CF) [20] is to produce effective recommendations from implicit feedback (e.g., clicking, rating, buying, and so on). The interaction data can be viewed as a user-item bipartite graph. Based on modeling such bipartite graphs, Graph Neural Networks (GNNs) [8,9,26] can learn the effective node representations of users and items for personalized recommendations. GNN model effectively utilizes the high-order graph structure information through the message-passing scheme and has achieved the-state-of-the-art results.

Although GNN models have achieved remarkable success, they still suffer from data sparsity issues. To overcome the difficulties, Graph Contrastive Learning (GCL) in a self-supervised manner, is introduced to improve the recommendation performance. Since no labeled data is required, GCL is considered a good solution for data sparsity issues in recommender systems [15,27]. GCL has two important components: data augmentation and contrastive loss. For data augmentation, the past GCL approaches in recommendation [27] generate hand-crafted graph augmentations by edge-dropping. After data augmentation, GCL uses graph neural network models to get the node representations over multiple views. The contrastive loss (e.g., InfoNCE) leverages the mutual information maximization principle (**InfoMax**) that aims to *maximize* the correspondence between the representations of the nodes in its different augmented graphs. However, improper data augmentation can hinder the performance of contrastive learning [33]. How to find the proper data augmentation is a promising research problem. Tian et al. [23] investigate the research problem of what makes for good views in contrastive learning (CL). Inspired by Information Bottleneck (IB) [24], they proposed the **InfoMin** principle that the good set of views shares the *minimal* information necessary to perform well at the downstream task. They find that stronger data augmentation indeed leads to decreasing mutual information and improves downstream tasks. **InfoMin** principle offers the guideline for us to find optimal data augmentation for GCL in the recommendation.

In this paper, we propose a new graph data augmentation in recommender systems (i.e., edge-operating including edge-adding and edge-dropping). First, to avoid randomly adding noisy edges that harm the graph structural information, we use a pre-trained model to predict the possible edges (*a.k.a.* link between user u and the item that is most likely to interact). Second, we constitute the original edges and the added edges as edge candidates. Our augmentation strategy can better explore the diversity of graph structure than traditional edge dropping. Third, we propose to a new adversarial Learnable Data Augmentation for Graph Contrastive Learning (LDA-GCL) framework in recommendation. LDA-GCL uses an edge operator model to learn the augmented graph from the candidate edges instead of randomly sampling. Our learning data augmentation (LDA) process follows the **InfoMin** principle, while GCL framework follows the **InfoMax** principle. LDA-GCL optimizes an adversarial loss function to get effective embeddings for recommendation tasks. Compared with heuristic design, our approach can automatically generate efficient graph data augmentation with **InfoMin** principle. We analyze the effectiveness of our approach on several public benchmark datasets. To the best of our knowledge, it's the first time to introduce **InfoMin** principle into GCL in recommendation.

The major contributions of this paper are summarized as follows:

- We propose a new data augmentation method (i.e., edge-operating) for contrastive learning, which includes edge-adding from a pre-trained GNN model and automatic edge-dropping from the edge candidates.
- Based on **InfoMin** and **InfoMax** principles, we proposed a new adversarial framework for learning efficient data augmentation, called LDA-GCL. LDA-GCL consists of learning data augmentation and graph contrastive learning.
- We conduct extensive experiments on four real-world public benchmark datasets. Experimental results demonstrate the effectiveness of the proposed LDA-GCL.

2   Preliminary

2.1   Bipartite Graph in Recommendation

As the fundamental recommender system, collaborative filtering (CF) can be modelled as a user-item bipratite graph as $G = (\mathcal{U}, \mathcal{I}, \mathcal{E})$, where $\mathcal{U}$ is the user set, $\mathcal{I}$ is the item set and $\mathcal{E} \subseteq \mathcal{U} \times \mathcal{I}$ is the inter-set edges. $\mathcal{E}$ can be denoted as the user-item interaction matrix $\mathbf{R} \in \{0,1\}^{|\mathcal{U}| \times |\mathcal{I}|}$. The adjacency matrix $\mathbf{A} = \begin{bmatrix} \mathbf{0} & \mathbf{R} \\ \mathbf{R}^\top & \mathbf{0} \end{bmatrix}$ is also widely used in [8].

2.2   GNN-Based Collaborative Filtering

Based on the bipartite graph definition, the general GNN-based collaborative filtering methods follow the message-passing scheme to generate informative representations for users and items:

$$z_w^l = f_{\text{aggregate}} \left(\left\{ z_v^{l-1} \mid v \in \mathcal{N}_w \cup \{w\} \right\} \right), z_w = f_{\text{update}} \left(\left[z_w^0, z_w^1, \ldots, z_w^L \right] \right), \quad (1)$$

where $\mathcal{N}$ denotes the neighbor set of node w in bipartite graph G and L denotes the number of GNN layers. z^0 is the learnable initial embeddings. $f_{\text{aggregate}}$ and f_{update} are aggregate function and update function designed by different models. Specifically, the state-of-the-art method (i.e., LightGCN [8]) removes nonlinear activation and feature transformation of NGCF [25] and applies a simple weighted sum aggregator:

$$Z^{l+1} = \left(\mathbf{D}^{-\frac{1}{2}} \mathbf{A} \mathbf{D}^{-\frac{1}{2}} \right) Z^l, Z = \frac{1}{L+1} (Z^0 + Z^1 + \cdots + Z^L), \quad (2)$$

where $\mathbf{D}_{ii} = \sum_j \mathbf{A}_{ij}$ is the diagonal matrix and Z^0 is initial trainable embeddings. After obtaining the final embedding Z, the inner product is used to predict how likely user u would adopt item i by $\hat{y}_{ui} = z_u^T z_i$. Most GNN-based CF methods (e.g., NGCF [25], DGCF [26], and LightGCN [8]) use the pairwise Bayesian Personalized Ranking (BPR) loss function for the model training:

$$\mathcal{L}_{\text{BPR}} = \sum_{(u,i,j) \in \mathcal{O}} - \log \sigma \left(\hat{y}_{ui} - \hat{y}_{uj} \right), \quad (3)$$

where $\mathcal{O} = \{(u,i,j) | (u,i) \in \mathcal{O}^+, (u,j) \in \mathcal{O}^-\}$, $\mathcal{O}^+$ and $\mathcal{O}^-$ are the observed and unobserved interactions, respectively.

2.3   Graph Contrastive Learning in Recommendation

To overcome the data sparsity issues, Graph Contrastive Learning (GCL) is introduced into recommender systems. GCL first applies data augmentation and then contrasts the two augmented samples. Common data augmentation is the perturbation of the graph structure due to the absence of node features. Specifically, SGL [27] proposes edge-dropping, node-dropping, and random walk data augmentation strategies. After data augmentation, the augmented views of the same user node are treated as the positive pairs (i.e., $\{(z'_u, z''_u)\}$), and the views of different user nodes are treated as the negative pairs (i.e., $\{(z'_u, z''_v)\}$)). Following SimCLR [1], the contrastive loss (i.e., InfoNCE [7]) is adopted to maximize the agreement of positive pairs and minimize that of negative pairs by

$$\mathcal{L}^{\mathcal{U}}_{\mathrm{NCE}} = \sum_{u \in \mathcal{U}} - \log \frac{\exp\left(sim\left(\mathbf{z}'_u, \mathbf{z}''_u\right)/\tau\right)}{\sum_{v \in \mathcal{U}} \exp\left(sim\left(\mathbf{z}'_u, \mathbf{z}''_v\right)/\tau\right)}, \qquad (4)$$

where τ is the temperature hyper-parameters and sim is the similarity function (e.g., cosine function). Analogously, contrastive loss is also adopted on the item side (i.e., $\mathcal{L}^{\mathcal{I}}_{\mathrm{NCE}}$). The final contrastive loss is the combination of two losses as $\mathcal{L}_{\mathrm{NCE}} = \mathcal{L}^{\mathcal{U}}_{\mathrm{NCE}} + \mathcal{L}^{\mathcal{I}}_{\mathrm{NCE}}$. It is worthwhile mentioning that the contrastive learning [15,27] in recommender systems usually adopts the joint learning strategy to train their model instead of pre-training and fine-tuning strategies [11]. In other words, both pretext tasks and downstream tasks are optimized jointly [28]. Wu et al. [27] demonstrate that joint training will achieve better performance, the pretext tasks and downstream tasks are mutually enhanced with each other.

3   Methodology

In this section, we introduce our framework for learning data augmentation for recommender systems. In Fig. 1, our methods include pre-trained edge candidate generation, edge operating, and min-max objective functions.

3.1   Graph Data Augmentation with Edge Operating

As mentioned in Sect. 2, the existing data augmentation in recommender systems is generally edge-dropping. The other node dropping and random walk dropping can also be regarded as different strategies of edge-dropping. Edge-dropping can be elaborated as follows:

$$s_1(G) = \mathbf{A}_1 = \mathbf{A} \odot \mathbf{M}_1, \quad s_2(G) = \mathbf{A}_2 = \mathbf{A} \odot \mathbf{M}_2, \qquad (5)$$

where $\odot$ is the Hadamard product and $\mathbf{M}_1, \mathbf{M}_2 \in \{0, 1\}^{|V| \times |V|}$ are two masking matrices to be applied on the original graph G to generate two augmented graph adjacency matrix $\mathbf{A}_1$ and $\mathbf{A}_2$. The node dropping and edge random walk strategies just use different masking matrices (e.g., the row or column sum of a masking matrix with node dropping will be zero). In practice, sampling edges follow a

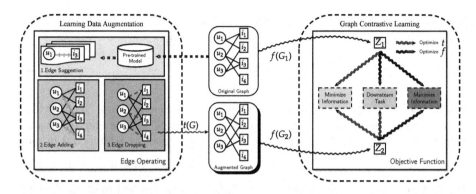

Fig. 1. Illustration of our framework LDA-GCL. LDA-GCL includes learning data augmentation and graph contrastive learning.

uniform distribution to keep $(1 - \rho) \times |\mathcal{E}|$ edges, where ρ is the edge-dropping ratio. ρ is usually set to a small value (e.g., 0.1).

However, the only edge-dropping strategies will suffer the data sparsity issues [27]. We propose a new data augmentation in recommender systems (i.e., **edge-operating** including both edge-adding and edge-dropping). Compared with the edge perturbation in self-supervised on graphs [28], edge-operating in recommendation systems faces some challenges. First, the complexity of randomly sampling edges from $\mathbf{A}$ is $\mathcal{O}((|V|)^2)$ and it is not acceptable in large-scale recommender systems, which have million-level users and items. Second, randomly adding edges to $\mathbf{A}$ will introduce noises. Therefore, we propose to first construct edge candidates and then sample from these edge candidates. As shown in Fig. 1, we firstly train a GNN model (e.g., LightGCN) to predict the item preference of user u (i.e., Edge Suggestion). We choose the top-K_u items for our candidate, where K_u is the degree of the user u in the user-item interaction matrix $\mathbf{R}$. After adding edges, our final edge candidates consist of original edges $\mathcal{E}_0$ and suggested edges $\mathcal{E}_1$ ($|\mathcal{E}_0| = |\mathcal{E}_1|$). We can sample edges from edge candidates and conduct graph contrastive learning (see Sect. 4.4).

3.2 Learning Data Augmentation

After introducing edge-operating augmentation, we further propose to use a learnable edge operator model t to generate informative data augmentation instead of random sampling. In order to learn the graph data augmentation, we use a Multi-layer Perception (MLP) to learn the weight for every edge candidate $e_{u,i}$ as follows:

$$\omega_{u,i} = \mathrm{MLP}\left([z_u \odot z_i] \| \mathbb{1}_{\mathcal{E}}(e_{u,i})\right), \qquad (6)$$

where $\odot$ is the Hadamard product, z_u and z_i are the embeddings for user u and item i, $\|$ is the concatenation operator and $\mathbb{1}_{\mathcal{E}}(e_{u,i})$ indicates if edge $e_{u,i}$

belongs to original or added edges. Then, we use the Gumbel-Max reparameterization [10] to get the probability $p_{u,i}$ for edge $e_{u,i}$ by

$$p_{u,i} = \text{sigmoid}(\frac{(\log \delta - \log(1 - \delta) + \omega_{u,i})}{\tau}), \tag{7}$$

where $\delta \sim \text{Uniform}(0, 1)$ and τ is the temperature hyperparameter. This style of edge learning has also been used in parameterized explanations and adversarial attacks of GNNs [16,22]. Further, we use $p_{u,i}$ to construct augmented graphs $t(G) = \mathbf{A'} = \begin{pmatrix} \mathbf{0} & \mathbf{P} \\ \mathbf{P}^\top & \mathbf{0} \end{pmatrix}$, where $\mathbf{P} \in R^{|\mathcal{U}| \times |\mathcal{I}|}$ is the probability matrix from Eq. 7. We can apply GNN model (i.e., LightGCN in this paper) to original graph $G_1 = G$ with the adjacency matrix $\mathbf{A}$ and augmented graph $G_2 = t(G)$ with the adjacency matrix $\mathbf{A'}$, and get the embeddings Z_1 and Z_2 by Eq. 2.

3.3   Objective Function

Inspired by related works on graph contrastive learning [21,29], we use an adversarial loss function to find good graph augmentations to enhance GCL in recommendation. The objective functions are defined as follows:

$$\begin{aligned} \min_{t} \ & \lambda_t I(f(G); f(t(G))) + \mathcal{L}(f(t(G)), y) \\ \max_{f} \ & I(f(G); f(t(G))) - \mathcal{L}(f(G), y), \end{aligned} \tag{8}$$

where $I(X_1; X_2)$ is the mutual information between two random variables X_1 and X_2, t is the data augmentation learner, f is the GNN encoder (i.e., LightGCN in this paper) and $\mathcal{L}$ is the task relevant supervised loss function (i.e., BPR loss function in this paper). λ_t is used to control the influence of I for t.

In order to estimate the Mutual Information (MI), we choose InfoNCE Estimator [1,7], which is one of the most popular lower-bound to the mutual information. Based on InfoNCE Estimator in Eq. 4, we can map $I(f(G), f(t(G))$ in Eq. 8 to:

$$I(f(G), f(t(G))) \rightarrow -\mathcal{L}_{\text{NCE}} = \frac{1}{B} \sum_{i=1}^{B} \log \frac{\exp (sim (z_{i,1}, z_{i,2}))}{\sum_{i'=1, i' \neq i}^{B} \exp (sim (z_{i',1}, z_{i',2}))}, \tag{9}$$

where sim is the cosine similarity to measure the agreement between two representations, z is the node representation encoded by $f(G)$ and $f(t(G))$, and B is the batch size. Equation 8 is the min-max optimize problem, we use the iterative training approach used in adversarial training [6].

When we fix t, we have following loss function:

$$\mathcal{L}_f = \mathcal{L}_{\text{BPR}}(f(G), y) + \lambda_{ssl}\mathcal{L}_{\text{NCE}} (f(G), f(t(G))) + \lambda_{reg}\|f\|_2^2, \tag{10}$$

where λ_{ssl} and λ_{reg} are the hyper-parameters to control the weights of the InfoNCE loss function and the regularization term. Equation 10 is also used in

---

Algorithm 1: LDA-GCL Training Algorithm

Input: Original bipartite graph $G(\mathcal{U}, \mathcal{I}, \mathcal{E})$; Pre-trained GNN encoder f_0; GNN encoder f; Edge operator model t; Epoch T;

Output: Node representation Z

1: Generate added edges $\mathcal{E}_1$ from pre-trained model f_0.
2: Merge added edges $\mathcal{E}_1$ and original edges $\mathcal{E}$ into edge candidates $\mathcal{E}_2$.
3: Initialize the parameters of edge operator model t and GNN encoder f
4: **for** *epoch* $= 1, ..., T$ **do**
5:   **for** each mini-batch interactions $B = \{(u_1, i_1, i_2)\}$ **do**
6:     Get node set V with user set U and item set I in mini-batch data
       /* Optimize t */
7:     Freeze GNN encoder f; unfreeze edge operator t
8:     Apply t on $\mathcal{E}_2$ to get augmented graph $t(G)$ and Apply f to get the embeddings Z_1, Z_2 for node V from G
9:     Compute loss in Eq. 11 with Z_1 and Z_2; Back propagation, update t.
       /* Optimize f */
10:    Freeze edge operator t; unfreeze of GNN encoder f
11:    Apply t on $\mathcal{E}_2$ to get augmented graph $t(G)$ and Apply f to get the embeddings Z_1, Z_2 for node V from G
12:    Compute loss in Eq. 10 with Z_1 and Z_2; Back propagation, update f.
       /* Judge early stopping condition   */
13:    **if** Z_1 match the early stopping condition **then**
14:      Stop training algorithm; Return the best GNN encoder f_{opt}
15:    **end if**
16:   **end for**
17: **end for**
18: **return** $Z = f_{opt}(G)$

---

self-supervised graph learning for recommendation [27]. We follow the setting of $\lambda_{ssl} = 0.1$ in [27]. Equation 10 leverages the mutual information maximization principle (**InfoMax**) to capture as much information as possible about the stimulus. When we fix f, we have following loss function:

$$\mathcal{L}_t = \mathcal{L}_{\text{BPR}}(f(t(G)), y) - \lambda_2 \mathcal{L}_{\text{NCE}}\left(f(G), f(t(G))\right) + \lambda_{reg}\|t\|_2^2, \qquad (11)$$

where $\lambda_2 = \lambda_t \times \lambda_{ssl}$ and λ_{reg} are the hyper-parameters to control the weights of the InfoNCE loss function and the regularization term. Contrary to Eq. 10, Eq. 11 follow the **InfoMin** principle to find augmented graph that share the minimal information necessary to perform well at the downstream task [23].

3.4 Training LDA-GCL

We briefly summarize LDA-GCL training process in Algorithm 1. As we discussed in Sect. 3.3, Algorithm 1 is iterative adversarial training procedure. We use two optimizers to optimize f and t, separately, which is similar to the training of GANs [5]. In this paper, we use the LightGCN as the pre-trained GNN, other pre-trained models were also tried but with worse results (see Sect. 4.4).

4   Experiments

In this section, we conduct experiments to evaluate the effectiveness of the proposed frameworks with comparison to state-of-the-art methods. Specifically, we aim to answer the following research questions:

- **RQ1**: How does LDA-GCL perform in recommendation tasks as compared with the state-of-the-art CF models and GCL models?
- **RQ2**: If LDA-GCL   performs well, what component benefits our LDA-GCL in collaborative filtering tasks?
- **RQ3**: What hyper-parameters affect the effectiveness of the proposed LDA-GCL?

Table 1. Statistics of the datasets used in this paper.

Datasets	#Users	#Items	#Interactions	%Density
Yelp	45,478	30,709	1,777,765	0.127
Gowalla	29,859	40,989	1,027,464	0.084
Amazon-Book	58,145	58,052	2,517,437	0.075
Alibaba-iFashion	300,000	81,614	1,607,813	0.007

4.1   Experimental Settings

Datasets. We conduct our experiments on four public benchmark datasets, which are widely used for recommender systems: Yelp [2], Gowalla [3], Amazon-Book [17] and Alibaba-iFashion [2]. For the Yelp and Amazon Books datasets, we filter out users and items with fewer than 15 interactions. The Alibaba-iFashion is a large and sparse dataset. The statistics of the datasets used in this paper are summarized in Table 1. Our experimental settings are close to [15], but we remove the duplicated user-item interactions on the Yelp dataset. For each dataset, we randomly select 80% of interactions as training data and 10% of interactions as validation data. The remaining 10% interactions are used for testing model performance. We run such training/validation/testing data split 5 times to report the average scores. We uniformly sample one negative item for each positive instance to form the training set.

Baselines. We compare the proposed method with the following different kinds of baseline methods including Matrix Factorization (MF) methods, Graph Neural Networks (GNN) methods, and Graph Contrastive Learning (GCL) methods:

- **BPRMF** [18]: It optimizes the BPR loss function (see Sect. 2.2) to learn the user and item representations with matrix factorization (MF) framework.
- **NeuMF** [9]: It is a generic matrix factorization model using a multilayer perceptron (MLP) to learn the user-item interaction function. NeuMF uses the pointwise binary cross-entropy loss function to optimize the model.

- **DMF** [30]: It is a matrix factorization model using a deep learning architecture to learn the representations of users and items. It uses a normalized cross entropy loss to optimize the DMF model.
- **NGCF** [25]: NGCF integrates the bipartite graph structure into the embedding learning process. It uses standard GCN [12] to enhance CF methods.
- **DGCF** [26]: DGCF is a GNN model to disentangle user intents factors and yield disentangled representations for user and item.
- **LightGCN** [8]: It simplifies the design of NGCF and devises a light graph convolution for training efficiency and generation ability. It can be viewed as the state-of-the-art GNN-based method.
- **SGL** [27]: SGL introduces several data augmentations (i.e., edge-dropping, node-dropping, and random-walk) and adopts GCL to enhance recommendation. In this paper, we adopt the most powerful SGL-ED as the instantiation of SGL.
- **SimGCL** [33]: It is a simple GCL method with uniform noises to the embedding space.
- **NCL** [15]: It explicitly incorporates the structural neighbors and semantic neighbors into contrastive pairs. It can be viewed as the most advanced GCL in recommendation.

Metrics. To evaluate the performance of top-N recommendation, we adopt two widely used metrics Recall@N and NDCG@N, where N is set to 10, 20 and 50 for consistency. Following [8], we adopt the full-ranking protocol [27], which ranks all the candidate items that the user has not interacted with.

4.2   Implementation Details

We implement the proposed model and all the baselines based on RecBole[1] [35], which is developed based on PyTorch for reproducing and developing recommendation algorithms in a unified, comprehensive, and efficient framework for research purpose. To be a fair comparison, we use Adam optimizer to optimize all the models. The embedding size is set to 64. The batch size is set to 4,096 and all the parameters are initialized by the default Xavier distribution. For NCL, we use the authors' released code from github[2]. We follow the authors' suggested hyper-parameter settings. We adopt early stopping with the patience of 10 epochs to prevent overfitting, and NDCG@10 is set as the early stopping indicator. All experiments run on an NVIDIA Tesla V100S GPU (32 GB) and Intel(R) Xeon(R) Silver 4310 CPUs (250 GB).

4.3   Performance Comparison (RQ1)

Table 2 shows the performance comparison of the proposed LDA-GCL and other baseline methods on four datasets. From Table 2, we can find that:

---

[1] https://recbole.io/.

[2] https://github.com/RUCAIBox/NCL.

- For MF-based methods, BPRMF outperforms NeuMF and DMF on all datasets (e.g., 17.9% in Recall@10 and 22.4% in NDCG@10 on Gowalla, compared with NeuMF). It can be due to the reason that it is difficult for MLP to learn the dot product [19] and the pointwise loss function (e.g., Binary Cross-Entropy loss function in NeuMF) is less effective than the pairwise loss function (e.g., BPR loss function in BPRMF).
- Compared to MF-based methods (e.g., BPRMF), GNN-based methods exhibit better performance on most datasets. GNN-based methods utilize the structural information of bipartite graphs into representations by introducing the powerful GNNs. Among all the GNN-based models, LightGCN performs best in most datasets, demonstrating the effectiveness of simplified architecture [8]. BPRMF performs even better than NGCF on Gowalla and Amazon-Book. Similar results are also reported in [15], again showing that the heavy GNN architecture will overfit and limit the performance. For the disentangled representation learning method DGCF, we limit the number of layers to 2 because of GPU memory limitations on the Amazon-Book dataset. DGCF is worse than LightGCN, especially on the sparse dataset (e.g., 11.2% in Recall@10 and 11.5% in NDCG@10 on Alibaba-iFashion). The low number of factors for disentanglement (default value is 4) may be the limit of DGCF.
- For the GCL-based baseline methods (i.e., SGL, SimGCL, and NCL), GCL-based methods consistently outperform other supervised GNN-based methods on all datasets, which shows the effectiveness of GCL for improving performance. With uniform noises, SimGCL outperforms SGL on most datasets. However, SimGCL performs poorly on some datasets (e.g., Alibaba-iFashion). The hyper-parameters in SGL and SimGCL (e.g., data augmentation strategy, edge dropout ratio, and contrastive ratio) will severely impact the performance. Compared with SGL and SimGCL, NCL achieves significant improvements on most datasets, which is consistent with the results reported in [15]. We consider NCL as the most competitive baseline. NCL incorporates the neighborhood-enriched contrastive learning objectives and achieves better results than SGL and SimGCL on most datasets.
- For our LDA-GCL, we can find that LDA-GCL outperforms all baselines and significantly performs better than the state-of-the-art NCL on most datasets. It is worth mentioning that our method achieves the largest improvement on the largest and sparsest dataset (i.e., Alibaba-iFashion). On Alibaba-iFashion, LDA-GCL outperforms NCL 23.5% and 25.28% in Recall@10 and NDCG@10, respectively. And it also surpasses the SGL on the Alibaba-iFashion dataset by about 10%. The experimental results show the effectiveness of the method proposed in this paper.

4.4   Benefits of LDA-GCL (RQ2)

Based on the previous experimental results, we analyze the advantages of our model in this subsection. We choose Gowalla and Alibaba-iFashion to analyze the reasons for the advantage in the previous subsection. We randomly resplit datasets as the previous subsection does.

Table 2. Performance Comparison of Different Baseline Models

Dataset	Metric	Matrix Factorization			Graph Neural Networks			Graph Contrastive Learning			
		BPRMF	NeuMF	DMF	NGCF	DGCF	LightGCN	SGL	SimGCL	NCL	LDA-GCL
Yelp	Recall@10	0.0499	0.0367	0.0372	0.0514	0.0606	0.0616	0.0664	<u>0.0743</u>	0.0713	**0.0751***
	Recall@20	0.0829	0.0629	0.0631	0.0857	0.0987	0.1001	0.1072	<u>0.1185</u>	0.1135	**0.1190***
	Recall@50	0.1549	0.1227	0.1215	0.1596	0.1798	0.1817	0.1928	<u>0.2068</u>	0.1997	**0.2101***
	NDCG@10	0.0335	0.0242	0.0248	0.0346	0.0412	0.0419	0.0456	<u>0.0515</u>	0.0489	**0.0518***
	NDCG@20	0.0438	0.0324	0.0327	0.0453	0.0530	0.0538	0.0581	<u>0.0652</u>	0.0619	**0.0653***
	NDCG@50	0.0622	0.0477	0.0476	0.0642	0.0738	0.0748	0.0801	<u>0.0878</u>	0.0841	**0.0886***
Amazon-Book	Recall@10	0.0619	0.0442	0.0313	0.0575	0.0787	0.0783	0.0844	0.0872	<u>0.0947</u>	**0.0975***
	Recall@20	0.0971	0.0726	0.0522	0.0920	0.1191	0.1210	0.1281	0.1251	<u>0.1395</u>	**0.1456***
	Recall@50	0.1676	0.1331	0.0984	0.1624	0.1965	0.2055	0.2117	0.1934	<u>0.2201</u>	**0.2346***
	NDCG@10	0.0431	0.0295	0.0216	0.0400	0.0563	0.0553	0.0606	0.0643	<u>0.0685</u>	**0.0699***
	NDCG@20	0.0537	0.0382	0.0280	0.0505	0.0681	0.0682	0.0739	0.0758	<u>0.0822</u>	**0.0845***
	NDCG@50	0.0721	0.0539	0.0400	0.0688	0.0887	0.0902	0.0956	0.0936	<u>0.1034</u>	**0.1078***
Gowalla	Recall@10	0.1040	0.0882	0.0634	0.0992	0.1343	0.1355	0.1386	0.1487	<u>0.1496</u>	**0.1505**
	Recall@20	0.1525	0.1307	0.0945	0.1462	0.1917	0.1969	0.1969	0.2123	<u>0.2131</u>	**0.2144**
	Recall@50	0.2476	0.2161	0.1559	0.2383	0.2972	0.3093	0.3055	0.3208	<u>0.3228</u>	**0.3284***
	NDCG@10	0.0738	0.0603	0.0450	0.0703	0.0963	0.0961	0.0999	0.1078	<u>0.1081</u>	**0.1085**
	NDCG@20	0.0878	0.0727	0.0540	0.0838	0.1127	0.1136	0.1166	0.1259	<u>0.1263</u>	**0.1268**
	NDCG@50	0.1109	0.0935	0.0692	0.1062	0.1384	0.1411	0.1431	0.1525	<u>0.1534</u>	**0.1547**
Alibaba-iFashion	Recall@10	0.0297	0.0157	0.0138	0.0355	0.0361	0.0402	<u>0.0518</u>	0.0450	0.0490	**0.0605***
	Recall@20	0.0458	0.0264	0.0229	0.0565	0.0549	0.0612	<u>0.0774</u>	0.0651	0.0729	**0.0882***
	Recall@50	0.0784	0.0501	0.0443	0.0994	0.0910	0.1015	<u>0.1258</u>	0.1029	0.1178	**0.1381***
	NDCG@10	0.0158	0.0079	0.0071	0.0185	0.0194	0.0216	<u>0.0280</u>	0.0252	0.0267	**0.0335***
	NDCG@20	0.0199	0.0106	0.0094	0.0237	0.0241	0.0269	<u>0.0344</u>	0.0303	0.0328	**0.0405***
	NDCG@50	0.0264	0.0152	0.0137	0.0323	0.0313	0.0350	<u>0.0440</u>	0.0378	0.0417	**0.0504***

The best result is **bolded** and the second result is <u>underlined</u>. * indicates the statistical significance for $p < 0.05$.

Sparsity Analysis. In general, graph contrastive learning on recommender systems can alleviate the problem of data sparseness commonly found in recommender systems [15,27]. To further verify the proposed LDA-GCL can alleviate the sparsity of interaction data, we evaluate the performance of the different groups of users. We split all the users into five groups based on their interaction numbers. Then, we compare the performance of LightGCN, SGL, and LDA-GCL on these groups. From Fig. 2, we can find that the performance improvements of SGL mainly come from recommending the items with sparse interactions, which is consistent with findings in [15]. On the Gowalla dataset, the performance of SGL is not even as good as LightGCN on the groups of users with a high number of interactions (e.g., $Group3, Group4$). For our LDA-GCL, we can find the performance of LDA-GCL is consistently better than or comparable with LightGCN and SGL on all user groups. Besides, as the number of interactions decreases, LDA-GCL achieves greater improvements. For instance, on the Alibaba-iFashion dataset, LDA-GCL outperform LightGCN 27.9%, 45.5%, and 51.8% on $Group3$, $Group2$, and $Group1$, respectively. In conclusion, LDA-GCL can further alleviate the problem of data sparsity compared to SGL.

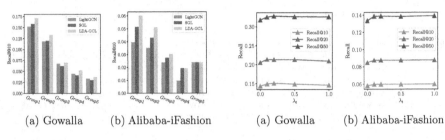

(a) Gowalla   (b) Alibaba-iFashion   (a) Gowalla   (b) Alibaba-iFashion

Fig. 2. Performance analysis over different users groups. G_1 is the group of users with the *lowest* interaction number.

Fig. 3. Parameter Analysis of λ_t. The list of recommended sites

Table 3. Performance comparison of different variants of LDA-GCL.

Method	Gowalla		Alibaba-iFashion	
	Recall@10	NDCG@10	Recall@10	NDCG@10
LightGCN	0.1342	0.0962	0.0395	0.0212
DA-GCL(0.0,0.0)	0.1488	0.1085	0.0497	0.0274
DA-GCL(0.1,0.0)	0.1492	0.1083	0.0529	0.0289
DA-GCL(0.0,0.1)	0.1487	0.1067	0.0544	0.0299
DA-GCL(0.1,0.1)	0.1479	0.1063	0.0553	0.0303
DA-GCL(0.0,0.5)	0.1412	0.1010	0.0533	0.0290
DA-GCL(0.1,0.5)	0.1409	0.1003	0.0542	0.0296
DA-GCL(0.0,1.0)	0.1369	0.0973	0.0520	0.0282
DA-GCL(0.1,1.0)	0.1359	0.0963	0.0526	0.0285
LDA-GCL (w NGCF)	0.1488	0.1078	0.0589	0.0322
LDA-GCL (w/o EA)	0.1499	0.1087	0.0579	0.0319
LDA-GCL	0.1512	0.1090	0.0599	0.0330

Ablation Study. As we discussed before, LDA-GCL consists of two parts: learning data augmentation and graph contrastive learning. We do ablation studies to investigate the role of each part.

First, since our edge candidates include original edges and added edges, we can remove the added edges from edge candidates. In other words, we use only edge-dropping data augmentation. We mark it as LDA-GCL (w/o EA). Second, we use NGCF as the pre-trained GNN to generate edge candidates, which is marked as LDA-GCL (w NGCF). Third, we can use randomly added and dropped edges for data augmentation and remove the module for learning data augmentation. We mark such data augmentation methods as DA-GCL(p_A, p_D), where p_A is the ratio of added edges, and p_D is the ratio of dropped edges. DA-GCL is close to SGL, the difference is that SGL only includes the edge-dropping data augmentation, which happens on two views. In contrast, DA-GCL has both edge-dropping and edge-adding on only one view. The other view in DA-GCL is the original graph. We vary p_D from $\{0.0, 0.1\}^3$ and vary p_A from $\{0.0, 0.1, 0.5, 1\}$. Note that, $p_A = 0.0$ and $p_D = 0.0$ mean that no data augmentation is employed, but the contrastive loss is still employed.

The results are reported in Table 3. From Table 3, we can find that: (1) All the variants of LDA-GCL outperform LightGCN by large margins, which demonstrates the effectiveness of contrastive learning in improving performance. (2) When we do not adopt the edge-adding strategy, the performance of LDA-GCL (w/o EA) shows a degradation compared to LDA-GCL. On Recall@10, LDA-GCL outperforms LDA-GCL (w/o EA) by 0.87% and 3.5% on the dataset Gowalla and Alibaba-iFashion, respectively. It demonstrates that the strategy of adding edges achieves greater improvement on more sparse datasets. (3) Compared with LDA-GCL, there is a drop in performance for LDA-GCL (w NGCF). It shows that better edge candidates have an impact on the results. (4) We can find that learnable data augmentation methods (i.e., LDA-GCL) show better performance than randomly sampling data augmentation methods (i.e., DA-GCL) on both datasets. (5) When we remove the learning data augmentation module, we can find that the edge-adding strategy (e.g., $p_A = 0.1$) performs better on the sparser Alibaba-iFashion dataset compared to the edge-dropping strategy (e.g., $p_A = 0.0$). Performance decreases as the ratio of added edges increases. We speculate that this is due to the introduction of more noise. (6) Surprisingly, when the graph augmentation is detached (i.e., DA-GCL(0.0, 0.0)), the performance gains are still so remarkable. This result was also reported in [33], which suggests the potential of a proper graph augmentation.

4.5   Parameter Analysis (RQ3)

We analyze the hyper-parameters of λ_t in Eq. 8. We vary it from $\{0.0, 0.1, 0.2, 0.5, 1.0\}$. Note that, $\lambda_t = 0.0$ means there is no $\min_t I(f(G); f(t(G)))$. Due to the page limits, we report the Recall in Fig. 3. The NDCG metrics have similar findings.

In Fig. 3, the best settings of λ_t on the Alibaba-iFashion dataset and the Gowalla dataset are 1.0 and 0.2, respectively. When we adopt $\min_t I(f(G); f(t(G)))$ (i.e., $\lambda_t > 0.0$), the hyper-parameter λ_t makes less of a difference. We can find that the performance is worst when $\lambda_t = 0.0$ on both datasets. More specifically, the settings of $\lambda_t = 0.1$ outperform the settings of $\lambda_t = 0.0$, 2.75%, 3.30 % and 3.82% on Recall@10, Recall@20, and Recall@50 on the Alibaba-iFashion dataset. The settings of $\lambda_t = 0.1$ outperform the settings of $\lambda_t = 0.0$, 3.82%, 4.13 % and 2.42% on Recall@10, Recall@20, and Recall@50 on the Gowalla dataset. It demonstrates the effectiveness of LDA-GCL adopting the InfoMin principle guided data augmentation learning.

---

[3] Large p_D will harm the performance according to the report in [27].

5   Related Work

5.1   GNN-Based Recommendation

Nowadays, GNNs are also widely used in recommender systems. Different from traditional CF methods, such as matrix factorization (MF) methods [9,13,18] and auto-encoder (AE) methods [14], Graph Neural Networks (GNN) are used to model interaction data into a bipartite graph and learn users and items effective representations from the graph structure information [8,25,26]. Most GNN methods in recommender system follow the message-passing scheme [4] to utilize the bipartite graph structure, including the design of aggregate functions and update functions for recommendations. Most representatively, LightGCN [8] removes the self-connection, feature transformation, and nonlinear activation of standard GCNs, and leverages the high-order graph structure to enhance the recommendation performance. LightGCN has also been used as a backbone model for many further works (e.g., GCL [27]). Although the GNN models are effective, they still suffer from data sparsity issues in recommender systems.

5.2   Contrastive Learning in Recommendation

Contrastive Learning (CL) as a self-supervised manner [34], has been applied in Recommender Systems (RS) [15,31,32]. In recommender system scenarios, Graph Contrastive Learning (GCL) is often used to alleviate the data sparsity problem [15,27]. For example, SGL [27] constructs the contrastive pairs by random sampling. The methods aforementioned for data augmentation are manually designed and require expertized knowledge. And improper data augmentation will hinder the performance of GCL [33]. The proper way of data augmentation requires guiding principles instead of the heuristic design.

Inspired by **InfoMin** principle proposed by [23], AD-GCL [21] optimizes adversarial graph augmentation strategies to train GNNs to avoid capturing redundant information during the training. However, AD-GCL is designed to work on unsupervised graph classification with lots of small graphs, under the pre-training & fine-tuning scheme. But in graph collaborative filtering in recommendation (link prediction with a large graph, under the joint learning scheme), how to design a learnable data augmentation method for better performance improvement is an open and promising research problem.

6   Conclusion and Future Work

In this work, we develop a theoretically motivated learnable data augmentation model for GCL in recommendation, instead of heuristic designs. Guided by both **InfoMin** and **InfoMax** principles, our model is an adversarial framework that can better enhance the effect of GCL in the recommendation. Via learning better data augmentation in GCL, our model achieves state-of-the-art performance on several public benchmark datasets. Further experiments on each component demonstrate the effective design of LDA-GCL. Our methods open the door to designing learnable data augmentation methods instead of heuristic augmentation methods in recommendation.

The main limitation of our model is the high complexity of learning data augmentation, which may cause low training efficiency, compared with simple data augmentation [33]. We can make improvements on the efficiency in future work. A potential boosting scheme is the pre-trained edge operator models.

Acknowledgement. This work is funded by the National Natural Science Foundation of China under Grant Nos. 62102402, U21B2046, 62272125, and the National Key R&D Program of China (2022YFB3103701). Huawei Shen is supported by Beijing Academy of Artificial Intelligence (BAAI). Junjie Huang is supported by Tencent Rhino-Bird Elite Training Program.

References

1. Chen, T., Kornblith, S., Norouzi, M., Hinton, G.: A simple framework for contrastive learning of visual representations. In: ICML, pp. 1597–1607. PMLR (2020)
2. Chen, W., et al.: POG: personalized outfit generation for fashion recommendation at alibaba ifashion. In: KDD, pp. 2662–2670 (2019)
3. Cho, E., Myers, S.A., Leskovec, J.: Friendship and mobility: user movement in location-based social networks. In: KDD, pp. 1082–1090 (2011)
4. Gilmer, J., Schoenholz, S.S., Riley, P.F., Vinyals, O., Dahl, G.E.: Neural message passing for quantum chemistry. In: ICML, pp. 1263–1272. PMLR (2017)
5. Goodfellow, I., et al.: Generative adversarial nets. In: NeurIPS 27 (2014)
6. Goodfellow, I., Shlens, J., Szegedy, C.: Explaining and harnessing adversarial examples. In: ICLR (2015)
7. Gutmann, M., Hyvärinen, A.: Noise-contrastive estimation: a new estimation principle for unnormalized statistical models. In: AISTATS, pp. 297–304 (2010)
8. He, X., Deng, K., Wang, X., Li, Y., Zhang, Y., Wang, M.: LightGCN: simplifying and powering graph convolution network for recommendation. In: SIGIR, pp. 639–648 (2020)
9. He, X., Liao, L., Zhang, H., Nie, L., Hu, X., Chua, T.S.: Neural collaborative filtering. In: WWW, pp. 173–182 (2017)
10. Jang, E., Gu, S., Poole, B.: Categorical reparameterization with gumbel-softmax. In: ICLR (2017)
11. Jin, W., et al.: Self-supervised learning on graphs: deep insights and new direction. arXiv preprint arXiv:2006.10141 (2020)
12. Kipf, T.N., Welling, M.: Semi-supervised classification with graph convolutional networks. In: ICLR (2017)
13. Koren, Y., Bell, R., Volinsky, C.: Matrix factorization techniques for recommender systems. Computer **42**(8), 30–37 (2009)
14. Liang, D., Krishnan, R.G., Hoffman, M.D., Jebara, T.: Variational autoencoders for collaborative filtering. In: WWW, pp. 689–698 (2018)
15. Lin, Z., Tian, C., Hou, Y., Zhao, W.X.: Improving graph collaborative filtering with neighborhood-enriched contrastive learning. In: WWW, pp. 2320–2329 (2022)
16. Luo, D., et al.: Parameterized explainer for graph neural network. In: NeurIPS 33, pp. 19620–19631 (2020)

17. McAuley, J., Targett, C., Shi, Q., Van Den Hengel, A.: Image-based recommendations on styles and substitutes. In: SIGIR, pp. 43–52 (2015)
18. Rendle, S., Freudenthaler, C., Gantner, Z., Schmidt-Thieme, L.: BPR: Bayesian personalized ranking from implicit feedback. arXiv preprint arXiv:1205.2618 (2012)
19. Rendle, S., Krichene, W., Zhang, L., Anderson, J.: Neural collaborative filtering vs. matrix factorization revisited. In: RecSys, pp. 240–248 (2020)
20. Sarwar, B., Karypis, G., Konstan, J., Riedl, J.: Item-based collaborative filtering recommendation algorithms. In: WWW, pp. 285–295 (2001)
21. Suresh, S., Li, P., Hao, C., Neville, J.: Adversarial graph augmentation to improve graph contrastive learning. In: NeurIPS 34 (2021)
22. Tao, S., Cao, Q., Shen, H., Huang, J., Wu, Y., Cheng, X.: Single node injection attack against graph neural networks. In: CIKM, pp. 1794–1803 (2021)
23. Tian, Y., Sun, C., Poole, B., Krishnan, D., Schmid, C., Isola, P.: What makes for good views for contrastive learning? In NeurIPS 33, pp. 6827–6839 (2020)
24. Tishby, N., Pereira, F.C., Bialek, W.: The information bottleneck method. arXiv preprint physics/0004057 (2000)
25. Wang, X., He, X., Wang, M., Feng, F., Chua, T.S.: Neural graph collaborative filtering. In: SIGIR, pp. 165–174 (2019)
26. Wang, X., Jin, H., Zhang, A., He, X., Xu, T., Chua, T.S.: Disentangled graph collaborative filtering. In: SIGIR, pp. 1001–1010 (2020)
27. Wu, J., et al.: Self-supervised graph learning for recommendation. In: SIGIR, pp. 726–735 (2021)
28. Wu, L., Lin, H., Tan, C., Gao, Z., Li, S.Z.: Self-supervised learning on graphs: contrastive, generative, or predictive. IEEE TKDE (2021)
29. Xu, D., Cheng, W., Luo, D., Chen, H., Zhang, X.: InfoGCL: information-aware graph contrastive learning. In: NeurIPS 34 (2021)
30. Xue, H.J., Dai, X., Zhang, J., Huang, S., Chen, J.: Deep matrix factorization models for recommender systems. In: IJCAI, Melbourne, Australia, vol. 17, pp. 3203–3209 (2017)
31. Yao, T., et al.: Self-supervised learning for large-scale item recommendations. In: CIKM, pp. 4321–4330 (2021)
32. Yu, J., Yin, H., Li, J., Wang, Q., Hung, N.Q.V., Zhang, X.: Self-supervised multi-channel hypergraph convolutional network for social recommendation. In: WWW, pp. 413–424 (2021)
33. Yu, J., Yin, H., Xia, X., Chen, T., Cui, L., Nguyen, Q.V.H.: Are graph augmentations necessary? simple graph contrastive learning for recommendation. In: SIGIR, pp. 1294–1303 (2022)
34. Yu, J., Yin, H., Xia, X., Chen, T., Li, J., Huang, Z.: Self-supervised learning for recommender systems: a survey. arXiv preprint arXiv:2203.15876 (2022)
35. Zhao, W.X., et al.: RecBole: towards a unified, comprehensive and efficient framework for recommendation algorithms. In: CIKM, pp. 4653–4664 (2021)

Sequential Hypergraph Convolution Network for Next Item Recommendation

Jiaxing Chen[1], Hongzhi Liu[2(✉)], Yingpeng Du[2], Zekai Wang[3], Yang Song[3], and Zhonghai Wu[4(✉)]

[1] Center for Data Science, AAIS, Peking University,
Beijing, People's Republic of China
`chenjiaxing@stu.pku.edu.cn`
[2] School of Software and Microelectronics, Peking University,
Beijing, People's Republic of China
`{liuhz,dyp1993}@pku.edu.cn`
[3] BOSS Zhipin NLP Center, Beijing, People's Republic of China
`{wangzekai,songyang}@kanzhun.com`
[4] National Engineering Research Center of Software Engineering, Peking University,
Beijing, People's Republic of China
`wuzh@pku.edu.cn`

Abstract. Graph neural networks have been widely used in personalized recommendation tasks to predict users' next behaviors. Recent research efforts have attempted to use hypergraphs to capture higher-order information among items. However, the existing methods ignore the sequential patterns between different sequences, which can be considered as collaborative signals at the sequential level. Moreover, the multifaceted nature of user behaviors is not sufficiently exploited in existing methods, which may mislead the representation learning of items. In addition, the high computational cost limits the application of hypergraph-based methods.

To address these challenges, we propose a novel architecture called the sequential hypergraph convolution network (SHCN) for next item recommendation. First, we design a novel data structure, called a sequential hypergraph, that accurately represents the behavior sequence of each user in each sequential hyperedge. Second, a well-designed node-hyperedge propagation method based on the sequential hypergraph is proposed to effectively capture the sequential patterns between different sequences and utilize the multifaceted nature of user behaviors. Third, we propose a hypergraph sampling strategy to improve the computational efficiency of hypergraph convolution. Extensive experiments on four real-world datasets demonstrate that the proposed method consistently outperforms state-of-the-art methods.

Keywords: Recommendation systems · Sequential recommendation · Hypergraph · Graph neural network

X. Wang et al. (Eds.): DASFAA 2023, LNCS 13944, pp. 389–405, 2023.
https://doi.org/10.1007/978-3-031-30672-3_26

1   Introduction

Next item recommendation is dedicated to predicting users' next behaviors based on their historical behavior sequences and has been widely used in online information systems, such as e-commerce and news systems [18]. The key to this task is to mine and utilize the sequential patterns in users' historical behaviors to capture each user's current intent for prediction.

Recently, graph natural networks (GNN) have achieved state-of-the-art performance in next item recommendation, which adopt the information diffusion with the user-to-item edge in the graph and transform the behavior sequences of all users into pairwise relations to make predictions [5]. However, the relations among items can be triadic, tetradic, or higher orders [1], which is difficult to capture with pairwise relation modeling. For example, a user's purchase of pumpkins and candy at the same time is more indicative of Halloween than the purchase of pumpkins or candy individually. Recently, several studies [16,17,20] tried to utilize the hypergraph to capture the beyond-pairwise relations among items for sequential recommendation. Although these studies have shown promising results, they suffer from some limitations.

One of the main challenges in next item recommendation is how to capture the collaborative signals between different sequences. Specifically, users with similar behaviors will exhibit similar sequential preferences on items [18]. For example, a user purchased a bed, a mattress, and a bedsheet successively, whose sequential preference on items can be shared by other users with similar purchasing behaviors. However, these important signals are not considered in existing methods. Existing data structures, such as graphs and hypergraphs, have difficulty accurately representing the behavior sequences of users, which limits the utilization of the sequential patterns between different sequences.

Another problem is that the multifaceted nature of user behaviors is not sufficiently exploited in existing methods [3]. Each hyperedge in a hypergraph is, in fact, the behavior sequence of a user, which tends to show more than one interest of the user. For instance, a user buys a camera bag, a video camera, an iPhone, and AirPods. Obviously, the camera and the camera bag are related to photography, while the iPhone and the AirPods are related to Apple. On the one hand, it is not reasonable to roughly and indiscriminately aggregate different facets of user behaviors, which may mislead the representation learning of items. On the other hand, the different facets of users' behaviors can reveal the semantic relations among items, especially when side information, such as brand or category, is not available.

The computational cost of GNN-based models limits their applications, as efficient recommender systems are more expected in the real world. Hypergraph-based models in existing work can effectively capture beyond-pairwise signals, but suffer from computational inefficiencies because the structure of hypergraphs is more complex than general GNN-based methods. Moreover, the implementation of complex hypergraph convolution strategy will result in a drastic increase in complexity.

To address these challenges, we propose a novel sequential model named the *Sequential Hypergraph Convolution Network* (SHCN) for next item recommendation. To accurately represent the behavior sequences of users among items, we design a novel data structure named *sequential hypergraph*. Each user sequence is represented as a sequential hyperedge and the items in the sequence are represented as nodes in the hyperedge. Based on the sequential hypergraph, we design a sequential hypergraph convolution module with a well-designed node-hyperedge propagation method consisting of a sequential pattern extractor and a user interest aggregator. The sequential pattern extractor is used to extract sequential patterns between different sequences and the user interest aggregator uses an attention mechanism to mine the semantic relations between items. To improve computational efficiency, we propose a hypergraph sampling strategy for convolution.

The main contributions of this work are summarized as follows:

- We design a data structure called a sequential hypergraph to accurately represent the behavior sequences of different users among items.
- We propose a sequential hypergraph convolution module that captures the sequential patterns between different sequences and exploits the multifaceted interests of users to capture the semantic relations between items.
- We propose a hypergraph sampling strategy to improve the computational efficiency of the sequential hypergraph convolution.
- Extensive experiments on four real-world datasets show that our proposed SHCN model significantly outperforms the state-of-the-art methods.

2 Related Work

2.1 Sequential Recommendation

Compared with traditional collaborative filtering (CF) methods [8], which consider users to be static, Sequential Recommendation (SR) approaches dynamically update users' embedding based on the users' historical interactions and generate a list of recommendations together with item embeddings. Early methods [13] used Markov chains to model sequence information. With the development of deep learning, researchers have used deep neural networks to model user preferences via sequential behaviors. Recurrent neural network (RNN)-based models use GRUs [9] or long short-term memory (LSTM) [2] to model users' sequential behaviors. Caser [15] encoded a user interaction sequence with convolutional filters to capture recent interests. As transformers have achieved state-of-the-art performance in machine translation tasks, many attention-based methods have been proposed and have achieved considerable success in recommendation. SASRec [10] uses a self-attention mechanism to find relevant items from the historical interactions of users and make recommendations. On this basis, [21] introduced time and position information into an attention mechanism to obtain more related items for recommendation purposes. However, these methods only use the historical interaction sequences of users to make recommendations, ignoring the high-order relations of items between different users.

2.2   GNN for Recommendation

Due to the advantages of GNNs in terms of constructing structural data and capturing high-order connectivity [5], many GNN-based recommenders construct input data as a graph structure and give predictions via graph learning. Some methods [22] utilize graph convolutional network (GCN) to make predictions. Graph attention network (GAT)-based models [19] utilize attention mechanisms to aggregate neighborhood features. Some works have combined a GNN with an RNN for prediction [25].

Although these methods have achieved some results, it is difficult to extract higher-order relations among items due to the limitation of the graph structure. Hypergraphs have certain advantages in terms of encoding high-order data correlations. In social recommendation, MHCN [23] applies hypergraphs to model social networks and enhances social recommendation by using high-order user relations. In session-based recommendation, DHCN [20] models each session as a hyperedge and all sessions as a hypergraph to obtain higher-order relations among items, while SHARE [17] models a hypergraph for each session and uses sliding windows to model hyperedges to capture user intents.

However, existing hypergraph-based methods, which express each user sequence as a hyperedge, ignore the sequential patterns between different sequences and the multifaceted nature of user behaviors. In addition, they rarely consider the computational inefficiency of hypergraph convolution.

3   Problem Definition

Let $U = \{u_1, u_2, \cdots, u_{|U|}\}$ and $I = \{i_1, i_2, \cdots, i_{|I|}\}$ represent the set of users and items, respectively. For each user u, we use $S_u = \{i_{1,u}, i_{2,u}, \cdots, i_{|S_u|,u}\}$ to represent the item sequence that user u has interacted with in chronological order.

The goal of next item recommendation is to predict the next item that user u will be interested in after S_u. More specifically, given the interaction data $\{S_u | u \in U\}$, our purpose is to construct a recommendation model that can estimate the probability that a user u will interact with an item i in the next action based on S_u.

4   Methodology

In this section, we propose an end-to-end next item recommendation framework empowered by the sequential hypergraph (Fig. 1). It consists of three main parts, including sequential hypergraph construction, sequential hypergraph convolution, and target sequence modeling.

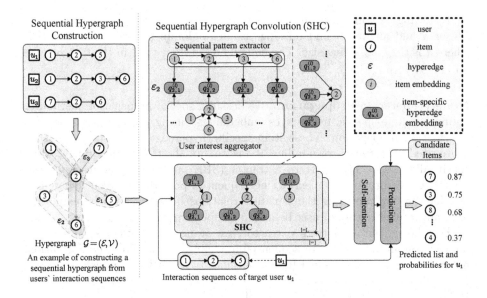

Fig. 1. The framework of our proposed SHCN.

4.1 Sequential Hypergraph Construction

Hypergraph provides a natural way to capture complex high-order relations. However, the traditional hypergraphs in existing methods have difficulty accurately representing the behavior sequences of users. To this end, we propose a data structure named *sequential hypergraph* to further capture the sequential patterns between different sequences.

Given the behavior sequences of all users $\{S_u | u \in U\}$, a sequential hypergraph is defined as $\mathcal{G} = (\mathcal{E}, \mathcal{V})$, where $\mathcal{V} = I$ and $\mathcal{E} = \{\varepsilon_u | \varepsilon_u = S_u, u \in U\}$ denote the node set and the sequential hyperedge set, respectively. Specifically, a sequence of items that user $u \in U$ has interacted with in chronological order is expressed as a sequential hyperedge, and each item in the sequence is expressed as a node in the sequential hypergraph.

Compared with the hypergraphs in existing methods, in which each hyperedge is the collection of items, the sequential hypergraph we constructed not only retains the beyond-pairwise relations among items but also preserves the sequential information of users' behaviors, making it possible to model the sequential patterns between different sequences.

4.2 Sequential Hypergraph Convolution (SHC)

To capture the sequential patterns between different sequences and sufficiently utilize the multifaceted nature of user behaviors, we develop a sequential hypergraph convolutional network with an elaborate node-hyperedge propagation strategy based on the sequential hypergraph.

Following existing methods [10], we use $e_i \in \mathbb{R}^d$ and $e_u \in \mathbb{R}^d$ to describe a user u and an item i and initialize them randomly according to their IDs, where d is the embedding size. The embedding of item $i \in I$ in the 0-th layer is initialized with e_i, i.e. $e_i^{(0)} = e_i$.

Node-Hyperedge Propagation. We design a special propagation method to obtain the item-specific hyperedges embeddings during the node-to-hyperedge information propagation. These item-specific hyperedge embeddings are then aggregated into the corresponding items to update their embeddings. More specifically, a sequential pattern extractor (SPE) and a user interest aggregator (UIA) are proposed for extracting sequential patterns between hyperedges and for directly capturing the semantic relations between items by exploiting the multifaceted nature of user behaviors, respectively.

For the sequential pattern extractor, we argue that both past and future information are important for sequential modeling. To capture the information, naturally, we use bidirectional RNNs, specifically Bi-GRU, to model the hyperedges. Given the hyperedge ε_u and the node $i \in \varepsilon_u$, we denote the output of the Bi-GRU network at node i as $h_{u,i}$ as follows:

$$h_{u,i}^{(l)} = \mathrm{BiGRU}(H_u^{(l)}, i), \tag{1}$$

where l refers to the l-th layer and $H_u^{(l)}$ is the embedding matrix consisting of the embeddings of all items in the hyperedge ε_u in the l-th layer.

For the user interest aggregator, an attention mechanism is used to aggregate items related to the user's interests. Given the hyperedge ε_u, for each item $i \in \varepsilon_u$, we calculate the similarity $\alpha_{j,i,u}^{(l)}$ between item i and every other item $j \in \varepsilon_u$ in the l-th layer with the normalized inner product as follows:

$$\alpha_{j,i,u}^{(l)} = \frac{\exp\left(e_j^{(l)} \cdot e_i^{(l)}\right)}{\sum_{k \in \varepsilon_u} \exp\left(e_k^{(l)} \cdot e_i^{(l)}\right)}. \tag{2}$$

We argue that the greater the similarity $\alpha_{j,i,u}^{(l)}$ between item i and item j, the greater the probability that these two items belong to the same user interest, which also indicates the greater the semantic similarity between the two items. Then, we use the similarity $\alpha_{j,i,u}^{(l)}$ as the attention weight to aggregate all the items together to obtain the output of the user interest aggregator in the l-th layer denoted as $g_{u,i}^{(l)}$ as follows:

$$g_{u,i}^{(l)} = \sum_{j \in \varepsilon_u} \alpha_{j,i,u}^{(l)} e_j^{(l)}. \tag{3}$$

After obtaining the outputs of the sequential pattern extractor and the user interest aggregator, we aggregate them together to obtain item-specific hyperedge embedding $q_{u,i}^{(l)} \in \mathbb{R}^d$ in the l-th layer as follows:

$$q_{u,i}^{(l)} = W^{(l)}(h_{u,i}^{(l)} \| g_{u,i}^{(l)}), \tag{4}$$

where $\|$ denotes the concatenation operation and $W^{(l)} \in \mathbb{R}^{d \times 2d}$ is the encoding matrix for information aggregation in the l-th layer.

After obtaining the item-specific hyperedge embeddings with sequential patterns as well as semantic relations between items from the node-to-hyperedge information propagation, we aggregate them into the corresponding items through the hyperedge-to-node information propagation. Noting that the training process becomes unstable when the convolutional module goes deeper, we apply a residual connection [6] to the aggregation as follows:

$$e_i^{(l+1)} = e_i^{(l)} + \frac{1}{|N_i|} \sum_{u \in N_i} q_{u,i}^{(l)}, \tag{5}$$

where N_i is the set of users that have interacted with item i and e_i^{l+1} is the embedding of item i in the $(l+1)$-th layer. This aggregation operation aggregates sequential patterns of different sequences into one item, thus achieving the purpose of extracting sequence features between different sequences.

After propagating through L layers, we obtain the final representation $e_i^{(L)}$ for each item i.

Hypergraph Sampling Strategy. To improve the computational efficiency of hypergraph convolution, we sample the nodes and hyperedges in the hypergraph before each sequential hypergraph convolution, which allows us to perform complex convolution operations on the hypergraph without concern for computational complexity. The idea is natural, namely, sampling some node-hyperedge connections during information propagation from nodes to hyperedges and hyperedges to nodes, respectively.

For node-to-hyperedge information propagation, we sample the nodes based on their relative positions to each other in each hyperedge. More specifically, we extract the item-centric sequence ε_u^i for each item $i \in \varepsilon_u$ to generate the item-specific hyperedge embedding as follows:

$$\varepsilon_u^i = \{ \underbrace{...}_{n_1}, i, \underbrace{...}_{n_1} \}, \tag{6}$$

where $|\varepsilon_u^i| = 2n_1 + 1$ and n_1 is a hyperparameter to control the length of ε_u^i. The computational complexity of node-to-hyperedge information propagation is linearly related to n_1. We pad zeroes to the item-centric sequence so that the length of the sequence is $2n_1 + 1$ for each item with fewer than $2n_1$ neighbors. Since Eq. (1) is computed only once in each hyperedge during each node-hyperedge propagation, we keep its content unchanged. For the user interest aggregator, we replace the hyperedge ε_u in Eq. (2) and Eq. (3) with ε_u^i to reduce the calculations.

For hyperedge-to-node information propagation, we sample the hyperedges to reduce the computational cost. In the real world, a popular item i may be interacted with by many users, which results in Eq. (5) having a worst-case complexity of $O(|U|)$. To reduce complexity, we sample N_i and denote the sampled result as N_i' as follows:

$$N_i' = \begin{cases} N_i & \text{if } |N_i| \leq n_2 \\ sample(N_i, n_2) & \text{if } |N_i| > n_2 \end{cases},\tag{7}$$

where n_2 is a hyperparameter to control the size of $|N_i'|$ and $sample(N_i, n_2)$ is a sampling function that selects n_2 elements from N_i. To enhance the robustness of the model, we use random sampling as the sampling function and leave other sampling methods for future exploration. Then we replace N_i in Eq. (5) with N_i'.

To implement the sampling strategy, we first sample the hyperedges according to Eq. (7) and then sample the items on the sampled hyperedges according to Eq. (6) since the opposite sampling order introduces many unnecessary calculations. It is worth noting that our hypergraph sampling strategy does not change the structure of the original hypergraph but samples a simpler hypergraph before each node-hyperedge propagation.

4.3   Target Sequence Modeling

To capture the short-term preference of the target user u, we adopt the self-attention mechanism as [10] since it achieved competitive performance in next item recommendation.

For the target user u, we have the corresponding SHC-based representation M_u as follows:

$$M_u = \{e_{1,u}^{(L)}, e_{2,u}^{(L)}, ..., e_{|S_u|,u}^{(L)}\},\tag{8}$$

where $e_{t,u}^{(L)}$ denotes the SHC-based embedding of the item that user u interacted with at time step t. The self-attention mechanism aggregates the user's recent engaged items to model his/her short-term preference. Following [10], we convert M_u into three matrices via linear projection and feed them into the self-attention layer as follows:

$$Z_u = \text{softmax}(\frac{(M_u W^Q)(M_u W^K)^T}{\sqrt{d}})M_u W^V$$
$$= \{e_{1,u}, e_{2,u}, ..., e_{|S_u|,u}\},\tag{9}$$

where the projection matrices $W^Q, W^K, W^V \in \mathbb{R}^{d \times d}$ are trainable parameters, Z_u is the output embedding matrix of the self-attention layer, and $e_{t,u}$ is the output embedding at time step $t \in \{1, 2, ..., |S_u|\}$ to predict the behavior of user u at time step $t + 1$.

4.4   Prediction Layer

Inspired by [24], we model users' hybrid preferences by aggregating their short-term and long-term preferences. After obtaining the short-term embedding $e_{t,u}$ of the target user u in Eq. (9) at time step t, we use e_u to represent the long-term preference. Then, the relevance of item i to the target user u at time step t can be calculated as follows:

$$\hat{y}_{i,u}^t = \sigma((e_{t,u} + e_u)^T \cdot e_i), \tag{10}$$

where $\hat{y}_{i,u}^t$ denotes the relevance score of candidate item i at time step t and $\sigma(\cdot)$ is the sigmoid function. $\hat{y}_{i,u}^t$ can also be considered the probability that user u interacts with item i after time step t. By sorting the scores of all the candidate items, we can obtain a list of recommendations for user u.

Algorithm 1. SHCN

Input: Users' interaction sequences $\{S_u | u \in U\}$
Parameter: L2 normalization coefficient λ, number of SHC layers L, number of neighboring nodes n, and initial learning rate η
Output: The set of trainable parameters Θ

1: Randomly initialize the parameters Θ
2: Construct the sequential hypergraph $G = \{E, V\}$ from input sequence set $\{S_u | u \in U\}$
3: **repeat**
4:    **for** $l = 0$ to $L - 1$ **do**
5:       Calculate e_i^{l+1} according to Eq. (5)
6:    **end for**
7:    Calculate loss function $\mathcal{L}$ with Eq. (11)
8:    $\theta \leftarrow \theta + \eta \frac{\partial \mathcal{L}}{\partial \theta}, \theta \in \Theta$
9:    Adjust the learning rate η by Adam
10: **until** Convergence or iteration number reaches the threshold
11: **return** Θ

4.5 Model Training

Given the behavior sequence $\{i_{1,u}, i_{2,u}, \cdots, i_{|S_u|,u}\}$ of the target user u, we take $i_{t,u}$ as the expected output after $i_{t-1,u}$. Then we construct a training set $C = \{(u, t, i, y_{i,u}^t)\}$, where $y_{i,u}^t \in \{0, 1\}$ indicates whether user u interacts with item i at time step t in each training sample $(u, t, i, y_{i,u}^t)$.

To train the model, we adopt cross-entropy loss as the loss function:

$$\mathcal{L} = \sum_{(u,t,i,y_{i,u}^t) \in C} \text{crossentropy}(\hat{y}_{i,u}^t, y_{i,u}^t) + \lambda ||\Theta||^2, \tag{11}$$

where Θ denotes the set of trainable parameters in our proposed model and λ is the L2 normalization coefficient to prevent overfitting. The pseudocode of the proposed model is shown in Algorithm 1.

4.6   Complexity Analysis for Hypergraph Convolution

For node-to-hyperedge information propagation, the computational complexity of generating item-specific hyperedge embedding $q_{u,i}^{(l)}$ is $O(n_1 d^2)$ for each item i and user u that have interacted with item i. For hyperedge-to-node information propagation, the computational complexity is $O(n_2|I|)$ since we randomly sample n_2 hyperedges for each item i if $|N_i| > n_2$. Then, we can derive the complexity of the sequential hypergraph convolution as $O(n_1 n_2 |I| d^2)$. Compared with existing methods that use the user-item interaction matrix for hypergraph convolution, such as DHCN [20] and HyperRec [16] with computational complexity $O(|I|^2 d)$, our proposed strategy greatly improves the computational efficiency of hypergraph convolution.

5   Experiments

In this section, we aim to evaluate the performance and effectiveness of the proposed method. Specifically, we conduct several experiments to study the following research questions:

- **RQ1:** Whether the proposed method outperforms state-of-the-art methods for next item recommendation?
- **RQ2:** Whether the proposed method benefits from the SPE and UIA in the sequential hypergraph convolution module to capture the sequential patterns between different sequences and the multifaceted nature of user behaviors?
- **RQ3:** (a) How do different configurations of key hyperparameters impact the performance of SHCN framework? (b) How does the hypergraph sampling strategy affect the recommendation performance?

5.1   Experimental Settings

Datasets. We evaluate the performance of the proposed approach on four real-world datasets that are all publicly available. We discard users and items with fewer than 5 interactions in each dataset following [4,10]. To prevent information leakage, we split the datasets following a real-world scenario. More specifically, the data located before the cutting timestamp are used to train the model. The first interaction after the cutting timestamp is used for validation and the second interaction is used for testing. The basic statistics of the four datasets are summarized in Table 1.

Table 1. Statistics of the datasets

Dataset	Users	Items	Interactions	Density
Taobao	4,378	22,774	42,651	0.0428%
Clothing	15,431	78,012	167,908	0.0139%
Beauty	5,865	27,570	83,012	0.0513%
Dianping	10,399	53,810	309,420	0.0553%

Table 2. Performance comparison. * indicates statistically significant improvement of the proposed method to every baseline model by an independent-samples t-test ($p < 0.01$).

Dataset	Taobao		Clothing		Beauty		Dianping	
Metrcis	NDCG	HR	NDCG	HR	NDCG	HR	NDCG	HR
BPR	0.046	0.056	0.070	0.070	0.055	0.065	0.042	0.049
NCF	0.033	0.058	0.033	0.056	0.036	0.054	0.031	0.054
LightGCN	0.085	0.118	0.072	0.104	0.094	0.134	0.143	0.212
GRU4Rec	0.053	0.076	0.085	0.134	0.075	0.107	0.105	0.158
Caser	0.058	0.087	0.087	0.128	0.084	0.127	0.119	0.167
SASRec	0.082	0.111	0.098	0.142	0.097	0.139	0.130	0.191
SHARE	0.034	0.058	0.032	0.053	0.037	0.058	0.036	0.061
HyperRec	0.072	0.107	0.106	0.154	0.088	0.141	0.115	0.176
SINE	<u>0.088</u>	<u>0.138</u>	<u>0.122</u>	<u>0.164</u>	0.098	0.147	<u>0.161</u>	<u>0.238</u>
DHCN	0.070	0.102	0.080	0.132	<u>0.115</u>	<u>0.198</u>	0.120	0.186
SHCN	**0.103***	**0.166***	**0.132***	**0.196***	**0.132***	**0.212***	**0.186***	**0.270***
Improv.	17.05%	20.29%	8.20%	19.51%	14.78%	7.07%	15.65%	13.49%

- **Taobao.** This dataset contains sequences of user actions from Taobao[1], including user behaviors from November 25 to December 3, 2017. We select the purchase behavior to conduct our experiment, and set the cutting timestamp to December 2, 2017, for this dataset.
- **Amazon.** This is a series of datasets comprising large corpora of product reviews crawled from Amazon[2]. We consider two challenging categories: Beauty and Clothing (Clothing, Shoes and Jewelry), which are famous for their sparsity. We set the cutting timestamp to January 1, 2014, for these datasets.
- **Dianping.** This dataset contains the user reviews as well as the detailed business metadata information crawled from a famous Chinese online review webset DianPing.com[3]. We set the cutting timestamp to January 1, 2010, for this dataset.

The sequential hypergraphs used in our experiments are constructed based on the training dataset. In other words, each hyperedge only contains the user behaviors before the cutting timestamp.

Evaluation Metrics. We use two widely adopted metrics, including the hit rate (HR) and the normalized discounted cumulative gain ($NDCG$) within the

---

[1] https://tianchi.aliyun.com/dataset/dataDetail?dataId=649.
[2] http://jmcauley.ucsd.edu/data/amazon/links.html.
[3] http://yongfeng.me/dataset.

top-5 list, to evaluate the performance of different models. We repeated each experiment five times and reported the average results.

Baselines. We compare the proposed model with several competitive baseline methods as follows.

- BPR [12] is a popular recommendation model that uses a pairwise ranking loss to optimize the MF model.
- NCF [8] combines matrix factorization and multilayer perceptions to predict user interactions.
- LightGCN [7] is a simplified GCN-based model that leverages user-item interaction diagrams to learn node representations and generate recommendations.
- GRU4Rec [9] uses RNNs to model user interaction sequences for session-based recommendation.
- Caser [15] applies convolutional operations on the embedding matrix to capture high-order Markov chains.
- SASRec [10] uses an attention mechanism to identify relevant items for predicting the next item.
- HyperRec [16] adopts hypergraphs to represent the short-term item collections to predict the next interesting item for each user.
- SHARE [17] uses sliding windows to build hypergraphs for each session to capture user intent.
- SINE [14] is a state-of-the-art model that captures user intentions by learning numerous interests and dynamically predicts the next intentions of users.
- DHCN [20] is a state-of-the-art method that captures beyond-pairwise relations via hypergraphs.

Parameter Settings. For a fair comparison, the embedding size and the maximum sequence length are both set as 50 for all methods. For the baseline models, we either follow the reported optimal parameter settings in their original papers or tune them to the best using the validation set. For the proposed method, we train our model on four datasets with an Adam optimizer [11], and the initial learning rate is set to $\eta = 1e^{-4}$. Our experiments are conducted on servers with a 16-core CPU, NVIDIA 2080Ti GPU, and 32 GB memory.

5.2   Experimental Results

Overall Performance (RQ1). Table 2 shows the recommendation performance of difficult methods on the four benchmark datasets. To make the table more notable, we bold the best results and underline the best baseline results in each case. In terms of different evaluation metrics, the proposed SHCN model significantly outperforms all the baselines on each of the datasets. More specifically, SHCN improves on average by 13.9% and 15.09% in terms of $NDCG$ and HR compared with the best baseline method, respectively. It confirms the effectiveness of our proposed model.

Among the baselines, SINE and DHCN achieve the best performance on different datasets. In most cases, sequential methods work better than CF-based methods on different datasets, which means that sequential patterns are important to predict users' next behavior. In certain situations, however, the non-sequential model LightGCN exceeds some sequential methods. This indicates that while extracting sequential patterns, structural information behind the graph still plays an important role in the performance of recommendations. Nevertheless, these baselines all fail to outperform our proposed SHCN methods, which makes use of both sequence patterns and collaborative information with the sequence hypergraph to make recommendations.

Table 3. Ablation study. * indicates statistically significant improvement on an independent-samples t-test ($p < 0.01$).

	Taobao		Clothing		Beauty		Dianping	
	NDCG	HR	NDCG	HR	NDCG	HR	NDCG	HR
SHCN	0.103*	0.166*	0.132*	0.196*	0.132*	0.212*	0.186*	0.270*
-SPE	0.080	0.125	0.126	0.188	0.094	0.149	0.142	0.208
-UIA	0.091	0.151	0.125	0.190	0.102	0.164	0.102	0.164

Ablation Study (RQ2). We conduct a series of ablation studies by removing the essential components in the SHCN to evaluate their effectiveness. The experimental results of removing different components are shown in Table 3.

-SPE. The sequential pattern extractor in sequential hypergraph convolution is removed in this study. Not surprisingly, the performance is worse than the default. This illustrates the effectiveness of SHCN model to capture the sequential patterns between different user sequences in sequential hypergraph convolution.

-UIA. We remove the user interest aggregator in sequential hypergraph convolution in this study. This implementation is worse than the default, which indicates the effectiveness of SHCN model to capture the multifaceted nature of user behaviors in sequential hypergraph convolution.

5.3 Hyperparameter Analysis

Parameter Analysis (RQ3). We conduct a series of experiments to evaluate the effects of different parameter settings on the performance of SHCN. The results are shown in Fig. 2.

- **Impact of Model Depth.** We compare the performance of the SHCN model with different numbers of convolutional layers in Fig. 2(a). When the number

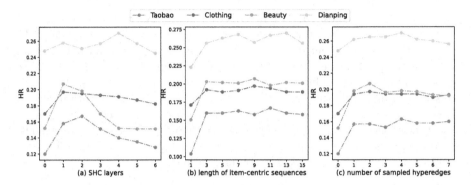

Fig. 2. The impact of the different numbers of SHC layers (a), the length of item-centric sequences (b) and different numbers of sampled hyperedges.

of convolution layers is zero, the SHC layers are removed and only the sequential module works. It was observed that SHCN achieved the best performance when there was only one convolution layer in the Beauty and Clothes datasets. On the Taobao and Dianping datasets, the best performance is achieved when the number of convolution layers is 2 and 4, respectively. Compared with the performance when the number of convolution layers is 0, the optimal performance of SHCN is greatly improved, which proves the effectiveness of SHC layers. After achieving optimal performance, the increase of the number of convolution layers will lead to the degradation of SHCN performance.

- **Impact of Length of Item-Centric Sequences.** To explore the influence of the node-to-hyperedge sampling strategy, we compare the performance of the SHCN by varying the length of the item-centric sequences $(2n_1 + 1)$, where $n_1 = \{0, 1, 2, 3, 4, 5, 6, 7\}$. The result is shown in Fig. 2(b). When the length is 1, the sequential hypergraph degenerates into a general graph. It is not surprising that this implementation is worse than others. It is observed that the best results were obtained for the Clothing and Beauty datasets at length 9, while the best results were obtained for the Taobao and Dianping datasets at lengths 11 and 13, respectively. Considering both effectiveness and efficiency, we propose to set the moderate length of the item-centered sequence to 11.

- **Impact of Number of Sampled Hyperedges.** To explore the influence of the hyperedge-to-node sampling strategy, we compare the performance of the SHCN by varying the number of sampled hyperedges (n_2), where $n_2 = \{0, 1, 2, 3, 4, 5, 6, 7\}$. When $n_1 = 0$, the sequential hypergraph convolution is actually interrupted, and only the target sequence modeling works. It is observed in Fig. 2(c) that on the Beauty and Clothing datasets the best performance is obtained when the number is 2 while on the Taobao and Dianping datasets the sampling number of 4 is more appropriate. When $n_2 > 0$, the performance of SHCN model is relatively stable for the hyperparameter

n_2. We propose to sample two hyperedges as a moderate setting, taking into account both effectiveness and efficiency.

6   Conclusion

In this work, we explore the role of sequential relations in hypergraph structures and propose a novel next item recommendation framework empowered by a sequential hypergraph. It can capture sequential patterns between different sequences and the semantic relations between items with the multifaceted nature of user behaviors. To improve the computational efficiency of sequential hypergraph convolution, we propose a hypergraph sampling strategy for hypergraph convolution, which makes it possible to apply complex convolution methods. Extensive empirical studies have shown the overwhelming superiority of our model, and ablation studies have validated the effectiveness and rationality of hypergraph convolution and hypergraph sampling strategies.

Acknowledgement. This work was supported by The National Key Research and Development Program of China (Grant No.2022YFB2703300).

References

1. Bai, S., Zhang, F., Torr, P.H.: Hypergraph convolution and hypergraph attention. Pattern Recogn. **110**, 107637 (2021)
2. Devooght, R., Bersini, H.: Long and short-term recommendations with recurrent neural networks. In: Proceedings of the 25th Conference on User Modeling, Adaptation and Personalization, pp. 13–21 (2017)
3. Du, Y., Liu, H., Wu, Z.: Modeling multi-factor and multi-faceted preferences over sequential networks for next item recommendation. In: Oliver, N., Pérez-Cruz, F., Kramer, S., Read, J., Lozano, J.A. (eds.) ECML PKDD 2021. LNCS (LNAI), vol. 12976, pp. 516–531. Springer, Cham (2021). https://doi.org/10.1007/978-3-030-86520-7_32
4. Du, Y., Liu, H., Wu, Z.: M 3-ib: A memory-augment multi-modal information bottleneck model for next-item recommendation. In: Database Systems for Advanced Applications: 27th International Conference, DASFAA 2022, Virtual Event, April 11–14, 2022, Proceedings, Part II. pp. 19–35. Springer (2022). https://doi.org/10.1007/978-3-031-00126-0_2
5. Gao, C., Wang, X., He, X., Li, Y.: Graph neural networks for recommender system. In: Proceedings of the Fifteenth ACM International Conference on Web Search and Data Mining, pp. 1623–1625 (2022)
6. He, K., Zhang, X., Ren, S., Sun, J.: Deep residual learning for image recognition. In: Proceedings of the IEEE Conference on Computer Vision and Pattern Recognition, pp. 770–778 (2016)
7. He, X., Deng, K., Wang, X., Li, Y., Zhang, Y., Wang, M.: Lightgcn: Simplifying and powering graph convolution network for recommendation. In: Proceedings of the 43rd International ACM SIGIR Conference on Research and Development in Information Retrieval, pp. 639–648 (2020)

8. He, X., Liao, L., Zhang, H., Nie, L., Hu, X., Chua, T.S.: Neural collaborative filtering. In: Proceedings of the 26th international conference on world wide we,. pp. 173–182 (2017)
9. Hidasi, B., Karatzoglou, A., Baltrunas, L., Tikk, D.: Session-based recommendations with recurrent neural networks. arXiv preprint arXiv:1511.06939 (2015)
10. Kang, W.C., McAuley, J.: Self-attentive sequential recommendation. In: 2018 IEEE International Conference on Data Mining (ICDM), pp. 197–206. IEEE (2018)
11. Kingma, D.P., Ba, J.: Adam: A method for stochastic optimization. arXiv preprint arXiv:1412.6980 (2014)
12. Rendle, S., Freudenthaler, C., Gantner, Z., Schmidt-Thieme, L.: Bpr: Bayesian personalized ranking from implicit feedback. In: Proceedings of the Twenty-Fifth Conference on Uncertainty in Artificial Intelligence, pp. 452–461 (2009)
13. Rendle, S., Freudenthaler, C., Schmidt-Thieme, L.: Factorizing personalized markov chains for next-basket recommendation. In: Proceedings of the 19th International Conference on World wide Web, pp. 811–820 (2010)
14. Tan, Q., et al.: Sparse-interest network for sequential recommendation. In: Proceedings of the 14th ACM International Conference on Web Search and Data Mining, pp. 598–606 (2021)
15. Tang, J., Wang, K.: Personalized top-n sequential recommendation via convolutional sequence embedding. In: Proceedings of the Eleventh ACM International Conference on Web Search and Data Mining, pp. 565–573 (2018)
16. Wang, J., Ding, K., Hong, L., Liu, H., Caverlee, J.: Next-item recommendation with sequential hypergraphs. In: Proceedings of the 43rd international ACM SIGIR Conference on Research and Development in Information Retrieval, pp. 1101–1110 (2020)
17. Wang, J., Ding, K., Zhu, Z., Caverlee, J.: Session-based recommendation with hypergraph attention networks. In: Proceedings of the 2021 SIAM International Conference on Data Mining (SDM), pp. 82–90. SIAM (2021)
18. Wang, S., Hu, L., Wang, Y., Cao, L., Sheng, Q.Z., Orgun, M.: Sequential recommender systems: Challenges, progress and prospects. In: Proceedings of the Twenty-Eighth International Joint Conference on Artificial Intelligence. pp. 6332–6338. International Joint Conferences on Artificial Intelligence Organization (2019)
19. Wang, Z., Wei, W., Cong, G., Li, X.L., Mao, X.L., Qiu, M.: Global context enhanced graph neural networks for session-based recommendation. In: Proceedings of the 43rd international ACM SIGIR Conference on Research and Development in Information Retrieval, pp. 169–178 (2020)
20. Xia, X., Yin, H., Yu, J., Wang, Q., Cui, L., Zhang, X.: Self-supervised hypergraph convolutional networks for session-based recommendation. In: Proceedings of the AAAI Conference on Artificial Intelligence. vol. 35, pp. 4503–4511 (2021)
21. Ye, W., Wang, S., Chen, X., Wang, X., Qin, Z., Yin, D.: Time matters: Sequential recommendation with complex temporal information. In: Proceedings of the 43rd International ACM SIGIR Conference on Research and Development in Information Retrieval, pp. 1459–1468 (2020)
22. Ying, R., He, R., Chen, K., Eksombatchai, P., Hamilton, W.L., Leskovec, J.: Graph convolutional neural networks for web-scale recommender systems. In: Proceedings of the 24th ACM SIGKDD International Conference on Knowledge Discovery & Data Mining, pp. 974–983 (2018)
23. Yu, J., Yin, H., Li, J., Wang, Q., Hung, N.Q.V., Zhang, X.: Self-supervised multi-channel hypergraph convolutional network for social recommendation. In: Proceedings of the Web Conference 2021, pp. 413–424 (2021)

24. Yu, Z., Lian, J., Mahmoody, A., Liu, G., Xie, X.: Adaptive user modeling with long and short-term preferences for personalized recommendation. In: Proceedings of the 28th International Joint Conference on Artificial Intelligence, pp. 4213–4219 (2019)
25. Zhu, T., Sun, L., Chen, G.: Graph-based embedding smoothing for sequential recommendation. In: IEEE Transactions on Knowledge and Data Engineering, pp. 1–1 (2021)

Multi-granularity Item-Based Contrastive Recommendation

Ruobing Xie[✉], Zhijie Qiu, Bo Zhang, and Leyu Lin

WeChat, Tencent, Beijing, China
ruobingxie@tencent.com

Abstract. Contrastive learning (CL) has shown its power in recommendation. However, most CL-based models build their CL tasks merely focusing on the user's aspects, ignoring systematically modeling the rich and diverse information in items. In this work, we propose a novel Multi-granularity item-based contrastive learning (MicRec) for the matching stage in recommendation, which systematically introduces multi-aspect item-related correlations to representation learning via CL. Specifically, we build three item-based CL tasks as a set of plug-and-play auxiliary objectives to capture item correlations in feature, semantic and session levels. In experiments, we conduct both offline and online evaluations on real-world datasets, verifying the effectiveness and universality of three proposed CL tasks. Currently, MicRec has been deployed on a real-world recommender system of WeChat Top Stories, affecting millions of users.

Keywords: Recommendation · Contrastive learning · Matching

1 Introduction

Real-world recommender systems aim to provide items according to user preferences from million-level item candidates [4]. Practical recommender systems usually adopt the classical two-stage architecture [2] to balance effectiveness and efficiency, which consists of both ranking and matching modules. The *matching* module [18] (i.e., candidate generation) tries to retrieve a small subset of (usually hundreds of) item candidates from the million-level large corpora.

Matching should jointly consider recommendation accuracy, efficiency, and diversity. The two-tower architecture is the mainstream matching method widely used in large-scale systems [16]. However, existing models mainly concentrate on the CTR-oriented objectives, struggling with the *data sparsity* and *popularity bias* issues [19,21]. These issues are even more serious in matching, since matching should deal with million-level item candidates and the supervised user-item labels are more sparse. Recently, with the blooming of contrastive learning (CL) [1], plenty of CL-based recommendation models are proposed, which alleviate the sparsity and popularity bias issues via more sufficient and diverse training

---

R. Xie and Z. Qiu—Authors have equal contributions.

X. Wang et al. (Eds.): DASFAA 2023, LNCS 13944, pp. 406–416, 2023.
https://doi.org/10.1007/978-3-031-30672-3_27

from additional self-supervised learning (SSL) signals [22]. However, most CL-based models only depend on different *user augmentations* from various aspects such as user behaviors [11,17] and user attributes [13], ignoring the rich multi-granularity information in *items*. Furthermore, the item representation often has a more direct influence on matching results due to the two-tower architecture, which multiplies the significance of learning good item representations via CL.

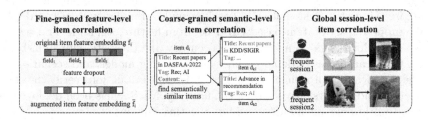

Fig. 1. Multi-granularity item correlations in real-world recommender systems.

In this work, we attempt to systematically conduct **item-based contrastive learning tasks** for matching. The key point is how to build more *effective and diverse item augmentations*. We find that an item could be described from multiple granularities, such as its features, semantics, and behavioral information. Figure 1 shows three different aspects of item correlations. (a) The *fine-grained feature correlations* of items reflect the implicit feature-level similarities between items and their augmentations. (b) The *coarse-grained semantic correlations* highlight items' semantic similarities extracted from titles, tags or contents. Differing from feature correlations, semantic correlations measure the semantic relevances between two items. (c) The *global session-level correlations* capture items' behavioral similarities via users' sequential behaviors in all sessions. For example, the [diaper→beer] in Fig. 1 are frequent sessions that reveal surprising but reasonable item associations beyond similar features and semantics. These item correlations help to build a comprehensive understanding of all items from diverse aspects.

Therefore, we propose a new **Multi-granularity Item-based Contrastive Recommendation (MicRec)** framework, aiming to encode the under-explored item correlations into representation learning via CL tasks. Specifically, we design three item-based CL tasks. (1) The ***feature-level item CL*** focuses on fine-grained feature correlations. It builds item augmentations via feature element/field dropouts as positive samples, improving the alignment and uniformity of item representations that are beneficial as [10]. (2) The ***semantic-level item CL*** attempts to model item similarities from the coarse-grained semantic aspect. Relevant items sharing similar titles/tags/contents are viewed as the positive item pairs. (3) The ***session-level item CL*** captures more implicit item behavioral similarities from the global view. Item pairs with high co-occurrences counted from all sessions are considered as positive samples. We conclude the advantages of MicRec as follows: (1) the multi-granularity item-based CL tasks bring in additional training signals beyond clicks to fight against sparsity. (2) These CL tasks encode more diverse item correlations into item representations beyond clicks, which is also beneficial for diversity. (3) MicRec is a plug-and-play framework that could be easily deployed on almost all matching models. Its universality and simplicity are welcomed in the industry. The contributions are:

- We highlight the importance of item-based CL in recommendation beyond lots of widely-explored user-based CL models. To the best of our knowledge, we are the first to systematically consider multi-granularity item-based CL.
- We propose a set of item-based CL tasks from feature, semantic, and session aspects, which could jointly improve accuracy and diversity. Our item-based CL tasks can be easily deployed on most matching models and scenarios.
- The significant improvements in offline and online evaluations verify the effectiveness and universality of MicRec. Currently, MicRec has been deployed on a real-world recommendation system, affecting millions of users.

2 Related Work

Matching in Recommendation. The two-tower architecture is the mainstream method in practical matching, which learns user and item representations separately in two towers [2,19]. Different feature interaction models [4,8] could be used in user/item towers for feature interaction. Some two-tower models [9,16] concentrate on user behavior modeling in the user tower. Other architectures such as graph-based methods [14] are also explored in matching. In this work, we deploy MicRec on three classical matching models.

Contrastive Learning in Recommendation. Recently, SSL and CL have been verified in recommendation [6,21]. Many of these models rely on user augmentations on user attributes and behaviors [12,13,22]. [21] adopts disentangled CL-based model to candidate generation with multi-intention queues. Other CL-based model focuses on graph learning [20], cross-domain recommendation [15], and multi-behavior recommendation [11]. [19] is the most related work. It proposes two random and correlated feature masking (RFM/CFM) strategies, building augmentations with feature masking and dropout. Differing from existing works that mainly focus on user-based augmentations or simply consider one aspect for item augmentation, we design a novel set of item-based CL tasks.

3 Methodology

3.1 Background and Overall Framework

Two-Tower Architecture. The two-tower architecture is the most widely-used matching model in practice [2,16]. A typical two-tower model contains two neural networks (i.e., towers) to construct user and item representations separately. Formally, we build the user and item representations u_i and d_k as $u_i = \mathrm{DNN}^u(f_i^u)$ and $d_k = \mathrm{DNN}^d(f_k^d)$, where $\mathrm{DNN}^u(\cdot)$ and $\mathrm{DNN}^d(\cdot)$ are deep neural networks and $\boldsymbol{f}_i^u$ and $\boldsymbol{f}_k^d$ are the input feature embeddings of user and item. The similarity between $\boldsymbol{u}_i$ and $\boldsymbol{d}_k$ is regarded as the click probability of (u_i, d_k). The classical sampled softmax is often used as the training objective. In the two-tower architecture, item representations have a direct impact on the final matching scores.

Multi-granularity Item Information. We introduce the item information used in MicRec as: (1) *Item features.* We follow classical models [8] that divide item features into feature fields. Each field (e.g., ID, tag) is allocated with an embedding. (2) *Item semantics.* Semantics is more coarse-grained information compared to specific feature fields. Semantic similarities are often reflected by similar item titles, contents, or taxonomies. (3) *Sessions.* Items sequentially clicked by a user in a session (within 1 h in this work) usually imply implicit and reasonable behavioral patterns (e.g., [diaper→beer]). All three types of item information widely exist and can be easily collected in real-world systems.

Overall Framework. MicRec adopts the classical two-tower architecture with three item-based CL tasks besides the original CTR loss: (a) The feature-level item CL considers an item and its feature-based augmentations as positive pairs. (b) The semantic-level item CL regards two distinct items having similar semantics as positive pairs. (c) The session-level item CL provides a new train of thought for modeling item similarities via the global item co-occurrences in all sessions. Note that MicRec is effective and universal, which is convenient to be deployed on different datasets with most embedding-based matching models.

3.2 Feature-Level Item Contrastive Learning

In MicRec, the i-th input item embedding $\boldsymbol{f}_i^d$ is the concatenation of the item's field embeddings. For each $\boldsymbol{f}_i^d$, we conduct a feature-level dropout to get the item's augmented embedding $\bar{\boldsymbol{f}}_i^d = \text{Dropout}(\boldsymbol{f}_i^d)$. We define the similarity score between item d_i and its augmented d_j as: $g_f(\boldsymbol{f}_i^d, \bar{\boldsymbol{f}}_j^d) = (\text{MLP}_f(\boldsymbol{f}_i^d))^\top \text{MLP}_f(\bar{\boldsymbol{f}}_i^d)$. $\text{MLP}_f(\cdot)$ is a multi-layer perception as the projector. Here, we want items to be closer to their augmentations than other items'. $d_j \in N_i$ are negative samples that $d_j \neq d_i$. Different from [1] that conducts in-batch negative sampling, we directly utilize randomly sampled $d_j \neq d_i$ as negative samples to avoid excessive suppression of popular items (effective in matching). Therefore, the feature-level item CL loss L_{fea} is defined following the classical InfoNCE [1] as:

$$L_{fea} = -\sum_{d_i} \log \frac{\exp(g_f(\boldsymbol{f}_i^d, \bar{\boldsymbol{f}}_i^d)/\tau)}{\sum_{d_j \in N_i} \exp(g_f(\boldsymbol{f}_i^d, \bar{\boldsymbol{f}}_j^d)/\tau)}. \tag{1}$$

τ is the temperature. We provide three possible feature dropout strategies to build item augmentations for different demands: (a) *Element-level dropout*, which randomly masks elements in any field of the raw feature embedding $\boldsymbol{f}_i^d$ with a certain mask ratio. (b) *Field-level dropout*, which randomly masks certain whole feature fields. (c) *Categorial feature dropout*, which randomly masks some tags when building the tag field embedding. In experiments, we find that the simple combination of the field-level and categorial feature dropout performs the best, with the mask ratio set as 0.5. It is also convenient to deploy this CL on $\boldsymbol{d}_i$.

3.3 Semantic-Level Item Contrastive Learning

The semantic-level item CL suggests that an item should be similar to its semantically similar items. For each item d_i, we first find its semantically similar item set P_i^t as the positive sample set in CL. We also randomly select other items not in P_i^t to form the negative set N_i^t. The similarity score $g_t(\boldsymbol{d}_i, \boldsymbol{d}_j)$ between d_i and d_j is calculated as $g_t(\boldsymbol{d}_i, \boldsymbol{d}_j) = (\mathrm{MLP}_t(\boldsymbol{d}_i))^\top \mathrm{MLP}_t(\boldsymbol{d}_j)$, where $\boldsymbol{d}_i = \mathrm{DNN}^d(\boldsymbol{f}_i^d)$ is the item representation. Formally, the semantic-level item CL loss L_{sem} is as:

$$L_{sem} = -\sum_{d_i} \sum_{d_k \in P_i^t} \log \frac{\exp(g_t(\boldsymbol{d}_i, \boldsymbol{d}_k)/\tau)}{\sum_{d_j \in N_i^t} \exp(g_t(\boldsymbol{d}_i, \boldsymbol{d}_j)/\tau)}, \tag{2}$$

L_{sem} highlights the intrinsic coarse-grained semantic correlations of items besides behavioral correlations. Item semantics could be represented by different attributes, such as item titles, taxonomies, and contents: (a) For *item titles*, we adopt their sentence embeddings learned from a pre-trained language model (e.g., we use BERT [3] in this work) as the title embeddings. The top-k nearest items retrieved by the cosine similarities between title embeddings are viewed as the semantically similar positive pairs in semantic CL. (b) For *taxonomies*, we regard items having the same taxonomies as positive item pairs of each other. (c) For *item contents*, we could use pre-trained visual/textual encoders to learn the content embeddings as item semantics and calculate the semantic similarities. These attributes have their own pros and cons in representing semantics. In our system, we find that using title similarities achieves the best results, since titles are sufficient to capture the main semantics without much noise.

3.4 Session-Level Item Contrastive Learning

Items appearing in a session are very likely to have implicit correlations, which is usually ignored in lots of recommendation models. To make full use of the sequential behavioral correlations in sessions, we propose a novel session-level item CL. Specifically, given a session $s = \{d_1, \cdots, d_n\}$ of n clicked items, we let any two items $d_i, d_j \in s$ have one co-occurrence relation. Next, we scan all sessions in the train set to learn the co-occurrence matrix of all item pairs. For the i-th item, we build its positive sample set P_i^s in the session view via a weighted sampling over its top-k co-occurred items proportional to the co-occurrence counts. The i-th item's negative sample set N_i^s are randomly selected from all items that have not co-occurred with d_i. We build L_{sess} with a similar session-level similarity $g_s(\boldsymbol{d}_i, \boldsymbol{d}_j) = (\mathrm{MLP}_s(\boldsymbol{d}_i))^\top \mathrm{MLP}_s(\boldsymbol{d}_j)$, noted as:

$$L_{sess} = -\sum_{d_i} \sum_{d_k' \in P_i^s} \log \frac{\exp(g_s(\boldsymbol{d}_i, \boldsymbol{d}_k')/\tau)}{\sum_{d_j \in N_i^s} \exp(g_s(\boldsymbol{d}_i, \boldsymbol{d}_j)/\tau)}, \tag{3}$$

L_{sess} assumes that items appearing together more times in a session should share the same user preferences and be more similar. It is the first attempt to encode the global session-level item correlations into item representations via CL.

3.5 Specifications of MicRec

Offline Training. MicRec is a universal and easy-to-deploy framework that can work with most embedding-based models. We follow a classical matching model ICAN [16] to build our MicRec's two-tower neural network. Precisely, we inherit the main neural networks and user features of ICAN and remove its multi-domain designs to get the final user and item representations u_i and d_k. We follow the classical sampled softmax matching loss L_S [16] as our loss with $d_k^\top u_i$ as the click score. Following [21], we jointly learn L_S with three item-based CL losses in a multi-task learning manner as: $L = L_S + \lambda_1 L_{fea} + \lambda_2 L_{sem} + \lambda_3 L_{sess}$.

Online Deployment and Complexity. We have deployed MicRec on a real-world recommender system, where MicRec functions as one of multiple matching channels with other modules unchanged. The online serving is the same as the base model: the learned user representation is used to fast retrieve top-K ($K = 500$) similar items via user-item similarities. MicRec adopts three additional CL losses, which inevitably brings in additional computation costs in offline. Nevertheless, the offline training time of MicRec is less than twice that of the base model. Note that MicRec has *equal online memory/computation costs* compared to its corresponding base model, since they have exactly the same online serving.

4 Experiments

4.1 Experimental Settings

Datasets. To simulate real-world scenarios, we build two large-scale datasets of different scenarios from WeChat Top Stories. The first Video-636M dataset contains nearly $25M$ users and $636M$ click instances on $2.4M$ videos. The second News-76M has nearly $6.4M$ users and $76M$ click instances on $1.1M$ news. These instances are then split into train and test sets in chronological order. We have $47M$ and $7.2M$ click instances as the test set in two datasets. All data are preprocessed via data masking to protect user privacy.

Competitors. In this work, we implement several competitive matching models and CL-based models as baselines, including FM [7], YoutubeDNN [2], DeepFM [4], AutoInt [8], ICAN* [16] (the same as the base model of MicRec), RFM* [19]. Other neural structures for matching (e.g., [14]) and user-focused CL tasks [22] can be easily deployed with MicRec, while it is not the core focus of this work.

Parameter Settings. Following [16], we use 20 most recent click behaviors as input. The dimensions of field embeddings and item/user representations are set as 64. The output dimensions of the 3-layer MLP are 128, 64, and 64 as the base model ICAN [16]. For fair comparisons, all models share the same features and

Table 1. Offline evaluation. All improvements are significant (t-test with $p < 0.01$).

Model	Video-636M				News-76M			
	hit@50	hit@100	hit@200	hit@500	hit@50	hit@100	hit@200	hit@500
FM	0.1731	0.2487	0.3219	0.4495	0.2466	0.3072	0.3871	0.4843
YoutubeDNN	0.1779	0.2576	0.3375	0.4719	0.2691	0.3307	0.4116	0.5188
DeepFM	0.1825	0.2587	0.3368	0.4776	0.2659	0.3268	0.4083	0.5171
AutoInt	0.1859	0.2626	0.3552	0.4857	0.2716	0.3431	0.4213	0.5316
ICAN*	0.1872	0.2627	0.3563	0.4873	0.2748	0.3438	0.4241	0.5403
RFM*	0.1967	0.2752	0.3671	0.5065	0.2911	0.3609	0.4413	0.5605
MicRec	**0.2047**	**0.2831**	**0.3825**	**0.5209**	**0.3018**	**0.3719**	**0.4543**	**0.5758**

training objective. In training, we use Adam with the learning rate $\alpha = 0.001$, and set the batch size as $4,096$. We conduct a grid search for parameter selection. We set the CL loss weights $\lambda_1 = 1.0$, $\lambda_2 = 0.3$, $\lambda_3 = 0.1$ for videos, and set $\lambda_1 = 1.0$, $\lambda_2 = 0.1$, $\lambda_3 = 0.1$ for news. We conduct 1-layer projectors with $\tau = 1$ for all CL tasks, and select up to 50 negative samples for efficiency.

4.2 Offline Evaluation on Matching

Following classical settings [16,21], we use HIT@N as our matching metric. We report HIT@N with different larger N such as 50, 100, 200, and 500 to simulate real-world matching. From Table 1 we find that:

(1) MicRec achieves significant improvements on all metrics compared with all baselines and ablation versions. The deviations of MicRec models are less than ± 0.002. It indicates that our multi-granularity item-based CL could well capture multiple item correlations from feature, semantic, and session levels.

(2) The improvements of MicRec mainly derive from the three CL tasks. Precisely, the feature-level CL improves the alignment and uniformity of items as classical CL models [10,19]. The semantic-level CL highlights the semantic item correlations besides user-item interactions. The session-level CL strengthens the sequential behavioral information to capture implicit but surprising item correlations that cannot be learned from L_S, L_{fea}, and L_{sem}. Considering various item correlations brings in diversity, which is also beneficial in matching.

(3) MicRec has consistent improvements in both video and news domains, implying its universality. Moreover, MicRec has more significant improvements with a larger N, which confirms its adaptability in real-world matching.

Results on Diversity. We also evaluate the diversity of MicRec via a classical aggregate diversity metric named *item coverage* [5,14]. It represents the percentage of items in a dataset that the recommendation model is able to provide predictions for in the test set, which focuses on the overall diversity of matching.

Table 2. Online A/B tests. All improvements are significant (t-test with $p < 0.05$).

Overall	ACC	ADT	ACC-c	ADT-c
MicRec	+3.287%	+1.389%	+11.517%	+6.472%

Video	ACC	ADT	ALR	AFR
MicRec	+5.643%	+2.824%	+4.138%	+5.165%

As in Table 3, MicRec outperforms its base model ICAN on item coverage by a large margin. The 29.3% diversity improvement shows that MicRec can alleviate popularity bias and improve diversity via our item-based CL.

4.3 Online Evaluation

We deploy MicRec on the video domain of WeChat Top Stories as Sect. 3.5 to show the power of MicRec in online. The online base model is ICAN. We focus on four overall metrics: (1) average click count per capita (ACC), (2) average dwell time per capita (ADT), (3) average click count of cold-start users (ACC-c), and (4) average dwell time of cold-start users (ADT-c). For the video domain, we further measure: (5) average like rate per capita (ALR), and (6) average follow rate per capita (AFR) besides ACC and ADT. We conduct the online A/B test for 7 days on 4 million users. From Table 2 we can find that:

(1) MicRec achieves significant improvements on all metrics, which confirms the effectiveness of MicRec in online. ACC focuses on the central click-related metric. ADT focuses on the user duration that reflects users' real satisfaction. The consistent improvements imply the effectiveness of MicRec in online. From ACC-c and ADT-c, we know that cold-start users get more improvements from MicRec thanks to the indirect influence of the additional item correlations.
(2) In the video scene, besides the consistent improvements of ACC and ADT, we also observe that MicRec outperforms the base model on ALR and AFR. The improvements on like- and follow- related metrics reconfirm users' satisfaction with our recommended items on diverse online metrics.

4.4 Ablation Study on Accuracy and Diversity

To verify the effectiveness of three item-based CL tasks of MicRec, we conduct four ablation versions without certain CL tasks (*fea, sem, sess*) on Video-636M. Table 3 shows the HIT@N and coverage results of different versions, we have:

(1) The feature-level item CL conducts feature dropouts to build fine-grained feature-level augmentations for items. It is an effective and classical way to introduce additional SSL signals for sufficient training, improving the *alignment* between items and their augmentations and the *uniformity* of all items.

Table 3. Impacts of three CL tasks on accuracy (HIT@N) and diversity (coverage).

Ablation version	HIT@50	HIT@500	item coverage
MicRec	0.2047	0.5209	16.1%
– Feature-level item CL	0.2008	0.5092	17.2%
– Semantic-level item CL	0.2033	0.5134	16.2%
– Session-level item CL	0.2012	0.5108	9.3%
– all item-based CL	0.1872	0.4873	12.4%

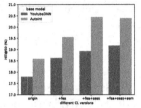 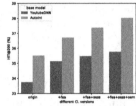 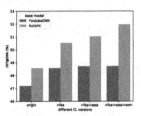

(a) Results on HIT@50.   (b) Results on HIT@200.   (c) Results on HIT@500.

Fig. 2. Improvements of three CL tasks with two matching models on Video-636M.

(2) We highlight the semantic information in CL since it is directly exposed to users and is one of the main factors largely impacting users. This CL also brings in significant improvements besides the other two CL tasks.

(3) The session-level item CL highlights the global behavioral correlations between items according to the "unconscious crowd-sourcing" by users' natural clicks in sessions. The session information is an essential supplement to the user-item click behaviors in discovering users' sequential behavioral preferences.

(4) From coverage, we find that the session-level item CL is the major source of diversity improvements. The diverse and less predictable sequential behaviors accumulated from the global sessions can bring in more high-quality occasionality for breaking the filter bubble and exploring more possibilities.

In contrast, models with feature-level and semantic-level CL tasks have comparable or even slightly worse diversity results. However, the diversity metric is only an indirect metric. Hence, we still conduct all three item-based CL tasks in MicRec, since it has the best accuracy results with relatively good diversity compared to other versions.

4.5   Universality of MicRec on Different Matching Models

To demonstrate MicRec's universality, we further deploy our item-based CL tasks on other matching models. Figure 2 shows the HIT@N results of different proposed CL combinations on YoutubeDNN [2] and AutoInt [8]. We find that in general, all feature-level, semantic-level, and session-level item CL tasks are effective even deployed with other models (while MicRec still has the best performances). The feature-level CL and session-level CL tasks contribute to the main part of improvements. It demonstrates the effectiveness and universality of MicRec and all proposed item-based CL tasks with different matching models.

5 Conclusions and Future Work

In this work, we propose a novel MicRec framework, which designs three effective, easy-to-deploy, and universal item-based CL tasks to improve the matching performances. In the future, we will design more effective item-based CL tasks, and jointly combine user-based CL with item-based CL tasks.

References

1. Chen, T., Kornblith, S., Norouzi, M., Hinton, G.: A simple framework for contrastive learning of visual representations. In: Proceedings of ICML (2020)
2. Covington, P., Adams, J., Sargin, E.: Deep neural networks for youtube recommendations. In: Proceedings of RecSys (2016)
3. Devlin, J., Chang, M.W., Lee, K., Toutanova, K.: BERT: pre-training of deep bidirectional transformers for language understanding. In: Proceedings of NAACL (2019)
4. Guo, H., Tang, R., Ye, Y., Li, Z., He, X.: DeepFM: a factorization-machine based neural network for ctr prediction. In: Proceedings of IJCAI (2017)
5. Herlocker, J.L., Konstan, J.A., Terveen, L.G., Riedl, J.T.: Evaluating collaborative filtering recommender systems. TOIS (2004)
6. Nie, P., et al.: MIC: model-agnostic integrated cross-channel recommender. In: Proceedings of CIKM (2022)
7. Rendle, S.: Factorization machines. In: Proceedings of ICDM (2010)
8. Song, W., et al.: AutoInt: automatic feature interaction learning via self-attentive neural networks. In: Proceedings of CIKM (2019)
9. Sun, F., et al.: BERT4Rec: sequential recommendation with bidirectional encoder representations from transformer. In: Proceedings of CIKM (2019)
10. Wang, T., Isola, P.: Understanding contrastive representation learning through alignment and uniformity on the hypersphere. In: Proceedings of ICML (2020)
11. Wu, Y., et al.: Multi-view multi-behavior contrastive learning in recommendation. In: Proceedings of DASFAA (2022)
12. Wu, Y., et al.: Personalized prompts for sequential recommendation. arXiv preprint arXiv:2205.09666 (2022)
13. Xiao, C., et al.: UPRec: user-aware pre-training for recommender systems. arXiv preprint arXiv:2102.10989 (2021)
14. Xie, R., Liu, Q., Liu, S., Zhang, Z., Cui, P., Zhang, B., Lin, L.: Improving accuracy and diversity in matching of recommendation with diversified preference network. TBD (2021)
15. Xie, R., Liu, Q., Wang, L., Liu, S., Zhang, B., Lin, L.: Contrastive cross-domain recommendation in matching. In: Proceedings of KDD (2022)
16. Xie, R., Qiu, Z., Rao, J., Liu, Y., Zhang, B., Lin, L.: Internal and contextual attention network for cold-start multi-channel matching in recommendation. In: Proceedings of IJCAI (2020)

17. Xie, X., Sun, F., Liu, Z., Wu, S., Gao, J., Ding, B., Cui, B.: Contrastive learning for sequential recommendation. In: Proceedings of ICDE (2022)
18. Xu, J., He, X., Li, H.: Deep learning for matching in search and recommendation. In: Proceedings of SIGIR (2018)
19. Yao, T., et al.: Self-supervised learning for large-scale item recommendations. In: Proceedings of CIKM (2021)
20. Yu, J., Yin, H., Xia, X., Chen, T., Cui, L., Nguyen, Q.V.H.: Are graph augmentations necessary? Simple graph contrastive learning for recommendation. In: Proceedings of SIGIR (2022)
21. Zhou, C., Ma, J., Zhang, J., Zhou, J., Yang, H.: Contrastive learning for debiased candidate generation in large-scale recommender systems. In: KDD (2021)
22. Zhou, K., et al.: S3-Rec: self-supervised learning for sequential recommendation with mutual information maximization. In: Proceedings of CIKM (2020)

Hyperbolic Mutual Learning for Bundle Recommendation

Haole Ke[1], Lin Li[1(✉)], PeiPei Wang[1], Jingling Yuan[1], and Xiaohui Tao[2]

[1] Wuhan University of Technology, Wuhan, China
{kehaole,cathylilin,ppwang07,yjl}@whut.edu.cn
[2] University of Southern Queensland, Toowoomba, Australia
Xiaohui.Tao@usq.edu.au

Abstract. Bundle recommendation aims to accurately predict the probabilities of user interactions with bundles. Most existing effective methods learn the embeddings of users and bundles from user-bundle interaction view and user-item-bundle interaction view. However, they seldom leverage the recommendation difference caused by the distinct learning trends of two views when modeling user preferences. Meanwhile, such two view interaction graphs are typically tree-like. If the graph data with this structure is embedded in Euclidean space, it will lead to severe distortion problem. To this end, we propose a novel Hyperbolic Mutual Learning model for Bundle Recommendation (HyperMBR). The model encodes the entities (user, item, bundle) of the two view interaction graphs in hyperbolic space to learn their accurate representations. Furthermore, a mutual distillation based on hyperbolic distance is proposed to encourage the two views to transfer knowledge for increasingly improving the recommendation performance. Extensive empirical experiments on two real-world datasets confirm that our HyperMBR achieves promising results compared to state-of-the-art bundle recommendation methods.

Keywords: Hyperbolic Space · Mutual Learning · Bundle recommendation

1 Introduction

Nowadays, many novel tasks have emerged in the field of recommendation system, such as group recommendation [30], sequential recommendation [12], etc. Bundle recommendation is one of them and is widely adopted by various platforms to produce a win-win effect. For users, it can reduce the blind search time and enjoy certain discounts; for the platform, this is a good way to increase sales profits and increase product exposure.

Bundle recommendation aims to accurately predict the probability of a user interacting with a bundle by learning the user's historical interaction information. Most existing effective methods learn user and bundle embeddings from the user-bundle interaction graph (i.e., bundle view) and the user-item-bundle interaction graph (i.e., item view) [3,5,6,11]. However, it is observed that distinct

X. Wang et al. (Eds.): DASFAA 2023, LNCS 13944, pp. 417–433, 2023.
https://doi.org/10.1007/978-3-031-30672-3_28

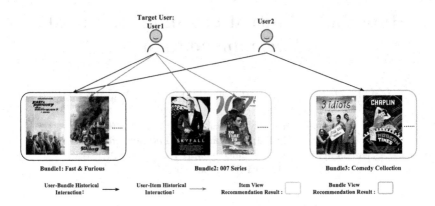

Fig. 1. User Interaction Cases in Real-World Scenarios.

learning trends of two views for user preference will cause different recommendation results. Taking Fig. 1 as an example, *User1* is our target user who has interacted with *Bundle1* and items "The Fast and Furious 9", "No Time to Die" and "Skyfall". We can see that *User1* and *User2* have the same behavior at the bundle view, i.e., they both interact with the *Bundle1* ("Fast & Furious"), which indicates that *User1* and *User2* may have similar preferences. Then if only bundle view is considered, *Bundle3* ("Comedy Collection") that *User2* has interacted with is likely to be recommended by the model to *User1*. But if at the item view, because *User1* interacts with items affiliated with *Bundle2* ("007 Series"), the model will be more inclined to recommend *Bundle2* to the *User1*. This indicates that the separate learning directions of two views will indeed create a difference in the final recommendation results. Bundle view tends to recommend the bundles that similar users have interacted with to the target user. Item view focuses more on the bundle-item similarity and utilizes items as a bridge to connect the target users with the bundles they may like.

How to make the two views learn the knowledge contained in their respective recommendation results will be the key to further improving recommendation performance. However, this important mutually reinforcing relationship is not well exploited in the existing work [3,5,6,11,19,28]. AttList [11] divides the two views separately and uses the interactive information and self-attention mechanism to learn entity embeddings independently. DAM [6] jointly models user-bundle and user-item interaction information, but only simply uses shared parameters to transfer information from denser user-item to user-bundle. Another method, BGCN [5], performs information propagation by graph convolution operations in two views to obtain the final representation and then takes the sum of inner products between representations at different views as the final prediction result. Although the above works perform better than a single view, loose modeling of two views is difficult to guarantee the effect of knowledge transformation. A recent work MIDGN [34] is aware of this problem. It abstractly divides user and bundle embeddings into chunks(intents) and applies contrastive learning framework to improve intents that are learned from different views. However, the above methods suffer from serious distance distortion problems since they uniformly embed the two views' interaction graphs with different tree-like degrees into Euclidean space [7,21]. A detailed analysis of the tree-likeness of interaction graphs can be found in Sect. 4.1.

To this end, we propose a novel Hyperbolic Mutual Learning model for Bundle Recommendation (HyperMBR), which enables two views to continuously extract complementary knowledge contained in soft labels through mutual learning framework [33]. And we observe that the interactive graph in the bundle recommendation has a tree-like structure (power-law distribution). It will incur large distortion if embedding graph data with such structure in the Euclidean space [16,24]. So in order to ensure that each view itself can learn a more accurate entity representation and reduce the error information passed to each other in mutual learning, we use hyperbolic graph convolution neural network (HGCN) [4] as an encoder to learn the representation of entities. To better combine the advantages of hyperbolic space and mutual learning, we also propose KL mimicry Loss based on hyperbolic distance, so that knowledge can be better transferred between two views.

The main contributions of this work are summarized as follows:

- We attempt to bridge mutual learning with hyperbolic GCN for bundle recommendation task and propose a simple but effective model HyperMBR. It encodes entity relations in hyperbolic space and further applies the mutual learning method to learn from each other for high-quality recommendation.
- Extensive experiments on two real-world datasets show that our proposed model outperforms existing state-of-the-art baselines by 9.26% to 30.94% in terms of Recall and NDCG.

2 Preliminaries

2.1 Problem Formulation

Let $\mathcal{U} = \{u_1, u_2, \ldots, u_M\}$, $\mathcal{B} = \{b_1, b_2, \ldots, b_N\}$ and $\mathcal{I} = \{i_1, i_2, \ldots, i_J\}$ be the set of users, bundles and items, where M, N and J represent the number of elements of the three entity sets, respectively. According to a user's interaction information and bundle composition information, we define two interaction matrices $X_{M \times N} = \{x_{ub} | u \in \mathcal{U}, b \in \mathcal{B}\}$, $Y_{M \times J} = \{y_{ui} | u \in \mathcal{U}, i \in \mathcal{I}\}$ and an affiliation matrix $Z_{N \times J} = \{z_{bi} | b \in \mathcal{B}, i \in \mathcal{I}\}$, in which $x_{ub} = 1$ or $y_{ui} = 1$ means that user u has interacted with bundle b or item i and $z_{bi} = 1$ mean bundle b includes item i. Based on the above definition, the task of bundle recommendation can be described as follows:

Input: Given U-I interaction matrix X, U-B interaction matrix Y and B-I affiliation matrix Z.

Output: A predictive function $\mathcal{F}$ that can estimate the probability $\hat{y}_{ub}$ of a user u interacting with bundle b based on the hyperbolic distance $d_{\mathbb{H}}(u, b)$.

2.2 Gromov's δ-hyperbolicity

HGCN [4] has shown that the benefits gain of hyperbolic space over Euclidean space is related to the degree of tree-likeness of the graph which can be measured by Gromov's δ-hyperbolicity. Here we take a simple example to describe the definition of δ-hyperbolicity. Suppose u_1, u_2, b_1, i_1 are vertices of interaction graph while the four nodes set is called quadruplet and let $S_1 = d_{\mathcal{G}}(u_1, u_2) + d_{\mathcal{G}}(b_1, i_1), S_2 = d_{\mathcal{G}}(u_1, b_1) + d_{\mathcal{G}}(u_2, i_1), S_3 = d_{\mathcal{G}}(u_1, i_1) + d_{\mathcal{G}}(u_2, b_1)$, where $d_{\mathcal{G}}$ denote the graph shortest distance. And then they are sorted in none-decreasing $S_1 \geq S_2 \geq S_3$. The δ-hyperbolicity of the *quadruplet* u_1, u_2, b_1, i_1 is defined as $\delta(u_1, u_2, b_1, i_1) = \frac{S_1 - S_2}{2}$. Traverse all the *quadruplets* (a, b, c, d), where the largest $\delta(a, b, c, d)$ is the final hyperbolicity, i.e.,

$$\delta(\mathcal{G}) = \max_{a,b,c,d \in \mathcal{V}} \delta(a, b, c, d). \tag{1}$$

The time complexity of the naive implementation of δ-hyperbolicity is in $O(n^4)$, which is a very costly operation. Computing the hyperbolicity of a large graph will take a long time. The works of [1,27] use sampling to approximate the average of $\delta(\mathcal{G})$ to decrease time cost and obtain robust results. We use the above two algorithms to calculate the hyperbolicity values of the Youshu [6] and NetEase [3], with the calculation results reported in Table 1.

3 Our Proposed Model

3.1 Hyperbolic Geometry

The first step in our work is to choose a suitable hyperbolic model. There are many famous hyperbolic models, including the Poincare ball model $\mathbb{D}$, the Klein model $\mathbb{K}$, the hyperboloid model $\mathbb{H}$, and so on [15]. Each of them has its own Riemannian metric and advantages. Among these models, the hyperboloid model (Lorentzian model) is a good choice due to its unique simplicity and numerical stability [14,18]. It also demonstrates strong embedding learning ability in the field of item recommendation [8]. We define the hyperboloid model here.

Suppose $x = \{x_0, x_1, \ldots, x_d\}$ is a vector in $d+1$ dimensional space, the hyperboloid model can be defined as follows:

$$\mathbb{H}^{d+1,K} = \{x \in \mathbb{R}^{d+1} : \langle x, x \rangle_{\mathbb{H}} = -K, x_0 > 0\}, \tag{2}$$

where $\mathbb{H}^{d+1,K}$ denotes hyperboloid manifold in $d+1$ dimensions with constant negative curvature $c = -1/K$ ($K > 0$) and $\langle \cdot, \cdot \rangle_{\mathbb{H}}$ is Minkowski inner product, $\langle x, y \rangle_{\mathbb{H}} = -x_0 y_0 + x_1 y_1 + \ldots + x_d y_d$. Another important definition in the hyperboloid model is the tangent space $T_x H^{d+1,K}$, which is a local approximation

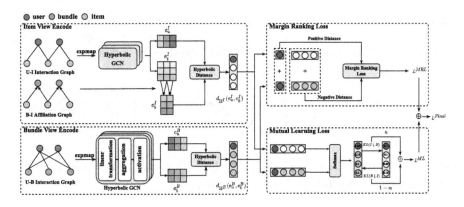

Fig. 2. The illustration of HyperMBR.

of hyperboloid at point x. In tangent space, we can perform GCN operations undefined in hyperbolic space [4]. It is defined by:

$$\mathcal{T}_x\mathbb{H}^{d+1,K} = \{v \in \mathbb{R}^{d+1} : \langle v, x \rangle_{\mathbb{H}} = 0\}. \tag{3}$$

In order to achieve the transformation of embedding between hyperbolic space and tangent space, we need exponential and logarithmic mappings. For arbitrary points x, y, z, $x,y \in \mathbb{H}^{d+1,K}(x \neq y)$, $z \in \mathcal{T}_x\mathbb{H}^{d+1,K}(z \neq 0)$, the exponential map $\exp_x^K(z) : \mathcal{T}_x\mathbb{H}^{d+1,K} \to \mathbb{H}^{d+1,K}$ and the logarithmic map $\log_x^K(y) : \mathbb{H}^{d+1,K} \to \mathcal{T}_x\mathbb{H}^{d+1,K}$ are given by:

$$\exp_x^K(z) = \cosh\left(\frac{\|z\|_{\mathbb{H}}}{\sqrt{K}}\right) x + \sqrt{K}\sinh\left(\frac{\|z\|_{\mathbb{H}}}{\sqrt{K}}\right)\frac{z}{\|z\|_{\mathbb{H}}}, \tag{4}$$

$$\log_x^K(y) = d_{\mathbb{H}}^K(x,y)\frac{y + \frac{1}{K}\langle x,y\rangle_{\mathbb{H}}x}{\left\|y + \frac{1}{K}\langle x,y\rangle_{\mathbb{H}}x\right\|_{\mathbb{H}}}. \tag{5}$$

It is worth mentioning that $d_{\mathbb{H}}^K(x,y)$ is the hyperbolic distance between x and y, which will be used to evaluate the probability of interactions between entities. The smaller the hyperbolic distance, the greater the probability of interaction. This is calculated as follows:

$$d_{\mathbb{H}}^K(x,y) = \sqrt{K}arcosh(-\langle x,y\rangle_{\mathbb{H}}/K). \tag{6}$$

3.2 Bundle View Encoder in Hyperbolic Space

To obtain bundle-view user preferences and bundle features, we use HGCN as an encoder to learn user and bundle embeddings based on user-bundle interaction graphs. As shown in the bottom left of Fig. 2, the encoding process can be divided into three steps: linear transformation, neighbor aggregation and nonlinear activation function. However, since the initial entity embedding is in

Euclidean space, it cannot be directly processed by hyperbolic graph convolution. We need to first map the entity embedding into the hyperbolic space using the exponential map function (Eq. 4). It can be defined as follows:

$$e_u^B = \exp_o^{K^B}\left((0, e_u^{\mathbb{E},B})\right) = \left(\sqrt{K^B}\cosh\left(\frac{\left\|e_u^{\mathbb{E},B}\right\|_2}{\sqrt{K^B}}\right), \sqrt{K^B}\sinh\left(\frac{\left\|e_u^{\mathbb{E},B}\right\|_2}{\sqrt{K}}\right)\frac{e_u^{\mathbb{E},B}}{\left\|e_u^{\mathbb{E},B}\right\|_2}\right),$$

(7)

where $e_u^{\mathbb{E},B}$ is origin Euclidean user embedding, e_u^B is user embedding in bundle view hyperbolic space $\mathbb{H}^B$, and K^B is curvature of $\mathbb{H}^{\mathbb{B}}$.

Linear Transformation. After transforming the Euclidean embeddings into hyperbolic space, we first perform the linear feature transformation operation for them. The operation is composed of multiplication of the embedding vector through a weight matrix and bias translation. Inspired by [4,10], we use the logarithmic map to project hyperbolic embeddings to tangent space and perform linear transformation operation on it. The matrix multiplication and bias addition in bundle view hyperbolic space $\mathbb{H}^B$ are defined as:

$$W^B \odot^{K^B} e_u^B = \exp_o^{K^B}(W^B \log_o^{K^B}(e_u^B)),$$
$$e_u^B \oplus^{K^B} b^B = \exp_{e_u^B}^{K^B}\left(b^B - \lambda\left(\log_o^{K^B}(e_u^B) + \log_{e_u^B}^{K^B}(o)\right)\right),$$

(8)

where $\lambda = \frac{\langle\log_o(e_u^B),b\rangle_{\mathbb{H}}}{d_{\mathbb{H}}^K(x,y)^2}$, W^B is the learned weigh matrix, b^B is learned bias, and parameters with superscript B indicates that the parameter is located in the bundle view hyperbolic space $\mathbb{H}^B$.

Hyperboloid Neighbor Aggregation. Following the linear transformation, aggregation is used to capture neighborhood structures and features. In this hyperbolic space, we exploit historical interaction information between users and bundles to capture user preferences and overall bundle features at the bundle view. While in hyperbolic space, the analog of mean aggregations is Fréchet mean which has no closed form solution [9]. We can map the hyperbolic embedding to tangent space through logmap (Eq. 5) and then complete the aggregation operation in tangent space. The aggregating process in bundle-based hyperbolic space can be formulated as follows:

$$e_u^{B,l+1} = \log_{e_u^{B,l}}^{K^B}(e_u^B) + \sum_{b\in\mathcal{N}(u)}\frac{1}{|\mathcal{N}(u)|}\log_{e_u^B}^{K^B}(e_b^B),$$
$$e_b^{B,l+1} = \log_{e_b^{B,l}}^{K^B}(e_b^{B,l}) + \sum_{u\in\mathcal{N}(b)}\frac{1}{|\mathcal{N}(b)|}\log_{e_b^{B,l}}^{K^B}(e_u^B),$$

(9)

where $\mathcal{N}(u),\mathcal{N}(b)$ represent neighbors of user u and bundle b.

Hyperbolic Non-linear Activation. In this part, we use non-linear activation in tangent space to learn non-linear transformations. As the last step of information propagation, it not only needs to process the data of this hyperboloid, but also needs to be responsible for mapping the data into the next hyperboloid. For $e_u^{I,l+1}$ and $e_b^{I,l+1}$, the hyperbolic activation is defined as:

$$
\begin{aligned}
e_b^{B,l+1} &= \exp_o^{K^B}(\sigma(\log_o^{K^B}(e_b^{B,l+1}))), \\
e_u^{B,l+1} &= \exp_o^{K^B}(\sigma(\log_o^{K^B}(e_u^{B,l+1}))),
\end{aligned}
\tag{10}
$$

where σ is non-linear activation function $Relu$. The updated embeddings $e_b^{B,l+1}$, $e_u^{B,l+1}$ have now learned one-hop neighbor information in the bundle-based hyperbolic space. By stacking multiple hyperbolic convolution layers, these embeddings will continue to learn information from high-order nodes.

3.3 Item View Encoder in Hyperbolic Space

To capture the user's preference and the item's own features in the item view, we perform a hyperbolic graph convolution operation on the user-item interaction graph in the item-based hyperbolic space. Similarly, we apply linear transformation (Eq. 8), neighbor aggregation (Eq. 9) and activation (Eq. 10) on the embeddings which is transformed by exponential map function, and finally we get the updated user embeddings e_u^I and item embeddings e_i^I.

Subsequently, since bundle is not an atomic unit, it is composed of multiple items which collectively reflect the characteristics of the bundle. We aggregate the item information into bundles according to their affiliation. This aggregation operation can be described as follows:

$$
e_b^I = \exp_o^{K^I}\left(\sum_{i \in \mathcal{N}(b)} \frac{1}{|\mathcal{N}(b)|} \log_o^{K^I}(e_i^I)\right),
\tag{11}
$$

where $\mathcal{N}(b)$ represents the set of items which are included in bundle b. During this information propagation, we utilize denser item information to synthesize bundled embeddings. This can not only obtain the characteristics of the bundle from a finer-grained level, but also alleviate the problem of bundle data sparsity, and finally improve the performance of the model.

3.4 Hyperbolic Mutual Learning for Bundle Recommendation

Mutual Leaning Loss. This module aims to utilize mutual learning loss to enable the representations of the two views can mutually enhance each other. After the embeddings of users and bundles (e.g., $e_u^B, e_b^B, e_u^I, u_b^I$) are obtained for the two hyperbolic spaces H^I and H^B, we use Eq. 6 to calculate the hyperbolic distance between these embeddings. And then we use softmax to convert the hyperbolic distances into probability scores for interactions. It can be described as follows:

$$
\begin{aligned}
p^I(u^I, b^I) &= softmax(d_{\mathbb{H}}^{K^I}(e_u^I, e_b^I)/\tau), \\
p^B(u^B, b^B) &= softmax(d_{\mathbb{H}}^{K^B}(e_u^B, e_b^B)/\tau),
\end{aligned}
\tag{12}
$$

where τ is a hyper-parameter known as the *temperature*. The posterior predictions (soft labels) p^I and p^B reveal the hidden relationships among entities that are not explicitly included in the training data and show the difference prone of the two views. By using each other's soft labels as their own additional training experience, the two views can learn knowledge that may not be learned with their own data alone. And then we use the KL divergence to encourage mutual learning between the two views. The KL distance can be formulated as follows:

$$KL(p^I \parallel p^B) = \sum_{j}^{C} p^I(x_j) \log \frac{p^I(x_j)}{p^B(x_j)}, \tag{13}$$

where C is the number of bundles randomly sampled [5] for each user. The final mutual learning loss function is as follows:

$$L^{ML} = \alpha KL(p^I \parallel p^B) + (1-\alpha)KL(p^B \parallel p^I), \tag{14}$$

where the α is a hyper-parameter that controls the degree of mutual learning between the two views. The higher the α, the more force the item view transmits knowledge to the bundle.

Margin Ranking Loss. The margin ranking loss which is widely used in distance model [17,26] is adopted as supervised loss. This loss function first uses the negative sample set N to estimate the pseudo-ranking of a positive sample j as follows:

$$rank(i,j) \approx r_{i,j} = \sum_{k \in N} \max(\mu + d_{\mathbb{H}}^2(u_i, b_j) - d_{\mathbb{H}}^2(u_i, b_k), 0), \tag{15}$$

where $\mu \in \mathbb{R}^+$ is safety margin parameter. Then the margin ranking loss is defined as:

$$L_{i,j}^{MRL} = log(1 + r_{i,j}). \tag{16}$$

And then we combine the Mutual Learning Loss and Margin Ranking Loss as the final loss. It is formulated as:

$$L^{Final} = L^{MRL} + \lambda L^{ML}, \tag{17}$$

where λ is the hyper-parameter to weight the mutual learning loss.

3.5   Prediction

In Euclidean space, the inner product between embeddings is widely used to compute matching scores, and models using the inner product usually regularize the norm of the embeddings to a limited scale. However, applying the inner product to hyperbolic space will make it difficult to push embeddings to the outer part of the hyperbolic space where has the largest capacity [32]. Therefore, in order to fully utilize the spatial expressive power of hyperbolic space, the

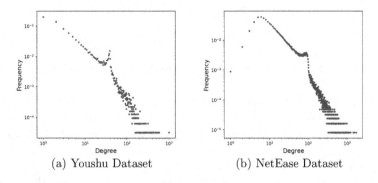

(a) Youshu Dataset   (b) NetEase Dataset

Fig. 3. Degree distribution of two real-world datasets.

hyperbolic distance between node embeddings is used to estimate the matching score. It is defined as follows:

$$\hat{y}_{ub} = d_{\mathbb{H}^I}(e_u^I, e_b^I) + d_{\mathbb{H}^B}(e_u^B, e_b^B), \tag{18}$$

which is composed of bundle view and item view hyperbolic distance. During the recommendation phase, K bundles closest to user u are selected as the final recommendation result of the model.

4 Experiments

We conduct extensive experiments on two real-world datasets to answer the following experiment questions (EQ): **EQ1:** Does HyperMBR have superior performance to state-of-the-art related models? **EQ2:** What are the contributions of the components in HyperMBR to its overall performance? **EQ3:** Whether mutual learning between two views works as we expect? **EQ4:** What is the effect of hyper-parameters (mutual learning coefficient α and temperature τ) on the result of HyperMBR?

4.1 Datasets and Metrics

Two public benchmarks, NetEase and Youshu, are used to evaluate our model. They differ in the number of entities, interaction density and hyperbolicity, allowing for a more complete picture of how the model performs under different conditions. This is also why they are widely used by other sota bundle recommendation models [3,6,28,31,34].

The statistics of the datasets are shown in the Table 1. And in order to verify that the interaction graphs in bundle recommendation indeed have tree-like (power-law distributed) [24] properties, we calculate the Gromov's δ-hyperbolicity value of the above two datasets in Table 1 and present their degree distributions in Fig. 3. We can conclude as follows: 1) These interaction graphs have a pretty low δ value (The lower the δ value, the more hyperbolic the graph)

Table 1. Statistics of two real-world datasets

Dataset	User	Bundle	Item	U-B	U-I	B-I	U-I δ_{avg}	B-I δ_{avg}	U-B δ_{avg}	δ_{max}	δ_{avg}
NetEase	18,528	22,864	123,628	302,303	1,128,065	1,778,838	0.3290	0.3406	0.2846	1.5	0.1934
Youshu	8,039	4,771	32,770	51,377	138,515	176,667	0.3366	0.3453	0.2921	2	0.2509

and the degree distribution does resemble a power-law distribution. 2) The δ values of the three types of interaction graphs are not exactly the same. The U-B interaction graph used in the bundle view has a lower δ value than U-I and B-I.

We evaluate the recommendation performance of our model with two widely used metrics, Recall@K and NDCG@K [5,6,28,31,34]. Recall@K measures the ratio of test bundles included in the top-K ranking list, while NDCG@K complements Recall by assigning higher scores to hits higher up on the list.

4.2 Baselines

To demonstrate the effectiveness of HyperMBR, we compared it with the following state-of-the-art models:

- DAM [6] applies a multi-task learning framework to jointly model user-item and user-bundle interactions and proposes a factorized attention network to learn bundle representations of affiliated items.
- Attlist [11] is an attention-based model that uses self-attention mechanisms and hierarchical structure of data to learn user and bundle representations, respectively.
- BGCN [5] is a graph-based model which first applies a graph neural network to bundle recommendation. It divides the entire interaction graph into bundle view and item view and then GCN operations are performed at the views to capture the different preferences of users.
- GRAM-SMOT [28] is another graph-based method that uses a graph attention mechanism to model the relative influence of user, bundle, and item nodes' neighbors.
- MIDGN [34] is the latest graph-based model in bundle recommendation task, which uses graph neighbor routing mechanism to disentangle the user-item/bundle-item graph and refines the user-item/bundle-item graph of different intents from multiple views.

4.3 Experimental Settings

For each baseline, we follow the optimal parameters given in the cited papers. The embedding dimension is fixed as 64. The optimizer is Adam with 4096 mini-batch. For the method using BPR loss [22], the number of negative samples is set to 1. In our model, we set the learning rate as 1e-3 and tune the hyper-parameters margin μ, mutual learning coefficient α, loss weight λ and temperature τ with the ranges of $\{0.5, 1, 1.5, 2\}$, $\{0.1, 0.3, 0.5, 0.7, 0.9\}$,

Table 2. Performance comparisons on experimental results on two real-world datasets, where the bold font indicates the best model and underline the second best.

Dataset	Model	Recall@5	NDCG@5	Recall@10	NDCG@10	Recall@20	NDCG@20	Recall@40	NDCG@40
Youshu	DAM	0.1098	0.0873	0.1460	0.0996	0.2082	0.1198	0.2890	0.1418
	AttList	0.0414	0.0370	0.0613	0.0424	0.0869	0.0499	0.1245	0.0599
	BGCN	0.0889	0.0848	0.1515	0.1032	0.2336	0.1268	0.3415	0.1559
	GRAM-SMOT	0.1129	0.1037	0.1673	0.1184	0.2408	0.1392	0.3451	0.1675
	MIDGN	0.1210	0.1076	0.1728	0.1210	0.2519	0.1440	0.3466	0.1696
	HyperMBR	**0.1321**	**0.1227**	**0.1974**	**0.1404**	**0.2813**	**0.1645**	**0.3798**	**0.1916**
	% Improv.	9.26%	14.00%	14.20%	16.08%	11.70%	14.18%	9.58%	12.94%
NetEase	DAM	0.0099	0.0077	0.0163	0.0102	0.0411	0.0210	0.0690	0.0281
	AttList	0.0194	0.0163	0.0325	0.0211	0.0548	0.0281	0.0898	0.0373
	BGCN	0.0211	0.0187	0.0361	0.0239	0.0581	0.0308	0.0934	0.0401
	GRAM-SMOT	0.0204	0.0174	0.0346	0.0225	0.0595	0.0303	0.0972	0.0402
	MIDGN	0.0225	0.0192	0.0374	0.0245	0.0607	0.0319	0.0968	0.0415
	HyperMBR	**0.0280**	**0.0251**	**0.0467**	**0.0316**	**0.0755**	**0.0406**	**0.1174**	**0.0517**
	% Improv.	24.32%	30.94%	24.80%	28.99%	24.42%	27.30%	20.75%	24.54%

$\{0.05, 0.1, 0.5, 1, 2\}$ and $\{0.8, 0.9, 1.0, 1.1, 1.2\}$. The curvature of $\mathbb{H}^I$ and $\mathbb{H}^B$ are selected in range $\{1, 2, 3, 4\}$. Note that the two curvatures are not necessarily the same. And due to exponential and logarithmic mapping, Adam Optimizer [13] is still used to optimize our model parameters because of its stability and slightly better performance than Riemannian optimization [2,4]. All experiments are conducted with Pytorch on a commodity machine equipped with 2 GTX 3090.

4.4 Performance Comparison (EQ1)

We compare the performance of our proposed model HyperMBR with other baselines and the results are presented in Table 2. As we can be seen HyperMBR obviously outperforms all other models by $9.26\% - 14.20\%$ and $20.75\% - 24.80\%$ on two datasets of Recall@K. As for the NDCG@K, HyperMBR achieves $12.94\% - 16.08\%$ and $24.54\% - 30.94\%$ improvements on Youshu and NetEase, respectively. We believe that this may be attributed to the following reasons: 1) **GCN's powerful high-order information encoding ability.** Thanks to this ability, Graph models (BGCN, GRAM-SMOT and MIDGN) show superior performance on bundle recommendation tasks compared to DAM and Attlist. 2) **Mutual learning between the two views.** Mutual learning successfully helps the two views learn a more comprehensive entity representation and finally improves the model's performance. This is also a reason why MIDGN is superior to other Graph Models. It not only obtains the user's intent representation from multiple views, but also utilizes contrastive learning for comparison to improve the learned intent representation. 3) **Hyperbolic space alleviates the distortion problem.** We fully consider the tree-like structure of interaction graphs in bundle recommendation, and exploit the strong embedding ability of hyperbolic space for this structure to learn more accurate entity representations.

Table 3. Ablation Study on Youshu and NetEase

Dataset	Metric	w/o IV	w/o BV	w/o Hyp	w/o ML	HyperMBR
Youshu	Recall@5	0.0640	0.1118	0.1206	0.1275	0.1321
	NDCG@5	0.0633	0.0993	0.1084	0.1149	0.1227
NetEase	Recall@5	0.0158	0.0156	0.0231	0.0239	0.0280
	NDCG@5	0.0151	0.0128	0.0199	0.0208	0.0251

4.5 Ablation Study (EQ2)

To show the impact of several key designs on the overall performance of the model, we conduct comprehensive ablation study with results shown in Table 3.

Impact of Item View. We set a variant, HyperMBR$_{w/o\ IV}$, which disable aggregating affiliated items into bundle. From Table 3, we can see that the performance of HyperMBR$_{w/o\ IV}$ drops dramatically, which indicates that aggregating affiliated items into a bundle is essential for performance improvement.

Impact of Bundle View. In this experiment, we disable the information propagation between user and bundle, denoted as HyperMBR$_{w/o\ BV}$. As shown in Table 3, for Youshu dataset, the performance of the model has dropped by 18.15%(0.1321 → 0.1118) and 23.53%(0.1227 → 0.0993) on Recall@5 and NDCG@5. For NetEase, Recall@5 and NDCG@5 are dropped by 79.69%(0.0280 → 0.0156) and 97.09%(0.0251 → 0.0128). The results indicate that considering user-bundle interaction information has an effect on performance improvement.

Impact of Hyperbolic Space. To examine the impact of the hyperbolic space, we let the model perform all operations in Euclidean space and ensure that other conditions are fully consistent with hyperbolic space. Based on the results reported in Table 3, the Euclidean version of the model produces worse results on both Recall@5 and NDCG@5 metrics by 21.20% (0.0280 → 0.0231) and 26.24%(0.0251 → 0.0199) for NetEase, respectively. As for Youshu, Recall@5 and NDCG@5 are dropped by 9.57%(0.1321 → 0.1206) and 13.15%(0.1227 → 0.1084). From this, we can conclude that embedding the tree-like hierarchy graph into hyperbolic space can learn better representations.

Impact of Mutual Learning. To evaluate whether mutual learning contributes to model performance, we design a variant of the model that invalidates the mutual learning loss in Eq. 17, called HyperMBR w/o ML. Compared with HyperMBR, HyperMBR w/o ML inevitably suffers a significant performance decline, which demonstrates the importance of encouraging the two views to learn from each other.

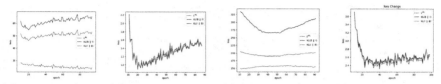

(a) Youshu w/o ML (b) Youshu with ML (c) NetEase w/o ML (d) NetEase with ML

Fig. 4. KL distance Change during training process of two real-world datasets.

4.6 Study of Mutual Learning (EQ3)

Although ablation study have demonstrated that mutual learning can indeed bring performance improvements to the model, in order to further investigate whether mutual learning works as we expect, we conducted some experiments to study the following two problems: 1)Whether the two models really distill knowledge from each other? and 2) whether the two views reinforce each other?

Knowledge Distillation Effect. To demonstrate that knowledge is indeed distilled between the two views, i.e., the two views output more similar logits. We visualize the change curve of the KL distance during training. It is worth noting that we give the two-views a warm-up process to give them time to learn a certain level of basic knowledge, rather than prematurely passing immature knowledge to each other. So here we start to calculate the KL distance change from epoch 10, and the result is shown in Fig. 4. We can see that mutual learning does greatly reduce the KL distance between the logits of the two views. This proves that as the training progresses, the two views are indeed learning from each other and slowly reaching a consensus on the recommendation result.

Mutual Reinforce Effect. To verify the two views are mutually reinforcing, rather than falling into the dilemma of "the blind lead the blind". We show the performance calculated based on single-view and both-view in Fig. 5. The bundle-view prediction uses $d_{\mathbb{H}^B}(e_u^B, e_b^B)$, and the item-view prediction uses $d_{\mathbb{H}^I}(e_u^I, e_b^I)$, both-view prediction is the same as Eq. 18. From the results, we can draw the following conclusions: 1)Mutual learning significantly improves the model performance for all three types of predictions. 2) The performance of a single view is far inferior to both-view, illustrates the strong complementarity of the knowledge which learned at the two views and justifies our motivation to use mutual learning to encourage them to reinforce each other.

4.7 Hyper-parameter Analyse (EQ4)

We also conduct experiments to further investigate the impact of some key hyper-parameters. To evaluate how the impact of the mutual learning coefficient α affects the performance, we select the α in the range $\{0.1, 0.3, 0.5, 0.7, 0.9\}$, with

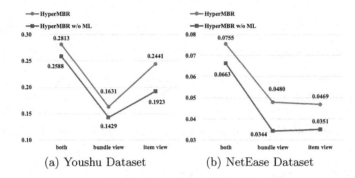

Fig. 5. Performance comparison of HyperMBR and HyperMBR w/o ML in three different cases in terms of Recall@20.

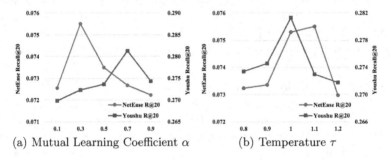

Fig. 6. The impact of mutual learning coefficient α and temperature τ .

the result shown in Fig. 6(a). We can see that for NetEase and Youshu, Hyper-MBR achieves the best performance on different mutual learning coefficients. This could be due to the fact that the model performance shown by the two views is not exactly the same during the training process, which can be seen from Fig. 5. For the temperature τ, we also conduct experiments to demonstrate the sensitivity of the model performance to τ. We gradually increase the value of τ from 0.8 to 1.2, the performance first peaks around 1.0, and then drops rapidly, which shows the change of τ has a relatively large impact on the model.

5   Related Work

5.1   Hyperbolic Space for Recommendation

Hyperbolic spaces and hyperbolic embeddings are being widely used in item recommendation systems due to their powerful modeling capabilities for hierarchical data. HGCF [25] combines hyperbolic space and GCN framework to develop a novel model for item recommendation that can effectively learn user-item embedding with marginal ranking loss. LGCF [29] proposes a model in which all GCN operations are finished in hyperbolic space to decrease the negative impact caused by implementing GCN operations in tangent space. In [32], the authors provide a detailed theoretical analysis of the role of hyperbolic space and hyperbolic embedding in the field of recommendation systems and show when and where hyperbolic can be used to improve embedding accuracy. However, there has been little work applying hyperbolic space to bundle recommendation.

5.2 Bundle Recommendation

With the increasing focus on bundle recommendation task, many advanced efforts have been made in this field [3,5,6,11,19,20,23,28,34]. Earlier work [23] doesn't treat bundle as a collection of items, but regards bundle as atomic unit similar to item. Later the difference between bundle and item began to be gradually paid attention to. In LIRE [19], the authors consider the user's historical interactions with bundles and items under the BPR [22] framework. On the basis of LIRE, EFM [3] regards bundle and item as sentence and word respectively, and word embedding algorithm is introduced to solve the problem of bundle recommendation. DAM [6] proposes a multi-task deep neural network model to jointly model user-item and user-bundle interactions and uses Factorized Attention Network to measure importance of items within a bundle. Attlist [11] leverages the user-bundle-item hierarchy and self-attention mechanism to capture user and list consistency. However, these methods fail to learn higher-order information in the user-bundle-item graph. Recently, graph neural networks have become the mainstream technology for recommendation due to their strong ability to encode high-order information. BGCN [5] is the first work to apply GCN to solve bundle recommendation task. The work [5] decouples the complete interactive graph of bundle recommendation into bundle view and item view, and then implements graph convolution operations to learn user preferences at different views. The following works [20,28,34] use attention mechanism or contrastive learning based on BGCN to obtain better entity representations. Inspired by the above work, our work fully learns different knowledge in hyperbolic space based on bundle view and item view and utilizes the respective predicted soft labels to enhance each other through mutual learning [33].

6   Conclusion and Future Work

In this paper, we apply mutual learning in hyperbolic space to encourage two views in bundle recommendation to reinforce each other and propose hyperbolic mutual learning model, HyperMBR, which achieves significant performance improvement on the two public benchmarks compared to other baselines. Our future work will consider that the user interaction behavior in both benchmarks is observational rather than experimental. This makes various biases widely exist in the data. It is a good direction to handle biases in bundle recommendation.

Acknowledgements. This work is supported in part by the National Natural Science Foundation of China (62276196) and the Key Research and Development Program of Hubei Province (2021BAA030).

References

1. Adcock, A.B., Sullivan, B.D., Mahoney, M.W.: Tree-like structure in large social and information networks. In: ICDM, pp. 1–10 (2013)
2. Bonnabel, S.: Stochastic gradient descent on riemannian manifolds. IEEE Trans. Autom. Control **58**(9), 2217–2229 (2013)
3. Cao, D., Nie, L., He, X., Wei, X., Zhu, S., Chua, T.S.: Embedding factorization models for jointly recommending items and user generated lists. In: SIGIR, pp. 585–594 (2017)
4. Chami, I., Ying, Z., Ré, C., Leskovec, J.: Hyperbolic graph convolutional neural networks. In: NeurIPS, pp. 4869–4880 (2019)
5. Chang, J., Gao, C., He, X., Jin, D., Li, Y.: Bundle recommendation with graph convolutional networks. In: SIGIR, pp. 1673–1676 (2020)
6. Chen, L., Liu, Y., He, X., Gao, L., Zheng, Z.: Matching user with item set: collaborative bundle recommendation with deep attention network. In: IJCAI, pp. 2095–2101 (2019)
7. Chen, W., Fang, W., Hu, G., Mahoney, M.W.: On the hyperbolicity of small-world and treelike random graphs. Internet Math. **9**(4), 434–491 (2013)
8. Chen, Y., et al.: Modeling scale-free graphs with hyperbolic geometry for knowledge-aware recommendation. In: WSDM, pp. 94–102 (2022)
9. Fréchet, M.: Les éléments aléatoires de nature quelconque dans un espace distancié. Annales de l'institut Henri Poincaré **10**(4), 215–310 (1948)
10. Ganea, O., Bécigneul, G., Hofmann, T.: Hyperbolic neural networks. In: NeurIPS, pp. 5350–5360 (2018)
11. He, Y., Wang, J., Niu, W., Caverlee, J.: A hierarchical self-attentive model for recommending user-generated item lists. In: CIKM, pp. 1481–1490 (2019)
12. Hu, K., Li, L., Xie, Q., Liu, J., Tao, X.: What is next when sequential prediction meets implicitly hard interaction? In: CIKM, pp. 710–719 (2021)
13. Kingma, D.P., Ba, J.: Adam: a method for stochastic optimization. In: ICLR (2015)
14. Law, M., Liao, R., Snell, J., Zemel, R.: Lorentzian distance learning for hyperbolic representations. In: ICML, pp. 3672–3681 (2019)
15. Levy, S.V.F.: Flavors of geometry, vol. 31 (1997)
16. Linial, N., London, E., Rabinovich, Y.: The geometry of graphs and some of its algorithmic applications. Combinatorica **15**(2), 215–245 (1995)
17. Liu, K., Natarajan, P.: WMRB: learning to rank in a scalable batch training approach. In: RecSys (2017)
18. Liu, Q., Nickel, M., Kiela, D.: Hyperbolic graph neural networks. In: NeurIPS, pp. 8228–8239 (2019)
19. Liu, Y., Xie, M., Lakshmanan, L.V.: Recommending user generated item lists. In: RecSys, pp. 185–192 (2014)
20. Ma, Y., He, Y., Zhang, A., Wang, X., Chua, T.: CrossCBR: cross-view contrastive learning for bundle recommendation. In: SIGKDD, pp. 1233–1241 (2022)
21. Ravasz, E., Barabási, A.L.: Hierarchical organization in complex networks. Phys. Rev. E **67**(2), 026112 (2003)
22. Rendle, S., Freudenthaler, C., Gantner, Z., Schmidt-Thieme, L.: BPR: bayesian personalized ranking from implicit feedback. In: UAI, pp. 452–461 (2009)
23. Rendle, S., Freudenthaler, C., Schmidt-Thieme, L.: Factorizing personalized Markov chains for next-basket recommendation. In: WWW, pp. 811–820 (2010)
24. Sala, F., De Sa, C., Gu, A., Ré, C.: Representation tradeoffs for hyperbolic embeddings. In: ICML, pp. 4460–4469 (2018)

25. Sun, J., Cheng, Z., Zuberi, S., Pérez, F., Volkovs, M.: HGCF: hyperbolic graph convolution networks for collaborative filtering. In: WWW, pp. 593–601 (2021)
26. Tay, Y., Anh Tuan, L., Hui, S.C.: Latent relational metric learning via memory-based attention for collaborative ranking. In: WWW, pp. 729–739 (2018)
27. Tifrea, A., Bécigneul, G., Ganea, O.E.: Poincare glove: hyperbolic word embeddings. In: ICLR (2019)
28. Vijaikumar, M., Shevade, S., Murty, M.N.: GRAM-SMOT: top-n personalized bundle recommendation via graph attention mechanism and submodular optimization. In: ECML/PKDD, pp. 297–313 (2020)
29. Wang, L., Hu, F., Wu, S., Wang, L.: Fully hyperbolic graph convolution network for recommendation. In: CIKM, pp. 3483–3487 (2021)
30. Wang, P., Li, L., Xie, Q., Wang, R., Xu, G.: Social dual-effect driven group modeling for neural group recommendation. Neurocomputing **481**, 258–269 (2022)
31. Wang, X., Liu, X., Liu, J., Wu, H.: Relational graph neural network with neighbor interactions for bundle recommendation service. In: ICWS, pp. 167–172 (2021)
32. Zhang, S., Chen, H., Ming, X., Cui, L., Yin, H., Xu, G.: Where are we in embedding spaces? In: SIGKDD, pp. 2223–2231 (2021)
33. Zhang, Y., Xiang, T., Hospedales, T.M., Lu, H.: Deep mutual learning. In: CVPR, pp. 4320–4328 (2018)
34. Zhao, S., Wei, W., Zou, D., Mao, X.: Multi-view intent disentangle graph networks for bundle recommendation. In: AAAI, pp. 4379–4387 (2022)

MixMBR: Contrastive Learning for Multi-behavior Recommendation

Zhicheng Qiao, Hui Yan[(✉)], and Lei Han

Nanjing University of Science and Technology, Nanjing, China
{zchengqiao,yanhui,hanl}@njust.edu.cn

Abstract. Contrastive learning has emerged as a dominant technique for unsupervised representation learning. Recent studies reveal that contrastive learning can effectively alleviate the limited supervision signals in Multi-behavior Recommendation(MBR). However, we argue that contrastive learning is not directly applicable to MBR due to (i) embedding different behaviors of the anchor close to positive samples violates the diversity of multiple behaviors; (ii) neglecting the effect of hard negative samples leads to sub-optimal performance. Inspired by mixup operation as a form of data argumentation in the field of computer vision, we propose MixMBR, a simple learning principle to alleviate these issues. Specifically, for each behavior, we execute the LightGCN to learn the user and item embeddings for a supervised task with dual type-specific feature transformation matrices. Meanwhile, we interpolate these learned transformation matrices to generate mixed views for a contrastive learning task. By doing so, parameter interpolation yields a parameter sharing contrastive learning, resulting in mining hard negative samples and preserving commonalities hidden in different behaviors. Extensive experiments on two real-world datasets indicate that our method outperforms state-of-the-art recommendation methods.

Keywords: Multi-behavior recommendation · Contrastive learning · Graph neural network

1 Introduction

Among various recommendation techniques, Graph Neural Networks (GNNs) has become an extremely important field for its advantages in processing the structural data and exploring structural information. This pipeline models user-item interactions as a graph and generate the user/item feature representations with the graph structure information preserved. The typical models include NGCF [1], LightGCN [2] and SGL [3]. However, most of the existing recommendation models are designed based on single behavior (*e.g.*, purchase). In real recommendation scenarios, there are multiple other types of auxiliary behaviors (*e.g.*, click and add to cart) between users and items.

Different types of behaviors may characterize user performance from different interaction dimensions and complement each other. However, there is a challenge

X. Wang et al. (Eds.): DASFAA 2023, LNCS 13944, pp. 434–445, 2023.
https://doi.org/10.1007/978-3-031-30672-3_29

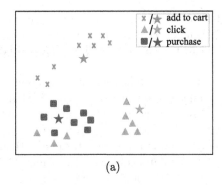 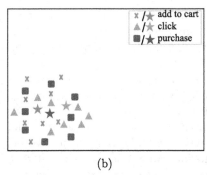

(a)   (b)

Fig. 1. An illustration of node embeddings for three behaviors. (a): Embedding of three behaviors without mixing; (b): Embedding of three behaviors with mixing. The target behaviors are represented by red squares, and auxiliary behaviors are represented by gray markers and blue triangles, respectively. We randomly select a node embedding of the user's target behavior as the anchor and denote it with a red pentagram. The positive samples under its auxiliary behaviors are represented by a gray pentagram and a blue pentagram, respectively. (Color figure online)

in multi-behavior that does not exist in single behavior, *i.e.* how to distinguish the semantic importance of different behaviors. To address this issue, recent works [4,5] have proposed different aggregation schemes to integrate the embedding of individual behaviors. For example, MBGCN [4] reconstructs multiple user-item interaction matrices into an unified graph to adequately model the different behaviors and preference strengths reflected by the various behavioral semantics. Inspired by the success of contrastive learning in CV, some recent works [6–8] introduce contrastive learning into multi-behavior recommendation (MBR). For example, S-MBRec [6] proposes a star-style contrastive learning task to capture the embedding commonalities between target and auxiliary behaviors. What's more, HMG-CR [8] constructs multiple hyper meta-graphs by combining multiple hyper meta-paths of a user, and performs contrastive learning between the constructed hyper meta-graphs to adaptively obtains complex dependencies between different behaviors and represents different behaviors embedding of patterns.

Despite the success of these methods, we argue that contrastive learning is not directly applicable to MBR. First, different types of behaviors embody diverse interaction patterns between users and items. Figure 1(a) shows the embedding of three behaviors (including one target and two auxiliary behaviors). As can be seen from the figure, the positive samples of different views are far away from the anchor, indicating that they have different semantic information. However, existing models based on contrastive learning directly close the distance between positive samples and anchors from different views, forcing them to present similar semantic information, which violates the diversity of multiple behaviors. Second, an important aspect of contrastive learning, the effect of hard negatives is ignored in the context of MBR. As shown in Fig. 1(a), in the node embeddings of auxiliary behaviors (grey markers and blue triangles), most are far away from the anchor and only a few are close to the anchor. This means that most of these nodes are easy negative samples and a few are hard negative samples, which have a limited contribution to the contrastive loss.

Based on these observations and inspired by the application of mixup operation in CV [9,10], we propose a mixing-based method for MBR, named MixMBR. The framework of MixMBR is shown in Fig. 2. Specifically, we first conduct LightGCN to learn the user and item embeddings and generate dual representations for each behavior using two different encoding matrices. Our model then splits into two branches: user preference prediction and contrastive learning. For user preference prediction, the embeddings of adjacent item nodes are propagated to user nodes via type-specific behavior message-passing, and an aggregation coefficient is learned for each behavior to aggregate all behaviors. For the branch of contrastive learning, we interpolate transformation matrices for different types of behaviors to generate mix-up augmented views for each behavior. And we perform contrastive learning between mixed views of target and auxiliary behaviors. As shown in Fig. 1(b), for the mixed anchor, positive samples have similar semantics to it, and the proportion of hard negative samples among all negative samples has also increased. In a nutshell, this work makes the following contributions:

- We introduce a mixup data augmentation method to generate mixed views, which interpolates two different transformation matrices for each type of behavior. In addition, we also adopt a parameter sharing strategy to extract common information among behaviors.
- We propose a simple but effective contrastive learning model for MBR named MixMBR, which performs constrastive learning on the mixed views of different behaviors to mine the commonalities instead of directly on the original views.
- We conduct extensive experiments on two real-world recommendation datasets to show that our framework improves recommendation performance compared to baselines.

2 Methodology

2.1 Problem Definition

Let $\mathcal{U} = \{u_1, ..., u_N\}$ and $\mathcal{I} = \{i_1, ..., i_M\}$ denote the set of users and items, respectively. N and M represent the numbers of users and items. In MBR scenarios, we define $\boldsymbol{A}^k \in \mathbb{R}^{N \times M}$ to denote the user-item interaction matrix under the k-th ($k = 1, 2, ..., K$) behavior type, where K denotes the number of behavior types. Among all behaviors, one is the target behavior and the others are auxiliary behaviors. The goal of MBR is to predict the possibility of interaction between user and item under target behavior with the help of all types of behaviors.

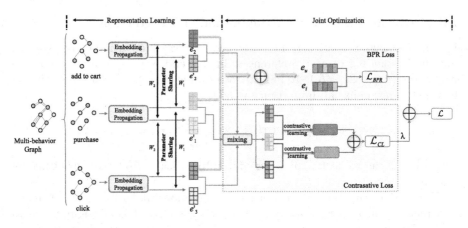

Fig. 2. Overall framework of our model. We take an example that $K=3$, and the purchase is target behavior and the rest two are auxiliary behaviors. We first generate two different views for each behavior (e.g., for purchase, we generate e_1 and e_1'). In the first branch, for the prediction of user preference, we only use one specific view for each behavior (e.g., e_1, e_2, e_3). In the second branch, in our proposed contrastive learning module, we perform contrastive learning of target and auxiliary behaviors between the mixed views (e.g., $\hat{e}_1$, $\hat{e}_2$, $\hat{e}_3$).

2.2 Type-Specific Behavior Message-Passing

In the initial step, we split K subgraphs of the multi-behavior graph, and adopt message-passing (a.k.a., graph convolution operation in LightGCN [2]) along each subgraph separately. It is formulated in a matrix form as follows:

$$\boldsymbol{X}_{\mathcal{U}}^{k,(l+1)} = \hat{\boldsymbol{A}}^k \boldsymbol{X}_{\mathcal{I}}^{k,(l)}, \quad \boldsymbol{X}_{\mathcal{I}}^{k,(l+1)} = (\tilde{\boldsymbol{A}}^k)^T \boldsymbol{X}_{\mathcal{U}}^{k,(l)}, \tag{1}$$

where $l \in \{0, 1, ..., L\}$. $\hat{\boldsymbol{A}}^k = (\hat{\boldsymbol{D}}_{\mathcal{U}}^k)^{(-\frac{1}{2})}(\boldsymbol{A}^k + \boldsymbol{I})(\hat{\boldsymbol{D}}_{\mathcal{U}}^k)^{(-\frac{1}{2})}$ is a normalized adjacency matrix with self-loop. The degree matrix for user $\hat{\boldsymbol{D}}_{\mathcal{U}}^k \in \mathbb{R}^{N \times N}$ is defined as a diagonal matrix where the i-th diagonal element is $\sum_j \boldsymbol{A}_{ij}^k$.

And $\tilde{\boldsymbol{A}}^k$ is similar to $\hat{\boldsymbol{A}}^k$. $X_{\mathcal{U}}^{k,(l)} \in \mathbb{R}^{N \times d}$ and $X_{\mathcal{I}}^{k,(l)} \in \mathbb{R}^{M \times d}$ are the embedding matrix of users and items under the k-th behavior after l layers propagation, respectively. Besides, d indicates embedding dimension.

Considering that information from different orders of neighbor are important, we concatenate the representations of users and items, respectively, obtained by all L layers, as below:

$$\boldsymbol{X}_{\mathcal{U}}^k = \boldsymbol{X}_{\mathcal{U}}^{k,(0)} \| \boldsymbol{X}_{\mathcal{U}}^{k,(1)} \| \dots \| \boldsymbol{X}_{\mathcal{U}}^{k,(L)}, \quad \boldsymbol{X}_{\mathcal{I}}^k = \boldsymbol{X}_{\mathcal{I}}^{k,(0)} \| \boldsymbol{X}_{\mathcal{I}}^{k,(1)} \| \dots \| \boldsymbol{X}_{\mathcal{I}}^{k,(L)}, \tag{2}$$

where $\|$ is the concatenation operation. For a particular user u (or item i), its embedding is the u-th (or i-th) row of $\boldsymbol{X}_{\mathcal{U}}^k$ (or $\boldsymbol{X}_{\mathcal{I}}^k$).

2.3   Behavior Embedding Combination and Prediction

In multi-behavior scenarios, different behaviors have different importance on the final understanding of user preferences. In this end, we assign a weight w_k to behavior k ($k = 1, 2, ..., K$). To limit the value of w_k as $(0, 1)$, we use the softmax function to normalize all importance weights:

$$w_k = softmax(w_k) = \frac{\exp(w_k)}{\sum_{m=1}^{K} \exp(w_m)}, \tag{3}$$

where w_k is the same for all users and automatically updated during model training.

In addition, the number of user-item interactions under different behaviors also reflects the user's behavior habits, which is also of great significance to the final realization of accurate recommendations to users. Based on the above considerations, we design an aggregation coefficient $\theta_{u,k}$ of user u under behavior k, as below:

$$\theta_{u,k} = \frac{w_k \cdot \boldsymbol{D}_{\mathcal{U}}^k(u, u)}{\sum_{m=1}^{K} \left(w_m \cdot \boldsymbol{D}_{\mathcal{U}}^k(u, u) \right)}, \tag{4}$$

where $\boldsymbol{D}_{\mathcal{U}}^k(u, u)$ is the number of user-item interactions of user u under behavior k.

Next, we design dual type-specific transformation matrices for target and auxiliary behaviors respectively to extract different features that correspond to the particular behavior, *i.e.*, W_1 and W_2. Different from our method, S-MBRec [6] uses the same encoding matrix for all types of behaviors, and different encoding matrices are used in different layers in MBGCN [4]. However, treating different types of behaviors indiscriminately inevitably loses some fine-grained and valuable information. The fine-grained differences among different user behaviors can further improve the performance of target behavior, especially when discriminating hard negative samples [7]. Furthermore, we need to generate a mixed view for each type of behavior in the later contrastive learning branch. So using two type-specific transformation matrices here is also beneficial to the later mixing operation.

For user u under target behavior, its representation can be extracted as:

$$e_u^{target} = W_1 \cdot (\theta_{u,target} \cdot x_u^{target}), \tag{5}$$

where W_1 is the encoding matrix for target behavior. Besides, for the auxiliary behaviors, we utilize another encoding matrix W_2 to help extract features. So the representation of user u under all auxiliary behaviors can be denoted as:

$$e_u^{k'} = W_2 \cdot (\sum_{k' \in \mathcal{N}_a} \theta_{u,k'} \cdot x_u^{k'}), \tag{6}$$

where $\mathcal{N}_a$ is the set of all auxiliary behaviors. It is worth mentioning that the parameter sharing strategy is involved here. By sharing parameters, MixMBR can capture common information among behaviors.

Following [4,6], we assume the contributions of different users to an item are the same. Thus, the representation of item i under target behavior is as follows:

$$e_i^{target} = W_1 \cdot x_i^{target}. \tag{7}$$

Similarly, item i merges all auxiliary behaviors as:

$$e_i^{k'} = W_2 \cdot \left(\sum_{k' \in \mathcal{N}_a} x_i^{k'} \right). \tag{8}$$

For the final prediction, we need to concatenate the representations of user u and item i separately for all types of behaviors:

$$e_u = e_u^{target} \| e_u^{k'}, \quad e_i = e_i^{target} \| e_i^{k'}. \tag{9}$$

The model prediction is the inner product of user and item final representations:

$$\hat{y}_{ui} = e_u^T \cdot e_i, \tag{10}$$

which is used as the ranking score for recommendation generation.

The conventional Bayesian Personalized Ranking (BPR) [11] loss is used as the main loss, which is a pairwise loss that encourages the prediction of an observed entry to be higher than its unobserved counterparts:

$$\mathcal{L}_{BPR} = \sum_{(u,i,j) \in O} -\log \sigma \left(\hat{y}_{ui} - \hat{y}_{uj} \right), \tag{11}$$

where $O = \{(u, i, j)\}$ consisting of observed positive and randomly-selected negative samples like (u, i) and (u, j) respectively. And $\sigma(\cdot)$ is the activate function such as sigmoid.

2.4   Multi-behavior Contrastive Learning

Inspired by the application of mixup in CV, we propose embedding-level mixing to generate data-augmented views for each type of behavior. Specifically, we generate data-augmented views by mixing two different embedding representations of each behavior, which are generated by the two different encoding matrices mentioned earlier. For user u and item i under behavior k $(k = 1, 2, ..., K)$, their representation can be represented as follows:

$$\hat{e}_u^k = x_u^k \cdot (\alpha \cdot W_1 + \beta \cdot W_2), \quad \hat{e}_i^k = x_i^k \cdot (\alpha \cdot W_1 + \beta \cdot W_2). \tag{12}$$

where α and β are mixing hyperparameters to control the proportion of encoding matrices. Actually, we can define $W = \alpha \cdot W_1 + \beta \cdot W_2$ as parameter interpolation. Instead of interpolating different classes of graphs in CV to generate mixed graphs [12], our proposed MixMBR interpolates two learned transformation matrices to generate mixed views of target and auxiliary behaviors for contrastive learning.

After building the contrastive view for each type of behavior, we leverage graph contrastive learning to construct an instance discrimination task that pulls together positive pairs (augmentation pairs of the same user under different behaviors) and pushes away negative pairs (augmentation pairs for different users). We assume the first behavior is the target behavior(e.g., purchase), and the rest are auxiliary behaviors (e.g., add to cart, view/click). For the established mixed view of user u under target behavior t, represented as $\hat{e}_u^t$, the positive and negative pairs are $\left\{\hat{e}_u^t, \hat{e}_u^{k'} \mid u \in \mathcal{U}\right\}$ and $\left\{\hat{e}_u^t, \hat{e}_v^{k'} \mid v \in \mathcal{U}, u \neq v\right\}$, where $k' = [2, .., K]$. Following works [3,6], we adopt the InfoNCE [13] loss to maximize the agreement of positive pairs and minimize that of negative pairs:

$$\mathcal{L}_{CL}^{user} = \sum_{k'=2}^{K} \sum_{u \in \mathcal{U}} - \log \frac{\exp\left(\varphi\left(\hat{e}_u^t, \hat{e}_u^{k'}\right) / \tau\right)}{\sum_{v \in \mathcal{U}} \exp\left(\varphi\left(\hat{e}_u^t, \hat{e}_v^{k'}\right) / \tau\right)} \tag{13}$$

where τ is the hyper-parameter, known as the temperature in softmax. And $\varphi(\cdot)$ is defined as the similarity function (such as inner-product) between two embeddings. Analogously, we obtain the contrastive loss of the item side $\mathcal{L}_{CL}^{item}$. We combine these two losses and get the objective function of contrastive task as $\mathcal{L}_{CL} = \mathcal{L}_{CL}^{user} + \mathcal{L}_{CL}^{item}$.

2.5   Joint Optimization

To optimize the proposed model, we combine the BPR loss $\mathcal{L}_{BPR}$, the contrastive loss $\mathcal{L}_{CL}$ and the L_2 regularization term $\|\Theta\|_2^2$, which is shown as :

$$\mathcal{L} = \mathcal{L}_{BPR} + \lambda_1 \mathcal{L}_{CL} + \lambda_2 \|\Theta\|_2^2, \tag{14}$$

where θ denotes all trainable parameters. λ_1 and λ_2 are hyperparameters to control the proportion of contrastive learning task and L_2 regularization strength, respectively.

Table 1. Statistics of datasets used in experiments.

Dataset	#Users	#Items	#Interactions	Interactive Behavior Type
Beibei	21,716	7,977	3.35×10^6	*View, Cart, Purchase*
Taobao	48,749	39,493	2.0×10^6	*Click, Cart, Purchase*

3   Experiment

3.1   Experimental Settings

Dataset. To evaluate the performance of our model, we experiment on two real-world e-commerce datasets: Beibei [5] and Taobao [5]. We summarize the statistics of two datasets in Table 1.

Baselines. We compare the performance of our model with the various recommendation methods, which can be divided into two categories: single-behavior models [1–3,11] that only utilize target behavior records, and multi-behavior models [2,4,6,14] that take all kinds of behavior into consideration. Notably, we apply LightGCN to multiple behaviors and name it LightGCN-MB. Different from LightGCN, we use all kinds of behavior to construct the user-item bipartite graph. And we treat all types of behavior as the same so there is only one kind of edge in the graph.

Parameters Settings. Our model is implemented in Pytorch. The batch size is fixed to 2048 and the learning rate is 0.002. The L_2 regularization is set to $1e^{-4}$. The hyperparameter α, β, λ_1 and temperature coefficient τ are searched in the range of 0.1 to 1, respectively. We fix the propagation layer numbers $L=4$. We also perform an early stop strategy to prevent overfitting, *i.e.*, premature stopping if recall@10 does not increase for 40 successive epochs. To evaluate the performance of each model, we use two widely used metrics called Recall@K and NDCG@K.

Table 2. Performance comparison of all compared methods on Beibei with the metrics of Recall@K and NDCG@K (K=10, 40, 80)

	Method	Recall@10	NDCG@10	Recall@40	NDCG@40	Recall@80	NDCG@80
Single-behavior	MF-BPR	0.0368	0.0172	0.1085	0.0331	0.1751	0.0444
	NGCF	0.0396	0.0191	0.1190	0.0366	0.1996	0.0503
	LightGCN	0.0424	0.0204	0.1223	0.0381	0.1987	0.0511
	SGL	0.0403	0.0216	0.1064	0.0360	0.1837	0.0491
Multi-behavior	NMTR	0.0317	0.0167	0.1116	0.0329	0.2260	0.0516
	LightGCN-MB	0.0613	0.0318	0.1530	0.0522	0.2276	0.0649
	MBGCN	0.0446	0.0232	0.1209	0.0399	0.2117	0.0553
	S-MBRec	0.0529	0.0148	0.1647	0.0429	0.2740	0.0615
	our model	**0.0950**	**0.0513**	**0.2140**	**0.0780**	**0.2983**	**0.0924**
	Improvement	54.97%	61.32%	29.93%	49.43%	8.87%	42.37%

3.2 Performance Comparison

The results are presented in Table 2 and Table 3. From the two tables, our model outperforms the baselines on almost all metrics. The average improvement of our model on the second best result is about 41.15% on the Beibei dataset and 13.82% on the Taobao dataset. We attribute the improved performance to the following reasons: 1) MixMBR adopts a parameter sharing strategy to extract common characteristics among different behaviors. 2) The mixed views generated by mixup augmentation provide more hard negative samples for contrastive learning, which is beneficial to the improvement of model performance.

Table 3. Performance comparison of all compared methods on Taobao with the metrics of Recall@K and NDCG@K (K=10, 40, 80)

	Method	Recall@10	NDCG@10	Recall@40	NDCG@40	Recall@80	NDCG@80
Single-behavior	MF-BPR	0.0333	0.0200	0.0579	0.0255	0.0720	0.0279
	NGCF	0.0311	0.0187	0.0580	0.0247	0.0748	0.0276
	LightGCN	0.0415	0.0246	0.0734	0.0320	0.0927	0.0353
	SGL	0.0420	0.0240	0.0767	0.0319	0.0975	0.0354
Multi-behavior	NMTR	0.0369	0.0237	0.0487	0.0305	0.0983	0.0336
	LightGCN-MB	0.0463	0.0253	0.0991	0.0371	0.1374	0.0436
	MBGCN	0.0305	0.0164	0.0665	0.0245	0.0925	0.0289
	S-MBRec	0.0608	**0.0391**	0.1027	0.0464	0.1647	0.0583
	our model	**0.0691**	0.0383	**0.1381**	**0.0538**	**0.1878**	**0.0623**
	Improvement	13.65%	−2.05%	34.47%	15.95%	14.03%	6.86%

The multi-behavior model performs better than the single-behavior model on both datasets, which reveals the importance of introducing multi-behavior data. Across all baselines, we find that the best multi-behavior model on the Taobao dataset outperforms the best single-behavior model by an average of 49.19% on recall and 57.69% on NDCG, while 28.07% and 39.97% on Beibei, which indicates that the auxiliary behaviors provide richer semantic information to fully reflect the user's preference.

3.3   Ablation Study

Effect of Auxiliary Behaviors. The results are presented in Fig. 3. Here, w/o view, w/o cart, and w/o click represent our MixMBR without the incorporation of individual *view*, *cart*, *click* behavior, respectively. From the results in the figure, it can be seen that deleting any type of auxiliary behavior causes the performance of the model to decrease, and the types of auxiliary behaviors have different effects on performance. This verifies the effectiveness of auxiliary behaviors in modeling user preferences.

Effectiveness of Mixing Operation. To evaluate the rationality of design mixing operation in our model, we compare it with two model variants including w/o-indist and w/o-mixing on two datasets. Here, w/o-indist means that the same encoding matrix is used for all types of behavior in Eq. 5–Eq. 8 to obtain the final representation. And w/o-mixing means that each type of behavior corresponds to an individual encoding matrix, and the representation used for contrastive learning is not mixed. According to the experimental result in Table 4, our model performs much better than the other two variants and has a huge gap with w/o-mixing. By mixing operation, there are more hard negative samples to be mined. This demonstrates the necessity and effectiveness of the mixing operation.

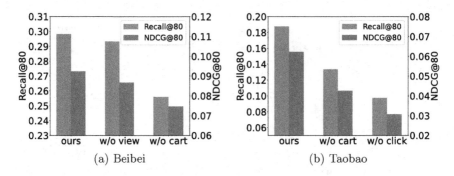

Fig. 3. Impact study of different types of auxiliary behavior.

3.4 Hyperparameter Analysis

Mixing Hyperparameter α and β. We evaluate α and β in the range of $0.1 \sim 1.0$ respectively. The experimental result on Taobao is shown in Fig. 4(a). When both α and β are set to large values (*e.g.* >0.5), our model outperforms with little difference in performance fluctuations. It shows that retaining more original information about the target and auxiliary behaviors is more conducive to mining harder negative examples in order to achieve better performance when performing mixing operation.

Contrastive Learning Task Weight λ_1. We start experimenting with different λ_1 on Beibei and Taobao to check its influence. We turn λ_1 in the range of $0.1 \sim 1.0$ and the result is presented in Fig. 4(b). As can be seen from the figure, $\lambda_1 = 0.2$ is the best choice for our model. As λ_1 keeps increasing, the impact of the contrastive learning module on the performance of the entire model keeps getting bigger. Contrastive learning becomes the main factor that dominates the model performance, but it hurts the performance of the model.

Table 4. Ablation study of mixing operation.

Dataset	Model	Recall@80	NDCG@80
Beibei	w/o-indist	0.1993	0.0674
	w/o-mixing	0.1649	0.0520
	ours	0.2983	0.0924
Taobao	w/o-indist	0.1631	0.0521
	w/o-mixing	0.1137	0.0356
	ours	0.1878	0.0623

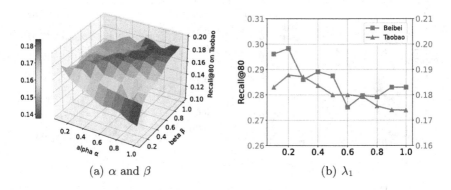

(a) α and β (b) λ_1

Fig. 4. Hyperparameter analysis of our model.

4   Conclusion

In this paper, we propose a new MBR model named MixMBR. It interpolates two different transformation matrices to generate the mixed views for different types of behaviors. Then we utilize contrastive learning and parameter sharing to learn the commonalities between target and auxiliary behaviors. MixMBR can effectively alleviate the shortcomings of directly applying contrastive learning to multi-behavior models in the past. We conduct comprehensive experiments and show that the proposed MixMBR improves the recommendation performance on both datasets.

References

1. Wang, X., He, X., Wang, M., Feng, F., Chua, T.-S.: Neural graph collaborative filtering. In: SIGIR, pp. 165–174 (2019)
2. He, X., Deng, K., Wang, X., Li, Y., Zhang, Y., Wang, M.: Lightgcn: simplifying and powering graph convolution network for recommendation. In: SIGIR, pp. 639–648 (2020)
3. Wu, J., et al.: Self-supervised graph learning for recommendation. In: SIGIR, pp. 726–735 (2021)
4. Jin, B., Gao, C., He, X., Jin, D., Li, Y.: Multi-behavior recommendation with graph convolutional networks. In: SIGIR, pp. 659–668 (2020)
5. Xia, L., Xu, Y., Huang, C., Dai, P., Bo, L.: Graph meta network for multi-behavior recommendation. In: SIGIR, pp. 757–766 (2021)
6. Gu, S., Wang, X., Shi, C., Xiao,D.: Self-supervised graph neural networks for multi-behavior recommendation. In: IJCAI, pp. 2052–2058 (2022)
7. Wu, Y., et al.: Multi-view multi-behavior contrastive learning in recommendation. In: DASFAA, pp. 166–182. Springer, Heidelberg (2022). https://doi.org/10.1007/978-3-031-00126-0_11
8. Yang, H., Chen, H., Li, L., Philip, S.Y., Xu, G.: Hyper meta-path contrastive learning for multi-behavior recommendation. In: ICDM, pp. 787–796. IEEE (2021)
9. Zhang, H., Cisse, M., Dauphin, Y.N., Lopez-Paz, D.: mixup: beyond empirical risk minimization. arXiv preprint arXiv:1710.09412 (2017)

10. Kim, S., Lee, G., Bae, S., Yun, S.-Y.: Mixco: mix-up contrastive learning for visual representation. arXiv preprint arXiv:2010.06300 (2020)

11. Rendle, S., Freudenthaler, C., Gantner, Z., Schmidt-Thieme, L.: Bpr: bayesian personalized ranking from implicit feedback. arXiv preprint arXiv:1205.2618 (2012)

12. Han, X., Jiang, Z., Liu, N., Hu, X.: G-mixup: graph data augmentation for graph classification (2022). arXiv preprint arXiv:2202.07179

13. Oord, A.V.D., Li, Y., Vinyals, O.: Representation learning with contrastive predictive coding. arXiv preprint arXiv:1807.03748 (2018)

14. Gao, C., et al.: Neural multi-task recommendation from multi-behavior data, pp. 1554–1557. In: ICDE. IEEE (2019)

CATCL: Joint Cross-Attention Transfer and Contrastive Learning for Cross-Domain Recommendation

Shuo Xiao[1], Dongqing Zhu[1], Chaogang Tang[1(✉)], and Zhenzhen Huang[1,2]

[1] School of Computer Science and Technology, China University of Mining and Technology, Xuzhou, China
cgtang@cumt.edu.cn
[2] Library School of Computer Science and Technology, China University of Mining and Technology, Xuzhou, China

Abstract. Cross-domain recommendation (CDR) improves recommendation accuracy by transferring knowledge from rich domains to sparse domains, which is a significant advancement in the effort to deal with data sparsity. Existing CDR works, however, still have some challenges, including (1) ignoring the user-item interaction long-tail distribution problem and (2) transferring only the domain-shared feature preferences of common users. In this paper, we propose a CDR framework named joint Cross-Attention Transfer and Contrastive Learning for Cross-Domain Recommendation (CATCL). We first add random uniform noise to the original representation to maximize the mutual information between the original representation and its augmented view, and then pre-train to obtain more uniformly distributed user/item representations, in order to address the issues of data sparsity and popularity bias within intra-domain. In addition, we introduce a cross-attention mechanism for extracting user domain-shared and domain-specific features in order to capture the relevance of user preferences between inter-domain. Then we employ an element-wise attention component that dynamically distributes weights between domain-specific and domain-shared features, allowing different features to exhibit different importance in rich and sparse domains. The final experimental results on three public datasets demonstrate the effectiveness of the proposed framework for many powerful state-of-the-art approaches.

Keywords: Cross-domain recommendation · Contrastive Learning · Long Tail Distribution

1 Introduction

Nowadays, recommender systems have been widely adopted in various businesses, including e-commerce, video streaming websites and social network [7].

S. Xiao and D. Zhu—Contributed equally to this research.
This work was funded by National Natural Science Foundation of China (No. 62271486 and No. 62071470).

The severity of the data sparsity and the long-tail problem of recommender systems has increased due to the richness of recommendation scenarios and the growing scale of recommendations [20]. Recent studies have demonstrated the effectiveness of cross-domain recommendation[9,10] in addressing the challenge of sparse data in recommender systems. By leveraging information from rich domains, this method effectively improves the accuracy of recommendations in sparse domains. The successful application of deep learning in the field of recommender systems has also greatly contributed to the development of CDR [1,18,26]. Researchers have proposed numerous DNN-based CDR methods to better capture user preferences [24]. However, sparse domains have a negative effect on the source domain, so typical transfer learning-based CDR models will suffer from negative transfer problems [15]. Dual-target CDR [32] avoids this problem, and the rich and sparse domains transfer to each other and promote each other. Nevertheless, most existing work on CDR still does not consider the problem of long-tailed distributions of user-item interaction. For example, on e-commerce platforms, it is common to find that a small percentage of popular products dominate the market, while a majority of lesser-known products are hardly recommended to users. User-item interactions on datasets with long-tailed item distributions are typically highly skewed in a power-law distribution [13]. It has been observed that recommendation models tend to overfit on a small proportion of popular items, resulting in suboptimal performance for the remaining long-tail items [8]. As a result, that these recommendation models usually retrieve popular items while ignoring more appropriate long-tail items.

Contrastive Learning (CL) has recently made significant advancements in various fields. It is considered as an emerging learning paradigm as a solution to the problem of data sparsity. The state-of-the-art CL-based recommendation model SGL [22] utilizes node dropout, edge dropout, and random walk on graph structure to create different views of nodes. Despite the promising results of CL in the recommendation domain, extensive experiments with SimGCL [28] have revealed that the uniformity of the representation distribution is the primary factor in performance improvement, while graph augmentation, which is thought to be crucial, has only played a minor role. Chen et al. [3] argue that learning a more uniformly distributed user/item representation can help to mitigate the prevalence bias. We leverage this property of CL to improve the long-tail problem in cross-domain by pre-training to generate more uniformly distributed user/item embedding for each domain in the CDR task.

In this paper, our goal is twofold: first, to improve the recommendation performance in both rich and sparse domains by means of cross-domain recommendation (CDR), and second, to provide users with diverse recommendations and reduce popularity bias. This task faces two main challenges. The first challenge is to generate a more uniform embedding of users and items within a single domain, which can be used to improve the long-tail distribution of items and thus reduce popularity bias in recommendations. Since long-tail items lack feedback to obtain a good feature representation, the feature distribution of long-tail items is inconsistent compared to feedback-rich popular items. The second challenge is how to perform knowledge transfer more effectively for both domains. Most CDR approaches consider only the domain-shared features of common users when transferring, and do not take into account the domain-specific preference features of users. Additionally, aggregating domain-specific and domain-shared features ignores the fact that they represent different semantic transfer knowledge in the training process.

To address these challenges, we propose a CDR framework named joint Cross-Attention Transfer and Contrastive Learning for Cross-Domain Recommendation (CATCL). First, we generate user and item embedding with more uniform distribution by CL during pre-training. Specifically, we construct positive samples for representation-level data augmentation by adding random uniform noise to the original representation based on the graph encoder. Then comparisons are made between different views to learn a more uniformity feature representation. Second, we propose a cross-attention transfer mechanism to extract user domain-shared and domain-specific features. Then, we dynamically assign weights to both domain-shared and domain-specific features by element-wise attention component, which allowing the model to learn the interest preferences of common users more effectively.

In this paper, we make the following contributions:

1. We propose a novel cross-domain recommendation framework that generates more uniform user/item representations for each of the two domains through a graph-free augmented CL-based approach. It not only alleviates the problem of non-uniform data distribution within intra-domain, but also promotes better knowledge migration between inter-domain.
2. We propose the cross-attention transfer mechanism in CDR in order to better capture the latent representation of user preferences, which extracts domain shared and domain specific features more efficiently and combined using the element-wise attention mechanism for dual-target recommendation.
3. We conduct experiments on three real-world CDR datasets to evaluate the model performance. Extensive results show that our model CATCL achieves consistent and significant improvements over a state-of-the-art baseline. In addition, we conducted a comprehensive ablation study and detailed analysis to investigate the effectiveness of our model components.

2   Related Work

2.1   Cross-Domain Recommendation

Cross-domain recommendation is an efficient method for dealing with the sparse data problem in recommendation systems. Most CDR approaches focus on transferring knowledge from one domain to another, for example, CMF [17], CDFM [11]. Dual-target CDR aims to improve performance on two domains simultaneously, compared to single-target CDR. DDTCDR [9] extends CoNet [6] by learning potentially orthogonal mapping functions to transfer user similarities across domains. GA-DTCDR [31] uses heterogeneous graphs in each domain to model the complex relationships between users and items, and applies element-wise attention networks to merge common user embeddings. BiTGCF [10] uses LightGCN [4] as an encoder to aggregate interaction information in each domain and further introduces a feature transfer layer to augment the two basic graph encoders. DisenCDR [2] uses two separation regularizers based on mutual information to separate domain-shared from domain-specific information. In recent work [12,23], the combination of cross-domain recommendation and self-supervised learning has also been developed. CCDR [23] designs an intra-domain CL and three inter-domain CL tasks to achieve better representation learning and knowledge transfer. Unlike existing work on dual-target CDR, our model takes into account not only the scarcity of user-item interaction data, but also the uneven distribution of data within the domain and the resulting popularity bias.

2.2 Contrastive Learning in Recommendation

Contrastive learning (CL) is a representative approach in self-supervised learning (SSL), which originates from metric learning and aims to learn models by comparing positive and negative samples. CL automatically constructs positive and negative samples by rules to embed positive sample closer and negative samples farther in the projection space. CL has been shown to be effective in dealing with the issue of data sparsity and reducing popularity bias in recommender systems. Some recent works have combined CL with recommender systems [22,27,28]. For example, Wu et al. [27] construct a multi-task self-supervised two-tower model to solve the label sparsity problem by learning the features of the item towers after training enhancement by comparison. SGL [22] performs data enhancement through node dropout, edge dropout, and random walk three ways of data augmentation as a way to generate multiple views of a node, so that the alignment between different views of the same nodes and the views of other nodes is maximized. In contrast, [28] improves SGL by discarding the graph augmentation approach and instead takes the approach of adding uniform noise for data augmentation. CL methods have also been used in various other directions, such as sequential recommendation [19] and cross-domain recommendation [12,23].

3 Proposed Model

3.1 Problem Definition

We consider two relevant domains and let D_A represent domain A and D_B represent domain B, respectively. Let $U = \{u_1, u_2, \ldots u_m\}$ and $V = \{v_1, v_2, \ldots v_n\}$ represent the set of users and the set of items, respectively, where m and n denote the number of users and items, the set of items for D_A and D_B is denoted as V_A and V_B, the set of users is denoted as U_A and U_B, and the set of overlapping users of the two domains is denoted as U_C. The $y_{ij} \in R$ represents the rated interactions between users and items, and the user-item interaction matrix is defined

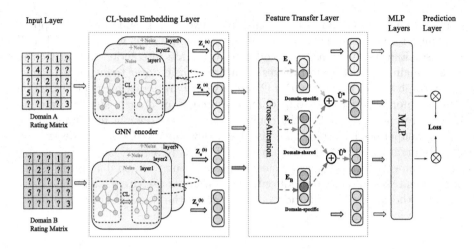

Fig. 1. The overview of CATCL

as $R_A \in R^{M \times N}$ in D_A and $R_B \in R^{M \times N}$ in D_B. The goal of this paper is to simultaneously improve the recommendation performance of both domains D_A (D_B). Specifically, a function is learned to predict the user's rating $\hat{y}_{ij}^a$ $(\hat{y}_{ij}^b)$ of the items to be interacted with based on the user's historical interaction records in both domains, which is used to rank the Top K items.

3.2   Model Overview

As shown in Fig. 1, our proposed model CATCL consists of five main modules:

(1) Input Layer
   User profiles and other auxiliary data are not readily available due to user privacy protection and other factors, so we chose more general and readily accessible scoring matrix data as the input of our model.
(2) CL-based Embedding Layer
   We perform representation-level data augmentation of nodes based on the graph encoder LightGCN, and obtain more uniformly distributed user/item embedding matrices by contrastive learning. These representations can implicitly mitigate the popularity bias.
(3) Feature Transfer Layer
   We consider both domain-shared and domain-specific features of users, and propose a cross-attention transfer mechanism to extract the common and specific features of users. And we use the element-wise attention component to combine the both for dual-target recommendation.
(4) MLP Layers
   In this component, we apply a fully connected neural network, the Multilayer Perceptron (MLP), to represent the nonlinear relationships between users and items on each domain.

(5) Prediction Layer

After propagation through the MLP layers, we can obtain feature representations of users and items for each domain. We use the cosine function to estimate the probability of user and target item interaction.

3.3 CL-Based Embedding Layer

A typical process for CL-based recommendation models is to first expand the user-item bipartite graph with structural perturbations and then maximize the consistency of node representations between different graph expansions. Inspired by the SimGCL [28] framework, we use a similar data augmentation approach to generate different contrastive views to generate a smoother and more uniform feature representation. The basic idea is to add random uniform noise to the embedding space to create contrastive views. Applying different random noise creates differences between the contrastive views, while retaining learnable invariance due to controlled magnitude. Based on the above CL, we generate user and item embedding matrices for each domain separately by pre-training. The CL-based Embedding Layer consists of three main components, namely Data Augmentation, Contrastive Learning, and Multi-task Training.

Data Augmentation:

Since the dropout-based graph augmentation is burdensome and inefficient, we choose a simpler and more effective graph-free augmentation method. Inspired by [28], specifically, we adds random uniform noise to the original representation for representation-level data augmentation, which will obtain a more uniform distribution of the embedding space.

Given a node i and its representation e_i in the d-dimensional embedding space, we can achieve the following representation-level augmentation:

$$\mathbf{z}_i' \leftarrow H\left(e_i, \delta'\right), \mathbf{z}_i'' \leftarrow H\left(e_i, \delta''\right), \tag{1}$$

where $H(\cdot)$ is the encoder function that encodes the connection information into a representation. δ' and δ'' are the randomly generated noise vectors. We use LightGCN as a graph encoder to perform convolution operations on the input graph data to capture node preferences by propagating information between neighboring nodes. At each layer, different scaled random noise is applied to the current node embedding. We consider the views of the same node as positive samples of each other (i.e., z_i' and z_i''), and the views of any different nodes as negative samples of each other (i.e., z_i' and z_j'').

The enhanced sample will lose the similar characteristics as the original sample if the added noise has an excessive impact. The direction of z_i and its augmented representations, z_i' and z_i'', must be kept within reasonable bounds. Therefore, we first restrict the noise δ size to a range of radius r size of a hypersphere. Secondly $\bar{\delta}$ keeps the direction of all three within a certain range, as shown in Eq. (2–3):

$$\delta = \bar{\delta} * r, \tag{2}$$

$$\bar{\delta} = \tilde{\delta} \odot \operatorname{sign}(\mathbf{z}_i), \tilde{\delta} \in \mathbb{R}^d \sim U(0, 1). \tag{3}$$

These restrictions do not allow for large deviations in z_i that lead to a reduction in the number of valid positive samples. The augmented representation retains most of the personalized feature information of the original representation since the noise added is light relatively. In the augmented embedding space, similar instances are closer together, and different instances are more uniformly distributed.

Contrastive Learning:
Formally, we use contrastive loss InfoNCE [14] to achieve consistency across views of the same node and divergence between different nodes. The InfoNCE is formulated as:

$$\mathcal{L}_{cl} = \sum_{i \in \mathcal{B}} -\log \frac{\exp\left(\mathbf{z}_i'^\top \mathbf{z}_i'' / \tau\right)}{\sum_{j \in \mathcal{B}} \exp\left(\mathbf{z}_i'^\top \mathbf{z}_j'' / \tau\right)}, \tag{4}$$

where i, j are users/items in a sampled batch $\mathcal{B}$, τ is the hyper-parameter, known as the temperature in softmax. The loss function above learns a robust embedding space such that similar items are close to each other after data augmentation, i.e., maximizing the consistency of z_i' and z_i'', and allowing random samples to be pushed further away, i.e., minimizing the consistency of z_i' and z_j''.

Multi-task Training: To learn robust representations from user item graphs, we use Bayesian personalized ranking (BPR) loss [16] as the main objective function:

$$\mathcal{L}_{\text{main}} = \sum_{(u,i,j) \in O} -\log\left(\sigma\left(\mathbf{z}_u^\top \mathbf{z}_i - \mathbf{z}_u^\top \mathbf{z}_j\right)\right), \tag{5}$$

where z_u is the user embedding, z_i and z_j denote the positive item embedding and negative item embedding, respectively, σ is the sigmoid function.

We jointly optimize the recommendation task and the contrastive learning task using a multi-task training strategy:

$$\mathcal{L} = \mathcal{L}_{\text{main}} + \lambda_1 \mathcal{L}_{cl} + \lambda_2 \|\Theta\|_2^2, \tag{6}$$

where Θ is the parameter of $\mathcal{L}_{\text{main}}$, λ_1 is the contrastive learning rate to adjust the loss ratio of $\mathcal{L}_{\text{main}}$ and $\mathcal{L}_{cl}$, and λ_2 is the hyperparameter to control the $\mathcal{L}_2$ regularization.

We can obtain the user and item embedding generated by graph contrastive learning after joint multi-task training, which is represented on domain A and domain B as Z_u^a, Z_v^a, Z_u^b, Z_v^b, respectively.

3.4   Feature Transfer Layer

Feature transfer is the key in cross-domain recommendation, and we propose a dual-attention transfer mechanism to achieve bidirectional user feature transfer between D_A and D_B. Our module effectively captures both domain-shared and domain-specific features for cross-domain recommendation. Unlike BiTGCF [10], our model dynamically assign the weights of user domain-shared and domain-specific features, allowing different features to exhibit different importance in rich and sparse domains. Our model is allowed to learn the interest preferences of public users more effectively. To improve the connectivity of items from different domains and capture the relevance of user performance across domains to facilitate the learning of domain-shared knowledge, we further adopt the cross-attention mechanism to learn the latent factors of users from both domains by using shared users with interaction histories.

In D_A, we use the user embedding vectors Z_u^a generated by graph contrastive learning to represent the projections to obtain three vectors, i.e., query $Q_u^a \in R^{N \times d_k}$, key $K_u^a \in R^{N \times d_k}$. d_k denotes their dimensions. They are calculated as follows:

$$Q_u^{(a)} = W_q^{(a)} Z_u^{(a)}, K_u^{(a)} = W_k^{(a)} Z_u^{(a)}, \tag{7}$$

where W_q^a, W_k^a are the learnable matrices of D_A. Similarly, we can get W_q^b, W_k^b on D_B, and consequently, we can get two vectors Q_u^b, K_u^b on D_B. The cross-attention mechanism is derived from the self-attention mechanism. The cross-attention module can be calculated as follows:

$$H_u^{(a)} = \text{softmax}\left(\frac{Q_u^{(a)} K_u^{(b)^T}}{\sqrt{d_k}}\right), \tag{8}$$

$$H_u^{(b)} = \text{softmax}\left(\frac{Q_u^{(b)} K_u^{(a)^T}}{\sqrt{d_k}}\right), \tag{9}$$

where the query Q_u^a of the cross-attention score H_u^a of domain A is from D_A, while the key K_u^b is from D_B. And the query Q_u^b of the cross-attention score H_u^b of domain B is from D_B, while the key K_u^a is from D_A.

Therefore, the embedding that is more similar to the query of Z_u^a will have a higher weight and will be more contributing to the output. In other words, the output of the cross-attention module manages to aggregate similar preferences of common users of both domains. It better connects users across domains and captures user-user correlations across domains to facilitate learning of domain sharable knowledge.

We get the common features E_C of users in D_A and D_B by Eq. (10). E_A and E_B are used to preserve the domain-specific features of the user in D_A and D_B, respectively. As shown in in the Eq. (11–12):

$$E_C = H_u^{(a)} Z_u^{(a)} + H_u^{(b)} Z_u^{(b)}, \tag{10}$$

$$E_A = \lambda^{(a)} Z_u^{(a)} + \left(1 - \lambda^{(a)}\right) Z_u^{(b)}, \tag{11}$$

$$E_B = \lambda^{(b)} Z_u^{(b)} + \left(1 - \lambda^{(b)}\right) Z_u^{(a)}, \tag{12}$$

where λ^a and λ^b are hyper-parameters within the range of [0, 1]. In Eq. (8–9) we can obtain the cross-attention scores H_u^a, H_u^b. We use the hyperparameters

λ^a and λ^b to control the retention ratio of user features in the corresponding domain. For example, when $\lambda^a = \lambda^b = 1.0$, it indicates that 100% of the user features in D_A and D_B are retained, so the user features have the greatest specificity in these two domains. When $\lambda^a = \lambda^b = 0.5$, the specificity of user features disappears and the same user has the same features in both domains.

Finally, we use the element-wise attention component to combine domain-shared and domain-specific features. It tends to have different proportions of common and specific domain features learned from the two domains. We denote the combination of user common and specific domain features embedding $\widetilde{U}^a$ and $\widetilde{U}^b$ of D_A and D_B as:

$$\widetilde{U}^a = W^{(a)} \odot E_A + \left(1 - W^{(a)}\right) \odot E_C, \tag{13}$$

$$\widetilde{U}^b = W^{(b)} \odot E_B + \left(1 - W^{(b)}\right) \odot E_C, \tag{14}$$

where $\odot$ is the element multiplication, $W^a \in R^{N \times k}$, $W^b \in R^{N \times k}$ are the weight matrices of the attention network on D_A and D_B, respectively.

3.5   Model Training

In the prediction layer, we obtain the predicted value of $\hat{y}^a$ for D_A by the cosine function:

$$\hat{y}_{ij}^a = \text{cosine}\left(U_i^a, V_j^a\right) = \frac{U_i^a \cdot V_j^a}{\|U_i^a\| \|V_j^a\|}, \tag{15}$$

where U_i^a and V_j^a are the user embedding and item embedding trained by MLP layer, $\|U_i^a\|$ and $\|V_j^a\|$ are the normalized embeddings, respectively. Similarly, we can also obtain the predicted value $\hat{y}^b$ the D_B.

We use the cross-entropy loss function as the objective function in D_A to train our model, which is defined as follows:

$$L\left(y^a, \hat{y}^a\right) = \sum_{y \in Y^{a+} U Y^{a-}} y^a \log \hat{y}^a + (1 - y^a) \log\left(1 - \hat{y}^a\right) + \lambda \left(\|U_{MLP}^a\|_F^2 + \|V_{MLP}^a\|_F^2\right), \tag{16}$$

where y^{a+} is the set of observed interaction histories, and y^{a-} is the set of randomly sampled non-observed interactions. λ is the hyperparameter that controls the regularization strength to avoid overfitting. $\|U_{MLP}^a\|_F^2 + \|V_{MLP}^a\|_F^2$ is the regularizer.

4   Experiments

In this section, we first describe the experimental setup, including the datasets, evaluation metrics, parameter settings, and baseline models. We then conduct

Table 1. Experimental datasets

Domain	Douban-Movie	Douban-Book	Douban-Music
#Users	2,711	2,110	1,672
#Items	9,555	6,777	5,567
#Interactions	1,133,420	96,041	69,709
Density	4.37%	0.67%	0.74%

Table 2. Experimental tasks

Tasks	Richer	Sparser	Overlap
Task1	Douban-Movie	Douban-Book	#Common Users = 2,106
Task2	Douban-Movie	Douban-Music	#Common Users = 1,666

an extensive analysis of the experimental results. The questions studied are as follows:

RQ1: How is the performance of CATCL compared to the baseline model?

RQ2: How do the different design modules (i.e. CL-based Embedding Layer, Feature Transfer layer) contribute to the performance of the model?

RQ3: Does CL-based Embedding help to improve the long tail problem?

RQ4: How does CATCL perform under different parameter settings?

4.1 Experimental Setup

Datasets. To evaluate our proposed model against the state-of-the-art baseline, we chose three real datasets in the movie, book and music domains of the Douban website (see Table 1) for our experiments. For the three datasets (douban_movie, douban_book and douban_music), we retain the users and items that had at least 5 interactions, and we transformed the user's rating to 0 or 1, indicating whether the user interacted with the item or not. This filtering strategy has been widely used in existing approaches [29]. With different combinations of domains, we constructed two CDR tasks, movie-book and movie-music (see Table 2). We validate the recommendation performance of our model in the CDR scenario with dual-target CDR experiments in rich and sparse domains.

Evaluation Metrics. In this paper we use a ranking-based evaluation strategy, the leave-one-out evaluation, which has been widely used in the literature [21,25]. For each test user, we selected the most recent interaction with the test item as the test interaction and randomly selected 99 unobserved interactions for the test user, and then ranked the test items among the 100 items. Hit Ratio (HR) and Normalized Discounted Cumulative Gain (NDCG) are the two evaluation metrics of this strategy. HR@N is the rate at which the test sample ranks in the N-th position of the ranked list, while NDCG@N measures the quality of a particular ranking that assigns high scores to the top-ranked hits. In this paper H@10 and N@10 represent HR@10 and NDCG@10 respectively.

Parameters Settings. For the pre-trained CL-based embedding, we used Adam with a learning rate of 5e-4 to optimize the model. We empirically set the temperature t to 0.2 and the contrastive learning rate to 1e-4. The batch size is set to 512. In addition, to train our model CATCL, we randomly select 7 negative samples for each observed positive sample to generate the training dataset, use Adam as the optimizer to train the neural network, and set the maximum number of training epochs to 50, the initial learning rate to 1e-4, the regularization factor λ to 1e-3, and the batch size to 4096. The L2 penalty is set to 10-3 to avoid overfitting. Furthermore, we set d = 128 for the embedding size of all the methods.

Comparison Methods. We take the following state-of-the-art methods as the baselines.

- NeuMF [5] is a neural network architecture that uses collaborative filtering methods to model the latent features of users and items
- DMF [25] is a neural network-based CF model that uses a deep architecture to learn low-dimensional factors of users and items
- DTCDR [30] performs dual target cross-domain recommendation by transferring the domain shared features of common users of both domains
- DDTCDR [9] introduces potential orthogonal mapping functions through dual learning mechanism to transfer user similarity across domains
- GA-DTCDR [31] generate graph embedding using rating and review information, and use attention mechanism to achieve cross-domain knowledge transfer
- BiTGCF [10] extends LightGCN to the CDR task. It first generates user/item representations for each domain using two linear graphical encoders, and then fuses the user representations using a feature transfer layer
- DisenCDR [2] separates domain-shared information from domain-specific information based on two separating regularizers of mutual information.

4.2 RQ1: Performance Comparisons

For RQ1, we conducted experiments with two CDR tasks. The experimental results are shown in Table 3. We have the following observations: First, we can see that in each task, CATCL significantly improves the recommended performance on both domains. Second, and notably, in most of cases, single-domain recommendation methods (e.g., NeuMF, DMF) are worse than cross-domain recommendation models. This proves that transferring knowledge from the source domain to the target domain can indeed improve recommendation performance. We can also note that GA-DTCDR performs best baseline in the movie domain in both tasks, while the performance is poor in the book domain in task1 and in the music domain in task2. The contrast is with DisenCDR, where HR and NDCG perform second best in the book and music domains, and mediocre in the other domain. The most dual-target CDR models fail to effectively improve the recommendation performance of both domains due to it is difficult to capture the dual transfer representations simultaneously. Our CATCL model alleviate the negative transfer problem through the cross-attention transfer mechanism.

Table 3. Experimental results of recommendation methods on evaluation datasets. Bold scores are the best in each column while underlined scores are the best among all baselines. Improvement denotes the improvement versus best baseline results.

Dataset	Movie & Book				Movie & Music			
	Movie		Book		Movie		Music	
Method	H@10	N@10	H@10	N@10	H@10	N@10	H@10	N@10
NeuMF	0.5512	0.3301	0.4012	0.2310	0.5512	0.3301	0.3314	0.1810
DMF	0.5776	0.3505	0.4046	0.2451	0.5776	0.3505	0.3301	0.1971
DTCDR	0.5991	0.3680	0.4317	0.2510	0.5792	0.3742	0.3580	0.2132
DDTCDR	0.5863	0.3589	0.4225	0.2439	0.5748	0.3762	0.3520	0.2117
BI-TGCF	0.5088	0.3491	0.4805	0.3169	0.5647	0.3719	0.4141	0.2583
GA-DTCDR	<u>0.6174</u>	<u>0.3729</u>	0.4241	0.2546	<u>0.6284</u>	<u>0.3733</u>	0.3325	0.1832
DisenCDR	0.5909	0.3478	<u>0.5085</u>	<u>0.3199</u>	0.5870	0.3728	<u>0.4563</u>	<u>0.2823</u>
CATCL	**0.6809**	**0.4229**	**0.5753**	**0.3601**	**0.6758**	**0.4221**	**0.5526**	**0.3402**
Improvement	10.29%	13.41%	13.14%	12.57%	7.54%	13.07%	21.10%	20.51%

4.3 RQ2: Ablation Study

To study the impact of each component of CATCL, we considered the following variants of CATCL: (1) CATCL-*Embed* is a variant of CATCL that uses ordinary embedding instead of CL-based embedding. To evaluate the quality of embedding, we compare this variant with CATCL. It is clear that the recommended performance using original embedding is lower in both experimental tasks. (2) CATCL-*LightGCN* replaces the normal embedding with an embedding generated by LightGCN, which performs neighborhood aggregation without distinguishing various relationships between nodes. The results demonstrate that our model component can effectively model heterogeneous relationships. (3) CATCL-*SGL* compares the components of our model (CL-based embedding) with the graph contrastive embedding generated by pre-training the SGL model, and the results demonstrate the superiority of CL-based embedding in terms of alignment and uniformity. (4) CATCL-*MLP* does not perform knowledge transfer for the source and target domains, but feeds the CL-based embedding directly into the MLP layer for training. From the experimental results, we can see that knowledge transfer can improve the recommendation performance. (5) CATCL-*Pooling* replaces the cross-attention transfer mechanism with a pooling operation for knowledge transfer, and the experimental results demonstrate that the cross-attention transfer mechanism effectively captures domain-shared features and domain-specific features recommended across domains. Table 4 summarizes the performance achieved by the different CATCL variants. We found that in both experimental tasks, our model components achieved a better performance.

Table 4. Performance achieved by different variants of CATCL

Dataset	Movie & Book				Movie & Music			
	Movie		Book		Movie		Music	
Method	H@10	N@10	H@10	N@10	H@10	N@10	H@10	N@10
CATCL-*Embed*	0.5695	0.3450	0.3118	0.2057	0.5651	0.3313	0.2266	0.1345
CATCL-*LightGCN*	0.6008	0.3605	0.4056	0.2469	0.6030	0.3645	0.3283	0.1792
CATCL-*SGL*	0.6402	0.3682	0.4374	0.2752	0.6395	0.3777	0.3068	0.1682
CATCL-*MLP*	0.6469	0.3941	0.5127	0.3148	0.6221	0.3982	0.5154	0.3125
CATCL-*Pooling*	0.6499	0.3974	0.5218	0.3262	0.6463	0.4039	0.5418	0.3302
CATCL	**0.6809**	**0.4229**	**0.5753**	**0.3601**	**0.6758**	**0.4221**	**0.5526**	**0.3402**

4.4 RQ3: Effectiveness of the Long Tail Recommendation

To verify the effectiveness of CL-based embedding, we divide the dataset into popular and long-tail items according to the popularity of the items. We define the top 20% of items as popular items and the rest as long-tail items. We divided the experimental data into three groups. The first group only contains long-tail items, the second group is the entire dataset, and the third group only includes popular items. Then we tested each of the two experimental tasks by examination of the values of Recall@10 and NDCG@10 for each group of experiments.

The results are shown in Fig. 2. From the experimental results, we can see that each model always performs much better on the popular items than on the long-tail items, which indicates that the distribution of popular items and long-tail items is non-uniform. Secondly, the results on both long-tail and entire datasets indicate that our model performs better than the other comparison models, and the outstanding advantage of our method in recommending long-tail items largely compensates for its lack in the hot group. Moreover, GA-DTCDR and LightGCN tend to recommend popular items and perform poorly in recommending long-tail items. Among them, GA-DTCDR achieves the highest recall on the movie domain of task1 and on the full dataset of task2, and then LightGCN achieves the best recall on the book domain of task1. The reason for this result is the popularity bias caused by the lack of uniformity in the feature representation learned by the model. Due to the superior performance on long-tail recommendations, we can conclude that learning CL-based embedding has a significant role in improving the long-tail problem.

4.5 RQ4: Impact Factor Analysis

In this set of experiments, we analyze the effect of the hyperparameter λ on the experimental performance. We set $\lambda = \lambda^a = \lambda^b$ with a step size of 0.1 and a range of $[0.5, 1.0]$. Figure 3 shows the experimental results of our model on the experimental tasks movie-book and movie-music. It can be seen that experimental task1 can achieve the best performance when λ is taken as 0.9, while experimental task2 can achieve better results overall when λ is 0.8. Secondly, the initial performance of the model is ordinary at $\lambda = 0.5$, and the performance

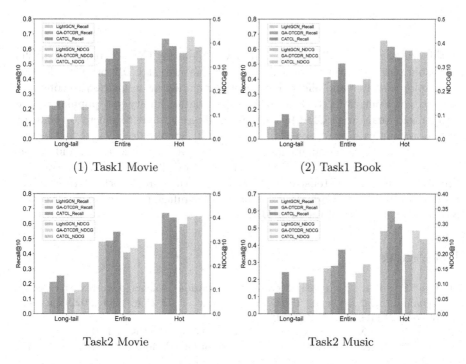

Fig. 2. Performance comparison over different item groups

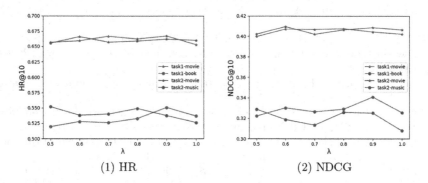

Fig. 3. Effect of λ

improves with the gradual increase of λ. However, at λ of 1, the performance of both tasks decreases. This can prove that the model needs to retain the specific features of both domains and also the proportion of feature retention needs to be considered.

5   Conclusion

In this work, we propose a CATCL framework jointing Cross-Attention Transfer and Contrastive Learning to deal with the dual-target CDR problem. Firstly, we employ a simple and effective graph-free augmentation method to construct an auxiliary contrastive task to mitigate the data sparsity and popularity bias problems in dual-target CDR by comparing different views to generate uniformly distributed embedding representations. Secondly, we also design a cross-attention mechanism to extract domain-shared and domain-specific features, and use an element-wise attention component to achieve more effective knowledge transfer. Finally, we conduct extensive experiments on three real-world datasets. The results show that our proposed model is a significant improvement over the baselines of the state-of-the-art CDR.

References

1. Cai, T., et al.: Incremental graph computation: Anchored vertex tracking in dynamic social networks. IEEE Trans. Knowl. Data Eng. 1–14 (2022). https://doi.org/10.1109/TKDE.2022.3199494
2. Cao, J., Lin, X., Cong, X., Ya, J., Liu, T., Wang, B.: Disencdr: Learning disentangled representations for cross-domain recommendation. In: SIGIR, pp. 267–277 (2022)
3. Chen, J., Dong, H., Wang, X., Feng, F., Wang, M., He†, X.: Bias and debias in recommender system: A survey and future directions. ACM Trans. Inf. Syst. **2022** (2022)
4. He, X., Deng, K., Wang, X., Li, Y., Zhang, Y., Wang, M.: Lightgcn: Simplifying and powering graph convolution network for recommendation. In: SIGIR, pp. 639–648 (2020)
5. He, X., Liao, L., Zhang, H., Nie, L., Hu, X., Chua, T.S.: Neural collaborative filtering. In: WWW, pp. 173–182 (2017)
6. Hu, G., Zhang, Y., Yang, Q.: Conet: Collaborative cross networks for cross-domain recommendation. In: CIKM, pp. 667–676 (2018)
7. Isinkaye, F.O., Folajimi, Y.O., Ojokoh, B.A.: Recommendation systems: Principles, methods and evaluation. Egyptian Inf. J. **16**(3), 261–273 (2015)
8. Krishnan, A., Sharma, A., Sankar, A., Sundaram, H.: An adversarial approach to improve long-tail performance in neural collaborative filtering. In: CIKM, pp. 1491–1494 (2018)
9. Li, P., Tuzhilin, A.: Ddtcdr: Deep dual transfer cross domain recommendation. In: WSDM, pp. 331–339 (2020)
10. Liu, M., Li, J., Li, G., Pan, P.: Cross domain recommendation via bi-directional transfer graph collaborative filtering networks. In: CIKM, pp. 885–894 (2020)
11. Loni, B., Shi, Y., Larson, M., Hanjalic, A.: Cross-domain collaborative filtering with factorization machines. In: de Rijke, M., et al. (eds.) ECIR 2014. LNCS, vol. 8416, pp. 656–661. Springer, Cham (2014). https://doi.org/10.1007/978-3-319-06028-6_72
12. Luo, S., Yang, Y., Zhang, K., Sun, P., Wu, L., Hong, R.: Self-supervised cross domain social recommendation. In: ICCAI, pp. 286–292 (2022)

13. Milojević, S.: Power law distributions in information science: Making the case for logarithmic binning. J. Am. Soc. Inform. Sci. Technol. **61**(12), 2417–2425 (2010)
14. Oord, A.v.d., Li, Y., Vinyals, O.: Representation learning with contrastive predictive coding. arXiv preprint arXiv:1807.03748 (2018)
15. Pan, S.J., Yang, Q.: A survey on transfer learning. IEEE Trans. Knowl. Data Eng. **22**(10), 1345–1359 (2009)
16. Rendle, S., Freudenthaler, C., Gantner, Z., Schmidt-Thieme, L.: Bpr: Bayesian personalized ranking from implicit feedback. In: UAI, pp. 452–461 (2009)
17. Singh, A.P., Gordon, G.J.: Relational learning via collective matrix factorization. In: SIGKDD, pp. 650–658 (2008)
18. Song, X., Li, J., Cai, T., Yang, S., Yang, T., Liu, C.: A survey on deep learning based knowledge tracing. Knowl.-Based Syst. **258**, 110036 (2022)
19. Sun, F., et al.: Bert4rec: Sequential recommendation with bidirectional encoder representations from transformer. In: CIKM, pp. 1441–1450 (2019)
20. Tao, Y., Gao, M., Yu, J., Wang, Z., Xiong, Q., Wang, X.: Predictive and contrastive: Dual-auxiliary learning for recommendation. arXiv preprint arXiv:2203.03982 (2022)
21. Wang, X., He, X., Cao, Y., Liu, M., Chua, T.S.: Kgat: Knowledge graph attention network for recommendation. In: SIGKDD, pp. 950–958 (2019)
22. Wu, J., et al.: Self-supervised graph learning for recommendation. In: SIGIR, pp. 726–735 (2021)
23. Xie, R., Liu, Q., Wang, L., Liu, S., Zhang, B., Lin, L.: Contrastive cross-domain recommendation in matching. In: SIGKDD, pp. 4226–4236 (2022)
24. Xu, K., Xie, Y., Chen, L., Zheng, Z.: Expanding relationship for cross domain recommendation. In: CIKM, pp. 2251–2260 (2021)
25. Xue, H.J., Dai, X., Zhang, J., Huang, S., Chen, J.: Deep matrix factorization models for recommender systems. In: IJCAI, Melbourne, Australia, vol. 17, pp. 3203–3209 (2017)
26. Yang, S., et al.: Robust cross-network node classification via constrained graph mutual information. Knowl.-Based Syst. **257**, 109852 (2022)
27. Yao, T., et al.: Self-supervised learning for large-scale item recommendations. In: CIKM, pp. 4321–4330 (2021)
28. Yu, J., Yin, H., Xia, X., Chen, T., Cui, L., Nguyen, Q.V.H.: Are graph augmentations necessary? simple graph contrastive learning for recommendation. In: SIGIR, pp. 1294–1303 (2022)
29. Yuan, F., Yao, L., Benatallah, B.: Darec: deep domain adaptation for cross-domain recommendation via transferring rating patterns. In: IJCAI, pp. 4227–4233 (2019)
30. Zhu, F., Chen, C., Wang, Y., Liu, G., Zheng, X.: Dtcdr: A framework for dual-target cross-domain recommendation. In: CIKM, pp. 1533–1542 (2019)
31. Zhu, F., Wang, Y., Chen, C., Liu, G., Zheng, X.: A graphical and attentional framework for dual-target cross-domain recommendation. In: IJCAI, pp. 3001–3008 (2020)
32. Zhu, F., Wang, Y., Chen, C., Zhou, J., Li, L., Liu, G.: Cross-domain recommendation: challenges, progress, and prospects. In: IJCAI 2021, International Joint Conferences on Artificial Intelligence, pp. 4721–4728. (2021)

Memory-Enhanced Period-Aware Graph Neural Network for General POI Recommendation

Tianchi Yang[1], Haihan Gao[2], Cheng Yang[1], Chuan Shi[1(✉)], Qianlong Xie[2], Xingxing Wang[2], and Dong Wang[2]

[1] Beijing University of Posts and Telecommunications, Beijing, China
{yangtianchi,shichuan}@bupt.edu.cn
[2] Meituan, Beijing, China
{gaohaihan,xieqianlong,wangxingxing04,wangdong07}@meituan.com

Abstract. As the popularity of Location-based Services increases, Point-of-Interest (POI) recommendations receive higher requirements to characterize the users, POIs and interactions. Although many recent graph neural network-based (GNN-based) studies have tried working on temporal and spatial factors, they still cannot seamlessly handle the temporal locality and spatial consistency. To tackle this issue, we propose a novel Memory-enhanced Period-aware Graph neural network for general POI Recommendation (MPGRec). Specifically, it exploits the advantages of the GNN module in characterizing user preferences. Moreover, we develop a period-aware gate mechanism after the GNN information propagation to characterize the temporal locality, and devise a dynamic memory module to extract, store and disseminate global information for spatial consistency. Furthermore, we propose a reading and writing strategy to merge the GNN module and memory module into a unified framework. Extensive experiments are conducted on four real-world datasets, and the experimental results demonstrate the effectiveness of our method.

Keywords: Recommendation systems · Graph neural network

1 Introduction

With the popularity of Location-based Services, Point of interest (POI) recommendation has already drawn lots of research attention [3,5,8,15].

There exist two research branches of POI recommendation. One focuses on sequential characteristics for temporal POI visit behavior mining, namely sequential POI recommendation [1,2,5,11]. The other focuses on the general characteristics of the users, POIs and interactions under temporal and spatial factors. Some existing works also refer to it as general POI recommendations [8]. In this work, we study the general POI recommendation problem. Owing to

X. Wang et al. (Eds.): DASFAA 2023, LNCS 13944, pp. 462–472, 2023.
https://doi.org/10.1007/978-3-031-30672-3_31

the rapid development of graph neural networks (GNNs), GNN-based recommendation approaches have attracted the interest of many researchers, which have advanced performance in capturing the general characteristics of interactions [4, 7, 14, 16–18]. Therefore, some recent studies [3, 8, 9] begin to explore GNN models for general POI recommendation, which introduce spatio-temporal contexts into the graph construction or message passing mechanism. In this way, GNNs can well characterize the user preferences under temporal and spatial factors.

Despite the great success of the existing methods, they still fail to consider the following two characteristics. (1) The interactions between users and POIs are not exactly the same among different time periods, e.g., morning and evening. We name this characteristic temporal locality. However, most GNN-based methods [3, 8] will aggregate neighbor information of all time periods and finally embed them into a single representation. It causes the information specific to different time periods to be confused, thus the temporal locality information also being ignored. (2) There is overlapping information among users and POIs that have not directly interacted or even are far away from each other, which can be treated as valuable and global information. We call this characteristic spatial consistency. However, existing GNN-based methods [3, 8] only pass messages for nodes on explicit interaction relationships (i.e., edges). It makes the nodes difficult to exploit valuable and global information from remote nodes, thus leading to difficulties in capturing spatial consistency.

Based on the above discussion, we propose to study the general POI recommendation based on the GNN framework to consider the temporal locality and spatial consistency. However, this is challenging due to the following questions. (1) *How to correctly distinguish information related to different time periods in the GNN, so as to consider temporal locality?* On the one hand, the GNN model cannot be trained on subgraphs divided by time period, since it separates the information related to different time periods into independent channels. On the other hand, it will still lead to a confusingly mixing of information to make time-specific transformations based on the whole graph like [3] for the neighbor information. Therefore, a special period-aware mechanism is needed that propagates the information of different periods, but outputs the information of a specific period for recommendation. (2) *How to make GNN not limited to the explicit interaction relationships in graph, thus further considering spatial consistency?* In the field of computer vision, the memory networks [10, 12] are recently explored to record global information across samples to improve few-shot learning tasks. Enlightened by it, we can transfer this idea to leverage a memory module to extract and store global node information that satisfies spatial consistency. Ultimately, it can serve as a springboard to propagate information to other nodes that are not explicitly connected. (3) *How to merge the GNN module and Memory module into a unified framework?* The existing memory approaches [6] are mainly implemented by a learnable parameter matrix, whose information is learned from the optimizer rather than the GNN. This results in an information gap between the memory module and the GNN module, which prevents them

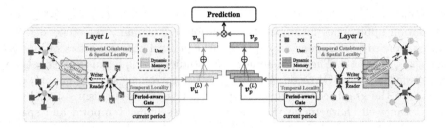

Fig. 1. Illustration of MPGRec.

from becoming a unified framework. Therefore, we propose a dynamic memory module, which stores the valuable and global information learned from GNN in memory and then feeds back into user/POI representations.

To address the aforementioned challenges, we propose a novel Memory-enhanced Period-aware Graph neural network to solve the task of general POI Recommendation (MPGRec). Specifically, first of all, a user-POI interaction graph is built to depict the user interaction history. Then, a novel memory-enhanced period-aware graph neural network is proposed to learn the user and POI embeddings. To further characterize the temporal locality, we develop a period-aware gate mechanism after information propagation to separate the mixed information of different periods. Moreover, a dynamic memory module is devised to extract and store the valuable and global information from the GNN module and accordingly send it back to all nodes, which takes the spatial consistency into account. Finally, we propose a writing strategy and a reading strategy to merge the GNN module and memory module into a unified framework, where the writing strategy is to extract global information learned by the GNN and store it in memory, and the reading strategy is to send the information that is valuable to a particular node. The main contributions are summarized as follows:

– We propose a novel Memory-enhanced Period-aware Graph neural network for POI recommendation (MPGRec). It designs a period-aware gate mechanism for temporal locality and a dynamic memory module for spatial consistency.
– We propose a writing/reading strategy for the dynamic memory module, which merges GNN module and memory module into a unified framework.
– Experimental results on four benchmark datasets show that our MPGRec significantly outperforms state-of-the-art baselines and demonstrate the effectiveness of the proposed method.

2 Methodology

In this section, we present the proposed MPGRec. Specifically, as illustrated in Fig. 1, based on a user-POI interaction graph, a novel memory-enhanced period-aware graph neural network is proposed to learn the user and POI embeddings.

In detail, a period-aware gate mechanism is designed for the temporal locality to filter out information related to other periods after the message passing process in GNN. Meanwhile, a dynamic memory module is introduced to store and disseminate the valuable and global information learned by the GNN module for all nodes, which takes advantage of spatial consistency. Finally, in order to merge the GNN module and memory module into a unified framework, we propose a writing strategy based on the principle of maximizing the expressiveness of memory and a reading strategy based on the principle of information correlation, which enables the memory to extract global information learned by the GNN and also enables the GNN to successfully benefit from the memory.

2.1 Problem Formulation and Graph Construction

In this paper, we focus on the task of general POI recommendation [3], which aims to recommend POIs to a given user at a specific time period. Formally, let $\mathcal{U} = \{u_1, \ldots, u_{|\mathcal{U}|}\}$ denote a set of users, $\mathcal{P} = \{p_1, \ldots, p_{|\mathcal{P}|}\}$ denote a set of POIs, and $\mathcal{T} = \{t_1, \ldots, t_{|\mathcal{T}|}\}$ denote a set of time periods which are obtained by dividing the timestamps into a specific range. Given a user $u \in \mathcal{U}$ and current time period $t \in \mathcal{T}$, the problem is defined to recommend POIs $p \in \mathcal{P}$ that u would be interested at t. We are aware of that explicit consideration for user/POI location might improve the POI recommendation [3,8], but we leave this extension in the future since it is not the key point focused in this work.

For modeling the user interaction history, similar to existing GNN-based recommendation methods [4], we construct a user-POI interaction graph, where each edge representing an interaction instance connects its corresponding user and POI. Formally, following [4], the user-POI interactions are modeled as a graph $\mathcal{G} = (\mathcal{V}, \mathcal{E})$. $\mathcal{V} = \mathcal{U} \bigcup \mathcal{P}$ represents the vertex set, where $\mathcal{U}$ is user set, and $\mathcal{P}$ is POI set. For each interaction instance that a user $u \in \mathcal{U}$ interacted a POI $p \in \mathcal{P}$, we build an edge $e = (u, p) \in \mathcal{E}$ to represent it. Note that following [4], we do not explicitly model the time period t into the graph.

2.2 Memory-Enhanced Period-Aware Graph Neural Network

Formally, the neighborhood aggregation and output rule of the l-th layer can be abstracted as follows,

$$\boldsymbol{v}_u^{(l+1)} = f_{\text{agg}}(\boldsymbol{v}_u^{(l)}, \{\boldsymbol{v}_p^{(l)} | p \in \mathcal{N}_u\}; \boldsymbol{M}_u^{(l)}), \quad \hat{\boldsymbol{v}}_u^{(l+1)} = g_{\text{out}}(\boldsymbol{v}_u^{(l+1)}; t), \quad (1)$$

$$\boldsymbol{v}_p^{(l+1)} = f_{\text{agg}}(\boldsymbol{v}_p^{(l)}, \{\boldsymbol{v}_u^{(l)} | u \in \mathcal{N}_p\}; \boldsymbol{M}_p^{(l)}), \quad \hat{\boldsymbol{v}}_p^{(l+1)} = g_{\text{out}}(\boldsymbol{v}_p^{(l+1)}; t), \quad (2)$$

where $\boldsymbol{v}_u$ and $\boldsymbol{v}_p$ are the aggregated embeddings of user u and POI p by aggregation function f_{agg} while $\hat{\boldsymbol{v}}_u$ and $\hat{\boldsymbol{v}}_p$ are the output embeddings by period-aware mechanism function g_{out}. $\mathcal{N}_u$ and $\mathcal{N}_p$ are the corresponding neighbors, and $\boldsymbol{M}_u^{(l)} \in \mathbb{R}^{K \times D}$ and $\boldsymbol{M}_p^{(l)} \in \mathbb{R}^{K \times D}$ are the memory matrices of K slots for user and POI vertices, respectively. D denotes the embedding size and $t \in \mathcal{T}$ denotes the current time period.

In the following paragraphs, Eq. (1) is taken as an example to illustrate the calculation procedure. Given a target node u and current time period t, suppose that the memory reader outputs a set of information denoted as $\hat{M}_u^{(l)} \in \mathbb{R}^{K \times D}$ for target user u and the corresponding information weight vector $\alpha \in \mathbb{R}^{K \times 1}$ after memory module calculations (which will discuss in detail in the next subsection). f_{agg} is defined as follows,

$$f_{\text{agg}}(v_u^{(l)}, \{v_p^{(l)} | p \in \mathcal{N}_u\}; M_u^{(l)}) = \sum_{p \in \mathcal{N}_u} \frac{1}{\sqrt{|\mathcal{N}_u|}\sqrt{|\mathcal{N}_p|}} v_p^{(l)} + \alpha^\top \hat{M}_u^{(l)}. \quad (3)$$

It is worth noting that in the above formula, the first term represents the neighbor information propagation of the GNN model. The second term represents the enhancement of the memory module.

Additionally, since the neighbors are of various time periods, the obtained representations are mixed up with different information specific to different time periods, thus causing other time-related information to become noise for the current time period t_u. Therefore, we apply a period-aware gating function g_{out} after the information propagation, which is calculated as follows,

$$\hat{v}_u^{(l+1)} = g_{\text{out}}(v_u^{(l+1)}; t) = v_u^{(l)} \| (v_u^{(l)} \odot g_t) \in \mathbb{R}^{2D}, \quad (4)$$

where g_t is the trainable gating vector towards the current time period t in the period-aware gating function $g_{\text{out}}(\cdot; \cdot)$. Symbol $\|$ denotes the concatenation operation to retain both the time period-specific information for temporal locality and the general information for other characteristics. In this paper, we use different groups of gating vectors for user and POI nodes, but share in different model layers. Consequently, there are a total of $2|\mathcal{T}|$ gating vectors.

Inspired by LightGCN [4], we do not explicitly integrate the self-connection information during message passing, but instead, combine this by summing the embeddings from each layer to form the final representation. Formally,

$$\hat{v}_u = \sum_{l=0}^{L} w_l \hat{v}_u^{(l)}, \quad \hat{v}_i = \sum_{l=0}^{L} w_l \hat{v}_i^{(l)}, \quad (5)$$

where $\hat{v}^{(0)} = v^{(0)} \| v^{(0)}$, L denotes the number of model layers, $w_l \geqslant 0$ denotes the importance of the l-th layer embedding in constituting the final embedding. Following [4], we set w_l uniformly as $1/(L+1)$.

2.3 Dynamic Memory Module

In this subsection, we present the proposed dynamic memory network designed for spatial consistency. Specifically, the core of the reading strategy is to determine which information in memory is relevant to a target node and how important it should be. Traditional memory methods mostly choose attention module to read memory and assign corresponding importance. We argue that the non-negativity of attention weights limits the capabilities of memory. Therefore, we propose a reader with a correlation-based reading strategy. Formally, given the current information stored in a K-slot memory $\boldsymbol{M} \in \mathbb{R}^{K \times D}$, the target node representation $\boldsymbol{v} \in \mathbb{R}^D$, the output messages $\hat{\boldsymbol{M}}$ and corresponding correlation weights $\boldsymbol{\alpha}$ in Eq. (3) are defined as follows,

$$\hat{\boldsymbol{M}} = \boldsymbol{M} \cdot \boldsymbol{W}_v^{(l)}, \quad \boldsymbol{\alpha} = \frac{1}{K}\mathrm{hardtanh}((\boldsymbol{M}\boldsymbol{W}_k^{(l)}) \cdot (\boldsymbol{v}\boldsymbol{W}_q^{(l)})^\top / D), \qquad (6)$$

where $\boldsymbol{W}_v^{(l)}$, $\boldsymbol{W}_k^{(l)}$, $\boldsymbol{W}_q^{(l)}$ is the parameter matrix for value, key, query, respectively. Different from the similarity-based readers that usually use the softmax function, we choose the hardtanh function to measure correlation, where positive weight $\boldsymbol{\alpha}$ represents positive correlation, while negative weight represents negative correlation. Thereby, the ability of memory enhancement increases since it relaxes the non-negative constraint on the weights.

As for writing strategy, the intuition is that more nodes can benefit from the memory only when information stored in memory should have the greatest possible coverage, which is named the principle of maximum memory expressiveness. Obviously, the K records with the maximum memory expressiveness should have the smallest sum of the pair-wise correlations with each other. To this end, the writing strategy finds a memory slot having the most overlapping information with others, and updates it with the least overlapping information. Formally, given the current information stored in the K-slot memory $\boldsymbol{M} \in \mathbb{R}^{K \times D}$ and the representation matrix $\boldsymbol{V} \in \mathbb{R}^{N \times D}$ of N nodes outputted by the GNN module,

$$\boldsymbol{m}^* = \arg\max_i \sum_{j \neq i} |\mathrm{corr}(\boldsymbol{m}_i, \boldsymbol{m}_j)|, \qquad (7)$$

where $\boldsymbol{m}_i \in \mathbb{R}^{1 \times D}$ denotes a slot of $\boldsymbol{M}$, and the correlation is measured by $\mathrm{corr}(\boldsymbol{x}, \boldsymbol{y}) = \boldsymbol{x}^\top \boldsymbol{y}$. Note that information with strong negative correlation is also worth storing, since our correlation-based reader allows negative weights. Finally, the information that is most worthy of being written in memory should have the least overlapping information with the existing slots, formally,

$$\boldsymbol{v}^* = \arg\min_i \sum_{\boldsymbol{m}_j \neq \boldsymbol{m}^*} |\mathrm{corr}(\boldsymbol{v}_i, \boldsymbol{m}_j)|, \qquad (8)$$

where $\boldsymbol{v}_i$ denotes a row of matrix $\boldsymbol{V}$. In order to ensure the updating convergence of memory, the above update operation will be performed if and only if

$$\sum_{\boldsymbol{m}_j \neq \boldsymbol{m}^*} |\mathrm{corr}(\boldsymbol{m}^*, \boldsymbol{m}_j)| > \sum_{\boldsymbol{m}_j \neq \boldsymbol{m}^*} |\mathrm{corr}(\boldsymbol{v}^*, \boldsymbol{m}_j)|. \qquad (9)$$

Besides, for the consideration of efficiency and stability, the writer updates $1 \leqslant k \ll K$ slots once a step according to the above rules. Note that for initialization, we randomly select K node representations from GNN module to build the initial memory, and then it will be trained by the above writing strategy.

2.4   Model Training

After propagating L layers, the user embeddings $\hat{\boldsymbol{v}}_u$ and POI embeddings $\hat{\boldsymbol{v}}_p$ have been obtained. In this section, for model training, we employ the Bayesian

Personalized Ranking (BPR) loss [13], which is a pairwise loss that encourages the prediction of an observed entry to be higher than its unobserved counterparts. Formally,

$$L_{BPR} = -\sum_{u \in \mathcal{U}; p \in \mathcal{N}_u; q \notin \mathcal{N}_u} \ln \sigma(\hat{y}_{up} - \hat{y}_{uq}), \text{ where } \hat{y}_{up} = \hat{v}_u^\top \hat{v}_p, \qquad (10)$$

Adam optimizer is adopted to train the model in a mini-batch manner.

3   Experiments

3.1   Experimental Setup

Datasets. To evaluate the effectiveness of MPGRec, we conduct experiments on four benchmark POI recommendation datasets: **Foursquare**, **Gowalla**, **Yelp** and **Meituan**[1]. For datasets Foursquare, Gowalla and Yelp,

Table 1. Statistics of datasets.

Dataset	# User	# POI	# Interaction	Density
Foursquare	830	1,090	21,367	2.4E−02
Gowalla	29,859	40,989	1,027,464	8.4E−04
Yelp	72,093	41,234	1,842,016	6.2E−04
Meituan	200,000	3,559	1,240,419	1.7E−03

we use the datasets released by a public tool RecBole [19], and eliminate those users with less than 10 check-in POIs, as well as those POIs with less than 10 visitors. We process their check-in timestamps at equal time intervals to obtain 4 periods. For dataset Meituan, we use the raw dataset without elimination. The statistics after preprocessed are reported in Table 1.

Evaluation Metrics. In the experiments, we randomly select 70%, 10% and 20% of interactions of each user for training, validation and testing, respectively. We choose the following widely used evaluation metrics: Recall@N, NDCG@N and HR@N for N = 1, 3, 5, 10 and 20. Following [4], all items that are not interacted with by a user are the candidates and a prediction is correct only when the item and the corresponding time period are both correct.

Baselines. We select the following baselines for POI recommendation to validate the effectiveness of our model: GNN-based general recommendation methods: LightGCN [4], SimGCL [18], HMLET [7]; GNN-based POI recommendation method: STGCN [3]; memory-based method: MMCF [6].

Implementation Detail. Following existing settings [4] for fair comparison, we set the dimension of node embeddings $D = 64$, the number of layers $L = 2$ for all datasets for all methods. For our model, we set the number of memory slots $K = 50$ and update $k = 2$ slots once a training step. L2 regularization coefficient is set $1e^{-5}$. The learning rate is set $1e^{-3}$ for Foursquare, $5e^{-4}$ for Gowalla and Yelp, $1e^{-4}$ for Meituan. The batch size is set 2048.

---

[1] https://www.biendata.xyz/competition/smp2021_1/.

Table 2. Overall recommendation performance on the four datasets. The best and second best results are bold and underlined, respectively. We also report improvement of MPGRec compared to the best baseline method.

Model	Recall@N					NDCG@N					HR@N				
	1	3	5	10	20	1	3	5	10	20	1	3	5	10	20
Dataset: Foursquare															
LightGCN	0.0214	0.0506	0.0700	0.0997	0.1563	0.0347	0.0456	0.0533	0.0645	0.0814	0.0347	0.0905	0.1222	0.1760	0.2634
SimGCL	0.0243	0.0495	0.0704	0.1153	0.1707	0.0372	0.0461	0.0548	0.0707	0.0874	0.0372	0.0860	0.1227	0.1926	0.2800
HMLET	0.0211	0.0475	0.0693	0.1158	0.1731	0.0362	0.0431	0.0522	0.0697	0.0866	0.0362	0.0804	0.1176	0.1946	0.2775
STGCN	0.0184	0.0421	0.0571	0.0966	0.1455	0.0327	0.0385	0.0440	0.0588	0.0736	0.0327	0.0719	0.0960	0.1664	0.2479
MMCF	0.0178	0.0513	0.0711	0.0997	0.1521	0.0341	0.0440	0.0526	0.0631	0.0773	0.0345	0.0877	0.1193	0.1697	0.2535
MPGRec	0.0301	0.0613	0.0840	0.1378	0.1954	0.0493	0.0571	0.0662	0.0857	0.1030	0.0493	0.1071	0.1438	0.2278	0.3172
Impr.	23.87%	19.49%	18.14%	19.00%	12.88%	32.53%	23.86%	20.80%	21.22%	17.85%	32.53%	18.34%	17.20%	17.06%	13.29%
Dataset: Gowalla															
LightGCN	0.0278	0.0555	0.0750	0.1087	0.1559	0.0592	0.0602	0.0663	0.0775	0.0916	0.0592	0.1159	0.1541	0.2178	0.2973
SimGCL	0.0243	0.0509	0.0697	0.1037	0.1490	0.0546	0.0558	0.0615	0.0727	0.0862	0.0546	0.1077	0.1431	0.2040	0.2762
HMLET	0.0258	0.0527	0.0712	0.1058	0.1535	0.0560	0.0573	0.0628	0.0745	0.0888	0.0560	0.1107	0.1474	0.2119	0.2906
STGCN	0.0178	0.0377	0.0514	0.0762	0.1101	0.0388	0.0408	0.0450	0.0532	0.0634	0.0388	0.0823	0.1114	0.1619	0.2207
MMCF	0.0031	0.0078	0.0115	0.0208	0.0366	0.0071	0.0084	0.0098	0.0131	0.0179	0.0071	0.0187	0.0279	0.0495	0.0852
MPGRec	0.0290	0.0583	0.0787	0.1167	0.1691	0.0613	0.0631	0.0695	0.0821	0.0978	0.0613	0.1221	0.1620	0.2300	0.3146
Impr.	4.32%	5.05%	4.93%	7.36%	8.47%	3.55%	4.82%	4.83%	5.94%	6.77%	3.55%	5.35%	5.13%	5.60%	5.82%
Dataset: Yelp															
LightGCN	0.0102	0.0232	0.0335	0.0532	0.0821	0.0196	0.0228	0.0268	0.0337	0.0426	0.0196	0.0446	0.0645	0.1018	0.1451
SimGCL	0.0105	0.0212	0.0284	0.0415	0.0590	0.0184	0.0205	0.0232	0.0278	0.0329	0.0184	0.0377	0.0505	0.0746	0.1060
HMLET	0.0117	0.0233	0.0314	0.0478	0.0719	0.0211	0.0232	0.0262	0.0320	0.0391	0.0211	0.0436	0.0599	0.0919	0.1364
STGCN	0.0018	0.0062	0.0096	0.0181	0.0353	0.0109	0.0111	0.0118	0.0144	0.0202	0.0109	0.0310	0.0445	0.0789	0.1353
MMCF	0.0103	0.0237	0.0337	0.0531	0.0751	0.0191	0.0234	0.0277	0.0332	0.0395	0.0190	0.0460	0.0648	0.1001	0.1441
MPGRec	0.0121	0.0250	0.0349	0.0543	0.0827	0.0222	0.0248	0.0286	0.0355	0.0439	0.0222	0.0470	0.0662	0.1033	0.1546
Impr.	3.42%	5.49%	3.56%	2.07%	0.73%	5.21%	5.98%	3.25%	5.34%	3.05%	5.21%	2.17%	2.16%	1.47%	6.55%
Dataset: Meituan															
LightGCN	0.2797	0.3346	0.3538	0.3770	0.4005	0.3039	0.3202	0.3284	0.3365	0.3428	0.3039	0.3605	0.3801	0.4034	0.4265
SimGCL	0.2699	0.2975	0.3065	0.3189	0.3345	0.2949	0.2934	0.2971	0.3014	0.3055	0.2949	0.3221	0.3316	0.3447	0.3613
HMLET	0.2813	0.3330	0.3519	0.3764	0.4014	0.3064	0.3198	0.3280	0.3364	0.3431	0.3064	0.3594	0.3786	0.4027	0.4273
STGCN	0.1840	0.2663	0.3002	0.3396	0.3737	0.1995	0.2377	0.2522	0.2657	0.2749	0.1995	0.2879	0.3237	0.3650	0.3997
MMCF	0.2760	0.3487	0.3629	0.3740	0.3843	0.3053	0.3293	0.3352	0.3391	0.3418	0.3053	0.3766	0.3899	0.4004	0.4103
MPGRec	0.3056	0.3620	0.3766	0.3921	0.4117	0.3357	0.3492	0.3554	0.3608	0.3659	0.3357	0.3894	0.4033	0.4182	0.4374
Impr.	8.64%	3.81%	3.78%	4.01%	2.57%	9.56%	6.04%	6.03%	6.40%	6.65%	9.56%	3.40%	3.44%	3.67%	2.36%

3.2 Overall Results

Table 2 shows the recommendation performance on the four datasets. The observations and conclusions are discussed as follows.

In all cases, the proposed MPGRec outperforms all the baselines across four datasets on all metrics. Specifically, MPGRec has achieved an average improvement of 20.5%, 5.4%, 3.7% and 5.3% on the four datasets, respectively, which validates the effectiveness and robustness of our model. In detail, MPGRec has a larger improvement ratio when the N (for top N) is smaller, e.g., in terms of Recall@N on Meituan, the performance has more than double improvement when $N = 1$ than $N = 3$. We attribute it to the consideration of both temporal locality and spatial consistency, which benefits MPGRec to make more accurate predictions. Besides, the GNN-based methods, especially the powerful LightGCN, SIMGCL and HMLET, have generally achieved sub-optimal results. However, STGCN, which should be a SOTA method for POI recommendation, has never achieved the expected performance. We analyze that this is due to the terrible information mixing caused by its propagation rules, which in turn indicates the effectiveness of our designs such as the period-aware mechanism. MMCF,

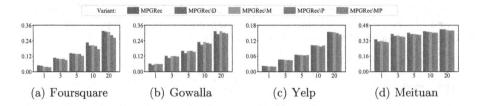

Fig. 2. Recommendation performance for ablation study.

the SOTA memory-based method, has generally performed sub-optimally on Meituan but poorly on Gowalla. This polarized phenomenon, on the one hand, verifies the enhancement capability of the memory mechanism, and on the other hand, reflects the insufficiency of memory that is implemented by a learnable parameter matrix. This validates our design for the dynamic memory module.

3.3 Ablation Study

In this subsection, we compare our MPGRec with 4 variants to validate the characteristics we have summarized and investigate the effectiveness of each proposed module in our model. Due to the limited space of the paper, we only report the performance on metric HR@N since the performances on other metrics are consistent. Specifically, $MPGRec_{\backslash D}$ is a variant that replace the proposed dynamic memory module with the simple memory implemented as a trainable parameter matrix like [6]. $MPGRec_{\backslash M}$ directly removes the entire memory module, $MPGRec_{\backslash P}$ removes the period-aware gate mechanism only, and $MPGRec_{\backslash MP}$ removes both. As reported in Fig. 2, the recommendation performance degrades with each module removed or modified, which verifies the effectiveness of each of the proposed modules. In detail, the performance of $MPGRec_{\backslash D}$ is sometimes even worse than $MPGRec_{\backslash M}$. This weird phenomenon is caused by the information gap between memory and GNN module since its expressiveness depends on the optimizer. It confirms the effectiveness of our reading and writing strategies in the dynamic memory module. Besides, the removal of either the dynamic memory module ($MPGRec_{\backslash M}$) or the period-aware mechanism ($MPGRec_{\backslash P}$) will reduce the performance, and removing both ($MPGRec_{\backslash MP}$) obtains the worst performance. This phenomenon on the one hand proves the necessity of the summarized characteristics; on the other hand, it also affirms the module effectiveness of each of our designs.

4   Conclusion

In this paper, we study the general POI recommendation based on the GNN framework and propose a novel Memory-enhanced Period-aware Graph neural network for general POI recommendation (MPGRec). In detail, it designs a period-aware gate mechanism for temporal locality, and a dynamic memory module for spatial consistency. Besides, we propose a correlation-based reading strategy and a writing strategy to maximize memory expressiveness, which merges the GNN module and memory module into a unified framework. Finally, we conduct extensive experiments to verify the effectiveness of our MPGRec.

Acknowledgments. This work is supported in part by the National Natural Science Foundation of China (No. U20B2045, 62192784, 62172052, 62002029, 62172052, U1936014) and also supported by Meituan.

References

1. Cui, Q., Zhang, C., Zhang, Y., Wang, J., Cai, M.: ST-PIL: spatial-temporal periodic interest learning for next point-of-interest recommendation. In: CIKM, pp. 2960–2964, October 2021
2. Cui, Y., Sun, H., Zhao, Y., Yin, H., Zheng, K.: Sequential-knowledge-aware next POI recommendation: a meta-learning approach. ACM Trans. Inf. Syst. **40**(2), 1–22 (2022)
3. Han, H., et al.: STGCN: a spatial-temporal aware graph learning method for POI recommendation. In: ICDM, pp. 1052–1057, November 2020
4. He, X., Deng, K., Wang, X., Li, Y., Zhang, Y., Wang, M.: LightGCN: simplifying and powering graph convolution network for recommendation. In: SIGIR, pp. 639–648, July 2020
5. Islam, M.A., Mohammad, M.M., Sarathi Das, S.S., Ali, M.E.: A survey on deep learning based Point-of-Interest (POI) recommendations. Neurocomputing **472**, 306–325 (2022)
6. Jiang, X., Hu, B., Fang, Y., Shi, C.: Multiplex memory network for collaborative filtering. In: ICDM, pp. 91–99 (2020)
7. Kong, T., et al.: Linear, or non-linear, that is the question! In: WSDM, pp. 517–525 (2022)
8. Li, Z., Cheng, W., Xiao, H., Yu, W., Chen, H., Wang, W.: You are what and where you are: graph enhanced attention network for explainable POI recommendation. In: CIKM, pp. 3945–3954, October 2021
9. Lim, N., et al.: STP-UDGAT: spatial-temporal-preference user dimensional graph attention network for next POI recommendation. In: CIKM, pp. 845–854, October 2020
10. Liu, X., et al.: Learn from concepts: towards the purified memory for few-shot learning. In: IJCAI, pp. 888–894, August 2021
11. Rahmani, H.A., Aliannejadi, M., Baratchi, M., Crestani, F.: A systematic analysis on the impact of contextual information on point-of-interest recommendation. ACM Trans. Inf. Syst. **40**(4), 1–35 (2022)
12. Ramalho, T., Garnelo, M.: Adaptive posterior learning: few-shot learning with a surprise-based memory module. In: ICLR (2019)
13. Rendle, S., Freudenthaler, C., Gantner, Z., Schmidt-Thieme, L.: BPR: Bayesian personalized ranking from implicit feedback. In: UAI, pp. 452–461 (2009)
14. Sun, L., Rao, Y., Zhang, X., Lan, Y., Yu, S.: MS-HGAT: Memory-enhanced sequential hypergraph attention network for information diffusion prediction. In: AAAI (2022)
15. Sánchez, P., Bellogín, A.: Point-of-interest recommender systems based on location-based social networks: a survey from an experimental perspective. ACM Comput. Surv. **54**, 3510409 (Jan 2022)

16. Wang, D., Wang, X., Xiang, Z., Yu, D., Deng, S., Xu, G.: Attentive sequential model based on graph neural network for next poi recommendation. In: WWW, vol. 24, pp. 2161–2184, November 2021
17. Wang, Z., Zhu, Y., Zhang, Q., Liu, H., Wang, C., Liu, T.: Graph-enhanced Spatial-temporal network for next POI recommendation. ACM Trans. Knowl. Discovery Data **16**, 1556–4681 (2022)
18. Yu, J., Yin, H., Xia, X., Chen, T., Cui, L., Nguyen, Q.V.H.: Are graph augmentations necessary? Simple graph contrastive learning for recommendation. In: SIGIR, pp. 1294–1303 (2022)
19. Zhao, W.X., et al.: RecBole: towards a unified, comprehensive and efficient framework for recommendation algorithms. In: CIKM (2021)

Attentive Hawkes Process Application for Sequential Recommendation

Shuodian Yu[1], Li Ma[1], Xiaofeng Gao[1(✉)], Jianxiong Guo[2], and Guihai Chen[1]

[1] Shanghai Key Laboratory of Scalable Computing and Systems, Department of
Computer Science and Engineering, Shanghai Jiao Tong University, Shanghai, China
{timplex233,mali-cs,gao-xf,gchen}@sjtu.edu.cn
[2] Advanced Institute of Natural Sciences, Beijing Normal University,
Zhuhai 519087, China
jianxiongguo@bnu.edu.cn

Abstract. Sequential recommendation systems which aim to predict
the users' future clicking items are essential in both research and industry services. Based on the contextual characteristics of users, a large
number of existing methods exploit abundant information from user
behavior sequences and gain users' interests through expressive sequential models like recurrent neural networks and self-attention mechanism
to predict the next item. However, the temporal generative process of
behavior sequences, which is crucial to the complex relationships and
distributions of users' dynamic preferences, is rarely considered. In this
paper, we propose HawRec (Hawkes process based sequential recommendation), which is a new representation learning approach to model
the interacted sequences of users from a temporal point process perspective. In particular, we leverage temporal self-attention mechanism
to make an analogy to the self-exciting mechanism of Hawkes process.
We bring expressive superiority of neural network into the Hawkes process with non-parametric conditional intensity kernel, resulting in more
accurate recommendation and inheriting robustness from the temporal
point process. Furthermore, extensive experiments conducted on four
public benchmarks demonstrate that our HawRec outperforms various
baselines of state-of-the-art sequential recommendation.

Keywords: Sequential Recommendation · Temporal Point Process ·
Self Attention

1 Introduction

Nowadays, with the rapid emergence of online websites, it is essential to model
the users' event sequences on these platforms so as to provide high-quality services for different users. For example, E-Commerce platforms on the Internet

This work was supported by National Key R&D Program of China [2020 YFB1707900];
National Natural Science Foundation of China [62272302, 62202055], Shanghai Municipal Science and Technology Major Project [2021 SHZDZX0102].

X. Wang et al. (Eds.): DASFAA 2023, LNCS 13944, pp. 473–488, 2023.
https://doi.org/10.1007/978-3-031-30672-3_32

tend to recommend proper products based on users' behavior sequences, which generally contain records of their interactions like click, favorite, purchase, etc. An accurate and effective personalized recommendation system is becoming extremely important for designing marketing strategies, promoting the users' purchase behavior, and improving the overall profit of platforms. Utilizing abundant information and discovering hidden intentions from users' temporal behavior sequences are the key point of recommendation.

Traditionally, sequential recommendation systems tend to predict the next clicked item by mining users' behavior sequences, which consist of their clicked items in history. Previous methods focus on mining the sequential pattern of users' interactions with sequential models, including Markov Chain based approaches [6,17], Recurrent Neural Networks [8,9], and Self-Attention based methods [10,14,18]. Temporal information has also been extensively studied to capture the temporal dynamics of users interests and item attributes [12]. These approaches incorporate the temporal patterns (change of item popularity, seasonality, user interests, etc.) into recommendation and achieve competitive performance. However, the mining and utilization of temporal information is far from sufficient, which yields inadequate modeling of temporal dependencies.

Temporal Point Process (TPP) [3] provides a solid mathematical framework based on prior knowledge of the real-world scenarios to model the occurrence of asynchronous event sequences. It is widely used in many forecasting applications where various types of events interact with each other and trigger the future events. The Hawkes process [7], a typical self-exciting TPP, was introduced to directly model the contribution of each event with the richer-get-richer phenomenon. Hawkes assumes that each historical event has independent influence on the future event and dynamically quantifies this influence with a parameterized function. Self-attention based model has also been proposed to model the parametric Hawkes process [26]. Although the parametric form of Hawkes process is too simplified to capture complex dependence of real-world temporal event sequences, TPP still intuitively points out a feasible direction of promoting the expression of sequential recommendation system. Each of clicked items in the user's history sequence implies the possibility that the user will click on the related items later. Modeling the dynamic influence between items collaboratively and efficiently is crucial to the performance for the recommendation services in real-world scenarios, which can be regarded as an instance of TPP, as shown in Fig. 1.

A number of attempts have been made to incorporate the idea of temporal intervals into sequential recommendations [12,22]. However, we point out that previous sequential recommendations still have the following limitations:

- They only take the timestamp together with item embedding as event embedding and feed the embedding into the self-attention block, which is not enough to fully capture the temporal characteristics. Inevitably, there methods suffer from poor interpretability to reflect the temporal influence.
- The complex distribution of temporal influence is not fully expressed. Previous TPP-based models [22] usually define an exponential exciting kernel to

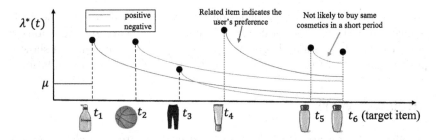

Fig. 1. An Example of diverse interests in user sequence, explained from Hawkes process perspective. Different items have dynamic effects on the target one over time. Related items indicating the user's preference can have a positive influence. The same cosmetic can decrease the possibility to buy it again in a short period.

describe the dynamic decay property of the influence between events. However, in real-world scenarios, the intensity is hard to be explicitly described. For example, it can even be seasonal that few human prior-based kernels have been considered. This complexity makes it infeasible to directly choose a suitable parametric kernel to cover all the cases.

To tackle those problems, we propose the **Haw**kes Process based Sequential **Rec**ommendation (**HawRec**), a novel sequential recommendation model, which fully exploits the expressive ability of Hawkes process to model the mutual interplay of actions at different temporal intervals and significantly enhances the results. In specific, we not only embed the time gap as a part of interaction representation, but also take advantage of Hawkes process by learning the mutual excitation between events in a non-parametric way without prior knowledge. Since Hawkes process well captures the exciting effects of history events on the current events, a non-parametric intensity kernel can be constructed to calculate the conditional intensity function with the given target item and timestamp for each user to predict the click probability, which fully integrates the expression capacity of neural networks into the traditional Hawkes model. We validate that our proposed model outperforms state-of-the-art baselines on four benchmarks for sequential recommendation. We also verify the effectiveness of our proposed Hawkes process based decoding module that contributes to the performance improvement through extensive ablation study and analysis.

Our contributions of this work can be summarized as follows:

- We claim that the interaction between users and items can be regarded as a temporal point event in the user behavior sequence, which leads to mutual excitation process between user actions. A Hawkes process based recommendation framework is proposed to fill the gap between user behaviors modeling and temporal point process.
- We design a novel non-parametric conditional intensity estimator, which fully takes advantage of Hawkes process and attention mechanism to decode the users' interest towards different items as well as the temporal dependencies.

– We conduct extensive experiments on four real-world datasets. The performance comparisons to several state-of-the-art approaches and variations validate the effectiveness and robustness of our proposed model, and show the positive impact of temporal point process on sequential recommendation.

2 Related Work

We would briefly discuss three lines of research that are related to our work: Sequential recommendation, temporal point process, and self-attention.

2.1 Sequential Recommendation

Sequential recommendation aims to model users' dynamic interests by leveraging their historical interactions, and predict the next item they would like to click. Typically, early recommenders [6,17] adopt Markov Chains, which naturally capture relationships of the adjacent items and use matrix factorization to learn the users' interests. Despite their success, with the vigorous development of neural networks, the deep sequential models like RNN are widely adopted in the next basket of recommendation [8,9,25]. SASRec [10] models the entire user sequence and takes advantage of self-attention mechanism to capture long-term semantics, thus achieving further success for prediction. Self-supervised learning methods [14,18] are also proposed to enhance the prediction.

While the methods based on deep learning have effectively modeled the relationship of user historical sequences with high expressive power, the researches still start to take an eye on the temporal features. In comparison to SASRec, TiSASRec [12] models the time intervals between items to explore the influence of different time intervals on next item prediction, which achieves higher performance. MHPE [22] uses multi-variable Hawkes process to simulate the user's temporal pattern, and assumes that the historical influence of items exponentially decays with time. However, up to now, the temporal recommendations mainly take the time information as a part of item embedding or explicitly aggregate the conditional intensity, of which the temporal dependencies between interactions are underutilized. Different from these existing methods, we combine a non-parametric temporal point process with a deep sequential model to obtain enhanced model performance and interpretability.

2.2 Temporal Point Process

Temporal Point Process (TPP) [3] is a stochastic process to model point process in continuous time space, and is usually simplified as a list of discrete events with multiple event time. TPP can be represented as a counting process $N(t)$, which denotes the number of events happened until timestamp t. Let $\mathcal{S} = \{(e_i, t_i)\}_{i=1}^{T}$ as the event sequences which denotes the i-th event happens at timestamp t_i with event type e_i. The conditional intensity function of a TPP is defined as the probability that an event e' happens at an infinitesimal time window $[t, t + dt)$ given the historical observed events $\mathcal{H}_t = \{(e_i, t_i) \mid t_i \leq t\}$:

$$\lambda^*(t)dt = P\left((e', t') : t' \in [t, t + dt) \mid \mathcal{H}_t\right) = \mathbf{E}\left(dN(t) \mid \mathcal{H}_t\right), \tag{1}$$

where $\mathbf{E}\left(dN(t) \mid \mathcal{H}_t\right)$ is the expected number of events in the time interval.

Hawkes process [7] is one of the most widely used TPP, which assumes that each previous event will influence the probability of future event temporarily by a time-dependent intensity function. It comes from the self-exciting effect of Poisson process. The probability of an event occurring in the future is calculated as the sum of the cumulative effects of historical events and a constant basic intensity. The definition of intensity function is given as:

$$\lambda^*(t) = \mu + \sum\nolimits_{j : t_j \leq t} \phi(t - t_j), \tag{2}$$

where μ is a base intensity which is independent of the history, while $\phi(t)$ is a kernel function that describes the influence of history events. In multivariate Hawkes process, $\phi(t)$ is also written as $\phi_{u,v}(t)$ to capture the impact of event u to a subsequent event v. Zhou et al. [11] proposes a parameterized form $\phi_{u,v}(t) = \alpha_{u,v}\kappa(t)$, which uses excitation parameter α to quantify the influence of different event types and uses kick function $\kappa(\cdot)$ to model the time decay effect, respectively. One of the most widely used kick function is the exponential function, i.e., $\kappa(t) = \exp(-\gamma t)$, which indicates that the influence of historical events decays exponentially with the time.

However, the assumptions about the self-exciting effect and the time decay effect fix the mutual excitation among events as parametric kernels based on human domain knowledge. In fact, the most common goal of TPP models is to predict the future, which is closely related to many applications of machine learning. TPP-based models have great potential in improving model performance and introducing interpretation ability into machine learning methods [2]. Du et al. [4] employs RNN to model the temporal point process and predicts when and which event will occur next. Zuo et al. [26] and Zhang et al. [24] predicts events using a combination of sequential self-attention and Hawkes process.

2.3   Self-attention

Attention mechanism has been widely used in a variety of fields and applications, e.g., image classification in computer vision [13] and translation in neural language processing [1]. Self-attention is a kind of attention mechanism that models the influences between items in the sequence [21], which calculates pairwise items' attention score between different positions in a sequence and learns the hidden representation of items and helps to achieve outstanding performance in a wide range of tasks, especially in sequential recommendation [10,18]. Fan et al. [5] propose stochastic self-attention to overcome uncertain issue in user behaviors. However, the fundamental form of self-attention only considers the absolute position in the sequence and is unable to model the temporal information directly.

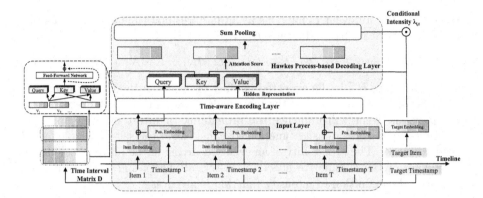

Fig. 2. Illustration of the proposed model HawRec which contains three layers. In the input layer, each interaction including the item and absolute position, as well as the time intervals, is mapped into low-dimensional embedding. The encoding layer employs a time-aware self-attention mechanism to obtain the hidden representations of interactions. In the Hawkes process-based decoding layer, the hidden representations are used to estimate the influence intensity through an attentive non-parametric kernel. A sum pooling is adopted to calculate the cumulative influence of historical interactions that excite the possibility of next-item.

3   Problem Definition

Let $\mathcal{U}$ and $\mathcal{V}$ denote the user set of m users and the item set of n items, respectively. For each user $u \in \mathcal{U}$, we define $\mathcal{S}^u = \{(v_1^u, t_1^u), (v_2^u, t_2^u), \dots, (v_{|\mathcal{S}^u|}^u, t_{|\mathcal{S}^u|}^u)\}$ as his/her event sequences in chronological order, where each interaction record (v, t) denotes that the user u interacts with item $v \in \mathcal{V}$ at timestamp t. The interaction event could be a click or a rating, while the item could be a movie or commodity depending on the dataset.

Problem 1 (Sequential Recommendation). *Given user set $\mathcal{U}$, item set $\mathcal{V}$, and all historical sequences $\{\mathcal{S}^u | u \in \mathcal{U}\}$, it is to predict the conditional probability $P(v_{|\mathcal{S}^u|+1}^u | \mathcal{S}^u, t_{|\mathcal{S}^u|+1}^u)$ that the user u will interact with target item $v_{|\mathcal{S}^u|+1}^u$, based on his historical behaviors at timestamp $t_{|\mathcal{S}^u|+1}^u$.*

4   Proposed Method

In this paper, we propose a Hawkes Process based Sequential Recommendation framework, *HawRec*. The architecture of HawRec is shown in Fig. 2, which consists of three components: an input layer, a time-aware encoding layer, and a Hawkes process-based decoding layer.

4.1   Input Layer

For the input sequence $\mathcal{S}^u = \{(v_1^u, t_1^u), (v_2^u, t_2^u), \dots, (v_{|\mathcal{S}^u|}^u, t_{|\mathcal{S}^u|}^u)\}$, following most of existing methods, we first convert it to a fixed length T. To be specific, if the sequence length is smaller than T, we add $T - |\mathcal{S}^u|$ special "padding" records at the beginning of the sequence while the padding timestamp is set as t_1^u. We choose the most recent T records for sequences longer than T. Thus the input of the model becomes $\mathcal{S} = \{(v_1, t_1), (v_2, t_2), \ldots, (v_T, t_T)\}$ (here we omit the subscript u for items and timestamps to avoid redundancy), which is used to predict the next clicked item at the given time t_{T+1}.

Item Embedding. We first transform the input items into embedding vectors with an embedding matrix $\mathbf{H}^V \in \mathbb{R}^{n \times d}$ by look-up operation, resulting in $\mathbf{q}_i \in \mathbb{R}^d$ for item v_i. As previous approaches shown in [12,18], absolute positions $i \in T$ in the interaction sequence is crucial for the prediction, while self-attention mechanism is not aware of the order of the inputs. A learnable positional embedding $\mathbf{p}_i \in \mathbb{R}^d$ is added as the state representation of i-th interaction:

$$\mathbf{x}_i = \mathbf{q}_i + \mathbf{p}_i. \tag{3}$$

We transform the input items into a sequence of embeddings $\boldsymbol{X} = (\mathbf{x}_i)_{T \times d}$.

Timestamp Embedding. Inspired by [12], we model the relative time intervals between interactions in user sequences to learn the information behind timestamps. For example, some users tend to frequently visit the shopping website to buy foods, while others may just buy clothes seasonally. These two kinds of people have distinctly different time patterns in their interaction sequences. Hence, we seek to embed the time intervals of two items within each sequence.

For two interactions (v_i^u, t_i^u) and (v_j^u, t_j^u) of user u, the absolute time interval is $t_i - t_j$. To accommodate different scenarios, we scale the intervals by dividing the minimal positive time interval in the dataset, which is denoted as $d_{\min} = \min\{t_i^u - t_j^u | u \in U, t_i^u - t_j^u > 0\}$. Previous work [12] clips the maximum relative interval to a specified upper bound to avoid sparsity encodings. But there are two drawbacks of such a clipping operation. First, it drops long-term intervals which plays an important role in user modeling. Second, it still suffers from sparsity in some shorter intervals. To address them, we bucketize the intervals through a log function to adaptively discretize the time gaps. For user u, the relative time interval between interactions i and j ($1 \le j \le i \le T$) is represented as:

$$d_{ij} = \left\lfloor \log_b \left(\frac{t_i - t_j}{d_{\min}} + 1 \right) \right\rfloor, \tag{4}$$

where b is a hyper-parameter as the base to scale the number of interval buckets. Each interval d_{ij} is also applied with a look-up operation by a learnable embedding matrix $\mathbf{H}^T \in \mathbb{R}^{\max(d_{ij}) \times d}$ to obtain the embedding form $\mathbf{t}_{ij}$. The input sequence is transformed into a matrix of time interval embeddings $\boldsymbol{T} = (\mathbf{t}_{ij})_{T \times T \times d}$.

4.2 Time-Aware Encoding Layer

With the embedding of historical sequences, we next present how to capture the mutual influence between each pair of interactions through the information of items and time intervals. Self-attention mechanism has been proved to be indispensable in sequential recommendations [10,12,18], which learns the hidden representations of items for further prediction. However, most existing methods only consider the item positions in the sequence. In order to enable the hidden embeddings jointly capture the dependencies between behaviors and the influence of time decay, we adopt a time-aware encoder $\Phi(\cdot)$ to leverage the representation capacity of self-attention and encode the sequence of item embeddings:

$$(\mathbf{h}_1, \mathbf{h}_2, \cdots, \mathbf{h}_T) = \Phi(\mathbf{X}, \mathbf{T}). \tag{5}$$

Time-Aware Self-attention Layer. Self-attention module takes the item embedding matrix $\mathbf{X}$ and the processed temporal intervals matrix $\mathbf{T}$ as input. We utilize the self-attention unit to compute the hidden representation $(\mathbf{h}_1, \mathbf{h}_2, \cdots, \mathbf{h}_T)$ considering the temporal information. In particular, each item embedding $\mathbf{x}_i$ is generated by a weighted sum of linearly embeddings, which includes its own and preceding interactions' embeddings:

$$\mathbf{z}_i = \sum\nolimits_{j=1}^{i} \alpha_{ij} \left(\mathbf{x}_j \mathbf{W}^V + \mathbf{t}_j \mathbf{W}^T\right), \tag{6}$$

where $\mathbf{W}^V, \mathbf{W}^T \in \mathbb{R}^{d \times d}$ are the projection matrices, and α_{ij} is the attention coefficient and is computed as Equ. 7:

$$\alpha_{ij} = softmax\,(e_{ij}) = \frac{\exp(e_{ij})}{\sum_{k=1}^{T} \exp(e_{ik})}. \tag{7}$$

The coefficient e_{ij} is calculated by the item embeddings and the time interval embeddings:

$$e_{ij} = \frac{\mathbf{x}_i \mathbf{Q}_1^V (\mathbf{x}_j \mathbf{K}_1^V + \mathbf{t}_{ij} \mathbf{K}_1^T)^\top}{\sqrt{d}}, \tag{8}$$

where the weight matrix $\mathbf{Q}_1^V, \mathbf{K}_1^V, \mathbf{K}_1^T \in \mathbb{R}^{d \times d}$ are used to project the embeddings into query and key vectors, respectively. $\sqrt{d}$ is a scale factor. The target item is regarded as the query and interacted with the keys which consist of historical items as well as the time intervals. In practice, we apply multiple heads following previous work [21].

Position-Wise Feed-Forward Network. After the time-aware self attention layer which fuses the item and time information along with their dependencies with previous interactions, we leverage a feed-forward network to learn the high-level representations of all the sequences, as shown in Eq. 9:

$$FFN(\mathbf{z}_i) = Softplus(\mathbf{z}_i \mathbf{W}_1 + \mathbf{b}_1) \mathbf{W}_2 + \mathbf{b}_2, \tag{9}$$

where $\mathbf{W}_1, \mathbf{W}_2 \in \mathbb{R}^{d \times d}$ and $\mathbf{b}_1, \mathbf{b}_2 \in \mathbb{R}^d$ are learnable parameters shared among all interactions. As most of self-attention based approaches [10,18], we adopt the residual dropout [21] at the end of feed-forward network to avoid over-fitting to get the hidden representation as the outputs of the encoder:

$$\mathbf{h}_i = \mathbf{z}_i + Dropout\,(FFN(LayerNorm(\mathbf{z}_i))). \tag{10}$$

4.3 Hawkes Process Based Decoding Layer

Since the sequential recommendation problem aims to predict whether the user will click on a given target item, we have to decode the learned hidden representation of historical interactions to calculate the click possibility with the target item at target time. Most existing sequential recommendation methods tend to directly calculate the similarity between the hidden user state and the embedding of target items, ignoring the specific interest of the user in the target item (i.e., the hidden state of the user should consider dependencies across clicked items together with the target item) and the influence of predicting time (i.e., the model is expected to recommend items based more on users' long-term preference if the time interval is long). In order to extract useful information from hidden states and model their relations with target interaction (v_{T+1}, t_{T+1}), the decoder needs to meet three principles. First, it should be able to decode the dependency relationship between the clicked iterm and the target one. Second, it is necessary to distinguish the influence of historical behavior of different time intervals on prediction time. Third, the interpretability of prediction results should be supported by the decoder. Inspired by the temporal point process, we incorporate a Hawkes process-based decoder to address the above concentrations.

Note that in self-exciting Hawkes process, the definition of conditional intensity function is given as Eq. 11, which models the influence between types of events for target item $v_i \in \mathcal{V}$ at position i:

$$\lambda_{v_i}^*(i) = \mu + \sum_{j=1}^{i-1} \phi_{v_i,v_j}(t_i - t_j). \tag{11}$$

Previous approaches [11,22] parameterize the endogenous components $\phi_{u,v}(i)$ as $\alpha_{u,v}\kappa(i)$. However, such predefined kernels of $\phi(\cdot)$ and $\kappa(\cdot)$ highly rely on prior domain knowledge. Moreover, the parametric form models the influence of event types and timestamps separately, which leads to poor expressiveness for complex relations among behavior sequences. To avoid drawbacks suffered by existing methods, we propose a non-parametric Hawkes process to model the intensity.

For the given item v_i, we also estimate the intensity function at time t_i as a weighted sum to summarizes the influence of all previous interactions:

$$\lambda_{v_i,t_i} = \sum_{j=1}^{i-1} \beta_{ij} \mathbf{h}_j(\mathbf{x}_i)^\top, \tag{12}$$

where the coefficient β_{ij} is calculated with another attention network:

$$\beta_{ij} = \sigma\left(\frac{\mathbf{x}_i \mathbf{Q}_2^V (\mathbf{x}_j \mathbf{K}_2^V + \mathbf{t}_{ij} \mathbf{K}_2^T)^\top}{\sqrt{d}}\right), \tag{13}$$

where σ is the sigmoid function and $\mathbf{Q}_2^V, \mathbf{K}_2^V, \mathbf{K}_2^T \in \mathbb{R}^{d \times d}$ are weight matrices. Instead of using softmax function like Eq. 8 for normalization, we relax the unnormalized attention weight β_{ij} since the intensity should not be limited by numerical scale of historical behaviors. It extends the vanilla self-attention to the "self-exciting" Hawkes process representation in a non-parametric way.

The hidden states $\mathbf{h}_j$ and the influence efficiency β_{ij} work collaboratively to learn the time intervals within previous observed user interactions without predefined time decay effect function. The non-parametric schema models the mutual excitation of different items by the context-aware attention score which extracts information related to the target item. Therefore, the output intensity λ_{v_i,t_i} captures abundant information of item relationship and temporal interval by incorporating Hawkes process into sequential recommendation.

4.4 Model Optimization

Based on our HawRec, the user's sequence $\mathcal{S} = \{(v_1,t_1),(v_2,t_2),\cdots,(v_T,t_T)\}$ is converted into the corresponding intensity with the target item. Following most paradigm of sequential recommenders [10,12], the expected output at timestamp i should be v_i given the observed history $\mathcal{H}_i = \{(v_1,t_1),(v_2,t_2),\cdots,(v_{i-1},t_{i-1})\}$, which is estimated based on the calculated intensity function.

The true value of intensity λ_{v_i,t_i} is hard to optimize since we have no explicit labels about user preference towards items. Instead, we estimate the probability with a sigmoid function $P(v_i|\mathcal{H}_i,t_i) = \sigma\left(\lambda_{v_i,t_i}\right)$. We adopt negative sampling for each positive item v_i and generate a negative sequence $\{v_1',v_2',\cdots,v_T'\}$. Such negative strategy can be regarded as a data augmentation to train the model on the whole sequence, while does not bring additional computation cost during inference stage. To optimize the model parameters, we choose Bayesian Personalized Ranking (BPR) loss [16] as our object, which is formulated as:

$$\mathcal{L} = \sum_{u \in \mathcal{U}} \sum_{i=1}^{T} \log \sigma\left(\lambda_{v_i,v_i^t} - \lambda_{v_i',v_i^t}\right) + \xi\|\Theta\|_2^2, \qquad (14)$$

which indicates that the recommendation system ranks the positive item v_i before the negative item v_i'. In practice, we add the BPR loss with a L2 regularization term to be the final objective, where Θ includes all the learnable weights and ξ is a hyper-parameter.

5 Experiments

In this section, we will apply HawRec on several datasets and evaluate the performance of HawRec comprehensively.

5.1 Setup

Datasets. In the experiments, we use four widely used datasets for recommendation, which are all publicly available:

- **Amazon Beauty**[1]. Amazon datasets contain a series of datasets comprising large corpora of product reviews from Amazon platform [15]. Among them we consider the category "beauty" which contains 52374 users and 121291 items after pre-processing.

---

[1] http://jmcauley.ucsd.edu/data/amazon/.

- **MovieLens-1M and MovieLens-20M**[2]. MovieLens (ML) is a widely used recommendation benchmark. We adopt two versions, ML-1M which has 6040 users and 3416 items, and ML-20M which has 138493 users and 18345 items.
- **Steam**[3]. Steam dataset was crawled from Steam website, which is a large online video game distribution platform . It was introduced by [10] and contains 281645 users and 13045 items.

Following the common pre-process methods in sequential recommendation [10,17,18], we treat all the user-item rating records as implicit feedbacks and group them by users to construct the interaction sequences. We filter out users and items with fewer than 5 records. For each user sequence, we sort the records in the order of timestamps and take the last (i.e., most recent) interaction for testing, the penultimate one for validation and the remaining for training.

Compared Methods. To evaluate the performance of our model, we compared it with several baselines including classical and state-of-the-art recommenders:

- **POP**: It is the simplest baseline that rank all items based on their popularity
- **BPR-MF** [16]: Bayesian Personalized Ranking is a classic matrix factorization (MF) based method for recommendation without sequential patterns
- **FPMC** [17]: FPMC introduces the first-order Markov Chains to MF method to capture users' long-term preferences
- **GRU4Rec+** [8]: GRU4Rec [9] models user interaction sequences for session-based recommendation with Gated Recurrent Unit. GRU4Rec+ is an improved version that adopts a different loss function and sampling strategy
- **Caser** [20]: Caser capture high-order Markov chains of item sequence by treating items as "images" and using CNN to model them
- **SASRec** [10]: It is a sequential recommendation method that model user behavior sequences with a left-to-right Transformer language model
- **BERT4Rec** [18]: BERT4Rec adopts self-supervised learning with bidirectional self-attention mechanism to train the recommendation model
- **TiSASRec** [12]: It introduces temporal information into SASRec and consider the time intervals between user behaviors
- **CL4SRec** [23]: It fuses contrastive self-supervised learning with a Transformer-based recommendation model to capture relationship among items.

Evaluation Metrics. To evaluate and compare the performance of the approaches described above, we use two widely used top-n measures: Hit Ratio (HR) and Normalized Discounted Cumulative Gain (NDCG). Following [18], we randomly sample 100 negative items for each user u according to popularity and rank them with the ground-truth one. HR@k counts the rates of the ground-truth items among the top k items based on the ranking results. NDCG@k assigns higher weights to higher position. We report HR and NDCG with $k = 5, 10$.

---

[2] https://grouplens.org/datasets/movielens/.
[3] https://cseweb.ucsd.edu/~jmcauley/datasets.html.

Table 1. Recommendation performance comparison of different methods. The bold scores mark the best, while underlined scores represent the second.

Datasets	Metric	POP	BPR-MF	FPMC	GRURec+	Caser	SASRec	CL4SRec	BERT4Rec	TiSASRec	HawRec	Improv.
Beauty	N@5	0.0241	0.0803	0.0921	0.1186	0.1054	0.1439	_0.1620_	0.1585	0.1411	**0.1687**	+4.1%
	H@5	0.0396	0.1209	0.1372	0.1791	0.1613	0.1929	**0.2207**	0.2107	0.1932	_0.2135_	-3.3%
	N@10	0.0337	0.1102	0.1215	0.1449	0.1361	0.1637	_0.1865_	0.1810	0.1761	**0.1943**	+4.2%
	H@10	0.0765	0.1997	0.2415	0.2658	0.2593	0.2656	**0.3043**	0.3029	0.2884	_0.2951_	-3.0%
Steam	N@5	0.0477	0.0744	0.0945	0.1613	0.1131	0.1727	_0.3298_	0.1842	0.3257	**0.3476**	+5.4%
	H@5	0.0805	0.1177	0.1517	0.2391	0.176	0.2559	_0.4258_	0.2710	0.4203	**0.4337**	+1.9%
	N@10	0.0665	0.1005	0.1026	0.1802	0.1484	0.2147	0.3541	0.2261	_0.3572_	**0.3972**	+6.1%
	H@10	0.1389	0.1993	0.2551	0.3594	0.2870	0.3783	0.5194	0.4013	_0.5271_	**0.5422**	+2.8%
ML-1m	N@5	0.0416	0.1903	0.2885	0.3705	0.3832	0.3980	_0.4620_	0.4454	0.4331	**0.4723**	+2.2%
	H@5	0.0715	0.2866	0.4297	0.5103	0.5353	0.5434	_0.5879_	0.5836	0.5725	**0.6187**	+5.2%
	N@10	0.0621	0.2365	0.3439	0.4064	0.4268	0.4368	_0.4930_	0.4818	0.4681	**0.5093**	+3.3%
	H@10	0.1358	0.4301	0.5946	0.6351	0.6692	0.6629	_0.7142_	0.6970	0.7003	**0.7164**	+0.3%
ML-20m	N@5	0.0511	0.1332	0.2239	0.3630	0.2538	0.4208	0.5162	0.4967	_0.5193_	**0.5491**	+5.7%
	H@5	0.0805	0.2128	0.3601	0.5118	0.3804	0.5727	_0.6631_	0.6323	0.6609	**0.6853**	+3.4%
	N@10	0.0695	0.1786	0.2895	0.4087	0.3062	0.4665	_0.5721_	0.5340	0.5480	**0.5804**	+1.4%
	H@10	0.1378	0.3538	0.5201	0.6524	0.5427	0.7136	_0.8062_	0.7473	0.7713	**0.8053**	+0.1%

Table 2. Ablation analysis in ML-1M and Beauty.

Variants	ML-1M		Beauty	
	HR@10	NDCG@10	HR@10	NDCG@10
HawRec-T	0.6741	0.4501	0.2759	0.1708
HawRec-H	0.6994	0.4827	0.2843	0.1795
HawRec-N	0.6430	0.4436	0.2605	0.1620
HawRec	**0.7164**	**0.5093**	**0.2951**	**0.1943**

Implementation Details. We implement HawRec with PyTorch and select hyper-parameters on validation data. All parameters, including weights of weight matrices, shared items, and temporal embeddings, are initialized by default and optimized via Adam with weight decay, of which $\beta_1 = 0.9$ and $\beta_1 = 0.999$. The learning rate is 0.0001 for all datasets while the dropout rate is 0.5. We set the regularization weight as 0.0001. The batch size is set as 128. We set the maximum sequence length $T = 100$ for two MovieLens datasets, $T = 50$ for Beauty and Steam. The embedding size is set as 40 with 2 attention heads for all the datasets. As for all baselines, we refer to the hyperparameters settings mentioned in the papers and fine-tune on different datasets.

5.2 Performance Comparison

Experiment results and comparisons of HawRec and other comparative baselines are demonstrated in Table 1. As we can see, Markov chain-based methods (i.e. Caser) outperforms non-sequential models that verifies the sequential information is beneficial. Self-attention based method SASRec distinctly performs better than GRURec+, suggesting that the powerful of such mechanism to model long-term sequential information within user behaviors. Furthermore, BERT4Rec [18] replaces the next-item prediction with a Cloze task to provide additional training signals to enhance representations. The additional time intervals information introduced in TiSASRec [12] further improve the performance. CL4SRec [23] achieve the best performance among baselines in different scenarios via contrastive self-supervised learning and sequence enhancement on an individual user level.

According to the results, our model consistently outperforms all the baselines in four datasets with most of the metrics. Concretely, it gains up to 6.1% improvements at NDGC@10 and up to 4.0% at HR@10 against the strongest baseline. HawRec not only takes advantage of the self-attention mechanism involving the temporal information, but also considers mutual excitation with time influence between items for prediction to achieve better performance. However, the remarkable results in CL4SRec also leaves us with the question of how to combine self-supervised comparative learning with point process, which is to be explored in future work.

6   Model Discussions

6.1   Ablation Study

In order to further validate the effectiveness of different units in our model and better understand their impacts, we design a series of ablation studies by comparing the performance of models without some components.

1. HawRec-T: Removing the temporal information in two attention layers. To be specific, we remove the key weights $\mathbf{K}_1^T$ and $\mathbf{K}_2^T$ in Eq. 8 and Eq. 13.
2. HawRec-H: Removing the Hawkes process-based decoder. We take inner product of the hidden and the target embeddings as the preference score.
3. HawRec-N: Replacing the non-parametric Hawkes process in Eq. 13 with an exponential time-decay kernel $\beta_{ij} = \exp\left(-\delta \times d_{ij}\right)$

We compare these variants on ML-1M and Beauty dataset. As we can see from Table 2, the temporal information plays a key role in the recommendation and the model performance would degrade significantly without it. The results of HawRec-H conform to the idea of our proposed method, that is, the temporal point process really helps the model to capture the tiem-dependent relationship among user behaviors. Moreover, the HawRec-N's frustrating performance demonstrates the failure of humanly parameterized time decay effect kernel.

6.2   Robustness

To validate the robustness of our proposed model, we corrupt the input user behavior sequences by adding synthetic noises. Specifically, we randomly remove several interactions in the sequence or replace some items and timestamps in the sequence with uniformly sampled items. We conduct the experiment on Beauty compared with the simplified version HawRec-H. The results reported in Fig. 3 shows that the recommendation performance of HawRec-H drops faster than our proposed model. The recommendation without Hawkes process-based decoder is more sensitive to the noises in the sequence while our method have the potential to ease the degradation and improve the robustness.

486   S. Yu et al.

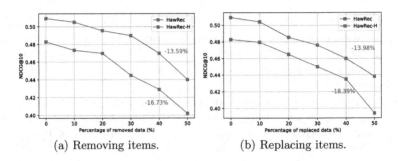

(a) Removing items.   (b) Replacing items.

Fig. 3. Performance comparison w.r.t. synthetic noises in sequences. The decline percentage (the colored numbers) of HawRec-H is greater than HawRec.

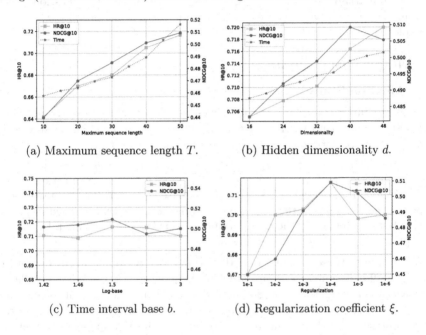

(a) Maximum sequence length T.   (b) Hidden dimensionality d.

(c) Time interval base b.   (d) Regularization coefficient ξ.

Fig. 4. Performance w.r.t. different hyper-parameters.

6.3 Hyper-parameter Sensitivity

In this section, we evaluate the model performance by studying the sensitivity to hyper-parameters. In Fig. 4, we present the metrics HR@10 and NDCG@10 with different parameter settings on dataset ML-1M.

We scale the maximum length T from 10 to 50 and find that the performance keeps increasing with longer sequences. We select the maximum length for each dataset as a trade off between the training time and the performance. We also show the performances when the range of hidden dimension is from 16 to 48 and find that the larger dimension leads to better prediction. As for the log-base for time intervals and the regularization coefficient, the model performances both increase first and then decrease, which demonstrates that a well-chosen hyperparameter will benefit the model remarkably.

6.4   Time Complexity and Scalability

We also discuss the computation cost of our proposed method w.r.t sequence length T and dimensionality d. In every iteration of the training phase, the time complexity is $O(T^2d + dT^2)$ given by the attention networks [19]. Here we omit the time cost during pre-process of relative time interval matrix d_{ij} which is $O(T^2)$. The overall complexity is dominated by the term $O(T^2d)$ and it is linear to traditional next item prediction model like SASRec. Moreover, as we can see from Fig. 4, the training time for each iteration increases in an approximately linear manner. We assume that the empirical time spending results from the parallel optimization of GPU.

7   Conclusion

In this paper, we propose HawRec (Hawkes process based sequential recommendation), which uses the self-exciting characteristics of temporal point process to predict the future interaction. This model incorporates time-dependent influence between user behaviors into the next item prediction with a non-parametric intensity kernel to fully take advantage of the theoretical foundation of Hawkes process and the expression power of neural networks. Extensive experiments on four datasets demonstrate the outstanding performance of our approach and the effectiveness of temporal influence on recommendation. For further study, we will focus on exploit the potentialities of temporal point process on more complicated areas in real-world to facilitate the temporal dependencies modeling.

References

1. Bahdanau, D., Cho, K., Bengio, Y.: Neural machine translation by jointly learning to align and translate. In: Bengio, Y., LeCun, Y. (eds.) ICLR (2015)
2. Choi, E., Bahadori, M.T., Sun, J., Kulas, J., Schuetz, A., Stewart, W.F.: RETAIN: an interpretable predictive model for healthcare using reverse time attention mechanism. In: NeurIPS, pp. 3504–3512 (2016)
3. Daley, D., Vere-Jones, D.: An Introduction to the Theory of Point Processes: Volume II: General Theory and Structure. Springer Science & Business Media (2007)
4. Du, N., Dai, H., Trivedi, R., Upadhyay, U., Gomez-Rodriguez, M., Song, L.: Recurrent marked temporal point processes: Embedding event history to vector. In: KDD, pp. 1555–1564 (2016)
5. Fan, Z., et al.: Sequential recommendation via stochastic self-attention. In: Proceedings of the ACM Web Conference 2022, pp. 2036–2047 (2022)
6. Feng, S., Li, X., Zeng, Y., Cong, G., Chee, Y.M., Yuan, Q.: Personalized ranking metric embedding for next new POI recommendation. In: IJCAI, pp. 2069–2075 (2015)

7. Hawkes, A.G.: Spectra of some self-exciting and mutually exciting point processes. Biometrika **58**(1), 83–90 (1971)
8. Hidasi, B., Karatzoglou, A.: Recurrent neural networks with top-k gains for session-based recommendations. In: CIKM, pp. 843–852 (2018)
9. Hidasi, B., Karatzoglou, A., Baltrunas, L., Tikk, D.: Session-based recommendations with recurrent neural networks. In: ICLR (2016)
10. Kang, W., McAuley, J.: Self-attentive sequential recommendation. In: ICDM, pp. 197–206 (2018)
11. Ke, Z., Hongyuan, Z., Le, S.: Learning social infectivity in sparse low-rank networks using multi-dimensional hawkes processes. In: Artificial Intelligence and Statistics, pp. 641–649. PMLR (2013)
12. Li, J., Wang, Y., McAuley, J.J.: Time interval aware self-attention for sequential recommendation. In: WSDM, pp. 322–330 (2020)
13. Liu, Z., et al.: Swin transformer: Hierarchical vision transformer using shifted windows. arXiv preprint arXiv:2103.14030 (2021)
14. Ma, J., Zhou, C., Yang, H., Cui, P., Wang, X., Zhu, W.: Disentangled self-supervision in sequential recommenders. In: KDD, pp. 483–491 (2020)
15. McAuley, J., Targett, C., Shi, Q., Van Den Hengel, A.: Image-based recommendations on styles and substitutes. In: SIGIR, pp. 43–52 (2015)
16. Rendle, S., Freudenthaler, C., Gantner, Z., Schmidt-Thieme, L.: BPR: bayesian personalized ranking from implicit feedback. In: UAI, pp. 452–461 (2009)
17. Rendle, S., Freudenthaler, C., Schmidt-Thieme, L.: Factorizing personalized markov chains for next-basket recommendation. In: WWW (2010)
18. Sun, F., et al.: Bert4rec: Sequential recommendation with bidirectional encoder representations from transformer. In: CIKM, pp. 1441–1450 (2019)
19. Tang, G., Müller, M., Rios, A., Sennrich, R.: Why self-attention? a targeted evaluation of neural machine translation architectures. arXiv preprint arXiv:1808.08946 (2018)
20. Tang, J., Wang, K.: Personalized top-n sequential recommendation via convolutional sequence embedding. In: WSDM, pp. 565–573 (2018)
21. Vaswani, A., et al.: Attention is all you need. In: NeurIPS, pp. 5998–6008 (2017)
22. Wang, D., Zhang, X., Xiang, Z., Yu, D., Xu, G., Deng, S.: Sequential recommendation based on multivariate hawkes process embedding with attention. IEEE Trans. Cybern. (2021)
23. Xie, X., et al.: Contrastive learning for sequential recommendation. In: 2022 IEEE 38th International Conference on Data Engineering (ICDE), pp. 1259–1273. IEEE (2022)
24. Zhang, Q., Lipani, A., Kirnap, O., Yilmaz, E.: Self-attentive hawkes process. In: International Conference on Machine Learning, pp. 11183–11193. PMLR (2020)
25. Zhao, J., Zhao, P., Zhao, L., Liu, Y., Sheng, V.S., Zhou, X.: Variational self-attention network for sequential recommendation. In: 2021 IEEE 37th International Conference on Data Engineering (ICDE), pp. 1559–1570. IEEE (2021)
26. Zuo, S., Jiang, H., Li, Z., Zhao, T., Zha, H.: Transformer hawkes process. In: International Conference On Machine Learning, pp. 11692–11702. PMLR (2020)

Modeling High-Order Relation to Explore User Intent with Parallel Collaboration Views

Xiangping Zheng[1], Xun Liang[1(✉)], Bo Wu[1], Jun Wang[2], Yuhui Guo[1], Sensen Zhang[1], and Yuefeng Ma[3]

[1] School of Information, Renmin University of China, Beijing, China
{xpzheng,xliang,wubochn}@ruc.edu.cn
[2] Swinburne University of Technology, Melbourne, Australia
junwang@swin.edu.au
[3] Qufu Normal University, ShangDong, China

Abstract. As an emerging paradigm, session-based recommendation (SBR) aims to predict the next item by exploiting user behaviors within a short yet anonymous session. Existing works focus on how to effectively model the information based on graph neural networks, which may be insufficient to capture the high-order relation for short-term interest. To this end, we propose a novel framework, named PacoHGNN, which models high-order relations based on **H**yper**G**raph **N**eural **N**etwork with **Pa**rallel **Co**llaboration views. Specifically, PacoHGNN learns two embedding views for the SBR task, respectively: (i) item-internal graph view, which is to learn the item embedding by modeling pairwise item connectivities among corresponding items; and (ii) session-external hypergraph view, which targets session embedding by learning beyond pairwise information from high-order relations across all sessions. These two types of graph modeling with data-driven can provide complementary information for each other while exhibiting collaboration to some degree. Additionally, we further propose Hyperedge-to-Node (H2N) to enhance supervised signals against the data sparsity problem for better graph representation. Extensive experiments on multiple real-world datasets demonstrate the superiority of the proposed model over state-of-the-art methods.

Keywords: Graph neural networks · High-order relations · Session-based recommendation

1 Introduction

With the ever-growing volume of online information, massive products, content, and services (uniformly described as items) are emerging every day [4,22,25,26,30]. Recommender systems (RS) have been widely adopted in various business applications (e.g., e-commerce services) to alleviate information

X. Wang et al. (Eds.): DASFAA 2023, LNCS 13944, pp. 489–504, 2023.
https://doi.org/10.1007/978-3-031-30672-3_33

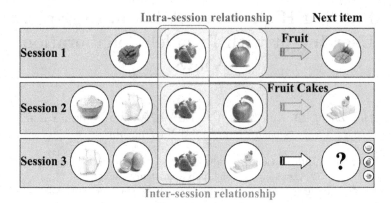

Fig. 1. An example of hidden user intents is revealed by groups of items. Three sessions are given, "Fruit" and "Fruit Cake" are user intents revealed by grouping different consecutive items in the session. Items on the right are possible items the user will click next under each intent.

overload, improve user experience, and boost business revenue. Most conventional recommender systems generally provide personalized recommendations based on user profiles, and long-term historical interactions are available as the basis of recommendation [8,15]. However, users' identity information is often unavailable in many real-world scenarios, e.g., unregistered users or those who are reluctant to log in for privacy concerns [11]. In these scenarios, only the interactions during an ongoing session are available for recommendation [8]. Given the limited interactions users engage with the portals, it is often hard for the systems to fully understand the users' behaviors, posing significant challenges to recommendation systems [6]. Thus there is an urgent practical need to provide recommendations with limited information [15].

Session-based recommendation (SBR) has emerged in time, encapsulating a range of consecutive user-item interactions as sessions for predicting the next items, gaining better performance [7]. Early efforts in this field mainly focused on discovering item-to-item relationships, such as co-occurrence relations [17,18]. Afterward, recurrent neural networks (RNNs) [9] exhibited a considerable advantage in modeling sequential data. However, these RNNs-based models overemphasize the order of items. Actually, there may be no such strict chronological order among user behaviors. For example, as shown in Fig. 1, if a user intends to buy ingredients for a cake, whether to purchase flour first or eggs later would not affect user preference. Instead, strictly and solely modeling the relative orders of items would be prone to overfitting, leading to suboptimal performance [25]. More recently, Graph Neural Networks (GNNs) [23,32,33] have been applied to capture complex item transitions by constructing sessions into graphs, which have effectively represented both item consistency and sequential dependency. However, due to the absence of long-term user behavioral information and very limited short-term interactions, GNNs-based models are more undermined by the problem of data sparsity. Additionally, sessions are normally of a small number of interactions. Learning from such short interaction sequences independently may be inherently inadequate to unveil users' real intents. We should attempt to explore collaborative information from other sessions for mining their high-order relations. In view of this, *a critical issue is how to model the high-order relations to unveil users' real intents*.

Individual items can reveal users' intent but only provide limited evidence. Consider the example in Fig. 1, a strawberry in different sessions can be viewed with different intent, i.e., as an option for fruits in session 1, or as part of the ingredients for fruit cake in session 2. In a sense, the meaning of an item (and what it reveals about users' intent) could be inferred from contextual sessions, each containing a set of consecutive items showing up together within a session. Hence, we suggest modeling session-wise representations that can robustly capture users' intent with only limited evidence available in short sessions. However, how to build a relationship beyond pairwise connections in a session? Obviously, simple graphs cannot depict such set-like relations since we need to consider connecting various numbers of items ranging from two to many. For instance, for session 1 in Fig. 1, the pairwise linkage between strawberry and apple is not enough to reveal that the user intends to buy various fruits. But such evidence could be inferred by analyzing the triadic relations among strawberries, apples, and pitaya as defined. To this end, we adopt the hypergraph [21,28,34] structure to model the correlations amongst items within each session. Hypergraph perfectly fits our assumption as hyperedge is set-like, emphasizing the involved items' coherence rather than relative orders. Then how to associate among sessions to mine the high-order relations by hypergraph. Figure 1 describes intra-session and inter-session relationships; some items (e.g., apples and strawberries) occur within a session and share their neighboring items for multiple sessions, implying potential item correlations. Technically, we first model each session as a hyperedge in which all the items in a session are connected. And different hyperedges are connected via shared items, constituting high-order correlations based on hypergraphs. We can borrow the strengths of the hypergraph convolution network to generate high-quality recommendation results.

Based on the analysis above, we propose a novel framework, named Paco-HGNN, which models high-order relations based on **HyperG**raph **N**eural **N**etwork with **Pa**rallel **Co**llaboration views. We exploit item transitions over sessions in a more subtle manner by modeling session-based data as a hypergraph for better inferring the user preference of the current session. Intuitively, PacoHGNN learns two embedding views for the SBR task, respectively: (i) item-internal graph view, which is to learn the item embedding by modeling pairwise item connectivities among corresponding items; and (ii) session-external hypergraph view, which targets session embedding by learning beyond pairwise information from high-order relations across all sessions. Moreover, SBR suffers more from the problem of data sparsity due to the very limited short-term interactions. Although Self-supervised learning [3,34] holds vast potential to tackle this problem. However, existing self-supervised recommendation models mainly rely on item dropout to augment data, which is not fit for SBR because the dropout leads to sparser data, creating unserviceable self-supervision signals. To circumvent this crucial issue, we further propose Hyperedge-to-Node (H2N) to enhance supervised signals against data sparsity for better graph representation. The main contributions of this work are summarized as follows:

- We propose a novel framework with parallel collaboration views, PacoHGNN, by modeling session-based data as a hypergraph, which can capture the beyond pairwise relations among items for better inferring the user preference of the current session.
- We further propose Hyperedge-to-Node (H2N) to enhance supervised signals against data sparsity problem for better graph representation.
- Extensive experiments show that the proposed framework is superior to the state-of-the-art baselines and achieves statistically significant improvements on benchmark datasets. datasets.

2   Related Work

In this section, we review three main topics of previous research: the session-based recommendation, hypergraph learning and self-supervised learning.

Session-Based Recommendation. Initial exploration of SBR mainly focuses on sequence modeling, where nearest neighbors [2,18] and Markov decision process [17] are the preferred technique at this phase. The boom of deep learning methods brings significant performance gain on SBR [9,12,14], such as GRU4Rec [9], NARM [12], STAMP [14]. Recently, GNNs adopt graph structure to model item-transition patterns, relaxing the strict order assumption over items by RNNs [4]. To capture the complex transitions over items within the entire session, SR-GNN [24] was perhaps the first to consider GNN for SBR. Other models [22,25,27] improved the performance by considering different aspects of GNN, such as SR-GNN [27], GCE-GNN [22]. Although significant progress has been achieved, previous methods have not considered users' multiple intents in learning session embeddings. They all fail to capture the complex and higher-order item correlations. Our proposed model employs graphs and hypergraphs to explore the higher-order dependency of sessions and items.

Hypergraph Learning. Recently, constructing hypergraphs to learn the data structure has become a popular approach. A hypergraph is a generalization of a simple graph in which a hyperedge can connect more than two nodes. Hypergraph provides a natural way to model complex data structures with high-order relations and has been extensively employed to tackle various problems [29]. Due to this flexibility of hypergraphs, some studies have recently tried combining the hypergraph with the recommender systems to improve their performances. Bu et al. [1] introduce hypergraph learning to music recommender systems, which is the earliest attempt. Li et al. [13] propose a novel architecture named hyperbolic hypergraph representation learning method for sequential recommendation (H^2SeqRec) with the pre-training phase. Unlike their works, our proposed model exploits inter-hyperedge information and designs for session-based scenarios.

Self-supervised Learning. Existing graph contrastive learning [3,34] is a class of self-supervised approaches, whose main idea of graph contrastive learning is to treat each sample as a distinct category and learn how to distinguish them [3]. For instance, [10] proposes Sub-con by utilizing the strong correlation between central nodes and their sampled subgraphs to capture regional structure information. [28] developed contrastive learning with augmentations to address the challenge of data heterogeneity in graphs. The theoretical analysis sheds light on the reasons behind their success [34]. Objectives used in these methods can be seen as designing different graph augmentation strategies to enhance the graph representation. As for our work, we combine graph contrastive learning with multi-view hypergraph learning for the recommendation, which gives us clues about applying training to the session-based recommendation.

3 Preliminaries

Notations. Given a set of sessions $\mathcal{S} = \left\{ s^{(1)}, s^{(2)}, \cdots, s^{(P)} \right\}$ over a set of items $\mathcal{I} = \{i_1, i_2, \cdots, i_N\}$, where N is the number of items. An arbitrary session $s \in \mathcal{S}$ is represented as a sequence $s = [i_{s,1}, i_{s,2}, \cdots, i_{s,k}, \cdots, i_{s,m}]$ ordered by timestamps with a length m. Here, $i_{s,k} \in \mathcal{I}$ represents an interacted item of an anonymous user within the session s. For simplicity, we embed each item i into the same space and let $x_i^l \in \mathbb{R}^{d^{(l)}}$ denote the representation of item i of dimension $d^{(l)}$ in the l-th layer of a deep neural network. By stacking P sessions, let $\mathbf{X} \in \mathbb{R}^{P \times d^{(l)}}$ denote a session-item interaction matrix, where P is the number of sessions. Each session s is represented by a vector $\mathbf{s}$.

Problem Formulation. Here we give a formulation of session-based recommendations problem: The goal of personalized session-based recommendation is to predict the next item, namely $i_{s,m+1}$, for any given session s. Specifically, given $\mathcal{I}$ and s, the output of recommendation model is trained to generate a ranked list $y = [y_1, y_2, \cdots, y_N]$ of top-K candidate items as the recommended next items, where y_i corresponds to the score of item i and the top-K items with highest score in y will be selected as the recommendations.

Hypergraph Definition. A hypergraph [25] is defined as $G_h = (V, E, \mathbf{W})$, which includes a vertex set V containing N unique vertices, a hyperedge set E containing M hyperedges. Each hyperedge contains two or more vertices and is assigned with a weight by $\mathbf{W}$, and all the weights formulate a diagonal matrix $\mathbf{W} \in \mathbb{R}^{M \times M}$. The hypergraph can be denoted by an incidence matrix $\mathbf{H} \in \mathbb{R}^{N \times M}$. For each vertex and hyperedge, their degree D_{vv} and $B_{\epsilon\epsilon}$ are respectively defined as $D_{vv} = \sum_{\epsilon=1}^{M} W_{\epsilon\epsilon} h_{v\epsilon}$; $B_{\epsilon\epsilon} = \sum_{v=1}^{N} h_{v\epsilon}$. $\mathbf{D}$ and $\mathbf{B}$ are diagonal matrices.

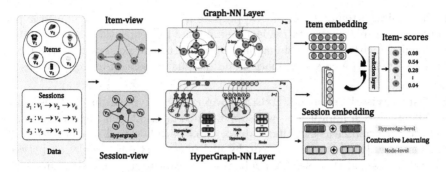

Fig. 2. An overview of the proposed framework. Firstly, PacoHGNN learns two levels of embeddings from the item graph view and session hypergraph view, respectively. Then we further propose Hyperedge-to-Node (H2N) to enhance supervised signals against data sparsity for better graph representation. Finally, we learn the contribution of each item to the next predicted item in the prediction layer.

4 The Proposed Method

We propose a novel framework with parallel collaboration views, PacoHGNN, for the session-based recommendation. The proposed method consists of two critical tasks: one is the main task for the session-based recommendation (SBR), and the other is the self-supervised learning (SSL) task that acts as the auxiliary task to boost the former. Firstly, we detail how to conduct the item embedding and session embedding for the SBR task, then we perform the SSL task to tackle the data sparse issue. Finally, we output the predicted probability of candidate items for recommendation. The motivation of this work is to effectively capture the high-order relations in items and exploit these relations to explore user intent for improving recommendation performance. As shown in Fig. 2, we now detail the proposed model as follows.

4.1 Item-Internal Graph View for Item Embedding

Graph Construction. We first construct a meaningful graph from all sessions for item embedding. Given a session $s = [i_{s,1}, i_{s,2}, \cdots, i_{s,k}, \cdots, i_{s,m}]$, we treat each item $i_{s,k}$ as a node and $(i_{s,k}, i_{s,k+1})$ as an edge that represents a user clicks item $i_{s,k}$ after $i_{s,k+1}$ in the session s. Subsequently, the graph view is educed by aligning all sessions. In other words, any two items ($i_{s,k}$ and $i_{s,k+1}$) which are connected in a session also get connected as nodes in the item view with a weighted directed edge, counting how many times they are adjacent in different sessions. Furthermore, if a node does not contain a self-loop, it will be added with a self-loop with a weight of 1. Based on our observation of our daily life and the datasets, it is common for a user to click two consecutive items a few times within the session.

Graph Neural Network Layer. An item may be involved in multiple sessions, from which we can obtain useful item-transition information to effectively help current predictions. We first convert each item $i \in \mathcal{I}$ into a unified embedding latent space and let $x_i^l \in \mathbb{R}^{d^{(l)}}$ denote the representation of item i of dimension $d^{(l)}$ in the l-th layer of a deep neural network. Then a simplified graph convolution layer for the graph view is defined as:

$$\mathbf{X}_g^{(l+1)} = \sigma(\hat{\mathbf{D}}_g^{-1}\hat{\mathbf{A}}_g\mathbf{X}_g^{(l)}\mathbf{W}_g^{(l)}) \tag{1}$$

where g means in the graph view and $\hat{\mathbf{D}}_{g,j,j} = \sum_{k=1}^m \hat{\mathbf{A}}_{g,j,k}$ is the degree matrix. $\hat{\mathbf{A}}_g = \mathbf{A}_g + \mathbf{I}$, $\hat{\mathbf{A}}_g$ and $\mathbf{I}$ are the adjacency matrix and the identity matrix, respectively. $\mathbf{W}_g^{(l)}$ denotes the parameter matrix and $\mathbf{X}_g^{(l)}$ represents the l-th layer's item embeddings. $\sigma(.)$ is a nonlinear activation function such as Sigmoid or ReLU. After passing L graph convolutional layers, we obtained the final item embeddings $\mathbf{X}_g$. In this way, the item view can capture the item-level connectivity information for better gaining item embeddings for the following process.

4.2 Session-External Hypergraph View for Session Embedding

Hypergraph Construction. To capture the beyond pairwise relations in the session-based recommendation, we adopt a hypergraph $G_h = (V, E, \mathbf{W})$ to denote each session as a hyperedge. For a more rigorous description, we define each item $i_{s,m} \in V$ and each hypergraph as $\epsilon = [i_{s,1}, i_{s,2}, \cdots, i_{s,k}, \cdots, i_{s,m}] \in E$. We transform the session's data structure into hypergraph construction. Compared with traditional graphs that mainly rely on pairwise items as linear sequences, hypergraphs can model higher relations in item interactions. For example, two items $i_{s,k}$ $i_{s,k+1}$ are connected only if a user interacted with item $i_{s,k}$ before $i_{s,k+1}$. Any two items clicked in a session are connected when we transform the session data into a hyperedge of a hypergraph. Besides, if we use a traditional graph, the graph is hard to reveal different item semantics in different sessions directly. Recall back to the previous instance that the strawberries in session 1 and session 2 reveal users' different intentions in Fig. 1, while a hypergraph is beneficial for dealing with the issue.

Hypergraph Neural Network Layer. After the hypergraph construction, we develop a hypergraph neural network to capture both the item-level high-order relations. Figure 2 illustrates the details of the hypergraph neural networks. Multiple hyperedge structure groups are constructed from the complex correlation of the multi-sessions. We concatenate the hyperedge groups to generate the hypergraph adjacent matrix $\mathbf{H}$. The hypergraph adjacent matrix $\mathbf{H}$ and the item embedding are fed into the hypergraph neural network to propagate high-order relations among items. Referring to the spectral hypergraph convolution proposed in [5], we can build a hyperedge convolutional layer in the following formulation:

$$\mathbf{X}_h^{(l+1)} = \sigma(\hat{\mathbf{D}}_h^{-1}\mathbf{H}\mathbf{W}\mathbf{B}^{-1}\mathbf{H}^T\mathbf{X}_h^{(l)}\Theta^{(l)}) \tag{2}$$

where h means in the hypergraph view and $\mathbf{X}_h^{(l)}$ represents the l-th layer's item embeddings. $\sigma(.)$ denotes the nonlinear activation function. $\Theta^{(l)}$ is the learnable filter matrix. Denote that $\hat{\mathbf{D}}_h^{-1}$ play a role of normalization.

Here, we further investigate item embedding in the property of exploiting high-order correlation among data. The hypergraph convolution can be viewed as a two-stage performing '**node-hyperedge-node**' feature transformation, which can better refine the features using the hypergraph structure. More specifically, at first, the initial item embedding $\mathbf{X}_h^{(l)}$ is processed by a learnable filter matrix $\Theta^{(l)}$. Then, the item features are gathered according to the hyperedge to form the hyperedge feature, which is implemented by the multiplication of $\mathbf{H}^T$. Finally, the output node feature is obtained by aggregating their related hyperedge feature, which is achieved by multiplying matrix $\mathbf{H}$. Thus, the hypergraph layer can efficiently propagate the high-order correlation on the hypergraph by the node-edge-node transform. After passing through the L hypergraph convolutional layer, we can get the final session embeddings $\mathbf{X}_h$ by aggregating representations of items contained in that session.

4.3   Enhancing SBR with SSL Task

Compared with other recommendation paradigms, SBR suffers more accessible from the problem of data sparsity due to the short-term interactions. Self-supervised learning [3,34] holds vast potential to tackle this problem. However, existing self-supervised recommendation models mainly rely on item dropout to augment data, which is unsuitable for SBR because the dropout leads to sparser data, creating unserviceable self-supervision signals. To solve this problem, we propose Hyperedge-to-Node (H2N) by employing the principle of deep graph infomax (DGI) [20]. Formally, suppose the features of hyperedge are denoted by $\mathbf{e}^L$ in the hypergraph layer L. Recall the hypergraph convolution, which has a two-stage performing *nodes-hyperedges-nodes* feature transformation. We can obtain hyperedges features in stage 1, while getting hypernode features in stage 2. The information propagation on the hypergraph encoder naturally aggregates information of homogeneous structural neighbors, making it convenient to extract the potential neighbors within sessions or items. In this way, we treat the hyperedge (or hypernode) representations and their corresponding whole hypergraph representations. For hyperedges features, to compute the final hyperedges representation s_e, we follow DGI to average all the hyperedges' representations and then apply a sigmoid activation function on the pooled result, i.e.,

$$\mathbf{s_e} = \sigma \left(\frac{1}{n} \sum_{j=1}^{n} \mathbf{e}_j^{(L+1)} \right) \tag{3}$$

where n is the number of hyperedges. Then the loss of hyperedge-level in H2N can be defined:

$$\mathcal{L}_{he} = -\frac{1}{2n} \sum_{i=1}^{n} \left(\mathbb{E}_g log \mathcal{D} \left(\mathbf{e}_i^{(L)}, \mathbf{s_e} \right) + \mathbb{E}_{\tilde{g}} log \left(1 - \mathcal{D} \left(\tilde{\mathbf{e}}_i^{(L)}, \mathbf{s_e} \right) \right) \right) \tag{4}$$

where $\mathcal{D}$ is a discriminator that outputs the affinity score of each local-global (i.e., hyperedge-hypergraph) pair. The hypergraph $\tilde{\mathcal{G}}$, generated by a row-wise shuffling of the initial feature matrix $\mathbf{e}_i^L$, provides the node representation $\tilde{\mathbf{e}}_i^{(L)}$ that can be paired with the hypergraph representation $\mathbf{s}$ as a negative sample. Likewise, we can get hypernode-level loss $\mathcal{L}_{hn}$. Consequently, we get two-scale self-supervised signals:

$$\mathcal{L}_{H2N} = \mathcal{L}_{he} + \mathcal{L}_{hn} \tag{5}$$

4.4 Prediction

We follow the strategy used in SR-GNN [24] to refine the embedding of session s. Considering the information in these embedding may have different levels of priority, we further adopt the soft-attention mechanism to better represent the embedding item in session. After obtaining the embedding of each session, we compute the score $\hat{z}_i$ for each candidate item $i \in \mathcal{I}$ by doing inner product between the item embedding $\mathbf{X}$. After that, a softmax function is applied to compute the probabilities of each item being the next one in the session: $\hat{\mathbf{y}} = softmax(\hat{\mathbf{z}})$. For each session graph, the loss function is defined as the cross-entropy of the prediction and the ground truth:

$$\mathcal{L}_t = -\sum_{i=1}^{N} \mathbf{y}_i log(\hat{\mathbf{y}}_i) + (1 - \mathbf{y}_i)log(1 - \hat{\mathbf{y}}_i) \tag{6}$$

where $\mathbf{y}$ is the one-hot encoding vector of the ground truth. Finally, we unify the recommendation task and this self-supervised task into a joint learning framework and combine Eq. (5) and Eq. (6) to get the total loss function:

$$\mathcal{L}_{total} = \mathcal{L}_t + \alpha\mathcal{L}_{N2H} \tag{7}$$

where $\alpha \in [0, 1]$ is a hyperparameter that controls the magnitude of the self-supervised task. Noted that, we jointly optimize the two throughout the training.

5 Experiments

5.1 Experimental Protocol

Datesets. We employ five real-world benchmark datasets, namely, *Tmall*[1], *Nowplaying*[2], *RetailRocket*[3], *Diginetica*[4], *Yoochoose*[5], which are often used in session-based recommendation methods. Particularly, *Tmall* dataset comes from IJCAI-15 competition, which contains anonymized user's shopping logs on Tmall

---

[1] https://tianchi.aliyun.com/dataset/dataDetail?dataId=42.
[2] http://dbis-nowplaying.uibk.ac.at/#nowplaying.
[3] https://www.kaggle.com/retailrocket/ecommerce-dataset.
[4] https://competitions.codalab.org/competitions/11161.
[5] http://2015.recsyschallenge.com/challege.html.

Table 1. Statistics of the datasets used in our experiments.

Statistics	Tmall	Nowplaying	RetailRocket	Diginetica	Yoochoose1/64	Yoochoose1/4
# Sessions (Training)	351,268	825,304	433,643	719,470	369,859	5,917,745
# Sessions (Testing)	25,898	89,824	15,132	60,858	55,898	55,898
# Items	40,728	60,417	36,968	43,097	16,766	29,618
Avg. Length of Sessions	6.69	7.42	5.43	5.12	6.16	5.71

online shopping platform. *Nowplaying* dataset comes from [31], which describes the music listening behavior of users. *RetailRocket* is a dataset on a Kaggle contest published by an e-commerce company, which contains the user's browsing activity within six months. *Diginetica* dataset contains sessions of product transaction data from an online retailer and was released as part of the 2016 CIKM Cup. *Yoochoose* dataset is obtained from the RecSys Challenge 2015, which contains a stream of user clicks on an e-commerce website within 6 months.

Data Preprocessing. For fair comparison, following [21,22,24], we conduct preprocessing step over the five datasets. More specifically, sessions of length 1 and items appearing less than 5 times were filtered across all the five datasets. Latest data (such as, the data of last week) is set to be test set and previous data is used as training set [22]. However, because the Yoochoose training set is quite large and training on the recent fractions yields better results than training on the entire fractions as per the experiments of [19]. As in [21,22,24], we sort all of the training sequences generated from Yoochoose, and retrieve the most recent 1/64 and 1/4 to be the training samples in Yoochoose1/64 and Yoochoose1/4. In addition, the training samples and testing samples in all of the five datasets are exactly the same as in [21,22,24,25]. The statistics of the six datasets are presented in Table 1, where dataset Yoochoose1 is sampled into Yoochoose1/64 and Yoochoose1/4.

Baseline Methods. The following models, including the state-of-art and closely related works, are used as representative baselines to evaluate the performance of our model. They are **Item-KNN** [18], **FPMC** [17], **GRU4Rec**[6] [9], **NARM**[7] [12], **STAMP**[8] [14], **SR-GNN**[9] [24], **FGNN**[10] [16], **GCE-GNN** [22], S^2-**DHCN**[11] [25]. More baseline details can be found in their corresponding original papers.

Evaluation Metrics. As recommender systems can only recommend a few items at once, the actual item a user might pick should be amongst the first few items of the list [9]. To keep the same setting as previous baselines, we adopt two widely used ranking based metrics: **P@K** and **MRR@K** by following previous

---

[6] https://github.com/hidasib/GRU4Rec.
[7] https://github.com/lijingsdu/sessionRec_NARM.
[8] https://github.com/uestcnlp/STAMP.
[9] https://github.com/CRIPAC-DIG/SR-GNN.
[10] https://github.com/RuihongQiu/FGNN.
[11] https://github.com/xiaxin1998/DHCN.

Table 2. Comparison of the state-of-art models and all the results are in percentage (%). The best performing method in each column is **boldfaced**, and the second-best method is marked with †.

Methods		Tmall				Nowplaying				RetailRocket			
		P@10	MRR@10	P@20	MRR@20	P@10	MRR@10	P@20	MRR@20	P@10	MRR@10	P@20	MRR@20
Traditional	Item-KNN	6.65	3.11	9.15	3.31	10.96	4.55	15.94	4.91	–	–	–	–
	FPMC	13.10	7.12	16.06	7.32	5.28	2.68	7.36	2.82	25.99	13.38	32.37	13.82
RNNs	GRU4Rec	9.47	5.78	10.93	5.89	6.74	4.40	7.92	4.48	38.35	23.27	44.01	23.67
	NARM	19.17	10.42	23.30	10.70	13.60	6.62	18.59	6.93	42.07	24.88	50.22	24.59
	STAMP	22.63	13.12	26.47	13.36	13.22	6.57	17.66	6.88	42.95	24.61	50.96	25.17
GNNs	SR-GNN	23.41	13.45	27.57	13.72	14.17	7.15	18.87	7.47	43.21	26.07	50.32	26.57
	FGNN	20.67	10.07	25.24	10.39	13.89	6.80	18.78	7.15	42.56	26.24	49.86	25.88
	GCE-GNN	28.01†	15.08†	33.42†	15.42†	16.94	8.03†	22.37	8.40†	–	–	–	–
	S^2-DHCN	26.22	14.60	31.42	15.05	17.35†	7.87	23.50†	8.18	46.15†	26.85†	53.66†	27.30†
	Ours	**32.17**	**17.22**	**38.21**	**18.01**	**18.69**	**8.32**	**24.12**	**8.89**	**47.31**	**28.46**	**54.47**	**38.71**

Table 3. Comparison of the state-of-art models and all the results are in percentage (%). The best performing method in each column is **boldfaced**, and the second-best method is marked with †.

Methods		Diginetica				Yoochoose 1/64				Yoochoose 1/4			
		P@10	MRR@10	P@20	MRR@20	P@10	MRR@10	P@20	MRR@20	P@10	MRR@10	P@20	MRR@20
Traditional	Item-KNN	25.07	10.77	35.75	11.57	–	–	51.60	21.81	–	–	52.31	21.70
	FPMC	15.43	6.20	26.53	6.95	-	-	45.62	15.01	-	-	51.86	17.50
RNNs	GRU4Rec	17.93	7.33	29.45	8.33	50.04	22.64	60.64	22.89	49.68	22.84	59.53	22.60
	NARM	35.44	15.13	49.70	16.17	57.50	27.97	68.32	28.63	57.83	28.10	69.73	29.23
	STAMP	33.98	14.26	45.64	14.32	58.07	28.92	68.74	29.67	59.62	29.24	70.44	30.00
GNNs	SR-GNN	36.86	15.52	50.73	17.59	60.21	30.13	70.57	30.94	61.28†	30.65†	71.36	31.89
	FGNN	37.72	15.95	50.58	16.84	60.97†	30.85	71.12†	31.68	61.25	30.48	71.97†	32.54†
	GCE-GNN	41.16†	18.15†	54.22†	19.04†	59.68	30.95†	70.58	31.12	59.66	30.12	69.28	30.35
	S^2-DHCN	40.21	17.59	53.66	18.51	59.38	29.17	69.87	29.86	60.54	29.05	69.95	29.85
	Ours	**42.28**	**18.91**	**55.03**	**19.67**	**61.90**	**31.49**	**71.85**	31.59†	**62.72**	**31.73**	**72.38**	**32.61**

works [16,22,25]. Specifically, we mainly choose to use top-10 and top-20 items to evaluate a recommender system.

5.2 Overall Comparison

To demonstrate the performance of the proposed model, we compare it with the state-of-the-art item recommendation approaches, illustrated in Table 2 and Table 3. We have the following observations: **1)** Overall, from the two tables, we can see that our proposed model consistently shows strong performance across all datasets, which ascertains our proposed method's effectiveness. **2)** Compared with traditional and RNN methods, our model has achieved remarkable performance. Those demonstrate that our proposed model converts the sequential item transitions into graph-structured data for capturing the inherent order of item-transition patterns. **3)** We also notice that our method achieves more competitive results compared with graph-based baselines. Particularly, it beats other models by a large margin on Tmall, showing the effectiveness of hypergraph modeling. **4)** Considering that S^2-DHCN and our method both have a hypergraph architecture, we think that the improvements mainly derive from the different SSL signals. The main reason is that S^2-DHCN uses the two types of hypergraph

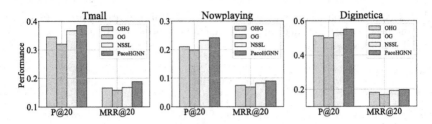

Fig. 3. Ablation Study.

to conduct session embedding, which may not be in the same representation space when two session embeddings, thus creating unsignificant self-supervision signals.

5.3 Ablation Study

In this section, we conduct experiments on three datasets (e.g., we choose Tmall, Nowplaying, and Diginetica as examples) to investigate the contribution of each component in our model. Specially, we design three variant versions of Paco-HGNN: **(1) OHG**: We only use the hypergraph view to conduct item embedding and then session embedding, removing the graph view and SSL task. **(2) OG**: We only use the graph view to encode item embedding and then session embedding, removing the hypergraph view and SSL task. **(3) NSSL**: We only remove the SSL task.

To compare them under different conditions, we show the results of these variants in Fig. 3 under two metrics (with $K=20$). We have the following observations: **(1)** The SSL task improves the base model, serving as the driving force of performance improvement. When removing the SSL task, we can observe a remarkable performance drop on both two metrics. **(2)** Furthermore, OHG can achieve better performance than OG, which shows the effectiveness of the hypergraph structure and the necessity of modeling the contextual relations for item representation learning. **(3)** Besides, the two views are effective in achieving better performance than the single view in three datasets, which indicates that two-view can cooperate and have mutual complementarity. **(4)** Overall, PacoHGNN can outperform all of its variants for $K=20$ values, indicating the effectiveness of its design for the session-based recommendation.

6 Conclusion

In this paper, we propose a novel framework, named PacoHGNN, by modeling session-based data as a hypergraph for better inferring the user preference of the current session. Moreover, we further propose Hyperedge-to-Node (H2N) to enhance supervised signals against data sparsity issue for better graph representation. Extensive empirical studies demonstrate the superiority of the proposed model over the current advanced methods, and the results validate the effectiveness of hypergraph modeling. Meanwhile, the research of hypergraph modeling for SBR remains in its infancy, and their application in GNNs has potential development, which is worthy of our further exploration.

Acknowledgement. This work was supported by National Natural Science Foundation of China (62072463, 71531012), Research Seed Funds of School of Interdisciplinary Studies of Renmin University of China, National Social Science Foundation of China (18ZDA309), and Opening Project of State Key Laboratory of Digital Publishing Technology of Founder Group. The computer resources were provided by Public Computing Cloud Platform of Renmin University of China. Xun Liang is the corresponding author of this paper.

References

1. Bu, J., et al.: Music recommendation by unified hypergraph: combining social media information and music content. In: Bimbo, A.D., Chang, S., Smeulders, A.W.M. (eds.) Proceedings of the 18th International Conference on Multimedia 2010, Firenze, Italy, 25–29 October 2010, pp. 391–400. ACM (2010)

2. Dias, R., Fonseca, M.J.: Improving music recommendation in session-based collaborative filtering by using temporal context. In: 25th IEEE International Conference on Tools with Artificial Intelligence, ICTAI 2013, Herndon, VA, USA, 4–6 November 2013, pp. 783–788. IEEE Computer Society (2013)

3. Fang, G., Song, J., Wang, X., Shen, C., Wang, X., Song, M.: Contrastive model inversion for data-free knowledge distillation. CoRR abs/2105.08584 (2021). https://arxiv.org/abs/2105.08584

4. Feng, L., Cai, Y., Wei, E., Li, J.: Graph neural networks with global noise filtering for session-based recommendation. Neurocomputing **472**, 113–123 (2022)

5. Feng, Y., You, H., Zhang, Z., Ji, R., Gao, Y.: Hypergraph neural networks. In: The Thirty-Third AAAI Conference on Artificial Intelligence, AAAI 2019, The Thirty-First Innovative Applications of Artificial Intelligence Conference, IAAI 2019, The Ninth AAAI Symposium on Educational Advances in Artificial Intelligence, EAAI 2019, Honolulu, Hawaii, USA, 27 January–1 February 2019, pp. 3558–3565. AAAI Press (2019)

6. Gong, S., Zhu, K.Q.: Positive, negative and neutral: Modeling implicit feedback in session-based news recommendation. In: Amigó, E., Castells, P., Gonzalo, J., Carterette, B., Culpepper, J.S., Kazai, G. (eds.) SIGIR 2022: The 45th International ACM SIGIR Conference on Research and Development in Information Retrieval, Madrid, Spain, 11–15 July 2022, pp. 1185–1195. ACM (2022)

7. Guo, J., et al.: Learning multi-granularity consecutive user intent unit for session-based recommendation. In: Candan, K.S., Liu, H., Akoglu, L., Dong, X.L., Tang, J. (eds.) WSDM 2022: The Fifteenth ACM International Conference on Web Search and Data Mining, Virtual Event/Tempe, AZ, USA, 21–25 February 2022, pp. 343–352. ACM (2022)

8. Han, Q., Zhang, C., Chen, R., Lai, R., Song, H., Li, L.: Multi-faceted global item relation learning for session-based recommendation. In: Amigó, E., Castells, P., Gonzalo, J., Carterette, B., Culpepper, J.S., Kazai, G. (eds.) SIGIR 2022: The 45th International ACM SIGIR Conference on Research and Development in Information Retrieval, Madrid, Spain, 11–15 July 2022, pp. 1705–1715. ACM (2022)

9. Hidasi, B., Karatzoglou, A., Baltrunas, L., Tikk, D.: Session-based recommendations with recurrent neural networks. In: Bengio, Y., LeCun, Y. (eds.) 4th International Conference on Learning Representations, ICLR 2016, San Juan, Puerto Rico, 2–4 May 2016, Conference Track Proceedings (2016)

10. Jiao, Y., Xiong, Y., Zhang, J., Zhang, Y., Zhang, T., Zhu, Y.: Sub-graph contrast for scalable self-supervised graph representation learning. In: Plant, C., Wang, H., Cuzzocrea, A., Zaniolo, C., Wu, X. (eds.) 20th IEEE International Conference on Data Mining, ICDM 2020, Sorrento, Italy, 17–20 November 2020, pp. 222–231. IEEE (2020)

11. Li, A., Cheng, Z., Gao, F.L.Z., Guan, W., Peng, Y.: Disentangled graph neural networks for session-based recommendation. CoRR abs/ arXiv: 2201.03482 (2022)

12. Li, J., Ren, P., Chen, Z., Ren, Z., Lian, T., Ma, J.: Neural attentive session-based recommendation. In: Lim, E., et al. (eds.) Proceedings of the 2017 ACM on Conference on Information and Knowledge Management, CIKM 2017, Singapore, 06–10 November 2017, pp. 1419–1428. ACM (2017)

13. Li, Y., et al.: Hyperbolic hypergraphs for sequential recommendation. In: Demartini, G., Zuccon, G., Culpepper, J.S., Huang, Z., Tong, H. (eds.) CIKM 2021: The 30th ACM International Conference on Information and Knowledge Management, Virtual Event, Queensland, Australia, 1–5 November 2021, pp. 988–997. ACM (2021)

14. Liu, Q., Zeng, Y., Mokhosi, R., Zhang, H.: STAMP: short-term attention/memory priority model for session-based recommendation. In: Guo, Y., Farooq, F. (eds.) Proceedings of the 24th ACM SIGKDD International Conference on Knowledge Discovery & Data Mining, KDD 2018, London, UK, 19–23 August 2018, pp. 1831–1839. ACM (2018)

15. Qiu, R., Huang, Z., Chen, T., Yin, H.: Exploiting positional information for session-based recommendation. ACM Trans. Inf. Syst. 40(2), 35:1–35:24 (2022)

16. Qiu, R., Li, J., Huang, Z., Yin, H.: Rethinking the item order in session-based recommendation with graph neural networks. In: Zhu, W., et al. (eds.) Proceedings of the 28th ACM International Conference on Information and Knowledge Management, CIKM 2019, Beijing, China, 3–7 November 2019, pp. 579–588. ACM (2019)

17. Rendle, S., Freudenthaler, C., Schmidt-Thieme, L.: Factorizing personalized markov chains for next-basket recommendation. In: Rappa, M., Jones, P., Freire, J., Chakrabarti, S. (eds.) Proceedings of the 19th International Conference on World Wide Web, WWW 2010, Raleigh, North Carolina, USA, 26–30 April 2010, pp. 811–820. ACM (2010)

18. Sarwar, B.M., Karypis, G., Konstan, J.A., Riedl, J.: Item-based collaborative filtering recommendation algorithms. In: Shen, V.Y., Saito, N., Lyu, M.R., Zurko, M.E. (eds.) Proceedings of the Tenth International World Wide Web Conference, WWW 2010, Hong Kong, China, 1–5 May 2001, pp. 285–295. ACM (2001)

19. Tan, Y.K., Xu, X., Liu, Y.: Improved recurrent neural networks for session-based recommendations. In: Karatzoglou, A., et al. (eds.) Proceedings of the 1st Workshop on Deep Learning for Recommender Systems, DLRS@RecSys 2016, Boston, MA, USA, 15 September 2016, pp. 17–22. ACM (2016)

20. Velickovic, P., Fedus, W., Hamilton, W.L., Liò, P., Bengio, Y., Hjelm, R.D.: Deep graph infomax. In: 7th International Conference on Learning Representations, ICLR 2019, New Orleans, LA, USA, 6–9 May 2019. OpenReview.net (2019)

21. Wang, J., Ding, K., Zhu, Z., Caverlee, J.: Session-based recommendation with hypergraph attention networks. In: Demeniconi, C., Davidson, I. (eds.) Proceedings

of the 2021 SIAM International Conference on Data Mining, SDM 2021, Virtual Event, 29 April–1 May 2021, pp. 82–90. SIAM (2021)

22. Wang, Z., Wei, W., Cong, G., Li, X., Mao, X., Qiu, M.: Global context enhanced graph neural networks for session-based recommendation. In: Huang, J., et al. (eds.) Proceedings of the 43rd International ACM SIGIR conference on research and development in Information Retrieval, SIGIR 2020, Virtual Event, China, 25–30 July 2020, pp. 169–178. ACM (2020)

23. Wu, B., Liang, X., Zheng, X., Wang, J., Zhou, X.: Reinforced sample selection for graph neural networks transfer learning. In: Adjeroh, D.A., et al. (eds.) IEEE International Conference on Bioinformatics and Biomedicine, BIBM 2022, Las Vegas, NV, USA, 6–8 December 2022, pp. 1281–1288. IEEE (2022)

24. Wu, S., Tang, Y., Zhu, Y., Wang, L., Xie, X., Tan, T.: Session-based recommendation with graph neural networks. In: The Thirty-Third AAAI Conference on Artificial Intelligence, AAAI 2019, The Thirty-First Innovative Applications of Artificial Intelligence Conference, IAAI 2019, The Ninth AAAI Symposium on Educational Advances in Artificial Intelligence, EAAI 2019, Honolulu, Hawaii, USA, 27 January– 1 February 2019, pp. 346–353. AAAI Press (2019)

25. Xia, X., Yin, H., Yu, J., Wang, Q., Cui, L., Zhang, X.: Self-supervised hypergraph convolutional networks for session-based recommendation. In: Thirty-Fifth AAAI Conference on Artificial Intelligence, AAAI 2021, Thirty-Third Conference on Innovative Applications of Artificial Intelligence, IAAI 2021, The Eleventh Symposium on Educational Advances in Artificial Intelligence, EAAI 2021, Virtual Event, 2–9 February 2021, pp. 4503–4511. AAAI Press (2021)

26. Xu, C., Zhao, P., Liu, Y., Sheng, V.S., Xu, J., Zhuang, F., Fang, J., Zhou, X.: Graph contextualized self-attention network for session-based recommendation. In: Kraus, S. (ed.) Proceedings of the Twenty-Eighth International Joint Conference on Artificial Intelligence, IJCAI 2019, Macao, China, 10–16 August 2019, pp. 3940–3946. ijcai.org (2019)

27. Xu, C., et al.: Graph contextualized self-attention network for session-based recommendation. In: IJCAI, vol. 19, pp. 3940–3946 (2019)

28. You, Y., Chen, T., Sui, Y., Chen, T., Wang, Z., Shen, Y.: Graph contrastive learning with augmentations. In: Larochelle, H., Ranzato, M., Hadsell, R., Balcan, M., Lin, H. (eds.) Advances in Neural Information Processing Systems 33: Annual Conference on Neural Information Processing Systems 2020, NeurIPS 2020, 6–12 December 2020, virtual (2020)

29. Yu, J., Yin, H., Li, J., Wang, Q., Hung, N.Q.V., Zhang, X.: Self-supervised multi-channel hypergraph convolutional network for social recommendation. In: Leskovec, J., Grobelnik, M., Najork, M., Tang, J., Zia, L. (eds.) WWW 2021: The Web Conference 2021, Virtual Event/Ljubljana, Slovenia, 19–23 April 2021, pp. 413–424. ACM / IW3C2 (2021)

30. Yu, Z., Zheng, X., Huang, F., Guo, W., Sun, L., Yu, Z.: A framework based on sparse representation model for time series prediction in smart city. Frontiers Comput. Sci. **15**(1), 151305 (2021)

31. Zangerle, E., Pichl, M., Gassler, W., Specht, G.: #nowplaying music dataset: Extracting listening behavior from twitter. In: Zimmermann, R., Yu, Y. (eds.) Proceedings of the First International Workshop on Internet-Scale Multimedia Management, WISMM 2014, Orlando, Florida, USA, 7 November 2014, pp. 21–26. ACM (2014)

32. Zheng, X., Liang, X., Wu, B., Guo, Y., Tang, H.: Adaptive attention graph capsule network. In: IEEE International Conference on Acoustics, Speech and Signal

Processing, ICASSP 2022, Virtual and Singapore, 23–27 May 2022, pp. 3588–3592. IEEE (2022)

33. Zheng, X., Liang, X., Wu, B., Guo, Y., Zhang, X.: Graph capsule network with a dual adaptive mechanism. In: Amigó, E., Castells, P., Gonzalo, J., Carterette, B., Culpepper, J.S., Kazai, G. (eds.) SIGIR 2022: The 45th International ACM SIGIR Conference on Research and Development in Information Retrieval, Madrid, Spain, 11–15 July 2022, pp. 1859–1864. ACM (2022)

34. Zhu, Y., Xu, Y., Yu, F., Liu, Q., Wu, S., Wang, L.: Graph contrastive learning with adaptive augmentation. In: Leskovec, J., Grobelnik, M., Najork, M., Tang, J., Zia, L. (eds.) WWW 2021: The Web Conference 2021, Virtual Event / Ljubljana, Slovenia, 19–23 April 2021, pp. 2069–2080. ACM / IW3C2 (2021)

Spatio-Temporal Position-Extended and Gated-Deep Network for Next POI Recommendation

Pengxiang Lan[1], Yihao Zhang[1(✉)], Haoran Xiang[1], Yuhao Wang[1],
and Wei Zhou[2]

[1] School of Artificial Intelligence, Chongqing University of Technology,
Chongqing, China
{pengxianglan,51202315124,yhwang}@2020.cqut.edu.cn, yhzhang@cqut.edu.cn
[2] School of Big Data and Software Engineering,
Chongqing University, Chongqing, China
zhouwei@cqu.edu.cn

Abstract. The next point of interest (POI) recommendation uses the user's check-in information on location-based social networks to make recommendations. The existing methods based on deep learning are evident in improving the performance of the recommendation model by capturing users' interests and preferences. However, the methods based on recurrent neural networks ignore the dependencies of non-continuous POIs for understanding users' behaviour under spatio-temporal factors. Most attention-based methods focus on the global POI sequence, which pays attention to all POIs in the users' check-in sequences, even if some attention has very little weight. To tackle these problems, we propose a novel spatio-temporal model based on the position-extended algorithm and gated-deep network (i.e., ST-PEGD) for next POI recommendation. Specifically, by combining spatio-temporal factors, we design a gated-deep network to capture the long-term behavioral dependencies of users by generating auxiliary binary gates. In addition, when capturing the short-term behaviour dependence of users, we use the position-extended algorithm to make the contextual interaction of RNNs more sufficient when performing POI sequence hopping selection. Extensive experiments on two real datasets prove that our model performs significantly better than state-of-the-art methods.

Keywords: Next Point-of-interest recommendation · User behaviour dependence · Spatio-temporal · Binary gate · Contextual interaction

1 Introduction

The emergence of the mobile internet has led to increasing demand for personalized recommendations in functional application areas (e.g., mobile forecasting, route planning, and location-based advertising) based on users' access interests.

The next POI recommendation is deeded as a particular item recommendation task that models the users' check-in records to recommend more personalized locations [11,22], and therefore has received a lot of attention in the development branch of the recommendation system.

Due to the rise of data computing resources, deep neural networks (DNNs) have blossomed in POI recommendation tasks and achieved superior performance. Numerous studies have confirmed that the target user's long-term and short-term behaviour dependencies coexist in his/her interaction sequence [6]. The long-term behaviour dependence represents the user's general interest preference with no noticeable change, while short-term behaviour dependence can better reflect the user's real-time interest preference [13]. Consequently, it is well understood that capturing general interest requires overall check-in sequences of the target user to be modeled, while real-time interest is more closely related to his/her current check-in sequence. The long short-term memory (LSTM) to be better applied to the POI recommendation task because it can learn long-term dependencies. The DAN-SNR implement two parallel channels to capture short- and long-term user preference by self-attention [8].

Simultaneously, it is conceivable that the behavioural information of users is generally periodic [5,21]. Firstly, users' check-in POIs around the company during the workday and are more likely to visit abundant and various venues in their leisure time. Secondly, during the day, users will check in the catering POIs during lunchtime and often visit the entertainment POIs such as bars and KTVs in the middle of the night. In a small number of modeling that considers spatio-temporal factors, the LSTPM computes temporal similarity and designs geo-nonlocal network and geo-dilated LSTM to integrate spatio-temporal context information [16]; the STMLA divides the time slots and integrates the distance relationship into the location-saltant algorithm to consider the impact of spatio-temporal factors on the user's long- and short-term interest preferences [20].

However, although these methods have undeniable contributions in the field of POI recommendation, they still have some limitations that cannot be ignored. (1) First, in the temporal dimension, RNN variants (like LSTM, GRU) are still considered serial pipelines, and their input and hidden states do not undergo any interaction before entering RNNs [14,18]. Lacking sufficient contextual interactions to capture behavioural dependencies may lead to poor performance of RNNs. (2) Second, the typical attention mechanism is generally global. It pays attention to all POIs in the user check-in sequence, even if some of the attention has very little weight. Some POIs in the check-in sequences may not reflect user preferences [19]. (3) Third, if the user's current check-in sequence has visited distant venues for non-subjective reasons (e.g., working), it is inappropriate to recommend POIs around these remote venues to the target user [12]. Vanilla RNNs do not fully integrate spatial information and cope with information sequentially, ignoring the indispensability of non-continuous POIs under spatio-temporal factors for understanding users' short-term behaviour dependence.

To address the aforementioned thorny issues, we propose a novel spatio-temporal model based on a position-extended algorithm and gated-deep network

(ST-PEGD) for next POI recommendation. Our ST-PEGD demarcates the over-all check-in sequences of each user into historical check-in sequences and current check-in sequence, which fuses multiple contextual information to capture behaviour dependencies better. For one thing, the ST-PEGD model leverages the gated-deep network generate binary gate to dynamically consider whether each POI needs attention in long-term behaviour dependence. For another thing, the ST-PEGD considers the indispensability of non-continuous POIs under spatio-temporal factors and makes the RNN interaction in the position-extended algorithm more sufficient, to understand users' short-term behaviour dependence better. In conclusion, the following are the primary contributions of this paper.

(1) We propose the ST-PEGD model to more fully capture users' behaviour dependencies. It integrates spatial (two methods) and temporal information and leverages the gated-deep network to generate binary gate to consider whether each POI needs attention dynamically.
(2) We propose a position-extended algorithm. It makes the RNN context more fully interactive and dynamically captures the users' short-term behaviour dependencies by combining spatio-temporal information.
(3) We carried out a comprehensive evaluation of ST-PEGD on two datasets. The experimental results verify the effectiveness of ST-PEGD and demon-strate the excellence of the gated-deep network and position-extended algorithm in capturing users' behaviour dependencies.

2   Related Work

The essence of the next POI recommendation technique is to recommend the next POIs that personalized users may visit based on multiple contextual information in the users' behavioural check-in sequence. Deep learning-based methods and machine learning-based methods can be used to separate the work of the next POI recommendation into two main directions.

Machine Learning-Based Methods. Rahmani et al. [15] proposed the STACP model, which incorporates a spatio-temporal activity-centers algorithm into the matrix factorization model to model users' behaviour more accurately. Cui et al. [4] designed a linear way and a non-linear way as preference encoders to combine sequential preference with spatial preference and use pair-wise Bayesian Personalized Ranking (BPR) to train it. Other variations of traditional machine learning-based methods are also proliferating in the field of POI recommendation. Wang et al. [17] proposed the GAIMC model, which consists of geographic feature extraction via a Gaussian mixture model (GMM) and inductive matrix completion for personalized POI recommendation. Baral et al. [1] proposed HiRecS, which formulates users' preferences as a hierarchical structure and models the locality trend using aggregated hierarchy.

Deep Learning-Based Methods. Chen et al. [2] modeled the relative ordering of locations with a one-dimensional convolution. The CEM learned a double-prototype representation for each location to eliminate the incorrect location transitions. Xiong et al. [18] used a network embedding method and dynamic factor graph model to jointly understand the effects of users' social connections, textual reviews, and the proximity of POIs in order to capture complicated spatio-temporal patterns of visiting behaviours. Luo et al. [12] proposed a STAN, which uses a bi-layer attention architecture that firstly aggregates spatio-temporal correlation within user trajectory and then recalls the target with consideration of personalized item frequency (PIF). Liu et al. [10] established a geographical-temporal attention network and a context-specific co-attention network to change user preferences by adaptively selecting relevant check-in activities from check-in histories. Chen et al. [3] designed a novel spatial-temporal transfer relation to intuitively capture users' transition patterns between neighbouring POIs. They used an end-to-end model to model long-term and short-term preferences. With the increasing scale of users' check-in data and the variability of users' behaviour, feature engineering construction has become more complicated, even if machine learning-based techniques may not necessitate intricate in-depth calculations. Whereas, even the prevailing methods based on deep learning still have some intractable problems. For instance, how to take into account the indispensability of non-continuous POIs under spatio-temporal factors for understanding users' behaviour dependencies; how to better balance the long-term and short-term behaviour dependencies connections. Consequently, we structure a fresh spatio-temporal model based on a position-extended algorithm and gated-deep network (ST-PEGD) for next POI recommendation and to understand user behaviour.

3   The ST-PEGD Method

This paper proposed the ST-PEGD model composed of a long-term behaviour dependence channel, a short-term behaviour dependence channel, and a balance and integration module for the next POI recommendation. The overall structure of ST-PEGD is shown in Fig. 1.

3.1   Problem Formulation

Users and POIs are denoted as $U = \{u_1, u_2, \cdots, u_{|U|}\}$, $L = \{l_1, l_2, \cdots, l_{|L|}\}$ respectively. Each POI's representation of spatial position (longitude and latitude) is geocoded by (lon_l, lat_l) tuple.

Definition 1. Overall check-in sequences. For each user $u \in U$, whose overall check-in sequences represented by $CS = \{CS_1, CS_2, \cdots, CS_{|n|}\}$, where n is the index of the current sequence. Each sequence $CS_i \in CS$ consist a battery of POIs accessed by the target user u in chronological order, i.e., $CS_i = \{l_1, l_2, \cdots, l_{|CS_i|}\}$ and $l \in L$.

Definition 2. Historical check-in sequences. In overall check-in sequences, each user's historical check-in sequences are expressed as $\{CS_1, CS_2, \cdots, CS_{|n-1|}\}$.

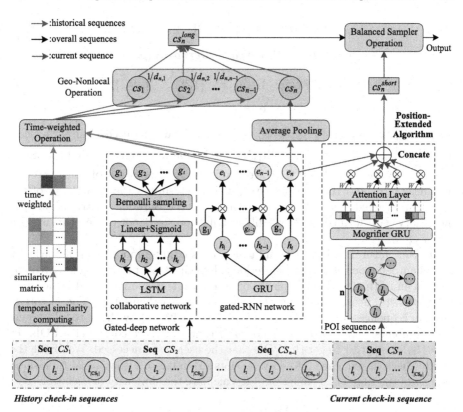

Fig. 1. The overview of ST-PEGD model. We utilize the overall check-in sequence to capture long-term behaviour dependencies. Users' short-term behaviour dependencies only utilize the current check-in sequence that has undergone different operations. Finally, the long-term and short-term behaviour dependencies are balanced and fused to obtain the predicted results. In the figure, the arrows refer to different information processing flow paths.

Definition 3. Current check-in sequence. The most recent continuous check-in is the current check-in sequence $CS_n = \{l_1, l_2, \cdots, l_{|L_{t-1}|}\}$, where the most recent POI that u has accessed at time $t-1$ is denoted by L_{t-1}.

3.2 Long-Term Behaviour Dependence Channel

The target user's long-term behaviour dependence can better map the user's general interest. Inspired by GA-Net [19] and LSTPM, we design the gated-deep network to capture users' long-term behaviour dependencies. As shown in Fig. 1, the collaborative network use LSTM to capture the target long-term user preferences roughly, generating a binary gate through Bernoulli distribution. The gated-RNN network to dynamically consider whether each POI needs attention. We construct the gated-deep network by the following equation:

$$\mathbf{h}_t' = LSTM\left(\mathbf{x}_t, \mathbf{h}_{t-1}'\right), t \in \{1, 2, \ldots, |CS_h|\}$$
$$p_t = sigmoid\left(\mathbf{U}\mathbf{h}_t'\right), g_t \sim Bernoulli\left(p_t\right) \tag{1}$$

$$\mathbf{h}_t = GRU\left(\mathbf{x}_t, \mathbf{h}_{t-1}\right), t \in \{1, 2, \ldots, |CS_h|\} \tag{2}$$

$$\mathbf{e}_t = g_t \odot \mathbf{h}_t \tag{3}$$

where $CS_h = \{CS_1, CS_2, \cdots, CS_{|n-1|}\}$ is user's historical check-in sequences, $\mathbf{h}' = \left[\mathbf{h}_1', \mathbf{h}_2', \ldots, \mathbf{h}_t'\right]$ and $\mathbf{h} = [\mathbf{h}_1, \mathbf{h}_2, \ldots, \mathbf{h}_t]$ be a sequence of hidden states generated by the LSTM and GRU, respectively. p_t is the result of going through the linear layer and the sigmoid layer, which determines the probability of the gate being opened, and utilizes to parameterize the Bernoulli distribution. g_t is the binary gate formed by Bernoulli distribution. $\odot$ denotes the element-wise product.

In the natural world, the influence of spatio-temporal factors on user behaviour dependence cannot be overlooked. Such as, people may look for restaurants at lunchtime and go to places of entertainment at night; under the same circumstances, users will prefer the place closest to themselves. The historical check-in sequences are encoded ($\{\mathbf{e}_1^{his}, \mathbf{e}_2^{his}, \cdots, \mathbf{e}_{|CS_h|}^{his}\}$) by the gated-deep network combines temporal factors. For each time slot i, we construct a POIs set $S_i = \{l_1, l_2, \cdots, l_{S_i}\}$ to get the POIs that the users have accessed. We divide the weekly check-in sequence into 48-time slots (Monday to Friday in 24 time slots, the rest of the time in 24 time slots.) to calculate the similarity of each time slot:

$$Sim_{i,j} = \frac{|S_i \cap S_j|}{\sqrt{|S_i| * |S_j|}}, (0 <= i < j <= 47) \tag{4}$$

$$cs_h = \sum_{t=1}^{|CS_h|} W_t \mathbf{e}_t^{his}, W_t = \frac{\exp\left(Sim_{c,S_t}\right)}{\sum_{i=1}^{|CS_h|} \exp\left(Sim_{c,S_i}\right)} \tag{5}$$

where in the same time slot S_t, the higher the similarity, the higher the weight W_t of POIs in the sequence. Based on the temporal factor, the historical check-in sequences are encoded as $cs_h \in \{cs_1, cs_2, \cdots, cs_{n-1}\}$.

The geographical distance is represented by $d_{n,h}$ (the haversine formula):

$$d_{n,h} = 2R\arcsin\left(\sqrt{A + B}\right), A = \sin^2\left(\frac{lon_{l_{t-1}} - lon_{cs_h}}{2}\right)$$
$$B = \cos\left(lon_{cs_h}\right)\cos\left(lon_{l_{t-1}}\right)\sin^2\left(\frac{lat_{l_{t-1}} - lat_{cs_h}}{2}\right) \tag{6}$$

where $\left(lon_{l_{t-1}}, lat_{l_{t-1}}\right) \in l_{t-1}$ is the POI location in the target user's previous moment, $CS_h \in \{CS_1, CS_2, \cdots, CS_{n-1}\}$ is the median coordinate (lon_{cs_h}, lat_{cs_h}) by average processing of the historical check-in sequences. R is the radius of the earth.

The current check-in sequence is encoded $\{\mathbf{e}_1^{cur}, \mathbf{e}_2^{cur}, \cdots, \mathbf{e}_{t-1}^{cur}\}$ by the gated-deep network retains all the information through the average pooling operation to better reflect the real-time behaviour dependencies of the users:

$$cs_n = \frac{1}{|CS_n|} \sum_{t=1}^{|CS_n|} \mathbf{e}_t^{cur} \tag{7}$$

where the current check-in sequence is represented as cs_n.

Incorporating the spatio-temporal factor into the processing of the overall check-in sequence by geo-nonlocal network is formulated as follows:

$$\mathbf{cs}_n^{long} = \frac{1}{\sum_h^{n-1} exp(\frac{1}{d_{n,h}} \widetilde{cs}_n^\top cs_h)} \sum_h^{n-1} exp(\frac{1}{d_{n,h}} \widetilde{cs}_n^\top cs_h) W_h cs_h \tag{8}$$

where $\widetilde{cs}_n = cs_h + cs_n$ is the overall check-in sequences, W_h is a weight matrix. Strikingly, $\mathbf{cs}_n^{long}$ is a representation of the user's long-term behaviour dependence.

3.3   Short-Term Behaviour Dependence Channel

The role of the target user's current check-in sequence CS_n in capturing the long-term behaviour dependence is unquestionable. However, compared with the overall check-in sequences CS, the target historical check-in sequences $CS_h \in \{CS_1, CS_2, \cdots, CS_{|n-1|}\}$ accounts for a large proportion, so the contribution of the current check-in sequence to capturing the user's long-term behaviour dependence is not as good as expected. On the contrary, in capturing short-term behaviour dependence, the personalized POI recommended to the target user for the next visit is more influenced by its current check-in venues. In addition, the vanilla RNNs can only process the current check-in sequence sequentially and do not blend the impact of spatial location on capturing short-term behaviour dependence, and dialed-RNN jump length is stationary. Therefore, we construct the current non-continuous POI sequence and propose a position-extended algorithm to determine the position jump length.

We initialize with $CS_n^{geo} = \phi$, and $k = 2$, where $CS_n^{geo} = \phi$ is the jump input sequence filtered by the distance factor, i is an index for the target user's current check-in sequence, and construct the POI geo-matrix $D \in \mathbb{R}^{(t-1) \times (t-1)}$ by two point distance formula (e.g., $d_{x,y} = d_{y,x}$, where $(lon_{l_x}, lat_{l_x}) \in l_x$, $(lon_{l_y}, lat_{l_y}) \in l_y$). The position-extended algorithm is manifested in Algorithm 1, where regaining to a basic GRU at $r = 0$. Meantime, we introduce the structure of mogrifier GRU in Fig. 2.

First, we utilize the geolocation of POIs to identify jump non-continuous input sequences of POIs. Second, inspired by a good improved LSTM, we construct mogrifier GRU to make GRU more fully interact with its contextual information. The current check-in sequence information, which is not rich, performs POI jumping again.

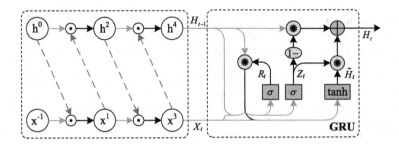

Fig. 2. The structure of mogrifier GRU through 4 iterations. The **green** dashed arrows refer to linear transformations. ⊙ is activation operation. (Color figure online)

Algorithm 1: Position-extended algorithm

Input: the target user's current check-in sequence $CS_n = \{l_1, l_2, \cdots, l_{|L_{t-1}|}\}$

Output: $\mathbf{cs}_n^{short}$:the representation of the user's short-term behaviour dependence

1 **while** $k \leq t - 1$ **do**

    set $j = k - 1$ and $Min = \infty$;

    while $j \geq 1$ **do**

        $Min = min(Min, d_{l_k, l_j})$;

        $j - -$;

    end

    find $d_{l_k, l_i} = Min$;

    $\{l_i, l_k\} \longmapsto CS_n^{geo}$

    $k + +$;

2 **end**

3 find $CS_n^{geo} = \{\{l_1, l_\mu\}, \ldots, \{l_\varphi, l_{t-1}\}\}$, where $1 < \mu < \cdots < \varphi < t - 1$;

4 $\{\{x_1, x_\mu\}, \ldots, \{x_\varphi, x_{t-1}\}\} \longmapsto CS_n^{geo}$ by embedding, where x_i is the vector;

5 $x^{-1} = x$ and $h_{t-1}^0 = h_{t-1}$, $W^i \in \mathbb{N}^{m \times n}$ and $D^i \in \mathbb{R}^{m \times n}$ are trainable matrices, cycle $r = 4$ round;

6 **for** $i = 1$ *in* r **do**

7   **if** $i \bmod 2 \neq 0$ **then**

        $h_{prev}^i \leftarrow 2\sigma\left(N^i x^{i-1}\right) \odot h_{prev}^{i-2}$;

8   **end**

9   **else**

10         $x^i \leftarrow 2\sigma\left(W^i h_{prev}^{i-1}\right) \odot x^{i-2}$;

11   **end**

12     $i + +$;

13 **end**

14 the input $x^\uparrow$ and $h_{prev}^\uparrow$ are the representation of the highest indexed x^j and h_{prev}^j, respectively;

15 $h_{t-1}^{geo} \leftarrow MogGRU(x_{t-1}^\uparrow, h_\phi^{geo\uparrow})$;

16 $\mathbf{cs}_n^{short} \leftarrow Attention(h_{t-1}^{geo})\alpha + \mathbf{e}_{t-1}^{cur}(1 - \alpha)$;

Therefore we use a general attention mechanism to select more available information. Finally, make contact with the output of $\mathbf{e}_{t-1}^{cur}$ and attention network. The position-extended algorithm was created by the above method to learns the user's short-term behavioural dependence $\mathbf{cs}_n^{short}$ combined with spatio-temporal factors.

3.4   Balance and Integration Module

The above modules combine multiple contextual information to capture the user's long-term behaviour dependence $\mathbf{cs}_n^{long}$ and short-term behaviour dependence $\mathbf{cs}_n^{short}$. In this module, we balance and integrate long-term and short-term behaviour dependence to generate rankings:

$$\mathbf{p} = softmax \left(\mathbf{W}_r \left(\eta \mathbf{cs}_n^{long} \oplus (1 - \eta) \mathbf{cs}_n^{short} \right) \right) \tag{9}$$

where the probability set $\mathbf{p}$ of each POI, $\mathbf{W}_r \in \mathbb{R}^{|L| \times 2d}$ is a projection matrix that can be trained. Obviously, the target user has the highest probability in $\mathbf{p}$ of accessing the POI at time t. Then the objective function (log likelihood) is formulated as:

$$\mathcal{L} = -\sum_{i=1}^{K} \log(p_i) \tag{10}$$

where K is the total number of training samples, p_i is the probability that the model will produce the i-th training sample's ground truth POI.

4   Experiments

In this section, we will analyze these research questions: **RQ1:** How does our ST-PEGD model performance compare with other cutting-edge recommendation algorithms? **RQ2:** How to understand the critical components of ST-PEGD and their impact on ST-PEGD? **RQ3:** How well does our proposed the Position-Extended algorithm perform in capturing short-term behaviour dependence?

4.1   Evaluation Datasets and Metrics

Evaluation Datasets. We conducted multi-faceted experiments on our proposed ST-PEGD model to show the universal applicability on two publicly available LBSNs datasets: Foursquare[1] (collected in New York from February 2010 to January 2011) [6] and Gowalla[2] (collected worldwide from February 2009 to October 2010) [16]. To lessen the effects of data sparsity, we expurgated some unpopular POIs (check-in by fewer than ten users), users (less than ten check-in POIs and checked in less than three times per day), and sequences (total check-in sequences of no more than five) [7]. The processed statistics of the selected dataset are shown in Table 1.

---

[1] https://developer.foursquare.com/places-api.

[2] http://snap.stanford.edu/data/loc-gowalla.html.

Table 1. Statistics of the datasets for evaluation

Datasets	#User	#POIs	#Data Collection	#Check-in	Density
Gowalla	5,802	40,868	World-Wide	301,080	0.13%
Foursquare	934	9296	New York	52,983	0.62%

Evaluation Criteria. We leverage two frequently-used evaluation metrics *Recall@K* and Normalized Discounted Cumulative Gain (*NDCG@K*) to evaluate the recommendation performance, where $K \in \{5, 10\}$. *Recall@K* measures the correct proportion of POIs predicted by top$K$, and *NDCG@K* cares whether these predicted POIs are placed in a more prominent position for users.

4.2 Experimental Settings

Parameter Configuration. In terms of public hyperparameters (hidden and embedding dimensions, batch size) and data partitioning (80% training set, 20% test set), we are consistent with the best hyperparameters in LSTPM [16]; using Adam to optimize all parameters with a learning rate to 0.0001 and a weight decay to 0.000001. Our implementation was written using PyTorch.

Baseline Recommenders. Our ST-PEGD model compares it with the eight baseline recommenders:

- **LSTM:** A variant of classical RNN that captures both long- and short-term behaviour dependencies.
- **ST-RNN:** [9] An RNN model fused with temporal and spatial transition matrix.
- **Time-LSTM:** [23] A variant LSTM incorporates temporal gates to account for the effect of time intervals on capturing long- and short-term behaviour dependencies.
- **DRCF:** [13] An RNN model with a pairwise ranking function combined with matrix factorization aims to model users' long- and short-term behaviour dependencies.
- **DeepMove:** [6] A combined attention and RNN network considers heterogeneous transition regularity and multi-level periodicity. Furthermore, it divides the user check-in sequence into two parts: historical and current.
- **STGN:** [22] This method proposes two pairs of time and distance gates to be integrated into the LSTM to obtain user preference updates.
- **LSTPM:** [16] This method designs spatio-temporal context-aware nonlocal networks and geo-dilated RNN while considering long- and short-term interests.
- **STMLA:** [20] This method uses the mogrifier LSTM to consider the contextual interaction of the RNN, and uses the location-saltant algorithm to understand the impact of non-continuous POIs.

Table 2. Recommendation performance comparison with baselines

Model	Gowalla				Foursquare			
	Rec@5	Rec@10	NDCG@5	NDCG@10	Rec@5	Rec@10	NDCG@5	NDCG@10
LSTM	0.0971	0.1259	0.0716	0.0809	0.2115	0.2639	0.1597	0.1768
ST-RNN	0.1245	0.1687	0.0885	0.1027	0.2291	0.2968	0.1615	0.1835
Time-LSTM	0.1500	0.1963	0.1082	0.1232	0.2868	0.3667	0.2117	0.2375
DRCF	0.1709	0.2192	0.1249	0.1406	0.2743	0.3466	0.1971	0.2206
DeepMove	0.1330	0.1656	0.1023	0.1128	0.3110	0.3831	0.2323	0.2556
STGN	0.1600	0.2041	0.1191	0.1333	0.2730	0.3547	0.1951	0.2217
LSTPM	0.1922	0.2442	0.1430	0.1597	0.3294	0.4065	0.2441	0.2688
STMLA	0.2520	0.3025	0.1978	0.2141	0.3440	0.4227	0.2594	0.2813
ST-PEGD	**0.3715**	**0.4191**	**0.3012**	**0.3166**	**0.4072**	**0.4856**	**0.3146**	**0.3387**
Improvement	23.61%	38.54%	52.27%	47.87%	18.37%	14.88%	21.28%	20.41%

4.3 Performance Analysis (RQ1)

We have summarized the results in $Recall@[5,10]$ and $NDCG@[5,10]$ of ST-PEGD and some influential POI recommendation methods as shown in Table 2. The best results we highlight are in bold.

Analyzing the results, we see that the ST-PEGD algorithm is comprehensive on both datasets and visibly exceeds other influential baseline recommenders. In particular, the average performance improvement of the ST-PEGD method on the Foursquare and Gowalla is 18.74% and 40.57%, respectively, compared to the most widely used baseline STMLA algorithm. Under the premise of uniformly setting all shared parameters, the results of ST-PEGD fully demonstrate the remarkable effectiveness of our algorithm.

Across overall baselines, models that capture users' long-term and short-term behaviour dependencies in different ways (STMLA, LSTPM, DeepMove, and DRCF) outperform than other baselines. So this fully indicates that while dynamically capturing short-term user behaviour dependencies, the role of long-term behaviour dependencies in recommending the user's next POI is needful. From a different perspective, ST-PEGD, STMLA, and LSTPM all consider some non-negligible multivariate contexts (e.g., time, distance), which is also the key to their high recommendation accuracy and better than other baselines. ST-PEGD, STMLA, LSTPM, DeepMove, and DRCF incorporate methods for capturing time periodicity, and the performance powerfully demonstrates the importance of the time factor. Obviously, STMLA and LSTPM are better in the baseline, but only a simple capture method is used in the long-term interest preference module. This is also why their performance is not as good as ST-PEGD.

Since Gowalla's data collection scope is all over the world, the ratio of POI to users is more significant, resulting in the sparseness of Gowalla being higher than that of the Foursquare dataset, so that the performance of all models on Gowalla is not as good as that on Foursquare. Nevertheless, from another perspective, it can be found that the performance improvement of our ST-PEGD on Gowalla is more prominent than that of Foursquare. The Gowalla can better

Table 3. Performance of the critical components in ST-PEGD

Model	Gowalla				Foursquare			
	Rec@5	Rec@10	NDCG@5	NDCG@10	Rec@5	Rec@10	NDCG@5	NDCG@10
STMLA	0.2520	0.3025	0.1978	0.2141	0.3440	0.4227	0.2594	0.2813
ST-GD	0.3659	0.4047	0.2939	0.3097	0.4058	0.4854	0.3106	0.3363
ST-PE	0.1994	0.2471	0.1442	0.1595	0.3307	0.4053	0.2490	0.2717
_L	0.3583	0.3814	0.2735	0.2807	0.3932	0.4735	0.2848	0.3108
_S	0.2346	0.2833	0.1812	0.2003	0.2509	0.3061	0.2007	0.2184
_outs	0.3536	0.3911	0.2794	0.2962	0.3932	0.4718	0.2890	0.3152
_nonlocal	0.3316	0.3783	0.2692	0.2846	0.3853	0.4527	0.2946	0.3185
ST-PEGD	**0.3715**	**0.4191**	**0.3012**	**0.3166**	**0.4072**	**0.4856**	**0.3146**	**0.3387**

record the check-ins of target users in different cities and countries, which leads to the relatively large spatial span of the locations visited. Therefore, it is more necessary to generate a binary gate to consider whether each POI needs attention dynamically and non-continuous POIs under spatio-temporal factors. It is the main reason that decides the ST-PEGD algorithm outperforms other baselines.

4.4   Analysis and Understanding of Critical Components of ST-PEGD (RQ2)

In order to verify how much each key component contributes to ST-PEGD performance, we separated the ST-PEGD model into six simplified variants from various angles for ablation studies.

- **ST-GD:** This version replaces the position-extended algorithm with a general GRU.
- **ST-PE:** This version replaces the gated-deep network with a general LSTM.
- **_L:** This version solely considers users' long-term behaviour dependence and leaves the component of capturing short-term behaviour dependence.
- **_S:** This version solely considers users' short-term behaviour dependence and leaves the component of capturing long-term behaviour dependence.
- **_outs:** This version that user short-term behaviour dependence is captured only with the output of the current check-in sequence in the position-extended algorithm.
- **_nonlocal:** This version omits geo-nonlocal operation and uses only non-local networks.

The experimental results of six different ST-PEGD algorithm variant structures on Foursquare and Gowalla are displayed in Table 3. We can observe the following finding by referring to Table 3.

Overall, the performance of each variant structure is lower than that of ST-PEGD. This illustrates volumes about the necessity of our design variant structures and the importance of these critical components. ST-GD obviously performs better than ST-PE, the performance of ST-PE is not satisfactory; on

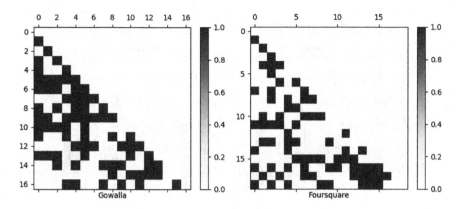

Fig. 3. Visualization of binary gates generated by Bernoulli distribution.

the other hand, the performance difference between ST-GD and ST-PEGD is not apparent, which strongly proves that the gated-deep network contributes more to our model. We believe that the binary gates generated by the collaborative network act on the gated RNN network to strengthen the weights further. As a result, the role of the gated-deep network is more pronounced. Since the models of ST-PE and LSTPM are similar, the performance of ST-PE is slightly better than that of LSTPM under the addition of the position-extended algorithm.

Comparing with the ST-PEGD, the experimental results of _L and _S show that it is essential to fuse both long-term and short-term behaviour dependencies. From another point of view, the improved performance of _L and _S also illustrates the effectiveness of the gated-deep network and position extended algorithm on the user's long-term and short-term behaviour dependencies, respectively.

The current check-in sequence through the gated-deep network captures the users' short-term behaviour dependencies without considering position jumps. The results of _outs show that it is crucial to consider the performance of the current check-in sequence in the users' long-term behaviour dependencies when capturing short-term behaviour dependencies. Combining the current check-in sequence without position jumps further handles negative noise when capturing short-term behaviour dependencies.

The geo-nonlocal operation combines geographic factors to select more helpful information. In ST-PEGD, we consider geographical factors' influence differently (geo-nonlocal operation and position-extended algorithm). The performance of the ST-PE and _nonlocal shows the importance of geographic factors for POI recommendation.

Interpretability. We visualized binary gates generated with Bernoulli distributions in Fig. 3. Noticeably, Fig. 3 effectively shows that we do not have a global consideration of weights like the general attention mechanism. The collaborative network facilitates the capture of behaviour dependencies in the users' check-in sequences.

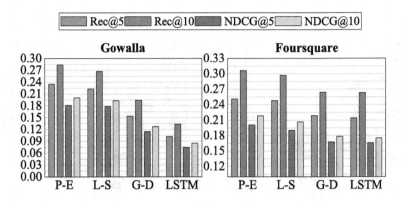

Fig. 4. The experimental results of our Position-Extended algorithm are compared with the performance of other baselines on Gowalla and Foursquare.

4.5 Analysis on Effectiveness of the Position-Extended Algorithm (RQ3)

In order to demonstrate how well our Position-Extended algorithm captures short-term user behaviour dependence, we compare it to the Location-Saltant algorithm [20], Geo-Dilated LSTM [16], and conventional LSTM processing approaches.

The results in Fig. 4 show that on the Gowalla and Foursquare datasets, the Position-Extended algorithm significantly outperforms the other three deep networks in capturing short-term user behaviour dependence. Obviously, the regular LSTM only sequentially process POI check-in sequences and does not take into account valuable spatial information, so it is not surprising that its performance is the worst. It is not enough to provide personalized recommendation services. Furthermore, the performances shown by the Position-Extended algorithm and the Location-Saltant algorithm demonstrate the indispensability of non-continuous POIs. In other words, the non-subjective visit noises in the geographic aspect of the user access sequence is reduced to a certain extent. Since the sparser Gowalla and the larger geographic span of users, it is more necessary to consider geographic factors in Gowalla. Therefore, the position-extended algorithm on Gowalla is more effective in denoising users' non-subjective visits.

According to the latent features captured by the current check-in sequence through the gated-deep network, the output of our proposed Position-Extended algorithm takes into account the different importance of each POI, which is also the critical factor that the performance of the Position-Extended algorithm is better than that of the Location-Saltant algorithm. Additionally, compared with the Geo-Dilated LSTM, the mogrifier LSTM context interaction is sufficient and is better at long-range information learning, so the Position-Extended Algorithm and Location-Saltant algorithm perform better. When capturing users' short-term behaviour dependencies, both LSTPM and ST-PEGD fuse the current check-in sequence output results (the long-term module). Nevertheless, the gated-deep network in our ST-PEGD is more comprehensive than the LSTPM, which only uses the ordinary LSTM in the long-term preference module to learn the current check-in sequence.

5 Conclusions

In this paper, we propose the ST-PEGD, a novel next POI recommendation algorithm. We design Gated-Deep network to address the case where typical attention mechanisms focus globally. Besides, we utilize the Position-extended algorithm to consider the non-continuous POIs and make the RNN contextual interactions to understand users' short-term behaviour dependence. Extensive experiments on Gowalla and Foursquare show that our model ST-PEGD achieves significant improvement in recommendation performance.

Overall, the collaborative module and the gated-RNN module of the gated-deep network are still based on the idea of RNNs. In the future, we will use other, more advanced technologies to improve.

Acknowledgements. The work is supported by the Natural Science Foundation of Chongqing (No. cstc 2019jcyj-msxmX0544), the Science and Technology Research Program of Chongq-ing Municipal Education Commission (No. KJZD-K202101105, KJ-QN202001136), the National Natural Science Foundation of China (No.61702063), the Graduate Innovation Foundation of Chongqing University of Technology (No. gzlcx20222135).

References

1. Baral, R., Iyengar, S.S., Zhu, X., Li, T., Sniatala, P.: HiRecS: a hierarchical contextual location recommendation system. IEEE Trans. Comput. Soc. Syst. **6**(5), 1020–1037 (2019)
2. Chen, M., Zuo, Y., Jia, X., Liu, Y., Yu, X., Zheng, K.: CEM: a convolutional embedding model for predicting next locations. IEEE Trans. Intell. Transp. Syst. **22**(6), 3349–3358 (2020)
3. Chen, W., et al.: Building and exploiting spatial-temporal knowledge graph for next poi recommendation. Knowl.-Based Syst. **258**, 109951 (2022)
4. Cui, Q., Tang, Y., Wu, S., Wang, L.: Distance2Pre: personalized spatial preference for next point-of-interest prediction. In: Yang, Q., Zhou, Z.-H., Gong, Z., Zhang, M.-L., Huang, S.-J. (eds.) PAKDD 2019. LNCS (LNAI), vol. 11441, pp. 289–301. Springer, Cham (2019). https://doi.org/10.1007/978-3-030-16142-2_23
5. Cui, Q., Zhang, C., Zhang, Y., Wang, J., Cai, M.: ST-PIL: spatial-temporal periodic interest learning for next point-of-interest recommendation. In: Proceedings of the 30th ACM International Conference on Information & Knowledge Management, pp. 2960–2964 (2021)
6. Feng, J., et al.: DeepMove: predicting human mobility with attentional recurrent networks. In: Proceedings of the 2018 World Wide Web Conference, pp. 1459–1468 (2018)
7. Feng, S., Li, X., Zeng, Y., Cong, G., Chee, Y.M., Yuan, Q.: Personalized ranking metric embedding for next new POI recommendation. In: Twenty-Fourth International Joint Conference on Artificial Intelligence (2015)

8. Huang, L., Ma, Y., Liu, Y., He, K.: DAN-SNR: a deep attentive network for social-aware next point-of-interest recommendation. ACM Trans. Internet Technol. (TOIT) **21**(1), 1–27 (2020)
9. Liu, Q., Wu, S., Wang, L., Tan, T.: Predicting the next location: a recurrent model with spatial and temporal contexts. In: Thirtieth AAAI Conference on Artificial Intelligence (2016)
10. Liu, T., Liao, J., Wu, Z., Wang, Y., Wang, J.: Exploiting geographical-temporal awareness attention for next point-of-interest recommendation. Neurocomputing **400**, 227–237 (2020)
11. Liu, Y., et al.: An attention-based category-aware GRU model for the next poi recommendation. Int. J. Intell. Syst. **36**(7), 3174–3189 (2021)
12. Luo, Y., Liu, Q., Liu, Z.: STAN: spatio-temporal attention network for next location recommendation. In: Proceedings of the Web Conference 2021, pp. 2177–2185 (2021)
13. Manotumruksa, J., Macdonald, C., Ounis, I.: A deep recurrent collaborative filtering framework for venue recommendation. In: Proceedings of the 2017 ACM on Conference on Information and Knowledge Management, pp. 1429–1438 (2017)
14. Melis, G., Kočiský, T., Blunsom, P.: Mogrifier LSTM. In: International Conference on Learning Representations (2020)
15. Rahmani, H.A., Aliannejadi, M., Baratchi, M., Crestani, F.: Joint geographical and temporal modeling based on matrix factorization for point-of-interest recommendation. In: Jose, J.M., et al. (eds.) ECIR 2020. LNCS, vol. 12035, pp. 205–219. Springer, Cham (2020). https://doi.org/10.1007/978-3-030-45439-5_14
16. Sun, K., Qian, T., Chen, T., Liang, Y., Nguyen, Q.V.H., Yin, H.: Where to go next: modeling long-and short-term user preferences for point-of-interest recommendation. In: Proceedings of the AAAI Conference on Artificial Intelligence, vol. 34, pp. 214–221 (2020)
17. Wang, W., Chen, J., Wang, J., Chen, J., Gong, Z.: Geography-aware inductive matrix completion for personalized point-of-interest recommendation in smart cities. IEEE Internet Things J. **7**(5), 4361–4370 (2019)
18. Xiong, X., Xiong, F., Zhao, J., Qiao, S., Li, Y., Zhao, Y.: Dynamic discovery of favorite locations in spatio-temporal social networks. Inf. Process. Manag. **57**(6), 102337 (2020)
19. Xue, L., Li, X., Zhang, N.L.: Not all attention is needed: gated attention network for sequence data. In: Proceedings of the AAAI Conference on Artificial Intelligence, vol. 34, pp. 6550–6557 (2020)
20. Zhang, Y., Lan, P., Wang, Y., Xiang, H.: Spatio-temporal Mogrifier LSTM and attention network for next poi recommendation. In: 2022 IEEE International Conference on Web Services (ICWS), pp. 17–26. IEEE (2022)
21. Zhao, P., et al.: Where to go next: a spatio-temporal gated network for next poi recommendation. IEEE Trans. Knowl. Data Eng. **34**(5), 2512–2524 (2020)
22. Zhou, X., Mascolo, C., Zhao, Z.: Topic-enhanced memory networks for personalised point-of-interest recommendation. In: Proceedings of the 25th ACM SIGKDD International Conference on Knowledge Discovery & Data Mining, pp. 3018–3028 (2019)
23. Zhu, Y., et al.: What to do next: modeling user behaviors by time-LSTM. In: IJCAI, vol. 17, pp. 3602–3608 (2017)

Leveraging Interactive Paths
for Sequential Recommendation

Aoran Li[1], Yalei Zang[1], Yani Wang[1], and Bohan Li[1,2,3]($\boxtimes$)

[1] College of Computer Science and Technology, Nanjing University of Aeronautics
and Astronautics, Nanjing 211106, China
`bhli@nuaa.edu.cn`
[2] Key Laboratory of Safety-Critical Software, Ministry of Industry and Information
Technology, Beijing, China
[3] National Engineering Laboratory for Integrated Aero-Space-Ground Ocean Big
Data Application Technology, Xi'an, China

Abstract. Sequential recommendation systems dynamically predict the
users' next behaviors from chronological historical records to provide
more accurate recommendations. However, in many real-world scenarios,
recommenders find it difficult to make recommendations for cold-start
users who have limited interactions with items. To address this issue,
existing methods utilize auxiliary information to capture user preferences, which have achieved performance improvement to some extent.
But with the gradual awareness of personal privacy, external information
becomes difficult to obtain. Therefore, we develop a novel model called
Path-based sequential **Rec**ommender for cold-start users (PathRec).
PathRec exploits the rich sequential dependencies within the interactive
paths to give reasonable recommendations for cold-start users without
additional external information. Specifically, (i) PathRec constructs the
global graph to learn the representations of the items. (ii) Then, PathRec
elaborately selects the sampled paths to derive the underlying dependencies of the cold-start sequence. (iii) Finally, PathRec provides rich
and complementary information to user-item interactions by strongly
coupling the paths to give suitable recommendations. Extensive experiments on three datasets demonstrate the effectiveness of PathRec. In
particular, our model can adapt to the cold-start users and has potential
interpretability for recommendations.

Keywords: Recommendation systems · Graph neural network ·
Cold-start · Path dependence

1 Introduction

Recommendation system (RS) is an information filtering technology that provides useful items for users. It predicts the content that users are most likely to
be interested in based on their preferences. As a new paradigm of recommendation system, sequential recommendation system (SRS) dynamically models

X. Wang et al. (Eds.): DASFAA 2023, LNCS 13944, pp. 521–536, 2023.
https://doi.org/10.1007/978-3-031-30672-3_35

users' chronological interactions to mine the behavioral preference patterns of users more accurately. Therefore, SRS is widely applied in various scenarios, like large-scale business systems and social networks.

Traditional methods of SRS mostly based on pattern mining or Markov chains, which can only model users' short-term preferences [18]. To model and capture the dependencies between different entities, the neural networks are introduced into SRS, such as RNN [4,13,22], CNN [16,19,20], GNN [23,24,26], and attention mechanism [6,12]. These methods model long-term and high-order sequential patterns and achieve superior performance.

The most critical task of SRS is how to mine user behavior patterns from sequences. Cold-start users have limited user-item interactions, recommenders will lose its effectiveness in this situation due to the lack of users' sequence behavior data. Given the sequence in Fig. 1(a). The cold-start user Alice clicks on two items successively, mobile phone and book. The correlation between the two items is weak, making it difficult for the recommender to infer Alice's next click. Considering this cold-start situation, the recommender will decrease users' stickiness if they get inappropriate recommendations. Consequently, the cold-start problem becomes an urgent and significant challenge in SRS.

To address the user cold-start problem, a straightforward method is to enrich the users' portrait with external information, such as user attribute information [8,14], knowledge graph [10,27] and social network information [11,28]. But, as people's awareness of privacy protection increases, such auxiliary information becomes difficult to gain. Some methods choose to directly ask users' suggestion to obtain their preferences, such as rating some items or filling out questionnaires. In this way, users can be recommended directly and effectively, but users usually do not want to spend too much time and energy during the process. Some other works utilize the idea of collaborative filtering [26]. They make recommendations for cold-start users by selecting items related to the sequence. However, they only consider the importance of items in the sequence, ignore the interactions between items.

We find that users' behavior sequences are chronological and have sequential dependence. So, we can regard the sequence as an explicit interactive path of the users' psychological activities. By modeling the path, we can capture the dependencies between items to obtain users' behavior patterns. But the interactive path for a cold-start user is too short, which is difficult to model. Therefore, we try to select some related paths from the original sequence data of all users whose starting and ending items are the same as the given sequence. This helps enrich the interaction between paths, and the recommender can more fully anticipate the user's intentions. To specifically understand why the dependency among paths is helpful, we take Fig. 1(b) as an example. In the graph built for cold-start user Alice, there are three object paths $p1$, $p2$, and $p3$. Each path represents a user's partial sequence and starts with the phone and ends with the book. For example, $p3$ clicked on the phone, then on the mask and sunglasses, and finally on the book. By considering the interactive paths of other users, we can more easily conjecture the user's main intention, such as making a journey. Then, the recommender will provide Alice with some items that may be used during the journey, such as a flashlight.

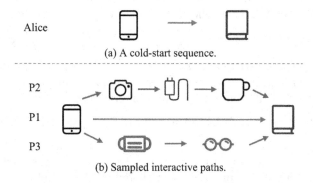

(a) A cold-start sequence.

P2

P1

P3

(b) Sampled interactive paths.

Fig. 1. Illustration of the cold-start sequence and sampled interactive paths.

In this paper, we propose a sequential recommender PathRec to provide advisable recommendations for cold-start users. Unlike previous work that only focused on items in the sequence, we focus more on the interaction between items. We utilize all sequences to construct the global graph and select object paths to add dependencies between items in the path to strongly couple the cold-start sequence. By employing graph neural networks, we learn the item embeddings. Then, we apply the attention mechanism to aggregate interactive paths' embedding outputs into a path representation. After enriching the interaction between paths, the recommender will provide more personalized suggestions for the cold-start users. The contributions are as follows:

- We view the interlinks between the users and the items as interactive paths. By strongly coupling the paths, we can obtain rich and complementary information to provide suitable recommendations without using any auxiliary information.
- We use the attention mechanism to obtain the strength of the relation in the sequence and propose an interactive path embedding method to better conjecture the user's intentions.
- Extensive experiments on three public datasets prove PathRec is well adapted to cold-start users and perform better than the baselines.

2   Preliminaries

In this section, we introduce the formal definition of the problem, then describe the specific steps to construct the global graph, and finally formulate a set of strategies for selecting sampled paths.

2.1   Problem Formulation

In SRS, we use $U = \{u_1, u_2, \ldots, u_m\}$ and $V = \{v_1, v_2, \ldots, v_n\}$ to represent the set of users and items, where m and n are the numbers of users and items respectively. For each user u, the historical record can be expressed as $S_u = \{v_{u,1}, v_{u,2}, \ldots, v_{u,l}\}$, where $v_{u,i}$ represents the i-th item clicked by the user u. In our task, each sequence can be viewed as a relation path. So the interaction sequence can also be written as $T_1 \xrightarrow{R_1} \ldots \xrightarrow{R_{\ell-1}} T_\ell$ where R_i represents the relation between item $v_{u,i}$ and item $v_{u,i+1}$. The model aims to recommend top-N items for the cold-start user (less than 5 interactions).

Input: The cold-start user's historical records $S_u = \{v_{u,1}, v_{u,2}, \ldots, v_{u,l}\}$, where $l \leq 5$.

Output: The top-N items that the user may click at step $l+1$, sorted in descending order according to the probability predicted by the model.

2.2   Global Graph Model

Previous work focuses only on the current sequence, ignoring the complex potential relations in other user sequences. Inspired by [23], we construct the global graph $G_g = \{V_g, \xi_g\}$ to capture similar user behavior patterns. The node set of the graph V_g is the item set V, the edge set is ξ_g where $\xi_i = (v_i, v_j)$ indicates that the user interacted with v_j and v_i successively.

For different items, we should give different weights. To extract the core interests and better capture users' historical interactions, we use the HITS algorithm to get the weight of each item from a global perspective. HITS is an algorithm for sorting web pages. It considers the authority of the linked node as well as the number of links of the node itself. For each node v, its value is characterized by two indicators, authority value, and hub value, which are interdependent calculated as follows,

$$\mathrm{auth}(v) = \sum_{v_i \in v_{to}} hub(v_i), \tag{1}$$

$$\mathrm{hub}(v) = \sum_{v_i \in v_{from}} auth(v_i), \tag{2}$$

where v_{to} are the nodes that point to v, v_{from} are the nodes that v points to. The initial authority and hub value of each node are 1. After a continuous iterative process, we can obtain the values for every node in the global graph.

2.3   Sampled Path Selection

We use the warm-start user sequence to get sampled paths, which can help us analyze the behavior pattern of the cold-start user. Given a cold-start user interaction sequence $\mathrm{Seq} = \{v_1, v_2, \ldots, v_n\}$, $n < 5$, which can also be written as $v_1 \xrightarrow{R_1} \ldots \xrightarrow{R_{l-1}} v_l$. For each relation R_i, we pick a set of sampled paths, we hope that by adding dependencies between v_i and v_{i+1} to strong coupling operation for R_i. With paths, we can enrich the interactions of paths and use users similar to cold-start users to help us make better recommendations. But selecting the proper sampled paths is a challenging problem. Previous work selects all paths for modeling, which will greatly increase the time cost, and bring some unnecessary redundant information [22]. Therefore, we propose a strategy for selecting the sampled path.

Take the relation R_i as an example. We first filter out the sequence that contains both v_i and v_{i+1}, requiring the order to be consistent with the cold-start user, that is, first click v_i, and then click v_{i+1}. There can be other interaction items between v_i and v_{i+1}. Since short paths are more expressive and too long paths may cause noise, we limit the length of the intercepted paths to be less than or equal to 5. Based on the sampled paths, we can build an interactive path structure between the two items, that is, a directed graph consisting of multiple paths. Each path should have a different impact on the target path, so we use the Path Ranking algorithm (PRA) to assign different weights to each path. The idea of PRA is to utilize the paths connecting two entities to predict the probability of a potential relation between them. Given the set of sampled paths $P = \{P1, P2, \cdots, Pn\}$ of node e. We regard each distribution w_{P_i} as the path feature, and rank the nodes by the linear model as follows,

$$\theta_1 w_{P_1} + \theta_2 w_{P_2} + \ldots \theta_n w_{P_n}, \tag{3}$$

where θ_i is the weight of path P_i, obtained by logistic regression. The probability score of the potential relation between nodes s and e is calculated by the following formula,

$$\text{score}(e, s) = \sum_{P_i \in P} w_{P_i} \theta_{P_i}. \tag{4}$$

Through the PRA algorithm, we can learn the weight of each sampled path, and get the strength of each relation in the original sequence. Then, we elaborate the proposed model.

3 Path-Based Sequential Recommender

We propose an effective sequential recommender for cold-start users named PathRec. PathRec regards recommendation as a path reasoning problem in the user-item interaction graph. PathRec makes recommendations for cold-start users by strongly coupling the paths to enrich the interaction between items. Figure 2 is the architecture of PathRec. It has four layers: 1) Global Graph Layer. We use warm-start users' sequences to construct the global graph (G_g) and employ PRA to capture users' core interests. 2) Path Representation Layer. After applying graph neural networks, we can learn the item embeddings, then further the representation of each sampled path. 3) Path Aggregation Layer. We enrich the cold-start sequence representation by aggregating the sampled paths. 4) Prediction Layer. It calculates the probability that the item will be interacted with the user. Next, we will introduce the four layers in detail.

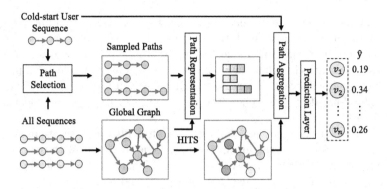

Fig. 2. Architecture of the PathRec model.

3.1   Global Graph Layer

Since GNN can utilize the rich underlying relations between nodes, it is suitable for sequential recommendation. Therefore, we utilize GNN to model the global graph to get items' latent vector representations. To fully use all sequences, we aggregate first-order neighbor information of the nodes. Since the correlation between each neighbor and the target node is not the same, it is unwise to aggregate information using average pooling. So, we apply the graph attention mechanism to calculate the attention weights. For item v_i, its first-neighbor information is linearly combined as follows,

$$h_{first(v_i)} = \sum_{v_j \in first(v_i)} h_{v_j} \cdot w(v_i, v_j), \tag{5}$$

where $first(v_i)$ represents v_i's first-order neighbors, $w(v_i, v_j)$ is the attention weight. Considering some items could appear many times in the sequences, we assign weights to each item $v_j \in first(v_i)$ computed as follows,

$$w(v_i, v_j) = \frac{n(v_i, v_j)}{\sum Out(v_i)} \cdot \frac{n(v_i, v_j)}{\sum In(v_j)}, \tag{6}$$

where $n(v_i, v_j)$ represents how many times edge $\xi = (v_i, v_j)$ appears. $\sum Out(v_i)$ is v_i's out-degree, $\sum In(v_j)$ is v_j's in-degree. Then we apply a softmax function to normalize the attention weight W,

$$w(v_i, v_j) = \text{softmax}(W). \tag{7}$$

After the processing, each item v_i in global graph G_g can be represented by aggregating its representation h_{v_i} and its neighbors' representation $h_{first(v_i)}$. W is the trainable parameter, and *LeakyRelu* is the activation function,

$$h'_{v_i} = \text{LeakyRelu}\left(W\left(h_{v_i} \| h_{\text{first}(v_i)}\right)\right). \tag{8}$$

3.2   Path Representation Layer

In the user-item interactions, the interaction paths provide important chronological order and dependency information. If we can improve the embeddings of
the interaction paths, a more effective representation learning method may be
raised. Since different items may have different importance in an interaction path.
We introduce the attention mechanism [17, 25] to obtain the attention weight of
items. Given a cold-start sequence Seq $= v_1 \xrightarrow{R_1} \ldots \xrightarrow{R_{l-1}} v_l$. For each relation
R_i, we pick a set of sampled paths P according to the strategy in Sect. 2.2. In
a path $P_u = \{v_{u_1}, v_{u_2}, \cdots, v_{u_m}\}$, we apply a two-layer attention mechanism to
learn the weight of item v_{u_i},

$$\alpha_{v_{u_i}}^{(1)} = \text{relu}\left(W_1 h'_{v_{u_i}} + b_1\right), \tag{9}$$

$$\alpha_{v_{u_i}}^{(2)} = \text{relu}\left(W_2 \alpha_{v_{u_i}}^{(1)} + b_2\right), \tag{10}$$

where W_1 and W_2 represent the transformation weight, b_1 and b_2 represent the
bias, $relu$ is the activation function. We adopt the softmax function to normalize
the attention weights to obtain the final item weights $\alpha_{v_{u_i}}$,

$$\alpha_{v_{u_i}} = \frac{\exp\left(\alpha_{v_{u_i}}^{(2)}\right)}{\sum_{v \in P_u} \exp\left(\alpha_v^{(2)}\right)}. \tag{11}$$

After we obtain the attention scores $\alpha_{v_{u_i}}$ for each item v_{u_i}, the embedding
for sampled path P_u can be expressed as the following sum,

$$h_{P_u} = \sum_{v_{u_i} \in P_u} \alpha_{v_{u_i}} \cdot h'_{v_{u_i}}, \tag{12}$$

where $h'_{v_{u_i}}$ are embedding for the item v_{u_i} learned in Eq. 8.

3.3   Path Aggregation Layer

After getting a set of interactive paths embedding for relation R_i, we hope to
enrich the interaction of paths through strong coupling operations on sampled
paths, to obtain additional structural information. By modeling each path independently, we can more comprehensively understand how different users establish
the relation. Considering that each sampled path may represent different semantics, we use PRA algorithm introduced in Sect. 2.3 to get the weight θ_u of path
P_u. θ_u can be calculated by logistic regression, which can be interpreted as P_u's
contribution to relation R_i. We normalize the weight by softmax function to get
the final weight θ'_u,

$$\theta'_u = \frac{\exp(\theta_u)}{\sum_{i=1}^{n} \exp(\theta_i)}. \tag{13}$$

After we gain the weight of each path, we aggregate all interactive path embeddings into a single path vector representation through a linear model. The new embedding is given as follows, which is also the embedding of relation R_i,

$$h_{R_i} = \sum_{P_u \in P} \theta'_u \cdot h_{P_u}. \tag{14}$$

Since the importance of each relation in users' interaction is different, we use HITS algorithm to get the strength of each relation in the original sequence. Given the relation R_i, its importance can be calculated as follows,

$$w_{R_i} = \sum_{x=i}^{i+1} \text{auth}(v_x) + hub(v_x). \tag{15}$$

Since the attention weights are generated for each relation, the whole cold-start sequence can be represented as follows,

$$h_{Seq} = \sum_{i=1}^{l-1} h_{R_i} \cdot \text{softmax}(w_{R_i}). \tag{16}$$

3.4   Prediction Layer

After the path aggregation, we obtain the updated cold-start sequence representation. We define $\hat{y}_i$ as the dot product of the initial representation h_{v_i} with the new sequence representation h_{Seq},

$$\hat{y}_i = \text{softmax}(h_{v_i} \cdot h_{Seq}), \tag{17}$$

where $\hat{y}_i$ denotes the probability that the user clicks item v_i in the next step. And we apply the cross-entropy of the prediction results $\hat{y}$ as the loss function,

$$\mathcal{L}(\hat{\mathbf{y}}) = -\sum_{i=1}^{m} \mathbf{y}_i \log(\hat{\mathbf{y}}_i) + (1 - \mathbf{y}_i) \log(1 - \hat{\mathbf{y}}_i), \tag{18}$$

where y is the one-hot encoding vector of the ground truth item.

4   Experiments

In this section, we conduct experiments on three public datasets to evaluate the performance of PathRec by answering the following four questions:

- RQ1: How does the method PathRec we proposed perform compared state-of-the-art sequential recommendation models?
- RQ2: Can the performance of PathRec be improved by learn item embedding from a global perspective?
- RQ3: Is it effective to alleviate the cold-start user problem by enriching the cold-start sequences?
- RQ4: What are the effects of different hyper-parameter settings on PathRec's accuracy?

4.1 Datasets

We evaluate the recommender on three public datasets *ml-1m*, *Amazon*, and *LAST.FM*. The basic statistics of the datasets are shown in Table 1. The average length is the average of the user's historical interactions. We consider users with less than five interactions with items as cold-start users. *MovieLens*MovieLens is a widely used dataset that contains user ratings of movies. It has various versions for different sizes, and we choose the *ml-1m*. *LAST.FM* contains the music playback records of more than one hundred thousand users. *Amazon* is a dataset of book purchase records from Amazon.com.

For all datasets, we filter out the sequences of length less than 2, and the items with fewer than 5 interactions. And we sort items that the user clicks by timestamp. As the *Amazon* and *LAST.FM* dataset are too large, we filter out the items that are clicked repeatedly. According to the number of clicks, we select the most frequent 5,000 and 6000 items, respectively. In the experiment, we divide each dataset into the training set, test set, and verification set in a ratio of 7:2:1.

Table 1. Statistics of the datasets.

Dataset	ml-1m	Amazon	LAST.FM
#users	6,039	70,570	106,518
#items	3,702	4,999	5735
#actions	999,868	581,546	6,997,127
#sparsity	4.47%	0.16%	1.14%
#cold-start user	0	24,662	2810
#proportion of cold-start user	0%	34.95%	2.64%

4.2 Baselines

To demonstrate the performance of our model PathRec, we compare it with seven sequential recommenders.

- POP: It directly recommends the top-N items that are most frequent in the training set.
- FPMC [18]: It combines the matrix factorization and the Markov chain for sequential recommendation.
- GRU4REC [4]: It applies RNN to model the sequential dependencies and makes recommendations.
- Caser [19]: It is a CNN-based model with an attention mechanism, which views the items embedding as an "image". It learns sequential patterns to provide more accurate recommendations.
- STAMP [15]: It is a short-term attention/memory priority model to get the user's general interests.

- SR-GNN [24]: It employs GNN with the soft-attention mechanism to model sequences as graph data and combines current interests and long-term preferences to predict user behavior.
- MCRec [5]: It learns representations of users, items, and meta-path-based context to provide users with more accurate recommendations.
- PathRec: Our proposed model.

4.3   Experimental Setup

Evaluation Metrics. We employ ranking metrics MRR@N, NDCG@N, and HR@N to evaluate the model. In the experiment, we set N to 10 and 20.

Parameter Setting. Following previous work [13,23,24], we set the batch size to 100, and the dimension of the latent vector to 100. Furthermore, the L2 penalty is set to 10^{-5}, and the initial learning rate is set to 0.001 and decays 0.1 after every three epochs. The dropout rate is in 0, 0.2, 0.4, 0.6, 0.8. In case some sequences are too long, we limit the maximum sequence length of *Amazon*, *LAST.FM* and *ml-1m* datasets to 10, 30, and 40, respectively.

4.4   Overall Performance (RQ1)

Table 2 shows the overall results of our proposed model and the seven baselines on the three public datasets. The best method in each column is bolded, and the second-best is underlined. Regarding the results, we note that:

- **Our model PathRec performs best on most occasions.** We observed that our model can adapt to the datasets with a higher proportion of cold-start users. It proves that PathRec is efficient in dealing with sequential recommendations for cold-start users and has no negative effect on warm-start users (according to the results on the *ml-1m* dataset).
- **Traditional methods like POP and FPMC cannot tackle the cold-start problem.** POP only recommends popular items for users, the items with less frequency will never be recommended. The performance of FPMC is better than POP, but it assumes that the current state only relies on its previous state, which limits its accuracy.
- **Neural network-based methods outperform traditional methods.** Since the RNN is suitable for modeling sequence data, it is widely applied in SRS. GRU4REC and Caser model user-item interactions to capture users' preferences. For this reason, their performance is much better than traditional methods. Compared with them, SR-GNN further improves performance, which shows that it is more advisable to model sequences as graph. Besides, MCRec uses meta-path-based context to get better recommendations. However, its meta-path is obtained by random walking, resulting in unstable performance.

Table 2. Performance comparisons of all models on the three datasets.

Datasets	Metric	N	POP	FPMC	GRU4REC	Caser	STAMP	SR-GNN	MCRec	PathRec	Improvement
ml-1m	MRR	10	0.0104	0.0498	0.0441	0.0719	0.0856	<u>0.0892</u>	0.0557	**0.0902**	1.12%
		20	0.0120	0.0544	0.0487	0.0790	0.0924	<u>0.0956</u>	0.0623	**0.0986**	3.14%
	NDCG	10	0.0114	0.0700	0.0618	0.1011	0.1141	<u>0.1185</u>	0.6900	**0.1203**	1.51%
		20	0.0159	0.0867	0.0789	0.1312	0.1390	<u>0.1421</u>	0.1048	**0.1469**	3.38%
	HR	10	0.0307	0.1366	0.0789	0.1975	0.2081	<u>0.2151</u>	0.1388	**0.2187**	1.67%
		20	0.0551	0.2025	0.1884	0.2889	0.3070	<u>0.3087</u>	0.2346	**0.3169**	2.66%
Amazon	MRR	10	0.0095	0.0594	0.0585	0.0422	0.0704	0.0716	<u>0.0787</u>	**0.0821**	4.32%
		20	0.0111	0.0629	0.0630	0.0466	0.0747	0.0758	<u>0.0821</u>	**0.0845**	2.92%
	NDCG	10	0.0109	0.0766	0.0779	0.0595	0.0899	<u>0.0909</u>	0.0845	**0.0986**	8.47%
		20	0.0151	0.0897	0.0945	0.0756	0.1060	<u>0.1064</u>	0.1057	**0.1165**	9.49%
	HR	10	0.0291	0.1332	0.1419	0.1169	0.1542	<u>0.1544</u>	0.1534	**0.1668**	8.03%
		20	0.0521	0.1852	0.2079	0.1812	0.2181	<u>0.2159</u>	0.1937	**0.2297**	6.39%
LAST.FM	MRR	10	0.0395	0.0544	0.0591	0.0649	0.0743	<u>0.0803</u>	0.0657	**0.0824**	2.62%
		20	0.0445	0.0623	0.0642	0.0729	0.0824	**0.0880**	0.0721	<u>0.0857</u>	−2.61%
	NDCG	10	0.0578	0.0824	0.0806	0.0976	0.1077	<u>0.1103</u>	0.8526	**0.1115**	1.09%
		20	0.0761	0.1115	0.0992	0.1267	0.1375	**0.1387**	0.1147	<u>0.1369</u>	−1.30%
	HR	10	0.1189	0.1764	0.1517	0.1862	0.2189	<u>0.2098</u>	0.1745	**0.2132**	1.62%
		20	0.1916	0.2318	0.2254	0.2918	0.2871	<u>0.3225</u>	0.2614	**0.3275**	1.55%

Table 3. Effects of the global graph.

Dataset	Model	MRR	NDCG	HR
Amazon	w/o Global Graph	0.0703	0.0859	0.1448
	w/o HITS	0.0786(+11.81%)	0.0920(+9.46%)	0.1592(+9.94%)
	w/ Global Graph	0.0814(+15.79%)	0.0973(+13.30%)	0.1661(+14.71%)
ml-1m	w/o Global Graph	0.0772	0.1034	0.1945
	w/o HITS	0.0831(+7.64%)	0.1109(+7.25%)	0.2079(+6.89%)
	w/ Global Graph	0.0895(+15.93%)	0.1189(+14.99%)	0.2172(+12.44%)

4.5 Ablation Study

Impact of Global Graph (RQ2). We construct the global graph to learn the representations of the items and apply HITS algorithm to mine the core interests of users. Now we investigate whether the global graph we constructed are necessary. Table 3 reports that the performance is better than the baseline under each condition, which shows the effectiveness of the global graph. It proves that capturing user preferences from a global perspective is significant.

Impact of Interactive Paths (RQ3). Our model exploits the rich sequential dependencies within the interactive paths to provide suitable recommendations for cold-start users. To evaluate the interactive paths, we use four models for comparison. As shown in Table 4, we notice that the performances of the model with the PRA algorithm and the model with the attention mechanism are both improved, which demonstrates that distinguishing the importance of items in the

Table 4. Effects of the interactive paths.

Dataset	Model	MRR	NDCG	HR
Amazon	w/o Sampled Paths	0.0689	0.0821	0.1437
	w/ PRA	0.0744(+7.98%)	0.0887(+8.04%)	0.1536(+6.89%)
	w/ Attention Mechanism	0.0791(+14.80%)	0.0943(+14.86%)	0.1621(+12.80%)
	w/ Sampled Paths	0.0817(+18.58%)	0.0977(+19.00%)	0.1663(+15.73%)
ml-1m	w/o Sampled Paths	0.0764	0.1024	0.1906
	w/ PRA	0.0819(+7.20%)	0.1091(+6.54%)	0.2007(+5.30%)
	w/ Attention Mechanism	0.0857(+12.17%)	0.1136(+10.94%)	0.2079(+9.08%)
	w/ Sampled Paths	0.0896(+17.28%)	0.1193(+16.50%)	0.2181(+14.43%)

path and obtaining the strength of each relation in the sequence is necessary. The model with the interactive paths works best, which means enriching rich and complementary information by strongly coupling the paths is significant.

4.6   Hyper-parameter Study (RQ4)

Effectiveness of Embedding Size. The embedding size is associated with the specific dataset and generally needs to be evaluated by specific tasks. Figure 3(a) shows how different embedding sizes affect the model. The embedding sizes range from 40 to 120. We notice that the low dimensions do not perform well due to their poor representation while the high dimensions may lead to overfitting problems. All four datasets achieve the best performance when the embedding size is 100.

Effectiveness of the Number of Sampled Paths. We select some sampled paths to enrich the interactive dependencies and control the number of them by limiting their length. To prove its effectiveness, we study the impact of the number of sampled paths on performance. As shown in Fig. 3(b), too few paths provide little performance improvement due to lacking of dependencies. Too many paths will greatly increase the time cost and introduce some unnecessary redundant information.

Effectiveness of the Proportion of Cold-Start Users. To prove that our model can adapt to the cold-start users, we discuss the performance of PathRec and some baselines under different cold-start scenarios on *Amazon* dataset. The results in Fig. 3(c) confirm that PathRec consistently outperforms the competition under these scenarios, which proves that PathRec can effectively alleviate the user cold-start problem. The performance of all models decreases as the proportion of cold-start users increases, but PathRec's performance decline is relatively small compared to other models.

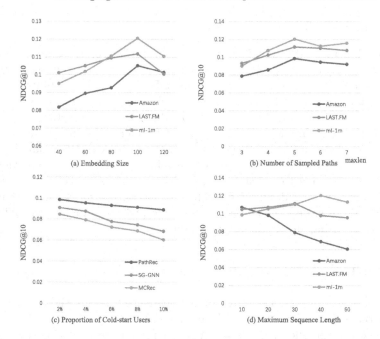

Fig. 3. Performance comparison of the PathRec using different parameter settings.

Effectiveness of the Maximum Sequence Length. To better handle the sequences, we restrict the maximum length of the sequence. The results are shown in Fig. 3(d). Since the data characteristics of each dataset are different, we have to treat it differently. In the comparison experiment, we set the suitable length for each dataset for optimal performance.

5   Related Work

5.1   Sequential Recommendation

As traditional sequential recommendation methods, Markov Chains-based methods achieve remarkable performance but ignore the long-term dependencies [18]. Lately, deep neural networks is introduced to SRS. Due to its natural advantages in modeling sequential dependency, many RNN-based methods are proposed. GRU4REC [4] utilizes RNN-based Gate Recurrent Units for sequential recommendation. KPRN [22] uses LSTM to model long-term dependences. CNN has been applied in SRS because of its success in other domain [7]. [20] models user behavior sequences based on CNN. Considering the advantages of GNN in graph representation learning, SR-GNN [24] applies GNN to capture users' preferences. GCE-GNN [23] explores item transitions in sequences for inferring users' interests and achieves good results. Therefore, We employ GNN to better capture the user-item interactions in sequences.

5.2 Cold-Start Recommendation

The main difficulty with the cold-start problem is the lack of information required for recommendations. Depending on how information is collected, methods can be divided into explicit and implicit methods. Explicit methods obtain information by interacting directly with the user, such as asking users to fill out questionnaires or rate several given items [1,9]. But users are reluctant to put too much effort into it. Implicit methods attempt to infer users' intentions by using auxiliary information like social media [2,3]. However, with the gradual awareness of personal privacy, it is hard to obtain valid personal information. MetaTL utilizes meta-learning to capture the transition patterns in the sequences [21]. But the model is only designed for cold-start users, not for warm-start users. Therefore, we propose a sequential recommender, which can adapt to cold-start users without using any auxiliary information.

6   Conclusion

In this work, we propose a novel model for sequential recommendation without any external information. We view the interlinks between the users and the items as interactive paths and give suitable recommendations by exploring the sequential dependencies within the paths. To better provide rich and complementary information, we learn the item embeddings through graph neural networks and strongly coupled the paths. Extensive experiments on three public datasets confirm our model is well adapted to cold-start users and outperforms the baseline methods.

Acknowledgements. This work was supported in part by the National Natural Science Foundation of China under Grant 62172351, the "14th Five-Year Plan" Civil Aerospace Pre-Research Project of China under Grant D020101, the Fund of Prospective Layout of Scientific Research for NUAA(Nanjing University of Aeronautics and Astronautics, and the Practice Innovation Program of NUAA (Grant No. xcxjh20221605).

References

1. Christakopoulou, K., Radlinski, F., Hofmann, K.: Towards conversational recommender systems. In: Proceedings of the 22nd ACM SIGKDD International Conference on Knowledge Discovery and Data Mining, pp. 815–824 (2016)
2. Fernández-Tobías, I., Braunhofer, M., Elahi, M., Ricci, F., Cantador, I.: Alleviating the new user problem in collaborative filtering by exploiting personality information. User Model. User Adap. Inter. **26**(2), 221–255 (2016)
3. Herce-Zelaya, J., Porcel, C., Bernabé-Moreno, J.: New technique to alleviate the cold start problem in recommender systems using information from social media and random decision forests. Inf. Sci. **536**, 156–170 (2020)
4. Hidasi, B., Karatzoglou, A., Baltrunas, L., Tikk, D.: Session-based recommendations with recurrent neural networks. Computer Science (2015)

5. Hu, B., Shi, C., Zhao, W.X., Yu, P.S.: Leveraging meta-path based context for top-N recommendation with a neural co-attention model. In: Proceedings of the 24th ACM SIGKDD International Conference on Knowledge Discovery & Data Mining, pp. 1531–1540. ACM (2018)
6. Kang, W.C., McAuley, J.: Self-attentive sequential recommendation. In: 2018 IEEE International Conference on Data Mining (ICDM), pp. 197–206. IEEE (2018)
7. Karpathy, A., Toderici, G., Shetty, S., Leung, T., Sukthankar, R.: Large-scale video classification with convolutional neural networks. In: Proceedings of the IEEE Conference on Computer Vision and Pattern Recognition, pp. 1725–1732 (2014)
8. Lee, H., Im, J., Jang, S.: MeLU: meta-learned user preference estimator for cold-start recommendation. In: Proceedings of the 25th ACM SIGKDD International Conference on Knowledge Discovery & Data Mining, pp. 1073–1082 (2019)
9. Lei, W., et al.: Interactive path reasoning on graph for conversational recommendation. In: Proceedings of the 26th ACM SIGKDD International Conference on Knowledge Discovery & Data Mining, pp. 2073–2083 (2020)
10. Li, B.H., Liu, Y., Zhang, A.M., Wang, W.H., Wan, S.: A survey on blocking technology of entity resolution. J. Comput. Sci. Technol. **35**(4), 769–793 (2020)
11. Li, C., Wang, F., Yang, Y., Li, Z., Zhang, X.: Exploring social network information for solving cold start in product recommendation. In: Wang, J., et al. (eds.) WISE 2015. LNCS, vol. 9419, pp. 276–283. Springer, Cham (2015). https://doi.org/10.1007/978-3-319-26187-4_24
12. Li, J., Wang, Y., McAuley, J.: Time interval aware self-attention for sequential recommendation. In: Proceedings of the 13th International Conference on Web Search and Data Mining, pp. 322–330 (2020)
13. Li, J., Ren, P., Chen, Z., Ren, Z., Lian, T., Ma, J.: Neural attentive session-based recommendation. In: Proceedings of the 2017 ACM on Conference on Information and Knowledge Management, pp. 1419–1428 (2017)
14. Li, J., Jing, M., Lu, K., Zhu, L., Yang, Y., Huang, Z.: From zero-shot learning to cold-start recommendation. In: Proceedings of the AAAI Conference on Artificial Intelligence, vol. 33, pp. 4189–4196 (2019)
15. Liu, Q., Zeng, Y., Mokhosi, R., Zhang, H.: STAMP: short-term attention/memory priority model for session-based recommendation. In: Proceedings of the 24th ACM SIGKDD International Conference on Knowledge Discovery & Data Mining, pp. 1831–1839 (2018)
16. Liu, Y., Li, B., Zang, Y., Li, A., Yin, H.: A knowledge-aware recommender with attention-enhanced dynamic convolutional network. In: Proceedings of the 30th ACM International Conference on Information & Knowledge Management, pp. 1079–1088 (2021)
17. Phan, M.C., Sun, A., Tay, Y., Han, J., Li, C.: NeuPL: attention-based semantic matching and pair-linking for entity disambiguation. In: Proceedings of the 2017 ACM on Conference on Information and Knowledge Management, CIKM 2017, Singapore, 06–10 November 2017, pp. 1667–1676. ACM (2017)
18. Rendle, S., Freudenthaler, C., Schmidt-Thieme, L.: Factorizing personalized Markov chains for next-basket recommendation. In: Proceedings of the 19th International Conference on World Wide Web, pp. 811–820 (2010)
19. Tang, J., Wang, K.: Personalized top-n sequential recommendation via convolutional sequence embedding. In: Proceedings of the 11th ACM International Conference on Web Search and Data Mining, pp. 565–573 (2018)
20. Tuan, T.X., Phuong, T.M.: 3d convolutional networks for session-based recommendation with content features. In: Proceedings of the 11th ACM Conference on Recommender Systems, pp. 138–146 (2017)

21. Wang, J., Ding, K., Caverlee, J.: Sequential recommendation for cold-start users with meta transitional learning. In: Proceedings of the 44th International ACM SIGIR Conference on Research and Development in Information Retrieval, pp. 1783–1787 (2021)

22. Wang, X., Wang, D., Xu, C., He, X., Cao, Y., Chua, T.S.: Explainable reasoning over knowledge graphs for recommendation. In: Proceedings of the AAAI Conference on Artificial Intelligence, vol. 33, pp. 5329–5336 (2019)

23. Wang, Z., Wei, W., Cong, G., Li, X.L., Mao, X.L., Qiu, M.: Global context enhanced graph neural networks for session-based recommendation. In: Proceedings of the 43rd International ACM SIGIR Conference on Research and Development in Information Retrieval, pp. 169–178 (2020)

24. Wu, S., Tang, Y., Zhu, Y., Wang, L., Xie, X., Tan, T.: Session-based recommendation with graph neural networks. In: Proceedings of the AAAI Conference on Artificial Intelligence, vol. 33, pp. 346–353 (2019)

25. Xu, K., Ba, J., Kiros, R.: Show, attend and tell: neural image caption generation with visual attention. In: Proceedings of the 32nd International Conference on Machine Learning, ICML, vol. 37, pp. 2048–2057 (2015)

26. Zang, Y. et al.: GISDCN: A graph-based interpolation sequential recommender with deformable convolutional network. In: Database Systems for Advanced Applications, DASFAA 2022. Lecture Notes in Computer Science, vol. 13246, pp. 289–297 (2022). Springer, Cham. https://doi.org/10.1007/978-3-031-00126-0_21

27. Huang, Y., Zhao, F., Gui, X., Jin, H.: Path-enhanced explainable recommendation with knowledge graphs. World Wide Web **24**(5), 1769–1789 (2021). https://doi.org/10.1007/s11280-021-00912-4

28. Li, C., Zhou, B., Lin, W., et al.: A personalized explainable learner implicit friend recommendation method. Data Sci. Eng. **8**, 23–35 (2023). https://doi.org/10.1007/s41019-023-00204-z

A Topic-Aware Graph-Based Neural Network for User Interest Summarization and Item Recommendation in Social Media

Junyang Chen[1,2(✉)], Ge Fan[3], Zhiguo Gong[4], Xueliang Li[1,5(✉)], Victor C. M. Leung[1], Mengzhu Wang[1], and Ming Yang[4]

[1] College of Computer Science and Software Engineering, Shenzhen University, Shenzhen, China
{junyangchen,lixueliang}@szu.edu.cn
[2] Guangdong Laboratory of Artificial Intelligence and Digital Economy (SZ), Shenzhen, China
[3] Tencent Inc., Shenzhen, China
[4] State Key Laboratory of Internet of Things for Smart City, Department of Computer Information Science, University of Macau, Macau, China
[5] National Engineering Laboratory for Big Data System Computing Technology, Shenzhen University, Shenzhen, China

Abstract. User-generated content is daily produced in social media, as such user interest summarization is critical to distill salient information from massive information. While the interested messages (e.g., tags or posts) from a single user are usually sparse becoming a bottleneck for existing methods, we propose a topic-aware graph-based neural interest summarization method (UGraphNet), enhancing user semantic mining by unearthing potential user relations and jointly learning the latent topic representations of posts that facilitates user interest learning. Experiments on two datasets collected from well-known social media platforms demonstrate the superior performance of our model in the tasks of user interest summarization and item recommendation. Further discussions also show that exploiting the latent topic representations and user relations are conductive to the user automatic language understanding.

1 Introduction

Social networking platforms enjoy a great popularity for individuals to share opinions and exchanging information [4,5]. Then, millions of user-generated posts are produced

---

This work was supported by National Natural Science Foundation of China under Grant No. 62102265, by the Open Research Fund from Guangdong Laboratory of Artificial Intelligence and Digital Economy (SZ) under Grant No. GML-KF-22-29, by the Natural Science Foundation of Guangdong Province of China under Grant No. 2022A1515011474, by the Science and Technology Development Fund, Macau SAR, China, under Grant No. (0068/2020/AGJ, SKL-IOTSC(UM)-2021-2023), and by Shenzhen Talents Special Project - Guangdong Provincial Innovation and Entrepreneurship Team Supporting Project under Grant No. 2021344612).
X. Li—unique corresponding author

X. Wang et al. (Eds.): DASFAA 2023, LNCS 13944, pp. 537–546, 2023.
https://doi.org/10.1007/978-3-031-30672-3_36

Table 1. POI: "The Vortex Bar And Grill" on Yelp. Posts are the short comments from different users for the POI. User interest summarization is the succinct description for these posts.

POI: The Vortex Bar And Grill - Midtown
Source posts from social media:
User 1: Best burgers in town.
User 2: The Yokohama Mama Burger for some Lunch!
User 3: Steakhouse Burger is one of the best.   Also, be sure to get onion rings as a side!
User 4: Yummy beer, and the beers aren't expensive either!   Comedy show!
User 5: Adding my vote as the best burger in ATL.
User 6: They have yamazaki whiskey.
User interest summarization: American (Traditional), Burgers, Restaurants,   Bars, Nightlife

daily, far outpacing the human being's reading and understanding capacity. As such, discovering gist information from large volume of posts becomes vital capability for current applications. Two characteristics can be summarized for current posts: short texts and word sparsity. As shown in Table 1, the source posts of one point-of-interest (POI) from social media (e.g., Yelp) usually only contain a few short sentences and their word co-occurrences are usually sparse. The user interest summarization can be considered as the keyphrases [22] to summarize the whole posts of a POI. These keyphrases can be further used for the downstream tasks, such as similar POI search [2,7], user sentiment analysis [6,20], POI recommendation [18,19], and so forth. Despite the widespread use of keyphrases, millions of posts are generated daily without summarization. Therefore, there exists a pressing need for automating user interest summarization for daily posts.

Most previous work focuses on extracting existed phrases from target posts. [20, 24] employ topic models to generate topical words as the keyphrases for a group of posts. These methods, ascribed to the limitation of most topic models, are incapable of generating non-existed keyphrases for each target post. More recently, [22] introduces a sequence generation framework that can generate keyphrases beyond the target post. It present a neural seq2seq model based on integrating more tweets related to the target post to generate keyphrases in a word-by-word training manner.

However, the aforementioned models encounter a common challenge: only limited number of relevant posts existed for one POI are encoded when processing posts from social media. To illustrate this challenge, we display Table 1 where a batch of posts are from Yelp. Such posts are the comments related to the POI "The Vortex Bar And Grill - Midtown". We can observe that each post only contains a few words, as such, this will inevitably encounter the sparsity problem. On way is to combine more relevant posts [22] to enrich the contents. However, even though all posts are focused on the same topic, it is difficult to summarize the keyphrases "American (Traditional), Burgers, Restaurants, Bars, Nightlife" due to the limited number and colloquial nature of social media language by looking at posts from 1 to 6 in Table 1.

To address the above challenges, we propose a novel graph-based neural interest summarization model (UGraphNet) that includes three complementary innovations. The first one is *user collaboration* that leverages neighboring information by construct the bipartite graph of user-post-user to enrich sparse contents. The second one is corpus-level *user latent topic modeling* with the constructed graph and the users interested posts. The last one is joint modeling the *latent topic embedding* of all users and the interest prediction of the target users. These approaches can effectively improve the accuracy and alleviate data sparsity in the tasks of user interest summarization and item recommendation. In general, the contributions of this work are as follows:

- To the best of our knowledge, our work is the first to study the benefit of leveraging user relations and latent topics on social media interest summarization and item recommendation. Also, our model enables an end-to-end training process.
- We propose there main components: a contrastive learning loss, a topic modeling loss, and a graph-based learning loss to achieve the above purposes by their jointly learning.
- We experiment on two newly constructed social media datasets. Our model can significantly outperform all the comparison methods. Ablation analysis also demonstrates the effectiveness of exploiting the latent topic representations and user relations in user automatic language understanding.

2 Proposed Model

In this section, we describe the proposed framework that how to leverage user collaboration and latent topics for the user interest summarization. Figure 1 shows the overall architecture consisting of three modules - a contrastive learning loss, a topic modeling loss, and a graph-based generative learning loss. Formally, given a collection D of social media posts, we process each post into bags-of-words word vector $[t_1, t_2, ..., t_{|V|}]$, which is a V-dim vector over the vocabulary and V denotes its size. Besides, each post consists of latent topics and we denote the topic size as $|K|$. Below we first introduce our three modules and then describe how they are jointly trained.

2.1 Contrastive Learning Loss

We exploit user collaboration by constructing the adjacent graph of users. Specifically, when two users are interested in the same posts, we make a connection between these users. Besides, it is difficult for every user has an unique embedding in the large-scale scenarios, which will inevitably make the number of parameters tremendous. As such, inspired by the work [8] that they represent the users by the terms of queries, we represent the users with a smaller number of tag embeddings. In other word, each user can be represented with a limited number of tags. Here we also use one-hot encoding to represent the tag (or call word) lexicon $(t_1, ..., t_{|V|})$. Then, we map the tags to d-dimensional vectors with a mapping function f to represent users as follows:

$$\mathbf{h}_{v_t} = f((t_1, ..., t_{|V|}), \mathbf{M}),\tag{1}$$

where $\mathbf{h}_{v_t} \in \mathbb{R}^d$ denotes the embedding of a user v_t, and $\mathbf{M} \in \mathbb{R}^{|V| \times d}$ is the transformation matrix. After that, we adopt an attention method to fuse the information of a

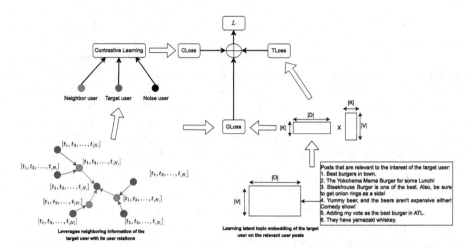

Fig. 1. Overview of the proposed UGraphNet model.

target user and its neighbors. First we perform the message propagation step for dealing with the messages passing from neighboring nodes, which is given by:

$$m_{v_i \leftarrow v_j} = \text{MLP}(n_{v_j v_i} \oplus h_{v_j}) \cdot h_{v_j}, \tag{2}$$

where $m_{v_i \leftarrow v_j} \in \mathbb{R}^d$ denotes the information passing from node v_j to v_i, $n_{v_j v_i}$ is one-hot encoded of the neighbor type (e.g., one-hop (0, 1) or multi-hop neighbors (1, 0)), $\text{MLP}(\cdot) \in \mathbb{R}^{d \times d}$ denotes a Multi-Layer Perception that takes as inputs both the neighbor type $n_{v_j v_i}$ and the representations of the user h_{v_j}, and $\oplus$ represents the concatenation.

Then, we aggregate the information of the target node and the messages passing from its neighbors in an attention way. The weight coefficient α_{v_i, v_j} between two nodes can be formulated by:

$$\alpha_{v_i, v_j} = \frac{\exp\left(\sigma(\mathbf{a}^T \cdot [\boldsymbol{W} h_{v_i} \| \boldsymbol{W} m_{v_i \leftarrow v_j}])\right)}{\sum_{v_k \in \mathcal{N}_{v_i}} \exp\left(\sigma(\mathbf{a}^T \cdot [\boldsymbol{W} h_{v_i} \| \boldsymbol{W} m_{v_i \leftarrow v_k}])\right)}, \tag{3}$$

where $\boldsymbol{W} \in \mathbb{R}^{d \times d}$ is a shared weight matrix for mapping nodes into the same embedding space, $\mathbf{a} \in \mathbb{R}^{2d}$ denotes a weight vector for learning the relations of the target node and its neighbors, and $\mathcal{N}_{v_i}$ is the set of neighbors of node v_i, and σ denotes the sigmoid function [11].

After that, with the learned weight coefficients α_{v_i, v_j} and the neighboring message information $m_{v_i \leftarrow v_j}$, the final representations of node v_i can be formulated by:

$$\mathbf{h}_{v_t}^L = \text{ReLU}\left(\sum_{v_j \in \mathcal{N}_{v_i}} \alpha_{v_i, v_j} \mathbf{W} m_{v_i \leftarrow v_j}\right), \tag{4}$$

where ReLU is an activation function [9], and L denotes the last layer of the network.

Finally, inspired by the recent advances in the contrastive learning work [12,13], we introduce a contrastive learning loss $\mathcal{L}_c$ formulated by:

$$\mathcal{L}_c = \sum_{(v_t, v_p, v_n) \in \mathcal{T}} [\sigma(v_t, v_p; \boldsymbol{h}) - \sigma(v_t, v_n; \boldsymbol{h}) + \nabla]_+, \tag{5}$$

where $\boldsymbol{h}$ denotes the hidden embeddings of users, v_t is the target user, v_p denotes its neighbor users, v_n is the negative users drawn from the whole set by using the alias table method [15] that only takes $O(1)$ time, ∇ is a margin hyper-parameter separating the positive pair and the corresponding negative one (we set it as 0.5 in the experiments), $\mathcal{T}$ denotes a training batch, and $[\cdot]_+$ denotes the positive part of the calculation. The above contrastive learning loss (Eq. (5)) explicitly encodes similarity ranking among node pairs into the embedding vectors.

2.2 Topic Modeling Loss

In this part, we refer to a matrix factorization [1] method to obtain the topic modeling loss $\mathcal{L}_t$. More concretely, given the document-word matrix $\mathbf{D}$, we decompose it into the product of the document-topic embedding matrix $\boldsymbol{\Theta}$ and the topic-word embedding matrix $\mathbf{T}$ with regularization as follows:

$$\mathcal{L}_t = \sum_{i \in \mathcal{T}} (\mathbf{D}_i - \boldsymbol{\Theta}_i \mathbf{T})^2 + \lambda(||\boldsymbol{\Theta}_i||_2^2 + ||\mathbf{T}||_2^2), \tag{6}$$

where $\mathbf{D} \in \mathbb{R}^{|D| \times V}$, D denotes the set of documents, V is the vocabulary size, $\boldsymbol{\Theta} \in \mathbb{R}^{|D| \times k}$, $\mathbf{T} \in \mathbb{R}^{k \times V}$, k is the dimension of the topic embedding, $|| \cdot ||_2^2$ is the l_2 norm regularization of the parameters, and λ is a harmonic factor for regularization. In Eq. (6), we explore the latent topics of the posts that are interested by the target user. Besides, the obtained document-topic embedding $\boldsymbol{\Theta}$ will be used in the generative learning in the following section.

2.3 Generative Learning Loss

With the target user embedding $\mathbf{h}_{v_t}^L$ from Eq. (4) that represents the user collaboration information, and the document-topic embedding $\boldsymbol{\Theta}_{v_t}$ from Eq. (6) that represents the interests of the target user, we can construct the generative learning loss $\mathcal{L}_g$ as follows:

$$\mathcal{L}_g = -\sum_{v_t \in \mathcal{T}} \log(\sigma([\mathbf{h}_{v_t}^L; \boldsymbol{\Theta}_{v_t}] \mathbf{W}_v)), \tag{7}$$

where $\mathbf{h}_{v_t}^L \in \mathbb{R}^{1 \times d}$, $\boldsymbol{\Theta}_{v_t} \in \mathbb{R}^{1 \times k}$, $\mathbf{W}_v \in \mathbb{R}^{(d+k) \times 1}$ are trainable weights, and $[;]$ denotes the concatenation operation. In Eq. (7), we aim to fuse the information of the two domains (i.e., the user relations and the interested latent topics) which exploits the assumption that relevant users may share similar interests.

2.4 Learning and Inference

In the training stage, we adopt stochastic gradient descent [14] to minimize the loss function of the total loss, which is given by:

$$\mathcal{L}_{total} = \mathcal{L}_c + \mathcal{L}_t + \mathcal{L}_g. \tag{8}$$

Table 2. The statistics of datasets.

Datasets	Delicious	Yelp
#Users	1847	7913
#Items	68755	12462
Avg. items interacted by per user	195.72	41.52
Avg. length of user summarization per item	3.67	6.36
Avg. length of description per item	7.09	12.15

With the above learning objective as shown in the Eq. (8), we can: (1) exploit the user collaboration information with the contrastive learning loss (Eq. (5)), (2) explore the latent topics of the semantic information to summarize user interests (Eq. (6)), (3) fuse the above information (Eq. (7)) to simultaneously learn them in an end-to-end way. **User Interest Inference:** Based on the concatenated embedding of user collaborative information $\mathbf{h}_{v_t}^L$ and user historical interest information Θ_{v_t}, we can conduct dot product with the topic-word embedding $\mathbf{T}$ to generate a ranking list of output words, where the top K ones serve as the user interest summarization in the evaluation. **Post Recommendation Inference:** Similarly, based on the $\mathbf{h}_{v_t}$ and $\Theta_{v_t}^L$ of the target user, we generate a ranking list with the document-topic embedding Θ of the output posts, where the top N ones serve as the post recommendation.

3 Experiments

3.1 Datasets

We adopt two real-world datasets to estimate the performance: *Delicious*[1] and *Yelp*[2] which are widely used in social recommendation [10,17]. The statistics of the datasets are shown in Table 2. Each dataset contains of users, items, the interactions including browse or access between users and items, user summarization of items, and item description. The "Avg. items interacted by per user" represents the average number of items that have been browsed or visited by users before. The "Avg. length of user summarization per item" denotes the average length of words that users summarize items. The "Avg. length of description per item" denotes the average length of words that are used to comprehensively describe the characters of items.

3.2 Comparison Methods

We include several traditional and state-of-the-art approaches that can be applied to user interest summarization, including probabilistic graph models and sequential learning models. Here are descriptions of selected methods: **GSDMM** [23] is a traditional and widely used probabilistic graph model which is designed for the short text modeling. The word and document representations are learned by combining Dirichlet and multinomial distributions. **DP-BMM** [5] is an another often used probabilistic graph model

[1] https://grouplens.org/datasets/hetrec-2011/.
[2] https://www.yelp.com/dataset.

Table 3. Main comparison results displayed with scores in %. Boldface scores in each column indicate the best results. The underlined scores denote the second best performance. Our model outperforms the strongest baselines with p-value < 0.05.

Model	Delicious				Yelp			
	HR@1	HR@5	HR@10	MAP	HR@1	HR@5	HR@10	MAP
Traditional models								
GSDMM	14.83	12.02	10.05	13.83	1.67	14.34	7.40	7.93
DP-BMM	16.56	14.75	14.38	17.15	4.16	17.68	11.12	12.41
State of the arts								
SEQ-TAG	20.16	22.03	21.79	21.03	12.21	27.49	21.62	20.56
SEQ2SEQ-CORR	23.21	29.59	24.27	23.10	15.17	32.70	22.31	22.96
TAKG	27.68	30.96	28.84	27.76	42.27	33.22	23.93	38.37
UGraphNet (Ours)	**34.56**	**35.51**	**33.07**	**34.63**	**55.99**	**35.09**	**25.30**	**48.20**
Improv.	24.86%	14.70%	14.67%	24.75%	32.46%	5.63%	5.73%	25.62%

which explicitly exploits the word-pairs constructed from each document to enhance the word co-occurrence pattern in short texts. It can deal with the topic drift problem of short text streams naturally. **SEQ-TAG** [24] is a state-of-the-art deep recurrent neural network model that can combines keywords and context information to automatically extract keyphrases from short texts. **SEQ2SEQ-CORR** [3] exploits a sequence-to-sequence (seq2seq) architecture for keyphrase generation which captures correlation among multiple keyphrases in an end-to-end fashion. **TAKG** [21] introduces a seq2seq based neural keyphrase generation framework that takes advantage of the recent advance of neural topic models [16] to enable end-to-end training of latent topic modeling and keyphrase generation.

Different from the above methods, we exploit the potential usefulness of user collaboration and the latent topics exhibited in the user interest and the item contents, which have been ignored in previous research and will be extensively studied here. We also present an ablation study to show the effectiveness of our proposed components.

3.3 User Interest Summarization Results

In this section, we examine our performance in user interest summarization for social media. The performance of the user summarization is accessed by calculating how many "hits" in an n-sized list of ranked words. To this end, we use popular information retrieval metrics *Hit Ratio* (HR) and *Mean Average Precision* (MAP) for evaluation. For the datasets Delicious and Yelp, most items are summarized by users with 3 to 6 on average (Table 2), thus HR@1, HR@5, HR@10 are reported. Besides, MAP is measured over the top 10 prediction for all datasets.

The main comparison results are shown in Table 3, where the highest scores are highlighted in boldface and the underlined ones denote the second best. The last row is the improvements of our method compared with the best baseline. In general, we can observe that:

Table 4. Ablation analysis. The highest scores are marked in boldface, and '_' denotes the second-best results.

Method	Delicious				Yelp			
	HR@1	HR@5	HR@10	MAP	HR@1	HR@5	HR@10	MAP
UGraphNet	**34.56**	**35.51**	**33.07**	**34.63**	**55.99**	**35.09**	**25.30**	**48.20**
w/o CLoss	26.36	12.30	6.88	10.80	53.02	33.56	22.23	46.99
w/o TLoss	4.92	5.63	4.44	5.28	52.07	25.79	15.81	14.29
w/o GLoss	28.64	27.22	26.44	24.76	12.54	11.15	9.71	3.25
Improv.	20.67%	30.46%	25.08%	39.86%	5.60%	4.56%	13.81%	2.58%

(1) Our model *UGraphNet* consistently outperforms other comparisons on all datasets under various metrics. This shows the usefulness of leveraging user neighboring information for their interest summarization. More concretely, *UGraphNet* achieves up to 24.86%, 14.70%, 14.67%, and 24.75% improvements over the second-best method *TAKG* in terms of HR@1, HR@5, HR@10 and MAP on Delicious. Besides, *TAKG* gains 32.46%, 5.63%, 5.73%, and 25.62% improvements on average against the second ones on Yelp. In general, the above improvements demonstrate the effectiveness of our method by jointly modeling user relations and user interests.

(2) Among the results of the baselines, the traditional methods including *GSDMM* and *DP-BMM* give poor performance. This indicates that user interest summarization is a challenging task. It is hard to rely on probabilistic graphical models to yield acceptable performance. On the contrary, seq2seq-based models consisting of *SEQ-TAG*, *SEQ2SEQ-CORR*, and *TAKG* yield better results than the traditional ones. Particularly, *TAKG* outperforms the other baselines, which suggests the helpful of exploiting latent topics in short texts. Interestingly, our model achieve larger improvements with a step further by exploring the user relations and their latent topics.

3.4   Ablation Analysis

To analyze the effectiveness of the proposed components on user interest summarization (introduced in Sect. 2) in our method, we conduct an ablation analysis as follows. In general, we have three ablated variants of our model:

    I. w/o CLoss (without constrastive learning loss): The CLoss (Eq. (5)) is used to exploit user relations that help to distinguish the target user from its neighboring users and negative users. We remove the CLoss and keep the TLoss and GLoss for comparison.

   II. w/o TLoss (without topic modeling loss): The TLoss (Eq. (6)) aims to exploit the latent topics in short texts which can alleviate the data sparsity in the user interest summarization.

  III. w/o GLoss (without generative learning loss): The TLoss (Eq. (7)) utilizes the assumption that relevant users share similar interests. We adopt it to generate keyphrases that is relevant to users' latent topics.

The results of the ablation tests are shown in Table 4. Our method *UGraphNet* outperforms the other variants. Specifically, *UGraphNet* achieves 20.67%, 30.46%, 25.08%, and 39.86% improvements over the second-best variant in terms of HR@1, HR@5, HR@10, and MAP on Delicious, and obtains 5.60%, 4.56%, 13.81%, and 2.58% gains on Yelp, respectively. These results validate that the user attention update gate is more appropriate to explore user interests. These results demonstrates the effectiveness of jointly learning different components. We observe that the influence levels of each component are presented as GLoss > CLoss > TLoss on Delicious. These results are identic with expectation that the generative loss contributes more than the topic modeling loss. This is because there are more dense interactions between users and items (i.e., Avg. items interacted by per user: 195.72) as shown in the statistics of datasets (Table 2), comparing with the sparsity of text information (i.e., Avg. length of user summarization per item (3.67)). Interestingly, the influence levels of components on Yelp are as: CLoss > TLoss > GLoss, which shows that the contrastive learning loss contributes the most while the generative loss contributes the lest. These results are also in accord with the characteristics in Table 2. There are more users information (#User: 7913) while less interactions between users and items (Avg. items interacted by per user: 41.52) in Yelp. Note that the above interesting points are observed by comparing results across the statistics of the datasets.

4   Conclusion

In general, we propose a topic-aware graph-based neural interest summarization method, called UGraphNet, that can enhance user semantic mining for user interest summarization and item recommendation in social media. The main innovations of our work include a contrastive learning loss, a topic modeling loss, and a graph-based learning loss that leverage user relations and latent topics on social media by jointly training. Experiments on two newly constructed social media datasets demonstrate that our model can significantly outperform all the comparison methods. Ablation analysis is also conducted to show the superiority of our proposed components.

References

1. Ahmed, N.K., et al.: Learning role-based graph embeddings. arXiv preprint arXiv:1802.02896 (2018)
2. Bansal, P., Jain, S., Varma, V.: Towards semantic retrieval of hashtags in microblogs. In: Proceedings of the 24th International Conference on World Wide Web, pp. 7–8 (2015)
3. Chen, J., Zhang, X., Wu, Y., Yan, Z., Li, Z.: Keyphrase generation with correlation constraints. arXiv preprint arXiv:1808.07185 (2018)
4. Chen, J., Gong, Z., Liu, W.: A nonparametric model for online topic discovery with word embeddings. Inf. Sci. **504**, 32–47 (2019)
5. Chen, J., Gong, Z., Liu, W.: A dirichlet process biterm-based mixture model for short text stream clustering. Appl. Intell. **50**(5), 1609–1619 (2020)
6. Davidov, D., Tsur, O., Rappoport, A.: Enhanced sentiment learning using twitter hashtags and smileys. In: Coling 2010: Posters, pp. 241–249 (2010)

7. Efron, M.: Hashtag retrieval in a microblogging environment. In: Proceedings of the 33rd International ACM SIGIR Conference on Research and Development in Information Retrieval, pp. 787–788 (2010)

8. Fan, S., et al.: Metapath-guided heterogeneous graph neural network for intent recommendation. In: Proceedings of the 25th ACM SIGKDD International Conference on Knowledge Discovery & Data Mining, pp. 2478–2486 (2019)

9. Glorot, X., Bordes, A., Bengio, Y.: Deep sparse rectifier neural networks. In: Proceedings of the 14th International Conference on Artificial Intelligence and Statistics, pp. 315–323 (2011)

10. Guo, Z., Wang, H.: A deep graph neural network-based mechanism for social recommendations. IEEE Trans. Industr. Inf. **17**(4), 2776–2783 (2020)

11. Han, J., Moraga, C.: The influence of the sigmoid function parameters on the speed of backpropagation learning. In: Mira, J., Sandoval, F. (eds.) IWANN 1995. LNCS, vol. 930, pp. 195–201. Springer, Heidelberg (1995). https://doi.org/10.1007/3-540-59497-3_175

12. Jaiswal, A., Babu, A.R., Zadeh, M.Z., Banerjee, D., Makedon, F.: A survey on contrastive self-supervised learning. Technologies **9**(1), 2 (2020)

13. Khan, A., AlBarri, S., Manzoor, M.A.: Contrastive self-supervised learning: a survey on different architectures. In: 2022 2nd International Conference on Artificial Intelligence (ICAI), pp. 1–6. IEEE (2022)

14. Kingma, D.P., Ba, J.: Adam: a method for stochastic optimization. arXiv preprint arXiv:1412.6980 (2014)

15. Li, A.Q., Ahmed, A., Ravi, S., Smola, A.J.: Reducing the sampling complexity of topic models. In: Proceedings of the 20th ACM SIGKDD International Conference on Knowledge Discovery and Data Mining, pp. 891–900 (2014)

16. Miao, Y., Grefenstette, E., Blunsom, P.: Discovering discrete latent topics with neural variational inference. In: International Conference on Machine Learning, pp. 2410–2419. PMLR (2017)

17. Wang, H., Wu, Q., Wang, H.: Factorization bandits for interactive recommendation. In: 31st AAAI Conference on Artificial Intelligence (2017)

18. Wang, W., Chen, J., Wang, J., Chen, J., Gong, Z.: Geography-aware inductive matrix completion for personalized point-of-interest recommendation in smart cities. IEEE Internet Things J. **7**(5), 4361–4370 (2019)

19. Wang, W., Chen, J., Wang, J., Chen, J., Liu, J., Gong, Z.: Trust-enhanced collaborative filtering for personalized point of interests recommendation. IEEE Trans. Industr. Inf. **16**(9), 6124–6132 (2019)

20. Wang, X., Wei, F., Liu, X., Zhou, M., Zhang, M.: Topic sentiment analysis in twitter: a graph-based hashtag sentiment classification approach. In: Proceedings of the 20th ACM International Conference on Information and Knowledge Management, pp. 1031–1040 (2011)

21. Wang, Y., Li, J., Chan, H.P., King, I., Lyu, M.R., Shi, S.: Topic-aware neural keyphrase generation for social media language. arXiv preprint arXiv:1906.03889 (2019)

22. Wang, Y., Li, J., King, I., Lyu, M.R., Shi, S.: Microblog hashtag generation via encoding conversation contexts. arXiv preprint arXiv:1905.07584 (2019)

23. Yin, J., Wang, J.: A dirichlet multinomial mixture model-based approach for short text clustering. In: Proceedings of the 20th ACM SIGKDD International Conference on Knowledge Discovery and Data Mining, pp. 233–242. ACM (2014)

24. Zhang, Q., Wang, Y., Gong, Y., Huang, X.J.: Keyphrase extraction using deep recurrent neural networks on Twitter. In: Proceedings of the 2016 Conference on Empirical Methods in Natural Language Processing, pp. 836–845 (2016)

MG-CR: Factor Memory Network and Graph Neural Network Based Personalized Course Recommendation

Yun Zhang[1] , Minghe Yu[2(✉)] , Jintong Sun[2] , Tiancheng Zhang[1] ,
and Ge Yu[1]

[1] School of Computer Science and Engineering, Northeastern University,
Shenyang, China
{tczhang,yuge}@mail.neu.edu.cn
[2] Software College, Northeastern University, Shenyang, China
yuminghe@mail.neu.edu.cn

Abstract. Course recommendations in universities select the most suitable courses for students according to their interests and academic requirements. However, existing works often focus on modeling course selection history, without paying sufficient attention to rich features contained in students' personal information and courses' teaching attributes, and various relationships between them. In order to realize the personalization and dynamics of course recommendation, we consider students and courses as two types of nodes to construct a heterogeneous information network (HIN), and propose a factor Memory network and Graph neural network based personalized Course Recommendation (MG-CR) on top of HIN. MG-CR captures multiple features of students and courses simultaneously in the input module, thereby maximizing the retention of personalized information. The feature transfer module optimizes course features by considering various types of relationships between courses in the HIN, and together with the input module, it ensures high recommendation accuracy in the cold-start scenario. The updating module explicitly stores the student's interest level for each factor in the state matrix, which is then used to dynamically recommend courses for students in the predicting module. Experimental results show that MG-CR provides accurate, dynamic and personalized recommendations, and has superior performance in the cold-start scenario.

Keywords: Course Recommendation · Heterogeneous Information Network · Factor Memory Network · Graph Neural Network

1 Introduction

Universities allow students a certain degree of freedom to select courses. Students choose optional courses according to their personal information and course teaching attributes. As shown in the left part of Fig. 1, course selection data constitutes a heterogeneous information network (HIN), which contains multiple types of relationships between courses. Taking the initiative to recommend suitable courses for students can help them contact new knowledge, improve abilities, and avoid failing the courses due to limited experience and weak autonomy. For colleges, course recommendation helps them understand students' preferences and requirements, which helps to design reasonable and effective training programs, so as to produce quality graduates.

Figure 1 illustrates the course recommendation process. For a student $S.1$, the model predicts his/her probability of choosing and successfully completing each course in the next semester based on course selection records, features contained in students' personal information and courses' teaching attributes, and relationships in the HIN, so as to make recommendations. Since student $S.1$'s knowledge level improves gradually over time, and his/her preference is affected by various factors, such as surrounding students, the content of the courses he has learned and credit requirements, course recommendation should be dynamic. In addition, $S.1$ tends to select courses that are related to each other, while modeling feature transfer between courses in HIN is helpful to optimize courses features and increase the probability of related courses being selected.

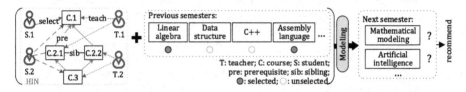

Fig. 1. Illustration of course recommendation process in course selection system.

Many efforts have been devoted to course recommendations. Some carry out a detailed analysis of data characteristics [14,21,33], demonstrating that the information of students and courses is very important for course recommendation. And works based on collaborative filtering (CF) [10,12], recurrent neural networks (RNN) [9],random walk [8,18–20], and attention mechanisms [35–37] have been proposed to provide accurate course recommendations. However, the current models still face a series of challenges. First, course information is studied in many works [3,4,8,14,17–19], but student features are often under-considered. In reality, students in one dormitory are more likely to select the same course, and gender has a great influence on students' selection of optional courses such as physical education. Therefore, we need to integrate various features for the recommendation. Second, existing research [8,18,19] models the student state as the hidden state of RNN, however explicitly modeling the student's preference for each factor that affects course selection will improve the explainability of the recommendation results. Third, there are multiple types of relationships between two courses, students have a greater probability of selecting courses associated with courses they have taken. However, few studies recommend students by measuring the relationships between courses, let alone consider multiple types of relationships.

To overcome the above challenges, we construct an HIN containing students, courses and various types of relationships, and propose a course recommendation model: factor Memory network and Graph neural network based personalized Course Recommendation (MG-CR) on the HIN. To realize personalization, MG-CR captures text description and attribute features of both students and courses, and represents them in the form of vectors. Memory-based methods like DKVMN [34] models students' knowledge mastery and the concept of problem through two matrixes. Inspired by it, we use a state matrix to store a student's interest states of different factors affecting course selection, and a factor matrix to store the vector of each factor. In addition, we construct a graph neural network (GNN) between courses in the HIN, and consider various relationships to model feature transfer. In order to dynamically realize course recommendations, MG-CR updates the state matrix of each student according to the courses they have selected. To summarize, the major contributions of our work are as follows:

- We build an HIN containing students, courses, teachers and various relationships to capture rich information contained in the course selection system.
- We integrate various features of students' personal information and courses' teaching attributes, as well as interactive information generated during the course selection process to ensure personalized recommendations.
- By introducing the memory network, we use a state matrix to store each student's interest state, and a factor matrix to store the vector corresponding to each factor, which is more explainable than just using the hidden state of RNN to represent each student. The state matrix will be dynamically updated according to the courses selected in the previous semester.
- We optimize the course representation by using message passing between courses. To the best of our knowledge, this is the first attempt to incorporate a GNN into a course recommendation task.
- Experimental results show that our model can provide accurate recommendations for students and has superior performance in cold-start scenarios.

2   Related Work

2.1   Knowledge Tracing Task

Knowledge tracing (KT) task predicts a student's performance by modeling his/her mastery of knowledge concepts. The RNN based model IEKT [13] introduces two additional estimation modules to explicitly represent students' cognitive level and knowledge acquisition sensitivity by vectors. LPKT [23] defines basic learning cells which are more capable to reflect the complete learning process and proposes a learning gate to control the transformation from gains to knowledge growth. Memory-based KT model DKVMN [34] proposed one static matrix that stores the knowledge concepts and one dynamic matrix that stores and updates corresponding concepts' mastery levels. Then SKVMN [1] improves DKVMN in two aspects. It uses a triangular layer to discover sequential dependencies between exercises and captures the long-term dependencies in exercise sequences by incorporating Hop-LSTM. We hope to recommend courses according to students' preferences for courses related factors. The above methods that model students' knowledge state have important reference value for our research.

2.2 Recommendation Task

GNN-Based Recommendation. The recommendation model based on GNN [5,15,16,25,26,28–32] updates nodes' features by aggregating information of neighbors. NGCF [29] integrate the user-item interactions into the embedding process to model high-order connectivity. Graph attention networks (GATs) [27], which are suitable for inductive tasks, use attention mechanisms to calculate the weight of relationships. MCCF [30] proposes two-layer attention on the bipartite graph for item recommendation. On the knowledge graphs (KGs), MKGAT [25] modifies GAT according to the KGs relation to increase the model representation ability, KCAN [26] uses attention mechanism to propagate embedding and refine the knowledge graph. AMR [31] uses GAT to aggregate users and their behaviors to make recommendations. However, there is no work focusing on the complex relationship between courses, which is still an exploratory research interest.

Course Recommendation. Recently, some works focus on analyzing data characteristics. HCR [14] analyzes the characteristics of students' course selection, and proposes a hybrid recommendation model considering factors including interest, time and grade. However, the importance of different factors is determined by fixed hyperparameters, which lack dynamics. [21] analyze characteristics of students and courses in detail. [33] recommends courses to students while explaining to them why they are recommended for specific courses, but it does not discuss how to improve the accuracy of recommendation results.

In addition, some progress has been made in course recommendation tasks based on various algorithms and frameworks. Classical CF algorithm is applied to course recommendation [10,12]. KPCR [10] creates knowledge graphs and user course level graphs, and combines embedding task and classification task to improve the recommendation performance. RNN is widely used for prediction and recommendation tasks, [9] chooses RNN as the framework to predict students' grades and recommend courses. Inspired by word2vec, course2vec is proposed to embed courses by applying a skip-gram to course enrollment histories [8,18,19]. Scholars Walk [20] captures the sequence relationship between courses through Markov chain. Attention mechanism is used to construct a deep learning model for course recommendation [35–37]. For example, [35] and [36] both calculate the attention coefficient of the course embeddings for recommendation in the MOOC platform. As a state-of-the-art research, PLAN-BERT [22] makes consecutive semester recommendations based on the Bidirectional Encoder Representations from Transformer (BERT) [2] model according to the past course histories and the future pre-specified courses of interest.

However, many studies have not comprehensively considered the factors that affect students' course selection, including the student's features, courses' features, course selection histories, the relationship between courses, etc. We hope that the course recommendation model can consider all factors and make personalized recommendations adaptively.

3   Preliminaries

In this section, we introduce the concepts of HIN and interest state that will be used in this paper, and clearly define the research problem.

Definition 1 (Heterogeneous Information Network (HIN)). *For an information network $H = (V, E)$, V is an object set and E is a link set, when the number of object types $|A| > 1$ or the number of link types $|R| > 1$, it is a heterogeneous information network.*

A toy example of the HIN consisting of a university course selection system is shown in the left part of Fig. 1. It comprises the object type set $A = \{Student, Teacher, Course\}$ and the link type set $R = \{Select, Prerequisite, Teach, Sibling\}$ (*Sibling* means different courses with the same code).

Definition 2 (Interest State). *Suppose there are N potential factors that influence a student's course selection, and each student has a different interest level for each factor. We define the interest levels of the student for all factors as interest state.*

Definition 3 (Problem Formulation). *Given the set C of all optional courses and the set $C^+ \subseteq C$ of courses that have been taken, our task is to predict the probability p_t of a student selecting and successfully completing each course in $C - C^+$ based on the information of the student and courses, and the course selection information, and give recommendations to students according to p_t.*

4   Course Recommendation Model

In this section, we propose a course recommendation model: factor Memory network and GNN based personalized Course Recommendation (MG-CR). The framework is shown in Fig. 2. We first capture rich information and build model inputs including student personal features, course teaching features, and interactive information. Second, considering various types of relationships on the HIN, a GNN based module is designed to fuse course features. Then, by improving the memory-based [1,34] architecture, the state matrix is designed to store the dynamically changing student interest state, thereby improving the explainability of course recommendation. Finally, we predict the student's course selection scores, which is the probability that the student selects each course and successfully completes it, and generates recommendations accordingly.

4.1   Features and Interactions

Course selection is a complex process that requires each student to weigh the pros and cons of many factors. Students generally tend to choose courses they enjoy and are likely to perform well. DKVMN [34] uses category vectors to represent exercises, making it difficult to distinguish nuances between different exercises. However, in the course recommendation task, the choice of courses is likely to be

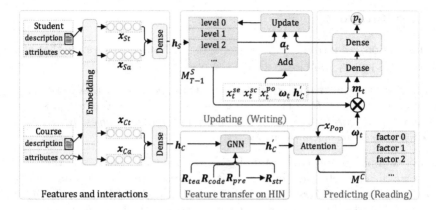

Fig. 2. Framework of MG-CR. We first capture the rich features of students' personal information, courses' teaching attributes and their interactive information. Courses' features will be optimized by GNN. The memory-based model contains two parts. The predicting (reading) process predicts the student's course selection score p_t based on the student's state matrix and course information. The updating (writing) process updates the state matrix according to the interactive information, the course information, the student's current state matrix, and the current prediction result.

affected by a very subtle factor. Students, courses, and *select* interactions in an HIN are informative. Based on this, we construct model inputs to fully reflect various external conditions that affect students' course selection:

- Text description of a student: It contains important professional information of a student, such as "*A student majoring in computer science and technology at the college of computer science and engineering*". We first perform word embedding on it, and feed the result x_{StRaw} into the GRU layer to get the student's text description embedding x_{St}.
- Attribute features of a student: Attributes include class, major, gender, dormitory, academic performance, etc. We represent each attribute with a one-hot vector and concatenate them to form the input x_{Sa}.
- Text description of a course: We treat the course name that can intuitively reflect its content as the text description, such as "*Game theory and social network*". Word embedding is also performed on it, and the result is fed into another GRU layer to get the course's text description embedding x_{Ct}.
- Attribute features of a course: Its attributes include department, credit, teacher, etc. We represent credits numerically and other attributes by one-hot vectors. Then, we get input x_{Ca} after connecting.
- Interactive information: Interactive information is generated during the course selection process. For example, the semester of course selection x_{Se}, the student's score x_{Sc}, and the popularity of this course x_{Pop}, that is, the number of students enrolled in the course. The above interactive information will be input into the model in the form of scalars, as detailed in Sect. 4.3.

Then we obtain the comprehensive features of students and courses:

$$h_S = \text{elu}(\boldsymbol{W}_S(GRU(\boldsymbol{x}_{StRaw}\|\boldsymbol{x}_{Sa}) + \boldsymbol{b}_S) \in \mathbb{R}^{d_S}, \tag{1}$$

$$h_C = \text{elu}(\boldsymbol{W}_C(GRU(\boldsymbol{x}_{CtRaw})\|\boldsymbol{x}_{Ca}) + \boldsymbol{b}_C) \in \mathbb{R}^{d_C}, \tag{2}$$

where $\boldsymbol{W}_S$, $\boldsymbol{b}_S$, $\boldsymbol{W}_C$ and $\boldsymbol{b}_C$ are trainable parameters, d_S and d_C are the dimensions of student and course vectors, elu$(\cdot)$ is the activation function, and $\|$ represent the connection operation.

4.2 Feature Transfer on HIN

In HIN, there are four types of relationships between courses. (i) Prerequisite. The content of some courses is based on the fundamentals contained in their prerequisite courses. We form the prerequisite relationship into matrix $\boldsymbol{R}_{pre} \in \mathbb{R}^{|C| \times |C|}$, where C is the course set and $|C|$ is the number of courses, and the corresponding prerequisite network is shown in Fig. 3(a). (ii) Co-instructor. Several different courses are taught by the same teacher, which constitutes the relationship matrix $\boldsymbol{R}_{coi}$. (iii) Sibling. Some courses with the same code are taught by different teachers in the same semester, we call these courses siblings, and form the relationship matrix $\boldsymbol{R}_{sib}$. (iv) Local structure similarity. This is a relationship extracted from the prerequisite network, which is detailed below.

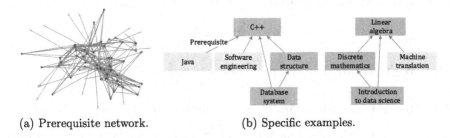

(a) Prerequisite network.   (b) Specific examples.

Fig. 3. Local structural relationship. Figure (a) is the prerequisite network. Figure (b) is two examples extracted from Figure (a), and the blue directed edges represent that the prerequisite relationships can form a specific local pattern.

Local structural pattern refers to the geometric relationship between nodes and neighbours in the network, that is, the structural features of nodes. Next, we extract the local structure pattern of the course from the prerequisite network, and obtain the local structural similarity relationship. As shown in Fig. 3(b), *C++* and *Linear algebra* have similar local structural patterns, so we consider them to be similar courses, which is consistent with the fact that they are both important foundation courses. In addition, *Data structure* and *Discrete mathematics*, as well as *Database system practice* and *Introduction to data science* have the same local structure patterns respectively. In Fig. 3(a), the edges involved in this local structural pattern are indicated in blue. In fact, these two courses will be selected by a student with a certain foundation, and they will not be selected in the first semester. Using anonymous random walk [7], we get the frequency $f_{c,w}$ of each possible walk path starting from the target course, where $c \in \{1, 2, \cdots, |C|\}, w \in \{1, 2, \cdots, |W_A|\}$, and W_A is the possible results set. Then use zero-mean normalization to process the frequency and get the local structure vector $\boldsymbol{v}_{str} = (f_{c,w} - \overline{f})/S_{str}$ of each course, where $\overline{f}$ is the mean and S_{str} is the standard deviation. By calculating the cosine similarity of $\boldsymbol{v}_{str}$, the local structure similarity matrix $\boldsymbol{R}_{str}$ is obtained.

To ensure that new courses can be added to the model, i.e. suitable for inductive tasks, we use a single-layer GAT [27], one of the GNN models, to fuse the input data and add relationships and network local structure to the course features. Specifically, for adjacent courses i and j under relationship $\boldsymbol{R}(\boldsymbol{R} \in \{\boldsymbol{R}_{pre}, \boldsymbol{R}_{coi}, \boldsymbol{R}_{sib}, \boldsymbol{R}_{str}\})$, the normalized attention coefficient is computed as:

$$\alpha_{i,j} = \frac{\exp(\text{LeakyReLU}(\boldsymbol{a}(\boldsymbol{W}_C \boldsymbol{h}_{Ci} \parallel \boldsymbol{W}_C \boldsymbol{h}_{Cj}) + b))}{\sum_{k \in N_i} \exp(\text{LeakyReLU}(\boldsymbol{a}(\boldsymbol{W}_C \boldsymbol{h}_{Ci} \parallel \boldsymbol{W}_C \boldsymbol{h}_{Ck}) + b))}, \tag{3}$$

where $\boldsymbol{W}_C$, $\boldsymbol{a}$ and b are weights for linear transformation, LeakyReLU($\cdot$) is the activation function, and N_i contains neighbors of course i. Then we update course features:

$$h_{Ci}^R = \text{ReLU}(\sum_{j \in N_i} \alpha_{ij} \boldsymbol{W} \boldsymbol{h}_{Ci}), \tag{4}$$

where $\boldsymbol{W}$ is the weight matrix and ReLU($\cdot$) is the activation function. To consider each relationship simultaneously, we fuse them into an aggregated feature:

$$h'_{Ci} = \sum_R \boldsymbol{W}_R h_{Ci}^R, \tag{5}$$

where $\boldsymbol{W}_R$ is the trainable weight matrix for each relationship.

4.3   Recommendation Model Based on Factor Memory Network

Memory-based models can generally be divided into predicting (reading) process and updating (writing) process. In this section, a factor memory network is designed to predict students' course selection score p_t, in which the state matrix $\boldsymbol{M}^S$ is used to store a student's interest state, and each vector $\boldsymbol{m}_n^S$ of $\boldsymbol{M}^S$ corresponds to an interest level. We first predict p_t based on the student's interest level for each factor. Then update $\boldsymbol{M}^S$ according to the prediction results and the course information.

A student's interest level for different factors is influenced, determined and revised by his/her original features and previous course selection. We generate the student's initial state matrix $\boldsymbol{M}^S$ by Gaussian distribution based on his/her feature vector $\boldsymbol{h}_S$. The distribution of the k-th ($k \in \{1, \cdots, d_F\}$, d_F is the dimension of $\boldsymbol{m}_n^S$) element of the n-th ($n \in \{1, \cdots, N\}$) vector in $\boldsymbol{M}^S$ is:

$$M_{n,k}^S \sim \mathcal{N}((\boldsymbol{W}_M \boldsymbol{h}_S)_n, (\lambda(\boldsymbol{W}_M \boldsymbol{h}_S)^T (\boldsymbol{W}_M \boldsymbol{h}_S))_n), \tag{6}$$

where $\mathcal{N}(\cdot, \cdot)$ refers to the Gaussian distribution, λ is the hyperparameter, and $\boldsymbol{W}_M \in \mathbb{R}^{N \times d_S}$ is the parameter matrix for linear transformation between student features and different factors.

The factor matrix $\boldsymbol{M}^C$ is a trainable parameter matrix consisting of N vectors $\boldsymbol{m}_n^C$ ($n \in \{1, \cdots, N\}$) representing factors. It is trained by the prediction sequence $\boldsymbol{S}^p$ and is used to measure the correlation between each course and each factor.

A student's course selection order is sorted by semester. Specifically, for each student, we construct a matrix $\boldsymbol{A} \in \mathbb{R}^{|C| \times |C|}$ for mapping, where $A_{i,j} = 1$ means the course i is at j after being arranged according to the course selection order. When updating the student's state matrix, we also take into account the scores of the courses they have selected, the semester, and the popularity of the corresponding courses, i.e., the data generated during the interaction between students and courses. Next we construct the sequence $\boldsymbol{S}^u$ for updating the state matrix. For the course t, the corresponding member $\boldsymbol{f}_t^u$ in $\boldsymbol{S}^u$ is:

$$\boldsymbol{f}_t^u = (\boldsymbol{A} \cdot \boldsymbol{h}_C')_t \| x_t^{po} \| x_t^{se} \| x_t^{sc}, \tag{7}$$

where x_t^{se} is the semester of course selection, x_t^{sc} is the student's score and x_t^{po} is the popularity of the course. In addition, for the predicted course, the student's score and the semester in which it was selected are unknown, so the corresponding member in the prediction sequence $\boldsymbol{S}^p$ is constructed as follows:

$$\boldsymbol{f}_t^p = (\boldsymbol{A} \cdot \boldsymbol{h}_C')_t \| x_t^{po}. \tag{8}$$

Similar to the reading process in memory-based KT models, we use an attention mechanism to estimate various factors included in the course. The correlation weight between course t and factor n is calculated by the inner product of the prediction sequence and the factor matrix:

$$\omega_{t,n} = \frac{\exp(\boldsymbol{f}_t^p \cdot \boldsymbol{m}_n^C)}{\sum_{k=1}^{N} \exp(\boldsymbol{f}_t^p \cdot \boldsymbol{m}_k^C)}. \tag{9}$$

The results of a student's course selection in semester T are affected by the interest state of each factor after the end of semester $T - 1$. Therefore, we use $\boldsymbol{M}_T^S$ to denote the state matrix after semester T, and the matrix $\boldsymbol{M}_t^S$ to record the temporary state during semesters. Then the interest vector $\boldsymbol{m}_t$ is extracted from $\boldsymbol{M}_t^S$ according to $\boldsymbol{\omega}_t$ considering various factors:

$$\boldsymbol{m}_t = \boldsymbol{\omega}_t \cdot \boldsymbol{M}_{T-1}^S. \tag{10}$$

The probability of a student selecting and completing courses is:

$$p_t = \text{Sigmoid}(\boldsymbol{W}_2(\tanh(\boldsymbol{W}_1(\boldsymbol{m}_t \| \boldsymbol{f}_t^p)) + \boldsymbol{b}_1) + \boldsymbol{b}_2), \tag{11}$$

where Sigmoid$(\cdot)$ and $\tanh(\cdot)$ are activation functions, and $\boldsymbol{W}_1$, $\boldsymbol{W}_2$, $\boldsymbol{b}_1$ and $\boldsymbol{b}_2$ are parameters to be trained.

Finally, during the updating process, the state matrix is updated according to the features of the selected courses, and the unselected courses will not affect the student's interest level for various factors. Therefore, the state is not updated when it is predicted that the student will not choose a course. We use the add vector $\boldsymbol{a}_t = \tanh(\boldsymbol{W}_a \boldsymbol{f}_t^u + \boldsymbol{b}_a)$ [1,34] to represent the impact of courses on a student's interests, where $\boldsymbol{W}_a$ and $\boldsymbol{b}_a$ are parameters to be trained. Since students' learning of courses is usually a long-term process throughout the semester and is not as easy to forget as answering questions, the decline of interest state is not considered, that is, the erase vector e_t [1,34] in traditional memory-based models is ignored. The temporary state update function of a student during the semester is designed as follows:

$$M_t^S = \begin{cases} M_{t-1}^S + \omega_t \cdot a_t, \, p_t \geq \xi \\ M_{t-1}^S, \qquad\quad p_t < \xi \end{cases}. \tag{12}$$

It should be noted that using the prediction results p_t to determine whether to update the state matrix enables MG-CR to predict the course selection in future multiple semesters, rather than only in the next semester. After one semester's course selection, the state matrix M_T^S is updated to:

$$M_T^S = M_t^S. \tag{13}$$

4.4   Objective Function and Model Training

For the course t selected by student s, we set the label $y_{s,t} = 1$, otherwise we set $y_{s,t} = 0$. We then use binary cross-entropy to calculate how well the student's course selection score $p_{s,t}$ matches $y_{s,t}$, and train the model:

$$\mathcal{O} = -\sum_{s \in S}\sum_{t \in C}(y_{s,t}\log p_{s,t} + (1 - y_{s,t})\log(1 - p_{s,t})) + \lambda_\Theta\|\Theta\|^2, \tag{14}$$

where Θ denotes all parameters of MG-CR and λ_Θ is the ℓ_2 regularization hyper-parameter.[1] And Dropout [24] is applied to ensure the robustness of the model.

5   Experimental Evaluation

5.1   Experimental Setup

Datasets. We use an anonymized private dataset from the course selection system for undergraduates at Northeastern University in China. All students are from the School of Computer Science and Engineering, the complete statistical information of the dataset can be found in Table 1. Due to less interaction in the fourth year, we extract data for the spring and autumn semesters of the first three years. Students are selected for training with probability r, and the rest are used for testing.

Table 1. Course selection system dataset details.

	Data	Number
Students	Student Id, Class, Major, Hometown, Gender, Dormitory, Text Description	1424
Courses	Course Id, Code, Name, Type, Credit, Text Description	629
Interactions	Student Id, Course Id, Semester, Score	104454

---

[1] The code is available on Github: https://github.com/hddyyyb/MGCR.

Baselines. We compare the recommendation results of our proposed MG-CR model, variants of this model, and three baselines. We designed several variants as follows to verify the effectiveness of each part of MG-CR one by one.

- MG-CR-GNN: Variant model without transfer module, in which courses' features will not be transferred to their neighbours.
- MG-CR-Up: The state matrix M^S is not updated.
- MG-CR-GU: Variant model without transfer module and update module.
- MG-CR-Stu: Variant model without considering student features. That is, M^S will be initialized randomly and will not be affected by h_S.

The baseline models are as follows:

- PopBased: We count the frequency p of each course being selected and the average number of courses k selected by each student in the training set, then recommend the most frequent k courses for students in the test set.
- LSTM [6]: A recurrent neural network model considering long-term dependence. The hidden state is updated according to the course selection history.
- DKVMN [34]: A Memory-based knowledge tracing model, and its input is course features h_C. The selected course is equivalent to the correct exercise, and the unselected course is equivalent to the wrong exercise.
- NGCF [29]: A GNN based CF algorithm, in which we construct a bipartite graph with *students* and *courses* as nodes and *select* relationships as edges.
- Course2vec [19]: It learns course representations using a skip-gram model and then makes recommendations. Hard-to-obtain students' favorite courses are replaced with 10% of the courses students take each semester.
- PLAN-BERT [22]: A state-of-the-art course recommendation model that takes user and project features as inputs. Since there is no pre-specified future reference course information in our dataset, we input the real course selection of last semester as such information into the model.

Evaluation Metrics and Configuration. We use the evaluation metrics including *Recall@n*, Root Mean Squared Error (*RMSE*), Area Under Curve (*AUC*) and Normalized Discounted Cumulative Gain (*NDCG@n*) in experiments to evaluate the recommendation results. Experimental results are reported as the average of five runs. For MG-CR, we train the Wiki corpus[2] using gensim.word2vec to obtain vectors x_{StRaw} and x_{CtRaw} for all words in text descriptions of students and courses. The Dropout rate is set to 0.3. λ is set to 0.01 and ξ is set to 0.5. We minimize the objective function $\mathcal{O}$ using the Adam optimizer [11].

5.2   Parameter Analysis

With other parameters fixed, we use *Recall@50* to compare the impact of the four hyperparameters on the recommendation results. It is worth noting that d_S is consistent with the dimension of the state matrix. We vary the dimension

---

[2] https://radimrehurek.com/gensim/corpora/wikicorpus.html.

d_S in $\{2, 4, 8, 16, 32, 64\}$. As shown in Fig. 4(a), the results are better when the dimension is large, and it reaches saturation when $d_S \geq 8$.

We change the dimension $d_C = \{2, 4, 8, 16, 32, 64\}$, and the result is shown in Fig. 4(b). The results also get better with the increase of dimension, and the model reaches saturation when $d_S \geq 32$.

Common factors that affect a student's course selection include course type, course difficulty, teaching style, etc. We change the parameter N and compare the results, as shown in Fig. 4(c). The recall first increases with factor number, reaches the best when $N = 20$, and then decreases with the increase of N. According to the above experimental results, we believe that there are 20 factors that affect students' course selection.

We change the proportion r of training and test samples, and the results are shown in Fig. 4(d). The results are relatively good $r = 0.7$. When $r = 0.7$, MG-CR achieves the optimal recommendation result. This is because too small r results in insufficient model training, and too large r will lead to overfitting. In the following experiments, the hyperparameters will be set according to the analysis results in this section: $d_S = 8$, $d_C = 32$, $N = 20$ and $r = 0.7$.

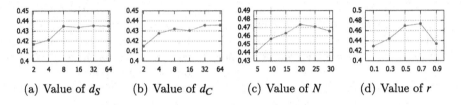

(a) Value of d_S (b) Value of d_C (c) Value of N (d) Value of r

Fig. 4. The influence of different hyperparameters evaluated by *Recall@*50.

5.3   Ablation Study

Impact of Each Component. To delve into the contribution of each component of the MG-CR, we conduct some ablation experiments to evaluate their impact on the recommendation. The results are shown in Table 2. We can see that the recommendation results of variants are not as good as MG-CR. The input of MG-CR-Stu has no student features, which affects the personalization of the model. MG-CR-GNN ignores the relationship between courses, while a series of courses that are systematically learned often have correlations. MG-CR-Up does not update the state matrix, it mistakenly assumes that the student's interest state will not change after completing the course, making recommendations less accurate than MG-CR in later semesters. Then MG-CR-GU with both updates and transfers removed obtained the worst results. According to the above analysis, each part of our model has indispensable significance.

Impact of Each Relationship. In addition, we evaluate the impact of different relationships of feature transfer modules on recommendation results by considering only a single relationship. As shown in Table 2, in all cases, MG-CR achieves the best recommendation results using all relationships. The results under R_{pre} and R_{coi} are more excellent, which indicates that students systematically study a series of courses according to the prerequisite relationship, and have a high probability to select courses taught by the same teacher.

Table 2. Results of ablation study. We use bold font to mark the best result.

Metrics	Recall@10	Recall@20	Recall@30	Recall@40	Recall@50	AUC	RMSE	NDCG@50
MG-CR-GU	0.1157	0.2062	0.2762	0.3459	0.4107	0.8450	0.3011	0.5800
MG-CR-Up	0.1254	0.2071	0.2725	0.3641	0.4295	0.8738	0.2850	0.5881
MG-CR-Stu	0.1314	0.2127	0.2920	0.3745	0.4531	0.8825	0.2815	0.6278
MG-CR-GNN	0.1326	0.2153	0.2866	0.3607	0.4464	0.8757	0.2876	0.6269
$-R_{str}$	0.1349	0.2108	0.2898	0.3685	0.4520	0.8700	0.2863	0.6342
$-R_{coi}$	0.1353	0.2082	0.2958	0.3701	0.4427	0.8789	0.2856	0.6436
$-R_{sib}$	0.1325	0.2055	0.2805	0.3637	0.4466	0.8902	0.2907	0.6310
$-R_{pre}$	0.1367	0.2231	0.2876	0.3671	0.4525	0.8831	0.2888	0.6475
MG-CR	**0.1376**	**0.2261**	**0.3054**	**0.3845**	**0.4734**	**0.8962**	**0.2804**	**0.6589**

5.4   Recommendation Result Compared to Several Baselines

We compare the recommended results of MG-CR with all baselines for evaluation. As shown in Table 3, the results of MG-CR significantly outperform all compared models ($p < 0.01$) on every metric, which demonstrate its effectiveness. Specifically, the superiority of MG-CR over DKVMN indicates that student features, relationships between courses and interactive information all contribute to the personalization and accuracy of recommendations. DKVMN takes into account the features of the course, and its memory module models the students' interest state on each factor, so its results are better than those of LSTM and NGCF. In addition, in most course selection systems, both pre-specified future reference course information in PLAN-BERT and favorite courses in Course2vec are hard to come by because students are often ignorant of future study content. According to the above analysis, MG-CR can provide students with accurate and personalized course recommendations without the need for additional reference information, which will help reduce students' learning pressure, improve their personal ability and successfully complete the learning tasks.

Table 3. Experimental results of MG-RC and baselines. Best results are in boldface and the second best is underlined. * denotes statistically significant improvements ($p < 0.01$) on paired t-tests. *Improv.* means improvement over the state-of-art methods.

Model	Recall@10	Recall@20	Recall@30	Recall@40	Recall@50	AUC	RMSE	NDCG@50
PopBased	0.0613	0.1415	0.2072	0.2826	0.2871	0.7026	0.3745	0.4889
LSTM	0.0998	0.1763	0.2570	0.3231	0.3917	0.7371	0.3645	0.5340
DKVMN	0.1100	0.2097	0.2874	0.3628	0.4258	0.8498	0.3011	0.6072
NGCF	0.1093	0.1857	0.2721	0.3491	0.4086	0.8502	0.3058	0.5916
Course2vec	0.1232	<u>0.2151</u>	<u>0.2896</u>	0.3459	0.4222	0.8668	0.2937	0.6091
PLAN-BERT	<u>0.1292</u>	0.2015	0.2863	<u>0.3674</u>	<u>0.4570</u>	<u>0.8714</u>	<u>0.2898</u>	<u>0.6130</u>
MG-CR	**0.1376***	**0.2261***	**0.3054***	**0.3845***	**0.4734***	**0.8962***	**0.2804***	**0.6589***
Improv.	6.50%	5.11%	5.45%	4.65%	3.58%	2.84%	3.24%	9.11%

5.5   Cold-Start Scenario Analysis

We hope to provide more accurate recommendations for new students. Freshmen have relatively little understanding of each course, and the lack of interactive data in the course selection system makes it more difficult to recommend. We classify this as a cold-start problem. According to statistics, students take an average of 12 courses in the first and the second semesters. Therefore, we believe that $Recall@10$ can be used to evaluate the recommendation results of different models in the first two semesters, as shown in Fig. 5.

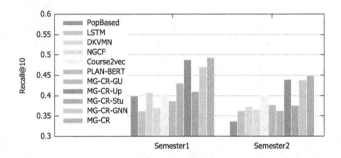

Fig. 5. $Recall@10$ of the first two semesters' recommendation results.

First, MG-CR achieves the best results because it captures features of the student and the course in the input part, and introduces a course feature transfer module on HINs, making it have good recommendation ability in cold-start scenarios. Second, MG-CR-Up performs relatively well in the first two semesters, because the module that updates the state matrix is more effective for the following semesters. MG-CR-Stu ignores the features of the student in the input process, and MG-CR-GNN and MG-CR-GU ignore the feature transfer of courses, both leading to a worse recommendation in the first semester. Third, PopBased outperformed the baselines NGCF, LSTM and course2vec in the first semester because freshmen do not know what they like and tend to follow the trend when selecting courses. Finally, baselines including NGCF and LSTM have suboptimal results in the first two semesters due to taking id as input, which illustrates the importance of student and course features in the recommendation process.

6   Conclusion

In this paper, we provided a focused study on the factor memory network and GNN based personalized course recommendation. We build an HIN to take full advantage of rich information contained in the course selection system. In the input module, we capture features of students and courses, as well as interactive information to maximize the retention of personalized information. For various types of relationships between courses, a GNN is used to optimize the feature vectors of courses. To achieve dynamics in the course selection process, we design a state matrix in the updating module to record the student's interest level for all factors, and update the matrix according to the selected courses. Finally, the model uses student state and course information to predict the probability that the student selects each course and successfully completes it, and recommends courses for each student. Experiments show that our model achieves the best results compared to other models and can handle the cold-start problem.

Acknowledgements. This research was funded by the National Natural Science Foundation of China (No. U1811261, 61902055, 62137001, 62272093), and Fundamental Research Funds for the Central Universities (No. N2117001).

References

1. Abdelrahman, G., Wang, Q.: Knowledge tracing with sequential key-value memory networks. In: SIGIR, pp. 175–184 (2019)
2. Devlin, J., Chang, M., Lee, K., Toutanova, K.: BERT: pre-training of deep bidirectional transformers for language understanding. In: NAACL-HLT, pp. 4171–4186 (2019)
3. Elbadrawy, A., Karypis, G.: Domain-aware grade prediction and top-n course recommendation. In: RecSys, pp. 183–190 (2016)
4. Farzan, R., Brusilovsky, P.: Social navigation support in a course recommendation system. In: Wade, V.P., Ashman, H., Smyth, B. (eds.) AH 2006. LNCS, vol. 4018, pp. 91–100. Springer, Heidelberg (2006). https://doi.org/10.1007/11768012_11
5. Feng, L., Cai, Y., Wei, E., Li, J.: Graph neural networks with global noise filtering for session-based recommendation. Neurocomputing **472**, 113–123 (2022)
6. Hochreiter, S., Schmidhuber, J.: Long short-term memory. Neural Comput. **9**, 1735–1780 (1997)
7. Ivanov, S., Burnaev, E.: Anonymous walk embeddings. In: ICML. Proceedings of Machine Learning Research, vol. 80, pp. 2191–2200 (2018)
8. Jiang, W., Pardos, Z.A.: Time slice imputation for personalized goal-based recommendation in higher education. In: RecSys, pp. 506–510 (2019)
9. Jiang, W., Pardos, Z.A., Wei, Q.: Goal-based course recommendation. In: LAK, pp. 36–45 (2019)
10. Jung, H., Jang, Y., Kim, S., Kim, H.: KPCR: knowledge graph enhanced personalized course recommendation. In: Long, G., Yu, X., Wang, S. (eds.) AI 2022. LNCS (LNAI), vol. 13151, pp. 739–750. Springer, Cham (2022). https://doi.org/10.1007/978-3-030-97546-3_60
11. Kingma, D.P., Ba, J.: Adam: A method for stochastic optimization. In: Bengio, Y., LeCun, Y. (eds.) ICLR (2015)
12. Li, J., Ye, Z.: Course recommendations in online education based on collaborative filtering recommendation algorithm. Complex **2020**, 6619249:1–6619249:10 (2020)
13. Long, T., Liu, Y., Shen, J., Zhang, W., Yu, Y.: Tracing knowledge state with individual cognition and acquisition estimation. In: SIGIR, pp. 173–182 (2021)
14. Ma, B., Taniguchi, Y., Konomi, S.: Course recommendation for university environment. In: EDM (2020)
15. Ojo, F., et al.: Visgnn: Personalized visualization recommendationvia graph neural networks. In: WWW, pp. 2810–2818 (2022)
16. Pang, Y., et al.: Heterogeneous global graph neural networks for personalized session-based recommendation. In: WSDM, pp. 775–783 (2022)

17. Parameswaran, A.G., Venetis, P., Garcia-Molina, H.: Recommendation systems with complex constraints: A course recommendation perspective. ACM Trans. Inf. Syst.**29**(4), 20:1–20:33 (2011)
18. Pardos, Z.A., Fan, Z., Jiang, W.: Connectionist recommendation in the wild: on the utility and scrutability of neural networks for personalized course guidance. User Model. User-Adap. Inter. **29**(2), 487–525 (2019). https://doi.org/10.1007/s11257-019-09218-7
19. Pardos, Z.A., Jiang, W.: Designing for serendipity in a university course recommendation system. In: LAK, pp. 350–359 (2020)
20. Polyzou, A., Nikolakopoulos, A.N., Karypis, G.: Scholars walk: A markov chain framework for course recommendation. In: EDM (2019)
21. Sánchez-Sánchez, C., González, C.R.J.: Course recommendation system for a flexible curriculum based on attribute selection and regression. In: IntelliSys, vol. 296, pp. 674–690 (2021)
22. Shao, E., Guo, S., Pardos, Z.A.: Degree planning with PLAN-BERT: multi-semester recommendation using future courses of interest. In: AAAI, pp. 14920–14929 (2021)
23. Shen, S., et al.: Learning process-consistent knowledge tracing. In: KDD, pp. 1452–1460 (2021)
24. Srivastava, N., Hinton, G.E., Krizhevsky, A., Sutskever, I., Salakhutdinov, R.: Dropout: a simple way to prevent neural networks from overfitting. J. Mach. Learn. Res. **15**(1), 1929–1958 (2014)
25. Sun, R., et al.: Multi-modal knowledge graphs for recommender systems. In: CIKM, pp. 1405–1414 (2020)
26. Tu, K., et al.: Conditional graph attention networks for distilling and refining knowledge graphs in recommendation. In: CIKM, pp. 1834–1843 (2021)
27. Velickovic, P., Cucurull, G., Casanova, A., Romero, A., Liò, P., Bengio, Y.: Graph attention networks. In: ICLR (2018)
28. Wang, L., Hu, F., Wu, S., Wang, L.: Fully hyperbolic graph convolution network for recommendation. In: CIKM, pp. 3483–3487 (2021)
29. Wang, X., He, X., Wang, M., Feng, F., Chua, T.: Neural graph collaborative filtering. In: SIGIR, pp. 165–174 (2019)
30. Wang, X., Wang, R., Shi, C., Song, G., Li, Q.: Multi-component graph convolutional collaborative filtering. In: AAAI, pp. 6267–6274 (2020)
31. Wei, Y., Ma, H., Wang, Y., Li, Z., Chang, L.: Multi-behavior recommendation with two-level graph attentional networks. In: DASFAA, vol. 13246, pp. 248–255 (2022)
32. Xu, J., Zhu, Z., Zhao, J., Liu, X., Shan, M., Guo, J.: Gemini: A novel and universal heterogeneous graph information fusing framework for online recommendations. In: KDD, pp. 3356–3365 (2020)
33. Yu, R., Pardos, Z.A., Chau, H., Brusilovsky, P.: Orienting students to course recommendations using three types of explanation. In: UMAP, pp. 238–245 (2021)
34. Zhang, J., Shi, X., King, I., Yeung, D.: Dynamic key-value memory networks for knowledge tracing. In: WWW, pp. 765–774 (2017)
35. Zhang, J., Hao, B., Chen, B., Li, C., Chen, H., Sun, J.: Hierarchical reinforcement learning for course recommendation in moocs. In: AAAI, pp. 435–442 (2019)
36. Zhao, Z., Yang, Y., Li, C., Nie, L.: Guessuneed: Recommending courses via neural attention network and course prerequisite relation embeddings. ACM Trans. Multim. Comput. Commun. Appl. **16**(4), 132:1–132:17 (2021)
37. Zhu, Q.: Network course recommendation system based on double-layer attention mechanism. Sci. Program **2021**, 7613511:1–7613511:9 (2021)

Revisiting Positive and Negative Samples in Variational Autoencoders for Top-N Recommendation

Wei Liu[1,2,3], Leong Hou U[1], Shangsong Liang[2,3,4], Huaijie Zhu[2,3(✉)], Jianxing Yu[2,3], Yubao Liu[2,3], and Jian Yin[2,3]

[1] University of Macau, Macau, China
liuw259@mail.sysu.edu.cn, ryanlhu@um.edu.mo
[2] Sun Yat-sen University, Guangzhou, China
{zhuhuaijie,yujx26,liuyubao,issjyin}@mail.sysu.edu.cn
[3] Guangdong Key Laboratory of Big Data Analysis and Processing, Guangzhou, China
[4] Mohamed bin Zayed University of Artificial Intelligence, Abu Dhabi, United Arab Emirates

Abstract. Top-N recommendation is a common tool to discover interesting items, which ranks the items based on user preference using their interaction history. Implicit feedback is often used by recommender systems due to the hardness of preference collection. Recent solutions simply treat all interacted items of a user as equally important positives and annotate all no-interaction items of a user as negatives. We argue that this annotation scheme of implicit feedback is over-simplified due to the sparsity and missing fine-grained labels of the feedback data. To overcome this issue, we revisit the so-called positive and negative samples for Variational Autoencoders (VAEs). Based on our analysis and observation, we propose a self-adjusting credibility weight mechanism to re-weigh the positive samples and exploit the higher-order relation based on item-item matrix to sample the critical negative samples. Besides, we abandon complex nonlinear structure and develop a simple yet effective VAEs framework with linear structure, which combines the reconstruction loss function for the positive samples and critical negative samples. Extensive experiments conducted on 4 public real-world datasets demonstrate that our VAE++ outperforms other VAEs-based models by a large margin.

Keywords: Variational AutoEncoders · Recommendation · Implicit Feedback · Collaborative Filtering

1 Introduction

Recommender Systems (RS) are generally used for ranking products and users, which are particularly helpful to discover favourite products on online shopping platforms. Collaborative Filtering (CF) is one of the most popular recommendation methods [3,11,15], which attempts to identify users of similar tastes and learns the user preference based on their interaction history data. Given the nature of user feedback, we can classify the data into *explicit feedback* and *implicit feedback*. The former can clearly indicate user preference, e.g., giving a like to a post, rating a movie, and commenting a hotel. However, the latter may provide some information but cannot fully reflect user preference, e.g., navigating a web page, purchasing a movie, and checking-in a location. Given the difficulty and the cost of collecting explicit feedback, some prior recommender systems [4] simply assume all interacted items (i.e., implicit feedback) of a user as positive samples. We argue that such assumption is over-simplified as the implicit feedback is easily affected by the first impression and the misconduct of users, which has been concluded in recent studies [1,14].

Recently, Variational Autoencoders (VAEs) [8,13] are recognized as an important CF technique in recommender systems due to its superior performance. To improve VAEs performance further, existing studies on VAEs attempt to enhance the Gaussian distribution of latent variables by a complex prior (e.g., variational mixture of posterior prior [7], composite prior [12]) or improve the encoder by a customized neural network design (e.g., hierarchical stochastic unit [7], dense CNN [12]). However, these studies do not directly examine the quality of implicit feedback data in VAEs. In details, they simply annotate interacted items of a user as 1 (a.k.a positive sample) and uninteracted items of a user as 0 (a.k.a negative sample). Accordingly, VAEs learn user latent variable distribution to reconstruct the user preference distribution over all items with a predefined distribution, e.g., multinomial distribution shown in Fig. 1, based on the implicit feedback. Obviously, the annotation quality of the feedback data plays a significant role in the learning process of VAEs. We argue that the annotation scheme in previous work is over-simplified, which generates two issues, (1) *over-confident on interacted items* and (2) *over-rigid on uninteracted items*.

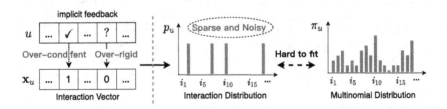

Fig. 1. The annotation scheme of implicit feedback (left), and the hard trained interaction distribution with multinomial distribution of user preference over all items assumed by VAEs (right).

(1) **Over-confident on interacted items.** The implicit feedback data is very sparse in practice. Due to the sparsity of the implicit feedback data and the big difference between the numbers of interacted and uninteracted items, the

learning process easily suffers from overfitting. To make things even worse, the interacted items are generally viewed as *positive samples*, which are handled equally in the learning process [8]. We argue that this over-simplified assumption is not good enough, since the user should have different satisfactory levels on the interacted items. For instance, giving a bad review of a movie, stop playing a song on Spotify, etc. Obviously, this over-simplified assumption may generate *false positive* or *noisy positive* labels, which is hard to train and would affect the learning performance.

(2) Over-rigid on items of no interaction. Similarly, recent CF-based recommender systems [3,10,11] simply treat the items without interaction as *negative samples*. However, an obvious fact is that not all these negative samples are the same due to the implicitness of the feedback. For example, an item i that has no interaction with user u_a but is interacted by many other users should not be simply taken as a negative sample for u_a. In other words, this item may be a *false negative* or *potential positive*, which is hardly distinguished from a positive sample. We call such samples as critical negatives in this work. Identifying important negatives could benefit for the learning since the optimization process of VAEs is unable to treat important negative samples differently from items of no interaction.

Our Solution. To address the issues mentioned above, we carefully review and re-design the label annotation mechanism, including two weighted loss functions for positives/negatives and a linear framework based on VAEs. We first consider the label credibility of interacted items in VAEs. Inspired by the supervised contrastive loss [6], a novel multinomial loss function is introduced and termed as Weighted Multinomial (WM) loss function, forming a self-supervised weight adjustment mechanism. Furthermore, based on our observation on the false positives and the credible positives (Sect. 3.1), we carefully reassign the weight for each *positive* depended on its predicted preference score. Besides the credibility of the positives, we also investigate the over-rigid issue on negative samples. According to our observations, the relationship or features of some negative samples are very close to those of positive samples. We formally define such negative samples as *critical negatives* (Sect. 3.2). Specifically, we exploit the critical negatives with high-order neighborhood information in the user-item interaction graph. Such high-order information is then considered in the weighted multinomial loss function such that the learning process is optimized with the critical negative information. Lastly, inspired by a recent study [2], we develop a simple yet effective linear encoder-decoder framework, which would prevent from overfitting caused by the noisy and sparse implicit feedback. The contributions of the paper can be summarized as follows:

- We observe the over-confident issue on the positive samples and the over-rigid issue on the negative samples in the model training of recommender systems.
- We then propose two weighted multinomial loss functions to adjust the credibility of the positives and criticality of the negatives, respectively.

- To better incorporate two enhanced loss functions, we devise a simple VAEs structure, called VAE++, which is shown to be more effective and efficient than existing VAE-based approaches.
- Extensive experiments on four public real-world datasets show that the proposed VAE++ framework achieves better performance compared with state-of-the-art CF methods (including VAEs-based CF methods).

2 Preliminary

In this section, we introduce the notations in the manuscript and then formally define our problem. We also give a brief introduction of Mult-VAE since it is the base method of our proposed solution.

Notations. We use bold lower-case and upper-case letters to denote vectors and matrices, respectively. $\mathbf{X} \in \mathbb{N}^{|\mathcal{U}| \times |\mathcal{I}|}$ represents the interaction matrix between users and items, where $\mathcal{U}$ and $\mathcal{I}$ denote user set and item set, respectively. For the binary matrix $\mathbf{X}$, a positive value of its entry indicates that there is an interaction between the user and the item, while 0 indicates no observed interaction.

Definition 1 (Problem Definition). *Given the implicit feedback and a sparse binary matrix (user implicit feedback)* $\mathbf{X}$, *the task of top-N recommendation is to recommend N items of interest from the candidate set for every user. In general, the candidate set of a user u refers to those items which have no interaction with u.*

Fundamental of Mult-VAE. As a generative model, Mult-VAE aims to reconstruct the generative process of the user-item interaction data. First, given a user $u \in \mathcal{U}$ historical interactions $\mathbf{x}_u$, the distribution of user u's latent vector $\mathbf{z}_u$ is inferred to comply with a Gaussian prior. Based on the Gaussian distribution, latent vector $\mathbf{z}_u$ is sampled and sent to a *non-linear* function $f_\theta(\cdot)$ that produces a preference probability $\boldsymbol{\pi}_u$ over $|\mathcal{I}|$ items. Finally, the user u's interaction vector, $\mathbf{x}_u$, is assumed to be drawn from the multinomial distribution parameterized by $\boldsymbol{\pi}_u$. The generative process can be formulated as follows.

$$\mathbf{x}_u \sim \text{Mult}(|\mathcal{I}_u^+|, \boldsymbol{\pi}_u), \tag{1}$$

where $\boldsymbol{\pi}_u = \text{softmax}(f_\theta(\mathbf{z}_u))$, $\mathbf{z}_u \sim \mathcal{N}(\mathbf{0}, \boldsymbol{I})$, and $|\mathcal{I}_u^+|$ denotes the cardinality of the interacted item set of user u. $\text{Mult}(|\mathcal{I}_u^+|, \boldsymbol{\pi}_u)$ represents the multinomial distribution parameterized by $|\mathcal{I}_u^+|$ and $\boldsymbol{\pi}_u$. The multinomial log-likelihood for user u is defined as

$$\mathcal{L}_u = \log p_\theta(\mathbf{x}_u \mid \mathbf{z}_u) \overset{def}{=} \sum_i \mathbf{x}_{ui} \log \pi_{ui}, \tag{2}$$

where $\mathbf{x}_{ui}$ and π_{ui} are the i's element in $\mathbf{x}_u$ and $\boldsymbol{\pi}_u$, respectively. The posterior distribution $p_\theta(\mathbf{z}_u|\mathbf{x}_u)$ with parameter θ is usually difficult to learn directly. A variational distribution $q_\phi(\mathbf{z}_u|\mathbf{x}_u)$ with parameter ϕ is then introduced to approximate $p_\theta(\mathbf{z}_u|\mathbf{x}_u)$. The objective function, called the Evidence Lower BOund (ELBO), is:

$$\mathcal{L}(\boldsymbol{\theta}, \boldsymbol{\phi}; \mathbf{x}_u) = \mathbb{E}_{q_\phi(\mathbf{z}_u|\mathbf{x}_u)}[\log p_\theta(\mathbf{x}_u|\mathbf{z}_u)] - \gamma \cdot D_{KL}(q_\phi(\mathbf{z}_u|\mathbf{x}_u)\|p(\mathbf{z}_u)), \qquad (3)$$

where $D_{KL}(\cdot\|\cdot)$ refers to the KL divergence between two distributions and $p(\mathbf{z}_u)$ refers to the prior distribution. $\log p_\theta(\mathbf{x}_u|\mathbf{z}_u)$ indicates the negative reconstruction error term, which is calculated using the multinomial log-likelihood in Eq. (2). γ is introduced to control the strength of the regularization, i.e., KL divergence term $D_{KL}(q_\phi(\mathbf{z}_u|\mathbf{x}_u)\|p(\mathbf{z}_u))$. Given a user-item interaction vector $\mathbf{x}_u$, we can infer user preference $\hat{\mathbf{x}}_u$ on unseen items with the learned Mult-VAE. In other words, the items with the top-N highest preference scores $\hat{\mathbf{x}}_{ui}$ are recommended to the user.

3   Credible Positives and Critical Negatives

In this section, we attempt to revisit the annotation scheme of the positive and negative samples so that we can reweigh the effect of the samples in the model training process.

3.1   Revisiting Positive Samples in VAEs

The implicit feedback may not fully reveal the preference of users. Given the uncertainty of the implicit feedback, a portion of annotated positives can be counted as *false positive*, which is not credible for the model training. To classify the positives, we give an observation of the predict scores ($\propto \hat{\mathbf{x}}_{ui}$) distribution of all items (including positives and negatives) in the model training phase as follows.

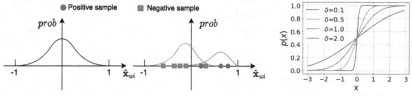

(a) Distributions of predicted scores over all items   (b) CDF of Gaussian Distribution

Fig. 2. (a) Distributions of predicted scores over all items (left) and positive/negative samples (right). Left figure shows the distribution of a user preference $\hat{\mathbf{x}}_{ui}$ over all the items, which are mostly distributed around 0. In right figure, the red line denotes the distribution of $\hat{\mathbf{x}}_{ui}$ for positive samples, the blue line represents the distribution of $\hat{\mathbf{x}}_{ui}$ for negative samples. (b) The cumulative probability function of Gaussian distribution has a similar shape with Sigmoid function.

Observation. Figure 2(a) illustrates the preference score distributions of a user u for positive and negative samples, respectively. Obviously, the preference values of both follow the Gaussian distributions since the prior distributions and the initialized model parameters (i.e., item latent vectors) are both based on Gaussian distributions. Even though a large portion of positives has better preference scores than negatives, there are still a few positives whose preference scores are indistinguishable from those of negatives. In other words, most positives of a user share similar latent features as compared with outlier positives. According to our analysis, we argue that the positives of relatively low scores are likely *false positives*, which should be *less-credible* to the learning process. Based on the observation, we estimate the credibility of a positive item i based on the user preference score $\hat{\mathbf{x}}_{ui}$ (computed by a VAEs-based model). For ease of our discussion, we denote the default positive set (i.e., items with interactions) and the default negative set (i.e., items with no interaction) of user u as $\mathcal{I}_u^+$ and $\mathcal{I}_u^-$, respectively. In this work, we define the *credible positives* as follows.

Definition 2 (Credible positive). *An item $i \in \mathcal{I}_u^+$ is a credible positive for a user u if and only if the preference score of i to u is not worse than the mean score of the positive set $j \in \mathcal{I}_u^+$, i.e., $\hat{\mathbf{x}}_{ui} \geq \frac{1}{|\mathcal{I}_u^+|} \sum_{j \in \mathcal{I}_u^+} \hat{\mathbf{x}}_{uj}$.*

Credible Positive Weighting. The credible positives should provide good impact for the training process as their preference scores are better than those of negatives with a visible difference. However, the other positives (a.k.a. less-credible positives) are likely noisy and should be lowered their influence in the learning process. We propose a *self-regulating* method to reweigh the sample probability of the positives based on the credibility level, denoted as the credibility weight α_{ui}. With the cosine similarity (Sect. 4), the preference score $\hat{\mathbf{x}}_{ui}$ for each sample is normalized into $[-1, 1]$. We say that an item is more likely credible if it is better than a large portion of other positives. The probability of a positive being credible is proportional to the cumulative distribution function of the Gaussian-like distribution (as shown in Fig. 2(a)).

Observing that the Sigmoid function with different temperature parameters δ has a similar shape to Gaussian distribution with different standard deviations (see Fig. 2(b)), we employ the Sigmoid function to approximate the credibility distribution of positives. For ease of the calculation, we set the mean of the user preference scores as the threshold (cf. Definition 2). To estimate the probability of true positive sample and transform the probability to a credibility weight, we leverage a transformation function $\alpha_{ui} = \sigma(\frac{\hat{\mathbf{x}}_{ui} - \mu_u^p}{\delta_u^p}) + 0.5$, where the Sigmoid activation function σ transforms the score to a probability in $(0, 1)$, μ_u^p denotes the mean of the preference scores of user u on all the positive samples, and δ_u^p represents the standard deviation of the preference scores of user u on all the positive samples. Note that, δ_u^p is adopted as a temperature hyperparameter, which automatically adapts the Sigmoid function to the positive distribution of different users. To adjust the credibility weights, a constant 0.5 is brought in, which keeps α_{ui} around 1.0 for average-credible samples. This adjustment prevents the weight from being too small or too large. With Sigmoid function σ, the credibility weight can better distinguish credible positives and less-credible positives.

Weighted Multinomial Loss. With the credibility weight, the multinomial log-likelihood in Eq. (2) for reconstructing a user interactions in VAEs can be revised as follows.

$$\mathcal{L}_u^{(p)} = \sum_{i \in \mathcal{I}_u^+} \alpha_{ui} \log \frac{\exp(\hat{\mathbf{x}}_{ui})}{\sum_{j \in \mathcal{I}} \exp(\hat{\mathbf{x}}_{uj})}, \tag{4}$$

where $\mathcal{L}_u^{(p)}$ means the weighted multinomial log-likelihood loss function for positive samples (i.e., user interactions) and $\mathcal{I}_u^+$ denotes the items interacted by user u. Note that we describe the loss function in terms of a user for clarity and the complete loss can be obtained by taking the sum of the loss for every user.

3.2 Revisiting Negative Samples in VAEs

As discussed in Sect. 1, previous work simply annotates all items of no interaction to a user as negative samples. This large set of negatives not only contains *true negatives* but also includes some potential positive items. Inspired by CF [15], users are interested in some *close* items of the positives, where the closeness can be defined by multi-hop relations between items and users. In the training process, a negative may be more *critical* than another negative if it is *closer* to those positives. In this work, we attempt to exploit the *critical negative* items and refine their influence based on the neighborhood relations. Though the neighborhood relations are commonly utilized in item-based CF and graph neural network (GNN)-based CF, they are seldom considered in VAEs before. For clarity, we formally define the critical negative as follows.

Definition 3 (Critical negative). *An item $j \in \mathcal{I}_u^-$ is a critical negative for a user u if and only if j is an item in the K-nearest neighbors (based on the item-item similarity matrix) of a positive item $i \in \mathcal{I}_u^+$.*

The K-nearest neighbor condition is to secure that every critical negative is more relevant to positives than non-critical negatives. In other words, a critical negative is more relevant (in terms of CF based relationship) and more similar (in terms of features) to some positives. We should increase the influence of these critical negatives in the model training.

Critical Negative Extraction. We normalize the interaction matrix $\mathbf{X}$ by $\widetilde{\mathbf{X}} = \mathbf{D}_U^{-\frac{1}{2}} \mathbf{X} \mathbf{D}_I^{-\frac{1}{2}}$, where $\mathbf{D}_U$ and $\mathbf{D}_I$ are diagonal matrices whose diagonal entries denote the number of interactions of users and items, respectively. Inspired by [5], we adopt the gram matrix (i.e., item-item co-occurrence matrix) $\widetilde{\mathbf{X}}^T \widetilde{\mathbf{X}} \in \mathbb{R}^{|\mathcal{I}| \times |\mathcal{I}|}$ as the item-item similarity matrix β. Thus, for each item i the user has interacted with, we obtain its K most similar negative items forming set $\mathcal{N}_i$, and gather all these K items of each interacted item to serve as the potential *critical negative samples* of the user.

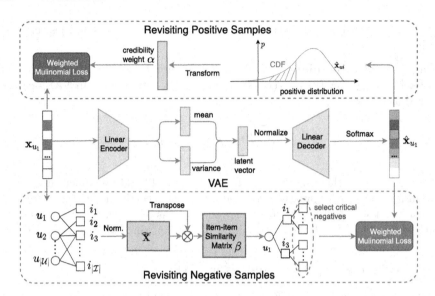

Fig. 3. The framework overview of VAE++. The middle part denotes the VAE with linear encoder and normalized decoder. The upper and bottom part represent the reconstruction process for positives and negatives, respectively.

Critical Negative Weighting. Similar to the credible positives, not every critical negatives are equally important for the training process. Thus, we consider the item-item co-occurrence matrix-based similarity as the weight of a critical negative, defined as $\beta_{ik} = \sum_{u \in \mathcal{U}} \mathbb{I}_{uik} \frac{1}{\sqrt{D_i D_k D_u}}$, where $\mathbb{I}_{uik}$ equals 1 when item i and k are commonly interacted by user u. D_i, D_k, and D_u denote the interaction number of item i, item k, and user u, respectively. The processing is illustrated in the bottom part of Fig. 3.

Weighted Multinomial Loss. Based on weight β, we optimize the loss for reconstructing user preference. Specifically, the weighted multinomial loss function $\mathcal{L}_u^{(c)}$ for reconstructing critical negative samples $k \in \mathcal{N}_i$ is formulated as follows.

$$\mathcal{L}_u^{(c)} = \sum_{i \in \mathcal{I}_u^+} \sum_{k \in \mathcal{N}_i} \beta_{ik} \log \frac{\exp(\hat{\mathbf{x}}_{uk})}{\sum_{j \in \mathcal{I}} \exp(\hat{\mathbf{x}}_{uj})} \tag{5}$$

4   Proposed Model VAE++

In this section, we discuss our framework, VAE++, based on the *weighted* multinomial losses of positives (cf. Sect. 3.1) and *critical negatives* (cf. Sect. 3.2). As shown in Fig. 3, VAE++ comprises three components, including the linear variational encoder/decoder for user interaction vector $\mathbf{x}_u$ (the middle part), the reconstruction for positives in $\mathbf{x}_u$ (the top part), and the reconstruction for critical negatives in $\mathbf{x}_u$ (the bottom part).

Encoder. Inspired by [2,4], we define the encoder of VAE++ as: $\mu_\phi(\mathbf{x}_u) = \mathbf{W}_\mu^T \mathbf{x}_u + \mathbf{b}_\mu$, $\sigma_\phi(\mathbf{x}_u) = \mathbf{W}_\sigma^T \mathbf{x}_u + \mathbf{b}_\sigma$, where $\mathbf{W}_\mu \in \mathbb{R}^{|\mathcal{I}| \times d}$ and $\mathbf{b}_\mu \in \mathbb{R}^d$ are weight and bias of mean function, and $\mathbf{W}_\sigma \in \mathbb{R}^{|\mathcal{I}| \times d}$ and $\mathbf{b}_\sigma \in \mathbb{R}^d$ are weight and bias of variance function, respectively. d refers to the dimension of latent vector.

Decoder. The decoder of VAE++ is simply defined as the dot product between the normalized latent representation of user u and a normalized matrix $\mathbf{W}_{dec}$, where each column corresponds an item i embedding, described as follows, $f_\theta(\mathbf{z}_u) = Norm(\mathbf{z}_u) \cdot Norm(\mathbf{W}_{dec})/\tau = [\hat{\mathbf{x}}_{u1}, ...\hat{\mathbf{x}}_{ui}, ...\hat{\mathbf{x}}_{u|\mathcal{I}|}]$, where $\hat{\mathbf{x}}_{ui} = \mathbf{e}_u \cdot \mathbf{e}_i/\tau$, the i-th column of the decoder weights $\mathbf{W}_{dec} \in \mathbb{R}^{d \times |\mathcal{I}|}$ is regarded as $\mathbf{z}_i \in \mathbb{R}^d$. Note that the dot product of $\mathbf{e}_u$ and $\mathbf{e}_i$ is equivalent to the cosine similarity between $\mathbf{z}_u$ and $\mathbf{z}_i$. τ is a hyperparameter to adjust the range $f_\theta(\mathbf{z}_u)$.

Loss Function. Based on the redesigned encoder, decoder, and the weighted multinomial log-likelihood for positives (cf. Eq. (4)) and critical negatives (cf. Eq. (5)), the loss function (cf. Eq. (3)) in vanilla VAEs can be now revised as follows.

$$\mathcal{L}_u(\boldsymbol{\theta}, \phi; \mathbf{x}_u) = \mathbb{E}_{q_\phi(\mathbf{z}_u | \mathbf{x}_u)} \sum_{i \in \mathcal{I}_u^+} [\alpha_{ui} \log \frac{\exp(\hat{\mathbf{x}}_{ui})}{\sum_{j \in \mathcal{I}} \exp(\hat{\mathbf{x}}_{uj})} + \lambda \sum_{k \in \mathcal{N}_i} \beta_{ik} \log \frac{\exp(\hat{\mathbf{x}}_{uk})}{\sum_{j \in \mathcal{I}} \exp(\hat{\mathbf{x}}_{uj})}]$$
$$- \gamma \cdot D_{KL}(q_\phi(\mathbf{z}_u | \mathbf{x}_u) \| p_\theta(\mathbf{z}_u)),$$

$$(6)$$

where the hyperparameter λ balances the loss between positives and critical negatives.

5 Experiments

Following [2], we conduct our experiments on four famous datasets to evaluate VAE++ for top-N recommendation task: **Gowalla**, **Yelp2018** (https:// www.yelp.com/dataset/), **Amazon-Book** and **Amazon-Video Games**. We answer the following research question: How does VAE++ perform in sparse setting as compared with state-of-the-arts VAEs-based models and other related models? Baselines include: **ItemPop** (ranking items based on popularity), **MF-BPR** [11], **Mult-VAE** [8], **EVCF** [7], **MacridVAE** [9], **RecVAE** [12], **Light-GCN** [2], **SimpleX** [4].

Experimental Result. Table 1 shows the experiment results of VAE++ and other methods on default settings. Observed from the table, VAEs-based methods mostly outperform MF-BPR, which shows that the probability generation list-wise models own better robustness than pair-wise models. Our method, VAE++, a list-wise ranking method, is superior to other methods due to our

Table 1. The comparison of performance with baselines. The best results are highlighted in bold. The second best ones are underlined.

Dataset	Gowalla		Yelp2018		Amazon-Book		Video-Games	
Method	recall	ndcg	recall	ndcg	recall	ndcg	recall	ndcg
ItemPop	0.0416	0.0317	0.0125	0.0101	0.0051	0.0044	0.0403	0.0188
MF-BPR	0.1475	0.1178	0.0485	0.0392	0.0351	0.0267	0.1120	0.0492
LightGCN	0.1840	0.1548	0.0659	0.0539	0.0417	0.0320	0.1362	0.0596
SimpleX	**0.1883**	0.1522	0.0704	0.0579	0.0553	0.0442	0.1463	0.0671
Mult-VAE	0.1541	0.1197	0.0615	0.0493	0.0436	0.0339	0.1407	0.0627
EVCF	0.1415	0.1077	0.0586	0.0474	0.0394	0.0302	0.1320	0.0574
RecVAE	0.1661	0.1321	0.0596	0.0488	0.0483	0.0379	0.1437	0.0633
MacridVAE	0.1750	0.1366	0.0713	0.0586	0.0598	0.0479	0.1484	0.0671
VAE++	0.1865	**0.1564**	**0.0740**	**0.0609**	**0.0662**	**0.0537**	**0.1601**	**0.0721**

enhanced loss functions based on denoised positives and critical negatives. The results show that VAE++ outperforms all VAEs-based CF methods and traditional methods, especially for Mult-VAE. Without applying the disentangle representations learning, VAE++ could still outperform MacridVAE. VAE++ also outperforms the two SOTAs in terms of NDCG in all 4 datasets, which verify the effectiveness of VAE++. Moreover, the simplified model structure makes VAE++ more efficient than other SOTAs. Compared with other VAEs-based CF methods, VAE++ avoids the sophisticated calculation in encoders. Compared with LightGCN and SimpleX, VAE++ is a list-wise learning approach, which ranks all items simultaneously and is more effective in top-N recommendation tasks. Besides, VAE++ avoids the aggregation computation of LightGCN at each iteration.

6 Conclusion

In this paper, we tackle with the common assumption in handling implicit feedback. We first analyze the effect of false positives and critical negatives in the implicit feedback, which are common overlooked in the prior studies. We then propose our sample re-weighting loss functions that enhance the model training performance. Besides, we design a simplified VAE training structure. Extensive experiments on four real-world datasets show the superior performance of VAE++ as compared with the state-of-the-arts, showing the re-weighting techniques in our model are effective. Moreover, we also demonstrate that the model performance can be improved by carefully handling the implicit feedback rather than merely devising complex model structure.

Acknowledgments. This work was supported in part by the National Natural Science Foundation of China (U1911203, 61902439, 61906219, 62002396), the Science and

Technology Development Fund Macau SAR (0015/2019/AKP, 0031/2022/A, SKL-IOTSC-2021-2023), the Research Grant of University of Macau (MYRG2022-00252-FST), Wuyi University Hong Kong and Macau joint Research Fund (2021WGALH14), Centre for Data Science University of Macau, Guangdong Basic and Applied Basic Research Foundation (2021A-1515011902), Macao Young Scholars Program (UMMTP2020-MYSP-016), and MBZUAI-WIS and MBZUAI start-up projects.

References

1. Gao, Y., et al.: Self-guided learning to denoise for robust recommendation. In: CoRR, p. abs/2204.06832 (2022)
2. He, X., Deng, K., Wang, X., Li, Y., Zhang, Y., Wang, M.: Lightgcn: simplifying and powering graph convolution network for recommendation. In: SIGIR, pp. 639–648 (2020)
3. Hu, Y., Koren, Y., Volinsky, C.: Collaborative filtering for implicit feedback datasets. In: ICDM, pp. 263–272. IEEE (2008)
4. Kelong, M., et al.: Simplex: a simple and strong baseline for collaborative filtering. In: CIKM, pp. 1243–1252 (2021)
5. Kelong, M., Jieming, Z., Xi, X., Biao, L., Zhaowei, W., Xiuqiang, H.: Ultra-gcn: Ultra simplification of graph convolutional networks for recommendation. In: CIKM, pp. 1253–1262 (2021)
6. Khosla, P., et al.: Supervised contrastive learning. In: NeurIPS, vol. 33 (2020)
7. Kim, D., Suh, B.: Enhancing vaes for collaborative filtering: flexible priors & gating mechanisms. In: RecSys, pp. 403–407 (2019)
8. Liang, D., Krishnan, R.G., Hoffman, M.D., Jebara, T.: Variational autoencoders for collaborative filtering. In: WWW, pp. 689–698 (2018)
9. Ma, J., Zhou, C., Cui, P., Yang, H., Zhu, W.: Learning disentangled representations for recommendation. In: NeurIPS, pp. 5711–5722 (2019)
10. Pan, R., et al.: One-class collaborative filtering. In: ICDM, pp. 502–511. IEEE (2008)
11. Rendle, S., Freudenthaler, C., Gantner, Z., Schmidt-Thieme, L.: BPR: Bayesian personalized ranking from implicit feedback. In: UAI, pp. 452–461. AUAI Press (2009)
12. Shenbin, I., Alekseev, A., Tutubalina, E., et al.: Recvae: a new variational autoencoder for top-n recommendations with implicit feedback. In: WSDM, pp. 528–536 (2020)
13. Truong, Q.T., Salah, A., Lauw, H.W.: Bilateral variational autoencoder for collaborative filtering. In: WSDM, pp. 292–300 (2021)
14. Wang, W., Feng, F., He, X., Nie, L., Chua, T.S.: Denoising implicit feedback for recommendation. In: WSDM, pp. 373–381 (2021)
15. Wu, S., Zhang, Y., Gao, C., Bian, K., Cui, B.: Garg: Anonymous recommendation of point-of-interest in mobile networks by graph convolution network. Data Sci. Eng. 5(4), 433–447 (2020)

Knowledge Graph

Question Answering over Knowledge Graphs via Machine Reading Comprehension

Weidong Han, Zhaowu Ouyang, Yifan Wang, and Weiguo Zheng$^{(\boxtimes)}$

School of Data Science, Fudan University, Shanghai, China
{wdhan20,ycwou20,zhengweiguo}@fudan.edu.cn, yifan_wang21@m.fudan.edu.cn

Abstract. Due to the representation gap between unstructured natural language questions and structured knowledge graphs (KGs), it is challenging to answer questions over KGs. The existing semantic parsing-based methods struggle for building structured queries that can be executed over the KG, and thus they are difficult to cover diverse complex questions. The information retrieval-based methods suffer from poor interpretability. In this paper, we present a novel approach powered by machine reading comprehension. To transform a subgraph of the KG centered on the topic entity into text, we sketch the subgraph through a carefully designed schema tree, which facilitates the retrieval of multiple semantically-equivalent answer entities. Instead of seeking answers from all the automatically generated paragraphs, we pick out the promising paragraphs containing answers by a contrastive learning module. Finally, it is straightforward to deliver the answer entities based on the answer span that is detected by the machine reading comprehension module. The results on benchmark datasets demonstrate that our method achieves significant improvement compared with the existing methods.

Keywords: Knowledge Graphs · Question Answering · Machine Reading Comprehension · Schema Tree

1 Introduction

By modeling the complex relationships among varieties of entities, knowledge graphs (shorted as *KGs*) have been widely used in many real-world applications, such as recommendation [30], fraud detection [33], and cyber security [18]. Generally, a knowledge graph is a structured repository that contains a collection of facts in the form (subject, relation, object). Knowledge graph question answering (KGQA) is an important task that aims at answering natural language questions based on a given knowledge graph, providing an easy and intuitive way to query knowledge graphs. Currently, how to understand and answer complex questions remains a challenging problem as they are often grounded to multiple facts in the underlying KG. For example, "Who won the prize at the spin-off of

X. Wang et al. (Eds.): DASFAA 2023, LNCS 13944, pp. 577–594, 2023.
https://doi.org/10.1007/978-3-031-30672-3_39

the 1885 Wimbledon Championships-Gentlemen's Singles?" is a multi-hop question involving two facts *(the 1885 Wimbledon Championships, spin-off, ?x)* and *(?obj, win, ?x)*. Moreover, the notorious ambiguity and variability of natural language further increase the difficulty of KGQA.

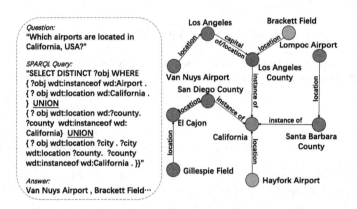

Fig. 1. An example question whose SPARQL query has multiple UNION operators. A subgraph of Wikidata is presented, where the red node denotes topic entity. (Color figure online)

The existing methods developed for KGQA can be roughly divided into three groups, i.e., rule-based/template-based (RT-based) methods, semantic parsing-based (SP-based) methods, and information retrieval-based (IR-based) methods. RT-based methods [34] exhibit superiority in precision but fail to handle the flexible and varied representation of the same semantic meanings. Existing SP-based methods try to transform natural language question q into symbolic logic form, e.g., SPARQL, which can be executed against the knowledge graph to get answers to q [6,12]. However, the existing SP-based methods are difficult to cover diverse complex queries (e.g., multi-hop reasoning, constrained relations, and numerical operations) [15]. For instance, let us consider the question in Fig. 1. To answer this question, the system should generate a complex SPARQL that is composed of three different basic graphs combined by two "UNION" operators. IR-based methods regard the process of finding answers as a classification task [21,24]. Most methods adopt deep neural networks like GNN to learn entity embeddings and rank candidate entities. One major challenge of these methods is lacking interpretability because they only take the ranking scores as objective. Moreover, the IR-based methods are not effective to deal with the multi-answer or aggregation questions as it is not known how many entities will be involved in the answers. For example, the question in Fig. 1 has 205 answers in total. It is difficult to predict the number of answers precisely.

To address the drawbacks of the existing methods, we propose a novel framework powered by machine reading comprehension (shorted as MRC) in this paper. Benefiting from the powerful pre-trained language model (PLM), MRC is promising to find the answers from plain text without complex structured queries. Thus we need to transform the knowledge graph or its subgraph into text (KG2Text). However, generating text for triples is a time-consuming task. To facilitate KG2Text, we present an effective approach based on a newly designed structure, namely *k-hop schema tree*, which sketches the subgraph through *aggregation nodes*. Meanwhile, it is natural to support the retrieval of multiple semantically-equivalent answers. Since generating text for irrelevant triples not only leads to noisy information but also is time-wasting, it is required to rule out the triples that are not relevant to the question. To the end, we perform relation linking to reduce the search space by excluding the paths that do not contain any desired relations. If the refined schema tree is still too large, multiple paragraphs will be generated for its decomposed subtrees respectively. For the automatically generated paragraphs, contrastive learning is applied to pick the target paragraph containing the answers. Powered by MRC, knowledge graph reasoning is regarded as the natural language understanding (NLU) task to obtain the aggregation answer that can be traced back to find all the answers along the schema tree. The contributions of the paper are summarized as follows:

- We develop a novel approach to KGQA based on MRC, migrating the KG reasoning to an NLU task that benefits from PLMs.
- We propose the schema tree to sketch the subgraph through aggregation nodes, facilitating KG2Text and supporting the retrieval of multiple answers.
- Contrastive learning is invoked to pick the target paragraph from all the generated paragraphs, over which MRC is conducted to find the answers.
- Empirical studies on benchmark datasets have demonstrated the effectiveness of the proposed method.

2 Problem Definition and Preliminary

2.1 Problem Definition

Definition 1 (Knowledge Graph). *A knowledge graph (KG), denoted by $\mathcal{G} = (\mathcal{E}, \mathcal{L}, \mathcal{R})$, is a directed graph consisting a set of triples (h, r, t), where $\mathcal{E}$, $\mathcal{L}$, and $\mathcal{R}$ represent the set of entities, literals, and relations (including predicates and properties), respectively. Specifically, $h \in \mathcal{E}$, $t \in \mathcal{E} \cup \mathcal{L}$, and $r \in \mathcal{R}$ represents the relation between two entities h and t.*

Definition 2 (k-hop Path Tree). *Given an entity $e \in \mathcal{E}$, the k-hop path tree rooted at e, shorted as k-PT(e), is a tree formed by combining all k-hop sequence of triples starting from e.*

Note that the "path" in the $k\text{-}PT(e)$ ignores the direction of triples. $k\text{-}PT(e)$ can be constructed through the extended breadth-first search by ignoring the direction and allowing the reusing of the nodes.

Definition 3 (Aggregation Node). *Given a k-PT(e), several entities are grouped as an aggregation node if they share the same father node with the identical edge labels.*

An aggregation node actually clusters the entities that act the same semantic role, making it feasible to return multiple semantically-equivalent answers. Specially, a single node in k-$PT(e)$ can be viewed as an aggregation node if it cannot be grouped with other nodes.

Definition 4 (k-hop Schema Tree). *Given an entity $e \in \mathcal{E}$, the k-hop schema tree rooted at e, shorted as k-ST(e), is a tree in which the starting node of each edge is an entity/aggregation node and each ending node is an aggregation node.*

Example 1. For the subgraph presented in Fig. 2(a), the corresponding 2-hop path tree rooted at the entity "California" is shown in Fig. 2(b). Figure 2(c) depicts the 2-hop schema tree, where the aggregation node in the second layer groups entities Hayfork Airport and Lake Tahoe Airport.

Definition 5 (Topic Entity). *Given a question q, its topic entity is the entity $e \in \mathcal{E}$ that is linked to a mention in q.*

Problem Statement 1. *Given a question q and a knowledge graph $\mathcal{G}$, the task is to find answers to q from the knowledge graph $\mathcal{G}$.*

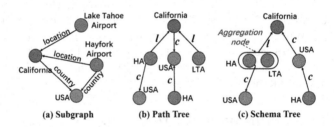

(a) Subgraph   (b) Path Tree   (c) Schema Tree

Fig. 2. An illustration of path tree and schema tree, where l denotes location, c denotes country, HA denotes Hayfork Airport, and LTA denotes Lake Tahoe Airport.

2.2 Preliminary

KG2Text. Given a subgraph g in $\mathcal{G}$, KG2Text aims at generating high-quality sentences to represent g. Recently, researchers mainly use encoder-decoder neural networks with attention [25,29] to address this problem. In this paper, we use the KG2Text tool JointGT [13] which presents three pre-training tasks on an encoder-decoder framework and a structure-aware self-attention layer.

Contrastive Learning. Contrastive Learning pulls semantically similar pair close and keeps dissimilar data away in continuous space, by using self-supervision signals. Inspired by SimCSE [8], we collect triples (q, p^+, p^-), where p^+ is called a positive instance corresponding to a paragraph containing the answers and p^- is a negative instance corresponding to a paragraph without any answer. Thus, the model picks out the sentences which are more likely to contain the final answers to the question.

Machine Reading Comprehension. Machine reading comprehension (MRC) requires a machine to answer questions based on a given textual context. It has attracted increasing attention with the incorporation of various deep-learning techniques over the past few years. The performance of the MRC task highly depends upon the ability to understand natural language for a machine. We notice that any MRC model could be applied to our framework and we choose a simple MRC model [26] in our experiments.

3   Overview of Our Approach

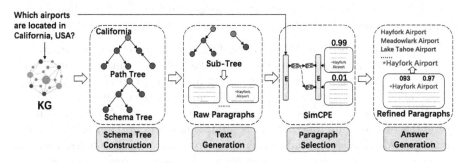

Fig. 3. Architecture of the proposed approach, consisting of four major steps. Red node indicates the topic entity. (Color figure online)

As depicted in Fig. 3, we propose a novel approach to KGQA, consisting of four components, i.e., schema tree construction, text generation, paragraph selection, and answer generation via MRC.

Schema Tree Construction. Motivated by fact that the answers to a question share the identical entity type and they appear in semantically-equivalent structures in the KG to represent the input question, we propose a novel algorithm to group and merge the semantically-equivalent nodes in a carefully designed *schema tree*. Specifically, for a question q, we extract the k-hop path tree rooted at the topic entity e_0 from the knowledge graph, aggregate the child nodes of one parent node connected by the same relation into an aggregation node, and refine the tree. In this way, we get a new subtree rooted at e_0, named schema tree, which simplifies the subgraph while preserving all necessary structure information in the knowledge graph.

Text Generation. The structure information of knowledge graphs is more difficult to understand and deal with for a computer than serialized text. It motivates us to generate texts for the built schema tree by employing KG2Text models and retrieve answers from the generated texts. On one hand, the tree structure almost complies with text writing. On the other hand, the generated text is more concise benefiting from the aggregate nodes, removing the repeated and redundant paths. Besides, just like the article describing objects from different aspects in segments, if the schema tree $k\text{-}ST(e)$ is too large, we will decompose it into smaller subtrees and generate a raw paragraph for each of them respectively.

Paragraph Selection. Since the amount of the generated texts from the knowledge graph may be still huge even if pruned by extracted relations from the question. Hence, we need to further filter out the unpromising generated texts. Motivated by *SimCSE*, we develop a paragraph selection module *SimCPE* to pick out the target paragraphs. In training stage, for each question q_i, we use the paragraph containing answers as a positive instance p_i^+. The paragraph having no answers or that is borrowed from the paragraphs of other questions will be taken as a negative instance p_i^-. Thus the triples (q_i, p_i^+, p_i^-) for contrastive learning can be collected to fine-tune the model. In test stage, we choose those paragraphs whose similarity to q_i goes beyond a predefined threshold.

Answer Generation via MRC. To find answers from the target paragraphs, we resort to the MRC model [26], because 1) the answers to questions are mostly nodes in schema tree, thus can be extracted from generated text. This process can be regraded as a sequence tagging problem, which is exactly what the MRC model elaborates. 2) the model can solve the questions which need arithmetic operations or aggregate functions, thus improving the performance. Specifically, we use this MRC model to find answer spans from the candidate texts. Once the aggregate nodes are labeled as answer spans, the final answers can be delivered straightforwardly by reporting the entities clustered in the aggregation node.

Advantages of the Proposed Framework. Compared to previous methods, our approach exhibits the following advantages:

(1) The novel schema tree has achieved lossless simplification of semantically-equivalent entities, which can alleviate the burden of text generation and improve the quality of generated texts.
(2) With the aggregation node, we can return an answer set that may contain multiple entities at once, thus overcoming the drawback of IR-based methods that they do not know how many entities will be involved in the answers.
(3) We can anchor the schema tree that generates the text containing answers, guaranteeing the interpretablity of our method.
(4) Our method utilizes the semantic correlation with the question to retrieve subgraphs from the underlying knowledge graph and finds answers in generated text, thus can handle the question with multiple answers containing different subgraph patterns respectively, which is difficult to address for both IR-based methods and SP-based methods.

Algorithm 1. Schema Tree Construction

Require: a natural language question q, a topic entity e_0, knowledge graph $\mathcal{G}$;
Ensure: a schema tree $k\text{-}ST(e_0)$.
1: $D \leftarrow$ triples within k hops from e_0, $Ents \leftarrow$ all entities in D
2: $Dict \leftarrow$ create an empty dict, an index structure where key is entity and value is (relation, aggregate node) pair
3: **for** ent in $Ents$ **do**
4:      $D_{ent} \leftarrow$ triples with one-hop relation of ent from D
5:      **for** each relation r_i in Adjacent(ent) **do**
6:          $agg\_ent \leftarrow$ merge all tail entities adjacent ent
7:          add (ent, r_i, agg_ent) into $Dict$
8: $k\text{-}ST(e_0) \leftarrow$ empty tree rooted at e_0
9: $frontier \leftarrow$ Queue(e_0)
10: **while** not IsEmpty($frontier$) **do**
11:     **if** $depth < k$ **then**
12:         $ent_p \leftarrow$ pop $frontier$
13:         $tmp \leftarrow$ retrieve $\{(r', aggnodes')\}$ from $Dict$ with key ent_p
14:         **for** r' and $aggnodes'$ in tmp **do**
15:             $k\text{-}ST(e_0) \leftarrow$ add an edge labelled with r' from ent_j to $aggnodes'$
16:             **for** $node$ in $aggnodes'$ **do**
17:                 **if** $node$ is not the same with its ancestor **then**
18:                     add $node$ to $frontier$
19: **return** $k\text{-}ST(e_0)$

---

4 Text Generation

4.1 Schema Tree Construction

Given a natural language question q as input, we first obtain the topic entity e_0 by using the existing method, e.g., spciy-entity-linker[1]. Then we extract and sketch the subgraph centered on the topic entity e_0 from KG. Algorithm 1 outlines the process of constructing a schema tree.

For each entity $e \in \mathcal{G}$, it may have outgoing and incoming edges as $\mathcal{G}$ is a directed graph. For ease of presentation, they are all called the edges starting from the entity e in the following sequel. We extract the set D of triples within h hops from e_0 and merge adjacent edges of the same relation, as shown in Algorithm 1 (lines 4–7). Then we merge tail entities with the same relation r starting from each $e \in D$ via breadth-first search, as shown in Algorithm 1 (lines 10–18). The combined entities form an aggregation node, taking one of the entities as its label. A special mark "*" can be assigned to indicate the role of the aggregation node. For example, as shown in Fig. 2, for the entity California, we merge the tail entities connected by the relation "location" into an aggregation node named "*Hayfork Airport". We can merge entities to form aggregation nodes iteratively. Finally, the schema tree $k\text{-}ST(e_0)$ is constructed.

---

[1] https://github.com/egerber/spaCy-entity-linker.

Algorithm 2. Schema Tree Refinement

Require: a schema tree $k\text{-}ST(e)$, a question q;
Ensure: a refined schema tree $k\text{-}ST(e)$.
  1: $R_l \leftarrow$ invoke relation linking methods for q, $R \leftarrow \{\}$
  2: **for** each path P_j in $k\text{-}ST(e)$ **do**
  3:      $R_p \leftarrow$ relations in path P_j
  4:      **if** $\exists rel \in R_p, rel \in R_l$ **then**
  5:          $R \leftarrow R \cup R_p$
  6: $k\text{-}ST(e) \leftarrow$ prune $k\text{-}ST(e)$ and maintain relations in R
  7: **return** $k\text{-}ST(e)$

To prevent introducing too many noisy triples, we perform relation linking-based path refinement as shown in Algorithm 2. Any relation linking methods, e.g., Falcon2.0 [23] and GenRL [22], can be applied to the question. From the topic entity e_0, we iteratively examine each path of the schema tree $k\text{-}ST(e)$ and maintain the path if and only if it contains a relation in the relation list R_t (line 2–7). We update the schema tree and remove the relation out of list R (line 9). In other words, we discard the path or sub-path which does not contain any relation in the linked relations. Though list R_l may do not contain golden relation r_{gold}, the schema tree could include r_{gold} via the other relation in R_l in the same path. In the end, we obtain a refined schema tree $k\text{-}ST(e)$.

4.2 Text Generation

In this process, we transform a schema tree into a set of paragraphs. JointGT is employed in our task. Since the number of triples in the whole schema tree $k\text{-}ST(e_0)$ is beyond model capacity, we decompose $k\text{-}ST(e_0)$ into subtrees $\{T_i\}_{i=1}^m$ based on relations $\{r_i\}_{i=1}^m$, $r_i \in \mathcal{R}_0 = \{r|(e_0, r, e_{1r}) \in k\text{-}ST(e_0)\}$:

$$T_i = (e_0, r_i, e_{1r_i}) \cup T'(e_{1r_i}),$$

where $T'(e_{1r_i})$ is the subtree rooted at e_{1r_i}. Then, we linearize T_i into a sequence $T_{linear} = \{\omega_1, \omega_2, \ldots, \omega_n\}$ consisting of n tokens. For instance, "$(A, r_1, B), (B, r_2, C)$" is transformed into "$< H > A < R > r_1 < T > B < H > B < R > r_2 < T > C$". We put T_{linear} into the encoder-decoder framework with a structure-aware self-attention layer proposed by JointGT framework: $x_{T_i} = \text{Encoder-Decoder}(T_{linear})$, where x_{T_i} is the representation of subtree T_i. Then, we use x_{T_i} to generate paragraph corresponding to subtree T_i.

5 Answer Selection

To retrieve answers from the text for the question, we first determine the target paragraphs in Sect. 5.1, based on which the final answers are generated powered by machine reading comprehension models in Sect. 5.2.

5.1   Paragraph Selection

Although we have refined the schema tree in the previous step, it is inevitable to generate irrelevant paragraphs that correspond to some subtrees T_i. Thus, we propose a Simple Contrastive Learning of Paragraph Embeddings, shorted as **SimCPE**, to pick out the most promising paragraphs that contain the answers.

As shown in Fig. 4, for each question q_i we use the paragraph which contains answers as positive instance p_i^+ and the paragraph which does not contain answers or the paragraph for other questions as negative instance p_i^-, thus obtaining triples (q_i, p_i^+, p_i^-). We apply RoBERTa$_{large}$ framework as a basic model and try to narrow the distance between q_i and p_i^+ and keep q_i and p_i^- away in continuous space. We minimize the objective $\mathcal{L}_i$ loss function:

$$-\log \frac{e^{\text{sim}(q_i, p_i^+)/\tau}}{\sum_{j=1}^{N}(e^{\text{sim}(q_i, p_j^+)/\tau} + e^{\text{sim}(q_i, p_j^-)/\tau})},$$

where the similarity measure $\text{sim}(x_1, x_2)$ is cosine similarity $\frac{x_1^T x_2}{||x_1||||x_2||}$, N is mini-batch size and τ is a temperature hyperparameter. In the test phase, we input previously generated paragraphs, output the probability of containing answers for each paragraph, and deliver the k paragraphs with the highest probability.

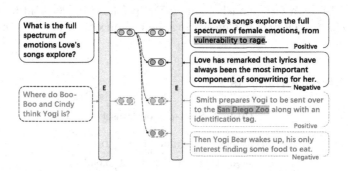

Fig. 4. Examples for contrastive learning. Orange circles denote representations of the question and ground truth text, and blue denotes noisy text. **E** is the encoder.

5.2   Answer Generation via MRC

Since an answer span may appear multiple times in the automatically generated paragraphs, we present a multi-span framework to tackle multi-span questions inspired by [26]. The paragraph containing at least one answer is assigned the possible-correct tagging. At training time, we maximize the marginal probability of possible-correct taggings:

$$\log p(\mathcal{S}|h) = \log \sum_{S \in \mathcal{S}} (\prod_{i=1}^{m} p_i[S_i]),$$

$$p_i = \text{softmax}(f(h_i))$$

where h is the representation of question and text generated by KG2Text module via BERT [4], f is a parameterized function, $\mathcal{S}$ is the set of possible-correct taggings and m is length of text. At test time, we output the tagging with the max probability: $\hat{S} = \arg\max_{S \in \mathcal{V}} \prod_{i=1}^{m} p_i[S_i]$, where $\mathcal{V}$ includes all possible taggings as for IO tags. We predict for each token whether it is part of the answer individually following the paper [4]. Then, we collect the predicted spans as answers. If the span contains the special mark "*" indicating an aggregation node, we can collect all the semantically-equivalent answers in the corresponding aggregation node. Meanwhile, the chain starting from e_0 to the detected aggregation node in the schema tree can be taken as the inferential chain of evidencing answers.

6   Experimental Evaluation

In this section, we evaluate the proposed method through extensive experiments and report the results.

6.1   Experimental Settings

In our experiments, we made the next settings. (1) **JointGT**: We choose JointGT(Bart), and fine-tune on WebNLG following [13]. (2) **MRC**: We choose $TASE_{IO}$+SSE as a base model, train on **Drop** following [26], and fine-tune on our datasets. (3) **Depth** k: Considering the complexity of datasets, k is set to 2.

6.2   Benchmark Datasets

Table 1. Statistics of datasets.

Datasets	Train	Dev	Test
LC-QuAD2.0	17,413	4,353	6,046
WebQSP	2,848	250	1,639

Table 1 lists the statistics of the used datasets.
WebQuestionsSP (WebQSP). [32] contains 4737 question-answer pairs with Freebase as the KG. We use the same splits and pre-processing as [24].

    LC-QuAD 2.0 [7] is a larger dataset based on Wikidata. The paraphrases and SPARQL queries are compatible with both Wikidata and DBpedia-2018.

6.3   Baselines

We choose the following methods as the baselines:
    QAnswer [5]: QAnswer is a system that relies on the simple template-based SPARQL queries that are then ranked and executed to get the answer.

Platypus [19]: Platypus is designed as a QA system driven by NLU, using grammar rules and template-based techniques for complex questions.

GRAFT-Net [28]: Graphs of Relations Among Facts and Text Networks (GRAFT-Net) classifies nodes in subgraphs containing KG entities and text.

PullNet [27]: PullNet uses self-learning to extract the subgraph related to the question. It combines the heterogeneous information, updates the subgraph, and finally uses GCN to perform representation learning on nodes.

UNIQORN [20]: UniQORN copes with questions by advanced graph algorithms for Group Steiner Trees, identifying the best answer candidates.

EmbedKGQA [24]: EmbedKGQA uses neutral network to get the representation of question and uses ComplEx embeddings of entities. Then it computes the similarity between question and candidate entities.

NSM [10]: NSM uses a teach-student framework to generate and learn supervision signal, thus enhancing the performance of reasoning against the KG.

6.4   Main Results

The results of our method and baselines are shown in Table 2. We can find that:

(1) The experimental results indicate that the existing methods, e.g. GRAFT-Net and PullNet, have poor performance on LC-QuAD2.0, while our approach achieves state-of-the-art performance on LC-QuAD2.0 and is much better than the above baselines. PullNet trains an effective subgraph retrieval module based on the shortest path between topic entities and answer entities. However, PullNet uses external corpus to enhance performance. UniQORN uses Group Stein Tree (GST) to reason against Wikidata, but ineffective to cover multi-hop queries with constraints. Our method exhibits a powerful ability to extract information from Wikidata that contains thousands of relations and billions of entities.

Table 2. Results on LC-QuAD2.0 and WebQSP. We copy the results on LC-QuAD2.0 in [20] and copy the results on WebQSP in [10].

Methods	LC-QuAD2.0		WebQSP
	Hits@1	Hits@5	Hits@1
QAnswer	–	31.8	–
Platypus	–	10.9	–
GRAFT-Net	26.5	41.8	66.4
PullNet	11.9	28.1	68.1
UniQORN	25.2	41.4	–
EmbedKGQA	–	–	66.6
NSM	–	–	74.3
ours	**42.6**	**53.7**	**75.2**

Table 3. Adaptive Ability of MRC model (Hits@1).

Methods	WebQSP
vanilla model (no fine-tune)	62.2
+ LC-QuAD2.0 train set	66.0
+ WebQSP train set	**75.2**

(2) Our method also achieves state-of-the-art performance on WebQSP. EmbedKGQA uses ComplEx embeddings to obtain the representation of entities and computes the similarity between the question and entities, ignoring the path information. NSM shows a competitive performance as it applies a teacher-student framework to learn intermediate supervision signals. However, it highly depends on the golden path, and is difficult to find all the semantically-equivalent answers. In contrast, we sketch the structure around the topic entity by schema trees and fit the refined paragraphs that are generated from decomposed schema subtrees into a multi-span MRC model. Thus, we can answer questions that even do not have explicit paths via natural language understanding.

6.5   Detailed Analysis

Adaptive Ability. We emphasize the importance of our proposed framework based upon MRC to select answers in Sect. 3. To further study its adaptive ability, we perform experiments to compare two variants including: (1) +LC-QuAD2.0 train set (fine-tune) using question-answer pair on LC-QuAD2.0 as train set. (2) +WebQSP train set (fine-tune) using WebQSP to fine-tune MRC model. As shown in Table 3, our proposed MRC model exhibits a certain inference ability even without fine-tuning. It has significant improvement when fine-tuned on the corresponding dataset. Thus, our framework based on MRC shows promising performance in adapting to other question-answering datasets.

Effect of KG2Text. To study the effect of KG2Text, we compare the performance of *w/o KG2Text* (without using KG2Text model to generate texts, concatenating the triples in schema tree instead) with our complete model equipped with the KG2Text module. The experimental results are shown in Table 4 (line 1 and line 3). We can see that the performance on both datasets decreases, indicating the contribution of KG2Text module. However, there is a significant difference on effect of this module on two datasets. Without KG2Text module, the decline on LC-QuAD2.0 is particularly severe, nearly three forth. While on WebQSP, the decline is relatively small, only 6.0%. The questions of LC-QuAD2.0 are more difficult to understand for the machine as they correspond to more complex subgraphs and more fluent and logical text. Hence, when the text is simply concatenated by strings, it is almost impossible for the model to answer the complex questions.

Table 4. Ablation study (Hits@1).

Methods	LC-QuAD2.0	WebQSP
fine-tuned model	**42.6**	**75.2**
w/o SimCPE	19.45	66.5
w/o KG2TEXT	10.77	69.18

Table 5. Effect of depth k (Hits@1).

Methods	LC-QuAD2.0	WebQSP
$k=1$	20.5	34.3
$k=2$	**42.6**	**75.2**

Effect of SimCPE. To study the effect of SimCPE, we compare *w/o SimCPE* (fine-tune MRC model without SimCPE) with our fine-tuned approach on LC-QuAD2.0 and WebQSP. As shown in Table 4 (lines 1–2), the performance on both datasets decreases when disabling SimCPE, which proves the efficiency and necessity of SimCPE. However, the degree of decline in two datasets has a different picture. Specifically, removing SimCPE causes more than half loss in Hits@1 on LC-QuAD2.0 dataset, while a slight decrease of 8.7% in WebQSP. The main reason for this result may be resulted from the size of the corresponding knowledge graph. The knowledge graph LC-QuAD2.0 is very large, thus the generated subgraph is also very large as well. With the help of SimCPE, we can quickly eliminate the unrelated texts and improve the accuracy. As for WebQSP, the knowledge graph is relatively small, so the truncation effect of SimCPE is much smaller than the former.

Effect of Depth k. We further study the influence of depth k. The results are reported in Table 5. Obviously, the Hits@1 drops sharply as there are many two-hop questions on both datasets. It reveals that equipped with higher-order relations our approach can handle more complex questions demanding multi-hop reasoning. Besides, since most questions have 2-hop structures in the knowledge graph, we set k to 1 and 2 in the experiments.

6.6   Case Study

As shown in Table 6, we provide several cases in LC-QuAD2.0 to further investigate and analyze our proposed approach.

Table 6. Case study on LC-QuAD2.0, where * means this entity is an aggregation node with its members in parentheses.

Case	Id	Question	Generated Text	Returned Answers	Ground Truth
1	14791	What has influenced the sculptors of Man in Shower in Beverly Hills?	The creator of Man in Shower in Beverly Hills is David Hockney, who was influenced by *Francis Bacon (Francis Bacon,Pablo Picasso).	Francis Bacon	Pablo Picasso
				Pablo Picasso	Francis Bacon
2	9652	What are the names of Keira Knightley's sibling and father?	The sister of Keira Knightley is Caleb Knightley who is also the son of Will Knightley.	Caleb Knightley	Caleb Knightley
				Will Knightley	Will Knightley
3	27990	Who is the curator of São Paulo Museum of Art?	The architect of São Paulo Museum of Art, which is located in Sao Paulo, Brazil, was Lina Bo Bardi. It is the home of the architect, Adriano Pedrosa and the museum 's director, Heitor Martins.	Adriano Pedrosa	Adriano Pedrosa
				Heitor Martins	
4	20699	Which is the rock band, member of which was Tom Petty?	Tom Petty is a member of Tom Petty and the Heartbreakers, a band that includes the band, Traveling Wilburys.	Tom Petty and the Heartbreakers	Tom Petty and the Heartbreakers
				Traveling Wilburys	
5	25382	Name an English written daily newspaper that starts with letter "T".	*the New York Times (The New York Times, Daily Times, ...) are all owned and operated by the company Nagaland Post.	The New York Times	The Times...
				Daily Times	The Dallas Morning News
				...	The Guardian

Both Case 1 and Case 2 are successful cases. The former shows that our approach can process questions with many answers. By merging all entities that influence the entity "David Hockney" into an aggregation node, the model is able to find all the answers. The latter one illustrates that our method can handle the complex question with multiple special interrogative words. Powered by the MRC model that targets at processing multi-span questions, our approach can find the sibling and father of "Keira Knightley" simultaneously.

Case 3 and Case 4 are both surprising cases. The answers are partially correct because some answers do not belong to ground truths. The main reason is that our method retrieves all relations related to the question, and the MRC model fails to distinguish "curato" from "manager" (or "rock band" from "musical group") quite clearly. However, from a perspective of real-world applications, the result may be pretty good. In most cases, the user is prone to use common words rather than professional words as she may not know the vocabulary of the huge knowledge graph. Moreover, it demonstrates that our methods can solve questions with semantically-equivalent subgraphs.

Case 5 is a failed case where our approach fails to handle character-level constraints. It returns all English written daily newspapers without considering the constraint "starts with letter 'T'". Although the MRC model can distinguish them in the text by dealing with the nodes in aggregation nodes separately, it is overwhelmed when encountering character-level constraints. That is because all character-level information has been hidden by aggregation nodes and is invisible in generated texts. If we present all entities in the aggregation node to the MRC model, the text will be long-winded for the model to understand.

7   Related Work

KGQA: Zheng et al. apply template-based methods including two steps, generating templates offline and understanding questions online [34]. Existing SP-based methods [2,6,12,17,31] try to transform natural language questions into a symbolic logic form which can be executed over the KG to return answers. Ding et al. introduce frequent query substructures to rank existing query structures or build new queries [6]. Kapanipathi et al. present a system that successfully utilizes a generic semantic parser, particularly AMR, to deal with a KGQA task for the first time [12]. Chen et al. propose a two-stage formal query building approach that automatically predicts the query structure and uses it as a constraint to avoid generating noisy candidate queries [2]. Liang et al. present a novel approach that can first identify the type of each question by training a Random Forest model and use it to guide different processes to generate SPARQL queries [17]. IR-based methods [3,10,11,21,24] regard answer selection as a classification task. Most studies use deep neural networks like GNN to score candidate entities. Qiu et al. introduce reinforcement learning to formulate multi-relation question answering as a sequential decision problem [21]. The proposed model performs an efficient path search on the knowledge graph to obtain answers and utilizes beam search to significantly reduce the number of candidates. Meanwhile, based on attention mechanism and neural network, policy network can enhance the unique influence of different parts of a given question on triple choice. He et al. propose a teacher-student framework to find a reasonable path via supervision signals at intermediate steps [10]. The main student model learns how to find answers to specific questions, while the teacher model strives to learn intermediate state supervision signals.

KG2Text: KG2Text is an important problem that generates natural language sentences from structured facts in the knowledge graph. Traditional methods [14] mainly focus on rule-based algorithms but subject to low adaptability. [1] proposes seq-to-seq and graph-to-seq framework to generate sentences from well-structured data and a large dataset KGText. KGText is a new pretrained corpus in which English sentences from Wikipedia are aligned with subgraphs from Wikidata for a total of about 1.8 million (subgraph, text) pairs. The method ensures that each subgraph and its paired sentences describe nearly the same facts. Based on that, [13] adds a structure-aware self-attention layer to better instruct sentence generation and proposes three pre-training tasks, including graph enhanced text reconstruction, text enhanced graph reconstruction, and graph-text embedding alignment, to explicitly promote the graph-text alignment.

MRC: Machine Reading Comprehension (MRC) requires a machine to answer questions based on a given textual context. Some pre-trained language models, like Bert [4] and BART [16], are trained on a large-scale corpus and fine-tuned on the downstream dataset, showing impressive performance on MRC. As for long-text, [9] proposes a model that learns to chunk text more flexibly via reinforcement learning and decides the next segment to process. However, forcing answers to a single span limits its ability to understand questions as some recent datasets contain multi-span questions, i.e., questions whose answers are a set of non-contiguous spans in the text. Naturally, models that return a single span cannot answer these questions. To answer multi-span questions, [26] proposes a new architecture to cast the problem as a sequence tagging task. In other words, the model predicts whether each token is part of an output.

8   Conclusion

We present a novel framework for KGQA based on machine reading comprehension. To facilitate KG2Text, we propose schema tree that sketches the subgraph centered on the topic entity. The search space is reduced by removing the irrelevant triples and paragraphs through relation linking and constrastive learning. The empirical results show that our approach outperforms the existing methods.

Acknowledgement. This work was supported by National Natural Science Foundation of China (Grant No. 61902074). Weiguo Zheng is the corresponding author.

References

1. Chen, W., Su, Y., Yan, X., Wang, W.: KGPT: knowledge-grounded pre-training for data-to-text generation. In: EMNLP (2020)
2. Chen, Y., Li, H., Hua, Y., Qi, G.: Formal query building with query structure prediction for complex question answering over knowledge base. In: IJCAI (2021)
3. Chen, Z.Y., Chang, C.H., Chen, Y.P., Nayak, J., Ku, L.W.: UHop: an unrestricted-hop relation extraction framework for knowledge-based question answering, pp. 345–356, June 2019
4. Devlin, J., Chang, M.W., Lee, K., Toutanova, K.: BERT: pre-training of deep bidirectional transformers for language understanding. In: NAACL, pp. 4171–4186 (2019)
5. Diefenbach, D., Migliatti, P.H., Qawasmeh, O., Lully, V., Singh, K., Maret, P.: Qanswer: A question answering prototype bridging the gap between a considerable part of the lod cloud and end-users. In: WWW, pp. 3507–3510 (2019)
6. Ding, J., Hu, W., Xu, Q., Qu, Y.: Leveraging frequent query substructures to generate formal queries for complex question answering. In: EMNLP-IJCNLP, pp. 2614–2622, November 2019
7. Dubey, M., Banerjee, D., Abdelkawi, A., Lehmann, J.: Lc-quad 2.0: a large dataset for complex question answering over wikidata and dbpedia. In: ISWC, pp. 69–78 (2019)
8. Gao, T., Yao, X., Chen, D.: SimCSE: simple contrastive learning of sentence embeddings. In: EMNLP, pp. 6894–6910 (2021)
9. Gong, H., Shen, Y., Yu, D., Chen, J., Yu, D.: Recurrent chunking mechanisms for long-text machine reading comprehension. In: ACL, pp. 6751–6761 (2020)
10. He, G., Lan, Y., Jiang, J., Zhao, W.X., Wen, J.R.: Improving multi-hop knowledge base question answering by learning intermediate supervision signals. In: WSDM, pp. 553–561 (2021)
11. Jain, S.: Question answering over knowledge base using factual memory networks. In: Proceedings of the NAACL Student Research Workshop, pp. 109–115 (2016)

12. Kapanipathi, P., Abdelaziz, I., Ravishankar, S., Roukos, S., Yu, M.: Leveraging abstract meaning representation for knowledge base question answering. In: Findings of the Association for Computational Linguistics: ACL-IJCNLP 2021 (2021)
13. Ke, P., et al.: JointGT: graph-text joint representation learning for text generation from knowledge graphs. In: Findings of ACL-IJCNLP, pp. 2526–2538 (2021)
14. Kukich, K.: Design of a knowledge-based report generator. In: ACL, pp. 145–150. Association for Computational Linguistics, USA (1983)
15. Lan, Y., He, G., Jiang, J., Jiang, J., Zhao, W.X., Wen, J.R.: A survey on complex knowledge base question answering: Methods, challenges and solutions. In: IJCAI (2021)
16. Lewis, M., et al.: BART: denoising sequence-to-sequence pre-training for natural language generation, translation, and comprehension. In: ACL, pp. 7871–7880, July 2020
17. Liang, S., Stockinger, K., de Farias, T.M., Anisimova, M., Gil, M.: Querying knowledge graphs in natural language. J. Big Data **8**(1), 1–23 (2021). https://doi.org/10.1186/s40537-020-00383-w
18. Liu, K., Wang, F., Ding, Z., Liang, S., Yu, Z., Zhou, Y.: A review of knowledge graph application scenarios in cyber security. CoRR abs/2204.04769 (2022)
19. Pellissier Tanon, T., de Assunção, M.D., Caron, E., Suchanek, F.M.: Demoing platypus - a multilingual question answering platform for Wikidata. In: The Semantic Web: ESWC 2018 Satellite Events, pp. 111–116 (2018)
20. Pramanik, S., Alabi, J., Roy, R.S., Weikum, G.: UNIQORN: unified question answering over RDF knowledge graphs and natural language text. CoRR (2021)
21. Qiu, Y., Wang, Y., Jin, X., Zhang, K.: Stepwise reasoning for multi-relation question answering over knowledge graph with weak supervision. In: WSDM, pp. 474–482 (2020)
22. Rossiello, G., et al.: Generative relation linking for question answering over knowledge bases. In: ISWC, pp. 321–337 (2021)
23. Sakor, A., Singh, K., Patel, A., Vidal, M.E.: Falcon 2.0: an entity and relation linking tool over Wikidata. In: CIKM, pp. 3141–3148 (2020)
24. Saxena, A., Tripathi, A., Talukdar, P.: Improving multi-hop question answering over knowledge graphs using knowledge base embeddings. In: ACL, pp. 4498–4507 (2020)
25. Scarselli, F., Gori, M., Tsoi, A.C., Hagenbuchner, M., Monfardini, G.: The graph neural network model. IEEE Trans. Neural Networks **20**(1), 61–80 (2009)
26. Segal, E., Efrat, A., Shoham, M., Globerson, A., Berant, J.: A simple and effective model for answering multi-span questions. In: EMNLP, pp. 3074–3080 (2020)
27. Sun, H., Bedrax-Weiss, T., Cohen, W.: PullNet: open domain question answering with iterative retrieval on knowledge bases and text. In: EMNLP-IJCNLP, pp. 2380–2390 (2019)
28. Sun, H., Dhingra, B., Zaheer, M., Mazaitis, K., Salakhutdinov, R., Cohen, W.: Open domain question answering using early fusion of knowledge bases and text. In: EMNLP, pp. 4231–4242 (2018)
29. Vaswani, A., et al.: Attention is all you need. In: NIPS, pp. 6000–6010 (2017)
30. Wang, X., Liu, K., Wang, D., Wu, L., Fu, Y., Xie, X.: Multi-level recommendation reasoning over knowledge graphs with reinforcement learning. In: WWW, pp. 2098–2108 (2022)
31. Xiao, G., Corman, J.: Ontology-mediated SPARQL query answering over knowledge graphs. Big Data Res. **23**, 100177 (2021)

32. Yih, W.t., Chang, M.W., He, X., Gao, J.: Semantic parsing via staged query graph generation: Question answering with knowledge base. In: ACL, pp. 1321–1331 (2015)
33. Zhang, L., Wu, T., Chen, X., Lu, B., Na, C., Qi, G.: Auto insurance knowledge graph construction and its application to fraud detection. In: IJCKG, pp. 64–70 (2021)
34. Zheng, W., Yu, J.X., Zou, L., Cheng, H.: Question answering over knowledge graphs: question understanding via template decomposition. Proc. VLDB Endow. **11**(11), 1373–1386 (2018)

MACRE: Multi-hop Question Answering via Contrastive Relation Embedding

Man Xu, Weiguo Zheng[✉], and Deqing Yang

School of Data Science, Fudan University, Shanghai, China
manxu21@m.fudan.edu.cn, {zhengweiguo,yangdeqing}@fudan.edu.cn

Abstract. Multi-hop question answering over knowledge graphs (KGs) is a crucial and challenging task as the question usually involves multiple relations in the KG. Thus, it requires elaborate multi-hop reasoning with multiple relations in the KG. Two existing categories of methods, namely semantic parsing-based (SP-based) methods and information retrieval-based (IR-based) methods, either suffer from complicated logic forms for covering diverse questions or fail to offer traceable reasoning. In this paper, we propose a novel approach for multi-hop question answering over KGs via contrastive relation embedding (MACRE), powered by contrastive relation embedding and context-aware relation ranking. An adaptive beam search is developed to deliver the answer by identifying the optimal weighted inferential chain, boosting the searching efficiency and alleviating error propagation. The proposed method offers both interpretable reasoning and powerful answering ability, unifying the strengths of SP-based methods and IR-based methods. Extensive experimental results on several benchmark datasets demonstrate the effectiveness of our method.

Keywords: Contrastive Learning · Knowledge Graph · Question Answering · Multi-hop Reasoning

1 Introduction

Answering natural language questions over knowledge graphs (KGQA) has been a hotspot research in both academic and industrial communities. In contrast with simple questions that just map single relations in the underlying KG, multi-hop questions are difficult to answer as they are usually grounded to multiple relations in sequence [12,29]. Currently, the approaches can be roughly divided into semantic parsing-based (SP-based) methods [12,24] and information retrieval-based (IR-based) methods [7,23]. SP-based methods heavily rely on the design of the logic form, while the staged parsing paradigm is easy to exclude the desired entities or relations too early, leading to error propagation [18]. The IR-based approaches benefit from advanced Graph Neural Network (GNN) architectures to conduct message passing but fail to offer adequate interpretability [11]. A question arises "Can we develop an approach that offers user-friendly interpretability while shaking off the complicated logic form and error propagation?"

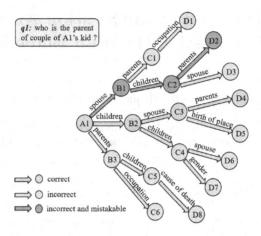

Fig. 1. Example of multi-hop reasoning over the KG.

It is critical to identify an inferential chain [24] which provides evidence of the answer and yet a challenging task especially for multi-hop questions since picking the correct one out of a huge number of candidate chains is difficult and requires elaborate multi-hop reasoning. Let us consider the question in Fig. 1, the system has to detect the three relations that are semantically related to and logically consistent to the question, which means the desired relations in the chain should be connected in the correct order. Otherwise, the system easily falls into the trap of confusing the chain $A_1 \xrightarrow{spouse} B_1 \xrightarrow{children} C_2 \xrightarrow{parents} D_2$ with the target inferential chain $A_1 \xrightarrow{children} B_2 \xrightarrow{spouse} C_3 \xrightarrow{parents} D_4$ since the two paths share the identical collection of relations.

Identifying the inferential chain demands elaborate multi-hop reasoning over the KG involving multiple relations, facing two major challenges: more powerful multi-relational linking and trade-off between searching efficiency and accuracy. Previous methods measure the similarity between the question and a single relation, essentially a classification task of relations in the KG [25], in which the same relation in different triples will get the same score, making it troublesome to perform multi-hop reasoning over the KG. And sentence embeddings from pre-trained language models (LMs) such as BERT [5] widely adopted for question-relation matching are poorly to capture semantic similarity out of a non-smooth anisotropic semantic space of sentences [14]. Another main challenge followed is the candidate chain space grows exponentially [7]. Methods [13] try to restrict the search space by beam search may suffer from excluding the right sub-chain too early, leading to cascading errors.

To handle the challenges mentioned above, we propose a new framework MACRE for interpretable multi-hop KGQA via contrastive relation embedding. As illustrated in Fig. 2, MACRE answers a multi-hop question q through three main modules. Firstly, a subgraph is extracted from KG by using the topic entity e_{topic} identified by entity linking and its neighbors within K hops, where K is inferred from q. Relation embeddings of the subgraph are obtained through

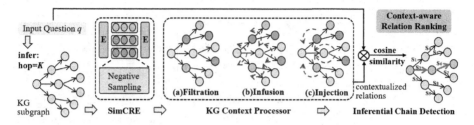

Fig. 2. Overview of MACRE. E denotes the encoder and s_i denotes the score of relation r_i in the weighted KG subgraph.

the contrastively trained **SimCRE**, and then the embeddings are fed into the **KG Context Processor** to get integrated representations, which are exploited to calculate contribution scores of edges in the subgraph. **Inferential Chain Detection** is finally conducted on the scored subgraph to deliver the answer to q.

To summarize, we make the following contributions in this paper:

- We propose SimCRE to generate effective relation embeddings based on contrastive learning, along with a representative negative sampling mechanism.
- We present a KG context processor that integrates the contrastive relation embedding, enabling multi-hop reasoning in a contextualized way.
- We develop an adaptive beam search with efficiency to find optimal weighted chains, alleviating error prorogation and supporting interpretability.
- Extensive experiments on benchmarks show that the proposed method outperforms the state-of-the-art methods.

2 Approach

2.1 SimCRE: Contrastive Relation Embedding

We propose SimCRE, a simple but effective relation embedding method based on contrastive learning with both question understanding and reasoning ability. With a representative negative sampling mechanism augmented, SimCRE encodes not only information of semantic similarity between relations and the question, but also logical structure hidden in the question text.

Idea of Contrastive Relation Embedding. Contrastive learning aims to learn effective representation by pulling semantically close neighbors together and pushing apart non-neighbors. Motivated by the example as shown in Fig. 1, we apply contrastive learning to get more informative relation representations, hoping that the encoder could capture the subtle differences between chains which are critical for multi-hop reasoning. By taking both semantic and logical similarity into consideration, the model could learn not only superficial semantic similarity, but also consistency with question structure and the underlying chain of relations in the process of contrastive learning.

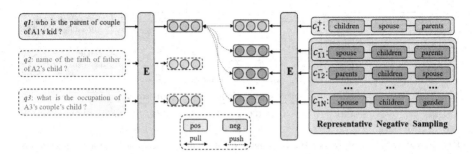

Fig. 3. Illustration of SimCRE with the representative negative sampling mechanism. E denotes the encoder.

Representative Negative Sampling Mechanism. Instead of just randomly selecting triples from KG, we propose a representative negative sampling mechanism to sample representative negative instances for each question. Negative instances more similar to the ground truth inferential chain are more representative, because highly deceptive and confusing instances can force the model to absorb important discriminative knowledge in the process of contrastive learning. As illustrated in Fig. 3, we use the full permutation of relations in the ground truth chain other than the ground truth chain as a strong negative augmentation. Meanwhile, in order to retain the general semantic comparison information, we also randomly sample a bunch of chains from the KG.

Model and Loss Function. SimCSE [8], one of the state-of-the-art methods for sentence embedding based on contrastive learning, extends (x_i, x_i^+) to (x_i, x_i^+, x_i^-), where x_i is the premise, x_i^+ and x_i^- are entailment and contradiction hypotheses. Following that setting, we take the NL question q as the premise q_i and the ground truth inferential chain as entailment hypothesis c_i^+, while contradiction chain c_i^- is generated following the representative negative sampling mechanism. We employ a pre-trained BERT as the backbone to encode all the text inputs (let $text_i$ denotes texts in q_i, c_i^+ or c_i^-):

$$h_i = f_{simcre}(text_i). \tag{1}$$

The parameters of f_{simcre} are fine-tuned with the objective l_i to pull questions and ground-truth chains together and push apart questions and negative chains:

$$l_i = -\log \frac{e^{sim(h_i, h_i^+)/\tau}}{\sum_{j=1}^{N}(e^{sim(h_i, h_j^+)/\tau} + e^{sim(h_i, h_j^-)/\tau})}, \tag{2}$$

where N is mini-batch size, τ is a temperature hyperparameter and $sim(h_1, h_2)$ is defined as $\frac{h_1^T h_2}{\|h_1\| \cdot \|h_2\|}$; h_i, h_i^+, and h_i^- are embeddings of q_i, c_i^+, and c_i^-, respectively.

2.2   Context-Aware Relation Ranking

Unlike previous methods [7,23], we directly rank relations instead of ranking entities. The intuition is that the intermediate entities are often hidden in multi-hop questions, while mentions of relation are always presented explicitly. A KG Context processor module that measures the contribution each relation can make to answer the question is designed to realize context-aware relation ranking, consisting of three components, namely filtration, infusion, and injection.

We define the k-hop KG context of relation r_i as its k-hop adjacent relation set R_i and entities set E_i, where the contextual relation $r_j \in R_i$ will be infused and injected into r_i and the contextual entity $e_j \in E_i$ is utilized for filtration. e_j is the tail entity connected by r_j, which can provide signal for filtration.

(a) Filtration. Considering that including all the adjacent relations would import too much noise, we utilize the inherent knowledge contained in LM for filtering. We use BERT's mask mechanism to predict the occurrence probability α_j of e_j given the question q, which is denoted as the co-occurrence rate of q and e_j. The higher α_j is, the more relevant r_j is to q, as a way of soft filtration.

$$\alpha_j = P_{BERT}(e_j|q). \tag{3}$$

(b) Infusion. We propose two alternative infusions to aggregate KG context, namely hard infusion and soft infusion. Co-occurrence rates of contextual entities are used as aggregating weights.

– **Hard Infusion:** Concatenate the relation with its k-hop ($1 \leq k < K$, K is the hop number of the question as predicted in Sect. 2.3) contextual relations and feed them into SimCRE to get an aggregated contextualized relation representation:

$$c_{r_i}^{hard} = \sum_{k=1}^{K-1} \sum_{j=1}^{m_k} \alpha'_{kj} \cdot f_{simcre}(r_i, r_{1j}, ..., r_{kj}), \tag{4}$$

$$\alpha'_{kj} = \frac{e^{\sum_{s=1}^{k} \alpha_{sj}}}{\sum_{j=1}^{m_k} e^{\sum_{s=1}^{k} \alpha_{sj}}}, \tag{5}$$

where m_k is the number of concatenated k-hop contextual relations for r_i.

– **Soft Infusion:** Directly aggregate the SimCRE representations of the relation's contextual relations:

$$c_{r_i}^{soft} = \sum_{k=1}^{K-1} \sum_{t=1}^{n_k} \alpha'_{kt} \cdot f_{simcre}(r_{kt}), \tag{6}$$

where $\alpha'_{kt} = \frac{e^{\alpha_{kt}}}{\sum_{t=1}^{n_k} e^{\alpha_{kt}}}$ and n_k is the number of r_i's k-hop relations.

(c) Injection. Next, we inject the infused KG context into the relation to get the final representation. λ is a hyperparameter to adjust the proportion of injection.

$$h_{r_i} = f_{simcre}(r_i), \tag{7}$$

Table 1. Statistics of benchmark datasets

Datasets	#Entity	#Relation	#Question
SimpleQuestions	131681	6837	>100k
PathQuestion-2/3H	2215	14	1908/5198
WorldCup-1/2H	1127	6	6482/1472

Table 2. Test accuracy on SimpleQuestions

Methods	Test accuracy
RNN-QA (2017)	0.883
Subgraph Ranking and Joint-Scoring (2019)	0.854
Relation-aware BERT (2020)	0.809
APVA-TURBO (2021)	0.751
MACRE(**Ours**)	**0.915**

$$h'_{r_i} = h_{r_i} + \lambda c_{r_i}. \tag{8}$$

The representation processed by the KG context processor of the relation is used to calculate the cosine similarity as its contribution score to answer the question:

$$s_{r_i} = sim(f_{simcse}(q_i), h'_{r_i}). \tag{9}$$

The higher the score, the greater the contribution it can make to answer q.

2.3   Inferential Chain Detection

With the hop number K of question q inferred by a simple text classification model, we implement the following K-hop inferential chain detection algorithm on the weighted subgraph, where the weights are scores given by context-aware relation ranking. To handle both the exponentially growing searching space and the risk of error propagation, we adopt an adaptive beam search by varying beam widths to expand the most promising chains.

A method similar to the elbow method for cluster analysis is adopted to choose the beam size for each searching step to reach a trade-off between searching cost and error propagation. Specifically, we plot the model performance as a function of the beam width for each step and pick the "elbow" of the curve as the beam width to use. The "elbow" is a point where cutting the beam width no longer maintains the performance level. Completing the adaptive beam search, the K-hop chain with the largest score will be taken as the predicted inferential chain, of which the end entity is inferred as the answer to the question.

Table 3. Test accuracy on PathQuestion and WorldCup2014

Methods	PQ-2H	PQ-3H	WC-1H	WC-2H
Subgraph Embed (2014)	0.744	0.506	0.448	0.507
Seq2Seq (2014)	0.899	0.770	0.537	0.548
KVMemN2N (2016)	0.937	0.879	0.870	0.928
IRN (2018)	0.960	0.877	0.843	0.981
SRN (2020)	**0.963**	0.892	0.989	0.978
MACRE-soft infusion (**Ours**)	0.938	0.841	**0.999**	0.930
MACRE-hard infusion (**Ours**)	0.944	**0.909**	**0.999**	**0.998**

3 Experiments

3.1 Datasets

We evaluate our model on three benchmark datasets, containing SimpleQuestions [2] for single-hop questions, PathQuestion [29] and WorldCup2014 [27] for multi-hop questions. Detailed statistics of the datasets are listed in Table 1.

3.2 Experimental Setups

We initialize the SimCRE model with pre-trained BERT and employ ADAM to optimize the parameters with batch size 128 and learning rate $5e^{-5}$. The temperature τ is set to be 0.05 and representations of token [CLS] are taken as embeddings for questions and relations. The hyperparameter λ (see Eq. 8) is set to 1. The training datasets are constructed following the proposed representative negative sampling mechanism (720k, 515k, and 683k for 1-hop, 2-hop, and 3-hop respectively) and splited into train/valid/test subset with a proportion of 8:1:1 [18,29]. For inferential chain detection, the beam widths are set as 8, 5, 3 for searching steps from 1 to 3. Following [18,29], we take the predicted answer ranked first to calculate accuracy.

3.3 Baselines

We compare our model with state-of-the-art methods for single-hop questions and multi-hop questions. With the identified topic entity, a single-hop question could be answered directly through SimCRE by a single-hop relation linking step in our setting. We include RNN-QA [21], Subgraph Ranking and Joint-Scoring [28], Relation-aware BERT [15], and APVA-TURBO [26] as baselines for single-hop questions in SimpleQuestions. For multi-hop questions, we compare with Subgraph Embed [1], Seq2Seq [20], KVMemN [17], IRN [29], and SRN [18].

Table 4. Ablation study of key modules

MACRE−SimCRE	Test accuracy
SQ	0.768 (−0.147)
PQ-2H	0.738 (−0.206)
PQ-3H	0.739 (−0.170)
WC-1H	0.820 (−0.179)
WC-2H	0.693 (−0.305)
MACRE−KG Context Processor	Test accuracy
PQ-2H w/1H KG context	0.548 (−0.396)
PQ-3H w/1H KG context	0.446 (−0.463)
PQ-3H w/2H KG context	0.479 (−0.430)
WC-2H w/1H KG context	0.630 (−0.368)

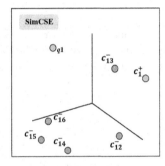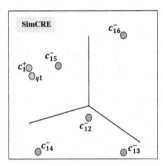

Fig. 4. Embedding visualization of SimCSE and SimCRE.

3.4   Main Results

We compare our model with state-of-the-art methods [1,15,17,18,20,21,28,29] focusing on single-hop questions and multi-hop questions. As reported in Table 2 and Table 3, MACRE outperforms all the competitors on SimpleQuestions and WorldCup2014, MACRE with hard KG context infusion module exhibits significant improvement over the baselines.

The method with hard infusion of KG context achieves better performance on both 2-hop and 3-hop questions, as a consequence of concatenating the relations first to form the context fits the sampling and training process of SimCRE better. Compared to 2-hop questions, 3-hop questions require multi-relational reasoning and integration of KG context at a higher level. Our method with hard infusion shows a more significant performance improvement (+0.017) on 3-hop questions, which demonstrates its superiority on multi-hop reasoning.

3.5   Analysis

Ablation Study. As summarized in Table 4, both SimCRE and KG context processor provide a significant boost to performance. Without SimCRE, the performance decreases on all hops of questions despite using the SOTA sentence embeddings SimCSE. It further confirms that SimCRE acquires better multi-relational reasoning ability by learning the subtle differences between positives and confusing negatives. Disabling the KG context processor, the model suffers an even sharper decline. Especially, performance on 3-hop questions falls to 0.479 in 2-hop KG context setting and 0.446 with only 1-hop KG context. This suggests the sufficient contextualization of relations in the KG is of great importance.

Representation Visualization. For the question illustrated in Fig. 1, we input $q1$ together with its ground truth inferential chain and representative negatives to SimCSE and SimCRE to get the embeddings. By conducting representation visualization through Embedding Projector[1], Fig. 4 depicts the distances between $q1$ and candidate inferential chains, all of the chains have the same relations as the ground truth. By SimCRE, the chain in the correct order is significantly the closest to the question, while SimCSE fails to encode the order information.

Model Interpretability
Intermediate entities and relations can be easily backtracked by the adaptive beam search to get the inferential chain, making it easy to evaluate whether the chain is consistent to the question or not. For question $q1$ illustrated in Fig. 1, MACRE finally picks the chain $A_1 \xrightarrow{children} B_2 \xrightarrow{spouse} C_3 \xrightarrow{parents} D_4$ correctly.

4   Related Work

Complex KGQA. Survey [11] summarizes the typical challenges and solutions for complex KGQA. Some advanced methods [3,10,19] are proposed to improve but challenges have not been well addressed, such as evolutionary KGQA and more robust models for general KGs.

Contrastive Learning for NLP. Contrastive Learning was first introduced by [16] and has shown significant improvement on NLP downstream tasks such as cross-lingual pretraining [4], language understanding [6], and textual representations learning [9].

Question Answering with KG and LM. A series of work [7,23] are proposed for question answering using knowledge from both KGs and pre-trained LMs. However, counting may play a crucial role in the knowledge-aware reasoning process [22], building more comprehensive complex reasoning modules for QA remains a challenging open problem.

---

[1] http://projector.tensorflow.org/.

5 Conclusion

Powered by contrastive relation embedding with a representative negative sampling mechanism and context-aware relation ranking, we develop a novel approach MACRE for multi-hop KGQA. An adaptive beam search is proposed to detect the inferential chain and get the answer entity, realizing the trade-off between efficiency and accuracy. Experiments on three benchmark datasets have demonstrated the significant improvement and effectiveness of our method.

Acknowledgement. This work was supported by National Natural Science Foundation of China (Grant No. 61902074) and Shanghai Science and Technology Innovation Action Plan (No. 21511100401).

References

1. Bordes, A., Chopra, S., Weston, J.: Question answering with subgraph embeddings. In: EMNLP, pp. 615–620 (2014)
2. Bordes, A., Usunier, N., Chopra, S., Weston, J.: Large-scale simple question answering with memory networks. arXiv preprint arXiv:1506.02075 (2015)
3. Cai, J., Zhang, Z., Wu, F., Wang, J.: Deep cognitive reasoning network for multi-hop question answering over knowledge graphs. In: Findings of ACL-IJCNLP 2021, pp. 219–229 (2021)
4. Chi, Z., et al.: InfoXLM: an information-theoretic framework for cross-lingual language model pre-training. In: NAACL-HLT, pp. 3576–3588 (2021)
5. Devlin, J., Chang, M.W., Lee, K., Toutanova, K.: BERT: pre-training of deep bidirectional transformers for language understanding. In: NAACL-HLT, pp. 4171–4186 (2019)
6. Fang, H., Wang, S., Zhou, M., Ding, J., Xie, P.: CERT: contrastive self-supervised learning for language understanding. arXiv preprint arXiv:2005.12766 (2020)
7. Feng, Y., Chen, X., Lin, B.Y., Wang, P., Yan, J., Ren, X.: Scalable multi-hop relational reasoning for knowledge-aware question answering. In: EMNLP, pp. 1295–1309 (2020)
8. Gao, T., Yao, X., Chen, D.: SimCSE: simple contrastive learning of sentence embeddings. In: EMNLP, pp. 6894–6910 (2021)
9. Giorgi, J.M., Nitski, O., Bader, G.D., Wang, B.: DeCLUTR: deep contrastive learning for unsupervised textual representations. In: ACL, pp. 879–895 (2021)
10. He, G., Lan, Y., Jiang, J., Zhao, W.X., Wen, J.R.: Improving multi-hop knowledge base question answering by learning intermediate supervision signals. In: WSDM, pp. 553–561 (2021)
11. Lan, Y., He, G., Jiang, J., Jiang, J., Zhao, W.X., Wen, J.R.: A survey on complex knowledge base question answering: methods, challenges and solutions. In: IJCAI, pp. 4483–4491 (2021)
12. Lan, Y., Jiang, J.: Query graph generation for answering multi-hop complex questions from knowledge bases. In: ACL, pp. 969–974 (2020)
13. Lan, Y., Wang, S., Jiang, J.: Multi-hop knowledge base question answering with an iterative sequence matching model. In: ICDM, pp. 359–368 (2019)
14. Li, B., Zhou, H., He, J., Wang, M., Yang, Y., Li, L.: On the sentence embeddings from pre-trained language models. In: EMNLP, pp. 9119–9130 (2020)

15. Luo, D., Su, J., Yu, S.: A BERT-based approach with relation-aware attention for knowledge base question answering. In: IJCNN, pp. 1–8. IEEE (2020)
16. Mikolov, T., Sutskever, I., Chen, K., Corrado, G.S., Dean, J.: Distributed representations of words and phrases and their compositionality. In: NIPS, pp. 3111–3119 (2013)
17. Miller, A., Fisch, A., Dodge, J., Karimi, A.H., Bordes, A., Weston, J.: Key-value memory networks for directly reading documents. In: EMNLP, pp. 1400–1409 (2016)
18. Qiu, Y., Wang, Y., Jin, X., Zhang, K.: Stepwise reasoning for multi-relation question answering over knowledge graph with weak supervision. In: WSDM, pp. 474–482 (2020)
19. Saxena, A., Tripathi, A., Talukdar, P.: Improving multi-hop question answering over knowledge graphs using knowledge base embeddings. In: ACL, pp. 4498–4507 (2020)
20. Sutskever, I., Vinyals, O., Le, Q.V.: Sequence to sequence learning with neural networks. In: NIPS, pp. 3104–3112 (2014)
21. Ture, F., Jojic, O.: No need to pay attention: Simple recurrent neural networks work! In: EMNLP, pp. 2866–2872, September 2017
22. Wang, K., Zhang, Y., Yang, D., Song, L., Qin, T.: GNN is a counter? Revisiting GNN for question answering. In: ICLR (2022)
23. Yasunaga, M., Ren, H., Bosselut, A., Liang, P., Leskovec, J.: QA-GNN: reasoning with language models and knowledge graphs for question answering. In: NAACL-HLT, pp. 535–546 (2021)
24. Yih, S.W.T., Chang, M.W., He, X., Gao, J.: Semantic parsing via staged query graph generation: question answering with knowledge base. In: ACL (2015)
25. Yu, M., Yin, W., Hasan, K.S., Santos, C.D., Xiang, B., Zhou, B.: Improved neural relation detection for knowledge base question answering. In: ACL, pp. 571–581 (2017)
26. Zhang, A., Lipton, Z.C., Li, M., Smola, A.J.: Dive into deep learning. arXiv preprint arXiv:2106.11342 (2021)
27. Zhang, L., Winn, J., Tomioka, R.: Gaussian attention model and its application to knowledge base embedding and question answering. arXiv preprint arXiv:1611.02266 (2016)
28. Zhao, W., Chung, T., Goyal, A., Metallinou, A.: Simple question answering with subgraph ranking and joint-scoring. In: NAACL-HLT, pp. 324–334 (2019)
29. Zhou, M., Huang, M., Zhu, X.: An interpretable reasoning network for multi-relation question answering. In: COLING, pp. 2010–2022 (2018)

AIR: Adaptive Incremental Embedding Updating for Dynamic Knowledge Graphs

Zhifeng Jia[1]($\boxtimes$), Haoyang Li[1], and Lei Chen[1,2]

[1] The Hong Kong University of Science and Technology, Hong Kong, China
{zjiaad,hlicg,leichen}@cse.ust.hk
[2] The Hong Kong University of Science and Technology (Guangzhou),
Guangzhou, China

Abstract. Recently, knowledge graphs (KGs) have been successfully applied on various downstream tasks, such as online recommendation and question answering. To better utilize the knowledge in KGs, recent researchers propose to learn low-dimension embeddings for entities and relations in KGs. Unlike static KGs, dynamic KGs evolve with knowledge events. Therefore, the embeddings of entities and relations should be updated to capture the information of latest knowledge events. However, when new events arrive, existing approaches update the whole knowledge graph or only update limited neighbors of entities directly involved in events. These approaches cannot maintain high-quality KG embedding in an efficient way. In this paper, we propose a framework AIR to adaptively update the embeddings for dynamic KGs. Specifically, AIR first measures the importance score of each triple. Then, AIR selects a set of triples that are most affected by knowledge events for updating. Besides, we apply the embedding propagation method to update the embedding to avoid retraining. Therefore, our proposed AIR can efficiently maintain high-quality KG embeddings. Extensive experiments on four real-world dynamic KGs demonstrate the effectiveness and efficiency of AIR against state-of-the-art baselines.

Keywords: Dynamic Knowledge graphs · Knowledge graph embedding

1 Introduction

Recently, with the development of information extraction techniques, numerous knowledge graphs [1,3,14,16], such as DBpedia [1] and Freebase [3], have been constructed and successfully applied on various downstream tasks, such as online recommendation [5] and question answering [30]. In these KGs, the knowledge is stored in the form (`subject`, `predicate`, `object`) triples, where each triple represents an real event existing in the real world. To better utilize the knowledge, recent researchers propose knowledge graph embedding approaches [4,29] to learn low dimension embeddings for entities and relations in KGs, and these embeddings are incapable of preserving triple semantics and KG structure.

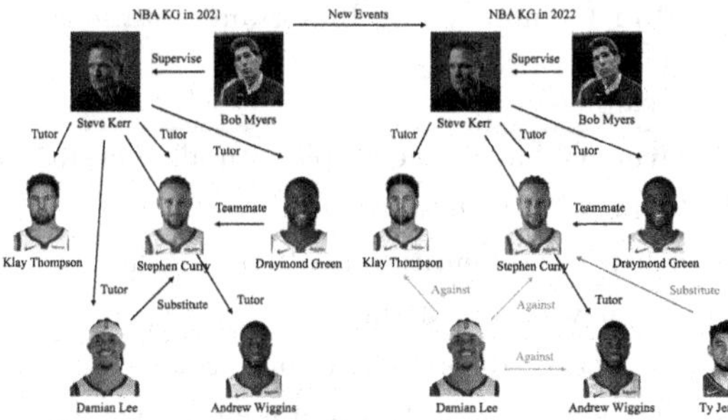

Fig. 1. An illustration of evolution in an NBA knowledge graph.

Unlike the above static KGs (e.g., DBpedia [1], Freebase [3]), dynamic KGs (e.g., GDELT [26], YAGO [21]) evolve with knowledge events. For example, in NBA knowledge graphs shown in Fig. 1, events occurred due to the trade of basketball players among the Warriors teams, and the dynamic KG (DKG) has been updated when events take place. As a result, these new knowledge events affect the embeddings of the other entities and relations in KGs. Therefore, to maintain high-quality KG embeddings, the embeddings of entities and relations should be updated to capture the information of latest knowledge events.

Based on the techniques employed to update KG embeddings, existing approaches can be categorized into two types, i.e., *neighbor-based* approaches [15] and *full-graph-based* approaches [7,17,20]. Specifically, given new knowledge events, neighbor-based approaches update the embedding of entities directly involved in these knowledge events and the neighbors of these entities. However, existing models restrict the update range to direct neighbors, but the entities out of the neighbors of these entities are also affected by new knowledge events. Consequently, such a strict range is not adaptive to different KGs, and thus these neighbor-based approaches fail to maintain high-quality embeddings for KGs. Unlike neighbor-based approaches, full graph-based approaches update the embedding of all entities and relations in the KGs. Such a way can enable the learned embeddings to capture the information of new knowledge events. However, these approaches that updating the whole graph are heavily time-consuming [7,17,20].

In this paper, we propose a framework to <u>A</u>daptive <u>I</u>ncremental update embedding for dynamic knowledge g<u>R</u>aphs, namely AIR, which updates KG embeddings in an efficient and effective way. Unlike existing works that update the whole graph or limited neighbors, our proposed AIR identifies and updates a set of triples that are most affected by knowledge events. Specifically, our proposed AIR first proposes a metric to measure the importance score of each triple based on KG structure and new knowledge events. Then, since the new information of new events is propagated through edges, we select a set of important triples that are connected, i.e., these triples can build a connected subgraph. Furthermore, we apply the embedding propagation method into the knowledge graph to further update the embedding without training. We also design perfor-

Table 1. Summary on Important Notations.

Symbols	Meanings
G	Dynamic knowledge graph with all timestamps
G_t	Snapshot at timestamp t
E_t, P_t, F_t	The set of all entities, relations and facts
$\triangle F_t$	The new events at time t
s, p, o	Subject, predicate, object of a fact
$\varepsilon_t^s, \varepsilon_t^p$	Embedding of entity s and relation p at timestamp t

mance recap mechanism to regularize model behavior in streaming events update to ensure practicability. In summary, we have made the following contribution.

- In this paper, we propose a framework to adaptive incremental update embedding for dynamic knowledge graphs, namely AIR. It first identifies the most important triples affected by updated facts and then update KG embedding based on performance recap mechanism. By these two steps, AIR is capable of efficiently maintaining high-quality KG embedding.
- We propose a novel metric to measure triple importance based on KG graph structure and new knowledge events. Based on this, we can identify the most important triple subgraph affected by new events for updating.
- We apply embedding propagation method into knowledge graph to fuse the updated embedding information with the rest entities in the DKG without training. We design a performance recap mechanism by dynamically combining incremental update and full batch retrain on the entire knowledge graph according to downstream task performance.
- Extensive experiments on four real-world dynamic KGs demonstrate the effectiveness and efficiency of AIR against state-of-the-art baselines.

2 Preliminary and Related Works

In this section, we first introduce the background of dynamic knowledge graphs (KG). Then, we present the preliminary of dynamic KG embedding models.

2.1 Dynamic Knowledge Graph

A dynamic knowledge graph (DKG) G can be expressed as a series of static knowledge graph snapshots $G = \{G_1, ..., G_T\}$, where T is the last timestamp. Each snapshot $G_t = \{E_t, P_t, F_t\}$ where E_t is the set of all entities, P_t is the set of all predicates (i.e., relations) and F_t is the set of all facts at time t. Each fact $f_t = \{s, p, o\} \in F_t$ consists of a subject entity s, a predicate (i.e., relation) p, and an object entity o. At each time $t \in [1, ..., T]$, the snapshot G_t is considered the static state of DKG G. For clarification and simplicity, we use $\triangle F_t = F_t \setminus F_{t-1}$ to denote the new events arriving at time t (Table 1).

2.2 Dynamic Knowledge Graph Embedding

Dynamic knowledge graph embedding is a method to convert entity and relation information in dynamic knowledge graphs to high dimensional vectors. Since knowledge graph stores data with the form of heterogeneous graph, knowledge graph embedding has a similar procedure to graph embedding models. A typical knowledge graph embedding model operates in 3 steps: 1) Knowledge graph; 2) Score function definition and 3) Representation learning [24]. With the emerging of graph neural network in the field of knowledge graph, researchers also use layer designs within neural networks to replace the score functions at step 2. Some also combine neural networks with other graph analysis technique including rule mining to improve the embedding performance [18].

We define the problem of incremental embedding update as follows: Given a dynamic knowledge graph $G = \{G_1, ..., G_T\}$, where T is the last timestamp. At a chosen timestamp $t \in [2, T-1]$ as the initial timestamp, the entity embedding ε_t^s and relation embedding ε_t^p are considered as static embedding ϵ_s^e, ϵ_s^r. The snapshot G_t is considered as the initial snapshot. When the initial snapshot G_t evolve into G_{t+1}, extract a set of triples $\mathcal{F}_{t+1}$ with minimum size k from snapshot G_{t+1} while guaranteeing performance in downstream tasks.

However, as discussed in Sect. 1, existing works [7,13,17,20] update the whole graph or only update limited neighbors when new events arrive. These approaches cannot maintain high-quality KG embedding in an efficient way.

3 Model Design

The AIR model consists of three steps: 1) Importance identification; 2) Important fact selection & critical subgraph formation; 3) Incremental embedding update.

3.1 Importance Identification

We propose to measure the importance of facts based on entity and predicate influence scoring in each knowledge graph snapshot.

Entity Influence. We measure the entity influence from the **structure score** and **feature score**. In brief, **structure score** is based on personal page rank (PPR) on the knowledge graph, while the **feature score** is based on KG historical embedding. The PPR calculation in a graph requires a specified set of target nodes, and PPR represents the relative influence of target nodes on other nodes in the graph. Given a knowledge graph G with E entities and P relations and a set of query entities Q, the personal page rank scores are in vector form. In DKG update problem, the updated entities substitute query entities in the PPR calculation. PPR score represents the influence of updated entities on existing graph structure. The PPR vector v on G represents the stationary distribution of a Markov process with the state transition probability [9]:

$$(1 - Tel) \cdot Av + Tel \cdot Pr \tag{1}$$

where $Tel \in (0,1)$ is the teleportation probability, and A is the column normalized transition probability matrix. Pr is the preference vector. Specifically, Pr has an $|V| \times 1$ shape with each element $Pr[n]$ calculated as follows:

$$Pr[n] = \begin{cases} \dfrac{b^{N_p}}{\sum_{p' \in P_t} b^{N_{p'}}}, & n \in Q \\ 0, & otherwise \end{cases} \tag{2}$$

where b is the message passing parameter, and N_p is the number of visits for the predicate type p at current timestamp t. We apply an approximation method [8] to calculate the structure score:

$$S_{st}(o, \Delta E_t) = PPR(o, \Delta E_t) \tag{3}$$

where ΔE_t is the set of updated entities. The method applies particle filtering to approximate PPR values. Particle filtering methods output ranks with high precision while exploiting local connections in knowledge graph structures. Training-based methods including PPRGO [2] are also efficient, but we avoid training in this step for higher efficiency. Similarity, we can compute the $S_{st}(s, \Delta E_t)$ for subject s in each triple.

Feature score represents the importance brought by semantic correlation between existing entities E_t and updated entities ΔE_t. We improve an approach of node influence calculation on homogeneous graphs to DKG [28]. The feature score incorporates input and hidden features of each entity, where we apply into importance identification. For the snapshot G_t where $t > 2$, ε_o^{t-2} is the embedding of entity o in previous snapshot G_{t-2}. h_o^{t-1} is the hidden embedding vector of entity u in last layer of the model. The Jacobian matrix J^t on vector pair ε^{t-2} and h^{t-1} in snapshot G_t is $J^t(s,o) \in \mathbb{R}^{|\varepsilon^{t-2}| \times |h^{t-1}|}$, where each element $J^t(s,o)[i][j]$ is $J^t(s,o)[i][j] = \frac{\partial h_s^{t-1}[i]}{\partial \varepsilon_o^{t-2}[j]}$. The feature influence score between two arbitrary entities s, o is calculated by the diagonal determinant of the Jacobian matrix to decouple other pairs of elements in the vectors:

$$I_o(s) = \frac{|\text{diag}(J^t(s,o))|}{\sum_{o \in \Delta E_t} |\text{diag}(J^t(s,o))|} = \frac{\Pi_{i=0}^{|\varepsilon^{t-2}|-1} J^t(s,o)[i][i]}{\sum_{o \in \Delta E_t} \Pi_{i=0}^{|\varepsilon^{t-2}|-1} J^t(s,o)[i][i]} \tag{4}$$

The feature score of an entity o is the average of feature influence scores between all updated entities ΔE_t and o:

$$S_{fe}(o) = \frac{1}{|\Delta E_t|} \sum_{s \in \Delta E_t} I_o(s) \tag{5}$$

The entity influence score of an entity o is calculated by linearly combining structure score $S_{st}(o)$ and feature score $S_{fe}(o)$:

$$EI(o) = (1 - \alpha) \cdot S_{st}(o) + \alpha \cdot S_{fe}(o) \tag{6}$$

where $\alpha \in [0,1]$ is a hyperparameter to adjust the weight of both scores.

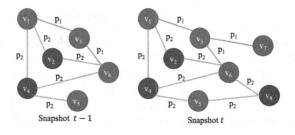

Fig. 2. Example of the importance calculation in knowledge graph.

Predicate Influence. The predicate influence score is based on edge betweenness centrality [6]. For a snapshot G_i, we compute all shortest paths between each pair of entities. Then for each predicate p and pair of entities s, o, we normalize the number of shortest paths that passes through p with regard to total number $n(s, o)$, and we denote it as $n_p(s, o)$. The predicate influence score of predicate p is calculated as $RI(p) = \sum_{s,o \in E_t} \frac{n_p(s,o)}{n(s,o)}$. The predicate influence score emphasizes by assigning scores to different.

Fact Importance. Within the RDF format of knowledge graph data, fact is the fundamental element of facts, and any multi-hop fact can be represented as a combination of consecutive facts [12]. Thus, it is vital to measure the importance of different facts to determine the choices during further model update. In a snapshot G_i, the fact importance score of an arbitrary fact $f = (s, p, o)$ is calculated as follows:

$$FI(s, p, o) = (1 - \beta) \cdot (EI(s) + EI(o)) + \beta \cdot RI(p) \tag{7}$$

where β is hyperparameter, $s, o \in E$ are head and tail entities in the fact, and $p \in P$ is the predicate of the fact.

Example. We use an example to demonstrate fact importance calculation. As shown in Fig. 2, the DKG snapshot at timestamp t is evolved from snapshot $t - 1$. Two entities v_7 and v_8 are added to this snapshot. We want to calculate the importance score of fact (v_1, p_1, v_3) The importance score of v_1 is calculated by the influence of v_7, v_8 on v_1.

$$EI(v_1) = (1 - \alpha_2) \cdot S_{st}(v_1) + \alpha_2 \cdot S_{fe}(v_1)$$
$$= (1 - \alpha_2) \cdot PPR(v_1, \{v_7, v_8\}) + \alpha_2 \cdot \frac{1}{2}(I_{v_7}(v_1) + I_{v_8}(v_1))$$

Importance score of v_3 is measured by a similar approach. On the other hand, the importance score of predicate type p_1 is calculated by the edge betweenness centrality at timestamp t as $RI(p_1) = \sum_{s,o \in \{v_1,...,v_8\}} \frac{n_{p_1}(s,o)}{n_{\{p_1,p_2\}}(s,o)}$. Thus, the fact importance score $FI(v_1, p_1, v_3)$ at timestamp t is calculated by:

$$FI(v_1, p_1, v_3) = (1 - \beta) \cdot (EI(v_1) + EI(v_3)) + \beta \cdot RI(p_1)$$

3.2 Important Facts Selection and Critical Subgraph Formation

The goal of subgraph formation at timestamp t is to select important facts from all static facts in snapshot G_{t-1}, and combine with updated facts ΔF_i to construct minimum subgraph $\mathcal{G}_i \subset G_i$ to guarantee connectivity between selected facts. The motivation is to further restrict input size. Since importance identification brings strong structure sensitivity to the importance between updated entities and static entities, we can omit the static entities with distance more than d from its nearest updated entity, so all choices can be found within a range of neighbors. Importance identification sets a score for every fact in G_{t-1}, and the selection of important fact set $\mathcal{F}_t \subset F_t$ follows the following rule:

$$\mathcal{F}_t = \{f = (s, p, o) | TI(f) > \sigma_t\} \tag{8}$$

where σ_t is the threshold determined by global fact importance score. The effect of σ_t is dependent of the distribution of importance scores of entities and predicates in each snapshot, so we design multiple functions for different distributions.

$$\sigma_t = \begin{cases} \gamma \max(FI(f))|_{f \in F_t} + (1 - \gamma) \min(FI(f))|_{f \in F_t}, \\ (1 - \gamma) \cdot \sqrt{\max(FI(f))|_{f \in F_t} \cdot \min(TI(f))|_{f \in F_t}}. \end{cases} \tag{9}$$

where $\gamma = R(G, \bar{S}_G, k)$. We also provide explorations into the selection of threshold functions and hyperparameter values $\gamma \in [0, 1]$ in the experiments by testing their impact on performance and efficiency. A brief conclusion is that the equation $\sigma_t = \gamma \max(FI(f))|_{f \in F_t} + (1 - \gamma) \min(FI(f))|_{f \in F_t}$ performs better on DKG with larger number of updated facts. The general process of minimum critical subgraph formation is listed in Algorithm 3.2. Step 4 extracts a minimum spanning tree from the whole knowledge graph, since the minimum subgraph of G_t containing certain important entities $\mathcal{E}$ is equivalent to the minimum spanning tree of G_t, which also satisfies the connectivity between entities. The subgraph formation step 5 satisfies the requirement that all important facts appear in the subgraph, while steps 6–10 prune the unimportant entities and predicates to the extent that only one cycle exists in the subgraph, or all facts on all cycles in the subgraph are critical. This ensures that the constructed graph has the smallest possible size, under restrictions that critical facts are contained.

3.3 Incremental Embedding Update

Training on subgraph The incremental training on the encoder-decoder framework of knowledge graph embedding model works on the encoder. At timestamp t, incremental training takes the generated subgraph $\mathcal{G}_t$ and the historical embedding of corresponding entities and predicates as the initialization of the training data, and outputs the updated embedding of important entities and predicates. Facts other than training data are not involved in training at current timestamp.

Algorithm 1. Minimum Subgraph Formation Algorithm

Input: Knowledge graph snapshots G_t, G_{t-1}
Output: Minimum subgraph $\mathcal{G}_t \subset G_t$
1: Calculate importance threshold σ_t.
2: Select important fact set $\mathcal{F}_{t-1}$.
3: Combine updated entities and predicates ΔG_i into $\mathcal{F}_{t-1}$ to form $\mathcal{F}_i$. Receive updated entity $\mathcal{E}_t$ and predicate set $\mathcal{P}_i$.
4: Construct Minimum Spanning Tree G'_t which all important entities $\{\mathcal{E}_t \in G'_t\}$.
5: Link all important predicates $\mathcal{P}_t$ to G'_t.
6: **while** Exists more than one cycle in $\mathcal{G}_t$ and exists entities that are not important: **do**
7:    Remove the unimportant entity $\mathcal{E}'$ in $\mathcal{G}_t$ with the lowest importance score.
8:    **if** No cycle exists in $\mathcal{G}_t$ **then**
9:       Undo the removal; mark the entity $\mathcal{E}'$ as important.
10:   **end if**
11: **end while**
12: Return $\mathcal{G}_t$.

Embedding propagation. After incremental training, the updated embedding is propagated to neighbor entities for better fusion on the embedding results. At timestamp t, an entity s_t has an updated embedding ε_t^s. The embedding of a neighbor entity $o \in N(s)$ is updated by weighted sum:

$$\varepsilon_t^o = (1 - \gamma)\varepsilon_{t-1}^o + \gamma\varepsilon_t^s \tag{10}$$

where $\gamma \in (0, 1)$ is a hyperparameter. The method helps updated facts spread their influence to farther entities and predicates while avoiding training on the whole graph.

Performance Recap. When DKGs continue to update, cumulative error occur in the performance of decoder in downstream tasks.

Thus, we set up a performance recap strategy to compare model performance between the most recent downstream task results p_{inc} and last static training p_{st}. The KGE base model is trained from scratch if incremental update performance $p_{inc} < \theta \cdot p_{st}$. θ is the performance control parameter. We discover that performance of a KGE model on the same streaming DKG falls below threshold after T timestamps in the experiments. AIR follows the period to retrain base model from scratch, and frees the strategy from downstream task feedback.

3.4   Complexity Analysis

Importance Identification. For a DKG snapshot $G_t = \{E_t, P_t, F_t\}$, $M = |E_t|$ is the number of entities, $N = |P_t|$ is the number of predicates, assume the average number of neighbor entities of each entity is k. The number of updated entities in the current timestamp is ΔM For each snapshot, the approximate

PPR calculation through particle filtering takes $O(\Delta M^2 k)$ [8]. The calculation of feature score on a basis of Jacobian matrix is $O(M)$ for all entities. For predicate influence score, the calculation of all edge betweennness centrality scores takes $O(N)$ time. The fact importance assignment takes $O(F)$ time, so the importance identification totally has a time complexity of $O(\Delta M^2 k + M + N + F)$.

Critical Subgraph Formation. Under the same setting in previous step, assume the selection range takes up $f(d) \in [0,1]$ portion of the total fact size, the selection of important facts takes $O(f(d)F)$ time. When subgraph is formed in step 3 in Algorithm 3.2, assume its size as $G'_t = \{E'_t, P'_t, F'_t\}$, $M' = |E'_t|$ is the number of entities, $N' = |P'_t|$ is the number of predicates. The predicates are sparse for G'_i, so Kruskal algorithm [10] is applied with time complexity $O(N' \log N')$. The repeat between steps 6 to 11 has a worst-case complexity when all predicates are on at least one cycle, and the traverse goes through all predicates in the G'_i, which is $O(N')$. Thus, the minumum subgraph formation algorithm has a worst case time complexity of $O(f(d)F + N' \log N' + N')$.

4   Experiments

4.1   Datasets

In the experiment, we use 4 common DKGs: ICEWS18 [26], WIKI12k [23], YAGO11k [21] and GDELT. The detailed information are listed in Table 2.

Table 2. Experiment dataset information

Dataset	$|E|$	$|R|$	$|F|$	Update Interval
ICEWS18	23,033	256	468,558	24 h
WIKI12k	12,554	24	669,934	1 year
YAGO11k	10,623	10	201,089	1 year
GDELT	7,691	240	2,278,405	15 min

4.2   Measurement Criteria

We use link prediction, which is a major downstream task of knowledge graph embedding, and compare the performance and efficiency between incremental training and static training of knowledge graph embedding models. The common criteria for measuring link prediction performance are Hit@1, Hit@10, mean rank (MR) and mean reciprocal rank (MRR). To give a better comparison, we design D@K to represent the difference between Hit@K values of incremental training and static training. Reciprocal rank difference is also valued [27]. We also compare the input training size N and the training time cost t.

4.3   Benchmarks

For the base models, we choose RGCN [19], KGAT [25] and CompGCN [22] as benchmarks of the AIR model. We compare the performance loss and efficiency improvement of AIR. We also choose typical GNN-based KGE models, including CyGNet [31], RE-NET [11] and RE-GCN [15]. CyGNet, RE-NET and RE-GCN models are trained with all snapshots in the time series, while AIR model is trained in the incremental update in the last time stamp $t = T$, where the static base models are trained on the previous timestamp $t = T - 1$ by static training.

We implement the AIR model with PyTorch, NetworkX, and DGL. The KGAT model is originally designed for knowledge graph based recommendation system, and we made simplification to keep a KGE model. AIR includes 3 hyper-parameters α, β and γ. We search the parameters $\alpha, \beta \in \{0, 0.2, 0.4, 0.6, 0.8, 1\}$, and the selection interval is set to 0.1. The hyperparameter γ for embedding propagation is set as $\gamma = 0.7$.

4.4   Performance Comparison

In each dynamic knowledge graph $G = \{G_1, ..., G_T\}$ which contains n snapshots separated by timestamp, the experiment is conducted in the last timestamp $t = T$. In this scenario, snapshot G_{T-1} is considered as a static snapshot from the DKG, and incremental update of graph information happens at timestamp $t = T$. The datasets are originally separated to training, validation and testing input, and each input dataset is further divided according to timestamp. We compare AIR and static training strategies on RGCN [19], KGAT [25] and CompGCN [22]. Static training method uses full training data as input to the base models; AIR strategy conducts selection prior to the training, and passes selected training data to the base model. The performance comparisons are listed in Table 3, separated with static model types. The bold numbers indicate the best experiment results. The notations in the table are:

1. **Update** refers to the update strategy. Full means using the full snapshot as input, and Sub means subgraph selection and using the subgraph as input. Ran is a random selection mechanism within the existing entities.
2. **Time** refers to process time. For full update, the time equals training time, while for subgraph update, time covers both subgraph selection and training.
3. **#Entity** refers to number of entities passed to the static models as input. This shows the efficiency difference of two methods.

The results show that the performance of AIR-RGCN can reach a guarantee of 85% of static RGCN model, while significantly reducing the time cost. The AIR-KGAT model can achieve a 90% guarantee of the static model. The AIR model has at least a training speed boost for around 12 times, and a maximum of 25 times acceleration in the GDELT dataset for RGCN model. Moreover, the subgraph selection mechanism guarantees the subgraph size to be small, which saves space resource during training. The subgraph selection results are the same for both models, so the Range values are the same with each other in the two

Table 3. Performance Comparison by Link Prediction

Dataset	Model	Update	H@1	H@10	MRR	Time	#Entity
ICEWS18	AIR-RGCN	Full	13.39	37.47	21.19	5323	42790
		Ran	9.43	23.81	12.42	379	**714**
		Sub	**12.09**	**34.18**	**19.47**	427	**714**
	AIR-KGAT	Full	15.77	43.98	24.03	7192	42790
		Ran	9.74	29.33	14.46	662	**714**
		Sub	**12.09**	**34.18**	**19.47**	684	**714**
	AIR-CompGCN	Full	15.29	41.19	22.86	5483	42790
		Ran	12.63	34.68	15.44	533	**714**
		Sub	**13.37**	**38.39**	**20.01**	576	**714**
WIKI12k	AIR-RGCN	Full	15.97	40.39	29.08	9218	12554
		Ran	13.61	35.02	20.21	682	**198**
		Sub	**15.24**	**38.76**	**27.59**	677	**198**
	AIR-KGAT	Full	18.14	44.76	32.54	12078	12554
		Ran	8.72	24.35	23.14	1259	**198**
		Sub	**17.10**	**41.39**	**30.08**	1235	**198**
	AIR-CompGCN	Full	16.29	44.76	32.54	9733	12554
		Ran	7.82	28.30	22.17	729	**198**
		Sub	**14.23**	**43.28**	**31.09**	802	**198**
YAGO11k	AIR-RGCN	Full	29.49	62.05	46.50	2847	10623
		Ran	16.06	29.60	39.94	**201**	**175**
		Sub	**28.39**	**58.09**	**42.17**	201	**175**
	AIR-KGAT	Full	33.00	65.52	47.10	3374	10623
		Ran	20.28	40.78	30.38	**301**	**175**
		Sub	**29.18**	**60.37**	**44.29**	307	**175**
	AIR-CompGCN	Full	31.19	61.03	39.90	3012	10623
		Ran	18.40	56.56	33.81	**221**	**175**
		Sub	**30.01**	**59.88**	**38.1**	278	**175**
GDELT	AIR-RGCN	Full	11.69	29.95	17.33	17338	7691
		Ran	10.43	15.10	11.87	**750**	**133**
		Sub	**10.81**	**28.76**	**15.69**	762	**133**
	AIR-KGAT	Full	11.69	33.15	19.04	19305	7691
		Ran	12.09	34.18	19.47	**1118**	**133**
		Sub	**11.05**	**30.76**	**17.80**	1104	**133**
	AIR-CompGCN	Full	11.29	32.08	18.19	16328	7691
		Ran	9.85	24.02	16.24	**832**	**133**
		Sub	**10.38**	**31.12**	**16.77**	912	**133**

tables. A comparing result is listed in Table 4, where each difference value is calculated by the corresponding pair of results from full update and subgraph update. The time and range difference are calculated by percentage between these pairs. The calculation follows the equations: $D@k = H@k_{Sub} - H@k_{Full}$, $Time\% = \frac{Time_{Sub}}{Time_{Full}} \times 100\%$, $Range\% = \frac{Range_{Sub}}{Range_{Full}} \times 100\%$.

The training time differences of AIR and base models in four DKGs are illustrated in Fig. 3. The yellow and orange bars show significant advantage in training time comparing to the original models.

Further comparisons between AIR-RGCN, AIR-KGAT and existing GNN-based DKG embedding models CyGNet, RE-NET and RE-GCN are also conducted on these four DKG datasets. The criteria also follow previous settings

Table 4. Performance Difference

Dataset	Model	D@1	D@10	RRD	Time%	Range%
ICEWS18	AIR-RGCN	−1.30	−3.29	−1.72	8.35%	1.67%
	AIR-KGAT	−1.52	−2.28	−2.7	9.51%	
WIKI12k	AIR-RGCN	−0.73	−1.63	−1.49	7.34%	1.58%
	AIR-KGAT	−1.04	−3.37	−2.46	10.23%	
YAGO11k	AIR-RGCN	−1.10	−3.96	−4.33	7.06%	1.65%
	AIR-KGAT	−3.82	−5.15	−2.81	9.09%	
GDELT	AIR-RGCN	−0.88	−1.19	−1.64	4.39%	1.73%
	AIR-KGAT	−0.64	−2.39	−1.24	5.72%	

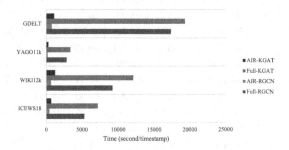

Fig. 3. Training time difference.

to compare Hit@1, Hit@10, MRR, training time and update range values. The experiment results are shown in Table 5.

The experiment results show the advantage of AIR-RGCN and AIR-KGAT with low training time cost and training entity size in incremental embedding update. The performance of AIR-RGCN is outperformed by AIR-KGAT in all datasets, while matching up with AIR-KGAT in YAGO11k. As a trade-off with efficiency, AIR-RGCN is outperformed by RE-NET, CyGNet and RE-GCN in Hit@1, Hit@10, MRR values, but AIR-KGAT has the same level of performance in link prediction, comparing with RE-NET and CyGNet. RE-GCN enjoys the best performance in all datasets, but the advantage is weak in GDELT dataset in comparison with AIR-KGAT. The results revealed the capability of AIR models to embed of frequently-updating dynamic knowledge graphs. With an update time interval of 15 min, it is not practical for the three DKG embedding models to catch up with the latest DKG information and incrementally update their models, while AIR-RGCN is fully capable of updating the embedding within each update period. AIR-KGAT can nearly catch up the latest information (with 3 min delay) and achieve reasonable performance.

Table 5. Performance Comparison of Dynamic Models

Dataset	Model	H@1	H@10	MRR	Time	Range
ICEWS18	RE-NET	16.79	43.56	25.82	8937	42790
	CyGNet	16.08	44.18	23.77	6155	42790
	RE-GCN	**17.20**	**45.45**	**27.13**	6798	42790
	AIR-RGCN	12.09	34.18	19.47	**427**	**714**
	AIR-KGAT	14.25	41.70	21.33	690	**714**
WIKI12k	RE-NET	17.35	41.73	30.87	10493	12554
	CyGNet	17.19	42.05	30.77	7591	12554
	RE-GCN	**19.92**	**53.88**	**39.84**	8313	12554
	AIR-RGCN	15.24	38.76	27.59	**677**	**198**
	AIR-KGAT	17.10	41.39	30.08	1179	**198**
YAGO11k	RE-NET	29.37	61.93	46.81	4128	10623
	CyGNet	28.58	61.52	46.72	2813	10623
	RE-GCN	**41.29**	**75.94**	**58.27**	3071	10623
	AIR-RGCN	28.39	58.09	42.17	**201**	**175**
	AIR-KGAT	29.18	60.37	44.29	313	**175**
GDELT	RE-NET	**12.03**	**33.89**	19.60	18304	7691
	CyGNet	11.13	31.50	18.05	15293	7691
	RE-GCN	11.92	33.19	19.15	14078	7691
	AIR-RGCN	10.81	28.76	15.69	**762**	**133**
	AIR-KGAT	11.05	30.76	17.80	1104	**133**

4.5   Performance Recap

We use DKG $G = \{G_1, ..., G_T\}$ by training on k consecutive snapshots $G_{n-k}, ..., G_T$ and observe the performance change with time. We also provide a comparison between the AIR model with and without performance recap to validate its affect in performance-efficiency trade-off. The performance decay graph with time is shown in Fig. 4. In the experiment, the performance trade-off

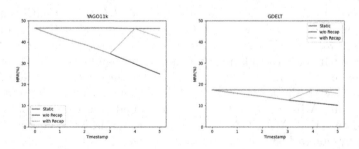

Fig. 4. Performance recap comparisons of AIR-RGCN. Blue lines show model performance decay without recap strategy, yellow lines show the effect of performance recap. (Color figure online)

ratio is set as $\theta = 0.75$, which means that the model requires full retrain using the static method, if the downstream task performance is below 0.75 of the last static performance. As we can see from the graph, the cumulative error drops below θ periodically, with different periods on different knowledge graphs.

The performance recap also effects the overall efficiency, since the recap require regular static training. The efficiency trade-off depends on the rate of performance decay. The general equivalent training time is measured by repeated experiments, and is listed in Table 6. **Avg. period** is the measured average period of static retraining the model, with the basic unit of one timestamp. **With Recap** and **w/o Recap** shows the average training time per timestamp with/without recap. The unit is seconds/timestamp.

Table 6. Equivalent Training Time by Performance Recap

Dataset	with Recap	w/o Recap	avg. period
ICEWS18	1364	431	6
WIKI12k	4933	681	6
YAGO11k	748	199	5
GDELT	3501	760	5

We can observe that the performance recap affects efficiency, but the overall efficiency still better than static training.

4.6   Threshold Function Selection

We display the experiment on ICEWS18 to investigate the choice of hyperparameter γ values. We train and test AIR-RGCN to measure the selected number of entities, performance in link prediction and training time cost with regard to γ and function selection. The results are shown in Fig. 5. The Hit@10 value is exaggerated for an easier view. Comparing from the figures, the scoring function varies in increasing trend due to their different designs and γ values, but the

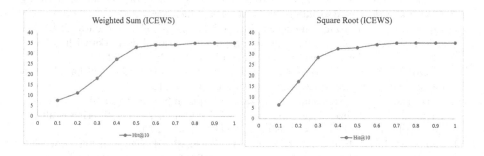

Fig. 5. Threshold function comparison.

performance growth is seemingly less relevant with the score function choices. Moreover, the increase of performance with γ has a turning point that, the growth will shrink when γ passes certain value. Thus, it is not worthy in the performance-efficiency trade off to achieve large *gamma* values to lower the threshold, while a high threshold is also not recommended.

5   Conclusion

In this paper, we propose an adaptive incremental embedding updating framework AIR for dynamic knowledge graphs. Specifically, we first propose a metric to measure the triple importance based on knowledge structure and semantics. Then, when new knowledge events arrive, we identify a set of the most valuable triples for updating. Besides, we propose to propagate the updated embeddings to other entities in KGs without retraining. Moreover, our proposed AIR is generalizable to accelerate the embedding updating for various knowledge graph embedding (KGE) models. Extensive experiments on four real-world dynamic KGs demonstrate that AIR can be deployed on various KGE models to efficiently maintain a high-quality embedding for KGs against state-of-the-art baselines.

Acknowledgements. Lei Chen's work is partially supported by National Science Foundation of China (NSFC) under Grant No. U22B2060, the Hong Kong RGC GRF Project 16213620, RIF Project R6020-19, AOE Project AoE/E-603/18, Theme-based project TRS T41-603/20R, China NSFC No. 61729201, Guangdong Basic and Applied Basic Research Foundation 2019B151530001, Hong Kong ITC ITF grants MHX/078/21 and PRP/004/22FX, Microsoft Research Asia Collaborative Research Grant and HKUST-Webank joint research lab grants and HKUST Global Strategic Partnership Fund (2021 SJTU-HKUST).

References

1. Auer, S., Bizer, C., Kobilarov, G., Lehmann, J., Cyganiak, R., Ives, Z.: Dbpedia: a nucleus for a web of open data (2007)
2. Bojchevski, A.: Scaling graph neural networks with approximate pagerank. In: KDD (2020)
3. Bollacker: Freebase: a collaboratively created graph database for structuring human knowledge. Association for Computing Machinery (2008)
4. Bordes, A.: Translating embeddings for modeling multi-relational data (2013)
5. Chen, Y.: Attentive knowledge-aware graph convolutional networks with collaborative guidance for personalized recommendation. In: ICDE (2022)
6. Cuzzocrea, A.: Edge betweenness centrality: a novel algorithm for QoS-based topology control over wireless sensor networks. J. NCA (2012)
7. Dettmers, T.: Convolutional 2d knowledge graph embeddings (2018)
8. Gallo, D.: Personalized page rank on knowledge graphs: particle filtering is all you need! Advances in Database Technology-EDBT 2020 (2020)
9. Gallo, D., Lissandrini, M., Velegrakis, Y.: Personalized page rank on knowledge graphs: Particle filtering is all you need! In: EDBT 2020, Copenhagen, Denmark, March 30 - April 02, 2020 (2020)

10. Guttoski: Kruskal's algorithm for query tree optimization. In: IDEAS (2007)
11. Jin, W.: Recurrent event network: autoregressive structure inference over temporal knowledge graphs (2019)
12. Lassila, O., Swick, R.R., et al.: Resource description framework (RDF) model and syntax specification (1998)
13. Li, H., Chen, L.: Cache-based GNN system for dynamic graphs. In: CIKM, pp. 937–946 (2021)
14. Li, H., Lin, X., Chen, L.: Fine-grained entity typing via label noise reduction and data augmentation. In: Jensen, C.S., et al. (eds.) DASFAA 2021. LNCS, vol. 12681, pp. 356–374. Springer, Cham (2021). https://doi.org/10.1007/978-3-030-73194-6_24
15. Li, Z.: Temporal knowledge graph reasoning based on evolutional representation learning. In: Proceedings of the 44th International ACM SIGIR (2021)
16. Lin, X., Li, H., Xin, H., Li, Z., Chen, L.: Kbpearl: a knowledge base population system supported by joint entity and relation linking. Proc. VLDB Endowment 13(7), 1035–1049 (2020)
17. Liu, Q.: Probabilistic reasoning via deep learning: Neural association models. arXiv preprint arXiv:1603.07704 (2016)
18. Sadeghian, A.: Drum: End-to-end differentiable rule mining on knowledge graphs. In: Advances in Neural Information Processing Systems (2019)
19. Schlichtkrull, M.: Modeling relational data with graph convolutional networks
20. Socher, R.: Reasoning with neural tensor networks for knowledge base completion. In: Advances in Neural Information Processing Systems (2013)
21. Suchanek, F.M.: Yago: A core of semantic knowledge. In: WWW (2007)
22. Vashishth, S.: Composition-based multi-relational graph convolutional networks
23. Vrandečić, D., Krötzsch, M.: Wikidata: a free collaborative knowledgebase. Commun. ACM 57(10), 78–85 (2014)
24. Wang, Q.: Knowledge graph embedding: a survey of approaches and applications. IEEE TKDE 29, 2724–2743 (2017)
25. Wang, X.: KGAT: knowledge graph attention network for recommendation. In: Proceedings of the 25th ACM SIGKDD (2019)
26. Ward, M.D., Beger, A., Cutler, J., Dickenson, M., Dorff, C., Radford, B.: Comparing GDELT and ICEWS event data. Analysis 21(1), 267–297 (2013)
27. Wu, J.: Tie: A framework for embedding-based incremental temporal knowledge graph completion. In: 44th International ACM SIGIR (2021)
28. Xu, K.: Representation learning on graphs with jumping knowledge networks. In: PMLR (2018)
29. Yang, B.: Embedding entities and relations for learning and inference in knowledge bases (2015)
30. Zheng, W.: Question answering over knowledge graphs: question understanding via template decomposition. Proc, VLDB Endow 11, 1373–1386 (2018)
31. Zhu, C.: Learning from history: modeling temporal knowledge graphs with sequential copy-generation networks. In: Proceedings of the AAAI (2021)

Class-Dynamic and Hierarchy-Constrained Network for Entity Linking

Kehang Wang[1], Qi Liu[1,2(✉)], Kai Zhang[1], Ye Liu[1], Hanqing Tao[1],
Zhenya Huang[1], and Enhong Chen[1]

[1] Auhui Province Key Laboratory of Big Data Analysis and Application,
University of Science and Technology of China, Hefei 230026, China
{wangkehang,kkzhang0808,liuyer,hqtao}@mail.ustc.edu.cn,
{qiliuql,huangzhy,cheneh}@ustc.edu.cn
[2] Institute of Artificial Intelligence, Hefei Comprehensive National Science Center,
Hefei, China

Abstract. Entity Linking (EL) aims to map mentions in a text to corresponding entities in a knowledge base. Existing EL methods usually rely on sufficient labeled data to achieve the best performance. However, the massive investment in data makes EL systems viable only to a limited audience. There is ample evidence that introducing entity types can provide the model prior knowledge to maintain the model performance in low-data regimes. Unfortunately, current low-data EL methods usually employ entity types by rule constraints, which are in a shallow manner. Furthermore, they usually ignore fine-grained interaction between mention and its context, resulting in insufficient semantic information of mention representation in low-data regimes. To this end, we propose a Class-Dynamic and Hierarchy-Constrained Network (CDHCN) for entity linking. Specifically, we propose a dynamic class scheme to learn a more effective representation for each entity type. Besides, we formulate a hierarchical constraint scheme to reduce the matching difficulty of the given mention and corresponding candidate entities by utilizing entity types. In addition, we propose an auxiliary task called mention position prediction (MPP) to obtain an informative mention representation in low-data regimes. Finally, extensive in-domain and out-of-domain experiments demonstrate the effectiveness of our method.

Keywords: Entity linking · Entity type · Hierarchical constraint

1 Introduction

Entity Linking (EL), also known as entity disambiguation, aims to link ambiguous textual mention to the correct entity in a particular knowledge base (KB). As a fundamental building block for many Natural Language Processing (NLP) applications, such as Question Answering [1] and Text Mining [2–4], this task has received increasing attention from researchers in recent years.

X. Wang et al. (Eds.): DASFAA 2023, LNCS 13944, pp. 622–638, 2023.
https://doi.org/10.1007/978-3-031-30672-3_42

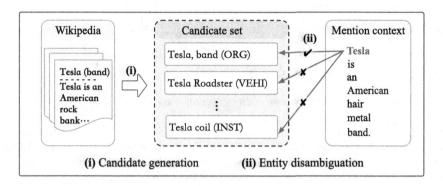

Fig. 1. A flowchart overview of the entity linking system.

Generally, as illustrated in Fig. 1, there are two main parts to the EL systems: the first part is the entity candidate generation module, which takes the given KB and selects a subset of entities that might be associated with mentions in the input text; the second part is the entity disambiguation module, which takes the given mentions and links them to the most corresponding entities in the candidates set. As Fig. 1 shows, in the sentence *"Tesla is an American hair metal band"*, the phrase *Tesla* is called a textual mention. Entity Linking seeks to map the mention to a real word entity from a specific KB. For instance, the mention *Tesla* in the above sentence should be mapped to *Tesla (band)* but not *Tesla (coil)* in Wikipedia. The entities to be linked may have similar word forms like *Tesla (band)* and *Tesla (coil)* which makes the EL task difficult.

Recently, learning better representations has been proposed as a promising direction for the EL tasks. For example, static word embeddings [5,6] and pre-trained models [8,9] have proved mention embedding's effectiveness for improving the performance of EL methods. Unfortunately, such methods often require a large amount of training data, and it is impractical for most researchers to repeatedly train EL models on massive data from KBs due to the constant updating of KBs (e.g., it took a week's time to train the BLINK [10] on 8-GPUs for $9M$ examples). For this reason, how to use external information to alleviate the problem of data dependence has received more attention in EL tasks [11–13] and other NLP tasks [14–16]. Specifically, Bhargav et al. [11] exploited external information about entity types by proposing an auxiliary task called entity type prediction in a low-data regime setting. Tedeschi et al. [13] proposed to improve the EL model trained on low amounts of labeled data by exploiting entity types.

However, the above approaches employ external knowledge (i.e., entity types) by rule constraints or simply splicing entity types and entity texts, which are in a shallow manner. For example, Tedeschi et al. [13] simply concatenated entity types (e.g., ORG) with entity description to enrich each candidate entity representation and use entity types to constrain the EL model's output. In this way, the models cannot make the best use of semantic information associated with entity types. Some deep semantic solutions (e.g., type embeddings) can pro-

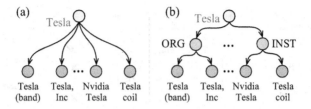

Fig. 2. The illustration of mention-entities selecting. (a) Traditional entity linking. These methods select the corresponding entity from a candidate set directly. (b) Hierarchical constraint entity linking. Our methods score the entity types for mention first and then use this score to select the corresponding entity.

vide more information for entity linking. Meantime, the previous models select the most suitable entity from the candidate set via similarity matching directly. However, these candidates are often highly similar. As shown in Fig. 2 (a), the candidate entity titles in Tesla's candidate set are similar, causing candidate entity representations generated to be similar, which is hard for models to select the mention's suitable entity. Therefore, it is necessary to exploit the semantic information of entity types to lower EL tasks difficulty.

With the above analysis, in this paper, we propose a **Class-Dynamic and Hierarchy-Constrained Network (CDHCN)** for effectively entity linking. Unlike traditional label embedding methods [17] embedded entity types statistically, we argue that the entity type representation should be dynamic as the meanings of the same entity type for different candidate entities maybe be different. For example, Tesla (band) and the European Union, the two mentions have the same entity type (ORG). However, their meanings are inconsistent, and we believe their entity type representations should be far away in the feature space. In response to this, we propose a dynamic class embedding generation method to learn a more effective representation for each entity type, which can ensure that the mention's unique characteristics can be retained as much as possible. Besides, different from traditional methods that match the mention and its candidate entities directly, we propose to use entity types help pre-classify of candidate entities and formulate a hierarchical scheme as shown in Fig. 2 (b). Specifically, the hierarchical constraint module in our method first classifies candidate entities according to entity types (e.g., ORG, INST) and then matches mention and entity types, as shown in Fig. 2 (b). In this way, we can substantially reduce the matching difficulty of the given mention and corresponding candidate entities. Furthermore, in order to solve the problem of the insufficient mention representation trained in low-data regimes, we propose an auxiliary task called *Mention Position Prediction (MPP)* to conduct multi-task learning. This task enables fine-grained interactions between mention and its context, which could capture more semantic information about the mention. In summary, the main contributions of our work could be summarized as follows.

- For the first time, we formulate a dynamic class scheme and hierarchical constraint scheme to fully utilize entity types in low-data regimes.

- We present an auxiliary task called *Mention Position Prediction* to obtain a more informative mention representation in low-data regimes.
- Extensive experiments on in-domain and out-of-domain datasets demonstrate the effectiveness of the proposed method. Our code is available at https://github.com/bigdata-ustc/CDHCN.

2 Related Work

2.1 Entity Linking

In recent years, the EL systems have made significant progress with the development of contextualized word embedding and representation methods [6,18,19]. There are a variety of different approaches proposed by researchers to tackle with EL task and its variants. For example, in order to maximize the similarity score between a mention embedding and its most corresponding entity embedding, Botha et al. [20] designed a dual-encoder architecture, consisting of two encoders for mentions and entities separately. GENRE, proposed by De Cao et al. [8], treats EL task as a generation problem by employing an auto-regressive formulation to teach a transformer-based model to produce a unique name for the mention. Gu et al. [21] proposed a machine reading comprehension framework for short text EL.

These methods described above require large amounts of training data, which is resource-intensive. This motivated researchers to make greater use of external information such as entity relations, entity types to provide the model with more prior information and reduce the scale of training data without performance degradation. Therefore, Raiman and Raiman [22] proposed DeepType, which makes use of type information to enhance the EL model's performance. Another noteworthy direction is how to fully leverage the entity types' commonalities between NER tasks and EL tasks. To take advantage of their relatedness, Martins et al. [23] performed joint learning of NER and EL, and obtain a robust system. Fine-grained NER labels are used to place limitations on the EL model's behavior in the most recent study [13]. Even though there is ample proof that introducing entity types (NER labels) into EL approaches is beneficial, current methods don't take fully utilize entity types information. For more than just labelling purposes: it can also be used to create informative embeddings that can be investigated with fine-grained features.

2.2 Name Entity Recognition

Most NER systems regard the task as sequence labelling and usually model it using conditional random fields (CRFs) or bi-directional Long Short Term Memory Networks (LSTMs). The performance of NER was further improved by recent large-scale pre-trained language models like BERT [9], which produced state-of-the-art results comparable to almost any other area in NLP. Thanks to NER classes, NER can cluster entities and solve the problem of intrinsic

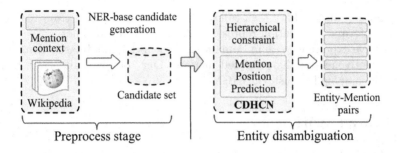

Fig. 3. The flowchart overview of our work.

sparsity in EL tasks. Besides, the relatedness of NER and EL has been proven to improve the performance of the entity linking model [23]. Nevertheless, there is a paucity of research on the effectiveness of enhancing EL models' capability by adding NER classes information or the relatedness of EL and NER. Previous approaches in this area have directly either used NER for mention detection or train a multi-task model that learns NER and EL jointly [23,24].

In this paper, we take the best of NER information. Following Tedeschi et al. [13], we add constraints to the candidate entities generation and inference phase. Additionally, we show using entity types embeddings is a feasible way to exploit NER for EL. Furthermore, we propose an auxiliary task called *mention position prediction*. This task is similar to NER but does not identify the type of detected mention, which can retain the mention's specific features for mention embedding generation as much as possible.

3   The Framework of CDHCN

In this section, we first present the problem statement of entity linking (EL), and then give an overview of our proposed CDHCN method. After that, we explain the technical details of CDHCN.

3.1   Problem Statement

The EL task can be formulated as a multi-label classification task where classes are represented by entities. Specifically, a text S contains a set of identified mentions $M = \{m_1, m_2, \dots, m_n\}$. The goal of the EL task is to find a mapping function that links each mention m_i to a unique entity e_i which is an unambiguous page in a given KB (e.g., Wikipedia).

Before entity disambiguation, we have a preprocessing step called entity candidate generation that chooses potential candidate entities $\Theta_i = \{e_1, e_2, \dots, e_K\}$ from a specific KB for each mention m_i to improve recall, where K is a pre-defined number to prune the candidate set. In this work, we adopt NER-enhanced candidate generation in line with Tedeschi et al. [13]. In a nutshell, by the use of NER classes, we can add constraints to the entity candidate set, which help decrease the size of the candidate set without reducing its recall.

3.2   An Overview of CDHCN

Figure 3 shows the framework of our method. In the preprocessing stage, our method makes use of mention's entity types (NER classes) to help generate a candidate set [13]. Note that each candidate has one pertinent description in the given KB. After obtaining the candidate set, we simultaneously input entities and mentions into the model. Through the hierarchical constraint module, we can obtain two similarity scores: one for the mention and entity type, and one for the mention and entity itself. Then, the former similarity score that we call it hierarchical constraint scores can be used to constrain the latter. Finally, we use the similarity score that is obtained after constraining for selecting the best mention-entity pair.

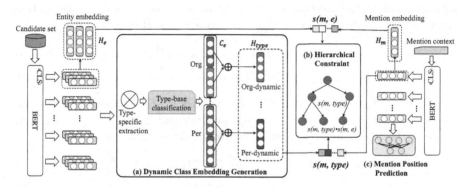

Fig. 4. Illustration of our proposed CDHCN framework for entity linking. (a) The dynamic class embeddings generation module can use candidate entities information to generate class embeddings for mention. (b) The hierarchical constraint module can use class embeddings to constrain the similarity scores between mention and its candidate entities. (c) The mention position prediction module can use auxiliary task to learn a more robust mention embedding.

3.3   Components of CDHCN

In this subsection, we will introduce the technical details of CDHCN. As shown in Fig. 4, CDHCN mainly contains three parts: (1) dynamic class embedding generation; (2) hierarchical constraint; (3) metion position prediction. We first obtain mention-specific entity type embeddings by dynamic class embedding generation module. Then the embedding can be used in hierarchical constraint module. Additionally, the mention position prediction module aims to obtain a more robust mention embeddings.

Input Representation. For a more accurate representation of mention and entity, we need to map each word in contextual mention and entity description into a low-dimensional vector. In this paper, we splice mention and mention's context as the input of transformer-base encoder to obtain initial mention embeddings. Specifically, we construct the input of each mention example as:

$$I_m = [CLS] \ lctx \ [M_s] \ mention \ [M_e] \ rctx \ [SEP], \tag{1}$$

where $lctx$ and $rctx$ denote context to the left of the mention and right of the mention, respectively. To address the problem of quadratic dependency in the transformer encoder when dealing with long texts, we limit the mention context length L to 128. Similarly, The entity input is as follows:

$$I_e = [CLS] \ title \ [SEP] \ description \ [SEP], \tag{2}$$

where *[CLS]* and *[SEP]* are special placeholders and *title* denotes entity. We consider the first token embedding, which is the output of the last hidden layer corresponding to the position of the *[CLS]*, as mention or entity representation. Both mention context and candidate entity are encoded into vectors:

$$H_m = BERT_{1[CLS]}(I_m) \in \mathbb{R}^h, \tag{3}$$

$$H_e = BERT_{2[CLS]}(I_e) \in \mathbb{R}^h, \tag{4}$$

where H_m and H_e are all h-dimensional vectors which represent mention and candidate entity respectively. Note that $BERT_1$ and $BERT_2$ are two different encoders. Besides, we also obtain every token representation as follows:

$$[H_{e_i^1}, H_{e_i^2}, ..., H_{e_i^L}]^T = BERT_1(I_{e_i}) \in \mathbb{R}^{L \times h}, \tag{5}$$

$$[H_{m^1}, H_{m^2}, ..., H_{m^L}]^T = BERT_2(I_m) \in \mathbb{R}^{L \times h}, \tag{6}$$

where H is an h-dimensional vector and L is text length.

Dynamic Class Representation. To capture more fine-grained semantic features of entity types (e.g., Organization), we need to encode entity types into low-dimensional vectors. This can be done quickly and easily by using BERT's built-in support for entity types as input text. However, due to the limited text (i.e., entity types are only one word in general), the embeddings that the model extract are insufficient to represent types features. Moreover, as mentioned above, even the meanings of the same entity type for different mentions are diverse. For example, *Tesla, the European Union* and *Paribas* have same entity type - Organization. However, there are clear distinctions between them, i.e., Tesla is a rock band, the European Union is a political and economic union and Paribas is a banking group. Fortunately, the candidate entity descriptions provide us with rich type information. We propose two simple yet effective methods to extract type-specific information for each candidate entity: one is context-aware type extraction and the other is mention-aware type extraction.

1) Context-aware Type Extraction. The local context in the candidate entity description provides most of the type-specific information. Therefore, we focus on how to extract these crucial local features. In this method, multiple

CNNs are used for generating entity types features because of the excellent ability of CNN in extracting local features [25]. Specifically, given a candidate entity description representation $[H_{e_i^1}, H_{e_i^2}, ..., H_{e_i^L}]^T$ generated in the above step, we need to splice them together:

$$H_{e_i^{1:L}} = H_{e_i^1} \oplus H_{e_i^2} \oplus ... \oplus H_{e_i^L}, \tag{7}$$

where $\oplus$ denotes concatenation operation.

Then, the local feature $C_{e_i^x}$ of candidate entity e_i can be generated by convolution operation with different convolution kernel sizes ks:

$$C_{e_i^x} = f(W_{CNN}^T \cdot H_{e_i^{x+ks-1}} + b), \tag{8}$$

where $W_{CNN} \in \mathbb{R}^{ks \times h}$ is a learnable matrix and b is a bias term. f is a nonlinear function. Particularly, the kernel is applied to each possible windows of word embddings $[H_{e_i^{1:ks}}, H_{e_i^{2:ks+1}}, ..., H_{e_i^{L-ks+1:L}}]$ in the sentence to produce a feature map:

$$\mathbf{C}_{e_i} = [C_{e_i^1}, C_{e_i^2}, ..., C_{e_i^{L-ks+1}}]. \tag{9}$$

After that, we use a max-pooling layer over the feature map to extract key information, and we can obtain the continuous fine-grained representations $[\mathbf{C}_{e_1}, \mathbf{C}_{e_2}, ..., \mathbf{C}_{e_K}]$ for each candidate entity.

2) Mention-aware Type Extraction. Lacking complex mention-entity interactions in the type-specific extraction stage might lead to the extracted features not being the ones that are required by the mention-entity disambiguation task. Therefore, we focus on how to utilize the mention information to assist in extracting type-specific features for each candidate entity. Some researches have proven that the associative attention can improve feature learning [26]. In this method, the multi-head attention mechanism [27] is used to help the model focus on features more relevant to the mention-entity disambiguation task. Specifically, we regard mention embedding H_{m_i} as *query*, candidate entity token embedding $[H_{e_j^1}, H_{e_j^2}, ..., H_{e_j^L}]^T$ as *key* and *value*. Then the mention-aware entity embeddings are computed by:

$$\begin{aligned} \mathbf{C} &= \{\mathbf{C}_{e_i} | i = 1, 2, ..., K\} \\ &= MultiHead(query * W_q, key * W_k, value * W_v) \end{aligned} \tag{10}$$

where W_q, W_k and W_v are learnable parameters. The output feature $\mathbf{C} = \{\mathbf{C}_{e_i} | i = 1, 2, ..., K\}$ is mention-aware representations for each candidate entity. For detailed implementation of MultiHead Attention Mechanism, please refer to Transformer [27].

After extracting type-specific information, we need to generate entity type representations by aggregating fine-grained candidate entity representations with the same entity type. In detail, we use entity type (e.g., Organization) to cluster candidate type representations, and the final mention-specific entity type representation $\mathbf{H}_{type}$ is computed by:

$$\mathbf{H}_{type_j} = \sum_{i=1}^{n} C_{e_{ij}}/n, \tag{11}$$

where e_{ij} ($i = 1, 2, \ldots, n$) denotes that candidate entity e_i type is $type_j$.

Hierarchical Constraint. As we discussed before, external knowledge can be used to improve model performance without requiring a larger data set. In the paper, we use entity types information as an EL task external knowledge. Unlike previous literature, we model entity type implicitly so that the proposed model can capture specific type information for each different mention. As shown in Fig. 2 (b), before calculating the similarity scores between mention and candidate entities, we need to obtain the similarity scores between mention and mention-specific entity types, which can also be called hierarchical constraint scores. The hierarchical constraint scores are given by the cosine similarity score:

$$s(m, type_i) = \frac{H_m^T \cdot \mathbf{H}_{type_i}}{||H_m||\ ||\mathbf{H}_{type_i}||}, i \in (1, 2, \ldots, 18), \tag{12}$$

where $type_i$ represents entity type, and we have 18 kinds of entity types. $\mathbf{H}_{type}$ are entity type embedding that is computed in equation (11).

After obtaining the hierarchical constraint scores between mention and entity types $s(m, type_i)$, we could use these scores to constrain the similarity scores between mention and candidate entities. Prior to matching mention and candidate entities, it is intended that the model have pertinent knowledge about candidate entity types. The similarity scores between mention and candidate entities are also calculated by cosine similarity score:

$$s(m, e_i) = \frac{H_m^T \cdot H_{e_i}}{||H_m||\ ||H_{e_i}||}, i \in (1, 2, \ldots, K). \tag{13}$$

And the final constrained score function is as follows:

$$s_{type_i}(m, e_j) = s(m, type_i) \cdot s(m, e_j), \tag{14}$$

where $s_{type_i}(m, e_j)$ denotes the similarity score between mention m and candidate entity e_j whose entity type is $type_i$. Then the network is trained to maximize the score of the gold mention-entity pairs. The training loss function of gold entity prediction is defined as cross-entropy:

$$\mathcal{L}_{EL} = -\frac{1}{N} \sum_{k=1}^{N} \sum_{j=1}^{K} y_k^j log(s_{type_i}(m_k, e_j)), \tag{15}$$

where y_k^j denotes gold entity of m_k. N is the number of examples. K is the number of candidate entities.

Mention Position Prediction. In order to enhance low-data mention representation through fine-grained interaction between mention and its context, we introduce *mention position prediction (MPP)* as an auxiliary task. We design an *MPP* task to teach the model to distinguish whether a token is a part of the mention. *MPP* is a token-level prediction task because we require tokens belonging to the correct mention span with the same position. In this way, the model can learn more token-level feature which is more compatible with the mention. This task is, in essence, a binary classification task that model predicts if a word is part of a mention or not. We feed mention context token embedding $[H_{m^1}, H_{m^2}, ..., H_{m^L}]$ obtained above into a feed-forward layer with Softmax function:

$$\tilde{y}_i^j = Softmax(W_{MPP}^T H_{m_j^i} + b_{MPP}), \tag{16}$$

where $W_{MPP} \in \mathbb{R}^h$ and $b \in \mathbb{R}^L$ are the learnable parameter and bias respectively. The training loss function of *MPP* is defined as cross-entropy:

$$\mathcal{L}_M = -\frac{1}{N} \sum_{j=1}^{N} \sum_{i=1}^{L} [y_i^j log(\tilde{y}_i^j) + (1 - y_i^j)log(1 - \tilde{y}_i^j)], \tag{17}$$

where $\tilde{y}_i^j$ denotes the prediction and y_i^j is the target indicating whether $H_{m_j^i}$ is mention embedding. N is the number of training examples.

3.4 Training Strategy

Our approach, in contrast to conventional methods, consists of two tasks: entity linking task and mention position prediction task. we need to optimize simultaneously for these two tasks by a single loss function. Given the losses defined in equation (15) and equation (17), our loss function is as follows:

$$\mathcal{L} = \lambda_{EL}\mathcal{L}_{EL} + \lambda_M \mathcal{L}_M, \tag{18}$$

where $\mathcal{L}$ is the final optimization target, λ_{EL} and λ_M are hyper-parameters that denote task weights for entity linking and mention position prediction respectively. In our experiment setup, λ_{EL} and λ_M are all set to 1.

In the same manner as Tedeschi et al. [13], we use NER classes (e.g., ORG, LOC) to constrain the model's output. In particular, for each mention, we require the model to output entities whose entity types is consistent with the prediction of an NER classifier.

4 Experiments

4.1 Datasets Preparation

For the reliability and authority of experimental results, we conduct experiments on both the in-domain and out-of-domain datasets. Specifically, we use AIDA-YAGO-CoNLL [28] as our in-domain dataset. This dataset contains AIDA-train for training, AIDA-A for development, and AIDA-B for testing. It is important to note that we train each of our models on only the AIDA-train (18k label instances). For out-domain datasets, we evaluated models on five popular datasets: MSNBC, AQUAINT, ACE2004 [29] and WNED-WIKI, WNED-CWEB [29,30]. The statistics of these datasets are shown in Table 1. For the KB, we use the November 2020 dump of the English Wikipedia labeled by Tedeschi et al. [13]. Each entity in Wikipedia was labeled by a new set of 18 fine-grained NER classes.

Table 1. Statistics of the datasets. It is important to note that MSNBC, AQUAINT, ACE2004, WNED-WIKI, and WNED-CWEB are out-of-domain datasets, meaning they do not have training data and development data. We train each of our models on only the AIDA-training.

Dataset	Train	Dev	Test
AIDA [28]	18,395	4,784	4,463
MSNBC [29]	–	–	656
AQUAINT [29]	–	–	727
ACE2004 [29]	–	–	257
WNED-WIKI [29,30]	–	–	6,821
WNED-CWEB [29,30]	–	–	11,154

4.2 Experiment Setup

In this paper, our goal is to demonstrate the superiority of our method in low-data regimes entity linking. For the low-data regimes scenario, we follow the setting of Tedeschi et al. [13]. A complete EL systems consists of mention detection, candidate generation and entity disambiguation. Here, we mainly focus on entity disambiguation methods. For mention detection and candidate generation, we follow the setting of Tedeschi et al. [13]. Note that, we do not have to figure out the right span of mention because it is provided in context. For candidate generation, we adopt NER-Enhance strategy to find a trade-off between recall and set size. Specifically, we train and employ an NER classifier to predict the entity type of the given mention in context and then try to choose the entities whose NER classes are consistent with the mention as candidate entities.

In our experiment, we use the pre-trained uncased BERT-based model with a 768 dimensions hidden representation as our backbone. We trained each model for 30 epochs, adopting an early stopping strategy with a patience value of 5. We adopt Adam [31] as optimizer with learning rate 1e−5 and maximum sequence length L 128. For each mention, the candidates' number K is set to 40. All experiments are performed on a NVIDIA GTX2080Ti with 11G GPU memory.

4.3 Baseline Methods

In our experiments, we compare our model with existing state-of-the-art baselines in entity linking:

- **Global-RNN** [32] is a method based on convolutional neural networks and recurrent neural networks.
- **Deep-ed** [6] is a joint document-level entity disambiguation method which leverage neural attention to reinforce context representations.
- **WNED** [29] is a greedy and global NED algorithm based on a sound information-theoretic notion of semantic relatedness derived from random walks on carefully constructed disambiguation graphs.
- **E-ELMo** [7] is a method to learn an entity-aware extension of pretrained ELMo [33]. The model obtains significant improvements.
- **DCA** [34] is a global method which sequentially accumulates context information to make efficient, collective inference.
- **WNEL** [19] is a two-stages method which exploits unlabelled documents.
- **GENRE** [8] exploits a sequence-to-sequence architecture to generate entity names in an autoregressive fashion conditioned on the context.
- **NER-EL** [13] is a method that use NER classes to improve performance in low-data regimes.
- **EXTEND** [12] is Transformer-based architectures framing EL task as a text extraction problem.

Most EL systems actually make use of massive additional data and information originating from Wikipedia at training time, i.e., GENRE benefits greatly from drastically increasing the size of the training set from 18K to 9000K labelled instances, gaining almost 5 points. Our focus, however, is to improve EL model performance in low-data regimes. Among the above methods, only **GENRE**, **NER-EL** and **EXTEND** have low-data regimes setting.

Table 2. Results (InKB accuracy) on the in-domain settings when training on the low-data regime, i.e., AIDA-training only (right) and when using additional resources coming from Wikipedia (left). We mark in **bold** the best scores.

Model(high-data)	AIDA-B	Model(low-data)	AIDA-B
Global-RNN	90.7	GENRE	88.6
Deep-ed	92.2	NER-EL	89.0
WNED	89.0	EXTEND(base)	87.9
E-ELMo	93.5	*CDHCN(w/o MPP, w CAT)*	89.6
DCA	**93.7**	*CDHCN(w/o MPP, w MAT)*	89.7
WNEL	89.6	*CDHCN(w MPP, w/o CAT)*	89.2
GENRE	93.3	*CDHCN(w MPP, w CAT)*	90.0
EXTEND(large)	92.6	***CDHCN(w MPP, w MAT)***	**90.3**

Table 3. Results (InKB accuracy) on the out-of-domain settings when training on the low-data regime, i.e., AIDA-training only. We mark in **bold** the best scores.

	Model	MSNBC	AQUAINT	ACE2004	CWEB	WIKI
	NER-EL	84.9	67.2	86.3	63.7	60.0
low-data	CDHCN(w/o MPP, w CAT)	85.4	67.3	86.7	64.0	60.4
	CDHCN(w/o MPP, w MAT)	**86.5**	67.1	85.9	64.1	60.0
	CDHCN(w MPP, w/o CAT)	86.0	67.2	86.7	64.0	60.0
	CDHCN(w MPP, w CAT)	84.2	**67.7**	**87.5**	64.2	60.2
	CDHCN(w MPP, w MAT)	84.6	67.4	86.7	**64.6**	**60.7**

4.4   Experimental Results

We present the results of our approaches for EL on an in-domain dataset known as AIDA-B and discuss the merits of each contribution in turn. Then, we conduct experiments on out-of-domain datasets named MSNBC, AQUAINT, ACE2004, WNED-WIKI and WNED-CWEB to prove that our contributions are robust and beneficial for EL task in low-data regimes. In order to demonstrate the effects of each contribution, we compare five CDHCN variants. Specifically, MPP denotes the mention position prediction module, CAT denotes the context-aware type extraction module and MAT denotes the mention-aware type extraction module.

In-domain Results. We present the in-domain entity linking eveluation results in low-data regimes in Table 2 (right). From Table 2 (right), we can observe that compared to other strong baseline methods, CDHCN (w MPP, w MAT) has state-of-the-art performance in low-data regimes and achieves 1.3% absolute improvement in terms of accuracy over the strong baseline [13]. On the AIDA-B dataset, CDHCN (w MPP, w CAT) performs better than other methods but slightly worse than CDHCN (w MPP, w MAT) in low-data regimes. Two CDHCNs (w/o MPP) all perform better compared to the strong baseline [13]. The performance of our model further increases to 90.0 and 90.3 respectively by combining auxiliary task MPP and two hierarchical constraint methods, indicating that our menthod enhances mention representation and makes use of entity type information to tackle EL task successfully. To further validate the effectiveness of our method, we also compare with baselines which are trained in high-data regimes in Table 2 (left). Although our method is trained in low-data regimes, it is still competitive and surpasses WNED [29] and WNEL [19] that are trained in high-data regimes.

Out-of-domain Results. Among all the baselines described in this paper, only GENRE, NER-EL and EXTEND have low-data regime settings. GENRE [8] and EXTEND [12] are trained based on Bart [35], and NER-EL is trained based on Bert [9]. Since Bart is more generalized than Bert, the models based on Bart can perform much better on out-of-domain datasets. Although our methods can also be realized on the Bart, in this paper, we only implemented on the Bert

Table 4. Per-class accuracy (%) of the entities on the AIDA-B and ACE2004.

Entity type	AIDA-B			ACE2004		
	NER-EL	**CDHCN**	Δ	NER-EL	**CDHCN**	Δ
PER	95.9	95.7	−0.2	88.9	88.9	+0.0
ORG	81.4	83.9	+2.5	84.5	85.9	+1.4
LOC	93.1	93.5	+0.4	87.7	89.0	+1.3
ANIM	100.0	100.0	+0.0	–	–	–
EVE	46.3	46.4	+0.1	–	–	–
VEHI	100.0	100.0	+0.0	100.0	100.0	+0.0

and left the implementation on the Bart as the future work. So we evaluate five CDHCNs on the out-of-domain setting and compare it to NER-EL which is the only model based on Bert in low-data regimes. Similarly to the in-domain evaluations, our method consistently improve the results across five out-of-domain test sets. Table 3 shows that CDHCN (w MPP, w MAT) presents consistent performance in term of accuracy by 0.7% points compared to baseline methond [13] on WIKI. CDHCN (w/o MPP, w MAT) presents consistent performance by 1.6% in term of accuracy compared to baseline methond [13] on MSNBC. On the ACE2004 and AQUAINT datasets, CDHCN (w MPP, w CAT) achieves a 1.2 and 0.5% point improvement over baseline approaches [13]. The performance on out-of-domain datasets reveals the robustness of our model.

4.5 Type-Base Results

To better evaluate the role of entity types in our method, we analyze the accuracy of each entity type. Table 4 shows the results of a comparison between NER-EL [13] and CDHCN (w MPP, w MAT) when dealing with partial entity types. We can observe that our approach improves the results of most entity types. Under the type ORG, CDHCN pushes performances up by 2.5 and 1.4% on the AIDA-B and ACE2004 respectively. Under the type LOC, CDHCN performs slightly better than NER-EL by 0.4% on the AIDA-B but far better than NER-EL by 1.3%. There is a main reason that our improvement in mainly focused on ORG and LOC: some entity types features are difficult to extract since the description information of these entities has low degree distinction (e.g., PER). These type-based results corroborate our hypothesis that using external knowledge (i.e., entity type) can boost entity linking performance.

4.6 Case Study

To better explain the CDHCN model's research results, we conduct a case study. As shown in Fig. 5, the model without hierarchical constraints incorrectly predicts "HSBC" as "The Hongkong and Shanghai Banking Corporation", predicts

"Yale University" as "Yale University Press". The model can learn partial mention's entity type according to the mention context, however, it does not dare to choose the gold candidate entity because of the lack of specific candidate entities' types information. After learning entity type by giving hierarchical constraints, the model will tend to select the correct candidate entity whose entity type is the same as mention.

Text	Gold Entity	Prediction
HSBC has moved its Shanghai branch ...	HSBC	NER-EL: The Hongkong and Shanghai Banking Corporation **CDHCN: HSBC**
... to head one of the departments at Yale University.	Yale University	NER-EL: Yale University Press **CDHCN: Yale University**
... International Airport in the southern Gaza Strip was ...	Gaza Strip	NER-EL: Gaza **CDHCN: Gaza Strip**
The New York Times on Wednesday reported that al Faroon told ...	The New York Times	NER-EL: Time (magazine) **CDHCN: The New York Times**

Fig. 5. Examples of sentences where CDHCN can correctly link entity in the KB, but NER-EL is not. "NER-EL" and "CDHCN" stand for the baseline system and CDHCN (w MPP, w MAT), respectively.

5   Conclusions

In this paper, we studied the problem of entity linking in a low-data regime and proposed the CDHCN model, which can take into account the information from entity types. Specifically, to better capture information from entity type, we proposed a method to generate a class embedding for each mention and then used the embedding to constrain the similarity score between the mention and its candidate entities. Additionally, we utilized an auxiliary task called *Mention Position Prediction* to generate a more robust embedding for mention. Experiments on in-domain and out-of-domain datasets verified the effectiveness of CDHCN. Since we proposed to study entity type representations in low-data regimes for the first time, we hope this work could lead to more research in the future.

Acknowledgement. This research was partially supported by grant from the National Natural Science Foundation of China (Grant No. 61922073), and the University Synergy Innovation Program of Anhui Province (GXXT-2021-002).

References

1. Dubey, M., Banerjee, D., Chaudhuri, D., Lehmann, J.: EARL: joint entity and relation linking for question answering over knowledge graphs. In: Vrandečić, D., et al. (eds.) ISWC 2018. LNCS, vol. 11136, pp. 108–126. Springer, Cham (2018). https://doi.org/10.1007/978-3-030-00671-6_7

2. Ji, H., Grishman, R.: Knowledge base population: successful approaches and challenges. In: Proceedings of the 49th Annual Meeting of the Association for Computational Linguistics: Human Language Technologies, pp. 1148–1158 (2011)
3. Liu, Y., et al.: Technical phrase extraction for patent mining: a multi-level approach. In: ICDM, pp. 1142–1147. IEEE (2020)
4. Huang, Z., et al.: Disenqnet: disentangled representation learning for educational questions. In: SIGKDD, pp. 696–704 (2021)
5. Le, Q., Mikolov, T.: Distributed representations of sentences and documents. In: ICML, pp. 1188–1196. PMLR (2014)
6. Ganea, O.-E., Hofmann, T.: Deep joint entity disambiguation with local neural attention. In: EMNLP, pp. 2619–2629 (2017)
7. Shahbazi, H., Fern, X.Z., Ghaeini, R., Obeidat, R., Tadepalli, P.: Entity-aware ELMO: learning contextual entity representation for entity disambiguation. arXiv preprint arXiv:1908.05762 (2019)
8. De Cao, N., Izacard, G., Riedel, S., Petroni, F.: Autoregressive entity retrieval. In: ICLR (2020)
9. Devlin, J., Chang, M.-W., Lee, K., Toutanova, K.: Bert: pre-training of deep bidirectional transformers for language understanding. In: Proceedings of NAACL-HLT, pp. 4171–4186 (2019)
10. Wu, L., Petroni, F., Josifoski, M., Riedel, S., Zettlemoyer, L.: Scalable zero-shot entity linking with dense entity retrieval. In: EMNLP, pp. 6397–6407 (2020)
11. Shrivatsa Bhargav, G.P., et al.: Zero-shot entity linking with less data. In: NAACL 2022, pp. 1681–1697 (2022)
12. Barba, E., Procopio, L., Navigli, R.: Extend: extractive entity disambiguation. In: ACL, pp. 2478–2488 (2022)
13. Tedeschi, S., Conia, S., Cecconi, F., Navigli, R.: Named entity recognition for entity linking: what works and what's next. In: EMNLP 2021, pp. 2584–2596 (2021)
14. Zhang, K., Zhang, H., Liu, Q., Zhao, H., Zhu, H., Chen, E.: Interactive attention transfer network for cross-domain sentiment classification. In: AAAI, vol. 33, pp. 5773–5780 (2019)
15. Zhang, K., et al.: Graph adaptive semantic transfer for cross-domain sentiment classification. In: ACM SIGIR, pp. 1566–1576 (2022)
16. Zhang, K.: Incorporating dynamic semantics into pre-trained language model for aspect-based sentiment analysis. In: ACL 2022, pp. 3599–3610 (2022)
17. Honglun Zhang, Liqiang Xiao, Wenqing Chen, Yongkun Wang, and Yaohui Jin. Multi-task label embedding for text classification. In EMNLP, pages 4545–4553, 2018
18. Le, P., Titov, I.: Improving entity linking by modeling latent relations between mentions. In: ACL, pp. 1595–1604 (2018)
19. Le, P., Titov, I.: Boosting entity linking performance by leveraging unlabeled documents. In: ACL, pp. 1935–1945 (2019)
20. Botha, J.A., Shan, Z., Gillick, D.: Entity linking in 100 languages. In: EMNLP, pp. 7833–7845 (2020)
21. Gu, Y., et al.: Read, retrospect, select: an MRC framework to short text entity linking. In: AAAI , vol. 35, pp. 12920–12928 (2021)
22. Jonathan Raiman and Olivier Raiman. Deeptype: multilingual entity linking by neural type system evolution. In AAAI, volume 32, 2018
23. Martins, Z., Marinho, P.H., Martins, A.F.T.: Joint learning of named entity recognition and entity linking. In: ACL, pp. 190–196 (2019)

24. Mrini, K., Nie, S., Gu, J., Wang, S., Sanjabi, M., Firooz, H.: Detection, disambiguation, re-ranking: autoregressive entity linking as a multi-task problem. In: ACL, pp. 1972–1983 (2022)
25. Chen, Y.: Convolutional neural network for sentence classification. Master's thesis, University of Waterloo (2015)
26. Tao, H., et al.: A schema-aware radical-guided associative model for Chinese text classification. IEEE TKDE, Learning from ideography and labels (2022)
27. Vaswani, A.. et al.: Attention is all you need. In: Advances in Neural Information Processing Systems, vol. 30 (2017)
28. Hoffart, J., et al.: Robust disambiguation of named entities in text. In: EMNLP, pp. 782–792 (2011)
29. Guo, Z., Barbosa, D.: Robust named entity disambiguation with random walks. Semantic Web **9**(4), 459–479 (2018)
30. Gabrilovich, E., Ringgaard, M., Subramanya, A.: Facc1: freebase annotation of clueweb corpora, version 1 (release date 2013–06-26, format version 1, correction level 0)
31. Kingma, D.P., Ba, J.: Adam: a method for stochastic optimization. In: ICLR (Poster) (2015)
32. Nguyen, T.H., Fauceglia, N.R., Muro, M.R., Hassanzadeh, O., Gliozzo, A., Sadoghi, M.: Joint learning of local and global features for entity linking via neural networks. In: COLING, pp. 2310–2320 (2016)
33. Peters, M., Neumann, M., Iyyer, M., Gardner, M., Zettlemoyer, L.: Deep contextualized word representations (2018)
34. Yang, X., et al.: Learning dynamic context augmentation for global entity linking. arXiv preprint arXiv:1909.02117 (2019)
35. Lewis, M., et al.: Bart: denoising sequence-to-sequence pre-training for natural language generation, translation, and comprehension. In: ACL, pp. 7871–7880 (2020)

Enhancing Knowledge Graph Attention by Temporal Modeling for Entity Alignment with Sparse Seeds

Chenchen Sun[1]($\boxtimes$), Yuyuan Jin[1], Derong Shen[2], Tiezheng Nie[2], Xite Wang[3],
and Yingyuan Xiao[1]

[1] School of Computer Science and Engineering, Tianjin University of Technology, Tianjin,
China
suncc_db@163.com, yuyuanjin202108@163.com, yyxiao@tjut.edu.cn
[2] School of Computer Science and Engineering, Northeastern University, Shenyang, China
{shendr,nietiezheng}@mail.neu.edu.cn
[3] College of Information Science and Technology, Dalian Maritime University, Dalian, China
wangxite@dlmu.edu.cn

Abstract. As a fundamental task of knowledge graph integration, entity alignment (EA) matches equivalent entities across knowledge graphs (KGs). Temporal knowledge graphs (TKGs) enhance static KGs with temporal information. Traditional EA approaches tackle alignments of static KGs, but cannot effectively deal with TKGs. Therefore, temporal EA solutions are called for. To this end, we propose a time-aware graph attention network for entity alignment (TGA-EA). Generally, we learn high-quality temporal-relational entity embeddings for alignments by systematically integrating temporal information into KG embeddings. We propose three temporal modeling methods to effectively represent and integrate temporal information into both entities and relations. Then we construct temporal enhanced graph attention to produce target temporal-relational entity embeddings with temporal entities and temporal relations. Thanks to the powerful design, TGA-EA achieves promising performances with sparse alignment seeds. Extensive experiments on five datasets demonstrate our approach's obvious advantages over previous works.

Keywords: Temporal Knowledge Graph · Entity Alignment · Graph Attention · Temporal Modeling · Data Integration

1 Introduction

Knowledge graphs (KGs), organizing knowledge in semi-structured forms, are applied to an increasing number of knowledge-driven tasks [1], such as web search, question answering, recommendation and machine translation. However, each KG is usually created separately with a single data source, which results in incompleteness of current KGs. To enrich KGs, KG integration is necessary. As a fundamental task in KG integration,

X. Wang et al. (Eds.): DASFAA 2023, LNCS 13944, pp. 639–655, 2023.
https://doi.org/10.1007/978-3-031-30672-3_43

Entity Alignment (EA) identifies equivalent entities in multiple KGs [2], and draws great attention from both research communities and industries. A number of EA approaches are proposed, and are categorized into translation based ones [3–5] and Graph Neural Network (GNN) based ones [6–9].

Recently, temporal knowledge graphs (TKGs), such as Wikidata, ICEWS and YAGO3, emerge for industry applications, and attract increasing research interests [10]. Current TKG researches mainly focus on TKG embeddings [11, 12] and completions [10], which show that temporal information really boosts performances in the two TKG tasks. Then an interesting arising problem is how to conduct temporal EA for TKGs. Most previous EA approaches consider alignments between static KGs [2], where relations are utilized but temporal information is ignored. Naturally, traditional EA approaches cannot produce satisfied performances in temporal EA.

The major challenge in temporal EA is how to effectively model and systematically integrate temporal information into alignment processes. Although there exist a few works with obvious progresses, we still find huge improvement space. As pioneer works, TEA-GNN [13] and TREA [14] directly treat timestamps as a property similar to relations in GNNs. Tem-EA [15] concatenates separated learned timestamp embeddings by recurrent neural networks (RNN) and structural embeddings by GNN. In TKGs, a timestamp τ is affiliated with a fact, denoted as a quadruple (h, r, t, τ), and constrains temporal existences of both entities h, t and relation r. First, previous works do not fully model timestamps according to their temporal characteristics [10]. Second, they do not systemically integrate temporal features into representations of entities and relations for alignments.

To tackle this challenge, we propose a Time-aware Graph Attention Network for Entity Alignment (TGA-EA). Essentially, we enhance graph attention with effective temporal modeling to learn temporal-relational entity embeddings for TKG alignments. First, we propose three temporal modeling methods to effectively temporalize both entities and relations. Then temporal entities and temporal relations are utilized to build time-aware graph attention (TGA), with which high-quality temporal-relational entity embeddings are generated for alignments. TGA-EA is able to achieve competitive performances with sparse alignment seeds, owing to effective temporal modeling and systematical temporal-relational integration into entity embeddings. Experimental evaluations show that TGA-EA outperforms previous works.

Our contributions are summarized as follows.

- We propose a novel time-aware graph attention network based entity alignment approach (TGA-EA) for TKGs. At the core, temporal entities and temporal relations are fully exploited to build a temporal enhanced graph attention network.
- We propose three temporal modeling solutions: temporal projection, temporal rotation and temporal encoding, which are able to effectively represent timestamps and then temporalize entities and relations.
- We evaluate TGA-EA on five TKG datasets with sparse alignment seeds. Experimental results show TGA-EA's advantages over existing works. Also, contributions of different components are tested in TGA-EA.

Organization of the rest. Section 2 presents preliminaries. Section 3 details our proposed time-aware graph attention network for entity alignment (TGA-EA). Section 4 evaluates our approach on five TKG datasets. Section 5 presents related work. Finally, Sect. 6 concludes the full paper.

2 Preliminaries

A **Knowledge Graph (KG)**, consisting of triples, is a directed graph, denoted as $KG = (\mathcal{E}, \mathcal{R}, \mathbb{T})$. $\mathcal{E}$, $\mathcal{R}$ and $\mathbb{T}$ are the set of entities, the set of relations, and the set of triples, respectively. Each fact in KGs is represented as a triple (h, r, t), where h, r and t refer to head entity, tail entity, and relation, respectively.

A **Temporal Knowledge Graph (TKG)**, consisting of quadruples, is a KG with temporal information, denoted as $TKG = (\mathcal{E}, \mathcal{R}, \mathcal{T}, \mathcal{Q})$. A quadruple essentially refers to a fact with a timestamp, denoted as $\mathcal{Q} = \{(h, r, t, \tau) \mid h, t \in \mathcal{E}, r \in \mathcal{R}, \tau \in \mathcal{T}\}$. $\mathcal{T}$ is the set of timestamps. Timestamps of TKGs have various forms, i.e., time point τ, start time τ_s & end time τ_e, and time interval $[\tau_s, \tau_e]$. We unify timestamps in the form of time intervals. For example, a time point τ is transformed into $[\tau, \tau]$. Besides, we use τ_0 to denote an unknown timestamp.

Entity Alignment for TKGs. Given two TKGs, $TKG_1 = (\mathcal{E}_1, \mathcal{R}_1, \mathcal{T}_1, \mathcal{Q}_1)$ and $TKG_2 = (\mathcal{E}_2, \mathcal{R}_2, \mathcal{T}_2, \mathcal{Q}_2)$, and $S = \{(e_i, e_j) \mid e_i \in \mathcal{E}_1, e_j \in \mathcal{E}_2, e_i \equiv e_j\}$ be the predefined alignment seeds, where $\equiv$ denotes two entities e_i, e_j refer to the same real-world entity, EA tries to identify all aligned entity pairs between TKGs. Timestamps $\mathcal{T}_1$ and $\mathcal{T}_2$ of two TKGs are aligned and unified, denotes as $\mathcal{T}^* = \mathcal{T}_1 \cup \mathcal{T}_2$. Note that the alignment relationship between two entities is fixed all the time.

3 Time-Aware Graph Attention Network for Entity Alignment

3.1 Overview

The key to entity alignment for TKGs is how temporal information is effectively exploited and integrated into the alignment process. To this end, we propose a time-aware graph attention network for EA (TGA-EA), as Fig. 1. Basically, we enhance graph attention with effective temporal modeling, and learn high-quality temporal-relational entity embeddings in a unified low-dimensional vector space, where potential aligned entities are located as near as possible. In particular, TGA-EA is supposed to perform well by fully utilizing temporal information even in the setting of sparse alignment seeds, which is experimentally verified in Sect. 4.

The TGA-EA framework is depicted in Fig. 1. Overall, two TKGs are put into the stacked time-aware graph attention network (TGA), where temporal-relational entity embeddings are learned; then entity embeddings are utilized to compute the distance matrix, according to which the alignment set is generated. As the core, each TGA layer contains two steps. (1) Relations and entities are temporalized with three technical options: temporal projection, temporal rotation and temporal encoding (Sect. 3.3). (2) Temporalized entity embeddings and relation embeddings in neighborhoods are aggregated via time-aware graph attention to update target entity embeddings (Sect. 3.2). Finally, the whole model is optimized with triple loss (Sect. 3.4).

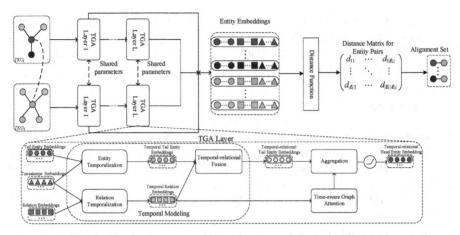

Fig. 1. The time-aware graph attention network for entity alignment.

3.2 Time-Aware Graph Attention Layer

Traditional Graph Attention Network (GAT) deals with ordinary graphs, but is not suitable for TKGs. In order to effectively process TKGs, we propose to enhance graph attention with temporal modeling. Following the classic GAT workflow, we first define time-aware graph attention, then linearly aggregate each entity's neighborhood with attention coefficients, and finally update each entity embedding with a nonlinear function. Besides, we consider bidirectional information propagations over directed edges in TKGs, where an inverse relation r^{-1} is created for each existing relation r. Each quadruple $(h, r, t, [\tau_s, \tau_e])$ is split into (h, r, t, τ_s) and (t, r^{-1}, h, τ_e).

Temporal Functions. To tackle temporal modeling issues in TKGs, we define two important temporal functions TM(·) and TRF(·). Later in Sect. 3.3, we introduce three implementations of TM(·) and TRF(·). As a core temporal function, TM(·) is able to integrate temporal information into entities and relations separately, as Eq. (1). Given an entity embedding $\mathbf{e}$ and a timestamp embedding τ, we get a temporal entity embedding $\bar{\mathbf{e}}$ with TM(·). Similarly, we generate a temporal relation embedding $\bar{\mathbf{r}}$ from a relation embedding $\mathbf{r}$ and a timestamp embedding τ.

$$\begin{aligned} \bar{\mathbf{r}} &= \text{TM}(\mathbf{r}, \tau) \\ \bar{\mathbf{e}} &= \text{TM}(\mathbf{e}, \tau) \end{aligned} \tag{1}$$

Given a quadruple (h, r, t, τ), we focus on tail entity t, relation r and timestamp τ here, which are utilized to generate head entity h's embedding with TGA later. Note that $\bar{\mathbf{t}}$ contains temporal information but lacks relational information. Then we propose another entity-centered temporal-relational fusion function TRF(·) based on TM(·), as Eq. (2). TRF(·) further fuses the temporal entity embedding $\bar{\mathbf{t}}$ and the temporal relation embedding $\bar{\mathbf{r}}$ into a temporal-relational entity embedding $\widetilde{\mathbf{t}}$.

$$\widetilde{\mathbf{t}} = \text{TRF}(\bar{\mathbf{t}}, \bar{\mathbf{r}}) = \text{TRF}(\text{TM}(\mathbf{t}, \tau), \text{TM}(\mathbf{r}, \tau)) \tag{2}$$

Time-Aware Graph Attention. We believe both relations and timestamps play important roles in learning entity embeddings. Thus, we define a light-weight graph attention with temporal relations, and then aggregate temporal-relational entities in neighborhoods to update target entities.

Although GAT defines graph attention with nodes, the recent EA work [8] shows that knowledge graph attention with relations is effective and light-weight. Thus, we design a time-aware graph attention with temporal relations. Given a quadruple (h, r, t, τ) and a target head entity h, its corresponding attention coefficient $\beta^l_{(h,r,t,\tau)}$ in the l-th TGA layer is defined as Eq. (3). $\mathbf{v}$ is a trainable shared weight vector, and $\bar{\mathbf{r}}$ is the temporal relation embedding of $\mathbf{r}$, computed with Eq. (1). We compute all attention coefficients around target head entity h, and normalize them with Eq. (4). $\alpha^l_{(h,r,t,\tau)}$ is the normalized attention coefficient of $\beta^l_{(h,r,t,\tau)}$. N_h is the tail neighbor set of h, and t' is a tail entity in N_h. $E_{h,t'} \subset \mathcal{R} \times \mathcal{T}$ is the set of relation-timestamp pairs from h to t', and $(r', \tau') \in E_{h,t'}$ denotes existence of a quadruple (h, r', t', τ').

$$\beta^l_{(h,r,t,\tau)} = \mathbf{v}^T \bar{\mathbf{r}} = \mathbf{v}^T \text{TM}(\mathbf{r}, \tau) \tag{3}$$

$$\alpha^l_{(h,r,t,\tau)} = \exp(\beta^l_{(h,r,t,\tau)}) \Big/ \sum_{t' \in N_h} \sum_{(r',\tau') \in E_{h,t'}} \exp(\beta^l_{(h,r',t',\tau')}) \tag{4}$$

Then, we aggregate target entity h's neighborhood information with attention coefficients, and apply a nonlinearity to get h's updated embedding $\mathbf{h}^{l+1}$ in the l-th TGA layer, as Eq. (5). ReLU($\cdot$) is an activation function. $\tilde{\mathbf{t}}^l$ is a tail neighbor's temporal-relational embedding from the $(l-1)$-th TGA layer, computed with Eq. (1) and (2).

$$\mathbf{h}^{l+1} = \text{ReLU}(\sum_{t \in N_h} \sum_{(r,\tau) \in E_{h,t}} \alpha^l_{(h,r,t,\tau)} \tilde{\mathbf{t}}^l) = \text{ReLU}(\sum_{t \in N_h} \sum_{(r,\tau) \in E_{h,t}} \alpha^l_{(h,r,t,\tau)} \text{TRF}(\tilde{\mathbf{t}}^l, \bar{\mathbf{r}}))$$

$$= \text{ReLU}(\sum_{t \in N_h} \sum_{(r,\tau) \in E_{h,t}} \alpha^l_{(h,r,t,\tau)} \text{TRF}(\text{TM}(\mathbf{t}^l, \tau), \text{TM}(\mathbf{r}, \tau)))$$

$$\tag{5}$$

We stack multiple TGA layers to aggregate more information from multi-hop neighborhoods. We concatenate the outputs of TGA layers to obtain the final output representation of entity h, $\mathbf{h}^{output}$, as Eq. (6). L is the total number of TGA layers. As Eq. (7), $\mathbf{h}^0$ denotes the initial embedding of entity h, and is the average embedding of entity h and its tail neighborhood.

$$\mathbf{h}^{output} = [\mathbf{h}^0 || \cdots || \mathbf{h}^L] \tag{6}$$

$$\mathbf{h}^0 = (1 / (|N_h| + 1)) \sum_{e \in N_h \cup h} \mathbf{e} \tag{7}$$

$$\mathbf{h}^{final} = \begin{cases} [\mathbf{h}^{output} || (1 / |N_h^r|) \sum_{r \in N_h^r} \mathbf{r} || (1 / |N_h^\tau|) \sum_{\tau \in N_h^\tau} \tau], & \text{temporal projection} \\ [\mathbf{h}^{output} || \text{StackedTGA}((1 / |N_h^r|) \sum_{r \in N_h^r} \mathbf{r}) || (1 / |N_h^\tau|) \sum_{\tau \in N_h^\tau} \tau], & \text{others} \end{cases} \tag{8}$$

Finally, we aggregate $\mathbf{h}^{output}$, temporal relation embeddings and timestamp embeddings around entity h together to generate h's final entity embedding $\mathbf{h}^{final}$, as Eq. (8). $N_h{}^r$ represents the set of relations outward from h, and $N_h{}^\tau$ represents the set of timestamps outward from h. For temporal projection, we simply concatenate these three types of embeddings. For other two temporal modeling methods, we put the average relation embedding into the stacked TGA network, and then concatenate its output with $\mathbf{h}^{output}$ and the average timestamp embedding. As we see, the final entity embeddings not only contain timestamp embeddings and relation embeddings explicitly, but also implicitly model timestamps and relations.

3.3 Temporal Modeling in Knowledge Graphs

For temporal modeling in KGs, there are two key issues: (1) timestamp representations; (2) temporal information integration into entities/relations. To this end, we introduce three methods: temporal projection (TP), temporal rotation (TR) and temporal encoding (TEn). For the first issue, TP and TR randomly initialize timestamp embeddings and learn them in the TGA pipeline; TEn computes fixed timestamp embeddings, which enables to represent time span information. For the second issue, TP temporalizes entities/relations with projections to hyperplanes; TR and TEn temporalize entities/relations with rotations in complex vector spaces. We implement three sets of TM($\cdot$) and TRF($\cdot$) with TP, TR and TEn.

Temporal Projection. We consider the timestamp τ as a hyperplane T_τ [11]. Basically, we get temporal entity embeddings and temporal relation embeddings by projecting entity embeddings and relation embeddings onto timestamp hyperplanes.

$$\begin{aligned} \bar{\mathbf{r}} = \mathrm{TM}_{\mathrm{tp}}(\mathbf{r}, \tau) = \mathbf{r} - (\tau^T \mathbf{r})\tau \\ \bar{\mathbf{e}} = \mathrm{TM}_{\mathrm{tp}}(\mathbf{e}, \tau) = \mathbf{e} - (\tau^T \mathbf{e})\tau \end{aligned}, \ s.t. \|\tau\|_2 = 1 \tag{9}$$

Since each normal vector identifies a hyperplane, we learn a timestamp embedding τ as the normal vector of its corresponding timestamp hyperplane T_τ. Each embedding vector can be decomposed into two mutually orthogonal components, one parallel to the normal vector τ and the other one falling onto the hyperplane T_τ (orthogonal to τ), where the latter is what we want. For instance, h, r, t are projected onto τ's hyperplane T_τ, in Fig. 2. We implement a temporal projection based TM function $\mathrm{TM}_{\mathrm{tp}}(\cdot)$ as Eq. (9), where an entity/relation embedding is projected onto the hyperplane T_τ. Timestamp embedding τ is randomly initialized, and is restricted with $\|\tau\|_2 = 1$. With $\mathrm{TM}_{\mathrm{tp}}(\cdot)$, we can compute the temporal entity embedding $\bar{\mathbf{e}}$ and the temporal relation embedding $\bar{\mathbf{r}}$ by projections.

Given a quadruple (h, r, t, τ), let us focus on relation r, tail entity t, and timestamp τ, which are used to generate head entity h's embedding with TGA. As Eq. (10), based on $\mathrm{TM}_{\mathrm{tp}}(\cdot)$, we define a temporal-relational fusion function $\mathrm{TRF}_{\mathrm{tp}}(\cdot)$ to fuse temporal entity embedding $\bar{\mathbf{t}}$ and temporal relation embedding $\bar{\mathbf{r}}$ to generate temporal-relational tail entity embedding $\tilde{\mathbf{t}}$, where $\bar{\mathbf{t}}$ is projected onto the hyperplane corresponding to $\bar{\mathbf{r}}$.

$$\tilde{\mathbf{t}} = \mathrm{TRF}_{\mathrm{tp}}(\bar{\mathbf{t}}, \bar{\mathbf{r}}) = \bar{\mathbf{t}} - (\bar{\mathbf{r}})^T \bar{\mathbf{t}}\bar{\mathbf{r}}, \ s.t. \|\bar{\mathbf{r}}\|_2 = 1 \tag{10}$$

Temporal Rotation. Complex vector spaces show great potential for temporal modeling [12]. We propose rotation based temporal functions in complex spaces. We first revisit rotation with the theoretical support of Euler's formula, and then introduce the proposed functions.

Any unit complex can be expressed by Euler's formula, as Eq. (11). Figure 3 presents a unit complex number $z = \exp(i\theta)$ in the complex plane, where the circle is the unit circle. When we want to rotate a number a in the complex plane by an angle θ, we only need to perform the multiplication operation between complex numbers z and a, expressed as $z*a = (x_z x_a - y_z y_a) + (y_z x_a + x_z y_a)i$, where $z = x_z + y_z i$ and $a = x_a + y_a i$.

Fig. 2. Projections of h,r,t onto a hyperplane T_τ.   **Fig. 3.** The representation of a unit complex number z with Euler's formula.

$$\exp(i\theta) = \cos(\theta) + i\sin(\theta) \tag{11}$$

$$\begin{aligned} \bar{\mathbf{r}} &= \mathrm{TM}_{tr}(\mathbf{r}, \tau) = \mathbf{r} \circ \tau \\ \bar{\mathbf{e}} &= \mathrm{TM}_{tr}(\mathbf{e}, \tau) = \mathbf{e} \circ \tau \end{aligned} \tag{12}$$

Given relation embedding $\mathbf{r} \in \mathbb{C}^d$, entity embedding $\mathbf{e} \in \mathbb{C}^d$, and timestamp embedding $\tau \in \mathbb{C}^d$ in complex spaces, temporalizations can be implemented with rotations. In Eq. (12), we implement a temporal rotation based TM function $\mathrm{TM}_{tr}(\cdot)$, where the temporal entity embedding $\bar{\mathbf{e}}$ and the temporal relation embedding $\bar{\mathbf{r}}$ are computed with rotations. The rotation operator $\circ$ is essentially element-wise complex number multiplication. Timestamp embedding τ is randomly initialized.

To further fuse temporal entity embeddings and temporal relation embeddings, we define a temporal-relational fusion function $\mathrm{TRF}_{tr}(\cdot)$ with a relational reflection transformation [7], as Eq. (13). By constructing an orthogonal reflection matrix $(I - 2\bar{\mathbf{r}}(\bar{\mathbf{r}})^T)$, norms of entities and their relative distances are guaranteed to be invariant during the transformation. The temporal-relational tail entity embedding $\tilde{\mathbf{t}}$ can be obtained without destroying the original distribution of the entity embedding space.

$$\tilde{\mathbf{t}} = \mathrm{TRF}_{tr}(\bar{\mathbf{t}}, \bar{\mathbf{r}}) = \bar{\mathbf{t}}(I - 2\bar{\mathbf{r}}(\bar{\mathbf{r}})^T), \quad s.t. \|\bar{\mathbf{r}}\|_2 = 1 \tag{13}$$

Temporal Encoding. Time spans between timestamps are important for timestamp representations. For instance, given a timestamp sequence "$\tau_1, \tau_2, \tau_3, \tau_4, \tau_5$", τ_1 should be more similar to τ_2 than τ_5. However, timestamp embeddings in temporal projection and temporal rotation fail to capture time span information, because they are randomly initialized in training pipelines. TP and TR are only able to tell that τ_1, τ_2 and τ_5 are different. Luckily, positional encoding in Transformer is able to effectively capture relative positions [16], which are similar to time spans for timestamps. Since time series are essentially timestamp sequences, we extend positional encoding to temporal encoding, which is defined in complex vector spaces. Our temporal encoding computes fixed timestamp embeddings, which do not change during training.

Following positional encoding [16], we define a real temporal encoding function $\text{TEn}_{\text{real}}(\cdot)$ in the real vector space, as Eq. (14), where timestamp τ is represented as a d'-dimensional embedding. $\text{TEn}_{\text{real}}(\tau, 2j)$ and $\text{TEn}_{\text{real}}(\tau, 2j + 1)$ represent the $2j$-th and $(2j + 1)$th elements in the real timestamp embedding of τ, respectively. According to Euler's formula, two adjacent elements ($2j$ and $2j + 1$) in a real timestamp embedding can be regarded as a unit complex. Inspired by this, we define a complex temporal encoding function $\text{TEn}_{\text{cplx}}(\cdot)$ to map a real timestamp embedding into a d-dimensional complex timestamp embedding in the complex vector space, as Eq. (15), where $d' = 2d$. With Eq. (15), we finally obtain a fixed complex d-dimensional timestamp embedding τ for timestamp τ, where $\text{TEn}_{\text{cplx}}(\tau, j)$ denotes the j-th element $\tau[j]$.

Then, with fixed timestamp embeddings, we apply rotations to get temporal entity embeddings, temporal relation embeddings and temporal-relational entity embeddings. Specifically, we define a TM function $\text{TM}_{\text{ten}}(\cdot)$ with rotations similar to Eq. (12), and define a temporal-relational fusion function $\text{TRF}_{\text{ten}}(\cdot)$ similar to Eq. (13).

$$\text{TEn}_{\text{real}}(\tau, 2j) = \sin(\tau \big/ 10000^{2j/d'}), j \in [0, d'/2 - 1]$$
$$\text{TEn}_{\text{real}}(\tau, 2j + 1) = \cos(\tau \big/ 10000^{2j/d'}), j \in [0, d'/2 - 1] \tag{14}$$

$$\begin{aligned}\text{TEn}_{\text{cplex}}(\tau, j) &= \text{TEn}_{\text{real}}(\tau, 2j + 1) + i\text{TEn}_{\text{real}}(\tau, 2j)\\ &= \cos(\tau \big/ 10000^{j/d}) + i \sin(\tau \big/ 10000^{j/d}) = \tau[j]\end{aligned}, j \in [0, d - 1] \tag{15}$$

3.4 Optimization

The prediction of EA relies upon distances of entity pairs in the embedding space. For entity e_i from TKG_1 and entity e_j from TKG_2, we use Eq. (16) to compute the L1 distance between e_i and e_j. The smaller the distance between two entities is, the more possibly the two entities are equivalent. We optimize the TGA-EA model with given alignment seeds, by minimizing the following triple loss function, as Eq. (17). S is the alignment seed set, S' is the set of negative examples, which are generated by randomly replacing e_i or e_j in alignment seeds, and γ is a hyper-parameter.

$$\text{D}(e_i, e_j) = ||\mathbf{h}_{e_i}^{final} - \mathbf{h}_{e_j}^{final}||_1 \tag{16}$$

$$Loss = \sum_{(e_i, e_j) \in S} \sum_{(e_{i'}, e_{j'}) \in S'} \text{Max}(0, \gamma + \text{D}(e_i, e_j) - \text{D}(e_{i'}, e_{j'})) \tag{17}$$

4 Experimental Evaluation

4.1 Experiment Setup

Datasets. We evaluate TGA-EA on five datasets: DGDELT7K, DGDELT-RM, DICEWS-20RM, DICEWS-RM and YAGOWIKI15K. They are constructed based on GDELT [10], ICEWS05–15 [10], and YAGO-WIKI20K [13]. Compared with previous TKG datasets [14], our datasets are more difficult to align for two reasons. (1) Our TKGs are sparser, where edges are reduced; (2) our alignment seeds are sparser, where seed sizes are set as 100 in default. The rest reference entity pairs are used for testing. Detailed statistics of the datasets are shown in Table 1. DGDELT7K and DGDELT-RM are dense in time density, DICEWS-20RM and DICEWS-RM are medium in time density, and YAGOWIKI15K is sparse in time density.

Table 1. Statistics of the datasets. ($\mathcal{P}$ is the set of reference entity pairs, where a reference entity pair is a real aligned entity pair.)

Datasets	DGDELT7K	DGDELT-RM	DICEWS-20RM	DICEWS-RM	YAGOWIKI15K
$|\mathcal{E}_1|$	7,203	7195	9,406	9,406	12,701
$|\mathcal{E}_2|$	7,410	7400	9,827	9,827	17,027
$|\mathcal{R}_1|$	236	235	248	248	7
$|\mathcal{R}_2|$	240	234	246	246	81
$|\mathcal{T}^*|$	44,625	44,625	4,017	4,017	404
time unit	15 min	15 min	1 day	1 day	1 year
time density	dense	dense	medium	medium	sparse
$|\mathcal{Q}_1|$	1,424,004	735,360	190,147	150,289	59,639
$|\mathcal{Q}_2|$	1,708,803	881,692	228,267	180,105	118,125
$|\mathcal{P}|$	6,922	6,909	8,745	8,745	12,625
$|\mathcal{S}|$	100	100	100	100	100

DGDELT7K. We copy GDELT into two TKGs $\mathcal{Q}_1$ and $\mathcal{Q}_2$, from which quadruples are randomly removed respectively. In the end, $\mathcal{Q}_1$ keeps 62.5% of all quadruples, $\mathcal{Q}_2$ keeps 75% of all quadruples, and the quadruple overlap between $\mathcal{Q}_1$ and $\mathcal{Q}_2$ covers 37.5% of all quadruples in GDELT.

DGDELT-RM. Based on DGDELT7K, we delete edges according to entity occurrence frequencies in quadruples. Entities are ordered into a list according to occurrence frequencies in quadruples. Edges connected to entities ranked top 10% in the list are deleted with a probability of 80%. Edges connected to entities ranked between top 30% and top 10% in the list are deleted with a probability of 40%. If a selected edge belongs to the last quadruple of any entity, its deletion is forbidden, which ensures the total entity number unchanged. After removals, the maximum connected components of Q_1 and Q_2 are retained to produce DGDELT-RM.

DICEWS-20RM. Based on ICEWS05–15, we generate an intermediate dataset DT_{20} in the same way as DGDELT7K. In DT_{20}, entities are ordered into a list according to occurrence frequencies in quadruples. Each edge connected to entities ranked top 20 in the list is removed unless the edge belongs to the last quadruple of any entity. After removals, rest Q_1 and Q_2 are all retained to generate DICEWS-20RM.

DICEWS-RM. Based on ICEWS05–15, we generate an intermediate dataset DT_{rm} in the same way as DGDELT7K. Then DT_{rm} is processed in the same way as DGDELT-RM, but rest Q_1 and Q_2 are all kept to get DICEWS-RM after removals.

YAGOWIKI15K. Based on YAGOWIKI20K, we randomly delete non-temporal edges with a probability of 60%. If a selected edge belongs to the last triple of an entity, its deletion is forbidden. Subsequently, the maximum connected components of Q_1 and Q_2 are retained to obtain YAGOWIKI15K. Note that YAGOWIKI15K also contains non-temporal edges.

Baselines. Our approach TGA-EA has three variants: TR-TGAEA with temporal rotation, TP-TGAEA with temporal projection, and TEn-TGAEA with temporal encoding. We compare our approach with six baselines, including four general EA approaches MTransE [3], AlignE [4], RREA [7], Dual-AMN [8], and two temporal EA approaches TEA-GNN [13], TREA [14], which are all detailed in related work.

Evaluation Metrics. According to [13, 14], CSLS is used for all approaches in the testing stage. We report Mean Reciprocal Rank (MRR) and Hits@k ($k = 1, 5, 10$) as our evaluation metrics. Hits@k measures the percentage of correct alignment ranked in top k. MRR calculates the average reciprocal of these ranks.

Experimental Settings. We implement our approach with Tensorflow and Keras. All baseline implementations are their publicly available source codes except MTransE, AlignE and TREA. Implementations of MTransE and AlignE in OpenEA [2] are used. We re-implement TREA since there are no available source codes. All experiments are conducted on a server with a single GPU RTX3090. All experiments are repeated three times, and average results are reported.

For all approaches, we perform grid searches for margin γ. For all baselines, rest hyper-parameters are set as the optimal parameters given in papers or source codes. Other hyper-parameters in our approach are in default set as $L = 2$, $d = 100$, dropout rate $dr = 0.3$, and learning rate $lr = 0.005$. Next, we list the optimal γ in TGA-EA. TH-TGAEA is set as $\gamma = 1$ on DGDELT7K, $\gamma = 0$ on DGDELTRM and $\gamma = 5$ on DICEWS-20RM, DICEWS-RM, and YAGOWIKI15K. TR-TGAEA is set as $d = 75$ (for memory restriction), $\gamma = 7$ on DGDELT7K, $\gamma = 3$ on DGDELT-RM and $\gamma = 10$ on DICEWS-20RM, DICEWS-RM and YAGOWIKI15K. TEn-TGAEA is set as $\gamma = 7$ on DGDELT7K and YAGOWIKI15K, $\gamma = 5$ on DGDELT-RM and $\gamma = 10$ on DICEWS-20RM and DICEWS-RM.

4.2 Main Results

Table 2, Table 3 and Table 4 show results of TGA-EA and all baselines on five datasets. It is observed that our approach achieves optimal results on all datasets. This demonstrates TGA-EA's effectiveness in the setting of sparse alignment seeds. In the following, we analyze our approach's performances specifically.

Table 2. EA results on DGDELT7K and DGDELT-RM. The best results are written in bold.

Models	DGDELT7K				DGDELT-RM			
	MRR	Hits@1	Hits@5	Hits@10	MRR	Hits@1	Hits@5	Hits@10
MTransE	0.002	0.02	0.10	0.16	0.001	0.02	0.07	0.15
AlignE	0.510	42.15	60.95	67.83	0.321	24.23	40.11	47.36
RREA	0.734	68.27	79.47	82.67	0.336	26.24	41.42	47.87
Dual-AMN	0.712	65.90	77.37	80.70	0.454	37.86	53.75	59.43
TEA-GNN	0.792	76.04	82.71	84.38	0.594	54.25	65.08	68.51
TREA	0.861	82.89	89.74	91.38	0.694	64.25	75.14	78.32
TP-TGAEA	0.894	87.44	91.57	92.72	0.778	73.49	82.72	84.78
TR-TGAEA	0.775	72.54	83.07	86.56	0.633	57.11	70.28	74.97
TEn-TGAEA	**0.960**	**94.84**	**97.31**	**97.76**	**0.873**	**84.56**	**90.52**	**91.65**

Table 3. EA results on DICEWS-20RM and DICEWS-RM.

Models	DICEWS-20RM				DICEWS-RM			
	MRR	Hits@1	Hits@5	Hits@10	MRR	Hits@1	Hits@5	Hits@10
MTransE	0.068	3.68	9.17	12.55	0.046	2.29	5.95	8.56
AlignE	0.321	24.03	40.72	47.69	0.214	14.48	28.04	35.04
RREA	0.703	63.69	78.15	82.43	0.497	41.01	59.63	65.90
Dual-AMN	0.707	64.13	78.28	82.71	0.501	41.49	59.80	66.22
TEA-GNN	0.883	85.09	92.08	93.05	0.743	69.12	80.48	83.13
TREA	0.833	80.01	87.08	89.09	0.694	63.50	76.45	79.55
TP-TGAEA	0.897	87.29	92.55	93.39	0.748	69.96	80.47	83.01
TR-TGAEA	**0.907**	**88.48**	**93.25**	**93.87**	**0.754**	**70.38**	**81.58**	**83.87**
TEn-TGAEA	0.878	84.96	91.23	92.72	0.747	69.95	80.61	83.59

DGDELT7K and DGDELT-RM. In Table 2, TEn-TGAEA achieves the best results on both datasets. Compared with the best baseline TREA, TEn-TGAEA improves Hits@1 by 11.95% and 20.31% on DGDELT7K and DGDELT-RM respectively. TP-TGAEA achieves the second best results on both datasets. Compared to TREA, TP-TGAEA improves Hits@1 by 4.55% and by 9.24% on DGDELT7K and DGDELT-RM respectively. TR-TGAEA performs worse than TREA and TEA-GNN on DGDELT7K, and performs worse than TREA on DGDELT-RM. As Table 1 shows, DGDELT7K and DGDELT-RM are dense in time density (44,625 timestamps for 14,595 to 14,613 entities, time unit = 15 min). Excellent performances of TEn-TGAEA come from temporal encoding, which computes fixed timestamp embeddings and represents time spans between timestamps. Temporal encoding is capable of effectively capturing fine-grained temporal features in datasets of dense time density, like DGDELT7K and DGDELT-RM. Temporal rotation is not able to well model dense timestamps, which results in TR-TGAEA's unsatisfactory performances. Temporal projection is able to represent dense timestamps, which enables TP-TGAEA to perform better than all baselines.

DICEWS-20RM and DICEWS-RM. In Table 3, TR-TGAEA achieves the best results on both datasets. Compared to the best baseline TEA-GNN, TR-TGAEA improves Hits@1 by 3.39% on DICEWS-20RM and by 1.26% on DICEWS-RM. The second best is TP-TGAEA, which improves Hits@1 by 2.20% on DICEWS-20RM and by 0.84% on DICEWS-RM compared to TEA-GNN. Compared to TEA-GNN, TEn-TGAEA performs slightly worse (Hits@1 by -0.13%) on DICEWS-20RM and slightly better (Hits@1 by 0.83%) on DICEWS-RM. In Table 1, DICEWS-20RM and DICEWS-RM are medium in time density (4,017 timestamps for 19,233 entities, time unit = 1 day). Both temporal rotation and temporal projection are able to well model medium timestamps. Therefore, TR-TGAEA and TR-TGAEA achieve the best and the second best performances. TEn-TGAEA cannot gain obvious advantages on such datasets, since temporal encoding prefers dense timestamps.

Table 4. EA results on YAGOWIKI15K.

Models	YAGOWIKI15K			
	MRR	Hits@1	Hits@5	Hits@10
MTransE	0.056	2.17	7.28	11.52
AlignE	0.018	0.81	2.45	3.61
RREA	0.453	35.94	55.43	63.41
Dual-AMN	0.502	42.17	58.91	65.64
TEA-GNN	0.620	54.60	70.43	76.12
TREA	0.686	62.91	75.01	79.02
TP-TGAEA	**0.728**	**67.16**	**79.38**	**82.97**
TR-TGAEA	0.638	57.24	71.53	76.35
TEn-TGAEA	0.492	41.12	58.13	64.74

YAGOWIKI15K. In Table 4, TP-TGAEA achieves the best performance, with a 4.25% Hits@1 improvement over the best baseline TREA. TR-TGAEA performs worse than TREA but better than other baselines. TEn-TGAEA has no advantage on YAGOWIKI15K. As Table 1 shows, YAGOWIKI15K is sparse in time density (404 timestamps for 29,728 entities, time unit = 1 year). Besides, it contains a certain amount of non-temporal facts, while other four datasets only contain temporal facts. Temporal projection is able to effectively model sparse timestamps, resulting in TP-TGAEA's good performance. Considering TP-TGAEA's general performances on all five datasets, we can say that temporal projection is of universal applicability. Temporal encoding cannot work well on sparse timestamps, where its generated fixed sparse timestamps embeddings cannot be well distinguished. For instance, given two close timestamps 1998 and 1999 (time unit = 1 year), temporal encoding computes with Eq. (15) to generate two similar embeddings (with the time span constraint), while temporal projection generates two different embeddings.

Overall Conclusions. According to overall performances on five datasets, we make following conclusions. TGA-EA generally demonstrates good overall performances with sparse alignment seeds. TEn-TGAEA is suitable for datasets with dense timestamps, as Table 2. TR-TGAEA is suitable for datasets with medium timestamps, as Table 3. TP-TGAEA is robust compared to TR-TGAEA and TEn-TGAEA, since it performs better than all baselines on all five datasets.

4.3 Ablation Study

We conduct ablation experiments for TR-TGAEA and TP-TGAEA respectively. We replace temporal-relational entity embeddings with temporal entity embeddings in Eq. (5) (TGA aggregation and update), where $TRF(TM(t^l, \tau), TM(r, \tau)) \rightarrow TM(t^l, \tau)$, denoted as "-R". We replace temporal-relational entity embeddings with relational entity embeddings in Eq. (5), where $TRF(TM(t^l, \tau), TM(r, \tau)) \rightarrow TRF(t^l, r)$, denoted as "-T". We replace temporal relation embeddings with concatenation of temporal head entity embeddings, temporal tail entity embeddings and temporal relation embeddings in Eq. (3) (time-aware graph attention computation), where $TM(r, \tau) \rightarrow [TM(h^l, \tau)||TM(t^l, \tau)||TM(r, \tau)]$, denoted as "w.hrt". "w.hrt" increases the vector dimension from d to $3d$, which results in parameter size increase and memory cost increase. The ablation results are reported in Table 5, Table 6 and Table 7.

Original approaches always greatly outperform their "-T" variants, which indicates that temporal modeling does play a key role in TGA aggregation and update. However, there are no clear winners between original approaches and their "-R" variants. Original approaches perform better in some scenarios, while their "-R" variants win in other scenarios. Recall that temporal relation embeddings are utilized in time-aware graph attention computation, according to Eq. (3), which explains why temporal relation information is not always necessary in TGA aggregation and update.

Original approaches and their "w.hrt" variatnts present similar performances overall, except TR-TGAEA(w.hrt). TR-TGAEA(w.hrt) is set as dimension $d = 75$ on DGDELT-RM and $d = 50$ on DGDELT7K, because TR-TGAEA(w.hrt) cannot run with the default

Table 5. Results for ablation experiments on DGDELT7K and DGDELT-RM.

Models	DGDELT7K				DGDELT-RM			
	MRR	Hits@1	Hits@5	Hits@10	MRR	Hits@1	Hits@5	Hits@10
TP-TGAEA	0.894	87.44	91.57	92.72	0.778	73.49	82.72	84.78
TR-TGAEA	0.775	72.54	83.07	86.56	0.633	57.11	70.28	74.97
TP-TGAEA(-R)	0.894	87.41	91.71	92.87	0.773	72.91	82.39	84.55
TP-TGAEA(-T)	0.891	86.96	91.52	92.79	0.773	72.83	82.39	84.45
TP-TGAEA(w.hrt)	0.895	87.44	91.94	93.09	0.774	72.97	82.66	84.54
TR-TGAEA(-R)	0.816	77.16	86.68	89.59	0.616	55.11	68.74	73.91
TR-TGAEA(-T)	0.642	58.41	70.67	74.72	0.627	56.94	69.28	73.25
TR-TGAEA(w.hrt)*	0.654	59.60	71.90	76.39	0.451	39.52	50.77	55.28

Table 6. Results for ablation experiments on DICEWS-20RM and DICEWS-RM.

Models	DICEWS-20RM				DICEWS-RM			
	MRR	Hits@1	Hits@5	Hits@10	MRR	Hits@1	Hits@5	Hits@10
TP-TGAEA	0.897	87.29	92.55	93.39	0.748	69.96	80.47	83.01
TR-TGAEA	0.907	88.48	93.25	93.87	0.754	70.38	81.58	83.87
TP-TGAEA(-R)	0.892	86.66	92.33	93.21	0.740	68.92	79.80	82.70
TP-TGAEA(-T)	0.883	85.59	91.52	92.50	0.731	68.02	78.99	81.84
TP-TGAEA(w.hrt)	0.892	86.68	92.04	93.06	0.749	69.99	80.52	83.19
TR-TGAEA(-R)	0.905	88.21	93.04	93.78	0.749	69.71	81.13	83.61
TR-TGAEA(-T)	0.823	78.31	86.87	89.25	0.672	61.37	73.74	78.00
TR-TGAEA(w.hrt)	0.907	88.46	93.42	94.01	0.748	69.53	81.06	83.55

dimension (memory exhaust). Dimension reductions lead to heavily performance declining. Therefore, TGA-EA's temporal relation based time-aware graph attention works well with a light-weight set of parameters, which reduces memory cost.

4.4 Influence of Seed Size

To explore seed size's influence on TGA-EA, we range the number of alignment seeds from 100 to 500, with 100 as the step length. Figure 4 and Fig. 5 present results with seed size ranging. TGA-EA has the greatest advantages over the baseline TREA when the seed size is 100. With seed size increasing, positive gaps between TGA-EA (partial or all variants) and TREA keep on narrowing. Especially, in some best cases, TGA-EA variants with $|S| = 100$ outperform TREA with $|S| = 500$: TEn-TGAEA on DGDELT7K and DGDELT-RM, TP-TGAEA and TR-TGAEA on DICEWS-20RM. In a nutshell,

Table 7. Results for ablation experiments on YAGOWIKI5K.

Models	YAGOWIKI15K			
	MRR	Hits@1	Hits@5	Hits@10
TP-TGAEA	0.728	67.16	79.38	82.97
TR-TGAEA	0.638	57.24	71.53	76.35
TP-TGAEA(-R)	0.722	66.65	78.55	82.28
TP-TGAEA(-T)	0.705	64.51	77.56	81.39
TP-TGAEA(w.hrt)	0.664	59.54	74.34	79.09
TR-TGAEA(-R)	0.651	58.55	72.81	77.25
TR-TGAEA(-T)	0.619	54.24	70.78	76.32
TR-TGAEA (w.hrt)	0.638	57.36	71.24	75.91

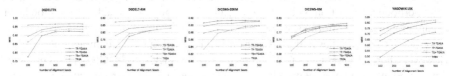

Fig. 4. MRR of TREA, TR-TGAEA, TP-TGAEA and TEn-TGAEA on entity alignment, w.r.t. number of alignment seeds $|S|$.

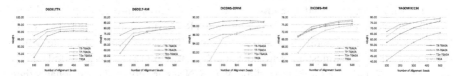

Fig. 5. Hits@1 of TREA, TR-TGAEA, TP-TGAEA and TEn-TGAEA on entity alignment, w.r.t. number of alignment seeds $|S|$.

TGA-EA is able to work well with sparse alignment seeds. In real scenarios, users expect to perform EA with sparse alignment seeds, and our approach is a good choice.

Specifically, TP-TGAEA and TEn-TGAEA consistently outperform TREA on DGDELT7K and DGDELT-RM. On DICEWS-20RM, TP-TGAEA and TR-TGAEA consistently outperform TREA; TEn-TGAEA outperforms TREA in the early stage, and they have similar performances in the late stage. On DICEWS-RM, TR-TGAEA, TP-TGAEA and TEn-TGAEA consistently outperform TREA. On YAGOWIKI15K, TP-TGAEA achieves similar overall performances to TREA, and slightly outperforms TREA with small seed sizes.

5   Related Work

Entity alignment approaches are divided into two main categories, translation based and GNN-based. Translation based approaches learn entity embeddings through TransE or its variants [1], which are used for alignment. MTransE [3] learns entity embeddings on each KG with TransE respectively, and then transforms entity embeddings into a unified embedding space. IPTransE [5] iteratively labels new alignment entities for training. BootEA [4] also adopts the iterative strategy, and proposes a ϵ-truncated uniform negative sampling method. GNN-based approaches usually learn entity embeddings in the same embedding space through parameter sharing. GCN-Align [9] is the first approach to apply GCN to EA, and is trained only with pre-aligned entities. MRAEA [6] explores and fuses three types of meta-relational knowledge (relation type, relation direction and inverse relation) to learn entity embeddings. RREA [7] proposes the relational reflection transformation for neighbor entity information aggregation. In this way, it learns relational entity embeddings, where the distribution of the original embedding space is kept. Dual-AMN [8] constructs a simplified relation attention layer to capture relation based neighborhood information within each KG, and builds an agent matching attention layer to capture cross-KG information.

Previous TKG studies majorly focus on TKG embeddings and completions [10–12]. However, there are relatively few studies on EA for TKGs. TEA-GNN [13] computes time based attention and relation based attention respectively, where orthogonal transformation matrices are utilized to process timestamps and relations. Then it aggregates neighborhood information with both attentions. TREA [14] computes time based attention similar to GAT, and aggregates neighborhood information with both time based attention and relation based attention. Also, it uses a margin-based full multi-class log-loss for efficient training. In TEA-GNN and TREA, timestamps are treated similar to relations, where temporal characteristics are not fully considered; also, interactions between relations and timestamps are not enough. Tem-EA [15] learns temporal embeddings via RNN and structural embeddings via GCN separately, and concatenates them for alignment in the end.

6   Conclusion

For TKG alignment, we propose a novel temporal enhanced graph attention network, TGA-EA. We propose three methods (temporal projection, temporal rotation and temporal encoding) to model timestamps and temporlize entities and relations. Then, with temporal entities and temporal relations, we build an attention based GNN to learn temporal-relational entity embeddings for alignment. Our approach is capable of well aligning TKGs with sparse seeds, owing to effective temporal modeling and systematical temporal integration into TKG embedding. We conduct sufficient experiments on five datasets, and demonstrate TGA-EA's effectiveness.

Acknowledgements. This work is supported by the National Natural Science Foundation of China (Grant Nos. 62002262, 62172082, 62072086, 62072084).

References

1. Wang, Q., Mao, Z., Wang, B., Guo, L.: Knowledge graph embedding: A survey of approaches and applications. IEEE Trans. Knowl. Data Eng. **29**(12), 2724–2743 (2017)
2. Sun, Z., et al.: A benchmarking study of embedding-based entity alignment for knowledge graphs. arXiv preprint arXiv:2003.07743 (2020)
3. Chen, M., Tian, Y., Yang, M., Zaniolo, C.: Multilingual knowledge graph embeddings for cross-lingual knowledge alignment. In: IJCAI, pp. 1511–1517 (2017)
4. Sun, Z., Hu, W., Zhang, Q., Qu, Y.: Bootstrapping entity alignment with knowledge graph embedding. In: IJCAI, pp. 4396–4402 (2018)
5. Zhu, H., Xie, R., Liu, Z., Sun, M.: Iterative entity alignment via knowledge embeddings. In: IJCAI, pp. 4258–4264 (2017)
6. Mao, X., Wang, W., Xu, H., Lan, M., Wu, Y.: MRAEA: an efficient and robust entity alignment approach for cross-lingual knowledge graph. In: WSDM, pp. 420–428 (2020)
7. Mao, X., Wang, W., Xu, H., Wu, Y., Lan, M.: Relational reflection entity alignment. In: CIKM, pp. 1095–1104 (2020)
8. Mao, X., Wang, W., Wu, Y., Lan, M.: Boosting the speed of entity alignment 10×: Dual attention matching network with normalized hard sample mining. In: WWW, pp. 821–832 (2021)
9. Wang, Z., Lv, Q., Lan, X., Zhang, Y.: Cross-lingual knowledge graph alignment via graph convolutional networks. In: EMNLP, pp. 349–357 (2018)
10. Cai, B., Xiang, Y., Gao, L., Zhang, H., Li, Y., Li, J.: Temporal knowledge graph completion: A survey. arXiv preprint arXiv:2201.08236 (2022)
11. Dasgupta, S.S., Ray, S.N., Talukdar, P.P.: HyTE: hyperplane-based temporally aware knowledge graph embedding. In: EMNLP, pp. 2001–2011 (2018)
12. Sadeghian, A., Armandpour, M., Colas, A., Wang, D.Z.: ChronoR: rotation based temporal knowledge graph embedding. In: AAAI, pp. 6471–6479 (2021)
13. Xu, C., Su, F., Lehmann, J.: Time-aware Graph Neural Networks for Entity Alignment between Temporal Knowledge Graphs. In: EMNLP, pp. 8999–9010 (2021)
14. Xu, C., Su, F., Xiong, B., Lehmann, J.: Time-aware Entity Alignment using Temporal Relational Attention. In: WWW, pp. 788–797 (2022)
15. Song, X., Bai, L., Liu, R., Zhang, H.: Temporal Knowledge Graph Entity Alignment via Representation Learning. In: DASFAA, pp. 391–406 (2022)
16. Vaswani, A., et al.: Attention is All you Need. arXiv preprint arXiv:1706.03762 (2017)

Learning Semantic-Rich Relation-Selective Entity Representation for Knowledge Graph Completion

Zenan Xu[1], Zexuan Qiu[2], and Qinliang Su[1,3(✉)]

[1] School of Computer Science and Engineering, Sun Yat-sen University, Guangzhou, China
`xuzn@mail2.sysu.edu.cn`, `suqliang@mail.sysu.edu.cn`
[2] The Chinese University of Hong Kong, Hong Kong SAR, China
`qzexuan@link.cuhk.edu.hk`
[3] Guangdong Key Laboratory of Big Data Analysis and Processing, Guangzhou, China

Abstract. Many existing knowledge graph embedding methods learn semantic representations for entities by using graph neural networks (GNN) to harvest their intrinsic relevances. However, these methods mostly represent every entity with one coarse-grained representation, without considering the variation of the semantics of an entity under the context of different relations. To tackle this problem, we propose a method to learn multiple representations of an entity, with each emphasizing one specific aspect of information contained in the entity. During training and testing, only the representation that is most relevant to the considered relation is selected and shown to the model, leading to the relation-selective representations. To enable the selection of representations according to the relation, we first propose to incorporate a relation-controlled gating mechanism into the original GNN, which is used to decide which and how much information can flow into the next updating stage of the GNN. Then, a mixture of relation-level and entity-level negative sample generation methods is further developed to enhance semantic information contained in relation-selective entity representations under the framework of contrastive learning. Experiments on three benchmarks show that our proposed model outperforms all strong baselines.

Keywords: Knowledge graph completion · Graph representation learning · Contrastive learning

Zenan Xu and Zexuan Qiu contribute equally. Zexuan Qiu did this work when he was a student at Sun Yat-sen University.

This work is supported by the National Natural Science Foundation of China (No. 62276280, U1811264), Key R&D Program of Guangdong Province (No. 2018B010107005), Natural Science Foundation of Guangdong Province (No. 2021A1515012299), Science and Technology Program of Guangzhou (No. 202102021205).

© The Author(s), under exclusive license to Springer Nature Switzerland AG 2023
X. Wang et al. (Eds.): DASFAA 2023, LNCS 13944, pp. 656–672, 2023.
https://doi.org/10.1007/978-3-031-30672-3_44

1   Introduction

Knowledge graphs (KGs) are widely used to capture the complex relations of entities in the real world. In practice, a KG is generally described by a huge amount of triplets of the format $< head, relation, tail >$, with each one indicating the existence of a link between the head and tail entities under the relation specified in the triplet. However, due to the inherent complexity and huge amount of links in KGs, it is unrealistic to expect a KG to contain almost all of the true links among entities. In fact, it has been reported that a large number of links that should exist between entities are missing, as observed in real-world KGs like NELL [2] and YAGO3 [13]. To enable KGs to serve downstream applications better, it is thus of practical importance to predict and complete the missing links for existing KGs. In practice, the KG completion (KGC) task is often reduced to predict the missing head or tail entity for an incomplete triplet of the form $<?, relation, tail >$ and $< head, relation, ? >$.

To complete the KGs, a widely used approach is to learn a low-dimensional meaningful representation for every entity and relation so that the relevance of entities and relations in KGs can be well captured by these low-dimensional representations. For instance, TransE [1] proposes to force the learned representations to satisfy the translation invariant property (*i.e.*, $\mathcal{H}(e_h) + \mathcal{H}(r) \approx \mathcal{H}(e_t)$) for all ground-truth triplets $< e_h, r, e_t >$, where $\mathcal{H}(\cdot)$ denotes the representation of an entity or relation. Differently, ConvE [5] proposes to estimate a matching score $s(e_h, r, e_t)$ for any triplet candidate $< e_h, r, e_t >$, with the score encouraged to be high for true triplets, while low for false ones. Recently, to take advantage of the broader neighborhood information beyond simple triplets in KGs, various kinds of graph neural networks (GNNs), like graph attention network (GAT) in [14], weighted graph convolutional network (WGCN) in [19], and heterogeneous relation attention network (HRAN) in [10], have been employed to tackle this task. However, in all these methods, only one low-dimensional representation is learned for an entity. For instance, in the task of predicting the tail entity in $< e_h, r, ? >$, no matter what the relation r is, the representation of head entity e_h is always kept identical. However, to facilitate the prediction, we argue that it is better to have the entity e_h maintaining multiple representations and show different representations to the model according to the specific value of r, *i.e.*, letting the representation of entity e_h shown to the model vary according to the value of relation r.

To see why it is preferable to have an entity represented by multiple vectors, let us take the following example as an illustration. The entity *MichaelJordan*, who was born in Brooklyn in 1963, contains information about the entity's date of birth and place of birth at the same time. In the task of predicting the tail entity for the missing triplet $< MichaelJordan, YearOfBirth, ? >$, if the entity maintains multiple representations, we can show to the model the representation that is most relevant to the relation *YearOfBirth*, which obviously could make the correct prediction for the missing entity *1963* easier. Thus, it is beneficial to have an entity showing different representations under the contexts of different relations, which, however, is not allowable in existing methods.

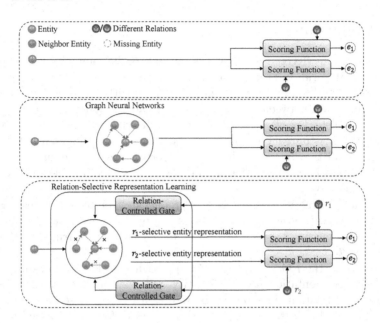

Fig. 1. A simple illustration of different kinds of KGE methods. From top to bottom are conventional methods, existing GNN-based approaches that learn one coarse-grained representation for each entity, and our model to learn relation-selective entity representations.

In this paper, we propose a method to learn **Re**lation-**se**lective **E**ntity representations (*i.e.*, ResetE), which enables an entity's representation shown to the model varying according to the specific value of the relation r. Figure 1 visually illustrates the difference between our proposed model and previous ones. Specifically, a relation-controlled gating mechanism is first developed, which can explicitly control which type of information is allowed to flow into the entity representation during the aggregation step of GNNs. Because of the key role played by the relation representation in the gating mechanism, it is important to ensure the quality of the learned representations of relations. To this end, a similarity-preserving relation representation learning method that encourages similar relations (*e.g.*, *PlaceOfBorn* and *PlaceOfResidence*) to share similar representations is developed. Furthermore, to improve the quality of entity representations, we further propose to use contrastive learning to enhance the semantic information in our learned relation-selective entity representation, in which a novel two-level negative sample generation method is developed. Extensive experiments are conducted on three benchmarks for the knowledge graph completion task. Experimental results on real-world datasets show that our ResetE outperforms all strong baselines, corroborating the effectiveness of proposed modules.

2 Preliminary

Thanks to the strong ability to learn commonalities of adjacent nodes for graph-structured data, graph neural networks (GNN) have been widely used to learn the entity representations of knowledge graphs in recent years [10,14,19]. The GNN-based models generally share the same architecture of using a GNN to learn the entity representation and then applying a score function to evaluate the matching score of a triplet $<$ head entity, relation, tail entity $>$ to be tested. Due to the similarities shared among these methods, here we take the SACN [19] as an example to illustrate the basic principles behind the GNN-based entity representation learning methods.

By viewing the KG as an entity graph G_e, with each node and edge representing an entity and a relation, respectively, SACN applies a L-layer weighted graph convolutional network onto the entity graph G_e and obtains the entity representations as

$$z_i^l = \sigma \left(\sum_{j \in \mathcal{N}_e(i)} \alpha_{i,j} z_j^{l-1} W^{l-1} + z_i^{l-1} W^{l-1} \right), \tag{1}$$

where $\ell = 1, 2, \cdots L$ denotes the ℓ-th layer of GNN; $\mathcal{N}_e(i)$ represents the neighbors of entity i in graph G_e; z_i^ℓ denotes the embedding of i-th entity e_i obtained at the ℓ-th layer, with the initial embedding $z_i^0 \in \mathbb{R}^{d_e}$ initialized from random Gaussian noise; $W^l \in \mathbb{R}^{d_e \times d_e}$ is the network parameter at ℓ-th layer; the coefficient $\alpha_{i,j}$ is used to control the interaction strength between node i and j; and $\sigma(\cdot)$ is the sigmoid activation function. z_i^L output from the L-th layer is then used to represent the final embedding of the i-th entity e_i, that is,

$$z_i = z_i^L. \tag{2}$$

Besides the entity embedding z_i, SACN also learns an embedding for every relation r. For the k-th relation r_k, its embedding $h_k \in \mathbb{R}^{d_e}$ is directly initialized with a random Gaussian vector.

With the entity embeddings z_i and relation embeddings h_k obtained above, for a given triplet $< e_i, r_k, e_j >$, the SACN computes a matching score for the triplet, with the scoring function defined as

$$\varphi(e_i, r_k, e_j) = CNN\left([z_i; h_k]\right) W^c z_j^T, \tag{3}$$

where $CNN(\cdot)$ denotes a convolutional network applied to a $2 \times d_e$ matrix $[z_i; h_k]$. The model then predicts the probability that the triplet $< e_i, r_k, e_j >$ is a true triplet as

$$p(True\ Triplet|e_i, r_k, e_j) = \sigma(\varphi(e_i, r_k, e_j)). \tag{4}$$

Obviously, the probability that the input $< e_i, r_k, e_j >$ is a false triplet is $p(False\ Triplet|e_i, r_k, e_j) = 1 - \sigma(\varphi(e_i, r_k, e_j))$. The model parameters and initial embeddings of relations can be optimized by minimizing the following cross-entropy loss

$$\mathcal{L}_c = -\frac{1}{N} \sum_{n=1}^{N} (y_n \log p_n + (1 - y_n) \log(1 - p_n)) \tag{5}$$

where p_n denotes the probability output from (4) when the n-th triplet $<e_h, r, e_t>$ is fed into the prediction model; and y_n denotes whether input triplet is true or false. Here, each triplet is either drawn from the ground-truth pool in KG or is fabricated intentionally for training.

Limitations of Existing Methods. From the methods introduced above, we can see that only one representation z_i is learned for the i-th entity e_i. To facilitate the prediction of the missing entity in $<e_h, r, ?>$ or $<?, r, e_t>$, we argue that it is better to have every entity e to be represented by multiple representations, letting each emphasizing on different aspects of the entity. When the representation is needed to use, the one that is most relevant to the relation r is selected and then fed into the model to predict the missing entity. Specifically, we propose to learn for every entity e_i a set of representations denoted by

$$\{z_i(r)\}_{r \in \mathcal{R}_i}, \tag{6}$$

with $z_i(r)$ denoting the representation of entity e_i under relation r, where $\mathcal{R}_i$ means the set of all possible relations associated with the entity e_i. Therefore, the entity e_i will maintain multiple representations, with each corresponding to a specific relation from the set $\mathcal{R}_i$, which makes a striking contrast with existing methods, in which only one representation z_i is learned for the i-th entity e_i. Notably, the number of possible relations associated with an entity is often small (several or tens at most), thus the number of representations that need to be learned for an entity is often limited.

3   Methodologies

In this section, we first present the learning framework of relation-selective entity representations, then introduce a novel contrastive learning method, featuring the use of the mixture of relation-level and entity-level negative samples to enhance the semantic information of entity representations.

3.1   Relation-Selective Entity Representation Learning Framework

To enable the entity representation shown to the model varying according to the specific relation r that the entity is interacting with, we propose to represent every entity with multiple vectors $\{z_i(r)\}_{r \in \mathcal{R}_i}$, in which the vector $z_i(r)$ denotes the representation of entity e_i under the relation r. To learn the representations $\{z_i(r)\}_{r \in \mathcal{R}_i}$, we propose to update the relation-selective representation under the GNN framework as

$$z_i^l(r) = f(\boldsymbol{h}_r, z_i^{l-1}(r))\boldsymbol{W}^{l-1} + \frac{1}{|\mathcal{N}_e(i)|} \sum_{j \in \mathcal{N}_e(i)} f(\boldsymbol{h}_r, z_j^{l-1}(r))\boldsymbol{W}^{l-1}, \tag{7}$$

where $z_i^{\ell}(r)$ denotes the intermediate representation of entity e_i from the ℓ-th layer under the relation r; $\boldsymbol{h}_r$ is the representation of relation r; and $f(\cdot, \cdot)$

denotes the gating function that controls how much information is allowed to flow into the next layer. Notably, only the representations corresponding to the relation r, i.e., $z_j^{\ell-1}(r)$, are incorporated into the process of updating $z_i(r)$, without utilizing the entity representations under other relation types. For the gating function $f(\cdot, \cdot)$ in (7), it is designed as

$$f(h_r, z_i^{l-1}(r)) = \sigma\left(W^f h_r + g^f\right) \odot z_i^{l-1}(r), \tag{8}$$

where $W^f \in \mathbb{R}^{d_e \times d_r}$, $g^f \in \mathbb{R}^{d_e}$ are parameters to be learned, $\odot$ is the feature-wise product. The final entity representation $z_i(r)$ is obtained by applying the sigmoid function σ to output of the last layer

$$z_i(r) = \sigma(z_i^L(r)). \tag{9}$$

It can be observed from the definition of gating function (8) that it can control which dimension's information in $z_i^{\ell-1}$ can flow into neighboring nodes. If the relevance between the relation and an entity is weak, the $\sigma(\cdot)$ function will output a value close to zero, cutting off the information flow to the entity's neighbors.

Given the entity and relation representations above, the matching score for the triplet $< e_i, r_k, e_j >$ is computed as

$$\varphi(e_i, r_k, e_j) = CNN([z_i(r_k); h_k]) W^c z_j^T(r_k), \tag{10}$$

where the only difference from the score in (3) is that the representations of head and tail entity e_i and e_j are selected from the representation sets $\{z_i(r)\}_{r \in \mathcal{R}_i}$ and $\{z_j(r)\}_{r \in \mathcal{R}_j}$ according to the provided relation r_k. With the score $\varphi(e_i, r_k, e_j)$, we can then use it to predict the truth probability of a triplet as in (4) and use the loss function in (5) to train the model. The reason why the scheme of using relation to select the representation is feasible is that the relation is always provided during the testing, no matter whether the problem is to predict the missing head or tail entity. Also, as mentioned before, the number of possible relations associated with an entity is often small (several or tens at most), making the number of representations that need to be learned for an entity also limited.

The motivation behind the introduction of the relation-controlled gating function above into the GNN is that KGs are usually densely connected [12], making a GNN-based encoder prone to aggregate from its neighbors much irrelevant information $w.r.t.$ the interest relation. Thus, to address this issue, we propose to let the relation filter the irrelevant information and only allow the information relevant to the relation to flow into the next layer. Notably, the proposed relation-selective representation learning framework is fundamentally different from the well-known relational graph convolutional network (R-GCN) in [18], in which there is no relation-controlled gating mechanism and only one representation is learned for every entity.

Similarity-Preserving Relation Representation Learning. The relation dependence in the proposed entity representations is achieved by incorporating the relation representations h_k into the entities' representation updating process

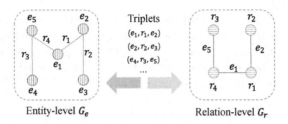

Fig. 2. The illustration of our graph construction methods of G_r and G_e.

through a gating mechanism. However, the relation representations used in the gating function (8) do not contain any correlation information among different relations as they are directly obtained from their initial embeddings without going through any information exchanging process. In practice, different relations are related, rather than isolated, to each other. For example, in KG, the relation *PlaceOfBorn* and *PlaceOfResidence* are both related to the city entity, suggesting they should share some common semantic information in their representations. To have the relation representations reflecting this kind of similarities, we propose to construct a relational graph G_r from the KG by representing every relation as a node and adding an edge between two relations if they refer to a common entity, as illustrated in Fig. 2. With the relation graph G_r, we can now apply the graph neural networks (*e.g.*, GCN) on the graph to obtain relation representations

$$h_r^l = \sigma \left(\sum_{j \in \mathcal{N}_r(r)} h_j^{l-1} W_r^{l-1} + h_r^{l-1} W_r^{l-1} \right), \qquad (11)$$

where $\ell = 1, 2, \cdots, L'$ denotes the ℓ-th layer of GCN; the initial embedding h_r^0 is initialized by random Gaussian noise; $\mathcal{N}_r(\cdot)$ denotes the set of the neighbors of relation r in G_r; and $W_r^l \in \mathbb{R}^{d_r \times d_r}$ is the GCN parameter. We set the output $h_r^{L'}$ from the last layer as the final relation representation, that is,

$$h_r = h_r^{L'}. \qquad (12)$$

Thanks to the message-passing process during the learning, the representation of a relation is not isolated anymore but is related to other relations that share common entities. In this way, the common information of different relations or their similar information can be manifested in the learned representations. By substituting the similarity-preserved relation representation (12) into entity representation updating equation (7), the final relation-selective entity representation updating method is obtained.

3.2 Enhancing Semantics of Entity Representation with Contrastive Learning

The link prediction task is to predict the missing head or tail entity given the other two components. Thus, similar to the classification tasks in images and texts, if more semantic information about entities is preserved in their representations, better prediction performance can be expected. Technically, contrastive learning can be understood as finding pairs of positive and negative instances and then trying to reduce the distance between positive pairs while enlarging that between negative ones under different contrast losses. Among them, the NT-Xent contrast loss below is used most widely

$$l = -\log \frac{\mathcal{D}(\boldsymbol{u}_i^{(1)}, \boldsymbol{u}_i^{(2)})}{\mathcal{D}(\boldsymbol{u}_i^{(1)}, \boldsymbol{u}_i^{(2)}) + \sum\limits_{j \neq i, m=1,2} \mathcal{D}(\boldsymbol{u}_i^{(1)}, \boldsymbol{u}_j^{(m)})},$$

where $\boldsymbol{u}_i^{(m)}$ represents the m-th view of the i-th instance. Different views from the same instances are generally treated as positive pairs, while views from different instances are considered negative pairs. The key to using contrastive learning lies in how to find effective positive and negative pairs, which can determine whether semantic information can be well preserved in the representations. For images, both positive and negative pairs can be easily obtained by applying transformations to the same or different images. However, generating effective positive and negative pairs is not that straightforward for graphs, especially for knowledge graphs that contain additional information of relation.

To generate positive pairs, inspired by the works that apply self-supervised learning on general graphs [27,32,36], we perturb the knowledge graph by randomly dropping some nodes and edges and then apply the aforementioned methods on the perturbed graph to obtain the entities' representations $\boldsymbol{z}_i'$. Then, the representations $\boldsymbol{z}_i$ and $\boldsymbol{z}_i'$ can be viewed as a positive pair. For convenience of presentation, the two representations $\boldsymbol{z}_i$ and $\boldsymbol{z}_i'$ are deemed as two views of entity i, and are denoted as $\boldsymbol{z}_i^{(1)}$ and $\boldsymbol{z}_i^{(2)}$.

As for the generation of negative pairs, a common method is to treat views of other entities as negative samples. However, in order to learn more meaningful semantic information in KG, we suggest collecting negative samples in two different levels, *i.e.*, the relation level and the entity level.

Relation-Level Negative Samples. For a relation-selective entity representation $\boldsymbol{z}_i(r)$, we hope it can retain discriminative semantic information of entity i under the specific relation of r. To strengthen the objective that the semantic information contained in $\boldsymbol{z}_i(r)$ is exclusive to the relation r, we propose to generate negative samples under the same entity by using different relations r' with $r' \neq r$. Specifically, for the representation of entity e_i under the relation r, *i.e.*, $\boldsymbol{z}_i(r)$, its relation-level negative samples is defined to be from the following set

$$\mathcal{Z}_i^{neg}(r) = \left\{ \boldsymbol{z}_i^{(1)}(r'), \boldsymbol{z}_i^{(2)}(r') \middle| r' \neq r \right\}. \tag{13}$$

Entity-Level Negative Samples. For an entity representation $\boldsymbol{z}_i(r)$, in addition, to include exclusive semantic information comparing to entity representations under other relations $\boldsymbol{z}_i(r')$ with $r' \neq r$, it should also contain exclusive

semantic information when compared with other entities. Therefore, we define
the entity-level negative samples of $\boldsymbol{z}_i(r)$ as

$$\widetilde{\mathcal{Z}}_i^{neg}(r) = \left\{ \boldsymbol{z}_j^{(1)}(r), \boldsymbol{z}_j^{(2)}(r) \middle| j \neq i \right\}, \tag{14}$$

where we require the relation in other entities to be the same as the considered
entity. In the implementation, the entity j can just be the other entities from
the same mini-batch.

With the two negative sample sets, we can define the final contrastive learning
loss as

$$\ell_i^{(1)} = -\log \frac{\mathcal{D}_{pos}}{\mathcal{D}_{pos} + \sum\limits_{\boldsymbol{u} \in \mathcal{Z}_i(r)} \mathcal{D}(\boldsymbol{z}_i^{(1)}(r), \boldsymbol{u})}, \tag{15}$$

where $\mathcal{Z}_i(r) \triangleq \mathcal{Z}_i^{neg}(r) \cup \widetilde{\mathcal{Z}}_i^{neg}(r)$; and $\mathcal{D}_{pos} \triangleq \mathcal{D}(\boldsymbol{z}_i^{(1)}(r), \boldsymbol{z}_i^{(2)}(r))$. Here,
$\mathcal{D}(\boldsymbol{z}_i^{(1)}(k), \boldsymbol{z}_i^{(2)}(k))$ is calculated as

$$\mathcal{D}(\boldsymbol{z}_i^{(1)}(k), \boldsymbol{z}_i^{(2)}(k)) = e^{sim(z_i^{(1)}(k), z_i^{(2)}(k))/\tau}, \tag{16}$$

where $sim(\cdot, \cdot)$ denotes the cosine similarity between vectors, and τ is a temperature parameter controlling the concentration level of the distribution [7]. By
averaging over a mini-batch of size N, the final contrastive loss $\mathcal{L}_{cl}$ is

$$\mathcal{L}_{cl} = \frac{1}{2N} \sum_{i=1}^{N} (\ell_i^{(1)} + \ell_i^{(2)}). \tag{17}$$

By minimizing $\mathcal{L}_{cl}$ with both the relation-level and entity-level negative samples, our ResetE can learn an entity representation preserving more meaningful
semantics. Finally, we unify the objective of the KGC task and the contrastive
learning as:

$$\mathcal{L} = \mathcal{L}_c + \lambda \mathcal{L}_{cl}, \tag{18}$$

where λ is a hyper-parameter used to control the trade-off between the loss
function.

4   Experiments

4.1   Experimental Setups

We evaluate the proposed ResetE model on three benchmark datasets from different domains. *1) FB15k-237* [22] contains the knowledge base relation triplets,
including real-world named entities and the relation. The FB15k-237 is the subset of the FB15K [1], which is originally collected from Freebase. Different from
the FB15K, the inverse relations are removed from FB15k-237. *2) WN18RR*
consists of English phrases and the corresponding semantic relations, which is
derived from the WN18 [1]. Similar to FB15k-237, the inverse relations and the

Table 1. The statistics of the three benchmark datasets.

Dataset	Entities	Relations	Train Edges	Dev Edges	Test Edges
FB15k-237	14541	237	272115	17535	20466
WN18RR	40943	11	86835	3034	652
UMLS	135	46	5216	652	661

leaky data are removed from the WN18RR. *3) UMLS* [9], named Unified Medical Language System, is a medical KG dataset. It contains 135 medical entities and 46 semantic relations. Statistics of these three datasets are listed in Table 1.

We evaluate the performance of our ResetE model on the link prediction task, *i.e.*, predicting the missing entity. Given an incomplete triplet, our model takes all the entities as the candidates in the inference phase and outputs the probabilities over all the candidates. Then each candidate is re-ranked according to their probabilities to calculate the Mean rank (MR), Mean reciprocal rank (MRR), and Hits@N. MR is the average of the rankings of entities predicted correctly over all triplets, while MRR targets the average of reciprocal rankings. Hits@N denotes the ratio of those predicted correctly entities which are ranked in top-N. Also, We follow [19] to use the filtered setting [1], which will filter out all valid triplets before ranking. In addition, we follow [21] to adopt the "RANDOM" protocol to handle the situation that the ground-truth triplets have the same scores as the negative triplets, which is caused by the float precision problem. Namely, the rankings of triplets with the same scores will be randomly determined. Especially, as indicated in [28], we avoid comparing with the previous methods (*e.g.*, KBAT, KGBAT, and GAATs), which adopt an inappropriate evaluation protocol and lead to mistaken very high results.

4.2   Experimental Results

The experimental results of our ResetE and the strong baselines on FB15k-237, WN18RR, and UMLS are shown in Table 2. From the table, the proposed ResetE outperforms the strongest baseline HRAN significantly, with relative MRR improvement of 4.5% and 2.3% on FB15K-237 and WN18RR, respectively.

Among all the baselines, SACN is the most similar to our model. SACN and our ResetE both utilize the Conv-TransE model to predict the missing entity, and the main difference is that SACN learns a unique representation for each entity while ResetE learns a relation-selective entity representation instead. It can be seen that our model outperforms SACN by 5.4% and 4.3% in MRR on FB15K-237 and WN18RR, respectively, showing the effectiveness of the relation-selective entity representation.

On UMLS, ResetE shows comparable performance with baselines. However, it is undeniable that ConvE outperforms our model on UMLS under the MR criterion. This may be due to the small size of UMLS, which leads to the over-fitting

Table 2. Performances on FB15k-237, WN18RR, and UMLS datasets. The performances of ConvE on UMLS are taken from the author's Github and are marked with *.

Model	FB15k-237			WN18RR			UMLS	
	Hits			Hits			Hits	
	@10	@1	MRR	@10	@1	MRR	@10	MR
TransE [1]	0.441	0.198	0.279	0.532	0.043	0.243	0.989	1.84
DistMult [33]	0.446	0.199	0.281	0.504	0.412	0.444	0.846	5.52
ComplEx [23]	0.450	0.194	0.278	0.530	0.409	0.449	0.967	2.59
ConvE [5]	0.497	0.225	0.312	0.531	0.419	0.456	0.990*	1.00*
ConvKB [15]	0.421	0.155	0.243	0.520	0.400	0.430	—	—
R-GCN [18]	0.300	0.100	0.164	0.207	0.080	0.123	—	—
RotatE [20]	0.533	0.241	0.338	0.571	0.428	0.476	—	—
SACN [19]	0.536	0.261	0.352	0.535	0.427	0.470	—	—
COMPGCN [25]	0.535	0.264	0.355	0.546	0.443	0.479	—	—
ATTH [3]	0.501	0.236	0.324	0.551	0.419	0.466	—	—
InteractE [24]	0.535	0.263	0.354	0.528	—	0.463	—	—
TorusE [6]	0.484	0.217	0.316	0.512	0.422	0.452	—	—
PairRE [4]	0.544	0.256	0.351	—	—	—	—	—
HRAN [10]	0.541	0.263	0.355	0.542	0.450	0.479	—	—
ResetE (Ours)	**0.562**	**0.275**	**0.371**	**0.555**	**0.460**	**0.490**	**0.993**	1.43
Improvements	**3.3%**	**4.2%**	**4.5%**	**2.4%**	**2.2%**	**2.3%**	—	—

issue when injecting the graph structure information into the entity representation. However, On FB15k-237 and WN18RR with the more complex graph structure, our ResetE outperforms ConvE by 18.9% and 7.5% under the MRR criterion.

4.3   Impacts of Different Components

This section gives a deep insight into how much improvement different components contribute to the model performance. To do this, we evaluate the performance of variants of ResetE that exclude one or more components that significantly impact the performance.

Specifically, three components included in ResetE are considered, and we follow our model's pipeline to describe the three components in turn: (1) Component C. It uses the relation to **C**ontrol the neighborhood information aggregation during the GCN-based encoding stage to generate the relation-selective entity representation. Without it, every entity will be assigned a unique representation instead. (2) Component R. It means the similarity-preserving **R**elation representation learning component, which obtains the relation representation by applying GCN on G_r. Without it, the relation representation degenerates the one ignoring its similar information. (3) Component D. The contrastive learning component with **D**ouble levels of negative samples is designed to enhance the semantics of the relation-selective entity representation. Dropping this com-

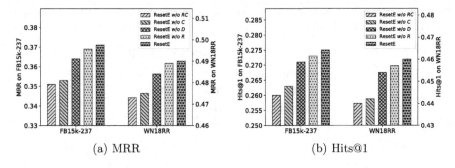

(a) MRR   (b) Hits@1

Fig. 3. Performances of variants of ResetE that exclude one or more components on FB15k-237 and WN18RR.

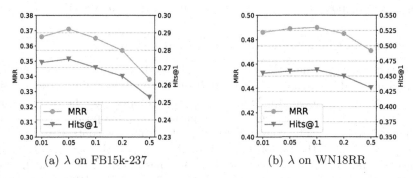

(a) λ on FB15k-237   (b) λ on WN18RR

Fig. 4. MRR and Hits@1 of ResetE under different values of λ on FB15k-237 and WN18RR.

ponent means that we remove the contrastive loss $\mathcal{L}_{cl}$. Please note that the D component is based on the C component; if we drop the C component, the D component will be dropped simultaneously. Based on the above-defined components, we propose four variants of ResetE: ResetE w/o R, ResetE w/o D, ResetE w/o C, and ResetE w/o RC. The four variants are compared with the original ResetE on FB15k-237 and WN18RR, and results are shown in Fig. 3.

From the result, we can have the following observations. First, ResetE w/o C, which removes the most fundamental component C, will induce a significant performance drop compared with the complete ResetE, suggesting the importance of taking the relation into account when learning the entity representation. Second, without using the proposed CL component (*i.e.*, ResetE w/o D), an immediate performance drop is observed on FB15k-237 and WN18RR, which demonstrates the necessity of utilizing the designed CL method to further improve our relation-selective entity representation. Third, ResetE w/o C is better than ResetE w/o RC, demonstrating that even if we solely learn a coarse-grained entity representation as previous methods do, improving the quality of the relation representation can still improve the performance. Also, ResetE w/o R works worse than ResetE, which indicates that similarity-preserving relation representations

Table 3. MRR when using one of the relation-level and entity-level negative samples on FB15-237 and WN18RR.

	FB15-237	WN18RR
w/o Entire $\mathcal{L}_{cl}$	0.364	0.484
+ Relation-Level	**0.369**	**0.488**
+ Entity-Level	0.366	0.486
+ Entire $\mathcal{L}_{cl}$	0.371	0.490

can better control the information aggregation from the entity's neighbors. Last but not least, if we remove all three components (*i.e.*, ResetE w/o *RC*), the performance is poorest, confirming the validity of the proposed ResetE.

4.4   Impacts of Different Levels of Negative Samples

In this section, we evaluate the influence of relation-level and entity-level negative samples in the denominator of (17). MRR on FB15-237 and WN18RR datasets when using one of these two kinds of negative samples are shown in Table 3. Note that no matter which negative samples we use, the positive samples are unchanged and always be considered when calculating the contrastive loss.

We can see that the performances brought by CL with solely relation-level negative samples are more excellent than the ones obtained by CL with solely entity-level negative samples on both FB15k-237 and WN18RR. In this paper, given an entity, CL with relation-level negative samples aims to increase the distances between different relation-selective entity representations of it among different relations. This is consistent with our motivation to learn an entity representation, in which semantics will vary depending on its relation. Therefore, it may explain why CL with relation-level negative samples achieves a more fantastic result. Also, the model performs best if two levels of negative samples are considered together simultaneously, indicating the credibility of the proposed CL method to enhance the entity's semantics.

4.5   Impacts of Parameter λ

In ResetE, we introduce the hyper-parameter λ, which controls the trade-off between the cross-entropy loss and the contrastive loss. In this section, we investigate the sensitivity of λ. We manually select the values of λ from {0.01, 0.05, 0.1, 0.2, 0.5}. MRR and Hits@1 w.r.t λ on FB15k-237 and WN18RR datasets are illustrated in Fig. 4.

It is shown that as λ grows up, the performance of ResetE first increases and reaches the peak when $\lambda = 0.05$ and 0.1 on FB15k-237 and WN18RR, respectively. Afterward, if λ is larger, the improvement is neutralized and lost. This phenomenon shows that the performance is sensitive to the hyper-parameter λ. And in practice, we suggest that the loss weight for the contrastive loss can be set to $[0.01, 0.1]$ to exploit the potentialities of the model.

5   Related Work

Knowledge graph embedding methods play an important role in KGC. Given a triplet $< e_1, r, e_2 >$, TransE [1] learns the representation of the entity and relation according to the translation-based constraint of $e_1 + r \approx e_2$. Later, TransH [31], TransR [11], and TransD [8] extend the translation-based constraint to model more complex features. To learn more expressive representation, ConvE [5] adopts multi-layer CNN architecture to capture the deeper correlation between e_1 and r. Then, ConvKB [15] further extends ConvE to consider the correlation between the entire triplet $< e_1, r, e_2 >$. InteractE [24] introduces more types of interactions between entity and relation in ConvE. In addition there are some other methods, such as DistMult [33], ComplEx [23], STransE [17], RotatE [20], and PairRE [4]. For more details, we refer interested readers to some surveys [16, 29].

However, these methods process each triplet independently, ignoring the neighborhood information inherent in the graph structure of a given entity. To address this, [14, 35] adopt the GAT and GCNs to aggregate neighborhood information, respectively. Similarly, SACN [19] adopts the weighted GCN, which assigns each relation a trainable weight. Further, COMPGCN [25] targets the directed multi-relational KG and proposes to systematically leverage entity-relation composition operations via a GCN-based approach. GAATs [30] integrate an attenuated attention mechanism to better preserve the neighborhood information according to different relation paths. Later, HRAN [10] divides the KG into sub-graph levels, where each sub-graph contains all the entities but only one relation to capture the heterogeneous relation features.

Another approach to KGE is to adopt the Transformer architecture [26], e.g., KG-BERT [34] and StAR [28]. These models adopt the deep Transformer architecture to learn a more meaningful representation and then advance the KGC. However, an obvious issue is that the computational complexity of those model are usually enormous, and they are urgent for substantial computing resources.

Our paper belongs to the category that considers neighborhood information. It can be concluded that the existing GNN-based methods represent each entity solely with one coarse-grained representation, which contains the entire miscellaneous neighborhood information. Unlike them, given incomplete triplets like $< entity, relation, ? >$, we use the *relation* as the guidance to selectively aggregate the neighborhood information into the entity representation. Moreover, we exclusively enhance the relation-semantic information using the proposed contrastive learning method with entity-level and relation-level negative samples.

6   Conclusion

In this paper, we proposed ResetE, a novel knowledge graph embedding method. We introduced the relation-controlled gate mechanism to control the information flow in the aggregation step of the graph neural network and thus obtained relation-selective entity representations. Further, we proposed a contrastive learning method with both relation-level and entity-level negative samples to enhance the meaningful semantic information of entities' representations. Extensive experiments have shown that ResetE significantly outperformed existing baselines.

References

1. Bordes, A., Usunier, N., García-Durán, A., Weston, J., Yakhnenko, O.: Translating embeddings for modeling multi-relational data. In: NIPS (2013)
2. Carlson, A., Betteridge, J., Kisiel, B., Settles, B., Hruschka, E.R., Mitchell, T.M.: Toward an architecture for never-ending language learning. In: AAAI (2010)
3. Chami, I., Wolf, A., Juan, D.C., Sala, F., Ravi, S., Ré, C.: Low-dimensional hyperbolic knowledge graph embeddings. In: Proceedings of the 58th Annual Meeting of the Association for Computational Linguistics, pp. 6901–6914 (2020)
4. Chao, L., He, J., Wang, T., Chu, W.: PairRE: knowledge graph embeddings via paired relation vectors. In: Proceedings of the 59th Annual Meeting of the Association for Computational Linguistics and the 11th International Joint Conference on Natural Language Processing (Volume 1: Long Papers), pp. 4360–4369. Online (Aug 2021)
5. Dettmers, T., Minervini, P., Stenetorp, P., Riedel, S.: Convolutional 2D knowledge graph embeddings. In: AAAI (2018)
6. Ebisu, T., Ichise, R.: Generalized translation-based embedding of knowledge graph. IEEE Trans. Knowl. Data Eng. **32**, 941–951 (2020)
7. Hinton, G.E., Vinyals, O., Dean, J.: Distilling the knowledge in a neural network. arXiv abs/1503.02531 (2015)
8. Ji, G., He, S., Xu, L., Liu, K., Zhao, J.: Knowledge graph embedding via dynamic mapping matrix. In: Proceedings of the 53rd Annual Meeting of the Association for Computational Linguistics and the 7th International Joint Conference on Natural Language Processing (Volume 1: Long Papers), pp. 687–696. Beijing, China (Jul 2015)
9. Kok, S., Domingos, P.M.: Statistical predicate invention. In: ICML '07 (2007)
10. Li, Z., Liu, H., Zhang, Z., Liu, T., Xiong, N.: Learning knowledge graph embedding with heterogeneous relation attention networks. IEEE Trans. Neural Netw. Learn. Syst. **33**(8), 3961–3973 (2021)
11. Lin, Y., Liu, Z., Sun, M., Liu, Y., Zhu, X.: Learning entity and relation embeddings for knowledge graph completion. In: AAAI (2015)
12. Lovelace, J., Newman-Griffis, D., Vashishth, S., Lehman, J.F., Rosé, C.: Robust knowledge graph completion with stacked convolutions and a student re-ranking network. In: Proceedings of the 59th Annual Meeting of the Association for Computational Linguistics and the 11th International Joint Conference on Natural Language Processing (Volume 1: Long Papers), pp. 1016–1029. Online (Aug 2021)
13. Mahdisoltani, F., Biega, J.A., Suchanek, F.M.: YAGO3: a knowledge base from multilingual wikipedias. In: CIDR (2015)
14. Nathani, D., Chauhan, J., Sharma, C., Kaul, M.: Learning attention-based embeddings for relation prediction in knowledge graphs. In: Proceedings of the 57th Annual Meeting of the Association for Computational Linguistics, pp. 4710–4723. Florence, Italy (Jul 2019)

15. Nguyen, D.Q., Nguyen, T.D., Nguyen, D.Q., Phung, D.: A novel embedding model for knowledge base completion based on convolutional neural network. In: Proceedings of the 2018 Conference of the North American Chapter of the Association for Computational Linguistics: Human Language Technologies, Volume 2 (Short Papers), pp. 327–333. New Orleans, Louisiana (Jun 2018)
16. Nguyen, D.Q.: A survey of embedding models of entities and relationships for knowledge graph completion. In: Proceedings of the Graph-based Methods for Natural Language Processing (TextGraphs) (2020)
17. Nguyen, D.Q., Sirts, K., Qu, L., Johnson, M.: STransE: a novel embedding model of entities and relationships in knowledge bases. In: Proceedings of the 2016 Conference of the North American Chapter of the Association for Computational Linguistics: Human Language Technologies, pp. 460–466. San Diego, California (Jun 2016)
18. Schlichtkrull, M., Kipf, T., Bloem, P., van den Berg, R., Titov, I., Welling, M.: Modeling relational data with graph convolutional networks. In: ESWC (2018)
19. Shang, C., Tang, Y., Huang, J., Bi, J., He, X., Zhou, B.: End-to-end structure-aware convolutional networks for knowledge base completion. In: AAAI (2019)
20. Sun, Z., Deng, Z., Nie, J.Y., Tang, J.: Rotate: Knowledge graph embedding by relational rotation in complex space. arXiv abs/1902.10197 (2019)
21. Sun, Z., Vashishth, S., Sanyal, S., Talukdar, P., Yang, Y.: A re-evaluation of knowledge graph completion methods. In: Proceedings of the 58th Annual Meeting of the Association for Computational Linguistics, pp. 5516–5522. Online (Jul 2020)
22. Toutanova, K., Chen, D.: Observed versus latent features for knowledge base and text inference. In: Proceedings of the 3rd Workshop on Continuous Vector Space Models and their Compositionality, pp. 57–66. Beijing, China (Jul 2015)
23. Trouillon, T., Welbl, J., Riedel, S., Gaussier, É., Bouchard, G.: Complex embeddings for simple link prediction. In: ICML (2016)
24. Vashishth, S., Sanyal, S., Nitin, V., Agrawal, N., Talukdar, P.P.: Interacte: improving convolution-based knowledge graph embeddings by increasing feature interactions. In: AAAI (2020)
25. Vashishth, S., Sanyal, S., Nitin, V., Talukdar, P.: Composition-based multi-relational graph convolutional networks. In: International Conference on Learning Representations (2019)
26. Vaswani, A., et al.: Attention is all you need. In: In Advances in Neural Information Processing Systems, pp. 5998–6008 (2017)
27. Velickovic, P., Fedus, W., Hamilton, W.L., Lio', P., Bengio, Y., Hjelm, R.D.: Deep graph infomax. In: ICLR (2019)
28. Wang, B., Shen, T., Long, G., Zhou, T., Chang, Y.: Structure-augmented text representation learning for efficient knowledge graph completion. In: Proceedings of the Web Conference 2021 (2021)
29. Wang, Q., Mao, Z., Wang, B., Guo, L.: Knowledge graph embedding: a survey of approaches and applications. IEEE Trans. Knowl. Data Eng. 29, 2724–2743 (2017)
30. Wang, R., Li, B., Hu, S., Du, W., Zhang, M.: Knowledge graph embedding via graph attenuated attention networks. IEEE Access 8, 5212–5224 (2020)
31. Wang, Z., Zhang, J., Feng, J., Chen, Z.: Knowledge graph embedding by translating on hyperplanes. In: AAAI (2014)
32. Xia, X., Yin, H., Yu, J., Wang, Q., Cui, L., Zhang, X.: Self-supervised hypergraph convolutional networks for session-based recommendation. In: AAAI (2021)
33. Yang, B., tau Yih, W., He, X., Gao, J., Deng, L.: Embedding entities and relations for learning and inference in knowledge bases. CoRR abs/1412.6575 (2015)

34. Yao, L., Mao, C., Luo, Y.: KG-BERT: BERT for knowledge graph completion. arXiv abs/1909.03193 (2019)
35. Yu, D., Yang, Y., Zhang, R., Wu, Y.: Knowledge embedding based graph convolutional network. In: Proceedings of the Web Conference 2021 (2021)
36. Yu, J., Yin, H., Li, J., Wang, Q., Hung, N.Q.V., Zhang, X.: Self-supervised multi-channel hypergraph convolutional network for social recommendation. In: Proceedings of the Web Conference 2021 (2021)

BiQCap: A Biquaternion and Capsule Network-Based Embedding Model for Temporal Knowledge Graph Completion

Sensen Zhang[1], Xun Liang[1(✉)], Zhiying Li[1], Junlan Feng[2], Xiangping Zheng[1], and Bo Wu[1]

[1] Renmin University of China, Beijing, China
{sensen0126,xliang,zhiyingli,xpzheng,wubochn}@ruc.edu.cn
[2] China Mobile Research Institute, Beijing, China
fengjunlan@chinamobile.com

Abstract. Temporal Knowledge Graphs (TKGs) provide a temporal context for facts, capturing temporal information and the dynamic nature of actual world facts. However, typical TKGs often suffer from incomplete dynamics with missing facts in real-world scenarios. Temporal Knowledge Graph Embedding (TKGE) is one of the critical approaches to tackling the challenge. However, the existing TKGE models are weak in simultaneously representing hierarchical semantics and other relation patterns. Therefore, embedding TKGs in a single space, no matter the Euclidean space, or hyperbolic space, cannot capture the complex structures of TKGs accurately. In addition, few existing models have a "deep" architecture for modeling the entries in a quadruple at the same dimension. In this paper, we propose a new TKGE model, **BiQCap**, which for the first time, combines biquaternion and capsule network in modeling to make up for the defects of existing TKGE models. BiQCap represents each temporal entity as a translation and each relation as euclidean rotation and hyperbolic rotation in biquaternions vector space. Further, we employ the embeddings of entities, relations, and temporal trained from biquaternions as the input to capsule networks. Experimental results on five well-known benchmark datasets show that our BiQCap achieves state-of-the-art performance.

Keywords: Temporal Knowledge Graph Embedding · Biquaternion · Capsule Network · Relation Patterns

1 Introduction

Knowledge Graph (KG) is a technical method to describe the relation between knowledge and modeling all things in the world with a graph model. It has essential applications in extensive data analysis, recommendation algorithms, assisted intelligent question answering, and other fields. Several receptive KGs

X. Wang et al. (Eds.): DASFAA 2023, LNCS 13944, pp. 673–688, 2023.
https://doi.org/10.1007/978-3-031-30672-3_45

have been constructed, including FreeBase [4], WordNet [23], conceptNet [28], and DBpedia [21]. However, the existing KGs are often incomplete due to the incompleteness and complexity of data collection.

Researchers have developed various methods to solve this problem to complete the fact that the existing KGs are missing. Knowledge Graph Embedding (KGE) model is one of the main techniques for Knowledge Graph Completion (KGC). It models and inferences various relation patterns in KG by learning the low-dimensional embedding of entities and relations. Currently, the mainstream KGE models [5,6,29,36] is mainly applied to the static KGC.

Different events and actions cause relations and entities to evolve over time. Static KGE models disregard temporal information, leading to an ineffectiveness of performing link prediction on KGs involving temporary relations. Therefore, some researchers have added a temporal factor to the original triplet, constituting Temporal Knowledge Graph (TKG), and built many Temporal Knowledge Graph Embedding (TKGE) models [8,11,27,33]. However, the TKGE models mentioned above use addition, subtraction, or simple multiplication operators and do not employ a "deep" architecture to model attributes of quadruple of the same dimension, thus only capturing the linear relationships between entities. Meanwhile, those models cannot capture the hierarchical relation due to the limitations of Euclidean space. For example, Pluto was a planet, but in 2006 it was removed from the group of nine planets and demoted to dwarf planet status, leaving the solar system with just eight planets. The relation between Pluto and the planets is a typical hierarchical relation that changes over time. Recently, HERCULES [24] projected TKGs into hyperbolic space. However, the hyperbolic models inevitably lose the basic properties of Euclidean space transformations and thus cannot avail themselves of these useful operations.

In this paper, we propose a new TKGE model, BiQCap, in which entity, relation, and temporal are embedded into the biquaternions of k-dimension, temporal is represented as a translation of the head entity, and the relation is represented as rotation in the biquaternions space after translation. The Hamilton product of biquaternions, at the core of BiQCap, imbues it with a solid geometric interpretation that combines both circular and hyperbolic rotations. Additionally, we employ the embeddings of entities, relations, and temporal trained from biquaternions as the input to capsule networks. The capsule network can not only extract features based on statistical information but also interpret features, thereby eliminating the constraint that existing TKGE models do not have the powerful ability to discern the position and direction of entities and relations. The capsule network requires the model to learn the feature variables in the capsule and retain valuable information to the maximum. We adopted a novel inverted dot-product attention routing algorithm for capsule networks in this work. This routing algorithm directly determines the routing probability by the agreements between parent and child capsules. Routing algorithms from prior work require child capsules to be explained by parent capsules. Therefore, our model can achieve superior performance by removing this constraint using a low number of parameters.

Our contributions are summarized as follows: **1)** For the first time, the concept of biquaternions is introduced into the construction of the TKGE model, and the superiority of biquaternions in the representation of the TKGE relation model is analyzed and proved. **2)** The first application of an inverted dot-product attention routing algorithm for capsule networks to construct the TKGE model. **3)** The BiQCap combines biquaternion and capsule network in modeling to compensate for existing TKGE models' defects. **4)** The experiment demonstrates that BiQCap significantly outperforms other models by inferring various relation patterns and encoding temporal information.

2   Related Work

Common KGE models can generally be divided into the static KGE and TKGE models. In this section, we mainly introduced two types of main models.

2.1   Static KGE

The static KGE models have been well studied, and many models have been constructed. TransE [5] is the earliest KGE model, which is a translational distance-based model. TransE [5] embeds the entities h and o, along with relation r, and maps them through the function: $h + r \approx o$. TransE [5] exhibits deficiencies when learning 1-n relations, so there are some models [17,22,31] that extend it. Adopting tensor factorization with a bilinear transformation, semantic matching models, e.g. DistMult [35] and ComplEx [30], capture the semantic relevance of entities. RotatE [29] expands TransE [5] to complex vector space to represent an asymmetric relation pattern, and QuatE [36] expands vector space to quaternions to realize 3D space rotation. DualE [6] further extends the dual quaternion into the embedding of relational entities and successfully represents the non-combinatorial and multiple relation patterns. Some researchers embed vectors into hyperbolic geometric space to express hierarchical relations according to the constructed model [3,7]. In addition, to express the deep communication between entities and relations, some researchers applied deep learning to construct KGE models, such as the ConvKB [18] of one-dimensional convolution and the ConvR [18] of adaptive convolution. CapsE [25] of capsule neural network and RSN [15] of Recurrent Neural Network(RNN). BiQUE [14] recently applied the concept of biquaternion to construct the KGE model for the first time, realizing the combination of rotation and hyperbolic geometry capable of modeling various relation patterns.

2.2   Temporal Knowledge Graph Embedding

In recent years, TKGE models have achieved remarkable results. Some models [9,20,26,34] are extended from TransE [5] and DistMult [35], but these models are just as flawed as models TransE [5] and DistMult [35]. DE-SimplE [13] incorporates temporal information into diachronic entity embeddings and has the capability of modeling various relation patterns. After that, TeRo [33] defines the temporal evolution of entity embedding as a rotation from the initial time to the current time in the complex vector space. ChronoR [27] learns a k-dimensional rotation transformation parametrized by relation and time and captures rich interaction between the temporal and multi-relational characteristics of a TKGE. RotateQVS [8] has introduced quaternions into the construction of TKGE models and defines the temporal evolution of entity embedding as a rotation from the initial time to the current time in the 3D space. Due to the limitations of Euclidean space, the model cannot capture the hierarchical relation. In addition, HERCULES [24] projected TKGs into hyperbolic space. TempCaps [11], the idea of a capsule network is first applied to the construction of the TKGE model, and good experimental results were obtained on the dataset. However, a major drawback of the model is the lack of interpretability.

3   Biquaternion

Endowed with rich algebraic properties, biquaternions represent the properties of geometry and hyperbolic geometry. Some widely used operations of biquaternions are introduced as follows:

1) **Complex Numbers**: The basic algebraic forms of a complex numbers c as follows:

$$c = c_r + c_i\mathbf{I}, \tag{1}$$

where c_r and $c_i \in \mathbb{R}$ are real numbers, $\mathbf{I}$ is the usual imaginary unit and $\mathbf{I}^2 = -1$, $c^* = c_r - c_i\mathbf{I}$ is the standard complex conjugate of a complex number c.

2) **Biquaternion**: The basic form of a biquaternion q as follows: $q = w + x\mathbf{i} + y\mathbf{j} + z\mathbf{k} = (w_r + w_i\mathbf{I}) + (x_r + x_i\mathbf{I})\mathbf{i} + (y_r + y_i\mathbf{I})\mathbf{j} + (z_r + z_i\mathbf{I})\mathbf{k} = q_r + q_i\mathbf{I}$, where $w, x, y, z \in \mathbb{C}$ are $q's$ coefficients, $w_r, x_r, y_r, z_r, w_i, x_i, y_i, z_i \in \mathbb{R}$.

3) **Multiplication**: Imaginary units $\mathbf{i}, \mathbf{j}, \mathbf{k}, \mathbf{I}$ have the following (non)commutative multiplication properties:

$$\mathbf{i}^2 = \mathbf{j}^2 = \mathbf{k}^2 = -1, \ \mathbf{ij} = -\mathbf{ji} = \mathbf{k}, \ \mathbf{jk} = -\mathbf{kj} = \mathbf{i},$$
$$\mathbf{ki} = -\mathbf{ik} = \mathbf{j}, \ \mathbf{iI} = \mathbf{Ii}, \mathbf{jI} = \mathbf{Ij}, \mathbf{kI} = \mathbf{Ik}. \tag{2}$$

4) **Conjugate**: $q = w + x\mathbf{i} + y\mathbf{j} + z\mathbf{k}$, the conjugate of q is denoted as $\overline{q} = w - x\mathbf{i} - y\mathbf{j} - z\mathbf{k}$, and the complex conjugate of q is denoted as $q^* = w^* + x^*\mathbf{i} + y^*\mathbf{j} + z^*\mathbf{k}$, and $\overline{q_1 q_2} = \overline{q_2} \ \overline{q_1}$.

5) **Add and Subtract**: Two biquaternions q_1 and q_2 are added and subtracted as: $q_1 \pm q_2 = (w_1 + x_1\mathbf{i} + y_1\mathbf{j} + z_1\mathbf{k}) \pm (w_2 + x_2\mathbf{i} + y_2\mathbf{j} + z_2\mathbf{k}) = (w_1 \pm w_2) + (x_1 \pm x_2)\mathbf{i} + (y_1 \pm y_2)\mathbf{j} + (z_1 \pm z_2)\mathbf{k}$.

6) **Hamilton Product**: The multiplication q_1, q_2 between q_1 and q_2 can be obtained via standard algebraic distributivity, and it is termed the Hamilton product between q_1 and q_2,

$$q_1 q_2 = w_1 w_2 - x_1 x_2 - y_1 y_2 - z_1 z_2 + (w_1 w_2 + x_1 x_2 + y_1 y_2 - z_1 z_2)\mathbf{i}$$
$$+ (w_1 w_2 - x_1 x_2 + y_1 y_2 + z_1 z_2)\mathbf{j} + (w_1 w_2 + x_1 x_2 - y_1 y_2 + z_1 z_2)\mathbf{k}. \tag{3}$$

7) **Multiplication**: The multiplication can be equivalently represented as a matrix-vector product: $\mathcal{V}(q_1q_2) = \mathcal{M}(q2)\mathcal{V}(q_1)$, or as a matrix-matrix product: $\mathcal{M}(q_1q_2) = \mathcal{M}(q_2)\mathcal{M}(q_1)$. The set of biquaternions and the set of quaternions are both closed under multiplication, and multiplication is associative but not commutative.

8) **Unit Biquaternion**: A unit biquaternion is one with unit norm, i.e.,$\|q\| = 1$, the norm of q is given by $\|q\| = \sqrt{\|q\bar{q}\|} = \sqrt{\|\bar{q}q\|} = \sqrt{w^2 + x^2 + y^2 + z^2}$. We can easily verify that $\mathcal{M}(\bar{q}) = \mathcal{M}(q)^T$ and $\mathcal{M}(q^*) = \mathcal{M}(q)^*$, where $\mathcal{M}(\cdot)^*$ refers to the complex conjugation of each element in the matrix.

We know from the basic form of a biquaternion, $q_r = w_r + x_r\mathbf{i} + y_r\mathbf{j} + z_r\mathbf{k}$, $q_i = w_i + x_i\mathbf{i} + y_i\mathbf{j} + z_i\mathbf{k}$, and $\mathbf{i}, \mathbf{j}, \mathbf{k}$ are imaginary units. We denote the scalar and vector parts of q respectively as $s(q)=w$, $v(q)=x\mathbf{i}+y\mathbf{j}+z\mathbf{k}$, quaternion, complex numbers and real numbers are both special cases of biquaternion.

The biquaternion q has several equivalent representations [14,16,32]: (1) as the vector $\mathcal{V}(q) = [w, x, y, z]^T$; (2) as $\|q\|(\cos\theta + \mu\sin\theta)$(where $\theta = \cos^{-1}(w/\|q\|)$, $\theta \in \mathbb{C}, \mu = v(q)/\|v(q)\|$, and $\|q\| = \sqrt{w^2 + x^2 + y^2 + z^2}$); and (3) as the matrix

$$\mathcal{M}(q) = \begin{bmatrix} w & -x & -y & -z \\ x & w & z & -y \\ y & -z & w & x \\ z & y & -x & w \end{bmatrix}. \tag{4}$$

Theorem 1. *A biquaternion unifies both circular and hyperbolic rotations in $\mathbb{C}^4$ space within a single representation.*

Proof. If $q = q_r + q_i\mathbf{I}$ is a unit biquaternion ($\|q\| = 1$),

$$\begin{aligned} \|q\|^2 &= (w_r + w_i\mathbf{I})^2 + (x_r + x_i\mathbf{I})^2 + (y_r + y_i\mathbf{I})^2 + (z_r + z_i\mathbf{I})^2 \\ &= (w_r^2 - w_i^2 + x_r^2 - x_i^2 + y_r^2 - y_i^2 + z_r^2 - z_i^2) \\ &\quad + 2(w_rw_i + x_rx_i + y_ry_i + z_rz_i)\mathbf{I}. \end{aligned} \tag{5}$$

Since $\|q\|=1$ is real, we know that the above imaginary part $(w_rw_i + x_rx_i + y_ry_i + z_rz_i =0)$.

$$\begin{aligned} s(\bar{q_r}q_i) &= s((w_r - x_r\mathbf{i} - y_r\mathbf{j} - z_r\mathbf{k}) \cdot (w_i + x_i\mathbf{i} + y_i\mathbf{j} + z_i\mathbf{k})) = \\ & w_rw_i + x_rx_i + y_ry_i + z_rz_i = 0, \end{aligned} \tag{6}$$

$$\begin{aligned} s(q_i\bar{q_r}) &= s((w_i + x_i\mathbf{i} + y_i\mathbf{j} + z_i\mathbf{k}) \cdot (w_r - x_r\mathbf{i} - y_r\mathbf{j} - z_r\mathbf{k})) \\ &= w_iw_r + x_ix_r + y_iy_r + z_iz_r = 0. \end{aligned} \tag{7}$$

Thus $\bar{q_r}q_i$ and $q_i\bar{q_r}$ are both pure quaternions.

$$\begin{aligned} \|q\| = \bar{q}q &= \overline{(q_r + q_i\mathbf{I})}(q_r + q_i\mathbf{I}) = (\bar{q_r} + \bar{q_i}\mathbf{I})(q_r + q_i\mathbf{I}) \\ &= \bar{q_r}q_r + \bar{q_i}q_i\mathbf{I}^2 + (\bar{q_r}q_i + \bar{q_i}q_r)\mathbf{I} = \|q_r\|^2 - \|q_i\|^2 + (\bar{q_r}q_i + \bar{q_i}q_r)\mathbf{I}. \end{aligned} \tag{8}$$

Since $\|q\|^2 = 1$ is real, $\overline{q_r}q_i + \overline{q_i}q_r = 0$ and $\|q_r\|^2 - \|q_i\|^2 = 1$. Since $\|q_r\|$ is real, there must exist $\phi \in \mathbb{R}$, so $\|q_r\| = \cosh\phi$ and $\|q_i\| = \sinh\phi$. It follows that $\|q_r\|^2 - \|q_i\|^2 = 1$ is the equation of a hyperbola($\cosh^2\phi - \sinh^2\phi = 1$).

We let $u = \frac{q_r}{\|q_r\|}, h = \|q_r\| + \mathbf{I}\frac{\overline{q_r}q_i}{\|q_r\|\|q_i\|}\|q_i\|$. Note that u is a unit quaternion and h is a biquaternion in which $\frac{\overline{q_r}q_i}{\|q_r\|\|q_i\|}$ is a quaternion that is both pure (using Eq. 6 and 7) of unit norm, it is a product of two unit quaternions $\frac{\overline{q_r}}{\|q_r\|} = \frac{\overline{q_r}}{\|q_r\|}$ and $\frac{\overline{q_i}}{\|q_i\|}$. If a and b are unit quaternions, then $\|ab\|^2 = \overline{ab}ab = \overline{b}\overline{a}ab = \overline{b}b = 1$. So, $uh = \frac{q_r}{\|q_r\|}(\|q_r\| + \mathbf{I}\frac{\overline{q_r}q_i}{\|q_r\|\|q_i\|}\|q_i\|) = q_r + \mathbf{I}\frac{q_r\overline{q_r}q_i}{\|q_r\|\|q_r\|} = q_r + \mathbf{I}\frac{\|q_r\|^2 q_i}{\|q_r\|^2} = q$. We get the following factorizations: $\mathcal{M}(q) = \mathcal{M}(h)\ \mathcal{M}(u)$. The quaternion $u = \frac{q_r}{\|q_r\|} = \frac{w_r + x_r\mathbf{i} + y_r\mathbf{j} + z_r\mathbf{k}}{\|q_r\|}$ can be represented equivalently as $u = \|u\|(\cos\theta + \frac{v(u)}{\|v(u)\|}\sin\theta) = \cos\theta + \frac{v(q_r)}{\|v(q_r)\|}\sin\theta$(where $\theta = \cos^{-1}\frac{w_r}{\|q_r\|}; \theta \in \mathbb{R}$ because w_r, $\|q_r\| \in \mathbb{R}$). Expanding $v(q_r)$ we get $u = \cos\theta + \frac{x_r\sin\theta}{\|v(q_r)\|}\mathbf{i} + \frac{y_r\sin\theta}{\|v(q_r)\|}\mathbf{j} + \frac{z_r\sin\theta}{\|v(q_r)\|}\mathbf{k}$. Then using the matrix representation given by Eq. 4, we get the form of $\mathcal{M}(u)$ as:

$$\mathcal{M}(u) = \begin{bmatrix} \cos\theta & -\frac{x_r\sin\theta}{\|v(q_r)\|} & -\frac{y_r\sin\theta}{\|v(q_r)\|} & -\frac{z_r\sin\theta}{\|v(q_r)\|} \\ \frac{x_r\sin\theta}{\|v(q_r)\|} & \cos\theta & \frac{z_r\sin\theta}{\|v(q_r)\|} & -\frac{y_r\sin\theta}{\|v(q_r)\|} \\ \frac{y_r\sin\theta}{\|v(q_r)\|} & -\frac{z_r\sin\theta}{\|v(q_r)\|} & \cos\theta & \frac{x_r\sin\theta}{\|v(q_r)\|} \\ \frac{z_r\sin\theta}{\|v(q_r)\|} & \frac{y_r\sin\theta}{\|v(q_r)\|} & \frac{x_r\sin\theta}{\|v(q_r)\|} & \cos\theta \end{bmatrix}$$

Since $\frac{\overline{q_r}q_i}{\|q_r\|\|q_i\|}$ is a pure unit quaternion, it can be represented as $a\mathbf{i} + b\mathbf{j} + c\mathbf{k}(a, b, c \in \mathbb{R})$. Thus, $h = \|q_r\| + \mathbf{I}(a\mathbf{i} + b\mathbf{j} + c\mathbf{k})\|q_i\|$, where $\frac{\overline{q_r}q_i}{\|q_r\|\|q_i\|} = a\mathbf{i} + b\mathbf{j} + c\mathbf{k}$. From Eq. 8, $\|q_r\| = \cosh\phi$ and $\|q_i\| = \sinh\phi$. Thus $h = \cosh\phi + (a\mathbf{I}\sinh\phi)\mathbf{i} + (b\mathbf{I}\sinh\phi)\mathbf{j} + (c\mathbf{I}\sinh\phi)\mathbf{k}$. Again, using the the matrix representation given by Eq. 4, we get the form of $\mathcal{M}(h)$ as :

$$\mathcal{M}(h) = \begin{bmatrix} \cosh\phi & -a\mathbf{I}\sinh\phi & -b\mathbf{I}\sinh\phi & -c\mathbf{I}\sinh\phi \\ a\mathbf{I}\sinh\phi & \cosh\phi & c\mathbf{I}\sinh\phi & -b\mathbf{I}\sinh\phi \\ b\mathbf{I}\sinh\phi & -c\mathbf{I}\sinh\phi & \cosh\phi & a\mathbf{I}\sinh\phi \\ c\mathbf{I}\sinh\phi & b\mathbf{I}\sinh\phi & -a\mathbf{I}\sinh\phi & \cosh\phi \end{bmatrix}.$$

Note that all elements in $\mathcal{M}(h)$, $\mathcal{M}(u)$ are derived from q. Hence, for any q, we can construct $\mathcal{M}(h)$, $\mathcal{M}(u)$ thus proving that $\mathcal{M}(q)$ can be factorized as: $\mathcal{M}(q) = \mathcal{M}(h)\mathcal{M}(u)$.

Next, we show that since both u and q are unit biquaternions, h is each a unit biquaternion. $q = uh \Rightarrow \overline{q}q = \overline{uh}uh = \overline{h}\overline{u}uh = \overline{h}h = \|h\|^2$. Since q is a unit biquaternion $\overline{q}q = 1 = \|h\|^2$. If $\mathcal{M}(q)$ is the matrix representation of a biquaternion q, then the matrix's determinant is given by det $\mathcal{M}(q) = \|q\|^4$ [16]. So, we can obtain det $\mathcal{M}(h) = \|h\|^4 = 1$ and det $\mathcal{M}(u) = \|u\|^4 = 1$.

$\overline{h}h = 1 \Rightarrow \mathcal{M}(h)\mathcal{M}(\overline{h}) = \mathcal{M}(h).\mathcal{M}(h)^T = \mathcal{M}(1) = \mathbb{II}$. Likewise $\overline{u}u = 1 \Rightarrow \mathcal{M}(u).\mathcal{M}(\overline{u}) = \mathcal{M}(u).\mathcal{M}(u)^T = \mathcal{M}(1) = \mathbb{II}$. Thus $\mathcal{M}(h)$, $\mathcal{M}(u)$ are orthogonal.

An orthogonal matrix with determinant 1 represents a rotation in the space in which it operates [1]. Both $\mathcal{M}(h)$ and $\mathcal{M}(u)$ are orthogonal and have determinants 1. $\mathcal{M}(h)$ represents a circular rotation and $\mathcal{M}(u)$ represents a hyperbolic rotation [14]. Biquaternion's matrix $\mathcal{M}(q)$ represents a circular rotation followed by a hyperbolic rotation. It is worth noting that the biquaternion's matrix $\mathcal{M}(q)$ can also represents a hyperbolic rotation followed by a circular rotation.(This theory has been proved in detail by [14], and this paper only briefly explains its proof process.)

4 Model Building

Each quadruple represents a fact that exists at a period in the real world. A TKG is referred to as a set of quadruples $\mathcal{TKG} = (h, r, o, t)$. Each quadruple consists of a head entity $h \in \xi$, a relation entity $r \in \mathcal{R}$, a tail entity $o \in \xi$, and a temporal entity $t \in \mathcal{T}$. ξ is the set of all entities, $\mathcal{R}$ is the set of all entities relations, and $\mathcal{T}$ represents the set of all possible time stamps. To predict missing links, TKGE maps entities, relations, and temporals to distributed representations and defines a score function to measure the plausibility of each quadruple. The BiQCap model constructed in this paper combines biquaternions and capsule networks to compensate for existing TKGE models' defects.

We represent the entities, relations and temporals in a $\mathcal{TKG}$ as vectors of biquaternions. Let $\mathbb{Q}$ be the set of biquaternions. Each element is a vector $\mathbf{Q}$ of k biquaternions, i.e., $\mathbf{Q} = [q_1, q_2, \ldots, q_k]^T$, where $q_1, q_2, \ldots, q_k \in \mathbb{Q}$, an entity , relation or temporal vector $\mathbf{Q}$ can be experssed as $\mathbf{Q} = w + x\mathbf{i} + y\mathbf{j} + z\mathbf{k}$ ($w, x, y, z \in \mathbb{C}^k$). We denote a head entity and a tail entity as $\mathbf{Q}_h$ and $\mathbf{Q}_o$, respectively. We denote a relation and a temporal as $\mathbf{Q}_r$ and $\mathbf{Q}_t$, respectively. The embedding $\mathbf{Q}_t$ applies a temporal-specific translation to a head entity's embedding $\mathbf{Q}_h$. We realize it by the element-wise addition of biquaternions (similar to what we do with real vectors for translation):

$$\mathbf{Q}_h' = \mathbf{Q}_t + \mathbf{Q}_h = (w_h + w_t) + (x_h + x_t)\mathbf{i} + (y_h + y_t)\mathbf{j}$$
$$+ (z_h + z_t)\mathbf{k} = w' + x'\mathbf{i} + y'\mathbf{j} + z'\mathbf{k}. \tag{9}$$

Then embedding $\mathbf{Q}_r$ applies a relation-specific multiplicative translation to the translated head entity $\mathbf{Q}_h'$. The multiplicative transformation is defined via the Hamilton product of biquaternions as follows:

$$\mathbf{Q}_{h,r}' = \mathbf{Q}_h' \boxplus \mathbf{Q}_r = (w' \otimes w_r - x' \otimes x_r - y' \otimes y_r - z' \otimes z_r)$$
$$+ (w' \otimes w_r + x' \otimes x_r + y' \otimes y_r - z' \otimes z_r)\mathbf{i}$$
$$+ (w' \otimes w_r - x' \otimes x_r + y' \otimes y_r + z' \otimes z_r)\mathbf{j} \tag{10}$$
$$+ (w' \otimes w_r + x' \otimes x_r - y' \otimes y_r + z' \otimes z_r)\mathbf{k},$$

where $\boxplus$ denotes the element-wise application of the Hamilton product between $\mathbf{Q}_h'$ and $\mathbf{Q}_r$, and $\otimes$ denotes the element-wise multiplication between vectors of complex numbers. Follower Rotate3D model [12], when evaluating the models, we find that multiplying a relation-specific bias can improve the performance. According to the above definition, for each quadruple (h, r, o, t), we define the objective function as follows:

Algorithm 1. Inverted dot-product attention routing

Output: r $= (r_1, r_2, \cdots r_i \cdots)$

1: **for** all capsule i in the first layer **do**
2:     $a_i \leftarrow 0$
3:     **for** $iteration = 1, 2, \cdots, k$ **do**
4:         $\mathbf{v}_i \leftarrow \mathbf{w}_i \cdot \mathbf{Q}_i$
5:         $r_i \leftarrow Softmax(\mathbf{a})$
6:         $\mathbf{s} \leftarrow \sum_{i=1}^{k} r_i \mathbf{v}_i$
7:         $\mathbf{e} = LayerNorm(\mathbf{s})$
8:         **for** all capsule i in the first layer **do**
9:             $a_i \leftarrow \mathbf{s}^T \mathbf{v}_i$
10:        **end for**
11:    **end for**
12: **end for**

$$f(\mathbf{h}, \mathbf{r}, \mathbf{o}, \mathbf{t}) = \left\| \mathbf{Q}'_{h,r} \otimes \mathbf{b} - \mathbf{Q}_o \right\|, \tag{11}$$

where $\mathbf{b} \in \mathbb{R}^k$. In the model, we use a loss function similar to the negative sampling loss [5] to effectively optimize distance-based models. The loss function is described as follows:

$$L = -\log \sigma(\gamma - f(\mathbf{h}, \mathbf{r}, \mathbf{o}, \mathbf{t})) - \frac{1}{k} \sum_{i=1}^{n} \log \sigma(f(\mathbf{h}_i', \mathbf{r}, \mathbf{o}_i', \mathbf{t}) - \gamma), \tag{12}$$

where σ is the sigmoid function, γ is a fixed margin, and (h_i', r, o_i', t) is the i-th negative quadruple and n is the number of negative samples. We use self-adversarial negative as with the RotatE [29] sampling, which has proven to be very effective, we sample negative quadruples from the following distribution:

$$P(h_j', r, o_j', t | \{(h_i, r_i, o_i, t_i)\}) = \frac{\exp \alpha f_r(h_j', o_j', t)}{\sum_i \exp f_r(h_i', o_i', t)}, \tag{13}$$

where α is the temperature of sampling. Therefore final loss function is shown below:

$$L = -\log \sigma(\gamma - f(\mathbf{h}, \mathbf{r}, \mathbf{o}, \mathbf{t})) - \sum_{i=1}^{n} P(h_i', r, o_i', t) \log \sigma(f(\mathbf{h}_i', \mathbf{r}, \mathbf{o}_i', \mathbf{t}) - \gamma).$$

We employ the entity, relation, and temporal embeddings trained from biquaternions as the input to our model. We define the embedding quadruple (h, r, o, t) as a matrix $\mathbf{B} = [\mathbf{h}, \mathbf{r}, \mathbf{o}, \mathbf{t}] \in \mathbb{Q}^{k \times 4}$. The i-th row of $\mathbf{B}$ is defined as $\mathbf{B}_i$ and filter ω is repeatedly operated over every row of $\mathbf{B}$ to generate a feature map $\mathbf{q} = [q_1, q_2, q_3, \ldots, q_k]$, as follows:

$$q_i = g(\omega \cdot \mathbf{B}_i + b), \tag{14}$$

where the filter $\omega \in \mathbb{Q}^{1\times4}$, $(\cdot)$ denotes a dot product, b is a bias term and g is a non-linear activation function such as ReLU or Sigmoid.

In the convolution layer, we use Eq. 14 to calculate the values of k capsules. We denote $\mathbf{N}$ as the number of filters, thus we have $\mathbf{N}$ k-dimensional feature maps, for which each feature map can capture one single characteristic among entries at the same dimension. The first capsule layer contains k capsules, for which each capsule $i \in \{1, 2, \ldots, k\}$ has a vector output $\mathbf{Q}_i \in \mathbb{Q}^{N\times1}$. Vector outputs $\mathbf{Q}_i$ are multiplied by weight matrices $\mathbf{w}_i \in \mathbb{Q}^{d\times N}$ to produce vectors $\mathbf{v}_i \in \mathbb{Q}^{d\times1}$ which are summed to produce a vector input $\mathbf{s} \in \mathbb{Q}^{d\times1}$ to the capsule in the second layer. Then, the capsule generates a vector output $\mathbf{e} \in \mathbb{Q}^{d\times1}$ by performing a operation of normalization. Finally, vector output $\mathbf{e}$ is multiplied by the weight matrix $\mathbf{W} \in \mathbb{Q}^{d\times1}$ to produce a score that is used to determine the correctness of a given quadruple. The detailed process is shown as follows:

$$\mathbf{v}_i = \mathbf{w}_i \cdot \mathbf{Q}_i, \; \mathbf{s} = \sum_i \mathbf{r}_i\mathbf{v}_i, \mathbf{e} = LayerNorm\,(\mathbf{s})\,, \; f = \mathbf{e} \bullet \mathbf{W}. \qquad (15)$$

The routing algorithm is shown in Fig. 1. We adopt Layer Normalization [2] as the normalization, which we empirically find to improve the convergence for routing, and r_i are coupling coefficients determined by the routing process as presented in Algorithm 1.

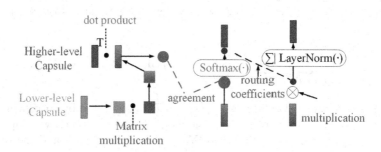

Fig. 1. Inverted Dot-Product Attention Routing.

We illustrate our proposed model in Fig. 2, where embedding size: $k = 6$, the number of filters: $N = 5$, the number of neurons within the capsules in the first capture layer is equal to N, and the number of neurons within the capsule in the second capture layer: $d = 4$. In summary, we define the score function for BiQCap as follows:

$$f(\mathbf{h}, \mathbf{r}, \mathbf{o}, \mathbf{t}) = \|caps(g([\mathbf{h}, \mathbf{r}, \mathbf{o}, \mathbf{t}] * \varOmega))\| \bullet \mathbf{W}, \qquad (16)$$

where the set of filters $\varOmega$ and $\mathbf{W}$ are shared hyper-parameters in the convolution layer, $*$ indicates a convolution operator, and the capsule network operator is denoted by $caps$, g represents the activation function ReLU. We use the Adam optimizer [19] to train BiQCap by minimizing the loss function [30] as follows:

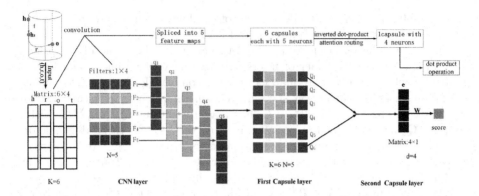

Fig. 2. An example explanation of our BiQCap with $k = 6$, $N = 5$, and $d = 4$. The input quadruple is trained with biquaternions as the output of the capsule network, and the score of the quadruple is obtained after one CNN layer and two capsule layers.

Table 1. Statistics for the various experimental datasets.

Datatset	#Entities	#Relations	#Training	#Validation	#Test	Time Span
ICEWS14	6869	230	72826	8941	8963	2014
ICEWS05-15	10094	251	368962	46275	46092	2005–2015
YAGO11k	10623	10	16406	2050	2051	−453–2844
Wikidata12k	12554	24	32497	4062	4062	1479–2018
GDELT	500	20	2735685	341961	341961	2015.03.31-2016.03.31

$$L = \sum_{(h,r,o,t)\in T} \sum_{o'\in T'} log(1 + \exp(y_o f(\mathbf{h}, \mathbf{r}, \mathbf{o}, \mathbf{t}))) + \lambda \|\mathbf{W}\|_2^2, \qquad (17)$$

where T is collections of valid quadruples and T' is generated by corrupting valid quadruples in T. In order to prevent the model from overfitting, we employ the L_2 regularization on the weight vector $\mathbf{W}$, and λ is the weight, and the value of y_o as follows:

$$y_o = \begin{cases} 1, & for(s,r,o,t) \in T \\ -1, & for(s,r,o,t) \in T'. \end{cases} \qquad (18)$$

5   Experiment

To validate BiQCap's effectiveness, we have conducted experiments on five TKG datasets ICEWS14, ICEWS05-15, YAGO11k, Wikidata12k, and GDELT. Table 1 is the summary of data sets information. ICEWS14 and ICEWS05-15 are the most commonly used datasets in TKGE embedding studies, and both are subsets of the ICEWS dataset. ICEWS14 data set records the facts in 2014, and each piece of data has the corresponding temporal information. Similarly, the ICEWS05-15 data set records the facts between 2005 and 2015, and each piece of data also records the corresponding temporal information. YAGO11k

Table 2. Link prediction results on ICEWS14 and ICEWS05-15. Best results are in bold.(All data are from the latest relevant papers.)

Modle	ICEWS14				ICEWS05-15			
	MRR	Hit@1	Hit@3	Hit@10	MRR	Hit@1	Hit@3	Hit@10
$TransE$	.280	.094	-	.637	.294	.090	-	.663
$DistMult$	.439	.323	-	.672	.456	.337	-	.691
$ComplEx$	.470	.350	.540	.710	.490	.370	.550	.730
$RotatE$	.418	.291	.478	.690	.304	.164	.355	.595
$QuatE$	.471	.353	.530	.712	.482	.370	.529	.727
$TTransE$	.255	.074	-	.601	.271	.084	-	.616
$TA - TransE$	.275	.095	-	.625	.299	.096	-	.668
$HyTE$	.297	.108	.416	.655	.316	.116	.445	.681
$TA - DistMult$	.477	.363	-	.686	.474	.346	-	.728
$DE - SimplE$	.526	.418	.592	.725	.513	.392	.578	.748
$ATiSE$	.550	.436	.629	.750	.519	.378	.606	.794
$TeRo$	.562	.468	.621	.732	.586	.469	.668	.795
$ChronoR$	.625	.547	.669	.773	.675	.593	.723	.820
$RotateQVS$	.591	.507	.642	.754	.633	.529	.709	.813
$TempCaps$	.489	.388	.544	.679	.521	.423	.576	.705
$BiQCap^-$	.635	.558	.681	.789	.685	.618	.731	.832
$BiQCap$	**.643**	**.563**	**.687**	**.798**	**.691**	**.621**	**.738**	**.837**

and Wikidata12k are subsets of YAGO3, and each fact of the two data sets contains temporal information, and temporal annotations are expressed in various forms. GDELT collects human societal-scale behaviors and events from April 1, 2015, to March 31, 2016, in news media.

5.1 Baselines and Evaluation Protocol

We compare our models with a number of baselines. For static KGE models, we report TransE [5], DistMult [35], ComplEx [30], RotatE [29], QuatE [36]. For TKGE models, we report TTransE [20], TA-TransE [26], HyTE [9], TA-DistMult [26], DE-SimplE [13], ATiSE [34], TeRo [33], ChronoR [27], TempCaps [11] and RotateQVS [8]. To explore the impact of the"deep" architecture employed by the capsule network on link prediction performance, we removed the capsule network part of our model, dubbed BiQCap$^-$.

We use standard evaluation metrics for the Temporal Knowledge Graph Completion (TKGC) task: Mean Reciprocal Rank (MRR) and Hits at $N(Hit@N)$. MRR is the average inverse rank for correct quadruples. $H@N$ measures the proportion of correct quadruples in the top N quadruples and $Hit@N$ with cutoff values $N \in \{1, 3, 10\}$. For both MRR and $Hit@N$, the larger the metric, the better the model's performance.

5.2 Implementation Details

For training, we restrict the iterations to 1000 and following the setup used in RoatE [29], the batch size $b = 512$ is kept for all datasets, the embedding dimension k is tuned in $\{100, 200, 500, 1000, 1500\}$, the learning rate r is chosen from 0.01 to 1, fixed margin $\gamma \in \{6, 9, 12, 24, 30\}$, self-adversarial sampling temperature $\alpha \in \{0.5, 1.0\}$. In the experiment, we defined the rotation angles of relation

Table 3. Link prediction results on YAGO11k , Wikidata12k and GDELT. Best results are in bold.(All data are from the latest relevant papers.)

Modle	YAGO11k				Wikidata12k				GDELT			
	MRR	Hit@1	Hit@3	Hit@10	MRR	Hit@1	Hit@3	Hit@10	MRR	Hit@1	Hit@3	Hit@10
$TransE$	.100	.015	.138	.244	.178	.100	.192	.339	.132	.000	-	.158
$DistMult$	.158	.107	.161	.268	.222	.119	.238	.460	.196	.117	.208	.348
$ComplEx$	.167	.106	.154	.282	.233	.123	.253	.436	-	-	-	-
$RotatE$	.167	.103	.167	.305	.221	.116	.236	.461	-	-	-	-
$QuatE$	.164	.107	.148	.270	.230	.125	.243	.416	-	-	-	-
$TTransE$	.108	.020	.150	.251	.172	.096	184	.329	.115	.000	.160	.318
$TA - TransE$	.127	.027	160	.326	.178	.030	.267	.429	-	-	-	-
$HyTE$	.105	.015	.143	.272	.180	.098	.197	.333	.118	.000	.165	.326
$TA - DistMult$	.161	.103	.171	.292	.218	.122	.232	.447	.206	.124	.219	.365
$DESimplE$	-	-	-	-	-	-	-	-	.230	.141	.248	.403
$ATiSE$	.170	.110	.171	.288	.280	.175	.317	.481	-	-	-	-
$TeRo$	.187	.121	.197	.319	.299	.198	.329	.507	.245	.154	.264	.420
$RotateQVS$	.189	.124	.199	.323	-	-	-	-	.270	.175	.293	.458
$TempCaps$	-	-	-	-	-	-	-	-	.258	.180	.277	.404
$BiQCap^-$	.192	.125	.201	.329	.301	.203	.331	.509	.262	.162	.291	.452
$BiQCap$	**.195**	**.126**	**.206**	**.332**	**.312**	**.216**	**.343**	**.517**	**.273**	**.183**	**.308**	**.469**

embeddings are uniformly initialized between $-\pi$ and π, and the same is true of the temporal embeddings.

Like CapsE [25], we use the embedding of entities, relationships, and temporals obtained from biquaternions to initialize BiQCap. The number of filters is set to $|\omega| = N \in \{50, 100, 200, 400\}$ and the regularization parameters of Eq. 17 $\lambda \in \{0.005, 0.01, 0.05, 0.1, 0.15\}$. The number of neurons within the capsule in the second capsule layer are set as b=10 and the batch size to 128. And the number of iterations in the inverted dot-product attention algorithm m is set to $\{1, 3, 5, 7, 9\}$.

5.3   Experimental Results

Tables 2 and 3 show that BiQCap outperforms all the baseline models over the five datasets across all metrics. In addition to the BiQCap, BiQCap$^-$ is the best performer on four of the five datasets, and the performance of the remaining dataset is close to the best results.

On ICEWS14 and ICEWS05-15 datasets, BiQCap and BiQCap$^-$ achieve new state-of-the-art results on all metrics and surpasses the second-best models by a clear margin. We believe these datasets have much more complex temporal information and quantitative relations (see Table 1). The excellent performance of BiQCap and BiQCap$^-$ on the three datasets demonstrates the importance of the biquaternions composition of circular and hyperbolic rotations for modeling relation patterns of rotational and hierarchical relations of TKG. BiQCap performs better than its closely related embedding model BiQCap$^-$ on all experimental datasets, especially on Wikidata12k and GDELT datasets and

Table 4. Link prediction results on ICEWS14 and ICEWS05-15, best results are in bold.

Models	Scoring Functions	ICEWS14				ICEWS05-15			
		MRR	Hit@1	Hit@3	Hit@10	MRR	Hit@1	Hit@3	Hit@10
TTransEN	$\mathbf{h} + LSTM([\mathbf{r};\mathbf{t}_s eq]) - \mathbf{o}$	.524	.503	.598	.720	.536	.523	.628	.795
TA-DistMultN	$\langle \mathbf{h}, LSTM([\mathbf{r};\mathbf{t}_s eq]),\mathbf{o}\rangle$	.526	.531	.627	.745	.568	.543	.673	.804
DE-SimplEN	$\frac{1}{2}(\langle \overrightarrow{\mathbf{h}}^t,\mathbf{r},\overleftarrow{\mathbf{o}}^t\rangle + \langle \overrightarrow{\mathbf{o}}^t,\mathbf{r}^{-1},\overleftarrow{\mathbf{o}}^t\rangle)$	.573	.538	.650	.759	.621	.589	.692	815
TeRoN	$\|\mathbf{h}\circ\mathbf{t}+\mathbf{r}-\overline{\mathbf{o}\circ\mathbf{t}}\|$	.571	.541	.540	.648	.646	.653	.705	813
RotateQVSN	$\|\mathbf{h}\otimes\mathbf{t}+\mathbf{r}-\overline{\mathbf{o}\circ\mathbf{t}}\|$	.595	.559	.667	.783	.659	594	.718	.837
BiQCap$^-$	-	.635	.558	.681	.789	.685	.618	.731	.832
BCovE	$g(vec(g(concat(\hat{\mathbf{h}},\hat{\mathbf{r}},\hat{\mathbf{t}}) * \Omega))\mathbf{W}) \cdot \mathbf{t}$	.638	.559	.679	.793	.689	.617	.733	.835
BConvKB	$concat(g([\mathbf{h},\mathbf{r},\mathbf{o},\mathbf{t}] * \Omega)) \cdot \mathbf{w}$	.638	.561	.683	795	.687	.619	.731	.833
BCapsE	$\|capsnet(g([\mathbf{h},\mathbf{r},\mathbf{o},\mathbf{t}] * \Omega))\|$	.641	.560	.684	.796	.689	**.621**	.736	.835
BiQCap	$\|caps(g([\mathbf{h},\mathbf{r},\mathbf{o},\mathbf{t}] * \Omega))\| \cdot \mathbf{W}$	**.643**	**.563**	**.687**	**.798**	**.691**	.621	**.738**	**.837**

demonstrating that the model's use of capsule network technology for deeper communication significantly improves the model's effectiveness.

5.4 Ablation Studies

In this work, we analyze the effect of the change in the number of routing iterations on the performance of our model. We train BiQcap for 1,000 epochs, using $k = 1000$, $b = 512$, $\alpha=1.0$, $\gamma=24$, the initial learning rate at 0.15, and the number of iterations in the inverted dot-product attention algorithm m is tuned in $\{0, 2, 4, 6, 8, 10\}$ on ICEWS14.

From Fig. 3, we can see how the number of routing iterations affects model performance and that the inverted dot-product attention routing aggregator outperforms the mean aggregator on the most commonly employed metrics of TKGE models.

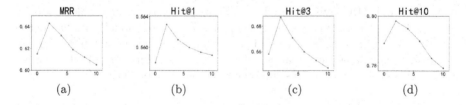

(a)   (b)   (c)   (d)

Fig. 3. Effects of the number of iterations. 0 indicates the model uses a mean aggregator,otherwise the model uses the inverted dot-product attention routing aggregator.

In order to investigate the number of filters on the performance of BiQCap, we record the performance of BiQCap on the validation set for each epoch. For Fig. 4. We observe that the performance of BiQCap improves as the number of filters increases since capsules can encode more valuable features for large embedding sizes.

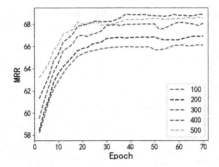

Fig. 4. The learning curves of the proposed BiQCap. We selected the number of filters among $\{100, 200, 300, 400, 500\}$, and the initial learning rate at 0.00005.

To verify the effectiveness of the biquaternion, we design four variants of BiQCap by replacing our score functions with previous score functions used in TA-TransE [26], TA-DistMult [26], DE-SimplE [13], TeRo [33], and RotateQVS [8], dubbed TTransEN, TA-DistMultN, DE-SimplEN, TeRoN, and RotateQVSN respectively. $LSTM\ (\cdot)$ denotes an $LSTM$ neural network. $[\mathbf{r}; \mathbf{t}_s eq]$ denotes the concatenation of the relation embedding and the sequence of temporal tokens. $\rightarrow$ and $\leftarrow$ denote the temporal part and untemporal part of a time-specific diachronic entity embedding. $\mathbf{r}^{-1}$ denotes the inverse relation embedding of r i.e., $(s, r, o, t) \leftrightarrow (o, r^{-1}, s, t)$, where $\circ$ denotes the Hermitian dot product between complex vectors, $\overline{\mathbf{n}}$ denotes the conjugate of $\mathbf{n}$. We reran the experiment using those score functions. It can be seen from Table 4 that on two benchmark datasets, ICEWS14 and ICEWS05-15, MRR, $Hit@1$, $Hit@3$ and $Hit@10$ indicate that the modeling the ability of biquaternion's model is much better than that of translation, rotation, and quaternion.

On data sets ICEWS14 and ICEWS05-15, we use the scoring function of CovE [10], ConvKB [18], and CapsE [25] to replace Eq. 16 to prove its validity, dubbed BCovE, BConvKB, and BCapsE, respectively. From Table 4, we can see that BiQcap performs better than its closely related models and has better experimental results. BiQcap uses biquaternion-embedded quadruples as input, so capsules have specific semantic information, providing important reference information for completing missing relationships between entities.

6   Conclusion

In this work, we introduce BiQCap, which combines biquaternion and capsule networks in modeling for the first time to make up for the defects of existing TKGE models. Our model employs biquaternions to integrate multiple geometric transformations, viz., Euclidean and hyperbolic rotations. We employ the embeddings of entities, relations, and temporal trained from biquaternions as the input to the capsule network, and inverted dot-product attention routing was used. Experimental results on four datasets demonstrate the superiority of our model. In future work, we will explore more unified mathematical models combined with capsule networks to model more temporal relation patterns.

References

1. Artin, E.: Geometric algebra. Courier Dover Publications (2016)
2. Ba, L.J., Kiros, J.R., Hinton, G.E.: Layer normalization. CoRR abs/1607.06450 (2016)
3. Balazevic, I., Allen, C., Hospedales, T.M.: Multi-relational poincaré graph embeddings. In: NeurIPS, pp. 4465–4475 (2019)
4. Bollacker, K.D., Evans, C., Paritosh, P.K.: Freebase: a collaboratively created graph database for structuring human knowledge. In: ACM, pp. 1247–1250 (2008)
5. Bordes, A., Usunier, N.: Translating embeddings for modeling multi-relational data. In: NIPS, pp. 2787–2795 (2013)
6. Cao, Z., Xu, Q., Yang, Z.: Dual quaternion knowledge graph embeddings. In: AAAI, pp. 6894–6902. AAAI Press (2021)
7. Chami, I., Wolf, A., Juan, D.: Low-dimensional hyperbolic knowledge graph embeddings. In: ACL, pp. 6901–6914. Association for Computational Linguistics (2020)
8. Chen, K., Wang, Y.: RotateQVS: representing temporal information as rotations in quaternion vector space for temporal knowledge graph completion. In: ACL, pp. 5843–5857 (2022)
9. Dasgupta, S.S., Ray, S.N., Talukdar, P.P.: HyTE: hyperplane-based temporally aware knowledge graph embedding. In: EMNLP, pp. 2001–2011 (2018)
10. Dettmers, T., Minervini, P., Stenetorp, P., Riedel, S.: Convolutional 2D knowledge graph embeddings. In: AAAI, pp. 1811–1818. AAAI Press (2018)
11. Fu, G., Meng, Z., Han, Z.: TempCaps: a capsule network-based embedding model for temporal knowledge graph completion. In: Proceedings of the Sixth Workshop on Structured Prediction for NLP, pp. 22–31. Association for Computational Linguistics (2022)
12. Gao, C., Sun, C., Shan, L., Lin, L., Wang, M.: Rotate3D: representing relations as rotations in three-dimensional space for knowledge graph embedding. In: CIKM, pp. 385–394. ACM (2020)
13. Goel, R., Poupart, S.M.K.P.: Diachronic embedding for temporal knowledge graph completion. In: AAAI, pp. 3988–3995. AAAI Press (2020)
14. Guo, J., Kok, S.: BiQUE: biquaternionic embeddings of knowledge graphs. In: EMNLP, pp. 8338–8351. Association for Computational Linguistics (2021)
15. Guo, L., Sun, Z., Hu, W.: Learning to exploit long-term relational dependencies in knowledge graphs. In: ICML, vol. 97, pp. 2505–2514. PMLR (2019)
16. Jafari, M.: On the matrix algebra of complex quaternions. TWMS J. Pure Appl. Math. (2016)
17. Ji, G., and S.H.: Knowledge graph embedding via dynamic mapping matrix. In: ACL, pp. 687–696. The Association for Computer Linguistics (2015)
18. Jiang, X., Wang, Q., Wang, B.: Adaptive convolution for multi-relational learning. In: NAACL-HLT, pp. 978–987. Association for Computational Linguistics (2019)
19. Kingma, D.P., Ba, J.: Adam: a method for stochastic optimization. In: ICLR (2015)
20. Leblay, J., Chekol, M.W.: Deriving validity time in knowledge graph. In: Companion of the The Web Conference 2018 on The Web Conference 2018, WWW 2018, Lyon, France, April 23–27, 2018, pp. 1771–1776. ACM (2018)

21. Lehmann, J., Isele, R., Jakob, M.: Dbpedia - A large-scale, multilingual knowledge base extracted from wikipedia. Semantic Web **6**(2), 167–195 (2015)
22. Lin, Y., Liu, Z., Sun, M.: Learning entity and relation embeddings for knowledge graph completion. In: AAAI, pp. 2181–2187. AAAI Press (2015)
23. Miller, G.A.: WordNet: a lexical database for english. ACM **38**(11), 39–41 (1995)
24. Montella, S., Rojas-Barahona, L.M., Heinecke, J.: Hyperbolic temporal knowledge graph embeddings with relational and time curvatures. In: ACL/IJCNLP, pp. 3296–3308 (2021)
25. Nguyen, D.Q., Vu, T., Nguyen, T.D.: A capsule network-based embedding model for knowledge graph completion and search personalization. In: NAACL-HLT, pp. 2180–2189 (2019)
26. Niepert, A.G.M.: Learning sequence encoders for temporal knowledge graph completion. In: Conference on Empirical Methods in Natural Language Processing, pp. 4816–4821 (2018)
27. Sadeghian, A., Armandpour, M., Colas, A., Wang, D.Z.: ChronoR: rotation based temporal knowledge graph embedding. In: AAAI, pp. 6471–6479 (2021)
28. Speer, R., Chin, J., Havasi, C.: Conceptnet 5.5: an open multilingual graph of general knowledge. In: AAAI, pp. 4444–4451. AAAI Press (2017)
29. Sun, Z., and Z.D.: Rotate: Knowledge graph embedding by relational rotation in complex space. In: ICLR (2019)
30. Trouillon, T., Welbl, J., Riedel, S., Gaussier, É., Bouchard, G.: Complex embeddings for simple link prediction. In: ICML, JMLR Workshop and Conference Proceedings, vol. 48, pp. 2071–2080. JMLR.org (2016)
31. Wang, Z., n, J.Z.: Knowledge graph embedding by translating on hyperplanes. In: AAAI, pp. 1112–1119. AAAI Press (2014)
32. Ward, J.P.: Quaternions and Cayley numbers: Algebra and applications, vol. 403. Springer Science & Business Media (2012)
33. Xu, C., Nayyeri, M., Alkhoury, F.: TeRo: a time-aware knowledge graph embedding via temporal rotation. In: COLING, pp. 1583–1593 (2020)
34. Xu, C., Nayyeri, M., Alkhoury, F., Lehmann, J.: Temporal knowledge graph embedding model based on additive time series decomposition. CoRR (2019)
35. Yang, B., Yih, W., He, X., Gao, J., Deng, L.: Embedding entities and relations for learning and inference in knowledge bases. In: ICLR (2015)
36. Zhang, S., Tay, Y.: Quaternion knowledge graph embeddings. In: NeurIPS, pp. 2731–2741 (2019)

CLNIE: A Contrastive Learning Based Node Importance Evaluation Method for Knowledge Graphs with Few Labels

Jingbo Wang, Yumeng Song, Yu Gu$^{(\boxtimes)}$, Xiaohua Li, and Fangfang Li

School of Computer Science and Engineering, Northeastern University,
Shenyang, Liaoning, China
wangjingbo@stumail.neu.edu.cn, ymsong94@163.com,
{guyu,lixiaohua,lifangfang}@mail.neu.cn

Abstract. Graph node importance estimation, which evaluates the importance of graph nodes, is an important graph mining problem and has been widely used in many fields such as search engines and recommender systems. Different from ordinary graphs, there are various types of nodes and relationships in knowledge graphs. The relationships in knowledge graphs encode different information, so the information of nodes in a knowledge graph is richer, which leads to the evaluation of the importance of nodes in knowledge graphs being more complicated. The existing research on the importance of nodes in knowledge graphs is mainly based on the assumption that there are enough labels, which even reach 70% of the dataset. However, there are usually few labels in reality. To better study the importance estimation of nodes in a knowledge graph when labels are sparse, we propose a node importance evaluation algorithm based on contrastive learning. First, an unsupervised contrastive loss is designed to generate rich node representations by maximizing the consistency of representations under different views of the same node. To utilize scarce but valuable labeled data for learning node importance, we design a semi-supervised contrastive loss, which solves the problem of failing to determine positive and negative examples in the task of node importance evaluation. In order to improve the effectiveness of contrastive learning, we propose a negative sampling strategy based on label similarity. Negative samples are constructed according to the label difference. Finally, the experimental results on real-world datasets confirm the effectiveness of CLNIE, which achieves a significant performance improvement over the state-of-the-art solutions.

Keywords: Contrastive learning · Node importance · Knowledge graph · Graph neural network

1 Introduction

Knowledge graphs, which can represent structural relationships between entities, have become an increasingly popular direction in cognitive and artificial intelligence [16,20]. Unlike normal graphs that have only one relationship, a knowledge graph is a multi-relational graph. The nodes in the knowledge graph correspond to entities, and edges correspond to relationships between entities. The nodes and edges in knowledge graphs contain rich structural and semantic information and play an important role in knowledge graph mining tasks.

Given a knowledge graph, how to estimate the importance of nodes in the graph is an important research topic, which has been widely used in recommender systems, web search, etc. [13,14]. At first, researchers use some characteristics based on the definition and direct description to measure the importance of nodes, such as degree centrality and betweenness centrality. Besides, there are also some algorithms derived from Internet link analysis, such as PageRank (PR) [19] and Personalized PageRank (PPR) [9]. However, these methods can not deal with various types of nodes and edges in knowledge graphs. With the development of graph deep learning, some machine learning algorithms such as GENI [21] and RGTN [12] have gradually emerged to estimate the importance of nodes in knowledge graphs. These methods have achieved significant success in estimating node importance, but they all rely on the assumption that there are enough dataset labels in the training process. In other words, these approaches do not study the importance of nodes when labels are scarce. However, there are usually few labels in practical applications, and node labeling usually requires huge labor and time costs. Therefore, how to evaluate node importance in knowledge graphs with few labels becomes a key challenge.

In recent years, contrastive learning has shown great potential in the absence or sparseness of labels as a self-supervised learning method [15,17]. Most of the existing contrastive learning methods are designed for other graph analysis tasks such as node classification. According to the characteristics of node classification tasks, nodes can be strictly divided into positive and negative samples by using class information because there are clear boundaries between node labels. Different from it, node importance evaluation aims to predict the importance values of nodes and the value is continuous. It leads to no clear boundary between positive and negative samples in the node importance evaluation task. Therefore, the existing contrastive learning methods are not suitable for it, which makes the task more challenging.

In this paper, we propose a node importance evaluation method based on contrastive learning according to the characteristics of the node importance evaluation task. Specifically, we first use the relational graph transformer network [12] to obtain two embeddings of nodes and treat them as two views. The relational graph transformer network allows us to take full advantage of the rich information in the knowledge graph. We then design a cross-view unsupervised contrastive loss to learn node representations containing structural and semantic information by minimizing the embeddings of the same node under the two views. In order to use sparse but valuable label information, we design a semi-supervised contrastive loss. Different from the practice of directly taking the same class of nodes as positive samples of each other in the node classification task, we take top k nodes as positive examples and the rest as negative examples to minimize the contrastive loss. Moreover, to enhance the effect of contrastive learning, we design a negative sampling strategy based on label similarity. Negative samples of contrast learning are selected according to the difference in node importance values. To our knowledge, we introduce contrastive learning into the task of node importance evaluation for the first time. Our contributions can be summarized as follows:

- We present CLNIE, a knowledge graph node importance evaluation algorithm based on contrastive learning, which can make full use of the structural and semantic information in the graph to estimate the node importance when the labels are sparse.
- We introduce contrastive learning into node importance tasks for the first time. A cross-view unsupervised contrastive loss is designed to learn node embeddings by contrasting node representations in different views, which is used for node importance calculation.
- To make full use of the scarce but valuable label information, we design the semi-supervised contrastive loss. At this time, the positive sample judgment condition is not whether it is the same node, but whether the node importance value is similar.
- Different from the traditional method of taking all other nodes as negative samples, we design a negative sampling strategy based on label similarity. The negative sampling strategy makes full use of the characteristics of node importance evaluation tasks and constructs high-quality negative samples to improve the contrastive learning effect.
- Experiment results exhibit that our model outperforms other methods on real-world datasets and demonstrate the effectiveness of CLNIE.

2   Related Work

2.1   Node Importance Estimation

Node importance is an attribute of the node itself, with which the characteristics of the graph can be deeply understood and mined. There are two main methods for evaluating the importance of nodes. One category is based on the characteristics of direct description by definition [2,3], such as degree centrality [25], betweenness centrality [18], and closeness centrality [5]. Another category is some important algorithms developed in recent years. PageRank propagates the importance of each node by traversing the graph structure or transmitting it to random nodes with a fixed probability. Personalized PageRank [9] is proposed to make random walks consider specific topics. These methods cannot handle different definitions of importance and various types of nodes and relationships in knowledge graphs. So they cannot be used for node importance evaluation tasks in knowledge graphs.

More recently, with the continuous study of graph data, models based on graph neural networks for node importance evaluation tasks in knowledge graphs have emerged. GENI [21] inputs the node representation into a scoring network, which converts node embedding into a one-dimensional importance value. MultiImport [22] effectively solves the challenge brought by multiple input signals to the task of evaluating the node importance in knowledge graphs. RGTN [12] proposes a representation learning-based framework, which fully utilizes the structural and semantic information of nodes. However, they can hardly maintain performance with insufficient labels. Therefore, we focus on node importance prediction with few labels on the knowledge graph. Moreover, there are studies considering the heterogeneous importance values of nodes [11] and the importance values of nodes on dynamic heterogeneous graphs [7]. These are different from our study of the importance of a single type of nodes on a static knowledge graph.

2.2   Graph Contrastive Learning

Contrastive learning is a self-supervised learning method that has been extensively studied in image classification, text classification, and visual question answering in recent years [4,6,10]. In the field of graph learning, graph data has the problem that labels are difficult to obtain, resulting in a lack of labels. More recently, researchers have focused on applying contrastive learning techniques to graph representation learning tasks, which have achieved good results. DGI [24] trains the encoder by maximizing local mutual information so that the node-level representations and the graph-level representations generated by the model are closer in the embedding space. GCC [23] introduces the contrastive learning method into the pre-training task of a graph, in which the training data and testing data come from different ranges. MVGRL [8] trains a graph encoder by maximizing mutual information between different structured view encoded representations of a graph. GRACE [26] generates two views through corruption and learns node representations by maximizing the consistency of node representations in these two views. Most of these graph contrastive learning methods are designed for tasks such as node classification without considering the characteristics of the node importance evaluation task. In this paper, we introduce contrastive learning into the task of node importance evaluation in knowledge graphs and design two kinds of contrastive losses to generate a rich representation of nodes with few labels.

3   Preliminaries

Definition 1. *Knowledge Graph.* *A knowledge graph can be defined as* $\mathcal{G} = (\mathcal{V}, \mathcal{R}, \mathcal{E}, \mathcal{X}, \mathcal{Z})$, *where nodes* $\mathcal{V}$ *correspond to entities, edges* $\mathcal{E}$ *correspond to relationships in* $\mathcal{R}$. $\mathcal{X}$ *and* $\mathcal{Z}$ *represent the features of nodes and relations, respectively.* (u, v, r) *represents that there is a relationship* $r \in \mathcal{R}$ *between node* u *and node* v.

Definition 2. *Node Importance Value.* *Given a node* i, *its importance value is expressed as a non-negative real number* $s_i \in R \geq 0$. *The larger the value, the higher the importance or popularity of the node.*

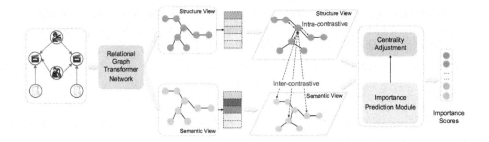

Fig. 1. The framework of the proposed model.

Definition 3. *Node Importance Estimation.* *Given a node i, its importance value is expressed as a non-negative real number $s_i \in R \geq 0$. The larger the value, the higher the importance or popularity of the node. Given a graph $G = ((\mathcal{V}, \mathcal{E}, \mathcal{X}_v))$ and the importance values of some of its nodes, node importance evaluation is to learn a function $S : \mathcal{V} \to [0, +\infty)$ to calculate the importance values of each node in the graph.*

4 Method

4.1 Framework

Figure 1 shows the framework of CLNIE. We introduce contrastive learning into the node importance evaluation task. Specifically, we first extract the structure and semantic features to construct different views of the KG and use the relational graph transformer network [12] to obtain two representations of nodes as two views for contrastive learning. Then we design an unsupervised contrastive loss and a semi-supervised contrastive loss to optimize the two representations. In order to select high-quality negative samples for contrastive learning, we propose a negative sampling strategy based on label similarity. Finally, we map the learned node representation to a one-dimensional importance value to get the final node importance score. Root mean squared error and learned ranking loss are used during the training of the model. The details of each component of the model are described next.

4.2 Unsupervised Contrastive Loss

Different from the traditional methods of contrastive learning between original graph and damaged graph or augmented graph, we use the structural and semantic embeddings obtained by the relational graph transform network as two views for collaborative supervision. With this approach, we design a novel dual-view contrastive mechanism, which can capture complex structural and semantic features in the knowledge graph at the same time. As the training progresses, these two views will guide each other and optimize together. In order to use two representations of nodes for contrastive learning, we first feed the embeddings obtained from two views into an MLP with a hidden layer and map them into the space where the contrastive loss is computed by:

$$z_v^{(s)} = W^2 \sigma \left(W^1 h_v^{(s)} + b^1 \right) + b^2 \tag{1}$$

$$z_v^{(c)} = W^2 \sigma \left(W^1 h_v^{(c)} + b^1 \right) + b^2 \tag{2}$$

where σ denotes the ELU nonlinear activation function, $h_v^{(s)}$ and $h_v^{(c)}$ are the structure representation and semantic representation generated from RGTN [12]. W^1, W^2, b^1 and b^2 are parameters shared by the two views. After transformation, we get the representation of structure and semantics in the contrastive space. Next, we need to contrast and optimize the two embeddings.

In order to make full use of the node structural information and semantic information in the knowledge graph, we first design an unsupervised contrastive learning method in CLNIE. The method makes the model distinguish the embedding of the same node from the embedding of other nodes in two views.

Given the embedding of node v in the structural view, we take its embedding in the semantic view as a positive sample. In this case, there are two types of negative samples. The first is the other nodes in the structural view, the second is the other nodes in the semantic view. Then the unsupervised contrastive loss of node v in the structural view is defined as:

$$\mathcal{L}_{v-un}^{(s)} = -\log \frac{e^{f\left(z_v^{(s)}, z_v^{(c)}\right)}}{e^{f\left(z_v^{(s)}, z_v^{(c)}\right)} + e^{f\left(z_v^{(s)}, z_{u'}^{(c)}\right)} + e^{f\left(z_v^{(s)}, z_{u''}^{(s)}\right)}} \tag{3}$$

where $u' \in N_{inter}$ denotes the nodes in a different view from node v, $u'' \in N_{intra}$ denotes the nodes in the same view as node v. $f(u,v) = \text{sim}(u,v)/\tau$, $\text{sim}(\cdot)$ denotes the cosine similarity of two vectors, and τ is the temperature parameter. Since the two views are symmetric, we can calculate the contrastive loss of node v under the semantic view similarly by:

$$\mathcal{L}_{v-un}^{(c)} = -\log \frac{e^{f\left(z_v^{(s)}, z_v^{(c)}\right)}}{e^{f\left(z_v^{(s)}, z_v^{(c)}\right)} + e^{f\left(z_v^{(s)}, z_{u'}^{(c)}\right)} + e^{f\left(z_v^{(c)}, z_{u''}^{(c)}\right)}} \tag{4}$$

Then the final unsupervised contrastive loss is obtained by:

$$\mathcal{L}_{un} = \frac{1}{2N} \sum_{i=1}^{N} \left(\mathcal{L}_{i-un}^{(s)} + \mathcal{L}_{i-un}^{(c)} \right) \tag{5}$$

where N denotes the number of nodes.

4.3   Semi-supervised Contrastive Loss

Unsupervised contrastive learning methods have achieved great success in various fields because they can use the rich information contained in the data itself to

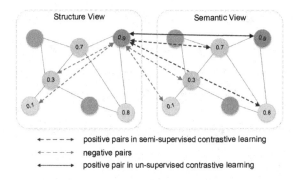

Structure View   Semantic View

◄ ─ ─ ─ ─►   positive pairs in semi-supervised contrastive learning
◄ ─ ─ ─ ─►   negative pairs
◄─────────►   positive pair in un-supervised contrastive learning

Fig. 2. Semi-supervised contrastive learning.

guide the process of representation learning. However, unsupervised contrastive
learning techniques do not exploit the scarce but valuable label information in
semi-supervised learning problems. Just as in the unsupervised contrastive learning model of the previous subsection, all nodes except the node itself are roughly
regarded as negative samples. This partitioning is obviously incomplete, which
may lead to the limitations of the learned node embedding. To tackle this issue,
as shown in Fig. 2, we treat nodes of similar importance as positive examples
and bring them close to each other in the embedding space, which facilitates
the hyperplane partitioning of CLNIE. For example, if two movies have similar
ratings, they are highly similar and closer in the embedding space.

To bring scarce but valuable label information into the model training, we
contrast the label-guided positive samples with the other negative samples.
Nodes with similar importance values will be mapped to adjacent points in
the low-dimensional hypersphere. Then the semi-supervised contrastive loss of
nodes in the structural view and the semantic view can be expressed as:

$$\mathcal{L}_{v-semi}^{(s)} = -\log \frac{\sum_{u \in N_{\text{inter-topk}}} e^{f\left(z_v^{(s)}, z_u^{(c)}\right)}}{e^{f\left(z_v^{(s)}, z_v^{(c)}\right)} + e^{f\left(z_v^{(s)}, z_{u'}^{(c)}\right)} + e^{f\left(z_v^{(s)}, z_{u''}^{(s)}\right)}} \qquad (6)$$

$$\mathcal{L}_{v-semi}^{(c)} = -\log \frac{\sum_{u \in N_{\text{inter-topk}}} e^{f\left(z_v^{(s)}, z_u^{(c)}\right)}}{e^{f\left(z_v^{(s)}, z_v^{(c)}\right)} + e^{f\left(z_v^{(s)}, z_{u'}^{(c)}\right)} + e^{f\left(z_v^{(c)}, z_{u''}^{(c)}\right)}} \qquad (7)$$

where $N_{inter-topk}$ denotes k nodes whose importance value is closest to that of
node v in different views. k is a hyperparameter, a threshold for the number
of positive samples that can be set to different values according to different
datasets. The semi-supervised contrastive loss of v, denoted by $\mathcal{L}_{semi}$, is then:

$$\mathcal{L}_{semi} = \frac{1}{2N} \sum_{i=1}^{N} \left(\mathcal{L}_{v-semi}^{(s)} + \mathcal{L}_{v-semi}^{(c)}\right) \qquad (8)$$

where N denotes the number of nodes.

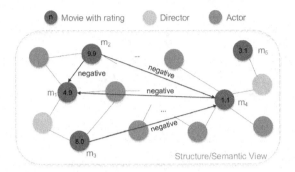

Fig. 3. Negative sampling strategy based on label similarity.

Finally, the overall contrastive loss of CLNIE $\mathcal{L}_{con}$ consisting of unsupervised and semi-supervised contrastive loss is as follows:

$$\mathcal{L}_{con} = \mathcal{L}_{un} + \mathcal{L}_{semi} \tag{9}$$

4.4   Negative Sampling Strategy Based on Label Similarity

The quality of negative samples is an important factor affecting the effect of contrastive learning. The existing negative sampling strategies can be summarized into two categories. One is to treat all other nodes in the current batch of the node as negative samples. However, the number of negative samples is limited by the size of the batch, and the size of the batch is limited by the size of GPU memory. Another negative sampling strategy is adversarial sampling. Adversarial negative sampling consists of a generator and a discriminator, where the generator acts as a sampler to generate samples to confuse the discriminator, and the discriminator needs to judge whether a given sample is a positive sample or a generated sample. However, this adversarial sampling strategy has a complex framework, which requires a long training time. Moreover, the game between the generator and the countermeasure may not converge to the ideal Nash equilibrium state, so its application scenarios are greatly limited.

Considering that samples with a greater difference from node importance values are more relevant to the results of the improved model, we propose a negative sampling strategy based on label similarity. The node with the least similar label value to the central node is added to the negative samples to contrast with the positive sample. Specifically, we calculate the label difference between the node and all other nodes at first, then sort all other nodes according to the label difference. After that, we select the percentage of the number of negative samples according to the size of the dataset. finally, the selected negative samples and positive samples will be used for contrastive learning. At this time, $u^{'}$ and $u^{''}$ in Eqs. (3), (4), (6) and (7) are no longer all other nodes except their own or label proximity nodes, but should be represented as $u^{'} \in N_{inter-w_k}$ and $u^{''} \in N_{inter-w_k}$. $N_{inter-w_k}$ and $N_{inter-w_k}$ represent the most dissimilar nodes to node v in different views and the same view respectively. Figure 3 shows the negative sampling strategy based on label similarity in a movie knowledge graph.

4.5   Model Training

After obtaining the representations of nodes through two different contrastive models, we use the importance prediction module in [12] to convert node representations to node importance values:

$$s_v^{(s)} = W_s^1 h_v^{(s)}, s_v^{(c)} = W_s^2 h_v^{(c)} \tag{10}$$

where W_s^1 and W_s^2 are projection matrices, $s_v^{(s)}$ is the structural importance of nodes and $s_v^{(c)}$ is the semantic importance of nodes. Considering the potential correlation between node importance and node degree, we adjust the structural importance of nodes centrally:

$$c(v) = \log(d(v) + \epsilon), s_v^{'(s)} = \sigma^* \left(s_v^{(s)} \cdot c(v) \right) \tag{11}$$

where ϵ is a small positive constant. Then the final importance is:

$$s_v^* = \text{LeakyReLU} \left(\gamma^{(s)} s_v^{'(s)} + \gamma^{(c)} s_v^{(c)} \right) \tag{12}$$

where $\gamma^{(*)}$ is attention weights, which can be calculated by:

$$\gamma^{(*)} = \frac{\exp\left(h_v^{(*)} \lambda^T \right)}{\exp\left(h_v^{(s)} \lambda^T \right) + \exp\left(h_v^{(c)} \lambda^T \right)} \tag{13}$$

where λ is a trainable attention vector. To train CLNIE, we use the mean squared error of the given importance of nodes and the importance predicted by the model. Then the total loss of the model can be calculated by:

$$\mathcal{L} = \alpha \mathcal{L}_{mse} + (1 - \alpha) \mathcal{L}_{con} + \mathcal{L}_{ltr} \tag{14}$$

where $\mathcal{L}_{mse}$ denotes the RMSE loss, $\mathcal{L}_{con}$ denotes the contrastive loss, $\mathcal{L}_{ltr}$ denotes the LTR loss [12]. α is a hyperparameter that controls the proportion of different parts of the loss function.

5   Experiments

5.1   Datasets

We selected three real-world public knowledge graph datasets for experiments, which were used in previous node importance studies [12,21,22]. We use the

Table 1. Statistics of datasets.

Dataset	# Nodes	# Edges	# Predicates
FB15K	14,951	592,213	1,345
TMDB5K	114,805	761,648	34
IMDB	1,124,995	9,729,868	30

datasets constructed in [12]. Table 1 summarizes the statistics of the datasets. FB15K is a subset of FreeBase [1], which contains rich text information of relational triples and entity pairs of the general knowledge base. FB15K is the dataset with the largest number of predicates used in our experiment. For each entity in the graph, we regard the pageview number in the last 30 d of its corresponding Wikipedia page as the node importance score. TMDB5K is a movie knowledge graph, which contains movie entities and other related information entities such as actors, directors, companies and casts. The original dataset provides the "popularity" information of movies, which is regarded as the node importance score. IMDB is also a movie knowledge graph created from the public IMDb dataset. IMDB contains related information such as movies, actors, genres and staff. This is the largest dataset we evaluated, which has millions of nodes. We take the number of votes a movie received provided by the original dataset as the node importance score.

5.2 Baselines

We compare the proposed model with a variety of the existing methods to evaluate the effectiveness of CLNIE, which can be classified into the three families. (1) **Non-Trainable Approaches.** PageRank [19] attaches an initial value to each node and obtains the PageRank value through multiple rounds of iterations until it finally converges. The PageRank value is the importance score of the node. Based on it, Personalized PageRank [9] calculates the correlation between a specific page and all other pages and takes it as the node importance value. (2) **Non-graph Supervised Approaches.** We select some representative supervised methods to explore their performance in node importance estimation, including Logistic Regression (LR), Random Forest (RF) and Multilayer Perceptron (MLP). These methods are all classical regression models. (3) **Graph Neural Network Based Approaches.** GENI [21] uses predicate-aware attention aggregation to consider the influence of predicates in the knowledge graph. RGTN [12] proposes a relational graph transformation encoder to obtain the structural and semantic information of nodes, with which the model achieves the state-of-art performance. These methods use graph neural networks to process graph data and have achieved excellent performance in the task of node importance estimation.

5.3   Evaluation Metrics

In order to evaluate the effectiveness of the model on the correlation of node importance and the quality of ranking, we use the Normalized Discounted Cumulative Gain (NDCG), Spearman correlation coefficient (SPEARMAN) and the overlap rate of the predicted nodes with the top k nodes (OVER). All indicators are that the higher the value, the better. We perform 5-fold cross validation on three metrics for all datasets and report the average and standard deviation values.

Table 2. The results of different methods over the three real-world datasets. The highest results are in bold and the second highest results are underlined.

Dataset	Method	NDCG@100	SPEARMAN	OVER@100
FB15K	PR	0.8334 ± 0.005	0.3457 ± 0.026	0.1099 ± 0.014
	PPR	$\underline{0.8341 \pm 0.005}$	0.3491 ± 0.004	$\underline{0.1101 \pm 0.014}$
	LR	0.8217 ± 0.011	$\underline{0.4325 \pm 0.005}$	0.0340 ± 0.004
	RF	0.8009 ± 0.017	0.3591 ± 0.018	0.0160 ± 0.010
	MLP	0.7910 ± 0.012	0.3890 ± 0.016	0.0240 ± 0.018
	GENI	0.7771 ± 0.021	0.2670 ± 0.050	0.0800 ± 0.022
	RGTN	0.8113 ± 0.013	0.2632 ± 0.133	0.0160 ± 0.010
	CLNIE	$\mathbf{0.8498 \pm 0.012}$	$\mathbf{0.4963 \pm 0.043}$	$\mathbf{0.1420 \pm 0.021}$
TMDB5K	PR	0.8391 ± 0.019	0.6210 ± 0.032	$\underline{0.4140 \pm 0.010}$
	PPR	0.8408 ± 0.009	0.6706 ± 0.002	$\mathbf{0.4233 \pm 0.019}$
	LR	0.8219 ± 0.019	0.6012 ± 0.012	0.2960 ± 0.030
	RF	0.8215 ± 0.016	0.5916 ± 0.025	0.3259 ± 0.018
	MLP	0.8345 ± 0.011	0.6434 ± 0.011	0.2917 ± 0.048
	GENI	0.8146 ± 0.068	0.6304 ± 0.139	0.2920 ± 0.086
	RGTN	$\underline{0.8422 \pm 0.021}$	$\underline{0.7160 \pm 0.019}$	0.3160 ± 0.058
	CLNIE	$\mathbf{0.8830 \pm 0.012}$	$\mathbf{0.7540 \pm 0.009}$	0.3880 ± 0.038
IMDB	PR	0.8587 ± 0.011	0.1541 ± 0.002	$\mathbf{0.3960 \pm 0.040}$
	PPR	0.8930 ± 0.014	0.4429 ± 0.001	0.2920 ± 0.017
	LR	0.8488 ± 0.020	0.5319 ± 0.002	0.1900 ± 0.018
	RF	0.8911 ± 0.008	0.4793 ± 0.002	0.1300 ± 0.021
	MLP	0.8738 ± 0.015	0.4979 ± 0.005	0.1080 ± 0.041
	GENI	0.9322 ± 0.011	0.5617 ± 0.116	0.2500 ± 0.036
	RGTN	$\underline{0.9348 \pm 0.011}$	$\underline{0.6880 \pm 0.011}$	0.2580 ± 0.042
	CLNIE	$\mathbf{0.9527 \pm 0.003}$	$\mathbf{0.6932 \pm 0.049}$	$\underline{0.3139 \pm 0.020}$

5.4   Node Importance

Following the setup of GENI [21], we treat the importance of the target entities as training, validation, and testing data. Other entities are only used to generate embeddings of the target nodes. For example, in the TMDB5K dataset, we only care about the importance of the movie nodes. The nodes of other entities such as actors are also fed into the encoder, but their importance is not predicted. We split the datasets into training, validation, and testing parts with the ratio

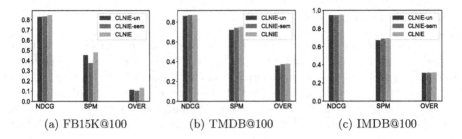

(a) FB15K@100   (b) TMDB@100   (c) IMDB@100

Fig. 4. The results of CLNIE and its variants.

of 1:2:7. Then we evaluate the performance of CLNIE in node importance estimation against several competitive baselines. Except for [12], all baselines use
the concatenation of structural features and semantic features as node input
features. Table 2 shows the evaluation results.

By analyzing the experimental results, we can see that: (1) The proposed
CLNIE achieves the best performance in most indicators of all datasets, which
proves the effectiveness of CLNIE. Moreover, we find that CLNIE obtains much
higher performance on NDCG@100 and SPEARMAN, which are more complicated than OVER@100. (2) When the labeling ratio is reduced to 10%, the
untrainable method can match or surpass the graph supervised methods GENI
and RGTN in some datasets. This phenomenon demonstrates the inevitable
drawback of the supervised method, that is, the model needs a large number of
labels to ensure its effectiveness. However, labels are usually difficult to obtain
in reality. (3) By observing the results of the baselines, we can see that PPR
has obtained suboptimal or optimal results on three datasets. Analysis of the
possible reasons is that the non-trainable method is not affected by the number
of node labels, so when the model effect of other methods decreases due to the
reduction of training labels, the non-trainable method can still maintain good
performance. Secondly, compared with the PR, PPR no longer randomly jumps
to any node during the random walk but determines the next node to jump
based on the user's preference, which can promote the model to evaluate the
importance of nodes better. In addition, LR obtains suboptimal results on some
datasets, which shows that when the ratio of node labels decreases, the simple
linear regression model can achieve good results instead. (4) Our model uses contrastive learning to utilize limited labels and achieves the best effect compared
with the baselines when the ratio of labels is few, which further illustrates the
effectiveness of contrastive learning when labels are scarce.

5.5   Ablation Study

In order to further measure the effectiveness of each component of our proposed model and explore the performance of the model more comprehensively,
we use CLNIE and its two variants to carry out further ablation studies on three
datasets. CLNIE-un denotes that CLNIE uses only unsupervised contrastive

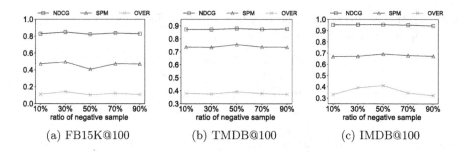

Fig. 5. Parameter sensitivity of CLNIE: Impact of w_k.

learning loss. CLNIE-sem denotes that CLNIE uses only semi-supervised contrastive learning loss. The experimental results are shown in Fig. 4. From the figure we can see that: (1) Compared with CLNIE, the effectiveness of the two variants in three evaluation indicators declined to a certain extent. (2) Observing the performance of the two variants in three datasets, we can find that the gain brought by semi-supervised contrastive learning loss is slightly greater than that of unsupervised contrastive learning loss, but this is not the case under a certain indicator of some datasets. This phenomenon is more obvious in TMDB5K, in which CLNIE-sem is superior to CLNIE-un in all three evaluation indexes. The finding shows that the positive sample of contrastive learning cannot be simply defined as the representation of the same node in another view. This simple way will wrongly lengthen the distance between two similar nodes in the embedding space, which will lead to the deviation of the model in evaluating the importance of nodes. (3) CLNIE achieves the best results under all evaluation metrics on all datasets, which demonstrates the effectiveness of using all components.

5.6   Parameter Sensitivity

Effect of Negative Samples Ratio. To observe how the negative sample ratio w_k in the negative sampling strategy affects CLNIE performance, we set up different w_k on three datasets for experiments. The results are displayed in Fig. 5, from which we can see that with the change of w_k, the performance of the model on all metrics changes to different degrees. For FB15K, with the increase of w_k, all three indexes reach the maximum when w_k reaches 30%, while TMDB5K and IMDB are both 50%. This finding demonstrates that if all nodes except the nodes with the closest labels are regarded as negative samples for contrastive learning, the model cannot distinguish nodes from negative samples well. Because among these negative samples, some are very different from the node and some are not so different from the node. However, if the label information is not used, the model will treat all negative samples equally. In addition, we can see that the three evaluation metrics have the same phenomenon of "low on both sides" in the figure. The phenomenon shows that when the ratio of negative samples is too low or too high, the effect of the model will be greatly reduced. Too many or too few negative samples make the model unable to distinguish positive and negative samples well. The results verify the effectiveness of the negative sampling strategy based on label similarity.

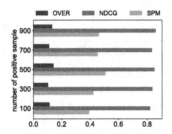

Fig. 6. Parameter sensitivity of CLNIE: Impact of k on FB15K.

Effect of Positive Samples Number. In semi-supervised contrastive learning, we take nodes with similar importance values as positive samples. Here we evaluate the effectiveness of CLNIE by setting different numbers of positive samples k on FB15K. The results are shown in Fig. 6. With the increasing number of positive samples, the accuracy of the model generally shows a trend of first increasing and then decreasing. When k reaches 500, all three metrics reach the maximum value. The phenomenon shows that if the number of positive samples is too small, the model will mistakenly pull away from the distance between the two samples that should be similar. However, if the number of positive samples is too large, the model will mistakenly regard the two samples that are not similar as positive samples, resulting in the degradation of model performance. To sum up, to make the model play its maximum role in each dataset, a different and appropriate number of positive samples should be set for different datasets.

5.7 Validity Analysis

In the previous experiment, we set the labeling ratio of the node to 10%. In order to observe the influence of the labeling ratio on the model, we decrease it from 50% to 10% in this section and display the experimental results of CLNIE and other trainable approaches. It can be seen from Fig. 7 that with the decrease in the ratio of the training set, the effectiveness of each method has declined to different degrees on the three metrics. When the labeling ratio is 50%, CLNIE has higher SPEARMAN and OVER@100 than the state-of-the-art solution, RGTN. In the process of changing the labeling ratio from 30% to 10%, CLNIE is always superior to other trainable approaches and the difference gradually widens. When the labeling ratio decrease to 10%, the effectiveness difference between CLNIE and other methods reaches the greatest. Experimental results prove the effectiveness of CLNIE in node importance evaluation with few labels.

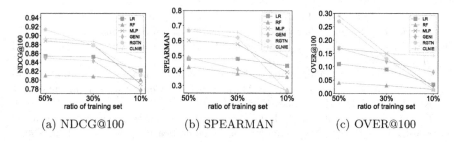

(a) NDCG@100   (b) SPEARMAN   (c) OVER@100

Fig. 7. Validity experiment of CLNIE on FB15K.

6   Conclusion

In this paper, we propose CLNIE for evaluating the importance of nodes in knowledge graphs based on graph neural networks. The model uses the idea of contrastive learning to make up for the limitation of the existing node importance evaluation methods that need to use a large number of labels. Firstly, unsupervised contrastive learning loss is used to minimize the embedding of the same node in two different views. When considering the influence of nodes with similar importance values on the central node, we use a semi-supervised contrastive loss to shorten the distance between the representations of two nodes with similar importance values in the embedding space. In order to construct high-quality negative samples, we propose a negative sampling strategy based on label similarity. The experimental results on three real-world datasets prove that the proposed model outperforms the state-of-art solutions and the improvement up to 4% on NDCG@100, 6.3% on SPEARMAN, and 3.19% on OVER@100.

Acknowledgements. This work is supported by the National Nature Science Foundation of China (62072083) and the Fundamental Research Funds of the Central Universities (N2216017)

References

1. Bollacker, K., Evans, C., Paritosh, P., Sturge, T., Taylor, J.: Freebase: a collaboratively created graph database for structuring human knowledge. In: ACM SIGMOD International Conference on Management of Data, pp. 1247–1250 (2008)
2. Borgatti, S.P.: Centrality and network flow. Soc. Netw. **27**(1), 55–71 (2005)
3. Borgatti, S.P., Everett, M.G.: A graph-theoretic perspective on centrality. Soc. Netw. **28**(4), 466–484 (2006)
4. Chen, T., Kornblith, S., Norouzi, M., Hinton, G.: A simple framework for contrastive learning of visual representations. In: International Conference on Machine Learning, pp. 1597–1607. PMLR (2020)

5. Cohen, E., Delling, D., Pajor, T., Werneck, R.F.: Computing classic closeness centrality, at scale. In: ACM Conference on Online Social Networks, pp. 37–50 (2014)
6. Gao, T., Yao, X., Chen, D.: SimCSE: simple contrastive learning of sentence embeddings. arXiv preprint arXiv:2104.08821 (2021)
7. Geng, H., et al.: Modeling dynamic heterogeneous graph and node importance for future citation prediction. In: ACM International Conference on Information & Knowledge Management, pp. 572–581 (2022)
8. Hassani, K., Khasahmadi, A.H.: Contrastive multi-view representation learning on graphs. In: International Conference on Machine Learning, pp. 4116–4126. PMLR (2020)
9. Haveliwala, T.H.: Topic-sensitive pagerank. In: International Conference on World Wide Web, pp. 517–526 (2002)
10. He, K., Fan, H., Wu, Y., Xie, S., Girshick, R.: Momentum contrast for unsupervised visual representation learning. In: IEEE/CVF Conference on Computer Vision and Pattern Recognition, pp. 9729–9738 (2020)
11. Huang, C., Fang, Y., Lin, X., Cao, X., Zhang, W., Orlowska, M.: Estimating node importance values in heterogeneous information networks. In: ICDE, pp. 846–858. IEEE (2022)
12. Huang, H., Sun, L., Du, B., Liu, C., Lv, W., Xiong, H.: Representation learning on knowledge graphs for node importance estimation. In: ACM SIGKDD, pp. 646–655 (2021)
13. Jung, J., Park, N., Lee, S., Kang, U.: Bepi: Fast and memory-efficient method for billion-scale random walk with restart. In: ACM International Conference on Management of Data, pp. 789–804 (2017)
14. Kleinberg, J.M.: Authoritative sources in a hyperlinked environment. J. ACM (JACM) 46(5), 604–632 (1999)
15. Liu, C., Fu, Y., Xu, C., Yang, S., Li, J., Wang, C., Zhang, L.: Learning a few-shot embedding model with contrastive learning. In: the AAAI Conference on Artificial Intelligence, vol. 35, pp. 8635–8643 (2021)
16. Liu, X., Tang, J.: Network representation learning: a macro and micro view. AI Open 2, 43–64 (2021)
17. Luo, X., Chen, Y., Wen, L., Pan, L., Xu, Z.: Boosting few-shot classification with view-learnable contrastive learning. In: IEEE International Conference on Multimedia and Expo, pp. 1–6. IEEE (2021)
18. Newman, M.E.: A measure of betweenness centrality based on random walks. Soc. Netw. 27(1), 39–54 (2005)
19. Page, L., Brin, S., Motwani, R., Winograd, T.: The pagerank citation ranking: bringing order to the web. Tech. Rep., Stanford InfoLab (1999)
20. Palmonari, M., Minervini, P.: Knowledge graph embeddings and explainable AI. Knowledge Graphs for Explainable Artificial Intelligence: Foundations, Applications and Challenges 47, 49 (2020)
21. Park, N., Kan, A., Dong, X.L., Zhao, T., Faloutsos, C.: Estimating node importance in knowledge graphs using graph neural networks. In: ACM SIGKDD, pp. 596–606 (2019)
22. Park, N., Kan, A., Dong, X.L., Zhao, T., Faloutsos, C.: Multiimport: inferring node importance in a knowledge graph from multiple input signals. In: ACM SIGKDD, pp. 503–512 (2020)
23. Qiu, J., Chen, Q., Dong, Y., Zhang, J., Yang, H., Ding, M., Wang, K., Tang, J.: GCC: graph contrastive coding for graph neural network pre-training. In: ACM SIGKDD, pp. 1150–1160 (2020)

24. Velickovic, P., Fedus, W., Hamilton, W.L., Liò, P., Bengio, Y., Hjelm, R.D.: Deep graph infomax. ICLR (Poster), vol. 2, no 3, p. 4 (2019)

25. Zhang, J., Luo, Y.: Degree centrality, betweenness centrality, and closeness centrality in social network. In: International Conference on Modelling, Simulation and Applied Mathematics, pp. 300–303. Atlantis Press (2017)

26. Zhu, Y., Xu, Y., Yu, F., Liu, Q., Wu, S., Wang, L.: Deep graph contrastive representation learning. arXiv preprint arXiv:2006.04131 (2020)

Block Decomposition with Multi-granularity Embedding for Temporal Knowledge Graph Completion

Lupeng Yue[1], Yongjian Ren[1], Yan Zeng[1], Jilin Zhang[1(✉)], Kaisheng Zeng[2], and Jian Wan[1]

[1] School of Computer Science and Technology, Hangzhou Dianzi University, Hangzhou, China
{lupengyue,yongjian.ren,yz,jilin.zhang,wanjian}@hdu.edu.cn
[2] Department of Computer Science and Technology, Tsinghua University, Beijing, China
zks19@mails.tsinghua.edu.cn

Abstract. Temporal knowledge graph (TKG) completion is the mainstream method of inferring missing facts based on existing data in TKG. Majority of existing approaches to TKG focus on embedding the representation of facts from a single-faceted low-dimensional space, which cannot fully express the information of facts. Furthermore, most of them lack the comprehensive consideration of both temporal and non-temporal facts, resulting in the inability to handle the two types of facts simultaneously. Thus, we propose BDME, a novel **B**lock **D**ecomposition with **M**ulti-granularity **E**mbedding model for TKG completion. It adopts multivector factor matrices and core tensor em-bedding for fine-grained representation of facts based on the principle of block decomposition. Moreover, it captures interaction information between entities, relationships, and timestamps in multiple dimensions. By further constructing a temporal and static interaction model, BDME processes temporal and non-temporal facts in a unified manner. Besides, we propose two kinds of constraint schemes, which introduce time embedding angle and entity bias component to avoid the overfitting problem caused by a large number of parameters. Experiments demonstrate that BDME achieves sub-stantial performance against state-of-the-art methods on link prediction.

Keywords: Temporal Knowledge Graph · Link Prediction · Knowledge Graph Completion

1 Introduction

Temporal knowledge graphs (TKGs) provide time-aware structural knowledge about the entities and relations in the real world by incorporating the facts' timestamps. Their powerful expressiveness ability has made them favorable for various applications over the last few years, e.g., social networks [3], and recommender systems [2]. Unfortunately, despite their successes and large scales, existing TKGs widely suffer from incompleteness which negatively affects their effectiveness for downstream applications. Each fact in TKGs is commonly represented by a quadruple, which is associated with a timestamp, e.g. *Barack Obama, is president of, The United States, [2008, 2017]*). TKG completion aims to predict missing entities in quadruples and alleviate the incompleteness, such as (*?, is president of, The United States, [2017, 2021]*) whose answer is Donald John Trump. Despite some methods have been successful, the following two critical issues still need further investigation.

How to provide the multi-faceted and fine-grained embedding representation for entities, relations, and timestamps in TKGs? To predict the missing entity, the main challenge is to fully mine the multi-faceted and fine-grained information of existing facts, which helps to fully understand the existing knowledge and thus make more comprehensive predictions for the missing facts. This perspective has led to the development of several embedding models. The SOTA methods, such as TcomplEx [1] and TIMEPLEX [4], introduce conjugate vectors for modeling entities from both real-valued and complex-valued dimensions. However, those methods still embed the entities in a single-faceted low-dimensional representation space, which has not only poor generalization and expressiveness, but also has low degrees of freedom for embedding and cannot model entities flexibly.

How to model the interactions in different types of facts for comprehensive knowledge fusion for TKG completion? Some facts in TKG may vary with time-wise dimension, such as *(Barack Obama, study at, Harvard University, [1988, 1991]*), while some facts in SKG are independent of time, such as (Barack Obama, was born in, America). It is crucial to consider different types of known facts for unknown fact prediction because Barack Obama was born and studied in America, both of which provide valuable information for the prediction of the fact *(Barack Obama, is president of, The United States, [2008, 2017]*). Existing methods only consider known temporal facts and ignore other types of known facts, that is, non-temporal facts that remain active all the time, such as HyTE [5] and RTGE [6]. These methods consider known facts at a coarse granularity level and thus bias the predictions of unknown facts.

To address the above two issues, we propose a novel model BDME for temporal completion over TKGs, which deals with the problems of the existing methods in these two aspects. **For multi-faceted and fine-grained embedding representation**, we represent quadruple *(subject, predicate, object, timestamp)* by a pair of basis vectors in 2-grade geometric space to learn the corresponding embedding representation. Furthermore, we design a geometric product to better integrate each other's fine-grained information. **For interactions in different types of facts**, we adopt the block decomposition (BD) [7] model with two component tensors, which address temporal and non-temporal facts. In particular, we present two kinds of constraint schemes to avoid the overfitting problem.

In summary, our contributions are as follows:

- We propose a new tensor decomposition model for TKG completion, which encodes facts' fine-grained information from two aspects. For multi-faceted and fine-grained embedding representation, we introduce multi-vector embeddings and a geometric product to model entities for TKG em-bedding. For interactions in different types of facts, we propose to encode the temporal and non-temporal facts through the two component tensors in BD.
- We propose two kinds of constraint schemes, which can effectively avoid the overfitting problem.
- Quantitative metrics on four datasets show that BDME outperforms several state-of-the-art baselines in link prediction.

2   Related Work

In this section, we introduce TKG completion methods proposed in recent years. Note that most of the related work is based on tensor decomposition.

2.1   Temporal Knowledge Graph Embeddings

It is also divided into three types according to the loss function.

Based on Translation Distance Models: HyTE extends TransH by embedding temporal information to a hyperplane. Then, it minimizes the translation distance to realize embedding learning of entities and predicates. RTGE extends HyTE by adding a smoothing factor to preserve the structural information and evolution pattern.

Based on Tensor Decomposition Models: TIMEPLEX extends ComplEx by constructing three time-related constraint schemes, and it incorporates prior knowledge into the model very well. TELM [8] extended RTGE and TIMEPLEX by introducing a randomly learned bias component. To a certain degree, it promotes the difference between time steps with different distances.

Based on Neural Network Models: RE-GCN [9] captures the structural dependencies of timestamps, sequential patterns of facts, and static properties through a relation-aware graph convolutional neural network. CEN [10] proposes length-diversity and time-variability in TKG sequences for the first time, and uses length-aware convolutional neural networks and online learning strategies to handle these two problems.

3   Proposed Method

3.1   Overview

We give a TKG consisting of observed facts as $G_{K_G} = (n, r, v, \tau)$, where $n \in N_{K_G}$, $r \in R_{K_G}$, $v \in V_{K_G}$, and $\tau \in T_{K_G}$. N_{K_G} and V_{K_G} are the set of entities,

R_{K_G} is the set of relations, and T_{K_G} is the set of timestamps associated with the relations, respectively. Then, given G_{K_G} and query $q = (?, r, v, \tau)$, TKG completion is formulated as predicting $n \in N_{K_G}$ as the most probable to fill in the query. Note that object entity prediction $q = (n, r, ?, \tau)$ can be evaluated in a similar way after adding an inverse edge to G_{K_G}. Finally, our goal is to maximize the score $G_{K_G} = (n, r, v, \tau)$.

We illustrate the overall framework of BDME, shown in (Fig. 1). BDME has three main components, namely **S**tatic and **T**emporal **I**nformation deep representation learning (STI) module, **T**imestamp **E**mbedding **E**volution constraint (TEE) module, and **T**ime-dependent **E**ntity **E**mbedding **I**nteraction constraint (TEEI) module. In the STI module, we manage temporal and non-temporal facts separately based on the BD of the two component tensors, and introduce the multivector factor matrix embedding representation in each part, which effectively solves the multi-faceted and fine-grained embedding representation and interactions in different types of facts. Then, in the TEE module, we consider adjacent and non-adjacent timestamps. Concretely, we confine the embedding angle to maximize valuable timestamp information. Finally, in the TEEI module, we propose bias embedding learned from the training process after random initialization to pro-mote the difference between entities with different distances.

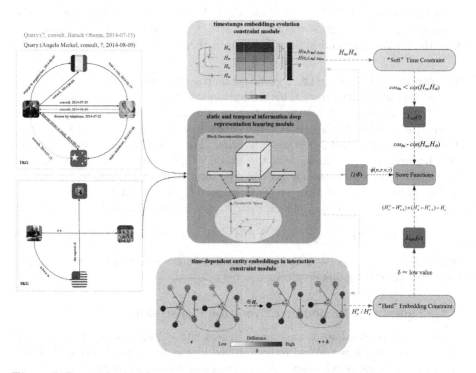

Fig. 1. Different types of facts in the KGs, including non-temporal and temporal facts.

3.2   Static and Temporal Information Deep Representation Learning

Block Decomposition. Static information in SKG can be considered as background knowledge for TKG. Thus, we merge the SKG into TKG representation to further obtain the plausibility score of their interactions as follows:

$$\phi(n, r, v, \tau) = \phi_t(n, r, v, \tau) + \phi_s(n, r, v, \tau) \tag{1}$$

where $\phi_t()$ and $\phi_s()$ encodes temporal and static properties, respectively. BD is more expressive and generalizable than CP and Tucker. Intuitively, BD introduces multiple core tensors and factor matrices. BD captures different aspects between four dimensions. Thus, BD is well suited for the TKG. We adopt BD with two component tensors for different types of facts as follows:

$$\chi \approx Z_1 \times_1 N_1 \times_2 R_1 \times_3 V_1 + Z_2 \times_1 N_2 \times_2 R_2 \times_3 V_2 \tag{2}$$

where $Z_1, Z_2 \in \mathbb{R}^{d_n \times d_r \times d_v}$ the core tensors. $N_1, N_2 \in \mathbb{R}^{n \times d_n}, R_1, R_2 \in \mathbb{R}^{r \times d_r}$ and $V_1, V_2 \in \mathbb{R}^{n \times d_v}$ represent the factor matrices. d_n, d_r, and d_v represent the embedding dimensions. $\times_n$ represents the n-mode product between tensors and matrices. Therefore, $\phi_t(n, r, v, \tau)$ can be defined as follows:

$$\phi_t(n, r, v, \tau) = Z_t \times_1 N_t^\tau \times_2 R_t^\tau \times_3 V_t + Z_s \times_1 N_s \times_2 R_s \times_3 V_s \tag{3}$$

where $N_t^\tau, R_t^\tau, V_t^\tau \in \mathbb{R}^{dn}, \mathbb{R}^{dr}, \mathbb{R}^{dv}$ and $N_s, R_s, V_s \in \mathbb{R}^{dn}, \mathbb{R}^{dr}, \mathbb{R}^{dv}$ represents the temporal and static representation. $Z_t \in \mathbb{R}^{d_n \times d_r \times d_v}$ and $Z_s \in \mathbb{R}^{d_n \times d_r \times d_v}$ can capture the interactions and influences between them.

Geometric Algebra. In the 2-grade geometric algebra $\mathbb{N}^2$, we refer to its members as 2-grade multivectors. This assumption is that there exists a pair of basis x_1, x_2, and $\mathbb{N}^2$ is based on two rules: $x_1 x_1 = x_2 x_2 = 1$ and $x_1 x_2 = x_2 x_1 = -1$. Then, the multivector space is 4-dimensional with basis: $0 - vector, scalars$, $1 - vector, vectors$ and $2 - vector, bivectors$. We introduce the geometric product denoted as $\bigotimes_n$, where n is the grade of multivectors embeddings. The geometric product represents the two vectors as a multiplication between a scalar, a vector, and a 2-dimensional vector. N in $\mathbb{N}^2$ can be represented as $N = a + bx_1 + cx_2 + dx_1 x_2$. Then, $\phi_t(n, r, v, \tau)$ is defined as follows:

$$\phi(n, r, v, \tau) = Z_t \bigotimes N_t^\tau \bigotimes R_t^\tau \bigotimes V_t + Z_s \bigotimes N_s \bigotimes R_s \bigotimes V_s \tag{4}$$

where $N_t^\tau = a_1 + a_2 x_1 + a_3 x_2 + a_{23} x_1 x_2$ and $R_t^\tau = b_1 + b_2 x_1 + b_3 x_2 + b_{23} x_1 x_2$ respectively. The geometric product of $\bigotimes$ is 2.

3.3   Timestamp Embedding Evolution Constraint

We consider both adjacent and non-adjacent timestamps and propose a "soft" time constraint scheme. We introduce embedding angles to make adjacent timestamps similar and non-adjacent timestamps as different as possible:

$$L_{soft(t)} = \sum_{\tau_a}^{T-1} \|max\langle[cos\theta_\alpha - cos(H_{\tau_b}, H_{\tau_a}), 0]\rangle\|_p^p \tag{5}$$

where T is the number of time steps. H_{τ_a} and H_{τ_b} are row a and row b. The embedding $cos(H_{\tau_b}, H_{\tau_a})$ between two timestamps should exceed $cos\theta_\alpha$, $\theta_\alpha = min(\sigma\alpha, 90°)$. Let σ denote the rate at which the angle climbs.

3.4   Time-Dependent Entity Embedding Interaction Constraint

In addition, We propose a "hard" entity embedding constraint scheme by adding a bias component between the entity embeddings of adjacent time steps:

$$L_{hard(\varepsilon)} = Mean[(H_\tau^n - H_{\tau-1}^n) + (H_\tau^v - H_{\tau-1}^v) - H_c], \tau \in [\tau_a, \tau_b] \qquad (6)$$

where H_c denotes the bias embedding learned from the training process after random initialization. This part promotes that the difference of adjacent entities is smaller than distant entities.

3.5   Model Training

For each relation r, we resort to data augmentation to add the opposite predicate in the form of (v, r^{-1}, n, τ) into the training set. Finally, the multiclass loss:

$$L(\phi) = -log\langle \frac{exp[\phi(n, r, v, \tau)]}{\sum_{n' \in N} exp[\phi(n', r, v, \tau)]} \rangle - log\langle \frac{exp[\phi(v, r^{-1}, n, \tau)]}{\sum_{v' \in N} exp[\phi(v', r^{-1}, n, \tau)]} \rangle \qquad (7)$$

where (n', r, v, τ) is the negative sample obtained by replacing the positive sample (n, r, v, τ). The final loss function:

$$L = L_\phi + \lambda_1 \cdot L_{soft(t)} + \lambda_2 \cdot L_{hard(\varepsilon)} \qquad (8)$$

where λ_1 and λ_2 are the tuning coefficients to balance the constraints.

4   Experiments

4.1   Experimental Setup

Table 1. Statistics of datasets.

DataSets	N	R	T	n_{train}	n_{valid}	n_{test}	Time Gap
ICEWS14	6869	230	365	72826	8941	8963	1 day
GDELT	7691	240	366	1734399	238765	305241	15 mins
Wikidata12k	12554	24	237	32497	4062	4062	1 year
Yago11k	10623	10	388	16406	2050	2051	1 year

Datasets. We evaluate our proposed model on four typical benchmark datasets for TKG completion. ICEWS14 is a subset of the ICEWS, each containing sociopolitical events in 2014. GDELT is the TKG that stores facts about human social relationships. This part of the dataset is processed similarly to RE-GCN. Then, we divide ICEWS14 and GDELT into training sets (80%), validation sets (10%), and test sets (10%). Wikidata12k is a subset of Wikidata, YAGO11k is a subset of YAGO3, and time representation is highly similar to Wikidata12k. The main statistics of the datasets are listed in Table 1 gives a summary of all heading levels.

Baselines. SKG: TransE, DisMult, ComplEx, SimplE, RotatE. **TKG:** HyTE, ATiSE, TeRo, TELM, RotateQVS-S [11], BTDG [12], TuckERTNT [13].

Parameter settings. We implement all the experiments in PyTorch and trained in GeForce RTX 3080. We use the Adgard optimizer with learning rate of 0.1 and batch size of 1000. σ is set to $10°$ in the experiments. The details of the parameters are presented in Table 2.

Table 2. Statistics of Parameter.

DataSets	λ_1	λ_2	$\triangle_1$	$\triangle_2$	d_n/d_v	d_r
ICEWS14	0.0075	0.001	1	-	300	100
GDELT	0.025	0.001	1	-	100	100
Wikidata12k	0.025	0.0025	-	1	100	50
YAGO11k	0.025	0.001	-	100	50	50

4.2 Experimental Results and Analysis

Table 3. Performance for link prediction task on ICEWS14 and GDELT.

DataSets	ICEWS14				GDELT			
Method	MRR	Hit@1	Hit@3	Hit@10	MRR	Hit@1	Hit@3	Hit@10
TransE(2013)[★]	0.280	0.094	-	0.637	0.113	0.0	0.158	0.312
DisMult(2014)[★]	0.439	0.323	-	0.672	0.196	0.117	0.208	0.348
ComPlEx(2016)[★]	0.467	0.347	0.527	0.716	-	-	-	-
SimplE(2018)[♣]	0.458	0.341	0.516	0.687	0.206	0.124	0.220	0.366
RotatE(2019)[★]	0.418	0.291	0.478	0.690	-	-	-	-
HyTE(2018)[★]	0.297	0.108	0.416	0.655	0.118	0.0	0.165	0.326
ATiSE(2019)[♦]	0.550	0.436	0.629	0.750	-	-	-	-
TeRo(2020)[★]	0.562	0.468	0.621	0.732	-	-	-	-
TELM(2021)[♦]	0.618	0.535	0.667	0.772	-	-	-	-
RQVS(2022)[♠]	0.575	0.489	0.625	0.737	0.259	0.165	0.270	0.428
BTDG(2022)[♠]	0.601	0.516	0.656	0.753	-	-	-	-
TuckT(2022)[♦]	0.625	0.544	0.673	0.773	0.275	0.190	0.298	0.440
BDME(ours)	**0.635**	**0.555**	**0.683**	**0.778**	**0.278**	**0.191**	**0.299**	**0.448**

Table 3 and Table 4 show the overall evaluation results of the BDME against baseline methods, where the best results are marked in bold, and the second best results are marked in underlined. The results that are not reported in the responding paper are marked in "-". Along with MRR, we report Hit@k k = 1, 3, 10 of the ground-truth entity. BDME outperforms all baseline methods on all the datasets. The results convincingly verify its effectiveness. Other results with [★] are from [14], [♣] is from [10], [♦] are from [4], and [♠] are from the original paper [11,12]. Note that RQVS and TuckT are shorthand for RotateQVS-S and TuckERTNT. We have the following observations.

Our proposed model outperforms all SKGE models because they do not consider temporal information on all the datasets. Especially, BDME surpasses all baseline models on MRR and Hits@1, 3, 10. The performance is close to TuckERTNT on ICEWS14 and GDELT. We argue that the main reason is that TuckERTNT uses Tucker decomposition, which is a special case of BD. Note that our proposed model even achieves the improvements of 3.2% on MRR, 3.3% in Hits@1, 3.0% in Hits@3, and 1.7% in Hits@10 over the second best baseline on YAGO11k. Then, the performance of BDME is close to that of TELM on Wikidata12k. Because the amount of training data on Wikidata12k and YAGO11k is smaller than GDELT, TELM and BDME can fit them better, it is worth noting that GDELT is substantially denser than Wikidata12k and YAGO11k. As seen in Table 1, the training data of GDELT is about 50 to 100 times larger than that of Wikidata12k and YAGO11k. BDME obtains a more promising result than TELM on GDELT.

Table 4. Performance for link prediction task on Wikidata12k and YAGO11k.

DataSets	ICEWS14				GDELT			
Method	MRR	Hit@1	Hit@3	Hit@10	MRR	Hit@1	Hit@3	Hit@10
TransE(2013)[★]	0.178	0.100	0.192	0.339	0.100	0.015	0.138	0.244
DisMult(2014)[★]	0.222	0.119	0.238	0.460	0.158	0.107	0.161	0.268
ComPlEx(2016)[★]	0.233	0.123	0.253	0.436	0.167	0.106	0.154	0.282
SimplE(2018)[♣]	-	-	-	-	-	-	-	-
RotatE(2019)[★]	0.221	0.116	0.236	0.461	0.167	0.103	0.167	0.305
HyTE(2018)[★]	0.105	0.015	0.143	0.272	0.180	0.098	<u>0.197</u>	<u>0.333</u>
ATiSE(2019)[♦]	0.280	0.175	0.317	0.481	0.170	0.110	0.171	0.288
TeRo(2020)[★]	0.187	0.121	0.197	0.319	-	-	-	-
TELM(2021)[♦]	<u>0.332</u>	<u>0.231</u>	<u>0.360</u>	<u>0.542</u>	<u>0.191</u>	<u>0.129</u>	0.194	0.321
RQVS(2022)[♠]	-	-	-	-	0.187	0.124	0.193	0.320
BTDG(2022)[♠]	0.314	0.214	0.351	0.523	-	-	-	-
TuckT(2022)[♦]	-	-	-	-	-	-	-	-
BDME(ours)	**0.339**	**0.241**	**0.371**	**0.548**	**0.223**	**0.162**	**0.227**	**0.350**

4.3   Ablation Studies

We further examine the effectiveness of each part in BDME, exploring how each component of BDME works. The results of all experiments are shown in Table 5.

Multivector Embeddings. We replaced the geometric product with the dot product for the experiment. This model is denoted as "-GA".

Timestamp Embedding Evolution Constraint. The TEE constraint penalty term is removed. This model is denoted as "-TEE".

Time-Dependent Entity Embedding Interaction Constraint. Similarly, The TEEI constraint penalty term is removed. This model is denoted as "-TEEI".

First, our proposed model still performs better than most baseline models under all ablation studies. Notably, "-GA" even achieves the deteriorations of 0.9% on MRR, 0.9% in Hits@1, 0.8% in Hits@3, and 0.8% in Hits@10 on ICEWS14. Second, the experimental results of "-TEE" are particularly influential on YAGO11k, with a maximum performance reduction of 2.2%. Third, the experimental results of "-TEEI" achieve average deteriorations of 1.0% on YAGO11k, while the effect is relatively mediocre in ICEWS14.

Second, the experimental results of "-TEE" are particularly influential on YAGO11k, with a maximum performance reduction of 2.2%. We speculate that the main reason is that YAGO11k has the long-tail characteristic, which makes it more challenging to collect temporal information.

Third, the experimental results of "-TEEI" achieve average deteriorations of 1.0% on YAGO11k, while the effect is relatively mediocre in ICEWS14. We argue that the facts distribution in YAGO11k is more dispersed, resulting in a high-er variability between two neighboring entities. Thus the bias component of the TEEI module will continue to learn during the training process, resulting in more penalties on YAGO11k compared to ICEWS14.

Table 5. Ablation study results on ICEWS14 and YAGO11k.

DataSets	ICEWS14				YAGO11k			
Method	MRR	Hit@1	Hit@3	Hit@10	MRR	Hit@1	Hit@3	Hit@10
BDME	**0.635**	**0.555**	**0.683**	**0.778**	**0.223**	**0.162**	**0.227**	**0.350**
-GA	0.626	0.546	0.675	0.770	0.218	0.160	0.223	0.340
-TEE	0.632	0.553	0.680	0.777	0.213	0.158	0.214	0.328
-TEEI	0.631	0.551	0.680	0.774	0.215	0.157	0.218	0.334

5   Conclusion

In this paper, we innovatively propose a model for TKG completion, named BDME, which explores multi-faceted and fine-grained embedding representation, and interactions in different types of facts. Unlike other existing models,

BDME effectively gathers useful information from the existing facts by adopting block decomposition and multivector embeddings for facts representation. Experimental results on four datasets show the significant merits and superiority of BDME. Additionally, we investigate the effect of the different time granularity and embedding dimensions on TKG completion. By combining these two kinds of constraint schemes with BDME, it outperforms baselines with an explicit margin on time overhead and memory costing compared to BTDG and ATiSE, the state-of-the-art baselines.

Acknowledgement. This work is supported by the National Natural Science Foundation of China under Grant No.62072146, The Key Research and Development Program of Zhejiang Province under Grant (No. 2021C03187, 2022C01125), National Key Research and Development Program of China 2019YFB2102100.

References

1. Lacroix, T., Obozinski, G., Usunier, N.: Tensor decompositions for temporal knowledge base completion. In: ICLR (2019)
2. Lee, D., Oh, B., Seo, S., Lee, K.H.: News recommendation with topic-enriched knowledge graphs. In: CIKM, pp. 695–704 (2020)
3. Molokwu, B.C., Shuvo, S.B., Kar, N.C., Kobti, Z.: Node classification in complex social graphs via knowledge-graph embeddings and convolutional neural network. In: Krzhizhanovskaya, V.V., Závodszky, G., Lees, M.H., Dongarra, J.J., Sloot, P.M.A., Brissos, S., Teixeira, J. (eds.) ICCS 2020. LNCS, vol. 12142, pp. 183–198. Springer, Cham (2020). https://doi.org/10.1007/978-3-030-50433-5_15
4. Jain, P., Rathi, S., Chakrabarti, S., et al.: Temporal knowledge base completion: new algorithms and evaluation protocols. In: EMNLP, pp. 3733–3747 (2020)
5. Dasgupta, S.S., Ray, S.N., Talukdar, P.: Hyte: hyperplane-based temporally aware knowledge graph embedding. In: EMNLP, pp. 2001–2011 (2018)
6. Zhu, F., Chen, S., Xu, Y., He, W., Yu, F., Zhang, X.: Temporal hypergraph for personalized clinical pathway recommendation. In: BIBM, pp. 718–725. IEEE (2022)
7. De Lathauwer, L.: A survey of tensor methods. In: 2009 IEEE International Symposium on Circuits and Systems, pp. 2773–2776. IEEE (2009)
8. Xu, C., Chen, Y.Y., Nayyeri, M., Lehmann, J.: Temporal knowledge graph completion using a linear temporal regularizer and multivector embeddings. In: NAACL, pp. 2569–2578 (2021)
9. Li, Z., et al.: Temporal knowledge graph reasoning based on evolutional representation learning. In: SIGIR, pp. 408–417 (2021)
10. Li, Z., et al.: Complex evolutional pattern learning for temporal knowledge graph reasoning. In: ACL, pp. 290–296 (2022)
11. Chen, K., Wang, Y., Li, Y., Li, A.: Rotateqvs: representing temporal information as rotations in quaternion vector space for temporal knowledge graph completion. In: AAAI, pp. 5843–5857 (2022)
12. Lai, Y., Chen, C., Zheng, Z., Zhang, Y.: Block term decomposition with distinct time granularities for temporal knowledge graph completion. Expert Systems with Applications, p. 117036 (2022)
13. Shao, P., Zhang, D., Yang, G.: Tucker decomposition-based temporal knowledge graph completion. Knowledge-Based Systems, p. 107841 (2022)
14. Xu, C., Nayyeri, M., Alkhoury, F.: Tero: a time-aware knowledge graph embedding via temporal rotation. In: COLING, pp. 1583–1593 (2020)

HIT - An Effective Approach to Build a Dynamic Financial Knowledge Base

Xinyi Zhu[1], Hao Xin[2], Yanyan Shen[3(✉)], and Lei Chen[1,2]

[1] Data Science and Analytics, HKUST (GZ), Guangzhou, China
{xzhu683,leichen}@connect.hkust-gz.edu.cn
[2] Computer Science and Engineering, HKUST, Hong Kong, China
hxinaa@cse.ust.hk
[3] Computer Science and Engineering, SJTU, Shanghai, China
shenyy@sjtu.edu.cn

Abstract. In recent years, due to their expertise and comprehensiveness in a specific domain, domain-specific knowledge bases (KBs) have attracted more and more attention from both academics and industries. Among these domain-specific KBs, financial KBs have become more and more popular and valuable due to their broad spectrum of downstream applications, such as quantitative investment analysis, financial risk analysis, and financial domain-based KBQA. However, due to their massive volume, high conflicts, and frequent volatile properties, it is pretty challenging to build an error-prone dynamic financial KB. To address these challenges, in this paper, we propose a dynamic financial KB construction pipeline that mainly consists of two fundamental modules, a **Human-Interacted (HI)** distant supervised evolved relation extraction module targets at obtaining the evolved knowledge with less manual annotations and high extraction accuracy, and a **Temporal (T)** duplication and conflict resolution module focus on applying a data fusion algorithm to the knowledge fusion task to select high-confidence knowledge without duplication and conflict by incorporating the temporal information. Through extensive experiments, we have demonstrated the effectiveness of **HIT**. Compared to state-of-the-art solutions, **HIT** can improve the accuracy by 10.6% on average for the relation extraction task and by 6.9% on average for the duplication and conflict resolution task, respectively.

Keywords: Dynamic knowledge base · Domain-specific · Distant supervision · Knowledge fusion

1 Introduction

The financial knowledge base (financial KB) has attracted significant attention since it provides more in-depth, professional, reasonable, and interpreted knowledge for academics and the industries. The financial KB has been widely utilized in many downstream applications, such as quantitative investment analysis [7],

X. Wang et al. (Eds.): DASFAA 2023, LNCS 13944, pp. 716–731, 2023.
https://doi.org/10.1007/978-3-031-30672-3_48

financial risk analysis [30], and KBQA. Current financial KBs are generally constructed by integrating knowledge from existing KBs, extracting knowledge from unstructured text data, or the combination of two methods [31]. Unstructured text data like public news or social events is time sensitive since information illustrated in it will expire after a while. It is difficult for the static financial knowledge graphs [7,10,11,32] to record knowledge flows, let alone to make the latent analysis based on the change of knowledge [17]. Thus, it is essential to build a dynamic financial KB to maintain the evolved knowledge of the timeline.

In order to better capture, store and maintain the knowledge, we propose to construct a dynamic financial KB, which incorporates knowledge's time property. There are three main challenges when applying such an algorithm. Firstly, extracting the structured knowledge from unstructured streaming text traditionally requires pre-specified ontology of relations and a heavy involvement of human efforts on annotations [10]. Apart from that, applying distant supervision saves much human efforts on data annotations but introduces noise in the training data. Secondly, considering the time dimension, conflicts are different from them in general KBs. For example, two pieces of extracted knowledge (Alibaba, Baidu, Cooperate) happened at 2022.02.01 and (Alibaba, Baidu, Rise) happened at 2022.02.05 should both be true and further be integrated into the dynamic financial KB. However, without temporal information, the different relations between the two same company entities "Alibaba" and "Baidu" are considered conflicted. Finally, there are multiple financial data sources, like different financial Websites. Various data sources may provide conflict knowledge along the timeline.

Existing automatic financial KB construction methods [7,10,11,17,32] do not pay enough attention to the time property of the news and events. To be specific, CFKG [10] and Yang et al. [32] both utilize the semantics in the financial data to extract the *(Subject, Predicate, Object)* or *(S, P, O)* triple, neither of them considers the temporal information, thus, leading to store outdated and incorrect knowledge. FR2KG [7] considers the publish date of reports, but it treats the timestamp as an entity related to only the "financial research report" entity, and thus its KB is still static. Miao et al. [17] build a dynamic KB incorporating the time property of knowledge. However, it cannot remove the conflict and duplication when it faces knowledge evolution.

In this paper, to address the above challenges, we propose our framework to build a dynamic financial KB, called **HIT**, with two main modules **H**uman-Interacted evolved relation extraction and **T**emporal duplication and conflict resolution. To be specific, firstly, we propose to use *(S, P, O, t, d)* to represent the financial knowledge where t is the timestamp, and d represents the data source. Secondly, to extract relations with high accuracy and less human workload, we propose a human-interacted distant supervised learning method. Finally, to integrate different time-granularity financial knowledge from multiple data sources, our framework applies a multi-truth discovery data fusion method DART [16] to knowledge fusion and utilizes the few-shot learning technique to reduce the parameter tuning workload. The temporal information together with the data source confidence knowledge helps further remove conflicts and duplication.

To summarize, our contributions made in this paper are as follows:

- We incorporate the time property of financial knowledge and its data source by representing it as *(S, P, O, t, d)*. Our method uses human-interacted distant supervised learning to do relation extraction from the unstructured text data.
- We apply a domain-aware multi-truth discovery data fusion method to our knowledge fusion task by introducing temporal information to help further remove the conflict and duplication in the evolved knowledge.
- We conduct experiments on the dataset based on A-share[1] listed companies, and the experimental results show that **HIT** outperforms the baselines. Moreover, we further show the effectiveness of the KB constructed by **HIT** compared to knowledge extracted by other solutions, such as the Open IE tool. Ours significantly improves downstream task, stock prediction accuracy.

2   Related Works

The related works section illustrates the financial knowledge graph's main constructing techniques from unstructured data.

2.1   Relation Extraction

The recognized entities extracted from each sentence can be retrieved as the nodes in the knowledge graph. To build up edges between different nodes in the knowledge graph, it is required to perform semantic relationship extraction between two entities from the unstructured text.

Relation extraction (RE) is one of the tasks in information extraction research. The different methods of RE can be divided into supervised, unsupervised, and semi-supervised.

Supervised Learning methods [6,15,19,34] take the pair of labeled entities and the labeled relationship between them as well as the sentence containing these as the input. The output is one of the pre-defined relation types [21]. Essentially, supervised relation extraction is regarded as a multi-classification task.

Unsupervised Learning RE methods require no manual annotations. There are some typical approaches like clustering and open information extraction. *Clustering based approaches* [14,29] tag the named entities in the text and then compute the similarity among the named entities pairs. Each cluster represents a relation. Later works make improvements on similarity measures and removing noise. Another widespread unsupervised learning relation extraction, *open information extraction* [2,28], also known as Open IE, can automatically extract possible relations in the different domains without any human involvement. However, there is a significant limitation in that multiple phrases represent the same semantic relation [21], which requires further processing.

---

[1] https://www.investopedia.com/terms/a/a-shares.asp.

Semi-supervised Learning tries to save manual workload while exploiting the unlabelled data. *Bootstrapping* [1,4,27], *active learning* [3,20,25], *distant supervision* [5,8,18], and so on are typical semi-supervised learning RE approaches. Among them, *distant supervision* tries to introduce the labels to the data from some existing structured knowledge bases (KBs). It assumes that if two entities have a relation, each sentence containing these two entities might represent the same relation [18]. In this way, distant supervised learning could generate a large number of positive instances of training data. In addition, in order to obtain the negative instances, entities not in a specific relation are randomly selected as the negative sample. However, there are several limitations, like the false negative instances introduced by the incompleteness of KBs. Later works [12,24] aggregate several sentences expressing the same relation between entities pair into one entity-pair bag. This idea is further utilized in RE tasks. Moreover, Snorkel [23] is proposed as a data annotation tool incorporating semi-supervised learning. It generates high training data for the downstream tasks through multiple labeling functions, such as crowdsourced labels [26] and distant supervision.

2.2 From Data Fusion to Knowledge Fusion

Knowledge fusion is defined as a task that assigns the truthfulness probability for each extracted triple to resolve conflicts further. The main difference between data fusion and knowledge fusion is that knowledge fusion has extra noise from the knowledge extraction process [9]. Thus, many recent works try to apply data fusion techniques to knowledge fusion scenarios.

DART [16], also known as Domain-Aware Multi-Truth Discovery from Conflicting Sources, is an unsupervised data fusion approach that introduces the reliability among domains and their correlations for different data sources. Moreover, it applies an unsupervised probability Bayesian-based approach to assign each instance a probability value for the multi-truth discovery goal.

To be specific, there are a source set $\mathcal{S} = \{s_1, s_2, \ldots, s_n\}$, a domain set $\mathcal{D}_a$, the objects set $\mathcal{O} = \left\{ o_1^{d_1}, o_2^{d_2}, \ldots, o_m^{d_D} \right\}$ where $d_1, d_2, \ldots, d_D \in \mathcal{D}_a$. DART tries to learn the probability of v being true as veracity score $\sigma(v)$ for each value v provided for each object o.

DART also utilizes a Bayesian probability model. To finally compute the veracity $\sigma(v)$ of a value v via the Bayes rule, it redefines the likelihood of $\psi(o)$ to compute the confidence score of value v as follows.

$$\Pr(\psi(o) \mid v) = \prod_{s \in S_{od}(v)} (\tau_d^{rec}(s))^{e_d(s)c_s(v)} \prod_{s \in S_{od}(\bar{v})} (1 - \tau_d^{sp}(s))^{e_d(s)c_s(v)} \quad (1)$$

$$\Pr(\psi(o) \mid \bar{v}) = \prod_{s \in S_{od}(\bar{v})} (\tau_d^{sp}(s))^{e_d(s)c_s(v)} \prod_{s \in S_{od}(v)} (1 - \tau_d^{rec}(s))^{e_d(s)c_s(v)} \quad (2)$$

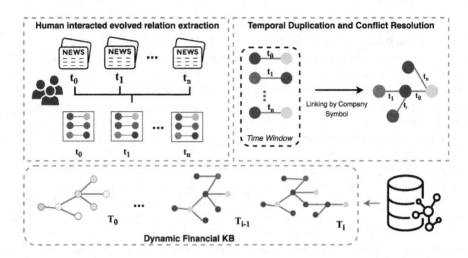

Fig. 1. Workflow of Building a Dynamic Financial KB

$V_s(o)$ is the set of values claimed for each object o from source s, and $\tau_d(s)$ is the trustworthiness value of each source s in domain d. $e_d(s)$ is the domain expertise score of source s in domain d. $c_s(v)$ stands for the confidence score of value v from source s. $\tau_d^{rec}(s)$ and $\tau_d^{sp}(s)$ are the trustworthiness of s in recall and specificity in domain d respectively, which defined as follows.

$$\tau_d^{rec}(s) = \frac{\sum_{o \in O^d(s)} \sum_{v \in V_s(o)} \sigma(v)}{\sum_{o \in O^d(s)} |V_s(o)|} \tag{3}$$

$$\tau_d^{sp}(s) = \frac{\sum_{o \in O^d(s)} \sum_{v' \in \bar{V}_s(o)} (1 - \sigma(v'))}{\sum_{o \in O^d(s)} |\bar{V}_s(o)|} \tag{4}$$

3   Methodology

This section describes the workflow of building a dynamic financial knowledge base, as illustrated in Fig. 1. It mainly consists of two modules. The human-interacted distant supervised module extracts a batch of temporal knowledge in a time interval T_i, which contains discrete timestamps $t_0, t_1, \dots$. Then, the temporal duplication and conflict resolution module takes the batch of temporal knowledge as input and generates the dynamic financial knowledge set without duplication and conflict. The database maintains the most up-to-date dynamic financial knowledge set at each time interval T_i.

3.1   Financial Entity Recognition and Linking

Before knowledge extraction, the essential information of the A-share listed companies and the raw financial news data is collected from multiple reliable data sources like "East Money", "Yuncaijing", and so on. We then show the further procedures as follows.

Financial Entities objectively exist and are closely related to the A-share listed companies [17]. Moreover, we contain the A-share listed companies as the financial entities in this paper.

Word Tokenization. We first apply a parser to divide the document-level text into sentences. Then, we use the word tokenization in HanLP² to split a sentence into words for further financial entity filtering.

Named Entity Recognition. After the word tokenization, we apply the named entity recognition technique in HanLP on each split word to filter the financial entities. This step results in many alias names of companies like "Nezha" and "360" which require further entity linking processing. For example, the recognized company entities "QQ" and"Tencent" actually refer to the same company "Tencent Holdings Limited" with the stock symbol "00700.HK".

Entity Linking. To link the different company alias names to the same mention, we use the fuzzy search API provided by *QICHACHA* to retrieve the unique company name and its stock symbol if it exists for each recognized company alias name.

3.2 Human-Interacted Distant Supervised Evolved Relation Extraction

As illustrated in the Sect. 2.1, distant supervised learning saves manual workload in generating labels by incorporating structured knowledge in the existing KBs. However, it introduces noise in the training data, which results in a low RE accuracy. Also, similar to the Open IE technique, multiple phrases in different KBs may link to the same semantic relation requiring further processing. Finally, the financial knowledge evolves along the timeline, which needs the dynamic financial KB to obtain the most up-to-date knowledge with the timestamp. To address these limitations, we put the human in the loop and incorporate the temporal information into the relation extraction process.

RE Architecture. Our RE is based on the sentence-level extraction module in OpenNRE [13] framework. Figure 2 shows the main RE architecture. The first input layer takes a piece of news with two recognized financial entities as the input. We denote x as the input text vector and two recognized entities as S and O. The second layer is the tokenization layer which tokenizes the input text into words as $x = [w_1, w_2, ..., w_n]$, where n is the length of the input text. The third encoder layer tries to convert the tokenization vector $[w_1, w_2, ..., w_n]$ into a low-dimensional semantic embedding $[e_1, e_2, ..., e_n]$ via the pre-trained Chinese whole word mask BERT. Finally, the fourth model layer is the softmax classifier which generates the extracted relation between two financial entities from the text and the news's timestamp and data source as (S, P, O, t, d).

Manual Pre-specified Ontology. Try to generate the formal representations of a set of entities and their relations in the financial domain. Our dynamic

---

² https://github.com/hankcs/HanLP.

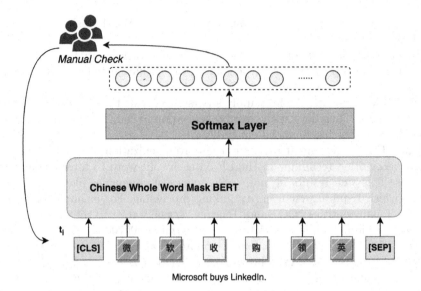

Fig. 2. Relation Extraction Architecture

financial KB is designed with professional and underlying ontology for the financial scenario. Especially for the dynamic company relations that evolve along the timeline, humans are asked to pre-define several relation categories. There are 16 types: "same industry", "rise", "compete", "cooperate", "increase holding", "fall", "supply", "reduce holding", "invest", "superior", "subordinate", "be increased holding", "be reduced holding", "be supplied", "be invested" and "dispute".

Manual Complementation and Check. From existing KBs, the distant supervision algorithm will initially obtain several structured data (S, P, O) as the training labels. For the *object* O in the structured data, which may have multiple semantics, humans are introduced to participate in the complementation. For example, the *object* "prosecute" and "argue" share a similar semantic meaning with our pre-specified ontology "dispute" which can be classified into the same *object*. Also, several pre-specified relations are directed like "supply" and "be supplied", which requires manual differentiation. Besides the manual complementation, there are also manual checks on instances with poor RE performance. For example, we test RE performance in a specific time interval and find that the relation "invest" has a low RE accuracy. Then, humans are asked to check and annotate several instances for its training augmentation.

3.3 Temporal Duplication and Conflict Resolution

With the evolution of knowledge along the timeline and the multiple time-granularity of social media information, duplication and conflict exist in a batch of extracted knowledge from the text data within a specific time interval. To address the knowledge evolution issue, we define the *temporal duplication* and *temporal conflict* when applying the multi-truth discovery data fusion method to the knowledge fusion task with temporal information incorporation. Considering the multiple time-granularity of social media information, the module supports a self-defined time window.

Data Model. There is a set of data source $S = \{s_1, s_2, \ldots, s_n\}$, a domain set $\mathcal{D}_a$ and a time window size T. When applying DART to a knowledge fusion scenario, our data model is defined as follows. Given a set of extracted knowledge tuples $K = k_1, k_2, \ldots$, where $k_i = (S_i, P_i, O_i, t_i, d_i)$ within the time window T, the original objects set mentioned in the last section is now a triple set defined as $\mathcal{T} = \left\{t_1^{d_1}, t_2^{d_2}, \ldots, t_m^{d_D}\right\}$. $t_i = (S_i, P_i, O_i)$ in the knowledge tuple k_i is extracted from the news set $\mathcal{N} = \{n_1, n_2, \ldots, n_l\}$, $d_1, d_2, \ldots, d_D \in \mathcal{D}_a$, and d_i is the domain of the news n_i that t_i extracted from. Our goal is to resolve the conflict of all incorrect triples within the time window.

Now we introduce some important definitions in this module.

Definition 1 *(Temporal Duplication). Given a set of extracted knowledge tuples $K = k_1, k_2, \ldots$, where $k_i = (S_i, P_i, O_i, t_i, d_i)$ and a time window T, temporal duplication is a pair of tuples which have the same (S, P, O) and both timestamp t are within the time window T.*

Definition 2 *(Temporal Conflict). Given a set of extracted knowledge tuples $K = k_1, k_2, \ldots$, where $k_i = (S_i, P_i, O_i, t_i, d_i)$ and a time window T, temporal conflict is a pair of tuples which have the same entities S, P but the different relation O between them and both timestamp t are within the time window T.*

For temporal duplication, the algorithm keeps the knowledge tuple with the smallest timestamp t and filters the others considering the timeliness of the news. For temporal conflict, we directly apply **DART** to assign a truthfulness probability for each triple within the time window. Triples with a truthfulness probability that is close to 0 are removed.

As knowledge fusion outputs a probability score between 0 and 1 for each input triple, it is natural to apply a Bayesian-based data fusion model, which also learns a probability of a given object being true [33]. The essential idea is using Bayes' rule to express the probability of the output value v is true as $\Pr(v \mid \psi(t))$ based on and for each observation $\psi(t)$ in Eq. 5.

$$\begin{aligned}
\Pr(v \mid \psi(t)) &= \frac{\Pr(\psi(t) \mid v)\Pr(v)}{\Pr(\psi(t))} \\
&= \frac{\Pr(\psi(t) \mid v)\sigma(v)}{\Pr(\psi(t) \mid v)\sigma(v) + \Pr(\psi(t) \mid \bar{v})(1 - \sigma(v))}
\end{aligned} \tag{5}$$

where $\psi(t)$ is the observed value of the knowledge triple t and $\Pr(v) = \sigma(v)$ is a prior probability that v is true.

In addition, as algorithm DART treats the initialization of recall in Eq. 3 and specificity in Eq. 4 as parameters to be tuned, it compares the performance regarding the parameter sensitivity in the paper. Different parameter settings lead to different performance results, which require much manual tuning workload. We would like to address this issue with few-shot learning. We define the initialization of recall and specificity in a standard way in reference to *PrecRec* [22]:

$$Recall = \frac{TP}{TP + FN}$$
$$specificity = \frac{TN}{TN + FP} \tag{6}$$

Given a randomly selected few-shot labeled data, the correctness of extracted knowledge triples in different sources' domains can be expressed in the Table 1. Thus, the initialization of recall and specificity can be computed by its standard definition in Eq. 6. Moreover, the initialization result of Table 1 is shown in Table 2.

Table 1. Knowledge Triples Derived from Sources Domains via the Extractor

ID	Knowledge Triple	Correct?	$s_1^{d_1}$	$s_1^{d_2}$	$s_1^{d_3}$	$s_2^{d_1}$	$s_2^{d_2}$
t_1	(002460.SZ, 600660.SH, fall)	Yes	✓	✓		✓	✓
t_2	(600009.SH, 600004.SH, invest)	No	✓	✓			
t_3	(002460.SZ, 601068.SH, rise)	Yes			✓		
t_4	(605358.SH, 002346.SZ, cooperate)	Yes		✓	✓	✓	✓
t_5	(300750.SZ, 600837.SH, unknown)	No		✓	✓		
t_6	(000066.SZ, 000748.SZ, superior)	Yes	✓			✓	✓
t_7	(002062.SZ, 600884.SH, cooperate)	Yes	✓	✓	✓		
t_8	(002284.SZ, 601633.SH, supply)	No	✓	✓		✓	✓
t_9	(603098.SH, 000786.SZ, rise)	No	✓	✓		✓	✓
t_{10}	(603098.SH, 000786.SZ, fall)	Yes	✓		✓	✓	✓

Table 2. Recall and Specificity of Each Source's Domain

	Recall	Specificity
$s_1^{d_1}$	0.67	0.25
$s_1^{d_2}$	0.5	0
$s_1^{d_3}$	0.67	0.75
$s_2^{d_1}$	0.67	0.5
$s_2^{d_2}$	0.67	0.5

4   Experiments

In this section, we will demonstrate the performance of each construction algorithm in our pipeline. In addition, to evaluate the effectiveness of the knowledge in the dynamic financial KB, we further test the effectiveness of the temporal knowledge in a downstream stock prediction task.

4.1   Dataset

We introduce two datasets, which are annotated based on Chinese financial texts. Their statistics information is listed in Table 3. The two datasets are separately constructed from financial news from online Websites and financial research reports proposed by different Securities Companies. In addition, we assess the performance of our two modules, human-interacted evolved relation extraction, and temporal duplication and conflict resolution, on the two datasets.

FinNews uses news crawled from online website like "Yuncaijing"[3], "East Money"[4] and "Hithink RoyalFlush"[5] as the corpus. We ask three master students to integrate similar semantic phrases from existing KBs, check several instances, and annotate the potential two entities and their relation in the raw text data. It consists 16 relations with an extra "Unknown" type for the relation extraction task. Moreover, each instance comes with two recognized A-share listed company entities, its data source, and the publishing timestamp of the news. Later, for the temporal duplication and conflict resolution task, three master students are asked to label whether the extracted temporal knowledge triple within the given time window ($T = 1$ days) should be kept (output 1) or removed (output 0).

FinReports is the dataset based on a collection of Chinese financial company research reports published by several Securities Companies. The three master students also annotate it in the same way as the dataset *FinNews* for two tasks. *FinReports* dataset does not have a training set, and it is used for the generalization test of models.

Table 3. Dataset Comparison

	# Relations	# Training set	# Test set
FinNews	17	1,811	453
FinReports	13	0	302

The schema of our processed dataset is defined as {head company stock symbol, tail company stock symbol, head company name, tail company name, relation, sentence, timestamp, source}, which targets the financial domain-specific

---

[3] https://www.yuncaijing.com/.

[4] https://www.eastmoney.com/.

[5] https://www.10jqka.com.cn/.

dynamic KB. In many distance supervision methods, benchmark dataset NYT is often used for testing. NYT is derived from the corpus annotated by the New York Times, and the named entities are recognized by the Stanford NER tool in conjunction with the Freebase knowledge base. The NYT dataset includes 53 relation types all from the Freebase KB, such as the relation "instance_of", and 522, 043 pieces of news sentences. The schema of an instance is defined as {head entity ID, tail entity ID, head entity, tail entity, relation, sentence, termination notation}. NYT has more affluent training instances; however, it does not target the financial domain and does not include the temporal property. Thus, we did not use NYT dataset in our experiments since it is not suitable for our financial evolved relation extraction.

4.2 Relation Extraction

We have obtained structured data from existing KBs like QICHACHA and the Github[6]. There are six relations between companies: cooperate (153 pairs), compete (114 pairs), invest (3, 415 pairs), acquisition (982 pairs), supplier (3, 016 pairs), and customer (2, 792 pairs). Moreover, we retrieve raw text data from the labeled entities. The raw text data is obtained by crawling official news websites such as "Yuncaijing", and "East Money" using the API provided by "Tushare". In total, we have collected 16, 612 pieces of financial news, including cooperate (2, 618 pieces), compete (7, 017 pieces), invest (3, 842 pieces), acquisition (632 pieces), supplier (1, 889 pieces), customer (614 pieces). Without the manual complementation and check, the distant supervision relation extraction accuracy is 63% using BiGRU+ATT model based on Zhou et al. [35]. Its poor performance is due to the strong assumption mentioned in Sect. 2.1, especially in the financial scenario. For example, there is a structured knowledge (General Motors Company, Ford Motor Company, compete) in the existing KBs. For a piece of crawled news text, "General Motors and Ford closed up about 2%', it is obscure for the model to learn their relation as "compete". In this example, the relation between "General Motors Company" and "Ford Motor Company" should be the pre-specified "rise" which means the two companies' shares rise together this time.

As shown in Table 4, we introduce our human-interacted relation extraction algorithm and train it by the *FinNews* training set. The prediction result is 89.4% on *FinNews* test set and 68.8% on *FinReports* test set. For the performance comparison, we also apply BiGRU + ATT model for the prediction with the same *FinNews* training set since BiGRU + ATT is a typical baseline relation extraction method applied in many works [7]. As we take the Chinese corpus as input, we use the Chinese Wiki GloVe pre-trained model with 100 dimensions to obtain the word embedding. The absolute accuracy of BiGRU + ATT model is 86.7% on *FinNews* test set and 50.3% on *FinReports* test set. The good performance of our HI-RE on *FinReports* test set demonstrates that ours has a much better generalization ability than Bi-GRU + ATT. Moreover,

---

[6] https://github.com/liuhuanyong/ChainKnowledgeGraph.

the performance gain is possibly due to the Chinese whole word mask BERT pre-trained model, which generates better word embedding than the pre-trained GloVe model.

Table 4. Accuracy Comparison on Relation Extraction

	Ours (HI-RE)	BiGRU + ATT
FinNews	**89.4%**	86.7%
FinReports	**68.8%**	50.3%

4.3 Temporal Duplication and Conflict Resolution

Table 5. Accuracy Comparison on Duplication and Conflict Resolution

	Ours (TDCR)	LTM	Majority Voting
FinNews	**88.0%**	82.0%	73.5%
FinReports	**86.4%**	78.5%	65.6%

We show the performance comparison of temporal duplication and conflict resolution task in Table 5. The best performance of our TDCR with the accuracy of 88.0% on *FinNews* and 86.4% on *FinReports*. The potential reason is that our algorithm design considers the temporal information for removing the conflict and duplication. We introduce LTM and Majority voting, two typical data fusion methods, as our compared baseline methods. Specifically, LTM is also an unsupervised data fusion method like DART. On the opposite, LTM [33] and Majority voting ignore it, which leads to their lower accuracy.

4.4 Effectiveness of the Financial Knowledge

To further evaluate the effectiveness of the knowledge in the KBs, we apply a downstream financial task. The downstream task takes a single stock or a portfolio of stocks to make the stock trend prediction. For the stock trend prediction model design, we treat the selected stock as the targeted company and retrieve all its related companies in the financial KB. We use LSTM as the prediction model, which concatenates these companies' time series data as input for the stock trend prediction. Given a single stock or a portfolio of stocks, the model utilizes their time series stock price data of the past year to train the model. The following backtesting module based on the Backtrader[7] framework will then give the corresponding returns based on the stock trend predictions as the trading signals as well as an investment strategy.

---

[7] https://github.com/mementum/backtrader.

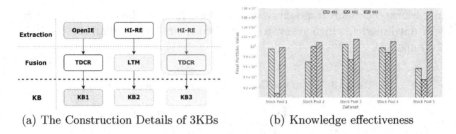

| (a) The Construction Details of 3KBs | (b) Knowledge effectiveness |

Fig. 3. Knowledge Effectiveness Result

To demonstrate the performance of our KB, we adopt three different ways to construct the Financial KB, which separately applies our model and other mainstream models in each step. In addition, three Financial KBs are constructed with the same raw data shown in detail in Fig. 3 (a). The knowledge in KB1 is extracted by the Open IE tool[8]. And we further apply our TDCR to remove the temporal conflict and duplication. Our HI-RE algorithm extracts KB2. LTM is applied to do duplication and conflict resolution. KB3 is constructed by our pipeline. After that, we evaluate three Financial KBs by the downstream task, stock trend prediction and backtesting. For a generally evaluated result, we randomly selected five different portfolios of stock data, and each portfolio (stock pool) contains several random stocks. The experimental results of each Financial KB are illustrated in Fig. 3 (b), which shows the backtesting results in the final portfolio value incorporating three KBs mentioned above in five randomly selected stock pools. Compared to KB1 and KB2, KB3 constructed by the HIT pipeline often generates a higher backtesting return due to the richer and more accurate financial knowledge maintained in it.

5   Conclusion

In this paper, we propose a pipeline **HIT** to construct a dynamic KB in the finance domain with high effectiveness and efficiency. Specifically, the human-interacted evolved relation extraction module works to extract the up-to-date knowledge along the timeline with the extraction accuracy of 89.4% on *FinNews* and 68.7% on *FinReports*. The following temporal duplication and conflict resolution module introduces the few-shot learning technique to save manual tuning workload for high-confidence temporal knowledge retrieval with the accuracy of 88.0% on *FinNews* and 86.4% on *FinReports*. Finally, we evaluate the effectiveness of the knowledge in our dynamic KB in the downstream stock prediction task and obtain higher backtesting returns in all five stock pools.

Acknowledgements. The authors would like to thank the anonymous reviewers for their insightful reviews. This work is supported by the National Key Research and

---

[8] https://github.com/PaddlePaddle/PaddleNLP/tree/develop/model_zoo/uie.

Development Program of China (2022YFE0200500), Shanghai Municipal Science and Technology Major Project (2021SHZDZX0102) and SJTU Global Strategic Partnership Fund (2021 SJTU-HKUST). Lei Chen's work is partially supported by National Science Foundation of China (NSFC) under Grant No. U22B2060, the Hong Kong RGC GRF Project 16213620, RIF Project R6020-19, AOE Project AoE/E-603/18, Theme-based project TRS T41-603/20R, China NSFC No. 61729201, Guangdong Basic and Applied Basic Research Foundation 2019B151530001, Hong Kong ITC ITF grants MHX/078/21 and PRP/004/22FX, Microsoft Research Asia Collaborative Research Grant and HKUST-Webank joint research lab grants.

References

1. Agichtein, E., Gravano, L.: Snowball: Extracting relations from large plain-text collections. In: Proceedings of the Fifth ACM Conference on Digital Libraries, June 2–7, 2000, San Antonio, TX, pp. 85–94. ACM (2000)

2. Banko, M., Etzioni, O.: The tradeoffs between open and traditional relation extraction. In: McKeown, K.R., Moore, J.D., Teufel, S., Allan, J., Furui, S. (eds.) ACL 2008, Proceedings of the 46th Annual Meeting of the Association for Computational Linguistics, June 15–20, 2008, Columbus, Ohio, pp. 28–36. The Association for Computer Linguistics (2008)

3. Blum, A., Mitchell, T.M.: Combining labeled and unlabeled data with co-training. In: Bartlett, P.L., Mansour, Y. (eds.) Proceedings of the Eleventh Annual Conference on Computational Learning Theory, COLT 1998, Madison, Wisconsin, July 24–26, 1998, pp. 92–100. ACM (1998)

4. Brin, S.: Extracting patterns and relations from the world wide web. In: Atzeni, P., Mendelzon, A., Mecca, G. (eds.) WebDB 1998. LNCS, vol. 1590, pp. 172–183. Springer, Heidelberg (1999). https://doi.org/10.1007/10704656_11

5. Bunescu, R.C., Mooney, R.J.: Learning to extract relations from the web using minimal supervision. In: Carroll, J.A., van den Bosch, A., Zaenen, A. (eds.) ACL 2007, Proceedings of the 45th Annual Meeting of the Association for Computational Linguistics, June 23–30, 2007, Prague. The Association for Computational Linguistics (2007)

6. Chan, Y.S., Roth, D.: Exploiting syntactico-semantic structures for relation extraction. In: Proceedings of the 49th Annual Meeting of the Association for Computational Linguistics: Human Language Technologies, pp. 551–560 (2011)

7. Cheng, D., Yang, F., Wang, X., Zhang, Y., Zhang, L.: Knowledge graph-based event embedding framework for financial quantitative investments. In: Proceedings of the 43rd International ACM SIGIR Conference on Research and Development in Information Retrieval, pp. 2221–2230 (2020)

8. Craven, M., Kumlien, J.: Constructing biological knowledge bases by extracting information from text sources. In: Lengauer, T., et al. (eds.) Proceedings of the Seventh International Conference on Intelligent Systems for Molecular Biology, August 6–10, 1999, Heidelberg, pp. 77–86. AAAI (1999)

9. Dong, X.L., et al.: From data fusion to knowledge fusion. PVLDB **7**(10), 881–892 (2015)

10. Elhammadi, S., et al.: A high precision pipeline for financial knowledge graph construction. In: Proceedings of the 28th International Conference on Computational Linguistics, pp. 967–977 (2020)

11. Guo, K., Jiang, T., Zhang, H.: Knowledge graph enhanced event extraction in financial documents. In: 2020 IEEE International Conference on Big Data (Big Data), pp. 1322–1329. IEEE (2020)
12. Han, X., Sun, L., Zhao, J.: Collective entity linking in web text: A graph-based method. In: Ma, W., Nie, J., Baeza-Yates, R., Chua, T., Croft, W.B. (eds.) Proceeding of the SIGIR, pp. 765–774 (2011)
13. Han, X., Gao, T., Yao, Y., Ye, D., Liu, Z., Sun, M.: Opennre: An open and extensible toolkit for neural relation extraction. In: Proceedings of the EMNLP-IJCNLP, pp. 169–174 (2019)
14. Hasegawa, T., Sekine, S., Grishman, R.: Discovering relations among named entities from large corpora. In: Scott, D., Daelemans, W., Walker, M.A. (eds.) Proceedings of the 42nd Annual Meeting of the Association for Computational Linguistics, 21–26 July, 2004, Barcelona, pp. 415–422. ACL (2004)
15. Kambhatla, N.: Combining lexical, syntactic, and semantic features with maximum entropy models for information extraction. In: Proceedings of the ACL Interactive Poster and Demonstration Sessions, pp. 178–181 (2004)
16. Lin, X., Chen, L.: Domain-aware multi-truth discovery from conflicting sources. Proceedings of the VLDB Endowment (2018)
17. Miao, R., Zhang, X., Yan, H., Chen, C.: A dynamic financial knowledge graph based on reinforcement learning and transfer learning. In: 2019 IEEE International Conference on Big Data (Big Data), pp. 5370–5378. IEEE (2019)
18. Mintz, M., Bills, S., Snow, R., Jurafsky, D.: Distant supervision for relation extraction without labeled data. In: Proceedings of the Joint Conference of the 47th Annual Meeting of the ACL and the 4th International Joint Conference on Natural Language Processing of the AFNLP, pp. 1003–1011 (2009)
19. Mooney, R., Bunescu, R.: Subsequence kernels for relation extraction. Adv. Neural Inf. Process. Syst. **18** (2005)
20. Muslea, I., Minton, S., Knoblock, C.A.: Selective sampling with redundant views. In: Kautz, H.A., Porter, B.W. (eds.) Proceedings of the Seventeenth National Conference on Artificial Intelligence and Twelfth Conference on on Innovative Applications of Artificial Intelligence, July 30–August 3, 2000, Austin, Texas, pp. 621–626. AAAI Press/The MIT Press (2000)
21. Pawar, S., Palshikar, G.K., Bhattacharyya, P.: Relation extraction: A survey. arXiv preprint arXiv:1712.05191 (2017)
22. Pochampally, R., Sarma, A.D., Dong, X.L., Meliou, A., Srivastava, D.: Fusing data with correlations. In: Dyreson, C.E., Li, F., Özsu, M.T. (eds.) International Conference on Management of Data, SIGMOD 2014, Snowbird, UT, June 22–27, 2014, pp. 433–444. ACM (2014)
23. Ratner, A., Bach, S.H., Ehrenberg, H., Fries, J., Wu, S., Ré, C.: Snorkel: Rapid training data creation with weak supervision. In: Proceedings of the VLDB Endowment. International Conference on Very Large Data Bases, vol. 11, p. 269. NIH Public Access (2017)
24. Riedel, S., Yao, L., McCallum, A.: Modeling relations and their mentions without labeled text. In: Balcázar, J.L., Bonchi, F., Gionis, A., Sebag, M. (eds.) ECML PKDD 2010. LNCS (LNAI), vol. 6323, pp. 148–163. Springer, Heidelberg (2010). https://doi.org/10.1007/978-3-642-15939-8_10
25. Sun, A., Grishman, R.: Active learning for relation type extension with local and global data views. In: Chen, X., Lebanon, G., Wang, H., Zaki, M.J. (eds.) 21st ACM International Conference on Information and Knowledge Management, CIKM'12, Maui, HI, October 29–November 02, 2012, pp. 1105–1112. ACM (2012)

26. Tong, Y., Yuan, Y., Cheng, Y., Chen, L., Wang, G.: Survey on spatiotemporal crowdsourced data management techniques. J. Softw. **28**(1), 35–58 (2017)
27. Vyas, V., Pantel, P., Crestan, E.: Helping editors choose better seed sets for entity set expansion. In: Cheung, D.W., Song, I., Chu, W.W., Hu, X., Lin, J. (eds.) Proceedings of the 18th ACM Conference on Information and Knowledge Management, CIKM 2009, Hong Kong, China, November 2–6, 2009, pp. 225–234. ACM (2009)
28. Weld, D.S., Hoffmann, R., Wu, F.: Using wikipedia to bootstrap open information extraction. SIGMOD Rec. **37**(4), 62–68 (2008)
29. Yan, Y., Okazaki, N., Matsuo, Y., Yang, Z., Ishizuka, M.: Unsupervised relation extraction by mining Wikipedia texts using information from the web. In: Su, K., Su, J., Wiebe, J. (eds.) ACL 2009, Proceedings of the 47th Annual Meeting of the Association for Computational Linguistics and the 4th International Joint Conference on Natural Language Processing of the AFNLP, 2–7 August 2009, Singapore, pp. 1021–1029. The Association for Computer Linguistics (2009)
30. Yang, S., et al.: Financial risk analysis for SMES with graph-based supply chain mining. In: Proceedings of the IJCAI, pp. 4661–4667 (2020)
31. Yang, Y., Miao, Z., Gao, J., Lu, J., Shi, G.: Automatic Chinese financial knowledge graph constructing framework. In: Proceedings of the ACAI, pp. 18:1–18:9 (2021)
32. Yang, Y., Miao, Z., Gao, J., Lu, J., Shi, G.: Automatic Chinese financial knowledge graph constructing framework. In: 2021 4th International Conference on Algorithms, Computing and Artificial Intelligence, pp. 1–9 (2021)
33. Zhao, B., Rubinstein, B.I.P., Gemmell, J., Han, J.: A bayesian approach to discovering truth from conflicting sources for data integration. Proc. VLDB Endow. **5**(6), 550–561 (2012)
34. Zhou, G., Su, J., Zhang, J., Zhang, M.: Exploring various knowledge in relation extraction. In: Proceedings of the 43rd Annual Meeting of the Association for Computational Linguistics (ACL'05), pp. 427–434 (2005)
35. Zhou, P., et al.: Attention-based bidirectional long short-term memory networks for relation classification. In: Proceedings of the ACL. The Association for Computer Linguistics (2016)

MulEA: Multi-type Entity Alignment of Heterogeneous Medical Knowledge Graphs

Mingxia Wang[1], Peng Tian[1], Yun Xiong[1(✉)], Jingwen Yue[1], Yao Zhang[1],
and Chunlei Tang[2]

[1] Shanghai Key Laboratory of Data Science, School of Computer Science,
Fudan University, Shanghai, China
{wangmx20,tianpeng,yunx,yuejw19,yaozhang}@fudan.edu.cn
[2] Brigham and Women's Hospital, Harvard Medical School, Boston, MA, USA

Abstract. The large-scale application of medical knowledge graphs has greatly raised the intelligence level of modern medicine. Considering that entity references between multiple medical knowledge graphs can lead to redundancy, knowledge graph alignment tasks are required to identify entity pairs or subgraphs of heterogeneous knowledge graphs pointing to the same elements in the real world. Existing medical knowledge graph alignment methods do not consider the multi-type features of medical entities. To tackle above challenge, we propose a **Mul**ti-type **E**ntity **A**lignment model of medical knowledge graphs based on attention mechanism named MulEA. Firstly, MulEA integrates the multi-type information features of medical entities to align various types of entities jointly through constructing an entity multi-type information embedding matrix for medical knowledge graphs. Secondly, MulEA designs a relationship collective aggregation module to fully utilize the features of different relationships to improve alignment accuracy. Finally, we evaluate the performance of MulEA on a real medical knowledge graph alignment dataset MED-OWN-15K, and it achieves the state-of-the-art performance on several metrics, showing the effectiveness of our method.

Keywords: Medical knowledge graph alignment · Entity alignment · Attention mechanism

1 Introduction

With the development of medical field, the number of medical texts has increased significantly including electronic health records, biomedical literature, and clinical trial reports. These texts contain rich knowledge. Existing studies focus on extracting knowledge from the above texts and presenting them in the form of knowledge graphs. Medical knowledge graph technology is the basis for medical knowledge retrieval and electronic medical record intelligence. The large-scale application of medical knowledge graphs [3] improve the intelligence of modern medicine, which can effectively serve intelligent medical decision-making systems, medical question answering systems, etc.

The advancement of medical big data technology has stimulated the emergence of many medical data sources. Since there is no unified norm to constrain the construction of knowledge graphs, entity references pointing to the same elements in reality in different medical knowledge graphs may be very different. When the intelligent medical system employs different knowledge graphs at the same time, multiple referencing of medical entities can easily lead to knowledge redundancy and knowledge noise.

To address these challenges, medical knowledge graph alignment has attracted a lot of attention [9,16], which aims to realize the logical integration of different medical knowledge graphs by aligning entities that point to the same elements in the real world, thereby constructing a more complete medical knowledge graph with higher quality.

In recent years, embedding-based methods have been dominant methods for knowledge graph alignment [5,8,13] without relying on manually constructed features or rules. They learn how to map equivalent entities between different knowledge graphs into a unified embedding vector space, and then measure whether they can be aligned by calculating the distance between embeddings. These methods are built on the assumption that entity pairs to be aligned between two knowledge graphs have similar neighborhood structures, so the learned embeddings of equivalent entities are similar. However, due to the incompleteness and heterogeneity of knowledge graphs, this assumption does not always apply in real-world scenarios. Sun et al. shows that most equivalent entities in the general neighborhood dataset DBP15K have different neighborhood structures [9].

In the medical filed, most existing knowledge graphs contain entities belonging to multiple types, for example, "Depression" is both a disease name and a complication. Previous researches on knowledge graph alignment have not considered this feature, and they need to train multiple models for multi-type entity annotation, which is complex and expensive. Besides, choosing a model among multiple models is difficult. This paper aims to tackle above challenges, and main contributions are as follows:

- We propose a novel Multi-type Entity Alignment model of medical knowledge graphs based on attention mechanism named MulEA, which can jointly align various types of medical entities.
- We design cross-graph attention and self-attention to alleviate the incompleteness of knowledge graphs, which can effectively aggregate positive information from the neighborhood and remain sensitive to negative information in order to distinguish similar but different entities.
- Experiments on a real medical knowledge graph alignment dataset MED-OWN-15K verify the effectiveness and superiority of MulEA.

2   Method

2.1   Problem Definition

$G = (\mathcal{V}, \mathcal{E}, \mathcal{T}, \mathcal{R})$ represents a knowledge graph, where $\mathcal{V}$ represents entities, $\mathcal{E}$ represents relationships, $\phi : \mathcal{V} \to \mathcal{T}$ is a mapping from an entity to its type, and $\psi : \mathcal{E} \to \mathcal{R}$ is a mapping from a relationship to its type. Given two medical knowledge graphs G and $G^{'}$, pre-aligned entity types and relation types in these two knowledge graphs are $\mathcal{T}^* = \left\{ \left(t, t^{'} \right) \in \mathcal{T} \times \mathcal{T}^{'} \middle| t \Longleftrightarrow t^{'} \right\}$, $\mathcal{R}^* = \left\{ \left(r, r^{'} \right) \in \mathcal{R} \times \mathcal{R}^{'} \middle| r \Longleftrightarrow r^{'} \right\}$, where $\Longleftrightarrow$ means an equivalence relation. Entity set $\mathcal{S} = \left\{ \left(v_i^{t^*}, v_{i^{'}}^{t^*} \right) \middle| t^* \in \mathcal{T}^* \right\}$ is training set, and $|\mathcal{S}| \ll |\mathcal{V}|$.

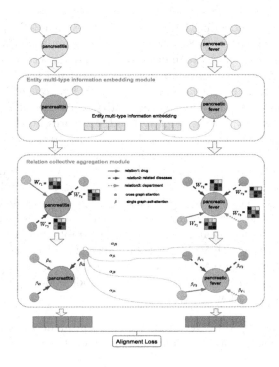

Fig. 1. MulEA architecture diagram

Medical Knowledge Graph Entity Alignment. Given two medical domain knowledge graphs $G = (\mathcal{V}, \mathcal{E}, \mathcal{T}, \mathcal{R})$ and $G^{'} = \left(\mathcal{V}^{'}, \mathcal{E}^{'}, \mathcal{T}, \mathcal{R} \right)$, the goal is to find as many aligned entity pairs $A_v = \left\{ \left(v, v^{'} \right) \in \mathcal{V} \times \mathcal{V}^{'} \middle| v \Longleftrightarrow v^{'} \right\}$ as possible. We take entity alignment problem as a classification problem and predict whether two nodes $i \in \mathcal{V}$ and $i^{'} \in \mathcal{V}^{'}$ represent the same entity in real world.

2.2   Model Overview

MulEA aligns multiple types of medical entities and captures useful neighborhood information through the relation collective aggregation module. Besides, it adds relation self-attention mechanism to prevent blindly aligning entities based on structural information.

As show in Fig. 1, MulEA follows a three-stage steps : (1) entity multi-type information embedding, (2) relation collective aggregation, (3) entity alignment. Through step (1) and (2), we can get the final representation $\left(h_i^K, h_{i'}^K\right)$ of each pair of entities $\left(i, i'\right)$ through K-layer GNN. Then we measure the distance between entity-entity pairs to determine whether they should be aligned based on entity embeddings, and the formula is as follows

$$d\left(h_i^K, h_{i'}^K\right) = \left|\left|h_i^K - h_{i'}^K\right|\right|_2.$$ (1)

We sample N negative entities. The loss function is:

$$\mathcal{L} = \sum_{(i,i')} \sum_{(i-,i-')} max\left(0, d\left(h_i^K, h_{i'}^K\right) - d\left(h_{i-}^K, h_{i-'}^K\right) + \gamma\right),$$ (2)

where $\gamma > 0$ and γ is the boundary hyperparameter.

2.3   Entity Multi-type Information Embedding Module

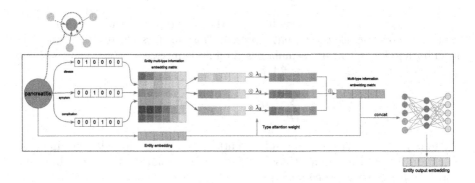

Fig. 2. Entity node information aggregation

As shown in Fig. 2, we combine all type information embeddings in two knowledge graphs to form a multi-type information embedding matrix. First, one-hot encoding is performed on multiple types of node i, and then multi-type information embedding matrix is multiplied to obtain embedding vector emb_{il} corresponding to type l. After that, we compute type information attention weights

λ_{il} with entity embedding vectors following

$$\lambda_{il} = \frac{exp\left(\sigma\left(a^T\left[h_i^{k-1}||emb_{il}\right]\right)\right)}{\sum_{n\in L_i} exp\left(\sigma\left(a^T\left[h_i^{k-1}||emb_{in}\right]\right)\right)}, \tag{3}$$

where a^T is weight vector, L_i is the type set of node i, and $\sigma(\cdot)$ is LeakyReLU. Then the final type information embedding vector emb'_{il} can be obtained by

$$emb'_{il} = \sum_{n\in L_i} \lambda_{in} \cdot emb_{in}. \tag{4}$$

Finally, the entity embedding and the type information embedding are concatenated, and the output embedding vector $h_i'^{k-1}$ of the final node fused with multi-type information is obtained by

$$h_i'^{k-1} = W_2\left(W_1\left[h_i^{k-1}\middle|\middle|emb'_{il}\right] + b_1\right), \tag{5}$$

where W_1 and W_2 are parameters.

2.4   Relation Collective Aggregation Module

We group the neighbor nodes of node i into $N_{i,r}$ by edge type r and apply different transformation matrices W_r. At layer k, the neighborhood information of the relationship between nodes i and j is transformed as follows,

$$z_{i,j}^k = W_r^k h_j'^{k-1}, j \in N_{i,r}. \tag{6}$$

Since an entity can belong to multiple overlapping types, the above transformation clearly distinguishes different representations of the same disease name. The neighborhood aggregation formula based on node intersection graph attention α and edge self-attention β is as follows,

$$z_i^k = \sum_{\bigcup N_{i,r}} \alpha_{ij}\beta_{ij}z_{i,j}^k, \Sigma_j\alpha_{ij}\beta_{ij} = 1, \tag{7}$$

where α_{ij} is the node's cross-graph attention, and β_{ij} is the relationship's self-attention. We will introduce these two attention mechanisms designed in neighborhood aggregation process later.

The output representation h_i^k is the combination of h_i^{k-1} and z_i^k as follows,

$$h_i^k = \sigma\left(\left[W_{self}^k h_i^{k-1}||z_i^k\right]\right). \tag{8}$$

In result, after node pairs $\left(i, i'\right)$ go through the K layer GNN, we can get the final representation $\left(h_i^K, h_{i'}^K\right)$.

In multi-layer GNN encoders, the input of layer k is the output representation of layer $k-1$. The layer 0 representation is the input node features, and different types of entities have features of the same length. Assuming that the hidden dimension of the model is m, the relationship matrix of first layer $W_r^1 \in R^{d_r \times \frac{m}{2}}$, where d_r is neighbor feature length of entities in group N_r. The output of each layer is the concatenation of hidden representation of the previous layer and the aggregated information representation of current layer, so the length of hidden representation is $\frac{m}{2} + \frac{m}{2} = m$.

Cross-Graph Attention. Existing GNN-based entity alignment methods reconcile the structural differences between two knowledge graphs in an implicit way, such as graph matching target [14] and rule base [2]. Structural differences are mainly caused by incompleteness in knowledge graph. In MulEA, this problem is solved by collective aggregation of confidence neighborhood information. Specifically, we assign higher weights to neighbors with corresponding weights in another graph through a cross-graph attention mechanism.

Given a node pair $\left(i, i^{'}\right)$, we take N_i and $N_i^{'}$ as neighbors of node i and node $i^{'}$ respectively. Soft decisions are made by computing the similarity of pairs $(p, q) \in N_i \times N_{i'}$,

$$\alpha_p = \frac{\sum_{q \in N_{i'}} exp^{-||z_p - z_q||_2}}{\sum_{p \in N_i} \sum_{q \in N_{i'}} exp^{-||z_p - z_q||_2}}, \tag{9}$$

$$\alpha_q = \frac{\sum_{p \in N_{i'}} exp^{-||z_q - z_p||_2}}{\sum_{q \in N_i} \sum_{p \in N_{i'}} exp^{-||z_q - z_p||_2}}, \tag{10}$$

where z_p and z_q are computed by Eq. 7. For node $p_1, p_2 \in N_i$, if the cumulative similarity between p_1 and neighbor $N_{i'}$ in graph $G^{'}$ is greater than p_2, then $\alpha_{p_1} > \alpha_{p_2}$. The weights α_p and α_q are vectors of the attention matrix $A_{i,i'} \in R^{|N_i| \times |N_{i'}|}$.

Single Graph Self-attention. Although cross-graph attention can improve the correctness of collective decision-making, it ignores negative evidence. If neighborhood aggregation only relies on cross-graph attention, it fails to predict negatives when only unimportant nodes are soft-aligned. Furthermore, the computation of cross-graph attention focuses on the similarity of neighbor nodes and considers each relationship equally important. In order to solve this problem, we refer to the graph attention network [10] and utilize edge-level self-attention considering triplet information to adjust the cross-graph attention. The edge-level self-attention calculation formula is as follows,

$$\beta_{ij} = \frac{exp\left(\sigma\left(a_r^T \left[z_i || z_j\right]\right)\right)}{\sum_{k \in N_i} exp\left(\sigma\left(a_r^T \left[z_i || z_k\right]\right)\right)} \tag{11}$$

where a_r is the weight vector, using self-attention to estimate the importance of the triplet edge composed of <head, relation, tail>. The $\sigma(\cdot)$ is the LeakyReLU function.

3   Experiment

3.1   Dataset

We use the real medical knowledge graph alignment dataset MED-OWN-15K in experiments. MED refers to the knowledge graph constructed through the medical encyclopedia network, and the entities and relationships for constructing the knowledge graph are obtained by fully crawling the medical encyclopedia network. OWN stands for OwnThink Medical Knowledge Graph. MED-OWN-15K contains rich node attributes and features with tens of thousands of nodes and different types of edges. Table 1 summarizes the entities, relations, and triples in MED-OWN-15K. Table 2 shows statistics of the number of various entities in dataset. After obtaining the Medbaike and OwnThink maps, medical seed collection and manual annotation are carried out. The reference is 9K medical seeds in [16].

Table 1. MED-OWN-15K dataset statistics

	Entity	Relation	Triple
OwnThink	24783	8903	47900
Medbaike	35827	10892	52729

Table 2. MED-OWN-15K entity statistics

	Disease	Symptom	Examination	Drug	Treatment	Food	Part	Operation
OwnThink	10783	7839	3900	7872	4728	1283	1834	2034
Medbaike	13827	8892	4729	7922	5334	2043	2109	2098

3.2   Comparison Methods

In order to evaluate the performance of MulEA on the alignment of heterogeneous medical knowledge graphs, we compare it with the following methods:

- JAPE [6] utilizes TransE [1] to jointly embed the structures of two knowledge graphs into a unified vector space, which is further refined by exploiting attribute correlations.
- GCN-Align [11] uses primitive graph convolutional networks [4] to model structural and attribute features.
- MuGNN [2] learns alignment-oriented knowledge graph embeddings by encoding two knowledge graphs through multiple channels.
- BootEA [7] iteratively labels possible entity alignments as training data for learning alignment-oriented knowledge graph embeddings.

- AliNet [8] introduces far neighbors to enlarge the overlap between their neighborhood structures, and uses a gating mechanism to control the aggregation of neighborhood information.
- MultiKE [15] establishes three representation views, namely entity name, relationship and attribute, and builds a model for each view to learn entity embeddings.
- RDGCN [12] builds a dual relation graph modeled by interaction with the original graph, and utilizes neural network gating to capture the neighbor structure.
- NMN [13] adopts a new graph sampling strategy to identify the most informative neighbors in entity alignment, and designs a matching mechanism to distinguish whether subgraphs match.

3.3   Performance Comparison

We present and analyze experimental results of MulEA and other benchmark algorithms on medical dataset MED-OWN-15K. From Table 3, it can be concluded that MulEA outperforms all other methods under Recall@Precision=0.95 (Rec@Pre=0.95 for short), PRAUC and F1 metrics, and Hit@1 score is second only to NMN. Compared with other benchmarks, MulEA can achieve joint alignment of multi-type entities. The results demonstrate the validity and superiority of our proposed model.

Table 3. Results of heterogeneous medical knowledge graph alignment

	Rec@Pre = 0.95	PRAUC	F1	Hit@1
JAPE	39.20	55.93	59.24	50.26
GCN-Align	45.42	59.45	50.33	51.34
MuGNN	47.41	61.63	55.36	52.15
BootEA	48.43	59.40	56.33	53.03
AliNet	50.42	55.40	56.43	59.90
MultiKE	53.49	59.45	56.94	60.40
RDGCN	63.43	61.45	59.49	64.35
NMN	66.42	62.25	61.33	**68.34**
MulEA	**70.90**	**63.20**	**62.50**	67.99

Furthermore, JAPE based on knowledge graph embedding performs worst on Rec@Pre = 0.95 and Hit@1 because it does not consider topology information. By considering structural heterogeneity, AliNet outperforms most other structure-based counterparts, showing the importance of addressing structural heterogeneity. RDGCN has a significant improvement over the structure-based model. It employs entity name information and structure information, which demonstrates that entity names provide positive effects for entity alignment. In addition, the performance of NMN with subgraph sampling method is only inferior to MulEA, indicating that entity alignment performance is greatly affected by information enhancement and a good neighborhood sampling strategy.

3.4 Ablation Experiment

We conduct ablation experiments on MulEA to analyze the effectiveness of entity multi-type information embedding module and relation collective aggregation module. Table 4 shows the alignment effect after removing entity multi-type information embedding module and two different attention mechanisms. The results show that entity multi-type information embedding module improves both metrics by nearly 6%. Cross-graph attention improves both metrics by nearly 4%, and self-attention improves by almost 2%. Therefore, the multi-type information of entities is the main part of performance improvement, indicating the multi-type features of entities in medical domain play a great role in entity alignment.

Table 4. Results of ablation experiment

	PRAUC	F1
w.o. multi-type	57.02	58.59
w.o. cross attention	60.58	59.99
w.o. self attention	61.98	60.05
MulEA	**63.20**	**62.50**

4   Conclusion

The medical knowledge graph alignment aims to eliminate the knowledge redundancy and knowledge noise problems, and alleviate the incompleteness of knowledge graph, which has great practical significance. The current heterogeneous knowledge graph alignment models do not take into account the multi-type characteristics of the medical domain entity relationship, and cannot utilize one model to align multi-type entities. In this paper, we propose a multi-type entity alignment method based on attention mechanism called MulEA for medical knowledge graphs. It can jointly align multiple types of entities through entity multi-type information embedding and relation collective aggregation. Furthermore, we design cross-graph attention and single-graph self-attention in relation collective aggregation module to alleviate the incompleteness of knowledge graphs. The results of experiments demonstrate that MulEA outperforms baselines on most metrics. In the meanwhile, the ablation study indicates that each individual component of MulEA benefits performance and their integration allows the model to achieve the optimal effect.

Acknowledgements. This work is partially supported by the National Key Research and Development Plan Project 2022YFC3600901CNKLSTISS, and the Shanghai Science and Technology Development Fund No. 19511121204.

References

1. Bordes, A., Usunier, N., Garcia-Duran, A., Weston, J., Yakhnenko, O.: Translating embeddings for modeling multi-relational data. Adv. Neural Inf. Process. Syst. **26**, 1–9 (2013)
2. Cao, Y., Liu, Z., Li, C., Li, J., Chua, T.S.: Multi-channel graph neural network for entity alignment. arXiv preprint arXiv:1908.09898 (2019)
3. Ehrlinger, L., Wöß, W.: Towards a definition of knowledge graphs. SEMANTiCS (Posters, Demos, SuCCESS) **48**(1–4), 2 (2016)
4. Kipf, T.N., Welling, M.: Semi-supervised classification with graph convolutional networks. arXiv preprint arXiv:1609.02907 (2016)
5. Pei, S., Yu, L., Hoehndorf, R., Zhang, X.: Semi-supervised entity alignment via knowledge graph embedding with awareness of degree difference. In: The World Wide Web Conference, pp. 3130–3136 (2019)
6. Sun, Z., Hu, W., Li, C.: Cross-lingual entity alignment via joint attribute-preserving embedding. In: d'Amato, C., et al. (eds.) ISWC 2017. LNCS, vol. 10587, pp. 628–644. Springer, Cham (2017). https://doi.org/10.1007/978-3-319-68288-4_37
7. Sun, Z., Hu, W., Zhang, Q., Qu, Y.: Bootstrapping entity alignment with knowledge graph embedding. In: IJCAI, vol. 18, pp. 4396–4402 (2018)
8. Sun, Z., et al.: Knowledge graph alignment network with gated multi-hop neighborhood aggregation. In: Proceedings of the AAAI Conference on Artificial Intelligence, vol. 34, pp. 222–229 (2020)
9. Sun, Z., et al.: A benchmarking study of embedding-based entity alignment for knowledge graphs. arXiv preprint arXiv:2003.07743 (2020)
10. Veličković, P., Cucurull, G., Casanova, A., Romero, A., Lio, P., Bengio, Y.: Graph attention networks. arXiv preprint arXiv:1710.10903 (2017)
11. Wang, Z., Lv, Q., Lan, X., Zhang, Y.: Cross-lingual knowledge graph alignment via graph convolutional networks. In: Proceedings of the 2018 Conference on Empirical Methods in Natural Language Processing, pp. 349–357 (2018)
12. Wu, Y., Liu, X., Feng, Y., Wang, Z., Yan, R., Zhao, D.: Relation-aware entity alignment for heterogeneous knowledge graphs. arXiv preprint arXiv:1908.08210 (2019)
13. Wu, Y., Liu, X., Feng, Y., Wang, Z., Zhao, D.: Neighborhood matching network for entity alignment. arXiv preprint arXiv:2005.05607 (2020)
14. Xu, K., et al.: Cross-lingual knowledge graph alignment via graph matching neural network. arXiv preprint arXiv:1905.11605 (2019)
15. Zhang, Q., Sun, Z., Hu, W., Chen, M., Guo, L., Qu, Y.: Multi-view knowledge graph embedding for entity alignment. arXiv preprint arXiv:1906.02390 (2019)
16. Zhang, Z., et al.: An industry evaluation of embedding-based entity alignment. arXiv preprint arXiv:2010.11522 (2020)

Analogy-Triple Enhanced Fine-Grained Transformer for Sparse Knowledge Graph Completion

Shaofei Wang, Siying Li, and Lei Zou$^{(\boxtimes)}$

Peking University, Beijing, China
{wangshaofei,lisiying,zoulei}@pku.edu.cn

Abstract. Sparse problem is the major challenge in knowledge graph completion. However, existing knowledge graph completion methods utilize entity as the basic granularity, and face the semantic under-transfer problem. In this paper, we propose an analogy-triple enhanced fine-grained sequence-to-sequence model for sparse knowledge graph completion. Specifically, the entities are first split into different levels of granularity, such as sub-entity, word, and sub-word. Then we extract a set of analogy-triples for each entity-relation pair. Furthermore, our model encodes and integrates the analogy-triples and entity-relation pairs, and finally predicts the sequence of missing entities. Experimental results on multiple knowledge graphs show that the proposed model can achieve better performance than existing methods, especially on sparse entities.

Keywords: Knowledge graph completion · Sequence-to-sequence model · Sparse knowledge graph

1 Introduction

Knowledge graphs (KGs) contain massive real world knowledge, and the knowledge in KGs is generally represented by structured triples in the form of (h, r, t), where h and t are head and tail entities respectively, and r denotes the relation between h and t. Due to the incompleteness of KGs, various knowledge graph completion methods are proposed to predict missing links based on existing data of KGs. However, as the long-tail distribution of entities in KGs, the sparse problem of entities is inevitable and becomes the major challenge for knowledge graph completion [27]. For example, in the open source KG Freebase [2], up to 58.2% of the entities appear lower than 10 times [6].

Previous knowledge graph completion methods can be divided into two streams. 1) One is embedding-based methods, these methods learn low-dimensional embeddings for entities and relations, and then score the candidate triples based on embeddings [3,17,28]. 2) The other line is rule-based methods, which learn logical rules form KGs, and then apply the rules on existing data to predict new triples [11,29]. However, although much remarkable progress has been achieved, existing methods still face the following **Semantic Under-transfer Problem**:

© The Author(s), under exclusive license to Springer Nature Switzerland AG 2023
X. Wang et al. (Eds.): DASFAA 2023, LNCS 13944, pp. 742–757, 2023.
https://doi.org/10.1007/978-3-031-30672-3_50

Semantic Under-Tansfer Problem. Current methods utilize entities as the basic granularity, and the granularity of entities is too coarse to transfer the semantics well (i.e., under-transfer problem), especially for sparse entities. For example, in Fig. 1, the entity `melbourne institute of technology` is a sparse entity, and it is difficult to predict its `location` and `field`. However, if the entity is split into the fine-grained components `melbourne` and `institute of technology`, whose semantics can be transferred from similar entities. As a consequence, the `location` and `field` can be predicted through learning from other analogy-triples (such as (`melbourne park, location, melbourne`) and (`massachusetts institute of technology, field, education`)).

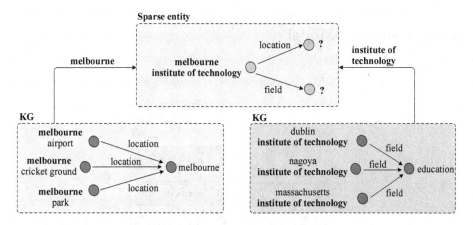

Fig. 1. A motivation example of our knowledge graph completion model on sparse entities. Considering a sparse entity `melbourne institute of technology`, the semantics of this entity is difficult to be modeled by traditional methods due to the data scarcity. While in our method, the entity is split into multiple fine-grained components (such as `melbourne` and `institute of technology`). Thus the semantics of these fine-grained components can be learnt from analogy-triples (showed in the left and right boxes respectively). Finally the `location` and `field` of the sparse entity `melbourne institute of technology` can be predicted.

In this paper, we propose an analogy-triple enhanced fine-grained knowledge graph completion model, the FineKGC, to alleviate the knowledge under-transfer problem. The main motivation of our model can be included as: 1) Traditional models are mainly data-driven methods, so it is difficult to model the semantics of sparse entities well due to the data scarce. While in our model, by splitting entities into fine-grained components, each component of spare entities could appear much more frequently; 2) Besides that, in order to alleviate the sparse problem, our model predicts entities not only based on the given entity-relation pair (such as (`melbourne institute of technology, location`)), but also incorporates the corresponding analogy-triples (such as (`melbourne airport, location, melbourne`)). Specifically, first the entities are split into fine granularities which are helpful to transfer semantics among entities. Then analogy-triples are extracted from KGs to enhance the modeling of entities. Furthermore, the knowledge graph completion is conducted by the sequence-to-sequence Transformer model and finally the predicted entities are directly generated.

The contribution of this paper can be summarized in the following:

- In order to alleviate the sparse problem, we propose a fine-grained knowledge graph completion method for sparse entities. The entities are split into fine granularities, which are helpful for semantic transfer among entities.
- The knowledge graph completion is completed by the sequence-to-sequence Transformer model. In the model, analogy-triples are extracted from KGs and are incorporated to enhance the modeling of entities.

2   Transformer Model

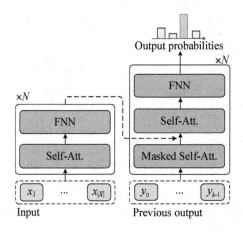

Fig. 2. The framework of Transformer model.

Transformer [21] is the state-of-the-art sequence-to-sequence model and achieves excellent performance in multiple fields due to its self-attention mechanism [24,32]. As illustrated in Fig. 2, the structure is composed of an encoder and a decoder.

Encoder. First, the input sequence $X = \{x_1, \cdots, x_{|X|}\}$ is first initialized to embeddings $\mathbf{X} = [\mathbf{x}_1; \cdots; \mathbf{x}_{|X|}]$ $(\mathbf{X} \in \mathbb{R}^{d \times |X|})^1$. Then $\mathbf{X}$ is feed into the encoder. The encoder is composed of N identical layers, and each layer is composed of two main sub-layers: self-attention sub-layer, and feed-forward sub-layer.

---

[1] In this paper, the bold characters represent the embeddings in the model, and d is the dimension of embeddings.

In the self-attention sub-layer, the input $\mathbf{X}$ is transformed to three matrices: query $\mathbf{Q}$, key $\mathbf{K}$, and value $\mathbf{V}$. Then they are encoded by the attention function, and the hidden states $\mathbf{M}_X$ is obtained. Formally,

$$M_X = Attention(\mathbf{Q}, \mathbf{K}, \mathbf{V}) = softmax(\frac{\mathbf{Q}\mathbf{K}^T}{\sqrt{g_k}}\mathbf{V}) \tag{1}$$

where $\frac{1}{\sqrt{g_k}}$ is the scaling factor.

In the feed-forward sub-layer, the hidden states $\mathbf{M}_X$ is processed by linear transformations and ReLU activation:

$$F_X = FNN(\mathbf{M}_X) = max(0, \mathbf{M}_X\mathbf{W}_1 + \mathbf{b}_1)\mathbf{W}_2 + \mathbf{b}_2 \tag{2}$$

where $\mathbf{W}_1$, $\mathbf{W}_2$, $\mathbf{b}_1$, and $\mathbf{b}_2$ are trainable parameters.

Decoder. The decoder is also composed of N identical layers, and each layer mainly contains three sub-layers. The first sub-layer is the masked self-attention sub-layer. In this sub-layer, the sequence generated by the decoder in previous steps (i.e., $\{y_0, \cdots, y_{k-1}\}$, where y_0 is a special token at the beginning) are processed by Eq. (1). The output hidden states of this sub-layer is denoted by $\mathbf{M}_Y$. Then in the self-attention sub-layer, the hidden states obtained by encoder F_X is integrated with the hidden states of the former sub-layer $\mathbf{M}_Y$. This procedure can be presented by:

$$F_Y = Attention(\mathbf{M}_Y, \mathbf{F}_X, \mathbf{F}_X) \tag{3}$$

Then F_Y is processed in the feed-forward sub-layer as Eq. (2). Finally, the decoder predicts the output probability of a token by:

$$P(y_k|X, Y_{<k}) = softmax(\mathbf{W}_3\mathbf{F}_Y + \mathbf{b}_3) \tag{4}$$

where $\mathbf{W}_3$ and $\mathbf{b}_3$ are trainable parameters. y_k denotes the output token in the k-th decode time, and $Y_{<k}$ is the output sequence in previous steps.

The procedure of predicting a sequence $Y = \{y_1, \cdots, y_{|Y|}\}$ can be represented by:

$$P(Y|X; \theta) = \prod_{k=1}^{K} P(y_k|Y_{<k}, X; \theta) \tag{5}$$

Note that all these sub-layers above are followed by residual connection and layer normalization, which are omitted for simplicity. More details can be seen in [21].

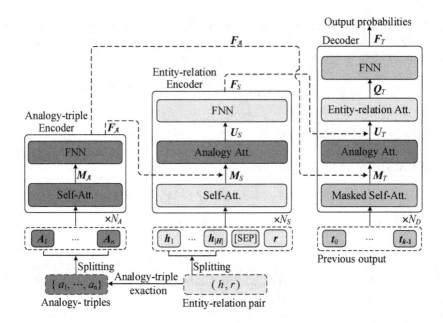

Fig. 3. The framework of our model. Given an entity-relation pair (h, r), first the entity h is split into fine granularities. Then corresponding analogy-triples are extracted from KG. Afterwards, the analogy-triples and entity-relation pair are encoded respectively, and then are integrated in the entity-relation encoder. Finally, in the decoder, previous output, hidden states of analogy-triples and entity-relation pair are all incorporated to finally generate the predicted entities.

3 Methodology

3.1 Problem Formulation

Given a KG $\mathcal{K} = \{\mathcal{E}, \mathcal{R}\}$, where $\mathcal{E}$ represents the set of entities, and $\mathcal{R}$ denotes the set of relations. KG is represented by a lot of triples (h, r, t), where entities h, $t \in \mathcal{E}$, and relation $r \in \mathcal{R}$. In this study we mainly focus on the link prediction task. Given an entity-relation pair (h, r), link prediction aims to predict the missing tail entity t.

3.2 Proposed Method

The proposed model is composed of five main modules: fine-grained splitting, analogy-triple extraction, analogy-triple encoder, entity-relation encoder, and decoder. First, all the entities are split into fine granularities. Then analogy-triples are extracted from KG based on the entity-relation pair. Afterwards, the sequences of entity-relation pair and analogy-triples are encoded and integrated. Finally, the sequences of predicted entities are generated by the decoder. The framework is illustrated in Fig. 3.

Fine-Grained Splitting. Fine-grained splitting is to split entities into fine granularities. In our model, we implement three different levels of granularity as follows:

- **Sub-entity granularity.** Under this granularity, each unit may contain multiple words. Take the entity `teahouse in chinatown` as an example, it can be split to {`teahouse in`, `chinatown`}.
- **Word granularity.** Word is the basic unit, and each entity is split by the words. Consider the entity `teahouse in chinatown` again, it is split to {`teahouse`, `in`, `chinatown`}.
- **Sub-word granularity.** Under this granularity, a word can be split into multiple sub-words. For example, the entity above can be split to {`tea`, `house`, `in`, `china`, `town`}.

To achieve fine-grained splitting, we introduce the text compression algorithm *byte pair encoding* (BPE) [15] in our model. In this algorithm, all the words are first split into characters, then two characters with the highest connection frequency are merged based on the statistics. Merging is conducted iteratively until the number of steps is reached or there is no more combination[2]. In this module, the BPE algorithm is modified to make it applicable for different levels of granularity on entities.

In this way, fine-grained splitting is applied on all the entities of KG. The entity h can be transformed into the sequence H, formally:

$$H = Splitting(h),$$
$$H = \{h_1, \cdots, h_{|H|}\} \tag{6}$$

Thus the entity-relation pair (h, r) is transformed to sequence $\{H, r\}$. Similarly, for a triple a, the head and tail entities of a are also split into sequences, and we denote the corresponding sequence of triple a by A.

Analogy-Triple Extraction. This module is to extract a set of analogy-triples for each entity-relation pair (h, r) based on the corresponding sequence $\{H, r\}$. Specifically, the head entities of analogy-triples are similar with H, and the relations contained in analogy-triples is same as r. For example, for the given entity-relation pair (`melbourne institute of technology`, `location`), the corresponding analogy-triple set may be {(`melbourne airport`, `location`, `melbourne`), (`melbourne park`, `location`, `melbourne`)}. The analogy-triples extraction can be formally represented by:

$$\mathcal{A} = \{A|Relation(A) = r, sim(Head(A), H) > \gamma\} \tag{7}$$

where $\mathcal{A}$ is a set which contains multiple sequences of analogy-triples, and A ($A \in \mathcal{A}$) is the sequence of an analogy-triple. $Head(A)$ and $Relation(A)$ are

---

[2] Similarly, for the sub-entity granularity, entities are first split into words, and then sub-entities can be obtained through the combination of words.

sequences of head entity and relation of A respectively. $sim(\cdot,\cdot)$ is the similarity function of entities, and the Levenshtein distance based similarity is used in this paper. γ is the similarity threshold. The maximum number of analogy-triples is set to n. In this module, a set of analogy-triples $\mathbb{A} = \{a_1, \cdots, a_n\}$ (a_i is the triple of KG $\mathcal{K}$) are extracted, and the corresponding sequences with the same level of granularity are denoted by $\mathcal{A} = \{A_1, \cdots, A_n\}$.

Analogy-Triple Encoder. The goal of the analogy encoder is to obtain the hidden states of analogy-triples and extract their features. Given the sequence of analogy-triples $\mathcal{A} = \{A_1, \cdots, A_n\}$, they are transformed to embeddings $\boldsymbol{\mathcal{A}} = [\boldsymbol{A_1}; \cdots; \boldsymbol{A_n}]$ ($\boldsymbol{\mathcal{A}} \in \mathbb{R}^{d \times |\mathcal{A}|}$). Then $\boldsymbol{\mathcal{A}}$ is feed into the analogy-triple encoder. The analogy-triple encoder is composed of N_A identical layers, and each layer contains the self-attention sub-layer and feed-forward sub-layer, which are same as Eq. (1) and Eq. (2). We denote the output hidden states by $\boldsymbol{F_A}$.

Entity-Relation Encoder. The entity-relation encoder first encodes the sequence of entity-relation pair, and then integrates the hidden states of analogy-triples with that of entity-relation pair. The structure of the entity-relation encoder is N_S identical layers, and each layer includes three sub-layers: self-attention sub-layer, analogy attention sub-layer, and feed-forward sub-layer. First, the sequence of entity-relation pair $\{H, r\}$ is separated by a special token [SEP], and the sequence is transformed to $S = \{H, [SEP], r\}$. Then they are initialized to embeddings $\boldsymbol{S}(\boldsymbol{S} \in \mathbb{R}^{d \times |S|})$.

In the self-attention sub-layer, $\boldsymbol{S}$ is encoded and transformed to the matrix $\boldsymbol{M_S}$ ($\boldsymbol{M_S} \in \mathbb{R}^{d \times |S|}$) according to Eq. (1).

The difference between the entity-relation encoder and the analogy-triple encoder is that there is an extra analogy attention sub-layer. In this sub-layer, the hidden states of analogy-triples $\boldsymbol{F_A}$ is attended by the hidden states $\boldsymbol{M_S}$. Formally:

$$\boldsymbol{U_S} = Attention(\boldsymbol{M_S}, \boldsymbol{F_A}, \boldsymbol{F_A}), \boldsymbol{U_S} \in \mathbb{R}^{d \times |S|} \tag{8}$$

Then the hidden states $\boldsymbol{U_S}$ is processed by the feed-forward sub-layer according to Eq. (2). Finally, the output of entity-relation encoder is denoted by $\boldsymbol{F_S}$ ($\boldsymbol{F_S} \in \mathbb{R}^{d \times |S|}$).

Decoder. The decoder aims to predict the missing entities for each entity-relation pair. The input of decoder includes three parts: the output sequences in previous decoding steps (a special beginning hidden states $\boldsymbol{t_0}$ ($\boldsymbol{t_0} \in \mathbb{R}^{d \times 1}$) is used in the first step), the hidden states of analogy-triples $\boldsymbol{F_A}$ and entity-relation pair $\boldsymbol{F_S}$. The decoder is composed of N_D layers, and each layer includes four main sub-layers: self-attention sub-layer, analogy attention sub-layer, entity-relation attention sub-layer, and feed-forward sub-layer. Finally the outputs of decoder are the predicted sequences of entities.

First, the sequence predicted in previous $k-1$ steps $\boldsymbol{T_{<k}} = \{t_0, t_1, \cdots, t_{k-1}\}$ is processed by the self-attention layer according to Eq. (1), and the output of this sub-layer is denoted by $\boldsymbol{M_T}$ ($\boldsymbol{M_T} \in \mathbb{R}^{d \times k}$).

Then in the analogy attention layer, the hidden states $\boldsymbol{M}_T$ is used to attend the hidden states of analogy-triples:

$$\boldsymbol{U}_T = Attention(\boldsymbol{M}_T, \boldsymbol{F}_\mathcal{A}, \boldsymbol{F}_\mathcal{A}), \boldsymbol{U}_T \in \mathbb{R}^{d \times k} \qquad (9)$$

Next the hidden states $\boldsymbol{U}_T$ is integrated with the entity-relation hidden states $\boldsymbol{F}_S$ in the entity-relation attention layer:

$$\boldsymbol{Q}_T = Attention(\boldsymbol{U}_T, \boldsymbol{F}_S, \boldsymbol{F}_S), \boldsymbol{Q}_T \in \mathbb{R}^{d \times k} \qquad (10)$$

Finally, the hidden states is processed in the feed-forward sub-layer and the final output of decoder is represented by $\boldsymbol{F}_T$.

The probability distribution at the k-th decode step is predicted by:

$$P(t_k|A, (h, r), T_{<k}; \theta) = softmax(\boldsymbol{W}\boldsymbol{F}_T + \boldsymbol{b}) \qquad (11)$$

where $\boldsymbol{W}$ and $\boldsymbol{b}$ are trainable parameters. Thus the sequence of predicted entity can be obtained by $T = \{t_1, \cdots, t_{|T|}\}$. Similar with [14], the output sequences of decoder are always exactly the entities, so the generated sequences are linked to entities through exactly string matching.

3.3   Training

During training, the parameters are optimized by minimizing the negative log-likelihood objective function:

$$NLL(\theta) = -\frac{1}{|T|} \sum_{k=1}^{|T|} logp(t_k|T_{<k}, \mathbb{A}, (h, r); \theta) \qquad (12)$$

where $\{\mathbb{A}\}$ denotes the analogy-triple sets, $\{(h, r)\}$ represents the entity-relation pairs, and $\{t\}$ is the set of corresponding tail entities. $T_{<k}$ is the decoded sequence in previous k steps. θ represents all the parameters in the model.

4   Experiment

4.1   Experiment Setup

Datasets. The experiments are conducted on three public datasets: 1) FB15k-237 [19] is a sparse dataset which is drieved from FB15k through removing the inverse triples. 2) YAGO3-10-dr [1] is a sparse dataset modified from YAGO3-10. Similar with FB15k-237, some duplicate relations are removed from YAGO3-10. 3) Wikidata5M [14, 25] is a large-scale benchmark dataset. The details of datasets are displayed in Table 1.

Table 1. Statistics of datasets.

	#entity	#relation	#train	#valid	#test
FB15k-237	14,541	237	272,115	17,535	20,466
YAGO3-10-dr	122,837	36	732,556	3,390	3,359
Wikidara5m	4,818,503	828	21,343,515	5,357	5,321

Baselines. We compare our method with three types of baselines as follows:

- **Embedding based methods.** TransE [3] and RotatE [17] are translation-based representation learning models. DisMult [28] and ComplEX [20] are tensor decomposition based methods.
- **Rule based methods.** Neural-LP [29] transforms the procedure of logic rule learning to be differentiable, and proposes an end-to-end differentiable model for rule learning. ProbCBR [5] extracts rules with probabilities from other entities and then apply the rules to the given entity-relation pair.
- **Pre-trained language model based methods.** KGT5 [14] adopts sequence-to-sequence pre-trained model T5 [13] to achieve knowledge graph completion. SimKGC [23] introduces negative samples and textual description of entities and relations in the pre-trained language model BERT.

Evaluating Metrics. We use mean reciprocal rank (MRR), and Hits@n ratio to evaluate the performance. These metrics are generally used to measure the quality of ranks. The results are evaluated under filtered settings[3].

Implementation Details. Our model is developed based on the papers [18,31], and the hyper-parameters of Transformer model are same as [18]. The hidden states d is set to 512, and the batch size is 4096. The iterations of layers in two encoders (N_A and N_S) and decoder (N_D) are set to 6. In the fine-grained splitting module, for sub-word granularity, the BPE algorithm is implemented for 8000 steps on FB15k-237, and 30000 steps on YAGO3-10-dr and Wikidata5M. In the analogy-triples extraction module, the similarity threshold γ is set to 0.5, and the maximum number of analogy-triples n is 3.

4.2 Performance Comparison with Baselines

Table 2 shows the link prediction results on three datasets by different methods. Lines 1–4 are results by embedding based methods[4]. Lines 5–6 are results of rule based methods[5]. Lines 7–8 show the performance obtained by pre-trained

---

[3] More details can be seen in paper [3].

[4] These results are quoted from papers [1,14].

[5] The results of these two models are obtained through our implementation using the open source codes. Note that the results on Wikidata5M is empty because it is difficult to extend to large-scale KGs for these two rule-based methods [4].

Table 2. The main results of link prediction.

#	Method	FB15k-237			YAGO3-10-dr			Wikidata5M		
		MRR	Hit@1	Hit@10	MRR	Hit@1	Hit@10	MRR	Hit@1	Hit@10
1	**TransE** [3]	0.288	0.198	0.441	0.190	0.136	0.323	0.253	0.170	0.392
2	**DisMult** [28]	0.238	0.199	0.446	0.192	0.133	0.307	0.253	0.208	0.334
3	**ComplEX** [20]	0.249	0.194	0.450	0.201	0.143	0.315	0.281	0.228	0.373
4	**RotatE** [17]	0.337	0.241	0.533	0.214	0.153	0.332	0.290	0.234	0.390
5	**Neural-LP** [29]	0.240	0.251	0.362	0.187	0.132	0.297	–	–	–
6	**ProbCBR** [5]	0.231	0.187	0.320	0.181	0.128	0.284	–	–	–
7	**SimKG** [23]	0.336	0.249	0.511	–	–	–	0.358	0.313	0.441
8	**KGT5** [14]	0.276	0.210	0.414	0.211	0.151	0.327	0.300	0.267	0.365
9	**FineKGC-Base**	0.379	0.289	0.556	0.254	0.186	0.387	0.369	0.306	0.443
10	**FineKGC-Ana.**	**0.389**	**0.299**	**0.567**	**0.273**	**0.204**	**0.404**	**0.387**	**0.321**	**0.452**

language model based methods[6]. In Line 9, our model is implemented without incorporating analogy-triples. In Line 10, the analogy-triples for each entity-relation pair are extracted and attended in the model. In our models, the entities are split into the sub-word granularity. The best performance is shown in bold fonts. We can reach the following conclusions:

1) Considering the overall results, our model `FineKGC-Ana.` (Line 10) achieves the best performance on three datasets, suggesting the superiority of the proposed model. Specifically, compared with `RotatE` (`RotatE` performs the best on 5 out of 9 metrics in baselines), our proposed model achieves the maximum improvement of 0.072 (Hits@10) on YAGO3-10-dr and 0.087 (Hits@1) on Wikidata5M. The results demonstrate the effectiveness of our model.
2) Compare the performance of our models in Lines 9–10, the performance in line 10 are better than that of line 9 on all three datasets. The maximum improvement is 0.019 on metric MRR of YAGO3-10-dr. The performance shows that it is helpful to incorporate analogy-triples in our model.

4.3   Performance with Different Frequencies

Our model mainly aims to improve the performance of knowledge graph completion on sparse entities. To further verify the effect of our model on sparse entities, in this experiment, we compare the performance of models on entities with different frequencies. Specifically, first the validation set of FB15k-237 is divided into four subsets according to the frequency of entities (the frequency of entities ranges in 1–10, 10–20, 20–30, and greater than 30, respectively). Then we randomly sample 500 test data from each subset. The results of different methods on these four test sets are plotted in Fig. 4.

---

[6] For SimKGC, the results on FB15k-237 and Wikidata5M is obtained from its original paper [23]. For KGT5, the results on FB15k-237 and Wikidata5M are quoted from [14].

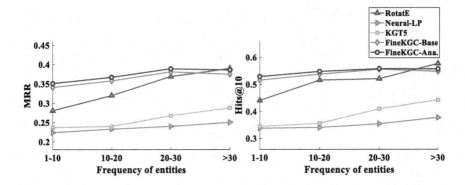

Fig. 4. The performance of models on entities with different frequencies.

In Fig. 4, the X-axises denote the frequencies of entities, and the Y-axises depict the MRR (left) and Hits@10 (right) respectively. We can see when the frequency of entities is low (when the frequency ranges in 1–10), the MRR and Hits@10 of three baselines are small, which demonstrates that the sparse problem is a challenge for knowledge graph completion. While our method can obtain remarkable progress than baselines on these sparse entities.

4.4 Performance with Different Levels of Granularity

As introduced in Sect. 3.2, the entities are split into different granularities. To test the corresponding performance, we conduct experiments on the validation set of YAGO3-10-dr with different levels of granularity. Table 3 shows the performance.

Table 3. Performance with different levels of granularity.

Gran.	Model	MRR	Hits@1	Hits@10
sub-word	FineKGC-Base	0.254	0.186	0.387
	FineKGC-Ana	0.273	0.204	0.404
word	FineKGC-Base	0.226	0.167	0.330
	FineKGC-Ana	0.238	0.178	0.350
sub-entity	FineKGC-Base	0.210	0.159	0.313
	FineKGC-Ana.	0.217	0.165	0.326

From the results, we achieve the following conclusions:

1) The comparison under different levels of granularity shows that the model FineKGC-Ana. always performs better than FineKGC-Base. The results indicate that incorporating analogy-triples in the model is effective under different levels of granularity.

2) Furthermore, the sub-word granularity beats the other two levels of granularity by both `FineKGC-Base` and `FineKGC-Ana`.. The reason is supposed that the sub-word granularity is finer than word and sub-entity granularities, which is more helpful for semantic transfer among entities.

4.5 Performance with Different Number of Analogy-Triples

In this experiment, we discuss the influence of the analogy-triple number on the performance of our model. Figure 5 plots the MRR (subfigure (a)), Hits@1 (subfigure (b)) and Hits@10 (subfigure (c)) of our model when the number of analogy-triples is 0, 1, 3, and 6 respectively.

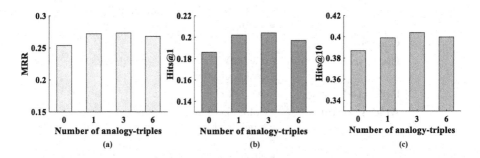

Fig. 5. The performance of our model with different numbers of analogy-triples.

From Fig. 5, we can see the results of model incorporating analogy-triples (the numbers of analogy-triples are 1,3, and 6) are greater than that of model without analogy-triples (the number of analogy-triples is 0). These results show that it is effective to introduce analogy-triples in our model. Moreover, our proposed model achieves the best performance when selecting 3 most similar analogy-triples.

4.6 Case Study

We choose three examples which are predicted correctly by our model on sparse entities (the frequencies of entities in these entity-relation pairs are all smaller than 10). The details are displayed in Table 4.

In Case 1, our model correctly predicts thes `nationality` of the entity `shakti kapoor`. As `kapoor` is the common seen surname of India, and in the analogy-triples, all the head entities which contain the fine-grained component `kapoor` has Indian nationality. The semantic of `kapoor` in analogy-triples can be transferred to the sparse entity `shakti kapoor`. As a consequence, the `nationality` can be correctly predicted.

In Case 2, all the entities in the entity-relation pair and analogy-triples contain the fine-grained component `zhou` and the tail entities are all in China.

Table 4. Case study of our model on sparse entities.

Case 1	
Entity-relation pair	(shakti kapoor, nationality)
Predicted entity	india
Analogy-triples	• (shammi kapoor, nationality, india) • (shashi kapoor, nationality, india) • (shobha kapoor, nationality, india)
Case 2	
Entity-relation pair	(chaozhou, location)
Predicted entity	china
Analogy-triples	• (shuozhou, is located in, china) • (chuzhou, is located in, anhui) • (changzhou, is located in, jiangsu)
Case 3	
Entity-relation pair	(new york university school of law, location citytown)
Predicted entity	new york city
Analogy-triples	• (johns hopkins university school of medicine, location citytown, baltimore) • (stanford university school of humanities and sciences, location citytown, stanford) • (boston university school of law, location citytown, boston)

As a consequence, the semantic of the fine-grained component **zhou** can be transferred, and the **location** of **chaozhou** may be China.

The analogy-triples in Case 3 indicate that the location information is usually implied in the fine-grained components of university schools. Thus the location of entity **new york university school of law** is predicted to be **new york city**.

5   Related Work

Embedding Based Methods. Embedding based methods aims at learning low-dimensional embeddings for entities and relations in KGs, and then predicting the missing triples through scoring candidate triples based on the learnt embeddings. TransE [3] is one of the representative methods which models the relations as translation operations from the head entities to tail entities. Then TransH [26], TransD [7] and TranSparse [8] extend TransE from different perspectives. DisMult [28] models the procedure of scoring triples as tensor factorization, and designs a bilinear formulation to score triples based on semantic matching. ComplEX [20] extends DisMult through mapping entities and relations to complex-valued space to model symmetric and anti-symmetric relations. RotatE [17] models relations as rotation operations from head to tail entities in the complex-valued space.

Besides that, several methods improve the representation learning through introducing extra resources, such as textual descriptions [10,16]. However, these methods utilize entity as the basic granularity, which face the semantic under-transfer problem. In our model, entities are split into different levels of granularity, thus the semantics of the fine-grained components in sparse entities can be transferred from other similar entities.

Rule Based Based Methods. Other stream of models aim to learn logic rules from KGs, and knowledge graph completion is conducted through applying the rules on existing data. AMIE [11] first extracts multiple candidate rules and then designs multiple statistic-based metrics (pca-confidence and head coverage) to evaluate the quality of candidate rules. Neural-LP [29] proposes an end-to-end differentiable model to learn logic rules. Rudik [12] extends the form of rules, and the obtained negative rules can be used for error recognition. ProbCBR [5] is a case-based method, the rules are obtained from entities in the same cluster, and the quality of rules are also obtained based on the statistic of the corresponding cluster. However, these models learn logic rules based on the frequency patterns of KG. For sparse entities, it is difficult to obtain qualified logic rules due to data scarcity.

Pre-trained Language Model Based Methods. Pre-trained language models (PLMs) have recently attracted enormous interest due to its large-scale background knowledge. KG-BERT [30] employs BERT [9] to complete knowledge graphs. First BERT is fine-tuned by knowledge graphs for relation prediction task and link prediction task respectively, and then the entity and relation is used as input (separated by token [SEP]), then the output embedding of special token [CLS] is used to predicate the other entity. StAR [22] also utilizes BERT to learn contextual embeddings of entities. Besides that, a scoring module is designed to learn both contextualized and structured knowledge of KGs. KGT5 [14] introduces the T5 [13] model in knowledge graph completion task and question answering task simultaneously. For KGC task, the input is the verbalized head/tail entities and the relations, then the output is the tail/head entities. SimKGC [23] incorporates contrastive learning on pre-trained language model BERT to implement knowledge graph completion. Besides that, textual descriptions of head entities and relations are also utilized to model the relation-aware semantics. Different from these methods, there is no need for our model to utilize the pre-trained language models and large amount of textual resources. The experimental results also demonstrate the superiority of our model. In the future we will explore to incorporate pre-trained models with our method.

6 Conclusion

In this paper, we propose an analogy-triple enhanced fine-grained Transformer model for sparse knowledge graph completion. First, entities are split with different levels of granularity. In this way, the semantics of sparse entities can be modeled and transferred from other similar entities. Then we introduce the analogy-triples to enhance the modeling of entities. Finally, knowledge graph completion task is conducted by the sequence-to-sequence model. In the model, the fine grained entity-relation pairs and analogy-triples are jointly attended to generate predicted tail entities. The experimental results demonstrate that our model outperforms existing methods on sparse knowledge graph completion.

Acknowledgements. This work was supported by NSFC under grant 61932001 and U20A20174.

References

1. Akrami, F., Saeef, M.S., Zhang, Q., Hu, W., Li, C.: Realistic re-evaluation of knowledge graph completion methods: An experimental study. In: SIGMOD, pp. 1995–2010 (2020)
2. Bollacker, K., Evans, C., Paritosh, P., Sturge, T., Taylor, J.: Freebase: A collaboratively created graph database for structuring human knowledge. In: SIGMOD, pp. 1247–1250 (2008)
3. Bordes, A., Usunier, N., Garcia-Duran, A., Weston, J., Yakhnenko, O.: Translating embeddings for modeling multi-relational data. NeurIPS **26** (2013)
4. Chen, X., Jia, S., Xiang, Y.: A review: Knowledge reasoning over knowledge graph. Expert Syst. Appl. **141**, 112948 (2020)
5. Das, R., Godbole, A., Monath, N., Zaheer, M., McCallum, A.: Probabilistic case-based reasoning for open-world knowledge graph completion. In: EMNLP, pp. 4752–4765 (2020)
6. Dong, X., et al.: Knowledge vault: A web-scale approach to probabilistic knowledge fusion. In: SIGKDD, pp. 601–610 (2014)
7. Ji, G., He, S., Xu, L., Liu, K., Zhao, J.: Knowledge graph embedding via dynamic mapping matrix. In: ACL, pp. 687–696 (2015)
8. Ji, G., Liu, K., He, S., Zhao, J.: Knowledge graph completion with adaptive sparse transfer matrix. In: AAAI (2016)
9. Kenton, J.D.M.W.C., Toutanova, L.K.: Bert: Pre-training of deep bidirectional transformers for language understanding. In: NAACL-HLT, pp. 4171–4186 (2019)
10. Kong, F., Zhang, R., Guo, H., Mensah, S., Hu, Z., Mao, Y.: A neural bag-of-words modelling framework for link prediction in knowledge bases with sparse connectivity. In: WWW, pp. 2929–2935 (2019)
11. Lajus, J., Galárraga, L., Suchanek, F.: Fast and exact rule mining with AMIE 3. In: Harth, A., et al. (eds.) ESWC 2020. LNCS, vol. 12123, pp. 36–52. Springer, Cham (2020). https://doi.org/10.1007/978-3-030-49461-2_3
12. Ortona, S., Meduri, V.V., Papotti, P.: Rudik: Rule discovery in knowledge bases. VLDB Endowm. **11**(12), 1946–1949 (2018)
13. Raffel, C., et al.: Exploring the limits of transfer learning with a unified text-to-text transformer. J. Mach. Learn. Res. **21**(140), 1–67 (2020)

14. Saxena, A., Kochsiek, A., Gemulla, R.: Sequence-to-sequence knowledge graph completion and question answering. In: ACL, pp. 2814–2828 (2022)
15. Sennrich, R., Haddow, B., Birch, A.: Neural machine translation of rare words with subword units. In: ACL, pp. 1715–1725 (2016)
16. Shi, B., Weninger, T.: Open-world knowledge graph completion. In: AAAI, vol. 32 (2018)
17. Sun, Z., Deng, Z.H., Nie, J.Y., Tang, J.: Rotate: Knowledge graph embedding by relational rotation in complex space. In: ICLR (2018)
18. Tan, Z., et al.: Thumt: An open-source toolkit for neural machine translation. In: AMTA, pp. 116–122 (2020)
19. Toutanova, K., Chen, D.: Observed versus latent features for knowledge base and text inference. In: CVSC, pp. 57–66 (2015)
20. Trouillon, T., Welbl, J., Riedel, S., Gaussier, É., Bouchard, G.: Complex embeddings for simple link prediction. In: ICML, pp. 2071–2080. PMLR (2016)
21. Vaswani, A., et al.: Attention is all you need. NeurIPS **30** (2017)
22. Wang, B., Shen, T., Long, G., Zhou, T., Wang, Y., Chang, Y.: Structure-augmented text representation learning for efficient knowledge graph completion. In: WWW, pp. 1737–1748 (2021)
23. Wang, L., Zhao, W., Wei, Z., Liu, J.: Simkgc: Simple contrastive knowledge graph completion with pre-trained language models. In: ACL, pp. 4281–4294 (2022)
24. Wang, S., Dang, D.: A generative answer aggregation model for sentence-level crowdsourcing task. IEEE Trans. Knowl. Data Eng. (2022)
25. Wang, X., et al.: Kepler: A unified model for knowledge embedding and pre-trained language representation. TACL **9**, 176–194 (2021)
26. Wang, Z., Zhang, J., Feng, J., Chen, Z.: Knowledge graph embedding by translating on hyperplanes. In: AAAI, vol. 28 (2014)
27. Xue, B., Zou, L.: Knowledge graph quality management: A comprehensive survey. IEEE Trans. Knowl. Data Eng. (2022)
28. Yang, B., Yih, S.W.t., He, X., Gao, J., Deng, L.: Embedding entities and relations for learning and inference in knowledge bases. In: ICLR (2015)
29. Yang, F., Yang, Z., Cohen, W.W.: Differentiable learning of logical rules for knowledge base reasoning. NeurIPS **30** (2017)
30. Yao, L., Mao, C., Luo, Y.: Kg-bert: Bert for knowledge graph completion. arXiv preprint arXiv:1909.03193 (2019)
31. Zhang, J., et al.: Improving the transformer translation model with document-level context. In: EMNLP, pp. 533–542 (2018)
32. Zhao, Y., Zhang, J., Zhou, Y., Zong, C.: Knowledge graphs enhanced neural machine translation. In: IJCAI, pp. 4039–4045 (2021)

Improving Hyper-relational Knowledge Graph Representation with Multi-grained Encoding

Ting Ma[✉], Longtao Huang, and Hui Xue

Alibaba Group, Beijing 100102, China
{mating.ma,kaiyang.hlt,hui.xueh}@alibaba-inc.com

Abstract. Learning hyper-relational knowledge graph (HKG) representation has attracted growing interest from research communities recently. HKGs are typically organized as structured triples associating with additional qualifiers (i.e., key-value pairs) to provide unambiguous hyper-relational facts. The main challenge of HKG is to disambiguate representations of entities and relations by addressing different qualifiers. However, most current models for HKG representation are trained based on single-grained encoders, usually with fine-grained entities, relations, keys, and values in qualifiers, which makes it hard to learn the precise impact of the qualifiers on the whole triple. To overcome this problem, we propose a novel model HEAT for HKG representation by exploring multi-grained encoding, including coarse-grained and fine-grained encoding. Specifically, HEAT performs a graph coarsening method to treat each triple as an integrated coarse-grained node, which satisfies the correlation constraint between the triple and its corresponding qualifiers. Then HEAT leverages a two-stage graph encoder to encode the fine-grained element nodes and coarse-grained triple nodes. The experimental results on three datasets demonstrate that the proposed HEAT consistently outperforms several state-of-the-art baselines.

Keywords: Hyper-relational knowledge graph · Multi-grained encoding · Graph Coarsening

1 Introduction

In recent years, research on knowledge graphs (KGs) has received considerable attention in both academia and industry communities. KGs usually store binary facts as triples in the form of (h, r, t), indicating that a specific binary relation r connects the head entity h and the tail entity t, e.g., (*O.Henry, Wrote, The Gift of the Magi*). Knowledge graph representation has developed as an effective way to represent these triples. The key idea is to embed components of KGs, including entities and relations, into continuous vector spaces. Because of its effectiveness, knowledge graph representation has been successfully applied to various knowledge-driven tasks, such as question answering [15,27], relation extraction [10,14], etc.

Traditional KGs have been explored profoundly, and many KG representation methods have been proposed, such as [2,3,22]. Recently, a new research focus on hyper relational knowledge graphs (HKGs) has drawn increasing attention [7,8,19], especially on how to utilize HKGs for link prediction. Different from traditional KGs, HKGs are

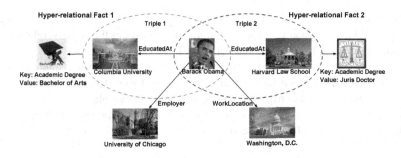

Fig. 1. An example of the hyper-relational knowledge graph.

organized as hyper-relational facts, which consist of triples associating with additional qualifiers. The qualifiers in the hyper-relational facts can provide auxiliary knowledge for disambiguating triples and improving prediction performance, thus better serving downstream applications. Figure 1 shows an example. When predicting the missing entity in the triple (*Barack Obama, EducatedAt, ?*), we cannot select a better prediction from *Columbia University* and *Harvard Law School* using the triple facts in the traditional KGs. After addressing qualifiers to the ambiguous entities, these two entities are disambiguated, and a more accurate prediction can be provided.

Over the past few years, some efforts have attempted to embed the hyper-relational facts [7–9,19,29]. However, these models are mainly trained based on single-grained encoders, usually with fine-grained entities, relations, keys, and values in qualifiers. As the example in Fig. 1, the entity *Barack Obama* and the relation *EducatedAt* are encoded separately, and their representations are updated with the message passing from the entities and qualifiers, respectively. In fact, the triple granularity encoding plays an important role in learning hyper-relational fact representations, because the qualifiers are related to the corresponding whole triples rather than fine-grained elements in the triples. That is, the qualifier (*Academic Degree: Bachelor of Arts*) is related to the *Triple 1*, not just the head entity *Barack Obama* or the relation *EducatedAt*. Although the existing methods have been proven effective, the single-grained encoding makes it hard to learn the precise impact of the qualifiers on the whole triple, and restricts the further enhancement of representation performance.

To solve the aforementioned problem, we propose **HEAT** (Hyper-relational knowledge graph representation), an end-to-end model for learning HKG representation. More specifically, HEAT explores multi-grained encoding, including coarse-grained and fine-grained encoding, to capture semantic correlations within the hyper-relational facts. A graph coarsening method is first devised to treat each triple as an integrated coarse-grained node, so as to satisfy the correlation constraints between the triples and their corresponding qualifiers. Then a two-stage directed graph encoder is leveraged to encode the fine-grained element nodes and the coarse-grained triple nodes. Finally, the multi-grained representations are fused to predict the probability of target fact. Experimental results on three benchmark datasets demonstrate that the proposed HEAT achieves improvements over several baselines.

The main contributions of this paper can be summarized as follows:

- We propose a novel multi-grained encoding model HEAT for learning hyper-relational knowledge graph representation. HEAT encodes the entities, relations, and qualifiers via graph convolutional networks in two stages.
- We devise a graph coarsening strategy to capture the impact of the qualifiers on the triples. To the best of our knowledge, this is the first work to explore the correlation constraints between the triples and the qualifiers in the graph.
- We conduct extensive experiments on three benchmark datasets, and the experimental results show the effectiveness of the HEAT over several baselines.

2   Related Work

According to the number of participating entities in the facts, the KGs can be divided into binary relational and hyper-relational KGs.

2.1   Binary Relational KG Representation

The traditional binary relational KG representation models usually learn low-dimensional vectors for entities and relations by transforming the prediction problem into simple vector operations, mainly including translational distance [3,4,6,13,16,21] and semantic matching [2,11,18,22,26,30] operations. TransE [3] and DistMult [26] are the representatives of these models and have derived a series of KG representation models. TransE assumes that the elements within the triples can be embedded via the additive function. DistMult represents each relation in the triples as a diagonal matrix. In addition, some researches [1,5,17,20,23,24] measure the plausibility of the facts via neural networks. ConvE [5] uses multi-layer CNNs with 2D reshaping to model the triples. To capture the graph structure information in the binary KGs, RGCN [20] and CompGCN [24] learn representations with GCN, which has a strong representation ability.

However, the binary relational KGs cannot describe some complex semantic scenarios well, such as the knowledge with attribute information, which limits the representation ability of the binary relational KG representation models.

2.2   Hyper-relational KG Representation

Hyper-relational KGs consist of triple facts and hyper-relational facts. The entities and relations not in the triples are called key-value pairs or qualifiers. m-TransH [25] and RAE [29] combine the relations and keys in the hyper-relational facts to obtain abstract combination relations. NaLP [9] transforms the hyper-relational facts into entity-relation pairs and defines the entities as the instances of the n-ary relations. However, the combination and transformation make the qualifiers lose their semantic information. To alleviate this problem, NeuInfer [8] and HINGE [19] attempt to represent the hyper-relational facts in the form of quintuple, which can retain both the structure information and the semantic information. To encode the arbitrary number of qualifiers, STARE [7] introduces GCN to learn hyper-relational facts and incorporates the representations of qualifiers into the representations of relations. Hy-Transformer [28] replaces the GCN module in STARE with layer normalization and dropout module.

However, these models mainly use fine-grained encoders and capture the correlations between the fine-grained elements in the hyper-relational facts. The coarse-grained correlations between triples and qualifiers, have received little attention.

Different from previous work, HEAT improves the models in two critical ways: (1) HEAT uses a graph coarsening method to integrate triples as coarse-grained nodes, which enables coarse-grained encoding without destroying the semantic information of the facts. (2) HEAT leverages a two-stage graph encoder to perform multi-grained encoding, so as to enhance fact representations and provide a more accurate prediction.

3 HEAT

The proposed HEAT for mainly includes three modules: Graph Coarsening, Multi-Grained Encoding, and Model Prediction, as illustrated in Fig. 2.

3.1 Problem Formulation

In this paper, the HKG representation problem is formalized as follows. Given a HKG $\mathcal{G}$ with the node set $\mathcal{V}$ and the edge set $\mathcal{E}$, the fine-grained graph $\mathcal{G}_{fine}(E, R)$ and the coarse-grained graph $\mathcal{G}_{coarse}(B, E, R)$ are organized as binary and hyper-relational facts, respectively, where $E \subset \mathcal{V}$, $R \subset \mathcal{E}$, B denotes the binary relational fact set. The binary relational and hyper-relational facts are denoted as $b : (e_1, r, e_2) \in \mathcal{G}_{fine}$ and $c : (b, k, v) \in \mathcal{G}_{coarse}$, where $e_1, e_2, v \in E$, $r, k \in R$, and k, v are the qualifier in c.

Given the HKG $\mathcal{G}$ and the target fact $o : (e_1, r, e_2, k, v)$, we reorganize the target fact o as the facts $b : (e_1, r, e_2) \in \mathcal{G}_{fine}$ and $c : (b, k, v) \in \mathcal{G}_{coarse}$. Then the prediction function for the target fact o can be defined as:

$$\hat{y}_o = f_\Omega(b, c | \mathcal{G}_{fine}, \mathcal{G}_{coarse}), \tag{1}$$

where $\hat{y}_o$ denotes the plausibility of the target fact o, and f_Ω is the model with the parameter set Ω.

3.2 Graph Coarsening

The triple granularity is exploited to satisfy the correlation constraint between the triples and their corresponding qualifiers. Instead of exploring the correlations between the qualifiers and each element in the triples individually, we devise a graph coarsening method to learn representations for triple granularity, so as to capture the correlations between the triples and their corresponding qualifiers. Given the coarsening function Γ, the graph coarsening can be performed via the triple coarsening: $\forall b_{e_1, r, e_2} \in B \Rightarrow b_{\Gamma(e_1), \Gamma(r), \Gamma(e_2)} \in \mathcal{V}$, where $e_1, e_2 \in E$, $r \in R$, b is the binary relational fact. After the binary relational fact (e_1, r, e_2) is coarsened to the node b, the facts of the form (b, k, v) are constructed as a graph, where k and v are the key-value pair.

3.3 Multi-grained Encoding

To learn the representations for the hyper-relational facts, we devise a two-stage graph encoder to encode the fine-grained graph $\mathcal{G}_{fine}$ and the coarse-grained graph $\mathcal{G}_{coarse}$.

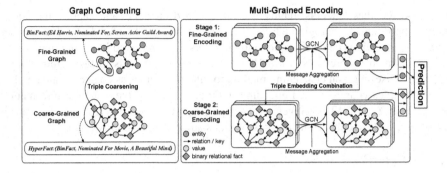

Fig. 2. Schematic illustration of our proposed HEAT.

Fine-Grained Encoding. Considering the considerable performance of GCN in representation learning, we use GCN as the encoder to learn the representations of $\mathcal{G}_{fine}$. Motivated by CompGCN [24], we leverage an entity-relation composition operation to combine the representations of neighbor entities and relations for each node, and aggregate neighbors to embed central nodes. The composition operation is defined as:

$$h_{er} = \psi(h_e, h_r), \tag{2}$$

where ψ is the composition operation, and we adopt the element-wise product in this paper. h_e and h_r denote the representations of e and r. For the node u in $\mathcal{G}_{fine}$, we adopt an aggregation function $f_{agg}(\cdot)$ to capture the influence of u's neighborhood. The representation of u is updated via aggregating the neighborhood information: $h_u = f_{agg}(\mathcal{N}_u)$. The aggregation function is defined as the weighted sum aggregator:

$$f_{agg}(\mathcal{N}_u) = \text{LeakyReLU}(\frac{1}{|\mathcal{N}_u|} \sum_{(e,r)\in\mathcal{N}_u} W_{\pi(r)} h_{er}), \tag{3}$$

where $\mathcal{N}_u = \{(e,r)|(b_{u,r,e} \in B)\}$ is the neighbors of u, LeakyReLU$(\cdot)$ is a nonlinear activation function, h_{er} is the combination representation of e and r. $W_{\pi(r)}$ denotes the relation-specific transformation matrix. Similar to CompGCN, the relation-pattern includes original, reverse and self-loop relations. Hence, the entities are encoded as the distribution of neighbors, and the relations are encoded via updating based on h_r.

Coarse-Grained Encoding. The coarse-grained encoding aims to learn representations for binary relational facts and qualifiers. For a binary relational fact b, its representation h_b is defined as the combination of the triple embedding: $h_b = W_b(h_u||h_r||h_e)$, where $||$ denotes the concatenation operation, W_b denotes the trainable weight matrix.

To update the value v, the fact-key pair is combined via the composition operation: $h_{bk} = \psi(h_b, h_k)$, where ψ denotes the element-wise product operation. For the node v, the neighborhood information $\mathcal{N}_v = \{(k,b)|(h_{b,k,v} \in \mathcal{G}_{coarse}\}$ is aggregated via $h_v = f_{agg}(\mathcal{N}_v)$. Similar to the fine-grained encoding, the aggregation function is defined as:

$$f_{agg}(\mathcal{N}_v) = \text{LeakyReLU}(\frac{1}{|\mathcal{N}_v|} \sum_{(b,k)\in\mathcal{N}_v} W_{\pi(k)} h_{bk}). \tag{4}$$

The representation h_b is updated in the same way as h_v. Then h_k is updated via $h_k = W_{rel}h_k$, where W_{rel} denotes the projection matrix associated with the relations. Finally, the representation for the fact (b, k, v) is obtained via the coarse-grained encoding.

3.4 Model Prediction

By stacking two-stage GCNs, HEAT encodes the binary as well as hyper-relational facts. Formally, the prediction for the hyper-relational facts is formulated as:

$$\hat{y}_o = \sigma(f_r(u, e) + f_k(b, v)), \tag{5}$$

where $f_r(u, e)$ and $f_k(b, v)$ are score functions for (u, r, e) and (b, k, v). Following CompGCN [24], we adopt ConvE [5] as the decoding function. $\sigma(x) = \frac{1}{1+\exp(-x)}$ is the activation function. The prediction score $\hat{y}_o$ is the plausibility of the target fact o.

3.5 Training and Optimization

In KGs, observed and unobserved facts are set as positive and negative samples, respectively. To predict the plausibility of the target facts, we employ the standard binary cross-entropy (BCE) loss function. The objective function can be formulated as:

$$\mathcal{L} = -\frac{1}{\mathcal{N}} \sum_{o \in \mathcal{G}} [y\ln\hat{y}_o + (1-y)\ln(1-\hat{y}_o)] + \lambda||\Omega||_2^2, \tag{6}$$

where $y \in \{0, 1\}$ denote the label of the fact o, $\mathcal{N}$ denotes the number of the training facts in the graph $\mathcal{G}$. λ and Ω are the regularization parameters of the model. The optimization of parameters is performed using Adam [12] with mini-batch.

4 Experiments

In this section, we evaluate the performance of our proposed HEAT on publicly datasets.

4.1 Experimental Settings

Datasets. We conduct experiments on three benchmark datasets: JF17K [25], WikiPeople [9], and WD50K [7]. JF17K is selected from Freebase, WikiPeople and WD50K are selected from Wikidata. The statistics of datasets are shown in Table 1.

Baselines. We evaluate the effectiveness of HEAT by comparing it with the following baseline models:
- **m-TransH** [25] combines the relation paths in the hyper-relational facts to obtain abstract combination relations.
- **RAE** [29] improves m-TransH by exploring the co-participating entities in the facts.
- **NaLP** [9] transforms the hyper-relational facts into entity-relation pairs and defines the entities as the instances of the relations.

Table 1. Statistics of three benchmark datasets.

Dataset	JF17K	WikiPeople	WD50K
#Entity	28,645	34,839	47,156
#Relation	322	375	532
#Total Fact	100,947	325,477	236,507
#Fact (Binary)	54,627	317,117	204,340
#Fact (Hyper)	46,320	8,360	32,167

- **HINGE** [19] utilizes two convolutional networks to encode hyper-relational facts with retaining the structure information of triples.
- **NeuInfer** [8] handles triple facts and hyper-relational facts separately for simple as well as flexible knowledge inference.
- **STARE** [7] learns hyper-relational facts via GCN and incorporates the embedded qualifiers into the embedded relations.
- **Hy-Transformer** [28] improves STARE by replacing the GCN module with layer normalization and dropout module.

Evaluation Protocols and Parameter Settings. To evaluate the performance of HEAT and baseline models, we adopt the widely-used evaluation protocols including MRR and Hit@k. To train our proposed model, we generate n_s negative samples by corrupting each positive sample in the batch randomly. In this paper, n_s is set to 20 for three datasets. The training epoch is set to 500, the learning rate to 0.0002. The batch size n is selected from $\{64, 128, 256\}$, the embedding dimension d from $\{100, 150, 200\}$, and the number of GCN layers l from $\{1, 2\}$ in the two-stage graph encoder.

4.2 Performance Comparisons

The performance comparisons include link prediction and entity prediction. The link prediction task aims to predict the head/tail entity with the given knowledge, and the entity prediction task refers to predicting the missing entity.

Link Prediction Performance. Table 2 shows the performance comparison results of the proposed HEAT and baselines in link prediction. The major observations from the experimental results are summarized as follows:

Among all baselines, the models that preserve the semantic structure usually generate more accurate predictions than models that destroy the semantic structure in most cases. The former includes HINGE, NeuInfer, STARE, and Hy-Transformer; the latter includes m-TransH, RAE, and NaLP. This indicates that preserving the complete semantic structure can help learn knowledge representation.

HEAT achieves better performance than the GCN-based model STARE, indicating that the performance improvements of HEAT are not only because of the graph convolution layers, but also the triple coarsening. This can verify the importance and effectiveness of the coarsening module in the HEAT to a certain extent.

Table 2. Link prediction performance comparisons.

Models	JF17K			WikiPeople			WD50K		
	MRR	Hit@1	Hit@10	MRR	Hit@1	Hit@10	MRR	Hit@1	Hit@10
m-TransH [25]	–	0.206	0.463	–	0.063	0.300	–	–	–
RAE [29]	–	0.215	0.467	–	0.059	0.306	–	–	–
NaLP [9]	0.245	0.185	0.358	0.420	0.343	0.556	0.177	0.131	0.264
HINGE [19]	0.449	0.361	0.624	0.476	0.415	0.585	0.243	0.176	0.377
NeuInfer [8]	0.431	0.342	0.611	0.342	0.272	0.463	–	–	–
STARE [7]	0.574	0.496	0.725	0.491	0.398	0.648	0.349	0.271	0.496
Hy-Transformer [28]	0.582	0.501	0.742	0.501	0.426	0.634	0.356	0.281	0.498
HEAT (ours)	**0.615**	**0.553**	**0.778**	**0.543**	**0.467**	**0.683**	**0.379**	**0.317**	**0.538**

Table 3. Entity prediction performance comparisons.

Fact Type	Methods	JF17K				WikiPeople			
		MRR	Hit@1	Hit@3	Hit@10	MRR	Hit@1	Hit@3	Hit@10
Binary Relational Facts	RAE	0.115	0.050	0.108	0.247	0.169	0.096	0.178	0.323
	NaLP	0.118	0.058	0.121	0.246	0.351	0.291	0.374	0.465
	NeuInfer	0.267	0.173	0.300	0.462	0.350	0.278	0.385	0.473
	HEAT	**0.362**	**0.317**	**0.479**	**0.616**	**0.553**	**0.472**	**0.570**	**0.692**
Hyper-relational Facts	RAE	0.397	0.294	0.434	0.618	0.187	0.126	0.198	0.306
	NaLP	0.477	0.394	0.512	0.637	0.283	0.187	0.322	0.471
	NeuInfer	0.628	0.554	0.666	0.770	0.349	0.303	0.364	0.439
	HEAT	**0.681**	**0.635**	**0.716**	**0.817**	**0.516**	**0.458**	**0.553**	**0.609**

The proposed model HEAT substantially outperforms the baselines by a margin. Compared with the best performance results of baselines, the absolute gains of HEAT are 4.85%, 7.73%, and 8.03% measured by Hit@10 on three datasets, respectively. The experimental results demonstrate the high effectiveness of our model and also verify the rationality of our motivation. We attribute such superior performance improvements to (1) leveraging two-stage GCN to encode the multi-grained graph, and (2) capturing the correlations among elements in HKGs.

Entity Prediction Performance. Table 3 shows the entity prediction performance comparisons of HEAT and baselines. The experimental results demonstrate that HEAT can perform more accurate predictions on both binary relational and hyper-relational facts.

Table 3 also demonstrates that the prediction results of the hyper-relational facts are better than those of the binary relational facts on JF17K when comparing absolute values, while the conclusion is the opposite on WikiPeople. We attribute that the proportion of hyper-relational facts in the WikiPeople is low (only 2.6%), and fewer qualifiers in the graph will limit the prediction performance, especially the value prediction.

4.3 Effects of Coarsening Granularity

To analyze the effect of different coarsening granularities, we compare the experimental results of three granularities, namely the single granularity (single entity or relation), the pair granularity (combining two elements in the triple), and the triple granularity. Figure 3 shows the performance of different coarsening granularities. We can observe

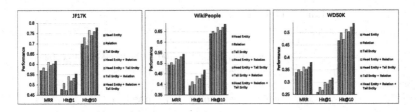

Fig. 3. Performance comparisons of different coarsening granularities on three datasets.

that removing any element in the triple degrades the performance. For the single granularity, treating the relation as a new node performs better than treating the head entity or the tail entity as a new node. Consistent with this, for the pair granularity, the model performance drops the most when the relation is removed. This illustrates the importance of the relation for prediction. The graph coarsening method is devised to satisfy the correlation constraints between the triples and the qualifiers. The key idea of the coarsening module is information fusion. The large coarsening granularity facilitates the completeness of triple representations, and the correlations between the triples and the qualifiers can be modeled more reasonably. HEAT with triple granularity achieves the best performance, which verifies the effectiveness of the graph coarsening module.

5 Conclusions

In this paper, we propose a novel model HEAT to learn representations for HKG. Specifically, we devise a two-stage directed graph encoder to learn the multi-grained nodes. To perform coarse-grained encoding, we use a graph coarsening method to treat the elements in the triples as the integrated nodes, so as to satisfy the correlation constraints between triples and qualifiers. The empirical experiments on three benchmark datasets demonstrate the effectiveness of the proposed HEAT.

References

1. Balazevic, I., Allen, C., Hospedales, T.M.: Hypernetwork knowledge graph embeddings. In: Proceedings of the ICANN, vol. 11731, pp. 553–565 (2019)
2. Balazevic, I., Allen, C., Hospedales, T.M.: Tucker: Tensor factorization for knowledge graph completion. In: Proceedings of the EMNLP-IJCNLP, pp. 5184–5193 (2019)
3. Bordes, A., Usunier, N., Garcia-Duran, A., Weston, J., Yakhnenko, O.: Translating embeddings for modeling multi-relational data. In: Proceedings of the NeurIPS, pp. 2787–2795 (2013)
4. Chami, I., Wolf, A., Juan, D.C., Sala, F., Ravi, S., Re, C.: Low-dimensional hyperbolic knowledge graph embeddings. In: Proceedings of the ACL, pp. 6901–6914 (2020)
5. Dettmers, T., Minervini, P., Stenetorp, P., Riedel, S.: Convolutional 2d knowledge graph embeddings. In: Proceedings of the AAAI (2018)
6. Ebisu, T., Ichise, R.: Toruse: Knowledge graph embedding on a lie group. In: AAAI (2018)
7. Galkin, M., Trivedi, P., Maheshwari, G., Usbeck, R., Lehmann, J.: Message passing for hyper-relational knowledge graphs. In: Proceedings of the EMNLP, pp. 7346–7359 (2020)

8. Guan, S., Jin, X., Guo, J., Wang, Y., Cheng, X.: Neuinfer: Knowledge inference on n-ary facts. In: Proceedings of the ACL, pp. 6141–6151 (2020)
9. Guan, S., Jin, X., Wang, Y., Cheng, X.: Link prediction on n-ary relational data. In: Proceedings of the WWW, pp. 583–593 (2019)
10. Hu, Z., Cao, Y., Huang, L., Chua, T.: How knowledge graph and attention help? A qualitative analysis into bag-level relation extraction. In: Proceedings of the ACL/IJCNLP, pp. 4662–4671 (2021)
11. Kazemi, S.M., Poole, D.: Simple embedding for link prediction in knowledge graphs. In: Proceedings of the NeurIPS, pp. 4284–4295 (2018)
12. Kingma, D., Jimmy, B.: Adam: A method for stochastic optimization. Comput. Sci. (2014)
13. Kolyvakis, P., Kalousis, A., Kiritsis, D.: Hyperkg: Hyperbolic knowledge graph embeddings for knowledge base completion. In: arXiv (2019)
14. Nadgeri, A., et al.: Kgpool: Dynamic knowledge graph context selection for relation extraction. In: Proceedings of the ACL/IJCNLP (Findings), pp. 535–548 (2021)
15. Naseem, T., et al.: A semantics-aware transformer model of relation linking for knowledge base question answering. In: Proceedings of the ACL/IJCNLP, pp. 256–262 (2021)
16. Nayyeri, M., Xu, C., Yaghoobzadeh, Y., Yazdi, H.S., Lehmann, J.: Toward understanding the effect of loss function on then performance of knowledge graph embedding. In: arXiv (2019)
17. Nguyen, D.Q., Nguyen, T.D., Nguyen, D.Q., Phung, D.: A novel embedding model for knowledge base completion based on convolutional neural network. In: Proceedings of the NAACL, pp. 327–333 (2018)
18. Nickel, Tresp, Kriegel: A three-way model for collective learning on multi-relational data. In: Proceedings of the ICML, pp. 809–816 (2011)
19. Rosso, P., Yang, D., Cudré-Mauroux, P.: Beyond triplets: Hyper-relational knowledge graph embedding for link prediction. In: Proceedings of the WWW, pp. 1885–1896 (2020)
20. Schlichtkrull, M., Kipf, T.N., Bloem, P., van den Berg, R., Titov, I., Welling, M.: Modeling relational data with graph convolutional networks. In: Gangemi, A., et al. (eds.) ESWC 2018. LNCS, vol. 10843, pp. 593–607. Springer, Cham (2018). https://doi.org/10.1007/978-3-319-93417-4_38
21. Sun, Z., Huang, J., Hu, W., Chen, M., Guo, L., Qu, Y.: Transedge: Translating relation-contextualized embeddings for knowledge graphs. In: Proceedings of the ISWC, pp. 612–629 (2019)
22. Sun, Z., Deng, Z.H., Nie, J.Y., Tang, J.: Rotate: Knowledge graph embedding by relational rotation in complex space. In: Proceedings of the ICLR (2019)
23. Vashishth, S., Sanyal, S., Nitin, V., Agrawal, N., Talukdar, P.P.: Interacte: Improving convolution-based knowledge graph embeddings by increasing feature interactions. In: Proceedings of the AAAI, pp. 3009–3016 (2020)
24. Vashishth, S., Sanyal, S., Nitin, V., Talukdar, P.P.: Composition-based multi-relational graph convolutional networks. In: Proceedings of the ICLR (2020)
25. Wen, J., Li, J., Mao, Y., Chen, S., Zhang, R.: On the representation and embedding of knowledge bases beyond binary relations. In: Proceedings of the IJCAI, pp. 1300–1307 (2016)
26. Yang, B., Tau Yih, W., He, X., Gao, J., Deng, L.: Embedding entities and relations for learning and inference in knowledge bases. In: Proceedings of the ICLR (2015)
27. Yasunaga, M., Ren, H., Bosselut, A., Liang, P., Leskovec, J.: QA-GNN: Reasoning with language models and knowledge graphs for question answering. In: Proceedings of the NAACL-HLT, pp. 535–546 (2021)
28. Yu, D., Yang, Y.: Improving hyper-relational knowledge graph completion. Arxiv (2021)
29. Zhang, R., Li, J., Mei, J., Mao, Y.: Scalable instance reconstruction in knowledge bases via relatedness affiliated embedding. In: Proceedings of the WWW, pp. 1185–1194 (2018)
30. Zhang, W., Paudel, B., Wang, L.: Iteratively learning embeddings and rules for knowledge graph reasoning. In: Proceedings of the WWW, pp. 2366–2377 (2019)

Temporal-Relational Matching Network for Few-Shot Temporal Knowledge Graph Completion

Xing Gong[1], Jianyang Qin[1], Heyan Chai[1], Ye Ding[2], Yan Jia[1,3], and Qing Liao[1,3(✉)]

[1] School of Computer Science and Technology, Harbin Institute of Technology,Shenzhen, China
{gongxing,22b351005,chaiheyan}@stu.hit.edu.cn,
{jiayan2020,liaoqing}@hit.edu.cn
[2] School of Cyberspace Security, Dongguan University of Technology, Dongguan, China
dingye@dgut.edu.cn
[3] Peng Cheng Laboratory, Shenzhen, China

Abstract. Temporal knowledge graph completion (TKGC) is an important research task due to the incompleteness of temporal knowledge graphs. However, existing TKGC models face the following two issues: 1) these models cannot be directly applied to few-shot scenario where most relations have only few quadruples and new relations will be added; 2) these models cannot fully exploit the dynamic time and relation properties to generate discriminative embeddings of entities. In this paper, we propose a temporal-relational matching network, namely TR-Match, for few-shot temporal knowledge graph completion. Specifically, we design a multi-scale time-relation attention encoder to adaptively capture local and global information based on time and relation to tackle the dynamic properties problem. Then, we build a new matching processor to tackle the few-shot problem by mapping the query to few support quadruples in a relation-agnostic manner. Finally, we construct three new datasets for few-shot TKGC task based on benchmark datasets. Extensive experimental results demonstrate the superiority of our model over the state-of-the-art baselines.

Keywords: Temporal knowledge graph completion · Few-shot learning · Link prediction

1 Introduction

Knowledge graphs (KGs) have proved their powerful strength in various downstream tasks, such as recommender system [18], information retrieval [12], concept discovery [6], and question answering [25], etc. KGs represent every fact with a triplet (s, r, o), where s, o are the subject entity and the object entity, and r is the relation between s and o. For example, $(Biden, President, USA)$

X. Wang et al. (Eds.): DASFAA 2023, LNCS 13944, pp. 768–783, 2023.
https://doi.org/10.1007/978-3-031-30672-3_52

represents that Biden is the president of the USA. However, many facts are not static but highly ephemeral. For instance, $(Biden,\ President,\ USA)$ is true only after $(Trump,\ President,\ USA)$. Thus, by incorporating temporal information into KGs, temporal knowledge graphs (TKGs) represent each fact with a quadruple (s, r, o, t), where t represents a temporal constraint specifying the temporal validity of the fact. Due to the incompleteness of TKGs, temporal knowledge graph completion (TKGC) becomes an increasingly important research task. The task is to infer missing facts at specific timestamps based on the existing ones by answering queries such as $(Biden,\ President,\ ?,\ 2022)$.

Most current TKGC models perform the completion task by proposing a distance-based scoring function that incorporates the time representations. For example, TTransE [11] modifies the distance formula of TransE [2] by adding the temporal information, ATiSE [22] projects the representations of TKGs into the space of multi-dimensional Gaussian distributions, TeRo [21] represents time as a rotation in complex vector space, etc. However, these methods still do not perform well because they all face the few-shot relations problem and dynamic entity properties problem:

(1) **Few-shot relations.** Few-shot relations problem widely exists in real-world TKGs, bringing difficulties to complete temporal knowledge graph due to the following twofold. On the one hand, most relations in TKGs only have a small number of quadruples. This brings difficulties for conventional TKGC models to learn discriminative embeddings of entities, relations and timestamps from few quadruples, because these models usually require a lot of quadruples for training. As a result, they cannot accurately compute the distance between embeddings for completion task. On the other hand, some new relations will be added into TKGs consistently, so that conventional TKGC methods should finetune to update the learned embeddings to fit new relations. Recently, some studies, such as GMatching [19], MetaR [4], FSRL [24] and FAAN [15], have been proposed to tackle the few-shot problem, but these methods elaborated for static KGs cannot be applied to temporal KGs.

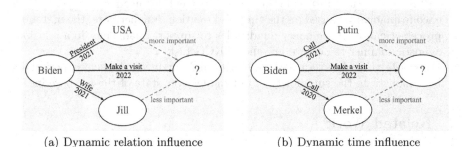

(a) Dynamic relation influence   (b) Dynamic time influence

Fig. 1. Two examples to illustrate the influence of neighbors on the target entity varying dynamically with the (a) relation and (b) time when completing the knowledge graph.

(2) **Dynamic entity properties.** Dynamic entity properties mean that the influence of neighbors on entity varies with the time and relation

of different completion tasks. Figure 1 illustrates two examples of dynamic entity properties problem. Specifically, when we perform the completion task (*Biden, Make a visit, ?, 2022*), the influence of entity *USA* on *Biden* is greater than the influence of entity *Jill* on *Biden*, as shown in Fig. 1(a). This is because the relation *President* shares the same work property as the relation *Make a visit*, while the relation *Wife* reflects a different family property than the relation *Make a visit*. Meanwhile, as shown in Fig. 3(b), the timespan affects the weights of different neighbors too. Despite that the entities *Markel* and *Putin* share the same relation *Call*, entity *Biden* weights more on its neighbor *Putin* than *Markel* due to the smaller timespan between *Biden* and *Putin*. Existing few-shot KGC methods [4,15,19,24] ignore the dynamic properties of entities, resulting in inaccurate encoding of entities. Thus, how to jointly exploit the relation and time for TKGC remains a challenging problem.

To address the above problems, we propose a **Temporal-Relational Match**ing network for few-shot temporal knowledge graph completion (TR-Match). Specifically, we firstly follow the few-shot settings [14,17] to split and generate each task with support and query quadruples based on relation. Secondly, we propose a multi-scale time-relation attention encoder to learn the representations of support quadruples with dynamic properties. The encoder adaptively aggregates local neighbor information based on time and relation to obtain quadruple representations, and interacts the representations with global relational information among all support quadruples via a multi-head attention mechanism. Thirdly, we introduce a matching processor to deal with the few quadruples and unseen relations existing in few-shot scenario. The processor utilizes an attention-based LSTM to generate the informative representation of each query quadruple, and maps the query to few support quadruples in a relation-agnostic manner to deal with new relation by ranking the similarity between quadruples. Main contributions of this paper are summarized as follow:

- We propose a novel few-shot TKGC model, namely TR-Match, to deal with the dynamic few-shot problem.
- In TR-Match, the multi-scale time-relation attention encoder can dynamically encode quadruples based on the time and relation. Furthermore, the matching processor can map the query quadruples to support quadruples in a relation-agnostic manner to achieve few-shot TKCG task.
- We create three few-shot TKG datasets and conduct extensive experiments to demonstrate the superiority of our model over state-of-the-art baselines.

2 Related Work

2.1 Temporal Knowledge Graph Completion Methods

Recently, many temporal knowledge graph embedding models have been proposed, which encode time information in their embeddings. TTransE [11] modifies the distance formula of TransE [2] to complete the temporal knowledge graph by adding the projection of temporal information and carrying out vector calculation. ATiSE [22] considers the temporal uncertainty during the evolution of entity/relation representations over time and projects the representations of TKGs into the space of multi-dimensional Gaussian distributions. TeRo [21] proposes scoring functions which incorporate time representations into a distance-based score function. DE-SimplE [5] uses diachronic entity embeddings to represent entities at different time steps and exploit the same score function as SimplE [8] to score the plausibility of a quadruple. Based on ComplEx [16], TComplEx [10] and TNTComplEx [10] analogously factorize the input TKG, which both models represent as a 4th-order tensor. TeLM [20] performs 4th-order tensor factorization of a TKG, using the asymmetric geometric product instead of complex Hermitian operator. These conventional TKGC methods do not consider the few-shot relations problem. Although FTMF [1] takes into account the few-shot relations problem, it focuses on temporal knowledge graph reasoning.

2.2 Few-shot Knowledge Graph Completion Methods

Recently, few-shot knowledge graph completion has attracted more and more research attention. GMatching [19] is the first one-shot knowledge graph completion model which consists an entity encoder to average ground aggregation of heterogeneous neighbors and a matching processor to measure the similarity between the support triple and the query triple. Based on GMatching, FSRL [24] uses an attention mechanism to aggregate neighbor information and a LSTM-based encoder to represent few-shot relations by support entity pairs. FAAN [15] is the first to propose a dynamic attention mechanism for one-hop neighbors adapting to the different relations which connect them. MetaR [4] focuses on transferring relation-specific meta to represent and fast update few-shot relations. MetaP [7] extracts the patterns effectively through a convolutional pattern learner and measures the validity of triples accurately by matching query patterns with reference patterns. GANA [13] puts more emphasis on neighbor information and accordingly proposes a gated and attentive neighbor aggregator. However, these models developed for static knowledge graphs cannot be applied to temporal knowledge graphs.

3 Preliminaries

In this section, we first present the notations of the temporal knowledge graph, then introduce the few-shot learning settings of our model, and finally define the few-shot temporal knowledge graph completion task in this work.

3.1 Temporal Knowledge Graph

Let $\mathcal{E}$, $\mathcal{R}$ and $\mathcal{T}$ represent a finite set of entities, relations and timestamps, respectively. A TKG is a collection of facts represented as a set of quadruples $G = \{(s, r, o, t)\} \subseteq \mathcal{E} \times \mathcal{R} \times \mathcal{E} \times \mathcal{T}$ in which $s \in \mathcal{E}$ and $o \in \mathcal{E}$ are subject entity and object entity respectively, $r \in \mathcal{R}$ is the relation and $t \in \mathcal{T}$ denotes the happened time of these facts.

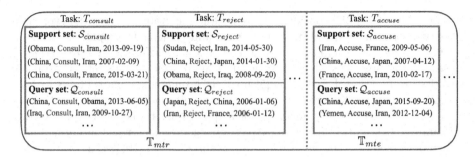

Fig. 2. A few-shot learning settings example, where few-shot size is 3.

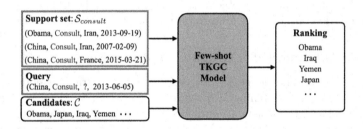

Fig. 3. A few-shot TKGC task $T_{consult}$ example.

3.2 Few-Shot Learning Settings

In this work, we classify the relations in TKG into two categories, frequent relations $\mathcal{R}_{freq}$ and sparse relations $\mathcal{R}_{sp}$, based on the frequency of their occurrence. Following GMamtching [19], the quadruples with the relations in $\mathcal{R}_{freq}$ construct background knowledge graph G' to get neighbor information. Each sparse relation corresponds to a few-shot relation. Following the standard few-shot learning settings [14,17], we consider the completion problem of quadruples with sparse relation $r \in \mathcal{R}_{sp}$ as a task, so as to access a set of tasks. In our problem, each task T_r corresponds to a sparse relation $r \in \mathcal{R}_{sp}$, and has its own support/query set: $T_r = \{\mathcal{S}_r, \mathcal{Q}_r\}$. Each support set $\mathcal{S}_r$ only contains few support quadruples $\{(s_1, r, o_1, t_1), (s_2, r, o_2, t_2), ..., (s_k, r, o_k, t_k)\}$, and $|\mathcal{S}_r| = k$ denotes the few-shot size. Besides, query set $\mathcal{Q}_r$ contains all query quadruples of relation r, including positive query quadruples $\mathcal{Q}_r^+ = \{(s_i, r, o_i^+, t_i)|(s_i, r, o_i^+, t_i) \in G, o_i^+ \in \mathcal{C}\}$ and corresponding negative query quadruples $\mathcal{Q}_r^- = \{(s_i, r, o_i^-, t_i)|(s_i, r, o_i^-, t_i) \notin G, o_i^- \in \mathcal{C}\}$. $\mathcal{C}$ is the candidate entity set, and the candidate entity set in this work is composed of all entities, i.e. $\mathcal{C} = \mathcal{E}$.

Moreover, we divide all the tasks into two sets, meta-train set $\mathbb{T}_{mtr}$ and meta-test set $\mathbb{T}_{mte}$. Notably, the relations in $\mathbb{T}_{mte}$ does not appear in $\mathbb{T}_{mtr}$. And we leave out a subset of relations in $\mathbb{T}_{mtr}$ as the meta-validation set $\mathbb{T}_{mtv}$. Figure 2 illustrates a few-shot learning settings example.

3.3 Few-Shot Temporal Knowledge Graph Completion

In this work, our purpose is to predict the object entity o given the subject entity, relation and timestamp: $(s, r, ?, t)$. In contrast to previous conventional TKGC methods that usually assume enough quadruples are available for training, this work studies the case where only few training quadruples are available. To be more specific, the goal is to rank the true object entity higher than other candidate entities in candidate entity set $\mathcal{C}$ to complete the quadruples in query set, given only few support quadruples. Figure 3 is an example of task $T_{consult}$.

Define $\ell_\Theta(\mathcal{Q}_r|\mathcal{S}_r)$ as the ranking loss of task T_r, Θ is the set of model parameters, the probabilistic optimization objective for this problem is given as:

$$\mathcal{L} = \arg\max_{\Theta} \mathbb{E}_{r\sim\mathcal{R}}\left[\ell_\Theta(\mathcal{Q}_r|\mathcal{S}_r)\right]. \tag{1}$$

4 Proposed Model

In this section, we give an introduction to our model in detail. This work aims to compute a similarity score $\mathcal{P}_\Theta((s_q, r, o_q, t_q), \mathcal{S}_r)$ for each query (s_q, r, o_q, t_q) given the support set. To achieve this purpose, we propose a completion network including an encoding and matching step, as shown in Fig. 4. Specifically, the encoding step utilizes multi-scale time-relation attention encoder to dynamically encode entity and time to obtain the representations of support and query quadruples, then the matching step uses matching processor to calculate the similarity between support and query quadruples for TKGC.

4.1 Multi-scale Time-Relation Attention Encoder

As shown in Fig. 4, multi-scale time-relation attention encoder is designed to obtain the temporal-relational representations of support and query quadruples. In this module, we design an adaptive neighbor aggregator and utilize a multihead attention to capture the local and global information in TKGs, respectively.

Adaptive Neighbor Aggregator. The influence of neighbors on one entity keeps changing based on the relevance of relation and the length of timespan. The neighbor with similar relation and smaller timespan put a higher weight on certain entity. However, existing few-shot KGC methods [15,19,24] cannot simultaneously consider the relation and timespan to obtain discriminative entity representations.

To tackle the above issue, we design a time-relation attention mechanism to dynamically assign neighbor weights. For every entity e, our model constructs the neighbors of e, i.e., $\mathcal{N}_e = \{(e_i', r_i', t_i')|, (e, r_i', e_i', t_i') \in G'\}$, by searching for the quadruples in background knowledge graph G' whose subject entity is e. e_i' is the object entity which is regarded as a neighbor of e, r_i' is the relation between

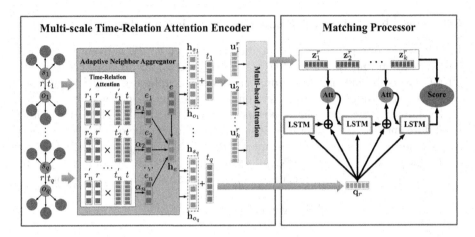

Fig. 4. The framework of TR-Match: it first obtains representations of the support and query quadruples by multi-scale time-relation attention encoder, then captures the similarity score between support and query quadruples by matching processor.

e and e_i', and t_i' is the time of fact (e, r_i', e_i'). Then, the weight α_i of the neighbor e_i' on entity e can be calculated as follows:

$$\alpha_i = \frac{\exp\left(\left(\mathbf{v}_r^\top \mathbf{W}_1 \mathbf{v}_{r_i'}\right) \times \langle \Phi(t), \Phi(t_i')\rangle\right)}{\sum_{(e_j', r_j', t_j') \in \mathcal{N}_e} \exp\left(\left(\mathbf{v}_r^\top \mathbf{W}_1 \mathbf{v}_{r_j'}\right) \times \langle \Phi(t), \Phi(t_j')\rangle\right)}. \tag{2}$$

It can be seen from Eq. (2) that we assign the weight α_i by jointly considering the relevance of different relations and the length of timespan. Specifically, the relevance of different relations can be calculated through $\mathbf{v}_r^T \mathbf{W}_1 \mathbf{v}_{r_i'}$. $\mathbf{v}_r$ denotes the embedding of current task relation r, which can be randomly initialized as a d-dimension vector, i.e. $\mathbf{v}_r \in \mathbb{R}^d$. $\mathbf{W}_1 \in \mathbb{R}^{d \times d}$ is a learnable parameter which is the similarity matrix to calculate the relevance between relations. The length of timespan can be measured via the inner product of paired time encoding, i.e., $\langle \Phi(t), \Phi(t_i')\rangle$. To ensure that the neighbors with smaller timespan have relatively higher weights, we refer to [23] to encode the time t as follows,

$$\Phi(t) = \sqrt{\frac{1}{d}}\left[\cos\left(\omega_1 t\right), \sin\left(\omega_1 t\right), \ldots, \cos\left(\omega_d t\right), \sin\left(\omega_d t\right)\right], \tag{3}$$

where $\{\omega_1, \omega_2, ..., \omega_d\}$ is a set of learnable parameters. With this encoding function, the neighbors with smaller timespan can receive higher weights on time factor, i.e., if $|t - t_i'| < |t - t_j'|$, then $\langle \Phi(t), \Phi(t_i')\rangle > \langle \Phi(t), \Phi(t_j')\rangle$.

Having obtained the weights of neighbors, we can learn entity e's representation by adaptively aggregating neighbor information and its own information.

$$\mathbf{h}_e = \sigma(\mathbf{W_2} \sum_{(e_i', r_i', t_i') \in \mathcal{N}_e} \alpha_i \mathbf{v}_{e_i'} + \mathbf{W_3} \mathbf{v}_e), \tag{4}$$

where $\mathbf{h}_e$ denotes e's entity representation obtained from the adaptive neighbor aggregator. $\mathbf{v}_e$ is the embedding of entity e, which can be randomly initialized as a d-dimension vector, i.e. $\mathbf{v}_e \in \mathbb{R}^d$. $\sigma(\cdot)$ denotes activation function of Relu. $\mathbf{W_2}, \mathbf{W_3} \in \mathbb{R}^{d \times d}$ are trade-off parameters learned by MLP, which balance the importance between neighbor information and entity itself information.

Multi-head Attention for Support Set. The entity representations obtained above consider only local neighbor information, but ignore global relational information in support set. Due to this, we use a multi-head attention module to generate informative representations of support set by refining the relational information of support set. We use the combination of the entity pair representations and time embedding as the input $\mathbf{U}_r$ of multi-head attention to better capture the temporal dependence between support quadruples:

$$\mathbf{u}_i^r = (\mathbf{h}_{s_i} \| \mathbf{h}_{o_i}) + \Phi(t_i), \quad \mathbf{U}_r = [\mathbf{u}_1^r, \mathbf{u}_2^r, ..., \mathbf{u}_k^r], \tag{5}$$

where $\mathbf{u}_i^r$ is the initial representation of support quadruple (s_i, r, o_i, t_i). $\mathbf{h}_{s_i}$ and $\mathbf{h}_{o_i}$ are the representations of entities s_i and o_i, respectively, obtained by the adaptive neighbor aggregator. $\|$ denotes concatenation operation to gather paired representations $\mathbf{h}_{s_i}$ and $\mathbf{h}_{o_i}$. Moreover, we take the time embedding $\Phi(t_i)$ as positional encoding to capture the temporal dependence of support quadruples. After that, we can obtain the support quadruple representations via the following multi-head attention layer,

$$head_i = \text{softmax}\left(\frac{\mathbf{Q}_i \mathbf{K}_i^\top}{\sqrt{2d}}\right) \mathbf{V}_i \tag{6}$$

$$\mathbf{Z}_r = [head_1, ..., head_N] \mathbf{W}^O \tag{7}$$

where $\mathbf{Z}_r = [\mathbf{z}_1^r, \mathbf{z}_2^r, ..., \mathbf{z}_k^r] \in \mathbb{R}^{k \times 2d}$ is the support quadruple representations. $\mathbf{W}^O \in \mathbb{R}^{2Nd \times 2d}$ is parameter matrix and N is the number of heads. $\mathbf{Q}_i = \mathbf{U}_r \mathbf{W}_i^Q$, $\mathbf{K}_i = \mathbf{U}_r \mathbf{W}_i^K$, $\mathbf{V}_i = \mathbf{U}_r \mathbf{W}_i^V$ are the 'queries', 'keys' and 'values' of the ith head attention. $\mathbf{W}_i^Q, \mathbf{W}_i^K, \mathbf{W}_i^V \in \mathbb{R}^{2d \times 2d}$ are the projection matrices of the ith head attention. The multi-head attention mechanism fully interacts with the global relational information of the support set.

4.2 Matching Processor

Conventional TKGC models are mainly encountered with two issues: (1)They require a large number of quadruples to train usually learn poor embeddings under few-shot scenario that most relations only have a few quadruples. (2)They should be finetuned to adapt newly added relations since the learned embeddings are unavailable to update after training. To this end, we propose a matching processor that includes an embedding step and a matching step to address the poor embeddings and new relation problems, respectively.

In the embedding step, we choose LSTM to improve the quality of query embedding because LSTM can selectively capture the feature impact of support set on query [17]. Moreover, we argue that different support quadruples have different weights for a query because of the semantic divergence existing in support quadruples. Therefore, we design an **Att** module based on attention mechanism to dynamically aggregate support quadruples. Finally, we combine LSTM and **Att** module to generate informative representation h_i of query quadruple as follows:

$$h_i = \hat{h}_i + \mathbf{q}_r, \ \hat{h}_i, c_i = \text{LSTM}\left(\mathbf{q}_r, p_{i-1}, c_{i-1}\right), \tag{8}$$

where $\text{LSTM}(\cdot)$ is a standard LSTM cell with input $\mathbf{q}_r$, hidden state p_{i-1} and cell state c_{i-1}, and $\mathbf{q}_r = (\mathbf{h}_{s_q} \| \mathbf{h}_{o_q}) + \Phi(t_q)$ is the initial representation of query (s_q, r, o_q, t_q). The hidden state p_{i-1} regarded as a hidden representation of query (s_q, r, o_q, t_q) can be calculated via the **Att** module as follows:

$$p_{i-1} = h_{i-1} + \sum_{\mathbf{z}_j^r \in \mathbf{Z}_r} \beta_j \mathbf{z}_j^r, \tag{9}$$

$$\beta_j = \frac{\exp(h_{i-1}^\top \mathbf{z}_j^r)}{\sum_{\mathbf{z}_m^r \in \mathbf{Z}_r} \exp(h_{i-1}^\top \mathbf{z}_m^r)}, \tag{10}$$

where β_j is the weight of support quadruple (s_j, r, o_j, t_j) and $\mathbf{z}_j^r$ is the representation of support quadruple (s_j, r, o_j, t_j) obtained by Eq. (7). After l layer LSTM, we can obtain the representation h_l of query (s_q, r, o_q, t_q).

In the matching step, our goal is to rank the query quadruples to find the best candidate object entity with respect to a certain relation, so as to complete the missing quadruples. Considering that the new relation will be added to TKGs, it is not suitable to complete quadruples by calculating the distance between subject entity, relation and object entity. This is because new relation unseen in the training process cannot be presented in testing process without finetuning. To address this, we aim to directly achieve the completion task in an end-to-end network via a relation-agnostic matching. Specifically, we rank the query quadruples via the similarity score between query and support set which can be defined as the sum of inner product between quadruple representations:

$$Score(\mathbf{q}_r, \mathbf{Z}_r) = \sum_{\mathbf{z}_j^r \in \mathbf{Z}_r} h_l^\top \mathbf{z}_j^r, \tag{11}$$

where h_l and $\mathbf{z}_j^r$ are query and support quadruple representations, respectively. It is worth noting that, the representations of query quadruples learned from Eq. (8) do not embed relation directly, meaning that each query representation is relation-agnostic. Thus, we match the quadruples only based on the similarity of entities and timespans between query and support quadruples. As a result, given a new relation, our matching process can find the best candidate object entity by selecting a query quadruple similar to support quadruples with new relation.

4.3   Loss Function and Training

We train the model on meta-training task set $\mathbb{T}_{mtr}$. We encourage high similarity scores for positive pairs and low similarity scores for negative pairs. The objective function is a hinge loss defined as follow:

$$\mathcal{L} = \sum_{r}^{\mathcal{R}} \sum_{q_r^+ \in \mathcal{Q}_r^+, q_r^- \in \mathcal{Q}_r^-} \left[\lambda + Score(\mathbf{q}_r^-, \mathbf{Z}_r) - Score(\mathbf{q}_r^+, \mathbf{Z}_r) \right]_+, \qquad (12)$$

where $[x]_+ = max(0, x)$ is standard hinge loss, and λ is a hyperparameter represents safety margin distance. The detail of the training process is shown in Algorithm 1.

Algorithm 1: TR-Match Training

Input: Meta-training task set $\mathbb{T}_{mtr}$; background knowledge graph $G^{'}$; randomly initial TKG embeddings; initial model parameters Θ.

1 **for** *epoch=0:M-1* **do**
2   Shuffle the tasks in $\mathbb{T}_{mtr}$;
3   **for** T_r *in* $\mathbb{T}_{mtr}$ **do**
4   Sample k quadruples as support set $\mathcal{S}_r$;
5   Sample a batch of positive query quadruples $\mathcal{Q}_r^+$;
6   Pollute the object entity in $\mathcal{Q}_r^+$ to get $\mathcal{Q}_r^-$;
7   Obtain the representations of entities in $\mathcal{S}_r$, $\mathcal{Q}_r^+$ and $\mathcal{Q}_r^-$ by Eq. (2)-(4);
8   Obtain the representations of support quadruples in $\mathcal{S}_r$ by Eq. (5)-(7);
9   Calculate the matching score for query in $\mathcal{Q}_r^+$ and $\mathcal{Q}_r^-$ by Eq. (8)-(11);
10   Calculate the batch loss $\mathcal{L}$ of task T_r by Eq. (12);
11   Update parameters Θ by Adam optimizer;
12   **end**
13 **end**
14 **return** Optimal model parameters Θ

---

5   Experiments

In this section, we begin with an introduction about how to construct datasets for few-shot settings. Then, we provide an overview of baselines and implement details. Finally, we conduct a series of experiments and provide an analysis of the experimental results.

Table 1. Dataset details of ICEWS14-few, ICEWS05-15-few and ICEWS18-few. $|\mathcal{E}|$, $|\mathcal{T}|$ and $|\mathcal{R}_{sp}|$ are the number of the entities, timestamps, and sparse relations, respectively. #Tasks is the number of tasks of $\mathbb{T}_{mtr}/\mathbb{T}_{mtv}/\mathbb{T}_{mte}$. #Frequency is the frequency interval of sparse relations $\mathcal{R}_{sp}$.

Dataset	$|\mathcal{E}|$	$|\mathcal{T}|$	$|\mathcal{R}_{sp}|$	#Tasks	#Frequency
ICEWS14-few	7121	365	114	92/11/11	(10,200)
ICEWS05-15-few	10471	4017	99	81/9/9	(50,500)
ICEWS18-few	24572	365	99	81/9/9	(50,500)

5.1   Datasets

Since conventional temporal knowledge graph datasets, such as ICEWS14, ICEW S05-15 and ICEWS18 [3], are not suitable for few-shot settings, we construct three new datasets ICEWS14-few, ICEWS05-15-few and ICEWS18-few based on conventional datasets for few-shot TKGC. The details of each dataset are illustrated in Table 1. Following GMatching [19], we construct each dataset by selecting appropriate number of sparse relations based on the size of corresponding conventional dataset as follows:

ICEWS14-few. To construct ICEWS14-few dataset, we select the relations with number less than 200 but greater than 10 quadruples as sparse relations $\mathcal{R}_{sp}$, and the relations with number more than 200 are considered as frequent relations $\mathcal{R}_{freq}$ from ICEWS14 dataset. Then, we use 92/11/11 task relations for $\mathbb{T}_{mtr}/\mathbb{T}_{mtv}/\mathbb{T}_{mte}$.

ICEWS05-15-few. To construct ICEWS05-15-few dataset, we select the relations with less than 500 but more than 50 quadruples as the sparse relations $\mathcal{R}_{sp}$ from ICEWS05-15 dataset, and the relations in greater than 500 as frequent relations $\mathcal{R}_{freq}$ from ICEWS05-15 dataset. Then, we use 81/9/9 task relations for $\mathbb{T}_{mtr}/\mathbb{T}_{mtv}/\mathbb{T}_{mte}$.

ICEWS18-few. ICEWS18-few dataset is constructed in the same way as ICEWS05-15-few. The frequency interval of $\mathcal{R}_{sp}$ and $\mathcal{R}_{freq}$ in ICEWS18-few is the same as ICEWS05-15-few. We use 81/9/9 task relations for $\mathbb{T}_{mtr}/\mathbb{T}_{mtv}/\mathbb{T}_{mte}$.

5.2   Baselines

Since there is no few-shot TKGC model focusing on the completion task for comparison, we select two kinds of baseline models for comparison in this experiment: few-shot KGC models and conventional TKGC models. (1) As for few-shot KGC models, we adopt the following state-of-the-art models as baselines: GMatching [19], MetaR [4], FSRL [24] and FAAN [15]. Since all few-shot KGC models are static and cannot be generalized into dynamic scenario, we provide these models with all the quadruples in the original datasets and neglect time information, i.e., neglecting t in (s, r, o, t). (2) As for conventional TKGC models, we adopt the following state-of-the-art models for comparison: DE-SimplE [5],

TNTComplEx [10], ATISE [22], TeRo [21] and TeLM [20]. Following GMatching, when evaluating these conventional TKGC models, we use the quadruples in background knowledge graph G', the quadruples in $\mathbb{T}_{mtr}$ and the quadruples in support set of $\mathbb{T}_{mtv}$ and $\mathbb{T}_{mte}$ as the train set.

5.3 Implementation

In our model, all entities and relations embeddings are initialized randomly with dimension of 100. The few-shot size k is set to 3 for the following experiments. We select the best hyperparameters that can achieve the highest MRR in validation set. The maximum number of local neighbors in adaptive neighbor aggregator is set to 50 for all datasets. In addition, we use LSTM in matching processor, the dimension of hidden state is 200 in our experiments. The number of layers of LSTM is set to 4 for ICEWS14-few and ICEWS05-15-few, and 5 for ICEWS18-few. The margin distance λ is set to 10. We implement all experiments with PyTorch and use Adam optimizer [9] to optimize model parameters with a learning rate of 0.001.

For models Gmatching, MateR, FSRL and FAAN, the few-shot size k is the same as our model, and the other parameters use the optimal parameters from the original papers. For models DE-SimpIE, TNTComplEx, ATiSE, TeRo, TeLM, we refer to the best hyperparameter settings of baseline methods reported in their original papers. We report two standard evaluation metrics: MRR and Hit@N. MRR is the mean reciprocal rank and Hits@N is the proportion of correct entities ranked in the top N, with N = 1,5,10.

5.4 Performance Comparison

Experimental Comparison with Baselines. We compare TR-Match with nine baselines on ICEWS14-few, ICEWS05-15-few and ICEWS18-few datasets, respectively, to evaluate the effectiveness of TR-Match. The performances of all models are reported in Table 2, where the best results are highlighted in bold, and the best performance of the two kinds of baselines on different datasets is underlined. It can be seen that our model outperforms all the baselines by achieving a higher MRR and Hits@1/5/10.

Compared to the few-shot KGC baselines, TR-Match consistently outperforms the best few-shot KGC baseline, i.e., FSRL, by achieving 3.0/13.4/18.1% improvement of MRR metric on ICEWS14-few/ICEWS05-15-few/ICEWS18-few datasets, respectively. This is because, compared to few-shot KGC models ignoring the temporal information, TR-Match can jointly and adaptively take the relation and time into consideration to aggregate local information, resulting in more accurate entity representations. Moreover, few-shot KGC models assume that different support quadruples are of equivalent importance to each query, while TR-Match can adaptively assign weights to support quadruples via matching process to capture the discriminative information.

Compared to the conventional TKGC baselines, TR-Match achieves significant improvements over the best results of conventional TKGC baselines by

Table 2. The overall results of all methods. The best results are highlighted in **bold**, and the best performance of the two kinds of baseline are marked as <u>underline</u>.

Model	ICEWS14-few				ICEWS05-15-few				ICEWS18-few			
	MRR	H@1	H@5	H@10	MRR	H@1	H@5	H@10	MRR	H@1	H@5	H@10
GMatching [19]	.213	.132	.286	.381	.222	.118	.318	.438	.184	.097	.269	.364
MetaR [4]	.224	.104	.352	.444	.215	.074	.333	.455	.132	.031	.240	.334
FSRL [24]	<u>.306</u>	<u>.205</u>	<u>.403</u>	<u>.490</u>	<u>.246</u>	<u>.149</u>	<u>.363</u>	<u>.467</u>	<u>.199</u>	<u>.113</u>	<u>.287</u>	<u>.376</u>
FAAN [15]	.241	.147	.343	.418	.200	.109	.291	.394	.155	.111	.177	.264
DE-SimplE [5]	.265	.163	.364	.464	<u>.208</u>	<u>.124</u>	.278	.382	.192	.109	.263	.371
TNTComplEx [10]	.218	.130	.317	.402	.097	.045	.138	.199	.138	.070	.195	.283
ATiSE [22]	.259	.153	.377	.479	.179	.087	.271	.378	.097	.049	.142	.196
TeRo [21]	.236	.131	.355	.469	.187	.086	<u>.292</u>	<u>.408</u>	.165	.087	.240	.336
TeLM [20]	<u>.270</u>	<u>.166</u>	<u>.364</u>	<u>.481</u>	.198	.108	.282	.383	<u>.203</u>	<u>.115</u>	<u>.280</u>	<u>.385</u>
TR-Match(Ours)	**.315**	**.220**	**.431**	**.529**	**.279**	**.176**	**.385**	**.497**	**.235**	**.150**	**.324**	**.408**

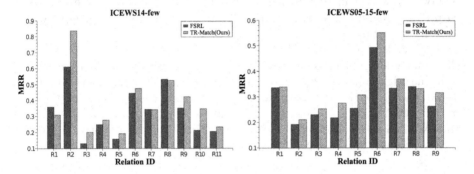

Fig. 5. The MMR of TR-Math and FSRL for each relation on ICEWS14-few and ICEWS05-15-few.

obtaining 16.7/34.13/15.8% gains of MRR metric on ICEWS14-few/ICEWS05-15-few/ICEWS18-few datasets, respectively. This is because conventional TKGC models requiring large-scale training data fail to achieve satisfying performance on such three datasets with few-shot relations. By contrast, our model is powerful to handle few-shot data by calculating the similarity between quadruples in matching processor.

Comparison over Different Relations. To demonstrate the superiority of our model in more detail, we set up comparative experiments on ICEWS14-few and ICEWS05-15-few with different relations. In this experiment, we compare our proposed TR-Match with the best few-shot static KGC baseline FSRL. The experimental results are shown in Fig. 5, where Relation ID represents a class of relation. On ICEWS14-few, our TR-Match outperforms FSRL in MRR metric with 8 out of 11 relations. On ICEWS05-15-few, our TR-Match outperforms FSRL in MRR metric with 8 out of 9 relations. Experimental results indicate that

Table 3. The results of ablation experiment.

Model	ICEWS14-few				ICEWS05-15-few				ICEWS18-few			
	MRR	H@1	H@5	H@10	MRR	H@1	H@5	H@10	MRR	H@1	H@5	H@10
TR-Match(v1)	.291	.197	.394	.497	.245	.135	.369	.483	.208	.129	.286	.362
TR-Match(v2)	.197	.104	.315	.373	.241	.145	.339	.434	.166	.102	.225	.304
TR-Match(v3)	.297	.203	.401	.488	.261	.152	.379	.486	.193	.114	.273	.363
TR-Match(Ours)	**.315**	**.220**	**.431**	**.529**	**.279**	**.176**	**.385**	**.497**	**.235**	**.150**	**.324**	**.408**

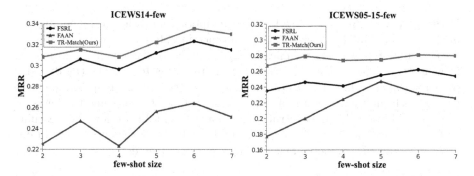

Fig. 6. Impact of few-shot size on ICEWS14-few and ICEWS05-15-few.

TR-Match is more powerful to learn discriminative quadruple representations by taking the advantage of time encoding in both encoding step and matching step.

5.5 Ablation Study

We perform experiments on all the datasets with several variants of our proposed model to provide a better understanding of the contribution of each module to our proposed model. The ablative results are shown in Table 3. In TR-Match(v1), we use the neighbor encoder proposed by GMatching [19] instead of our proposed adaptive neighbor aggregator to encode entities. Experiments demonstrate that dynamically aggregating neighbors based on relation and time can improve the model performance compared to aggregating neighbors with fixed weights. In TR-Match(v2), we remove multi-head attention from our model. The experimental results demonstrate that multi-head attention can stably boost the performance of our model by capturing the global relational information for support set. In TR-Match(v3), we replace **Att** in matching processor with mean-pooling. Experimental results show that **Att** can improve our model performance by adaptively aggregating support features compared to fixed support weights.

5.6 Impact of Few-shot Size

In this subsection, we study the impact of few-shot size k. We perform experiments on TR-Match, FSRL [24], and FAAN [15] models on ICEWS14-few and ICEWS05-15-few datasets, and set different k values from a subset

$\{2, 3, 4, 5, 6, 7\}$. Experimental results in Fig. 6 demonstrate that: (1) The performance of TR-Match is always better than the comparative models, indicating the capability of our proposed method in few-shot TKGC. (2) TR-Match obtains relatively stable boosts compared to FSRL and FAAN, which shows the robustness of TR-Match to few-shot size.

6   Conclusion

In this paper, we propose a new few-shot temporal knowledge graph completion model, i.e., TR-Match, which consists of an encoding step and a matching step. In the encoding step, we can dynamically aggregate the local and global information to generate temporal-relational representations, so as to capture the dynamic properties in completion task. In the matching step, we can map the query to few support quadruples in a relation-agnostic manner to overcome the few-shot problem. Additionally, we construct three datasets suitable for few-shot learning based on public datasets. The experimental results show the superiority of our model and the effectiveness of each component in our model.

Acknowledgements. This work was partially supported by the National Natural Science Foundation of China: 61976051, U19A2067, and the Major Key Project of PCL: PCL2022A03

References

1. Bai, L., Zhang, M., Zhang, H., Zhang, H.: FTMF: Few-shot temporal knowledge graph completion based on meta-optimization and fault-tolerant mechanism. In: World Wide Web, pp. 1–28 (2022)
2. Bordes, A., Usunier, N., García-Durán, A., Weston, J., Yakhnenko, O.: Translating embeddings for modeling multi-relational data. In: NeurIPS, pp. 2787–2795 (2013)
3. Boschee, E., Lautenschlager, J., O'Brien, S., Shellman, S., Starz, J., Ward, M.: ICEWS Coded Event Data (2015). https://doi.org/10.7910/DVN/28075
4. Chen, M., Zhang, W., Zhang, W., Chen, Q., Chen, H.: Meta relational learning for few-shot link prediction in knowledge graphs. In: EMNLP-IJCNLP, pp. 4217–4226 (2019)
5. Goel, R., Kazemi, S.M., Brubaker, M.A., Poupart, P.: Diachronic embedding for temporal knowledge graph completion. In: AAAI, pp. 3988–3995 (2020)
6. Jeon, I., Papalexakis, E.E., Faloutsos, C., Sael, L., Kang, U.: Mining billion-scale tensors: Algorithms and discoveries. VLDB J. **25**(4), 519–544 (2016). https://doi.org/10.1007/s00778-016-0427-4
7. Jiang, Z., Gao, J., Lv, X.: Metap: Meta pattern learning for one-shot knowledge graph completion. In: SIGIR, pp. 2232–2236 (2021)
8. Kazemi, S.M., Poole, D.: Simple embedding for link prediction in knowledge graphs. NeurIPS, pp. 4289–4300 (2018)
9. Kingma, D.P., Ba, J.: Adam: A method for stochastic optimization. CoRR (2015)
10. Lacroix, T., Obozinski, G., Usunier, N.: Tensor decompositions for temporal knowledge base completion. In: ICLR (2019)

11. Leblay, J., Chekol, M.W.: Deriving validity time in knowledge graph. WWW, pp. 1771–1776 (2018)
12. Liu, Z., Xiong, C., Sun, M., Liu, Z.: Entity-duet neural ranking: Understanding the role of knowledge graph semantics in neural information retrieval. In: ACL, pp. 2395–2405 (2018)
13. Niu, G., et al.: Relational learning with gated and attentive neighbor aggregator for few-shot knowledge graph completion. In: SIGIR, pp. 213–222 (2021)
14. Ravi, S., Larochelle, H.: Optimization as a model for few-shot learning. In: ICLR (2017)
15. Sheng, J., et al.: Adaptive attentional network for few-shot knowledge graph completion. In: EMNLP, pp. 1681–1691 (2020)
16. Trouillon, T., Welbl, J., Riedel, S., Gaussier, É., Bouchard, G.: Complex embeddings for simple link prediction. In: ICML, pp. 2071–2080 (2016)
17. Vinyals, O., Blundell, C., Lillicrap, T., Wierstra, D., et al.: Matching networks for one shot learning. NeurIPS, pp. 3637–3645 (2016)
18. Wang, X., Wang, D., Xu, C., He, X., Cao, Y., Chua, T.S.: Explainable reasoning over knowledge graphs for recommendation. In: AAAI, pp. 5329–5336 (2019)
19. Xiong, W., Yu, M., Chang, S., Guo, X., Wang, W.Y.: One-shot relational learning for knowledge graphs. In: EMNLP, pp. 1980–1990 (2018)
20. Xu, C., Chen, Y.Y., Nayyeri, M., Lehmann, J.: Temporal knowledge graph completion using a linear temporal regularizer and multivector embeddings. In: NAACL, pp. 2569–2578 (2021)
21. Xu, C., Nayyeri, M., Alkhoury, F., Yazdi, H.S., Lehmann, J.: Tero: A time-aware knowledge graph embedding via temporal rotation. In: COLING, pp. 1583–1593 (2020)
22. Xu, C., Nayyeri, M., Alkhoury, F., Yazdi, H., Lehmann, J.: Temporal knowledge graph completion based on time series gaussian embedding. In: ISWC, pp. 654–671 (2020)
23. Xu, D., Ruan, C., Körpeoglu, E., Kumar, S., Achan, K.: Inductive representation learning on temporal graphs. ArXiv (2020)
24. Zhang, C., Yao, H., Huang, C., Jiang, M., Li, Z.J., Chawla, N.: Few-shot knowledge graph completion. In: AAAI, pp. 3041–3048 (2020)
25. Zhang, Y., Dai, H., Kozareva, Z., Smola, A.J., Song, L.: Variational reasoning for question answering with knowledge graph. In: AAAI, pp. 1–8 (2018)

Distantly Supervised Entity Linking
with Selection Consistency Constraint

Haipeng Dai, Lei Meng, Hancheng Wang, Rong Gu, Siwen Chen, Feng Chen,
and Wei Hu[⊠]

State Key Laboratory for Novel Software Technology,
Nanjing University, Nanjing, Jiangsu 210023, China
{haipengdai,gurong,whu}@nju.edu.cn,
{hanchengwang,181250013,fengchen}@smail.nju.edu.cn

Abstract. Entity linking (EL) aims to find entities that the textual mentions refer to from a knowledge base (KB). The performance of current distantly supervised EL methods is not satisfactory under the condition of low-quality candidate generation. In this paper, we consider the scenario where multiple KBs are available, and for each KB, there is an EL model corresponding to it. We propose the selection consistency constraint (SCC), that is, for one sample, the entities selected from multiple KBs should be consistent if these selections are all correct. In this work, we aim to utilize the SCC to improve the performance of each EL model (not the combination of multiple EL models) under low-quality candidate generation. Specifically, we define an SCC model from two different aspects: minimizing probability and upper bound, which are used to introduce the SCC into the training of EL models. The experimental results show that our method, jointly training multiple EL models with the SCC model, outperforms the baseline which trains multiple EL models separately, and it has low cost.

1 Introduction

Entity linking (EL) is the task of finding out the corresponding entities of textual mentions from a knowledge base (KB). For example, the mention "Michael Jordan" in "Michael Jordan is a great basketball player" should be linked to entity Michael_Jordan in DBpedia. Most of the previous work focuses on EL under full supervision and requires annotated datasets [7, 12, 14, 28, 29]. Annotated datasets, such as webpages with Wikipedia hyperlinks and some manually annotated datasets (*e.g.*, AIDA-CoNLL [13]), are limited and exist only for a few domains. For new domains, annotated datasets are often not available and expensive to obtain. In recent years, there are some distantly supervised approaches [1, 15, 16, 18] that try to train an EL model without using mention-level annotations. Because the correct entity is not labeled, they get supervision from a candidate set which is often generated by a mention-entity dictionary [15, 18] or a surface matching heuristic [16]. Intuitively, the smaller size and the higher recall rate a candidate set has, the greater supervision it gives. High-quality candidate generation often relies on a mention-entity dictionary built from hyperlinks, such as Wikipedia hyperlinks, or hand curation. So, for the domains or languages that are not rich in Wikipedia resources, building high-quality candidates is not easy, and the performance of EL under low-quality candidate generation is not satisfactory (low-quality candidate generation, *i.e.*, the high recall rate and the small size cannot be satisfied at the same time). In this paper, we mainly consider the distantly supervised EL under low-quality candidate generation. We define an EL model, which consists of a noise detection model and an entity ranking model. The former is used to determine whether the candidate set contains the correct entity, while the latter is used to rank the entities in the candidate set according to their relevance to the mention.

In reality, in some domains, we often have multiple KBs to use. For example, we have DBpedia and YAGO for general encyclopedias, and Twitter and Weibo for public figures. We believe that with the progress of knowledge extraction technology, scenarios with multiple KBs will become more common. Obviously, the entities corresponding to the same textual mention and context in different KBs should be the same. We model this requirement as the selection consistency constraint (SCC). If the selections of multiple EL models do not conform to the SCC, there must be at least one wrong selection. The wrong prediction may be caused by the fact that the candidate set does not contain the correct entity or the wrong selection generated by entity ranking model. It can be seen that the occurrence of inconsistent selections is related to the performance of noise detection model and entity ranking model. In this paper, we explore whether the SCC can be leveraged to improve the performance of distantly supervised EL. Another more intuitive way to utilize multiple KBs is to treat extra KBs as extra features, but there are two problems: (1) the entities in different KBs need to be aligned first, and the alignment errors will be passed to EL task (two entities are said to be aligned if they denote the same entity); (2) the entity distributions of multiple KBs should have large overlap. However, our method can avoid these two problems.

First, we define an entity alignment (EA) model to determine whether two entities from different KBs are aligned. This is similar to the EA task that aims to find the aligned entities in different knowledge graphs (KGs) (*e.g.*, [4,27,30]). We use some seed alignment to construct the training data of the EA model. Note that this is the only labeled data used in this paper. However, our experiments show that our method can still work even if there is little or no seed alignment, because the EA model can get supervision from EL models (see Sect. 5.4). This ensures that considering SCC does not cost much on annotated data.

Second, we define an SCC model to introduce the SCC into the training of EL models. Specifically, the SCC model calculates the loss for inconsistent selections through the results of EL models and EA models. By reducing this loss, the training of EL models and EA models can get extra supervision. We train them using an end-to-end style.

Our contributions can be summarized as follows: (1) we explore the impact of SCC on the performance of distantly supervised EL under low-quality candidate generation, and to the best of our knowledge, we are the first to explore this problem; (2) we verified the effectiveness of SCC from two different aspects; and (3) we conducted sufficient experiments and the results show that our method significantly improves the performance compared with the baseline which trains multiple EL models separately.

2   Related Work

We briefly discuss previous work related to distantly supervised entity linking and entity alignment.

2.1   Distantly Supervised Entity Linking

Most EL approaches include two steps: candidate generation and entity ranking. In the test phase, candidate generation can greatly reduce the workload of entity ranking. In the training phase, candidate generation can provide high-quality negative samples under full supervision, while under distant supervision, it provides weak supervision. [15,18] generate candidates by a mention-entity dictionary built from Wikipedia hyperlinks and the latter further filters out the unlikely entities while maintaining a high recall rate. [18] extends the skip-gram model to jointly learn embeddings of words, mentions, and entities from shared textual contexts, and uses the cosine similarity between embeddings of entities and mentions to rank candidates. [15,16] frame the EL task as the multi-instance learning problem [10]. An EL model is trained to score at least one candidate in a candidate set higher than any negative example from a negative entity set. [16] considers the candidate generation without Wikipedia resources and uses a surface matching heuristic to construct candidate set. An EL model is jointly trained with a noise detection model to complete automatic noise detection. The idea is similar to the data quality detection [22]. SNERL [1] jointly trains the EL and relation extraction (RE) models so that EL model can get supervision from RE task. SNERL constructs the candidate set by the cosine similarity between TF-IDF vectors of mentions and entities. Overall, the setting of [16] is closer to ours and other methods do not consider the situation of low-quality candidate generation. Different from [16], our work focuses on the impact of SCC on EL models. Our method adds SCC to the training to improve performance.

2.2   Embedding-Based Entity Alignment

EA task aims to find aligned entities in different KGs. In recent years, embedding-based EA methods have been greatly developed. In terms of the way for KG embedding, previous work can be divided into: translation-based methods [5,6,20,23,30] and GCN-based methods [4,19,21,24,26,27]. Translation-based methods regard a relation as a translation vector from a head entity to a tail. GCN-based methods use graph convolutional networks [3,32] to encode entities in KG. Different from previous work, the goal of the EA task in this work is to determine whether two entities from different KBs denote the same entity, and we model this task as a binary classification problem.

3   Proposed Model

In this section, we formalize the problem and describe the proposed model. Let (m, c) denote a mention m and its context c (*i.e.*, the sentence that contains m). Given $N (> 1)$ KBs $KB^1, \cdots, KB^N$, a training sample consists of a mention-context pair (m, c), N candidate sets $E^{1+}, \cdots, E^{N+}$, and N negative entity sets $E^{1-}, \cdots, E^{N-}$, where E^{t+}

and E^{t-} are generated from KB^t ($t = 1, \cdots, N$). It can be assumed that there are no duplicate entities in each KB. We call a triple $\langle E^{t+}, m, c \rangle$ as a data point. Moreover, if E^{t+} contains the correct entity, we call the data point valid; otherwise, we call it noisy.

As mentioned earlier, an EL model includes a noise detection model and an entity ranking model. We define the noise detection model as $p_N^t \left(1|E^{t+}, m, c\right)$ which represents the probability that the data point $\langle E^{t+}, m, c \rangle$ is noisy, and the entity ranking model as $g^t \left(e_i^t, m, c\right)$ which scores the matching degree between (m, c) and entity e_i^t ($\in E^{t+}$). The EA model is defined as $p_a^{uv}(e_i^u, e_j^v)$ ($u \neq v$, $e_i^u \in KB^u$ and $e_j^v \in KB^v$) which represents the probability that e_i^u and e_j^v are aligned. Moreover, if e_i^u ($\in KB^u$) and e_j^v ($\in KB^v$) are aligned, we have $e_i^u = e_j^v$; otherwise, $e_i^u \neq e_j^v$. If e_i^t ($\in E^{t+}$) is the correct entity for (m, c), we have $e_i^t = \langle E^{t+}, m, c \rangle$; otherwise, $e_i^t \neq \langle E^{t+}, m, c \rangle$.

Definition 1 (Inconsistent Selections). *When both E^{u+} and E^{v+} contain the correct entity, if the entities selected by g^u and g^v from E^{u+} and E^{v+}, respectively, are not aligned, we call them inconsistent selections.*

Based on Definition 1, we define the SCC model between two EL models from two different aspects and describe how to extend the SCC model to the scenario with more than two KBs in Sect. 3.1. The trainings of EL and EA models are described in Sect. 3.2, and the joint training is described in Sect. 3.3. Last, the neural network implementation of $p_N^t (1|\cdot)$, $g^t (\cdot)$ and $p_a^{uv} (\cdot)$ is described in Sect. 3.4.

3.1 Selection Consistency Constraint Model

First, the probability that g^t selects entity $e_i^t (\in E^{t+})$ is defined as

$$p \left(e_i^t\right) = \operatorname*{softmax}_{e_i^t \in E^{t+}} \left(g^t \left(e_i^t, m, c\right) / T\right) \tag{1}$$

where T (> 0) is temperature. Apparently, there are no duplicate entities in E^{t+}. That is, for one entity in E^{u+}, there is at most one entity is aligned with it in E^{v+}. To prevent one entity from being aligned with multiple entities, we define $p_A \left(e_i^u, e_j^v\right)$ ($u \neq v$, $e_i^u \in E^{u+}$ and $e_j^v \in E^{v+}$) as the probability that (e_i^u, e_j^v) is one-to-one alignment. The $p_A \left(e_i^u, e_j^v\right)$ is given by

$$p_A \left(e_i^u, e_j^v\right) = p_a^{uv} \left(e_i^u, e_j^v\right) \cdot \prod_{e' \in E^{v+} \setminus \{e_j^v\}} \left(1 - p_a^{uv} \left(e_i^u, e'\right)\right)$$
$$\cdot \prod_{e' \in E^{u+} \setminus \{e_i^u\}} \left(1 - p_a^{uv} \left(e', e_j^v\right)\right) \tag{2}$$

Minimizing Probability (Min-P). An intuitive way is to minimize the probability of inconsistent selections during training. According to Definition 1, the probability of inconsistent selections for one sample is given by

$$p_{inc}^{uv} (m, c) = p_N^u \left(0|E^{u+}, m, c\right) \cdot p_N^v \left(0|E^{v+}, m, c\right)$$
$$\cdot \left(1 - p_{same}^{uv} (m, c)\right), \tag{3}$$

where

$$p_{same}^{uv}(m,c) = \sum_{e_i^u \in E^{u+}} \sum_{e_j^v \in E^{v+}} p\left(e_i^u\right) p\left(e_j^v\right) p_A\left(e_i^u, e_j^v\right) \tag{4}$$

is the probability that the entities selected by g^u and g^v are one-to-one aligned and $p_N^t\left(0|E^{t+}, m, c\right) = 1 - p_N^t\left(1|E^{t+}, m, c\right)$. We take $p_{inc}^{uv}(m,c)$ as the loss function directly: $l_{inc}^{uv}(m,c) = p_{inc}^{uv}(m,c)$.

Upper Bound (U-Bound). From another aspect, Event $\left(e_i^u = \langle E^{u+}, m, c \rangle\right) \bigcap$ Event $\left(e_j^v = \langle E^{v+}, m, c \rangle\right) \subset$ Event $\left(e_i^u = e_j^v\right)$. So, the probability that e_i^u ($\in E^{u+}$) and e_j^v ($\in E^{v+}$) are both correct should be no greater than the probability that they are one-to-one aligned.

The probability that e_i^t is correct is given by

$$p_c\left(e_i^t\right) = p_N^t\left(0|E^{t+}, m, c\right) \cdot p\left(e_i^t\right), \tag{5}$$

which means that the $\langle E^{t+}, m, c \rangle$ is valid and e_i^t ($\in E^{t+}$) is selected. Then, we define the loss of inconsistent selections for one sample as

$$l_{inc}^{uv}(m,c) = \sum_{e_i^u \in E^{u+}} \sum_{e_j^v \in E^{v+}} \left[p_c\left(e_i^u\right) p_c\left(e_j^v\right) - p_A\left(e_i^u, e_j^v\right)\right]_+, \tag{6}$$

where $[a]_+ = \max(a, 0)$. This is equivalent to using $p_A\left(e_i^u, e_j^v\right)$ as the upper bound of $p_c\left(e_i^u\right) p_c\left(e_j^v\right)$.

Total Loss of SCC Model. The total loss of SCC model between two EL models is

$$L_{inc}^{uv} = \eta_{inc} \cdot \frac{1}{|D|} \cdot \sum_{(m,c) \in D} l_{inc}^{uv}(m,c) + \eta \cdot L_{kl}^{uv}, \tag{7}$$

where

$$L_{kl}^{uv} = \text{KL}\left(\frac{\sum_{(m,c) \in D} p_N^u(0|E^{u+}, m, c) p_N^v(0|E^{v+}, m, c)}{|D|} \| P^*\right) \tag{8}$$

is used to prevent $p_N^u(0|E^{u+}, m, c) p_N^v(0|E^{v+}, m, c)$ from being zero and is estimated at the mini-batch level. P^* is a prior distribution manually set. η_{inc} and η are coefficients. D is the data set.

Extension. When N is greater than 2, one direct way is to add SCC between every two KBs, namely

$$L_{inc} = \sum_{u=1}^{N-1} \sum_{v=u+1}^{N} L_{inc}^{uv}, \tag{9}$$

and the time complexity is $O(N^2)$. Another way is to select a KB (such as KB^1) as intermediary and only add SCC between intermediary and other KBs, namely

$$L_{inc} = \sum_{t=2}^{N} L_{inc}^{1t}, \tag{10}$$

and the time complexity is $O(N)$. When $N = 2$, these two methods are equivalent.

3.2 Trainings of EL Model and EA Model

Training of EL Model. For the training of EL model, we directly use the method proposed in [16]. The loss function of the EL model is given by

$$L_{el}^t = \frac{1}{|D|} \cdot \sum_{(m,c) \in D} p_N^t \left(0 | E^{t+}, m, c\right) l^t (m, c)$$

$$+\eta \cdot \text{KL} \left(\frac{1}{|D|} \cdot \sum_{(m,c) \in D} p_N^t \left(1 | E^{t+}, m, c\right) || P_N^* \right), \tag{11}$$

where

$$l^t(m, c) =$$

$$\left[\max_{e_i^t \in E^{t-}} g^t \left(e_i^t, m, c\right) + \delta - \max_{e_i^t \in E^{t+}} g^t \left(e_i^t, m, c\right) \right]_+, \tag{12}$$

E^{t-} is obtained by random sampling, $\delta > 0$ is a margin, and P_N^* is a prior distribution manually set. The reader is referred to [16] for more details. In theory, the SCC model is applicable as long as the method conforms to our definition of the EL model.

Training of EA Model. We use S^{uv} to represent the set of all seed alignment between KB^u and KB^v. The wrong alignment set S_n^{uv} is obtained by negative sampling (see Sect. 4.2). Then, the loss function of the EA model is

$$L_{ea}^{uv} = -\frac{1}{|S^{uv}| + |S_n^{uv}|} \cdot \sum_{(e_i^u, e_j^v) \in S^{uv}} \ln(p_a^{uv}(e_i^u, e_j^v))$$

$$-\frac{1}{|S^{uv}| + |S_n^{uv}|} \cdot \sum_{(e_i^u, e_j^v) \in S_n^{uv}} \ln(1 - p_a^{uv}(e_i^u, e_j^v)) \tag{13}$$

3.3 Joint Training

We jointly train the EL, EA, and SCC models with the following loss function:

$$L = \sum_{t=1}^{N} L_{el}^t + \sum_{u=1}^{N-1} \sum_{v=u+1}^{N} \mathbb{I} \left(\mathbf{F}[u, v] = 1\right) \left(L_{inc}^{uv} + L_{ea}^{uv}\right), \tag{14}$$

where $\mathbf{F}[u, v] = 1$ denotes that the SCC between KB^u and KB^v needs to be added. In a sense, SCC acts as a bridge between EL and EA, so that they can supervise each other.

3.4 Structure of Neural Network

Encoding Mention and Context. Let $e^t_{mc}(m, c)$ denote the embedding of a mention-context pair (m, c) for KB^t. Like [16], we used a bidirectional LSTM to encode mention and context and the reader is referred to [16] for more details.

Encoding Entity. The embedding of an entity e^t_i is $\mathbf{e}^t_e (e^t_i) = [\mathbf{e}^t_{name} (e^t_i); \mathbf{e}^t_{TransE} (e^t_i)]$ where $\mathbf{e}^t_{name} (e^t_i) \in \mathbb{R}^{d_n}$ is entity name embedding and $\mathbf{e}^t_{TransE} (e^t_i) \in \mathbb{R}^{d_T}$ is TransE [2] embedding. TransE embeddings remain unchanged during training. Previous work [11,31] has shown that entity name features are useful for EL and EA tasks. We used a unidirectional LSTM to encode name (limited maximum length to 5) and used last hidden state as $\mathbf{e}^t_{name} (e^t_i)$. For UNK words in entity name, we used the average of char embeddings which are pre-trained with the CBOW model [17] as word embeddings.

$g^t(\cdot), p^t_N (1|\cdot)$ and $p^{uv}_a (\cdot)$. For $g^t(\cdot)$ and $p^t_N(1|\cdot)$, we used the same method as that in [16]. The details are as follows:

$$g^t(e^t_i, m, c) = \mathbf{FFN}^t_g([\mathbf{e}^t_e(e^t_i); \mathbf{e}^t_{mc}(m, c)]) \tag{15}$$

$$p^t_N \left(1|E^{t+}, m, c\right) = \\ \sigma \left(\mathbf{FFN}^t_{nd} \left([\mathbf{e}^t_{E+}(E^{t+}); \mathbf{e}^t_{mc} (m, c)]\right) /T\right) \tag{16}$$

where

$$\mathbf{e}^t_{E+}(E^{t+}) = \sum_{e \in E^{t+}} \alpha(e) \cdot \mathbf{e}^t_e(e). \tag{17}$$

Note that $\alpha(e)$ is attention weight and is given by

$$\alpha (e) = \operatorname*{softmax}_{e \in E^{t+}} \left(g^t \left(e, m, c\right) /T\right) \tag{18}$$

Last,

$$p^{uv}_a(e^u_i, e^v_j) = \sigma \left(\mathbf{FFN}^{uv}_{ea}([\mathbf{e}^u_e(e^u_i); \mathbf{e}^v_e(e^v_j)])\right) \tag{19}$$

$\mathbf{FFN}^t_g$, $\mathbf{FFN}^t_{nd}$ and $\mathbf{FFN}^{uv}_{ea}$ are one-hidden layer feedforward neural networks. Note that each EL or EA model does not share the parameters of the neural networks (include word embeddings and char embeddings) with other models.

4   Dataset

In this section, we introduce the KBs and datasets for EL and EA, and the method for candidate generation.

Table 1. The statistics of the datasets (What's in brackets is the proportion of the total).

Sets	# of sentences	# of mentions
Train	169,906 (99.94%)	586,051 (91.04%)
Dev	2,261 (97.96%)	4,497 (94.30%)
Test	2,403 (97.52%)	4,211 (94.50%)

4.1   Knowledge Bases and EL Datasets

In this paper, a single SCC model can only be used between two EL models. When $N > 2$, the essence of extension is to use multiple SCC models. Therefore, we believe that as long as a single SCC model is effective, then when $N > 2$, the extension is also effective. We used DBpedia[1] and YAGO[2] as our KBs ($N = 2$). There are 3,017,446 and 3,059,272 entities in DBpedia and YAGO, respectively. We used Fast-TransX[3] to train TransE models for generating entity embeddings (see Sect. 3.4).

We took EL-170k [16], which is extracted from New York Times corpus and not annotated, as training set, AIDA-CoNLL-testa as development set, and AIDA-CoNLL-testb as test set. The AIDA-CoNLL datasets are manually annotated. The statistics of the three datasets are shown in Table 1. We did not count the mentions without candidates in both KBs, nor the sentences without mentions. But in the testing phase, we took all the mentions into account. If the candidate set of a mention is empty, it is directly judged as noise (*i.e.*, $p_N^t \left(1|E^{t+}, m, c\right) = 1$).

Table 2. Oracle recall (%) as function of the size of E^{t+} on development set.

# of candidates	1	10	100	1,000	10,000	100,000
DBpedia	40.87	48.94	58.19	66.45	66.89	66.89
YAGO	43.20	51.18	61.42	68.38	69.13	69.15

Candidate Generation. For each mention m, we listed the entities whose names contain all the words that appear in m. Then, we sorted these entities in ascending order by the length of entity names and took the first C entities as candidates. For training, we set C as 20. For development and testing, we set C as 100. During training, we randomly sampled 20 entities from the remaining entities in KB^t as E^{t-}. The oracle recall of our method is shown in Table 2.

---

[1] http://downloads.dbpedia.org/2016-04/core-i18n/en/infobox_properties_en.ttl.bz2.

[2] http://resources.mpi-inf.mpg.de/yago-naga/yago3.1/yagoFacts.ttl.7z.

[3] https://github.com/thunlp/Fast-TransX.

4.2   Training Set for Entity Alignment

The seed alignment between DBpedia and YAGO can be obtained on the official web-sites of YAGO[4]. We randomly took 50,000 seed alignment. For each seed alignment $\left(e_i^u, e_j^v\right)$, we took 40 negative samples by replacing e_i^u or e_j^v with 1-to-1 ratio by another entity in KB^u or KB^v, respectively. From our candidate generation method, it is obvious that all the entities in E^{u+} and E^{v+} have overlap in their names, that is, words in the mention. Therefore, in order to make the distribution of $\left(e_i^u, e_j^v\right)$ ($\in S^{uv} \cup S_n^{uv}$) close to that of $\left(e_x^u, e_y^v\right)$ ($e_x^u \in E^{u+}$ and $e_y^v \in E^{v+}$), we did negative sampling by the following steps: first, arranged the entities in KB^u (KB^v) in descending order according to the number of words overlapped with the name of e_i^u (e_j^v); and second, replaced e_i^u (e_j^v) by the top 20 entities. 50,000 correct alignment and 2,000,000 wrong alignment make up the training set for EA model.

Table 3. Setting of hyper-parameters.

Hyper-parameter	Value
learning rate (Adam)	0.001
mini-batch size	256
d_w (word emb. dim.)	300
d_p (position emb. dim.)	5
d_n (name emb. dim.)	100
d_T (TransE emb. dim.)	50
BiLSTM hidden dim.	100
$\mathbf{FFN}_g^t, \mathbf{FFN}_{nd}^t, \mathbf{FFN}_{ea}^t$ hidden dim	300
δ (margin)	0.3
T (temperature)	1/3
η	5
η_{inc}	0.5
P_N^*	0.9
P^*	0.04

5   Experiments

5.1   Baselines

We compared our methods: Min-P and U-Bound with the following methods:

Separate (Sep) [16]: it trains each EL model independently, that is, the models are trained with loss function $\left(\sum_{t=1}^{N} L_{el}^t\right)$.

Zero-η_{inc}: in order to understand the effect of L_{kl}^{uv} on the results, in this model, we set $\eta_{inc} = 0$, i.e., $L_{inc}^{uv} = \eta \cdot L_{kl}^{uv}$.

---

[4] http://resources.mpi-inf.mpg.de/yago-naga/yago3.1/yagoDBpediaInstances.ttl.7z.

5.2 Test Tasks

We evaluated the four models in the following test tasks:

'ND': it uses the noise detection model to judge whether a data point is valid, so as to complete the noise reduction. This is a binary classification task and the AP (Average Precision) is used to compare the performance. AP is calculated by $AP = \sum_n (R_n - R_{n-1}) \cdot P_n$, where P_n and R_n are the precision and recall at the n_{th} threshold. We labeled valid data points as positive and noisy data points as negative and used $p_N^t (0|E^{t+}, m, c)$ as the score of each positive label.

'ER' [16]: it uses the entity ranking model to select an entity from E^{t+} for (m, c), so as to complete EL on the candidate sets which have a high recall rate but a large size. This task is tested on the all valid data points and the accuracy is used to compare the performance. We also report the performance of a fully supervised model whose loss function is $L = \sum_{t=1}^{N} \frac{1}{|D|} \sum_{(m,c)\in D} l^t (m, c)$. AIDA-CoNLL-train was used as the training data and the candidate set contains only the correct entity.

'ND&ER' [16]: it jointly tests the performance of noise detection model and entity ranking model. Specifically, if $p_N^t (1|E^{t+}, m, c) > \tau$, the data point is predicted as 'undecidable' (*i.e.*, the entity ranking model will not select an entity for this data point). The (micro) precision, (micro) recall, and (micro) F1 are used to compare the performance. We labeled the noisy data point as 'undecidable', so that the influence of candidate generation can be eliminated when calculating the recall. As P_N^* is close to the mean of $p_N^t (1|\cdot)$, we set the value of τ as P_N^*.

For each model, there are two options during testing:

IND: each EL model predicts independently.

REF: each EL model refers to the prediction of other EL models in order to reach an agreement. Specifically, we replace g^t with G^t:

$$ G^t \left(e_i^t, m, c\right) = p \left(e_i^t\right) \cdot \prod_{u \in B\setminus\{t\}} \left(\sum_{e_j^u \in E^{u+}} p \left(e_j^u\right) p_a^{tu}(e_i^t, e_j^u) \right) \tag{20} $$

where $B \subseteq \{1, \cdots, N\}$. For 'ER', B is the collection of u that E^{u+} contains the correct entity; and for 'ND&ER', B is the collection of u with $p_N^u (1|E^{u+}, m, c) < \tau$. Note that this does not affect the performance of noise detection model.

5.3 Results

We ran each model five times. And for each metric, we report mean and 95% confidence interval. We used the early stop strategy. Since the training convergence speed of EA model is slow, we first trained EA model alone, *i.e.*, with loss function $\sum_{u=1}^{N-1} \sum_{v=u+1}^{N}$ $\mathbb{I} \left(\mathbf{F}\left[u, v\right] = 1\right) L_{ea}^{uv}$, and then jointly trained models with Eq. (14). The settings of hyper-parameters are shown in Table 3. For parameters P_N^*, T and η, we used the same setting as [16]. We selected δ among $\{0.1, 0.3, 0.5\}$ for all the four models. Then, we selected P^* among $\{0.02, 0.04, 0.06, 0.08\}$ for Min-P and U-Bound. In order to avoid paying too much attention to the SCC model during training and misleading the learning of EL models, we set η_{inc} to 0.5 for Min-P and U-Bound.

Table 4. Results of 'ND' and 'ER' on the test set.

	ND		ER	
Model	DBpedia	YAGO	DBpedia	YAGO
Sep (IND)	65.39±1.73	65.65±1.49	64.19±1.26	66.65±0.34
Sep (REF)	-	-	67.75±0.44	67.81±0.86
Zero-η_{inc} (IND)	68.85±1.34	70.29±1.49	64.91±0.67	66.78±0.22
Zero-η_{inc} (REF)	-	-	67.64±0.36	67.70±0.33
Min-P (IND)	**74.15±1.37**	76.20±1.61	66.72±0.44	68.36±0.20
Min-P (REF)	-	-	**68.37±0.23**	**68.68±0.33**
U-Bound (IND)	73.70±0.87	**77.09±0.64**	66.85±0.29	67.83±0.19
U-Bound (REF)	-	-	68.34±0.15	68.27±0.14
fully supervised (IND)	-	-	82.13±0.56	79.68±0.45
fully supervised (REF)	-	-	84.21±0.28	83.83±0.41

Table 5. Results of 'ND&ER' on the test set.

	DBpedia			YAGO		
Model	P	R	F1	P	R	F1
Sep (IND)	49.47	35.98	41.65±0.35	48.51	34.57	40.31±0.78
Sep (REF)	50.13	**36.98**	42.53±0.60	49.09	**37.72**	42.64±0.47
Zero-η_{inc} (IND)	57.98	32.01	41.17±1.04	57.34	32.17	41.21±1.12
Zero-η_{inc} (REF)	59.41	33.08	42.40±0.78	58.73	32.64	41.93±1.02
Min-P (IND)	66.33	35.18	45.94±0.46	**65.34**	34.69	**45.30±0.34**
Min-P (REF)	66.31	35.24	**45.99±0.47**	64.25	34.94	45.26±0.34
U-Bound (IND)	66.33	34.39	45.27±0.87	64.17	34.45	44.79±0.51
U-Bound (REF)	**66.55**	34.72	45.62±0.67	64.22	34.48	44.83±0.52

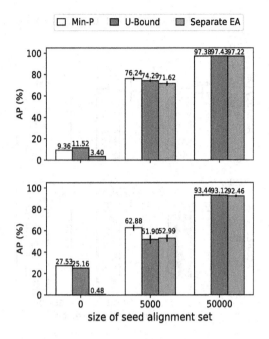

Fig. 1. Results on EA_{all} and EA_c with different size of seed alignment set (top: EA_{all}; bottom: EA_c).

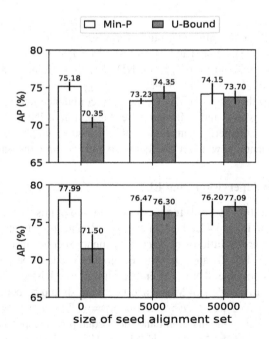

Fig. 2. AP of ND when the size of the seed alignment set changes (top: DBpedia; bottom: YAGO).

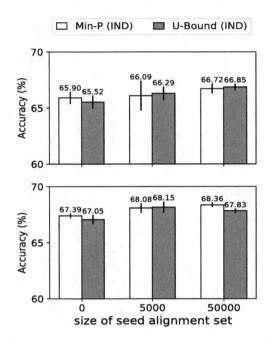

Fig. 3. Accuracy of ER when the size of the seed alignment set changes (top: DBpedia; bottom: YAGO).

Min-P and U-Bound significantly outperform Sep and Zero-η_{inc} on 'ND' (base on Table 4). Therefore, the precision and F1 scores of Min-P and U-Bound are significantly better than those of Sep and Zero-η_{inc} on 'ND&ER' (see Table 5), although the recall scores have declined. On 'ER' (see Table 4), Zero-η_{inc} is close to Sep, and Min-P and U-Bound are better than Sep (on YAGO, the 95% confidence interval of Sep does not overlap with that of Min-P and U-Bound under the option of IND). From the results of 'ER' and 'ND&ER', the performance under REF is better than that under IND in most cases. The experiments show that it is effective to add SCC into the training and testing.

5.4 Can EA Get Supervision from EL?

We answer this question by verifying the following two points. First, if EA model can get supervision from EL models, the Min-P and U-Bound should outperform a separate EA model when there is little or no seed alignment. Second, in order to reduce the loss of SCC model (Eq. (3) and Eq. (6)), EA model needs to be consistent with EL models as much as possible, that is, the entities that EL models consider correct should be aligned. So, the EA models of Min-P and U-Bound should be better at dealing with such entity pairs $\{(e_i^u, e_j^v) | e_i^u = \langle E^{u+}, m, c \rangle \wedge e_j^v = \langle E^{v+}, m, c \rangle\}$. We set two test sets EA_{all} and EA_c to verify the first point and the second point, respectively. Specifically, $EA_{all} = \{(e_i^u, e_j^v) | e_i^u \in E^{u+} \wedge e_j^v \in E^{v+}\}$ and $EA_c = \{(e_i^u, e_j^v) | e_i^u \in E^{u+} \wedge e_j^v \in E^{v+} \wedge e_i^u \neq e_j^v\} \cup \{(e_i^u, e_j^v) | e_i^u = \langle E^{u+}, m, c \rangle \wedge e_j^v = \langle E^{v+}, m, c \rangle\}$. We constructed EA_{all} and EA_c on the development set, and set the maximum sizes of E^{u+} and E^{v+} to

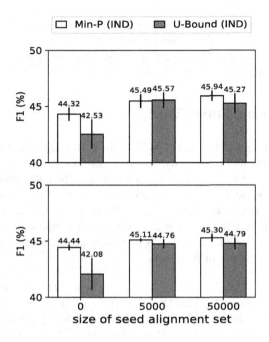

Fig. 4. F1 of ND&ER when the size of the seed alignment set changes (top: DBpedia; bottom: YAGO)).

20. There are 6,490 correct alignment and 185,413 wrong alignment in EA_{all}, and there are 938 correct alignment and 185,413 wrong alignment in EA_c. The performance of EA models on EA_{all} and EA_c are shown in Fig. 1.

The results verify the above two points. We can see from Fig. 1 that: (1) when there is little or no seed alignment, Min-P and U-Bound outperform the separate EA model, especially on EA_c; (2) the performance of Min-P and U-Bound on EA_c is better than that on EA_{all}. In Eq. (3), the larger value of $p_c\left(e_i^u\right)p_c\left(e_j^v\right)$ an entity pair (e_i^u, e_j^v) has, the greater gradient of $p_A(e_i^u, e_j^v)$ it has. Similarly, in Eq. (6), the larger value of $p_c\left(e_i^u\right)p_c\left(e_j^v\right)$ an entity pair (e_i^u, e_j^v) has, the larger lower bound of $p_A(e_i^u, e_j^v)$ it has. Last, it can be seen from Eq. (2) that p_A also gives negative samples when giving positive samples to the training of EA model. These are the reasons why EA model can get supervision from EL models.

5.5 Size of Seed Alignment Set

We tested the performance of Min-P and U-Bound with different size 0, 5,000, and 50,000 of seed alignment set. The results in Figs. 2, 3, and 4 fully show that our method does not need a lot of seed alignment. Because EA model can get supervision from EL models, SCC model can still work when there is little or no seed alignment. The performance of Min-P only decreases slightly on 'ER' and 'ND&ER' when the size of seed alignment set decreases. On 'ND', the performance of U-Bound is sensitive to the

798 H. Dai et al.

size of seed alignment set. This can be explained as follows: when the supervision of seed alignment becomes weak, the binding force of p_A as the upper bound (see Eq. (6)) would be weakened, which causes many gradients returned by Eq. (6) to become zero, thus making U-Bound degenerate to Zero-η_{inc}.

6 Conclusion and Future Work

In this paper, we discuss the impact of SCC on the performance of distantly supervised EL under the condition of low-quality candidate generation. We define the SCC model from two different aspects and the experimental results show that adding SCC is effective, especially in noise detection. Furthermore, the cost of our approach is low. For future work, we will explore the application of the proposed scheme to more real-world scenarios [8,9,25]. In addition, we plan to study how to use a learned entity ranking model to help candidate generation, so as to achieve high-quality candidate generation without abundant resources in Wikipedia.

Acknowledgements. This work was supported in part by the National Natural Science Foundation of China under Grant No. 62272219, 61872178, 62272223, 61832005, and 62072230, in part by the Jiangsu High-level Innovation and Entrepreneurship (Shuangchuang) Program, in part by the Postgraduate Research & Practice Innovation Program of Jiangsu Province No. KYCX22_0152,in part by the Fundamental Research Funds for the Central Universities No. 020214380089 and 020214380098, and in part by the Collaborative Innovation Center of Novel Software Technology and Industrialization, Nanjing University.

References

1. Bansal, T., Verga, P., Choudhary, N., McCallum, A.: Simultaneously linking entities and extracting relations from biomedical text without mention-level supervision. In: Proceedings of AAAI, pp. 7407–7414 (2020)
2. Bordes, A., Usunier, N., Garcia-Duran, A., Weston, J., Yakhnenko, O.: Translating embeddings for modeling multi-relational data. In: Proceedings of NIPS, pp. 2787–2795 (2013)
3. Bruna, J., Zaremba, W., Szlam, A., LeCun, Y.: Spectral networks and locally connected networks on graphs. In: Proceedings of ICLR, pp. 1–14 (2014)
4. Cao, Y., Liu, Z., Li, C., Liu, Z., Li, J., Chua, T.S.: Multi-channel graph neural network for entity alignment. In: Proceedings of ACL, pp. 1452–1461 (2019)
5. Chen, M., Tian, Y., Chang, K.W., Skiena, S., Zaniolo, C.: Co-training embeddings of knowledge graphs and entity descriptions for cross-lingual entity alignment. In: Proceedings of IJCAI, pp. 3998–4004 (2018)
6. Chen, M., Tian, Y., Yang, M., Zaniolo, C.: Multilingual knowledge graph embeddings for cross-lingual knowledge alignment. In: Proceedings of IJCAI, pp. 1511–1517 (2017)
7. Chen, S., Wang, J., Jiang, F., Lin, C.Y.: Improving entity linking by modeling latent entity type information. In: Proceedings of AAAI, pp. 7529–7537 (2020)
8. Dai, H., Li, M., Liu, A.X.: Finding persistent items in distributed datasets. In: Proceedings of INFOCOM, pp. 1403–1411 (2018)
9. Dai, H., Shahzad, M., Liu, A.X., Zhong, Y.: Finding persistent items in data streams. Proc. VLDB Endowm. **10**(4), 289–300 (2016)

10. Dietterich, T.G., Lathrop, R.H., Lozanoperez, T.: Solving the multiple instance problem with axis-parallel rectangles. Artif. Intell. **89**(1), 31–71 (1997)
11. Francis-Landau, M., Durrett, G., Klein, D.: Capturing semantic similarity for entity linking with convolutional neural networks. In: Proceedings of NAACL, pp. 1256–1261 (2016)
12. Ganea, O.E., Hofmann, T.: Deep joint entity disambiguation with local neural attention. In: Proceedings of EMNLP, pp. 2619–2629 (2017)
13. Hoffart, J., Yosef, M.A., Bordino, I., Fürstenau, H., Weikum, G.: Robust disambiguation of named entities in text. In: Proceedings of EMNLP, pp. 782–792 (2011)
14. Le, P., Titov, I.: Improving entity linking by modeling latent relations between mentions. In: Proceedings of ACL, pp. 1595–1604 (2018)
15. Le, P., Titov, I.: Boosting entity linking performance by leveraging unlabeled documents. In: Proceedings of ACL, pp. 1935–1945 (2019)
16. Le, P., Titov, I.: Distant learning for entity linking with automatic noise detection. In: Proceedings of ACL, pp. 4081–4090 (2019)
17. Mikolov, T., Chen, K., Corrado, G., Dean, J.: Efficient estimation of word representations in vector space. In: Proceedings of ICLR, pp. 1–12 (2013)
18. Newman-Griffis, D., Lai, A.M., Fosler-Lussier, E.: Jointly embedding entities and text with distant supervision. In: Proceedings of RepL4NLP, pp. 195–206 (2018)
19. Nie, H., et al.: Global structure and local semantics-preserved embeddings for entity alignment. In: Proceedings of IJCAI, pp. 3658–3664 (2020)
20. Pei, S., Yu, L., Hoehndorf, R., Zhang, X.: Semi-supervised entity alignment via knowledge graph embedding with awareness of degree difference. In: Proceedings of WWW, pp. 3130–3136 (2019)
21. Raiman, J.: Deeptype 2: Superhuman entity linking, all you need is type interactions. In: Proceedings of AAAI, pp. 8028–8035 (2022)
22. Rong, G., et al.: Sparkdq: Efficient generic big data quality management on distributed data-parallel computation. J. Parall. Distrib. Comput. **156**(1), 132–147 (2021)
23. Sun, Z., Hu, W., Zhang, Q., Qu, Y.: Bootstrapping entity alignment with knowledge graph embedding. In: Proceedings of IJCAI, pp. 4396–4402 (2018)
24. Sun, Z., et al.: Knowledge graph alignment network with gated multi-hop neighborhood aggregation. In: Proceedings of AAAI, pp. 222–229 (2020)
25. Wang, H., et al.: Bamboo filters: Make resizing smooth. In: Proceedings of ICDE, pp. 979–991 (2022)
26. Wang, Z., Lv, Q., Lan, X., Zhang, Y.: Cross-lingual knowledge graph alignment via graph convolutional networks. In: Proceedings of EMNLP, pp. 349–357 (2018)
27. Wu, Y., Liu, X., Feng, Y., Wang, Z., Zhao, D.: Neighborhood matching network for entity alignment. In: Proceedings of ACL, pp. 6477–6487 (2020)
28. Xue, M., et al.: Neural collective entity linking based on recurrent random walk network learning. In: Proceedings of IJCAI, pp. 5327–5333 (2019)
29. Yamada, I., Shindo, H., Takeda, H., Takefuji, Y.: Learning distributed representations of texts and entities from knowledge base. Trans. Assoc. Comput. Linguist. **5**(1), 397–411 (2017)
30. Yang, K., Liu, S., Zhao, J., Wang, Y., Xie, B.: Cotsae: Co-training of structure and attribute embeddings for entity alignment. In: Proceedings of AAAI, pp. 3025–3032 (2020)
31. Zhang, Q., Sun, Z., Hu, W., Chen, M., Guo, L., Qu, Y.: Multi-view knowledge graph embedding for entity alignment. In: Proceedings of IJCAI, pp. 5429–5435 (2019)
32. Zhaokang, W., Yunpan, W., Chunfeng, Y., Rong, G., Yihua, H.: Empirical analysis of performance bottlenecks in graph neural network training and inference with GPUS. NeuroComputing **446**(1), 165–191 (2021)

Sample and Feature Enhanced Few-Shot Knowledge Graph Completion

Kai Zhang[1,2], Daokun Zhang[3], Ning Liu[1], Yonghua Yang[4], Yonghui Xu[2(✉)], Zhongmin Yan[1], Hui Li[1], and Lizhen Cui[1,2]

[1] School of Software, Shandong University, Jinan, China
[2] Joint SDU-NTU Centre for Artificial Intelligence Research (C-FAIR), Shandong University, Jinan, China
xu.yonghui@hotmail.com
[3] Monash University, Melbourne, Australia
[4] Alibaba Group, Hangzhou, China

Abstract. Knowledge graph completion is to infer missing/new entities or relations in knowledge graphs. The long-tail distribution of relations leads to the few-shot knowledge graph completion problem. Existing solutions do not thoroughly solve this problem, with the few training samples still deteriorating knowledge graph completion performance. In this paper, we propose a novel data augmentation mechanism to overcome the learning difficulty caused by few training samples, and a novel feature fusion scheme to reinforce data augmentation. Specifically, we use a conditional generative model to increase the number of entity samples on both entity structure and textual content views, and adaptively fuse entity structural and textual features to get informative entity representations. We then integrate adaptive feature fusion and generative sample augmentation with few-shot relation inference into an end-to-end learning framework. We conduct extensive experiments on five real-world knowledge graphs, showing the significant advantage of the proposed algorithm over state-of-the-art baselines, as well as the effectiveness of the proposed feature fusion and sample augmentation components.

Keywords: Knowledge Graph Completion · Few-Shot Learning · Data Augmentation · Feature Fusion

1 Introduction

Knowledge graphs contain extensive world information about the entities, their descriptions, and mutual relations, with applications in various domains such as recommendation, medical data mining and question answering, etc. However, due to the labor-intensive construction process, knowledge graphs suffer from the incompleteness problem, limiting their further usages [15]. To complete the knowledge graphs automatically, some algorithms (e.g., TransE [1], Distmult [14], SimplE [4] and ComplEx [9]) apply machine learning-based methods to predict the missing relations, which have been verified with good performance.

© The Author(s), under exclusive license to Springer Nature Switzerland AG 2023
X. Wang et al. (Eds.): DASFAA 2023, LNCS 13944, pp. 800–809, 2023.
https://doi.org/10.1007/978-3-031-30672-3_54

The success of the traditional knowledge graph completion methods relies on large quantities of known relation facts as training samples. However, due to the long-tail distribution of relations in knowledge graphs, most relations do not have enough support entity pairs, which makes traditional knowledge graph completion methods fail to perform well. This gives rise to the *few-shot knowledge graph completion* problem: inferring missing/new facts for relations with few observed facts. The few-shot knowledge graph completion problem is faced with the following two main challenges: **(1) Few Training Samples**: The long-tail distribution property makes only few known relation facts can be leveraged to perform few-shot relation inference, which inevitably results in inaccurate inference. **(2) Insufficient Structural Evidence**: The few-shot relations are usually surrounded by sparse structure with heterogeneous semantics, which makes it difficult to infer few-shot relation facts from only the structure patterns.

In order to overcome the above two main challenges and effectively solve the few-shot knowledge graph completion problem, we propose a novel algorithm called **S**ample **A**nd **F**eature **E**nhanced knowledge graph completion (SAFE) algorithm. SAFE utilizes two key data augmentation strategies, i.e., adaptive feature fusion and generative sample augmentation, and makes the two components effectively collaborate with each other. First, to overcome the **Few Training Samples** challenge, we propose a novel generative sample augmentation strategy. Specifically, we use Conditional Wasserstein Generative Adversarial Networks (cWGAN) [7] to generate ground-truth-like entity samples on both entity textual content and structure views. With augmented relational facts, we can effectively capture the intrinsic semantics of few-shot relations. Second, to combat the **Insufficient Structural Evidence** challenge, we fuse entity structure features with rich textual content features to form a unified representation. Even if structural evidence is insufficient and the structural representations are not informative enough, the adaptive feature fusion scheme can compensate for this deficiency by adaptively weighting more textual content representations. In this way, we can build robust entity representations. Finally, to make the generative sample augmentation and adaptive feature fusion components well serve the few-shot knowledge graph completion task, we seamlessly integrate them with a relational few-shot learner into a unified end-to-end learning framework.

We conduct extensive comparison and ablation experiments on five real-world knowledge graphs with entity textual content features. Experimental results show that the proposed SAFE algorithm significantly outperforms state-of-the-art baselines. The results also demonstrate that the proposed adaptive feature fusion and generative sample augmentation components indeed greatly boost the few-shot knowledge graph completion performance.

2   Related Work

Knowledge graph completion aims to infer new/missing entities or relations in knowledge graphs. TransE [1] is the pioneer of the translational distance based methods, which models relations as transnational vectors from their head entities to tail entities. To make TransE able to capture the diverse entity semantics, a number of follow-up variants (e.g., TransH [11], TransR [5] and TransD [3]) are proposed. On the other hand, knowledge graph completion can be performed through measuring the semantic matching credibilities between head and tail entities, with relations defining the matching semantics. Representative algorithms include RESCAL [6], DistMult [14] and ComplEx [9].

To handle few-shot knowledge graph completion, some solutions have been proposed by adapting few-shot learning strategies from image and natural language domains to the knowledge graph domain, such as GMatching [13], FSRL [15], FANN [8] and REFORM [10]. Data fusion and augmentation have also been leveraged to solve the few-shot knowledge graph completion problem, like wRAN [16] and TCVAE [12]. However, they either augment samples on only one view or fail to well balance the contributions of structural features and textual content.

3   Problem Definition

A knowledge graph $\mathcal{G} = (\mathcal{E}, \mathcal{R}, \mathcal{TP})$ includes a set of entities $\mathcal{E}$, a set of relations $\mathcal{R}$, and is a set of triplets $\mathcal{TP} = \{(h, r, t)\} \subseteq \mathcal{E} \times \mathcal{R} \times \mathcal{E}$. Each entity $e \in \mathcal{E}$ has a content feature vector $\boldsymbol{x}_e^c \in \mathbb{R}^d$ described by the entity name or textual description, and a structure feature vector $\boldsymbol{x}_e^s \in \mathbb{R}^d$ obtained by TransE [1] and a neighborhood aggregation operation.

For few-shot knowledge completion, we aim to predict the unseen facts determined by few-shot relations. Formally, for each few-shot relation r, we have a positive support set $\mathcal{S}_r^+ = \{(h_i, t_i) \in \mathcal{E} \times \mathcal{E} | (h_i, r, t_i) \in \mathcal{TP}\}$ with a fixed size, and we are to predict the tail entity t given the head entity h_j of relation r: $(h_j, r, ?)$. Following practice, we cast the completion task into a triplet classification problem. We then construct the overall support set by augmenting a negative set: $\mathcal{S}_r = \mathcal{S}_r^+ \cup \mathcal{S}_r^-$. The negative support set is defined as $\mathcal{S}_r^- = \{(h_i, t_{i'}) \in \mathcal{E} \times \mathcal{E} | (h_i, r, t_{i'}) \notin \mathcal{TP}\}$, i.e., for each positive relation instance $(h_i, t_i) \in \mathcal{S}_r^+$, we sample a negative counterpart $(h_i, t_{i'})$ with $(h_i, r, t_{i'}) \notin \mathcal{TP}$ to form the negative support set. Through referring to the support set $\mathcal{S}_r$, few-shot knowledge graph completion aims to classify the relation instances in the query set $Q_r = \{(h_j, t_j) | (h_j, r, t_j) \in \text{ or } \notin \mathcal{TP}\}$.

We will train a general few-shot knowledge graph completion model from a training task set $\mathcal{T}_{train} = \{(\mathcal{S}_{r_i}, \mathcal{Q}_{r_i})\}$, formed by a set of training few-shot relations r_i, with known labels of relation instances in each $\mathcal{S}_{r_i}$ and $\mathcal{Q}_{r_i}$. We then evaluate the performance of the trained model on a test task set $\mathcal{T}_{test} = \{(\mathcal{S}_{r_j}, \mathcal{Q}_{r_j})\}$, i.e., we use the trained model to infer the labels of the relation instances in each query set $\mathcal{Q}_{r_j}$ according to the corresponding support set $\mathcal{S}_{r_j}$.

4   Our Approach

4.1   Overall Learning Framework

In Fig. 1, we illustrate the learning framework of the proposed SAFE approach for few-shot knowledge graph completion. Given a pair of support and query sets

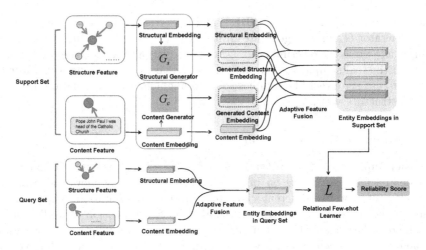

Fig. 1. The overall learning framework of the proposed SAFE approach for few-shot knowledge graph completion.

$(\mathcal{S}_{r_i}, \mathcal{Q}_{r_i}) \in \mathcal{T}_{train}$, we perform generative sample augmentation and adaptive feature fusion for the entities of the relation instances in the support set $\mathcal{S}_{r_i}$, while conduct adaptive feature fusion for the entities of the relation instances in the query set $\mathcal{Q}_{r_i}$. The obtained entity embeddings in the support and query sets are then used to train a relational few-shot learner.

4.2 Generative Sample Augmentation

For each given support relation instance $(h_j, t_j) \in \mathcal{S}_{r_i}$, we use two conditional generators G_s and G_c pre-trained on structure and content views respectively, to generate real-like new samples in structure and content views. To be specific, for entity h_j, we use the given structure feature $\boldsymbol{x}^s_{h_j}$ and content feature $\boldsymbol{x}^c_{h_j}$ and the corresponding conditional generators to respectively generate $\mathcal{K}$ samples on the structure and content views, $\{\tilde{\boldsymbol{x}}^{s,k}_{h_j}\}$ and $\{\tilde{\boldsymbol{x}}^{c,k}_{h_j}\}$:

$$\tilde{\boldsymbol{x}}^{s,k}_{h_j} = G_s(\boldsymbol{x}^s_{h_j}, \boldsymbol{z}^{s,k}_{h_j}), \quad \tilde{\boldsymbol{x}}^{c,k}_{h_j} = G_c(\boldsymbol{x}^c_{h_j}, \boldsymbol{z}^{c,k}_{h_j}), \tag{1}$$

where $k = 1, 2, \cdots, \mathcal{K}$ and $\boldsymbol{z}^{s,k}_{h_j}, \boldsymbol{z}^{c,k}_{h_j} \in \mathbb{R}^d$ are random noises sampled from standard normal distribution, i.e., $\boldsymbol{z}^{s,k}_{h_j}, \boldsymbol{z}^{c,k}_{h_j} \sim N(0, \mathbf{I}_d)$ with $\mathbf{I}_d$ being the $d \times d$ identity matrix. For entity t_j, $\mathcal{K}$ augmented samples on the structure view $\{\tilde{\boldsymbol{x}}^{s,k}_{t_j}\}$ and content view $\{\tilde{\boldsymbol{x}}^{c,k}_{t_j}\}$ are obtained in the same way.

The generators G_s and G_c are pre-trained with the conditional Wasserstein Generative Adversarial Network [7]. For example, the generator G_s is pre-trained through optimizing the following objective:

$$\min_{G_s} \max_{D_s} \mathop{\mathbb{E}}_{e \sim \mathcal{E}} [D_s(\tilde{\boldsymbol{x}}^s_e)] - \mathop{\mathbb{E}}_{e \sim \mathcal{E}} [D_s(\boldsymbol{x}^s_e)] + \eta \mathop{\mathbb{E}}_{e \sim \mathcal{E}} \left[(\|\nabla_{\hat{\boldsymbol{x}}^s_e} D_s(\hat{\boldsymbol{x}}^s_e)\|_2 - 1)^2 \right], \tag{2}$$

where $\tilde{x}_e^s = G_s(x_e^s, z_e^s)$, $z_e^s \sim N(0, \mathbf{I}_d)$, and $\hat{x}_e^s = \alpha x_e^s + (1 - \alpha)\tilde{x}_e^s$, $\alpha \sim U(0, 1)$. The generator G_s aims to generate entity structure sample $\tilde{x}_e^s$ similar to the real sample x_e^s, while the discriminator D_s tries to distinguish the generated sample $\tilde{x}_e^s$ from the real sample x_e^s. Through the adversarial pre-training, generator G_s is able to generate real-like samples so as to augment support relation instances on the structure view.

4.3   Adaptive Feature Fusion

For each support relation instance $(h_j, t_j) \in \mathcal{S}_{r_i}$, given the entity original and augmented feature vectors on both structure and content views, we can fuse entity features on two different views to a united representation, which is more informative to represent entities. To make the feature fusion more flexible to handle different cases where structure or content is more informative, we adopt an adaptive approach. For entity h_j, to obtain more real-like embedding vectors with larger diversity, we adopt an crossover fusion scheme, i.e., considering all possible fusion combinations of the four groups of features: original and generated structure features $\{x_{h_j}^s\}$ and $\{\tilde{x}_{h_j}^{s,k}\}$, as well as original and generated content features $\{x_{h_j}^c\}$ and $\{\tilde{x}_{h_j}^{c,k}\}$. Then $3\mathcal{K} + 1$ embedding vectors are generated:

$$
\begin{aligned}
h_j^{(0)} &= [1 - \lambda(x_{h_j}^c)] \cdot x_{h_j}^s + \lambda(x_{h_j}^c) \cdot x_{h_j}^c, \\
h_j^{(k)} &= [1 - \lambda(\tilde{x}_{h_j}^{c,k})] \cdot x_{h_j}^s + \lambda(\tilde{x}_{h_j}^{c,k}) \cdot \tilde{x}_{h_j}^{c,k}, \quad k = 1, 2, \cdots, \mathcal{K}, \\
h_j^{(k+\mathcal{K})} &= [1 - \lambda(x_{h_j}^c)] \cdot \tilde{x}_{h_j}^{s,k} + \lambda(x_{h_j}^c) \cdot x_{h_j}^c, \quad k = 1, 2, \cdots, \mathcal{K}, \\
h_j^{(k+2\mathcal{K})} &= [1 - \lambda(\tilde{x}_{h_j}^{c,k})] \cdot \tilde{x}_{h_j}^{s,k} + \lambda(\tilde{x}_{h_j}^{c,k}) \cdot \tilde{x}_{h_j}^{c,k}, \quad k = 1, 2, \cdots, \mathcal{K},
\end{aligned}
\tag{3}
$$

where $\lambda(\cdot)$ defines the adaptive trade-off weight, which is a function of entity content feature vector and implemented through a four-layer neural network. For the tail entity t_j, four groups of entity embedding vectors $t_j^{(0)}, \{t_j^{(k)}\}, \{t_j^{(k+\mathcal{K})}\}$ and $\{t_j^{(k+2\mathcal{K})}\}$ are also constructed in the same way.

For each query relation instance $(h_l, t_l) \in \mathcal{Q}_{r_i}$, without performing sample augmentation, we obtain the entity embeddings h_l and t_l for h_l and t_l, by adaptively fusing their original features on structure and content views:

$$
\begin{aligned}
h_l &= [1 - \lambda(x_{h_l}^c)] \cdot x_{h_l}^s + \lambda(x_{h_l}^c) \cdot x_{h_l}^c, \\
t_l &= [1 - \lambda(x_{t_l}^c)] \cdot x_{t_l}^s + \lambda(x_{t_l}^c) \cdot x_{t_l}^c.
\end{aligned}
\tag{4}
$$

4.4   End-to-End Few-Shot Learning

We adopt the gradient based meta-learning strategy [2] to perform relational few-shot learner training. Given the embedding pair $(h_j^{(k)}, t_j^{(k)})$ for the positive relation instance $(h_j, t_j) \in \mathcal{S}_{r_i}^+ \subset \mathcal{S}_{r_i}$, we calculate its relation meta as

$$
R(h_j^{(k)}, t_j^{(k)}) = g(h_j^{(k)} \| t_j^{(k)}),
\tag{5}
$$

where $\|$ is the concatenation operator and $g(\cdot)$ is a multi-layer neural network. The relation meta for the few-shot relation r_i is obtained through averaging over all positive relation instances with all entity embedding variants:

$$R_{r_i} = \frac{\sum_{(h_j, t_j) \in \mathcal{S}_{r_i}^+} \sum_{k=0}^{3\mathcal{K}} R(h_j^{(k)}, t_j^{(k)})}{(3\mathcal{K} + 1)|\mathcal{S}_{r_i}^+|}. \tag{6}$$

For each embedding pair $(h_j^{(k)}, t_j^{(k)})$ of the relation instance $(h_j, t_j) \in \mathcal{S}_{r_i}$, we calculate its reliability score as

$$s(h_j^{(k)}, t_j^{(k)}) = \|h_j^{(k)} + R_{r_i} - t_j^{(k)}\|_2. \tag{7}$$

We then build the following ranking loss on the support set $\mathcal{S}_{r_i}$:

$$L(\mathcal{S}_{r_i}) = \sum_{(h_j, t_j) \in \mathcal{S}_{r_i}^+} \sum_{k=0}^{3\mathcal{K}} \left[\gamma + s(h_j^{(k)}, t_j^{(k)}) - s(h_j^{(k)}, t_{j'}^{(k)})\right]_+, \tag{8}$$

where $[\cdot]_+$ denotes the function $\max(\cdot, 0)$ and γ is the margin hyperparameter. We then update the relation meta R_{r_i} by descending the gradient of $L(\mathcal{S}_{r_i})$ with regard to R_{r_i}:

$$R'_{r_i} = R_{r_i} - \beta \nabla_{R_{r_i}} L(\mathcal{S}_{r_i}), \tag{9}$$

where β is the step hyperparameter. For the embedding pair (h_l, t_l) of the relation instance $(h_l, t_l) \in \mathcal{Q}_r$, we calculate its reliability score as

$$s'(h_l, t_l) = \|h_l + R'_{r_i} - t_l\|_2. \tag{10}$$

Then the loss on the query set is defined as

$$L(\mathcal{Q}_{r_i}) = \sum_{(h_l, t_l) \in \mathcal{Q}_{r_i}^+} [\gamma + s'(h_l, t_l) - s'(h_l, t_{l'})]_+. \tag{11}$$

The overall loss to be minimized is defined as

$$L_{All} = \sum_{(\mathcal{S}_{r_i}, \mathcal{Q}_{r_i}) \in \mathcal{T}_{train}} L(\mathcal{Q}_{r_i}). \tag{12}$$

4.5   Overall Workflow and Time Complexity

The overall workflow of few-shot knowledge graph completion with the proposed SAFE algorithm is described in Algorithm 1. In Algorithm 1, the pre-training of the generators G_s and G_c in Steps 1–2 is also conducted in a batch mode, with a time complexity of $O(d^2 \cdot B_1 \cdot I_1)$, where d is embedding dimension, B_1 and I_1 are respectively the batch size and the number of iterations for pre-training; the time complexity for training SEAF model in Steps 3–9 is $O((n+m) \cdot d^2 \cdot B_2 \cdot I_2)$, where n and m are respectively the size of support and query sets for each task, B_2 and I_2 are respectively the batch size and the number of iterations for SAFE model training; the time complexity for relation inference for $\mathcal{T}_{test}$ in Step 10 is $O((n+m) \cdot d^2 \cdot |\mathcal{T}_{test}|)$, with $|\mathcal{T}_{test}|$ being the size of the test task set $\mathcal{T}_{test}$.

Algorithm 1. Few-shot knowledge graph completion with SAFE

Input: The given knowledge graph $\mathcal{G} = (\mathcal{E}, \mathcal{R}, \mathcal{TP})$, training tasks $\mathcal{T}_{train} = \{(\mathcal{S}_{r_i}, \mathcal{Q}_{r_i})\}$ and test tasks $\mathcal{T}_{test} = \{(\mathcal{S}_{r_j}, \mathcal{Q}_{r_j})\}$.

Output: Likelihood scores of relation instances in test query sets $\mathcal{Q}_{r_j}$.

 1: Pre-train the structure-view entity generator G_s;
 2: Pre-train the content-view entity generator G_c;
 3: **repeat**
 4:    Sample a batch of tasks $\{(\mathcal{S}_{r_i}, \mathcal{Q}_{r_i})\}$ from $\mathcal{T}_{train}$;
 5:    Augment entity samples on structure and content views for each support relation instance $(h_j, t_j) \in \mathcal{S}_{r_i}$ for each task in the batch;
 6:    Fuse entity features on structure and content views for each support relation instance $(h_j, t_j) \in \mathcal{S}_{r_i}$ for each task in the batch;
 7:    Fuse entity features on structure and content views for each query relation instance $(h_j, t_j) \in \mathcal{Q}_{r_i}$ for each task in the batch;
 8:    Update model parameters with gradient descent for reducing the loss in Eq.(12);
 9: **until** Converge or a maximum number of iterations expire;
10: Get the likelihood of being linked by the target few-shot relation for each query relation instance $(h_l, t_l) \in \mathcal{Q}_{r_j}$ at each test task $(\mathcal{S}_{r_j}, \mathcal{Q}_{r_j}) \in \mathcal{T}_{test}$ with Eq. (10).

Table 1. Summary of five knowledge graph benchmarks.

Dataset	# Entities	# Relations	# Triplets	# Tasks
Wiki	4,838,244	822	5,859,240	183
NELL	68,545	358	181,109	67
DBpedia	19,752	2,646	31,742	34
FB15K-237	14,541	474	310,116	67
SRPRS	15,000	221	36,508	61

5 Experiments

5.1 Benchmark Knowledge Graphs

To provide an extensive evaluation, we study on five real-world knowledge graphs with entity textual content features. The statistics of the five knowledge graphs are summarized in Table 1. We construct a varying number of 5-shot relation inference tasks for each knowledge graph, as is shown in Table 1. Then, we split the tasks into the training, validation and test sets according to the ratios: 133:34:16 for Wiki, 51:11:5 for NELL, 23:8:3 for DBpedia, 51:10:5 for FB15K-237 and 30:15:16 for SRPRS. The test set is used to evaluate the performance of the trained model, while the validation set is used to select the best training epoch.

5.2 Baseline Methods

We compare the proposed SAFE algorithm with two groups of knowledge graph completion algorithms: **1) Traditional Knowledge Graph Completion Algorithms**, which use translational modeling and semantic modeling to

Table 2. Few-shot knowledge graph completion performance comparison.

	Wiki			NELL			DBpedia			FB15K-237			SRPRS		
	MRR	Hits@3	Hits@1	MRR	Hits@3	Hits@1	MRR	Hits@3	Hits@1	MRR	Hits@3	Hits@1	MRR	Hits@3	Hits@1
TransE	.072	.105	.015	.046	.067	.010	.096	.181	.001	.105	.101	.100	.083	.129	.015
Distmult	.035	.039	.014	.030	.032	.015	.044	.051	.021	.038	.041	.013	.062	.068	.026
ComplEx	.058	.060	.031	.008	.010	0.010	.032	.023	.009	.017	.017	.002	.084	.099	.042
SimplE	.034	.034	.012	.028	.030	.020	.030	.032	.009	.023	.018	.003	.063	.066	.027
GMatching(MeanP)	.095	.084	.059	.076	.051	.036	.231	.271	.108	.320	.377	.203	.148	.186	.097
GMatching(MaxP)	.010	.096	.061	.080	.047	.040	.241	.274	.130	.207	.219	.131	.149	.187	.097
GMatching(Max)	.094	.085	.061	.079	,048	.039	.232	.272	.118	.244	.271	.151	.150	.186	.097
Meta-R	.148	.155	.105	.144	.142	.062	.286	.324	.209	.164	.185	.103	.180	.189	.103
FSRL	.270	.280	.249	.298	.316	.220	.273	.395	.153	.232	.308	.133	.096	.137	.023
FAAN	.193	.205	.148	.238	.259	.164	.250	.276	.151	.372	.378	.321	.284	.297	.193
REFORM	.134	.149	.076	.134	.149	.076	.256	.305	.150	.297	.304	.237	.216	.253	.093
TCVAE	.133	.135	.065	.200	.238	.122	.418	.424	.376	.319	.369	.237	.467	.459	.441
SAFE	.297	.300	.277	.387	.421	.236	.461	.505	.363	.671	.669	.664	.531	.501	.499

perform knowledge graph completion: TransE [1], DistMult [14], ComplEx [9] and SimplE [4]. **2) Few-Shot Knowledge Graph Completion Algorithms**, which use few-shot learning to do knowledge graph completion: GMatching [13] (with three different configurations: MeanP, MaxP and Max [8]), Meta-R [2], FSRL [15], FAAN [8], REFORM [10] and TCVAE [12].

5.3 Evaluation Metrics

We use the MRR and Hit@R metrics to evaluate the few-shot knowledge graph completion performance. MRR is the mean reciprocal rank to indicate the ranking position of the first true relation fact among all the relation fact candidates in the test query set. Hit@R represents the ratio of correct relation facts among the relation fact candidates ranking top R in the test query set. Here, R takes values 1 and 3. Averaged MRR and Hit@R over all test tasks are reported. Higher scores in both two metrics mean better performance.

5.4 Performance Comparison

In Table 2, we compare the proposed SAFE algorithm with all baseline methods on the five benchmark knowledge graphs, with the best and second best results respectively highlighted by **boldface** and underline.

From Table 2, we can find that the proposed SAFE algorithm significantly outperforms all baseline methods. This owes to SAFE's ability to effectively augment informative training samples with generative models, and fuse entity structural encodings with rich entity text features. TCVAE, another data augmentation based method, achieves the overall second best performance. The good performance of SAFE and TCVAE verifies the effectiveness of data augmentation for few-shot knowledge graph completion. However, the proposed SAFE augments samples on both textual content and structural encoding views, and adopts an adaptive crossover feature fusing scheme to augment more informative and diverse samples, making it superior to TCVAE.

Table 3. Performance comparison between SAFE and its ablated variants.

	Wiki			NELL			DBpedia			FB15K-237			SRPRS		
	MRR	Hits@3	Hits@1	MRR	Hits@3	Hits@1	MRR	Hits@3	Hits@1	MRR	Hits@3	Hits@1	MRR	Hits@3	Hits@1
w/o G_s	.158	.152	.139	.259	.264	.161	.298	.336	.233	.422	.431	.407	.165	.167	.156
w/o G_c	.183	.181	.144	.278	.266	.179	.405	.471	.317	.497	.503	.482	.369	.362	.362
w/o G_s & G_c	.260	.289	.152	.352	.405	.202	.387	.445	.239	.464	.510	.340	.505	.480	.469
w/o $\lambda(\cdot)$	.159	.152	.093	.339	.340	.204	.308	.342	.237	.468	.464	.456	.296	.285	.285
w/o CF	.099	.091	.031	.325	.391	.214	.357	.383	.229	.307	.301	.293	.433	.433	.415
full SAFE	**.297**	**.300**	**.277**	**.387**	**.421**	**.236**	**.461**	**.505**	**.363**	**.671**	**.669**	**.664**	**.531**	**.501**	**.499**

5.5 Ablation Study

We also conduct some ablation experiments to study the importance of generative sample augmentation adaptive feature fusion and crossover feature fusion components. Table 3 compares SAFE with different ablated variants on Wiki, NELL, DBpedia, FB15K-237 and SRPRS, with the best and second best results respectively highlighted by **boldface** and underline.

Ablation on Generative Sample Augmentation. We compare the full SAFE model, with the variants without the structure generator G_s for generating entity samples on structure view ("w/o G_s"), without the content generator G_c for generating entity samples on content view ("w/o G_c"), and without the use of both G_s and G_c ("w/o G_s & G_c"). The full SAFE outperforms all the three variants in all cases. From this, we can find that sample augmentation on structure and content views are both important.

Ablation on Adaptive Feature Fusion. We replace the adaptive weight function $\lambda(\cdot)$ with a fixed value 0.5 for feature fusion in Eq. (3–4) to generate the "w/o $\lambda(\cdot)$" variant. As is shown in Table 3, without the adaptive feature fusion mechanism, the SAFE model goes through a serious performance drop. As structure features and content features play different roles for different few-shot relation inference tasks, simply mixing the features on two views with a rigid weighting configuration cannot capture such flexibility.

Ablation on Crossover Fusion. For this variant, we respectively perform adaptive fusion on the original and augmented entity samples, with no crossovers between them, which is named as the "w/o CF" variant. Obviously, the "w/o CF" variant is inferior to the full SAFE model. Through the use of the crossover feature fusion scheme, SAFE can augment more informative and diverse training samples, and yield better performance.

6 Conclusion

In this paper, we propose a novel approach SAFE to solve the few-shot knowledge graph completion problem with the well designed sample augmentation and feature fusion components, which are integrated into an end-to-end relational few-shot learning framework. The experiments on five real-world knowledge graphs show the significant superiority of the proposed SAFE algorithm over state-of-the-art baselines, and the effectiveness of the proposed entity sample augmentation and adaptive feature fusion components.

Acknowledgements. This research is supported by the National Key R&D Program of China (No. 2021YFF0900800), NSFC No.62202279, Shandong Provincial Key Research and Development Program (Major Scientific and Technological Innovation Project) (No. 2021CXGC010108), Shandong Provincial Natural Science Foundation (No. ZR202111180007) and the Fundamental Research Funds of Shandong University.

References

1. Bordes, A., Usunier, N., Garcia-Duran, A., Weston, J., Yakhnenko, O.: Translating embeddings for modeling multi-relational data. NeurIPS (2013)
2. Chen, M., Zhang, W., Zhang, W., Chen, Q., Chen, H.: Meta relational learning for few-shot link prediction in knowledge graphs. In: EMNLP-IJCNLP, pp. 4216–4225 (2019)
3. Ji, G., He, S., Xu, L., Liu, K., Zhao, J.: Knowledge graph embedding via dynamic mapping matrix. In: ACL, pp. 687–696 (2015)
4. Kazemi, S.M., Poole, D.: Simple embedding for link prediction in knowledge graphs. In: NeurIPS, pp. 4289–4300 (2018)
5. Lin, H., Liu, Y., Wang, W., Yue, Y., Lin, Z.: Learning entity and relation embeddings for knowledge resolution. Procedia Comput. Sci. **108**, 345–354 (2017)
6. Nickel, M., Tresp, V., Kriegel, H.P.: A three-way model for collective learning on multi-relational data. In: ICML (2011)
7. Rawat, S., Shen, M.H.H.: A novel topology optimization approach using conditional deep learning. arXiv preprint arXiv:1901.04859 (2019)
8. Sheng, J., et al.: Adaptive attentional network for few-shot knowledge graph completion. In: EMNLP, pp. 1681–1691 (2020)
9. Trouillon, T., Welbl, J., Riedel, S., Gaussier, É., Bouchard, G.: Complex embeddings for simple link prediction. In: ICML, pp. 2071–2080 (2016)
10. Wang, S., Huang, X., Chen, C., Wu, L., Li, J.: Reform: error-aware few-shot knowledge graph completion. In: CIKM, pp. 1979–1988 (2021)
11. Wang, Z., Zhang, J., Feng, J., Chen, Z.: Knowledge graph embedding by translating on hyperplanes. In: AAAI, vol. 28, pp. 1112–1119 (2014)
12. Wang, Z., Lai, K., Li, P., Bing, L., Lam, W.: Tackling long-tailed relations and uncommon entities in knowledge graph completion. In: EMNLP-IJCNLP, pp. 250–260 (2019)
13. Xiong, W., Yu, M., Chang, S., Guo, X., Wang, W.Y.: One-shot relational learning for knowledge graphs. In: EMNLP, pp. 1980–1990 (2018)
14. Yang, B., Yih, W., He, X., Gao, J., Deng, L.: Embedding entities and relations for learning and inference in knowledge bases. In: ICLR (2015)
15. Zhang, C., Yao, H., Huang, C., Jiang, M., Li, Z., Chawla, N.V.: Few-shot knowledge graph completion. In: AAAI, vol. 34, pp. 3041–3048 (2020)
16. Zhang, N., Deng, S., Sun, Z., Chen, J., Zhang, W., Chen, H.: Relation adversarial network for low resource knowledge graph completion. In: WWW, pp. 1–12 (2020)

Author Index

Printed in the United States
by Baker & Taylor Publisher Services